Edited by
Günter Schmid

Nanoparticles

Related Titles

Gubin, S. P. (ed.)

Magnetic Nanoparticles

2009

ISBN: 978-3-527-40790-3

Amabilino, D. B. (ed.)

Chirality at the Nanoscale

Nanoparticles, Surfaces, Materials and more

2009

ISBN: 978-3-527-32013-4

Geckeler, K. E. / Nishide, H. (eds.)

Advanced Nanomaterials

2010

ISBN: 978-3-527-31794-3

Elaissari, A. (ed.)

Colloidal Nanoparticles in Biotechnology

2008

ISBN: 978-0-470-23052-7

Martin, J. M., Ohmae, N.

Nanolubricants

2008

ISBN: 978-0-470-06552-5

Astruc, D. (ed.)

Nanoparticles and Catalysis

2008

ISBN: 978-3-527-31572-7

Mirkin, C. A., Niemeyer, C. M. (eds.)

Nanobiotechnology II

More Concepts and Applications

2007

ISBN: 978-3-527-31673-1

Niemeyer, C. M., Mirkin, C. A. (eds.)

Nanobiotechnology

Concepts, Applications and Perspectives

2004

ISBN: 978-3-527-30658-9

Edited by
Günter Schmid

Nanoparticles

From Theory to Application

Second Edition

WILEY-VCH

WILEY-VCH Verlag GmbH & Co. KGaA

The Editor

Prof. Dr. Günter Schmid
Universität Duisbug - Essen
Inst. für Anorganische Chemie
Universitätsstr. 5-7
45117 Essen

First Edition 2004

All books published by Wiley-VCH are carefully produced. Nevertheless, authors, editors, and publisher do not warrant the information contained in these books, including this book, to be free of errors. Readers are advised to keep in mind that statements, data, illustrations, procedural details or other items may inadvertently be inaccurate.

Library of Congress Card No.: applied for

British Library Cataloguing-in-Publication Data
A catalogue record for this book is available from the British Library.

Bibliographic information published by the Deutsche Nationalbibliothek
The Deutsche Nationalbibliothek lists this publication in the Deutsche Nationalbibliografie; detailed bibliographic data are available on the Internet at http://dnb.d-nb.de.

© 2010 WILEY-VCH Verlag GmbH & Co. KGaA, Boschstr. 12, 69469 Weinheim

All rights reserved (including those of translation into other languages). No part of this book may be reproduced in any form – by photoprinting, microfilm, or any other means – nor transmitted or translated into a machine language without written permission from the publishers. Registered names, trademarks, etc. used in this book, even when not specifically marked as such, are not to be considered unprotected by law.

Cover Design Formgeber, Eppelheim
Typesetting Thomson Digital, Noida, India
Printing and Binding Strauss GmbH, Mörlenbach
Printed in the Federal Republic of Germany
Printed on acid-free paper

ISBN: 978-3-527-32589-4

Contents

List of Contributors XI

1 **General Introduction** 1
   Günter Schmid

2 **Quantum Dots** 3
   Wolfgang Johann Parak, Liberato Manna, Friedrich C.
   Simmel, Daniele Gerion, and Paul Alivisatos
2.1 Introduction and Outline 3
2.2 Nanoscale Materials and Quantum Mechanics 5
2.2.1 Nanoscale Materials as Intermediate Between Atomic and Bulk Matter 5
2.2.2 Quantum Mechanics 6
2.3 From Atoms to Molecules and Quantum Dots 7
2.4 Shrinking Bulk Material to a Quantum Dot 10
2.4.1 Three-Dimensional Systems (Bulk Material) 10
2.4.2 Two-Dimensional Systems 13
2.4.3 One-Dimensional Systems (Quantum Wires) 17
2.4.4 Zero-Dimensional Systems (Quantum Dots) 18
2.5 Energy Levels of a (Semiconductor) Quantum Dot 19
2.6 Varieties of Quantum Dots 22
2.6.1 Lithographically Defined Quantum Dots 23
2.6.2 Epitaxially Self-Assembled Quantum Dots 25
2.6.3 Colloidal Quantum Dots 26
2.7 Optical Properties of Quantum Dots 28
2.7.1 Absorption and Emission Spectra 28
2.7.2 Spectral Diffusion and Blinking 29
2.7.3 Metal Nanoparticles 31
2.7.4 Overview of Some Selected Applications 32

Nanoparticles: From Theory to Application. Edited by Günter Schmid
Copyright © 2010 WILEY-VCH Verlag GmbH & Co. KGaA, Weinheim
ISBN: 978-3-527-32589-4

2.8	Some (Electrical) Transport Properties of Quantum Dots 33
2.8.1	Coulomb Blockade: Basic Theory and Historical Sketch 33
2.8.2	Single-Electron Tunneling 35
2.8.3	Tunneling Transport: The Line Shape of Conductance Peaks 39
2.8.4	Some Applications 40
	References 42

3	**Syntheses and Characterizations** 49
3.1	Zintl Ions 49
3.1.1	Homoatomic and Intermetalloid Tetrel Clusters – Synthesis, Characterization, and Reactivity 49
	Sandra Scharfe and Thomas F. Fässler
3.1.1.1	Introduction 49
3.1.1.2	Homoatomic Clusters of Tetrel Elements 50
3.1.1.2.1	Discrete Clusters in Neat Solids and from Solutions 50
3.1.1.2.2	Cluster Shapes and Ion Packing 52
3.1.1.2.3	Linked E_9 Clusters 55
3.1.1.3	Intermetalloid Clusters of Tetrel Elements 56
3.1.1.3.1	Complexes of Zintl Ions 56
3.1.1.3.2	Ligand-Free Heteroatomic Cluster: Intermetalloids 58
3.1.1.4	Beyond Deltahedral Clusters 63
	References 65
3.2	Semiconductor Nanoparticles 69
3.2.1	Synthesis and Characterization of II–VI Nanoparticles 69
	Alexander Eychmüller
3.2.1.1	Historical Review 69
3.2.1.2	Thiol-Stabilized Nanoparticles 76
3.2.1.3	The "Hot-Injection" Synthesis 81
3.2.1.4	Core–Shell Nanocrystals 83
3.2.1.5	Quantum Dot Quantum Wells 88
	References 92
3.2.2	Synthesis and Characterization of III-V Semiconductor Nanoparticles 101
	Uri Banin
3.2.2.1	Introduction 101
3.2.2.2	Synthetic Strategy 103
3.2.2.3	InAs and InP Nanocrystals 105
3.2.2.3.1	Synthesis of InAs and InP Nanocrystals 105
3.2.2.3.2	Structural and Basic Optical Characterization of InAs and InP Nanocrystals 107
3.2.2.4	Group III–V Core–Shell Nanocrystals: Synthesis and Characterization 111
3.2.2.4.1	Synthesis of Core–Shell Nanocrystals with InAs Cores 113
3.2.2.4.2	Optical Characterization of the Core–Shell Nanocrystals 114
3.2.2.4.3	Chemical and Structural Characterization 117

3.2.2.4.4	Model Calculations for the Band Gap *122*	
3.2.2.4.5	Stability of Core–Shell Nanocrystals *124*	
	References *125*	
3.2.3	Synthesis and Characterization of Ib–VI Nanoclusters *127*	
	Stefanie Dehnen, Andreas Eichhöfer, John F. Corrigan, Olaf Fuhr, and Dieter Fenske	
3.2.3.1	Introduction *127*	
3.2.3.2	Chalcogen-Bridged Copper Clusters *128*	
3.2.3.2.1	Synthesis Routes *128*	
3.2.3.2.2	Sulfur-Bridged Copper Clusters *129*	
3.2.3.2.3	Selenium-Bridged Copper Clusters *136*	
3.2.3.2.4	Tellurium-Bridged Copper Clusters *165*	
3.2.3.3	Chalcogen-Bridged Silver Clusters *177*	
3.2.3.3.1	Sulfur-Bridged Silver Clusters *178*	
3.2.3.3.2	Selenium-Bridged Silver Clusters *186*	
3.2.3.3.3	Tellurium-Bridged Silver Clusters *196*	
3.2.3.4	Selenium-Bridged Gold Clusters *208*	
	References *210*	
3.3	Synthesis of Metal Nanoparticles *214*	
3.3.1	Noble Metal Nanoparticles *214*	
	Günter Schmid	
3.3.1.1	Introduction *214*	
3.3.1.2	History and Background *214*	
3.3.1.3	Stabilization of Metal Nanoparticles *215*	
3.3.1.4	Synthetic Methods *218*	
3.3.1.4.1	Salt Reduction *219*	
3.3.1.4.2	Controlled Decomposition *226*	
3.3.1.5	Shape Control *228*	
	References *232*	
3.3.2	Synthesis, Properties and Applications of Magnetic Nanoparticles *239*	
	Galyna Krylova, Maryna I. Bodnarchuk, Ulrich I. Tromsdorf, Elena V. Shevchenko, Dmitri V. Talapin, and Horst Weller	
3.3.2.1	Introduction *239*	
3.3.2.1.1	Reverse Micelles Technique *241*	
3.3.2.1.2	Sonochemical Synthesis *242*	
3.3.2.1.3	Colloidal Syntheses *242*	
3.3.2.2	Colloidal Synthesis of Magnetic Metal Nanoparticles *242*	
3.3.2.2.1	General Remarks on the Synthesis of Co and $CoPt_3$ Nanocrystals *243*	
3.3.2.2.2	Synthesis of Cobalt Nanoparticles with Different Crystalline Modification *244*	
3.3.2.2.3	Synthesis of $CoPt_3$ Magnetic Alloy Nanocrystals *247*	
3.3.2.2.4	Shape-Controlled Synthesis of Magnetic Nanoparticles *252*	
3.3.2.2.5	Other Metal Magnetic Nanoparticles Synthesized by Methods of Colloidal Chemistry *255*	
3.3.2.3	Iron Oxide-Based Magnetic Nanocrystals *259*	

3.3.2.3.1	Maghemite and Magnetite Nanocrystals *259*
3.3.2.3.2	Nanocrystals of Other Iron Oxides (Hematite, Wüstite, Goethite) *262*
3.3.2.3.3	Nanocrystals of Metal Ferrites *263*
3.3.2.4	Multicomponent Magnetic Nanocrystals *264*
3.3.2.4.1	Magnetic Core–Shell Nanoparticles *265*
3.3.2.4.2	Dumbbell-Like Nanoparticles *266*
3.3.2.4.3	Hollow Magnetic Nanocrystals *268*
3.3.2.5	Size- and Shape-Dependent Magnetic Properties of Magnetic Metal Nanoparticles *271*
3.3.2.6	Magnetic Nanocrystals for Data Storage Applications *281*
3.3.2.7	Biomedical Applications of Magnetic Nanoparticles *282*
3.3.2.7.1	Design of Magnetic Particles for Biomedical Applications *283*
3.3.2.7.2	Drug Delivery *285*
3.3.2.7.3	Gene Delivery *290*
3.3.2.7.4	Magnetic Separation *291*
3.3.2.7.5	Magnetic Hyperthermia *292*
3.3.2.7.6	Magnetic Resonance Imaging *295*
3.3.2.7.7	Tomographic Imaging *297*
3.3.2.7.8	The Role of Magnetic Nanoparticle-Based Contrast Agents in MRI *298*
	References *302*

4 Organization of Nanoparticles *311*
4.1 Semiconductor Nanoparticles *311*
Nikolai Gaponik and Alexander Eychmüller
4.1.1 Molecular Crystals and Superlattices *311*
4.1.2 Layers of Semiconductor Nanocrystals *315*
4.1.3 Coupling of Semiconductor Nanocrystals *322*
References *323*
4.2 Metal Nanoparticles *328*
Günter Schmid, Dmitri V. Talapin, and Elena V. Shevchenko
4.2.1 Three-Dimensional Organization of Metal Nanoparticles *328*
4.2.2 Two- and One-Dimensional Structures of Metal Nanoparticles *338*
4.2.2.1 Self-Assembly *338*
4.2.2.2 Guided Self-Assembly *348*
4.2.2.3 Aimed Structures *357*
References *368*

5 Properties *371*
5.1 Semiconductor Nanoparticles *371*
5.1.1 Optical and Electronic Properties of Semiconductor Nanocrystals *371*
Uri Banin and Oded Millo
5.1.1.1 Introduction *371*
5.1.1.2 Semiconductor Nanocrystals as Artificial Atoms *372*
5.1.1.3 Theoretical Descriptions of the Electronic Structure *379*

5.1.1.4	Atomic-Like States in Core–Shell Nanocrystals: Spectroscopy and Imaging *381*
5.1.1.5	Level Structure of CdSe Quantum Rods *385*
5.1.1.6	Level Structure and Band-Offsets in Heterostructured Seeded Quantum Rods *385*
5.1.1.7	Optical Gain and Lasing in Semiconductor Nanocrystals *388*
	References *390*
5.1.2	Optical and Thermal Properties of Ib–VI Nanoparticles *392*
	Stefanie Dehnen, Andreas Eichhöfer, John F. Corrigan, Olaf Fuhr, and Dieter Fenske
5.1.2.1	Optical Spectra of Selenium-Bridged and Tellurium-Bridged Copper Clusters *392*
5.1.2.2	Thermal Behavior of Selenium-Bridged Copper Clusters *397*
	References *400*
5.2	Electrical Properties of Metal Nanoparticles *401*
	Kerstin Blech, Melanie Homberger, and Ulrich Simon
5.2.1	Introduction *401*
5.2.2	Physical Background and Quantum Size Effect *402*
5.2.2.1	Single-Electron Tunneling *403*
5.2.2.2	The Single-Electron Transistor *406*
5.2.3	Thin Film Structures *408*
5.2.4	Single-Electron Tunneling in Metal Nanoparticles *409*
5.2.4.1	STM Configurations *411*
5.2.4.2	Chemical Switching and Gating of Current Through Nanoparticles *417*
5.2.4.3	Individual Particles and 1-D Assemblies in Nanogap Configurations *421*
5.2.5	Collective Charge Transport in Nanoparticle Assemblies *435*
5.2.5.1	Two-Dimensional Arrangements *438*
5.2.5.2	Three-Dimensional Arrangements *446*
5.2.6	Concluding Remarks *450*
	References *451*
6	**Semiconductor Quantum Dots for Analytical and Bioanalytical Applications** *455*
	Ronit Freeman, Jian-Ping Xu, and Itamar Willner
6.1	Introduction *455*
6.2	Water Solubilization and Functionalization of Quantum Dots with Biomolecules *458*
6.3	Quantum Dot-Based Sensors *462*
6.3.1	Receptor- and Ligand-Functionalized QDs for Sensing *462*
6.3.2	Functionalization of QDs with Chemically Reactive Units Participating in the Sensing *470*
6.4	Biosensors *472*
6.4.1	Application of QDs for Probing Biorecognition Processes *475*
6.4.2	Probing Biocatalytic Transformations with QDs *485*

6.4.3	Probing Structural Perturbations of Proteins with QDs	496
6.5	Intracellular Applications of QDs	499
6.6	Conclusions and Perspectives	503
	References	504

7 Conclusions and Perspectives 513
Günter Schmid, on behalf of all the authors

Index 517

List of Contributors

Paul Alivisatos
University of California, Berkeley
Department of Chemistry
B-62 Hildebrand Hall
Berkeley, CA 94720-1460
USA

Uri Banin
The Hebrew University of Jerusalem
Institute of Chemistry and the Center
for Nanoscience and Nanotechnology
Jerusalem 91904
Israel

Kerstin Blech
RWTH Aachen University
Institute of Inorganic Chemistry
and JARA-Fit
Landoltweg 1
52076 Aachen
Germany

Maryna I. Bodnarchuk
Department of Chemistry
Chicago, IL 60637
USA

John F. Corrigan
University of Western Ontario
Department of Chemistry
Chemistry Building, London
Ontario, N6A 5B7
Canada

Stefanie Dehnen
Universität Marburg
Fachbereich Chemie
Hans-Meerwein-Straße
35043 Marburg
Germany

Andreas Eichhöfer
Forschungszentrum Karlsruhe
Institut für Nanotechnologie
Postfach 3640
76021 Karlsruhe
Germany

Alexander Eychmüller
TU Dresden
Physikalische Chemie und
Elektrochemie
Bergstrasse 66b
01062 Dresden
Germany

Nanoparticles: From Theory to Application. Edited by Günter Schmid
Copyright © 2010 WILEY-VCH Verlag GmbH & Co. KGaA, Weinheim
ISBN: 978-3-527-32589-4

Thomas F. Fässler
Technische Universität München
Department of Chemistry
Lichtenbergstr.
85747 Garching
Germany

Dieter Fenske
Universität Karlsruhe
Institut für Anorganische Chemie
Engesserstraße
76128 Karlsruhe
Germany

Ronit Freeman
The Hebrew University of Jerusalem
Institute of Chemistry
Jerusalem 91904
Israel

Olaf Fuhr
Forschungszentrum Karlsruhe
Institut für Nanotechnologie
Postfach 3640
76021 Karlsruhe
Germany

Nikolai Gaponik
TU Dresden
Physikalische Chemie und
Elektrochemie
Bergstrasse 66b
01062 Dresden
Germany

Daniele Gerion
Lawrence Berkeley National Laboratory
Life Science Division
Berkeley, CA 94720
USA

Melanie Homberger
RWTH Aachen University
Institute of Inorganic Chemistry and
JARA-Fit
Landoltweg 1
52076 Aachen
Germany

Liberato Manna
Istituto Italiano di Tecnologia
Via Morego 30
16163 Genova
Italy

Oded Millo
The Hebrew University of Jerusalem
Institute of Physics and the Center for
Nanoscience and Nanotechnology
Jerusalem 91904
Israel

Wolfgang Johann Parak
Philipps-Universität Marburg
Fachbereich Physik
AG Biophotonik
Renthof 7
35032 Marburg
Germany

Galyna Krylova
Argonne National Laboratory
Center for Nanoscale Materials
Argonne, IL 60439
USA

Sandra Scharfe
Technische Universität München
Department of Chemistry
Lichtenbergstr.
85747 Garching
Germany

Günter Schmid
Universität Duisburg-Essen
Institut für Anorganische Chemie
Universitätstr. 5–7
45117 Essen
Germany

Elena V. Shevchenko
Argonne National Laboratory
Center for Nanoscale Materials
Argonne, IL, 60439
USA

Friedrich C. Simmel
Technische Universität München
Physics Department - E14
Biomolecular Systems and
Bionanotechnology
James-Franck-Strasse
85748 Garching
Germany

Ulrich Simon
RWTH Aachen University
Institute of Inorganic Chemistry and
JARA-Fit
Landoltweg 1
52076 Aachen
Germany

Dmitri V. Talapin
University of Chicago
Department of Chemistry and
James Franck Institute
Chicago, IL 60637
USA

Ulrich I. Tromsdorf
University of Hamburg
Institut of Physical Chemistry
Grindelallee 117
20146 Hamburg
Germany

Horst Weller
University of Hamburg
Institut of Physical Chemistry
Grindelallee 45
20146 Hamburg
Germany

Itamar Willner
The Hebrew University of Jerusalem
Institute of Chemistry
Jerusalem 91904
Israel

Jian-Ping Xu
Zhejiang University
Department of Polymer Science and
Engineering
Key Laboratory of Macromolecular
Synthesis and Functionalization
(Ministry of Education)
Hangzhou 310027
China

1
General Introduction
Günter Schmid

Six years after the publication of the First Edition of *Nanoparticles*, in 2004, the Second Edition became necessary due to the impressive developments in the important field of nanosciences and nanotechnology. Today, the predictions made in the "General Introduction" in 2004 have, more or less, all been confirmed. In other words, developments with regards to the study and application of nanoparticles have made decisive progress, and nanotechnology in the broader sense has today become a general expression for technical progress which, in public discussions, is often used in a scientifically incorrect sense. Nevertheless, the public have become very much aware of these new techniques, and have accepted them to a great extent.

This Second Edition of *Nanoparticles: From Theory to Application* is, of course, based on the construction of the First Edition, with most of the chapters having been considerably renewed, extended or even totally rewritten, largely as the result of scientific progress made during the past six years.

The changes in Chapter 2 on "Quantum Dots" are only marginal, as the original chapter contained mainly the basic physical facts regarding the nature of nanoparticles; however, some new relevant literature has been added. Chapter 3, on "Synthesis and Characterization" begins with a new Section 3.1 on "Homoatomic and Intermetalloid Tetrel Clusters," a contribution which contains details of the latest results in the field of the famous Zintl ions (especially of Ge, Sn, and Pb), although those with endohedral transition metal atoms are also considered. Particular importance is attached to inter-cluster relationships, to form oligomeric and polymeric nanostructures. The following Sections 3.21–3.23, on "Semiconductor Nanoparticles," have been adjusted to the development of literature. In particular, those sections on Group II–VI and Group Ib–VI semiconductor nanoparticles are now complemented by the latest published results. Both, Section 3.3.1 and Section 3.3.2, on the synthesis and characterization of noble metal and magnetic nanoparticles, respectively, have consequently also been renewed and extended, following the preconditions of literature. The same is valid for Chapter 4, which deals with the "Organization of Nanoparticles." The increase in knowledge concerning "Properties," in Chapter 5, differs depending on the systems to be considered. While the progress of "Optical and Electronic Properties of Group III–V and Group

Nanoparticles: From Theory to Application. Edited by Günter Schmid
Copyright © 2010 WILEY-VCH Verlag GmbH & Co. KGaA, Weinheim
ISBN: 978-3-527-32589-4

II–VI Nanoparticles" (Section 5.1.1) is obviously limited, that of Group Ib–VI nanoparticles (Section 5.1.2) is much more marked. There has also been a considerable increase in information concerning the "Electrical Properties of Metal Nanoparticles," as can be seen from the extended Section 5.2. Finally, it must be noted that nanoscience and nanotechnology have definitely arrived in the biosciences, including medicine. Therefore, the former Chapter 6 on "Biomaterial–Nanoparticle Hybrid Systems" has been quantitatively substituted by a new chapter "Semiconductor Quantum Dots for Analytical and Bioanalytical Applications." Semiconductor quantum dots, meanwhile, have acquired a decisive role as molecular sensors and biosensors, due to their photophysical properties. Fundamental studies conducted during the past few years have demonstrated the ability of semiconductor quantum dots to act as biosensors, not only as passive optical labels as in the past but, based on the progress of molecular and biomolecular modifications, as indicators of biocatalytic transformations and conformational transitions of proteins. Comparable progress has been achieved in the field of chemical sensors, such that specific recognition ligands are now capable of sensing for ions, molecules, and macromolecules.

Altogether, this Second Edition provides an actual insight into the present situation on the development of metal and semiconductor nanoparticles.

It should be mentioned at this point that not all aspects of the world of nanoparticles can be considered in a single volume. For instance, the rapidly developing field of nanorods and nanowires has again not been considered, as these species are indeed worthy of their own monographs. The terminus "Nanoparticles," as in the First Edition, is restricted to metal and semiconductor species. Numerous other materials exist as nanoparticles, while nonmetallic and oxidic nanoparticles exist and exhibit interesting properties, especially with respect to their applications. Nevertheless, from a scientific point of view, metal and semiconductor nanoparticles play perhaps the most interesting role, at least from the point of view of the Editor.

2
Quantum Dots

Wolfgang Johann Parak, Liberato Manna, Friedrich C. Simmel, Daniele Gerion, and Paul Alivisatos

2.1
Introduction and Outline

During the past decade, new directions of modern research, broadly defined as "nanoscale science and technology," have emerged [1, 2]. These new trends involve the ability to fabricate, characterize, and manipulate artificial structures, the features of which are controlled at the nanometer level. Such trends embrace areas of research as diverse as engineering, physics, chemistry, materials science, and molecular biology. Research in this direction has been triggered by the recent availability of revolutionary instruments and approaches that allow the investigation of material properties with a resolution close to the atomic level. Strongly connected to such technological advances are the pioneering studies that have revealed new physical properties of matter at a level which is intermediate between atomic and molecular level, and bulk.

Materials science and technology is a field that is evolving at a very fast pace, and is currently making the most significant contributions to nanoscale research. It is driven by the desire to fabricate materials with novel or improved properties. Such properties might include strength, electrical and thermal conductivity, optical response, elasticity or wear-resistance. Research is also evolving towards materials that are designed to perform more complex and efficient tasks; examples include materials with a higher rate of decomposition of pollutants, a selective and sensitive response towards a given biomolecule, an improved conversion of light into current, or a more efficient energy storage system. In order for such, and even more, complex tasks to be realized, novel materials must be based on several components, the spatial organization of which is engineered at the molecular level. This class of materials – defined as "nanocomposites" – are made from assembled nanosized objects or molecules, their macroscopic behavior arising from a combination of the novel properties of the individual building blocks and their mutual interaction.

In electronics, the design and the assembly of functional materials and devices based on nanoscale building blocks can be seen as the natural, inevitable evolution of the trend towards miniaturization. The microelectronics industry, for instance, is today fabricating integrated circuits and storage media, the basic units of which are approaching the size of a few tens of nanometers. For computers, "smaller" goes along with higher computational power at lower cost and with higher portability. However, this race towards higher performance is driving current silicon-based electronics to the limits of its capability [3–6]. The design of each new generation of smaller and faster devices involves more sophisticated and expensive processing steps, as well as requiring the solution of new sets of problems, such as heat dissipation and device failure. If the trend towards further miniaturization persists, silicon technology will soon reach the limits at which these problems become insurmountable. In addition to this, it has been shown that device characteristics in very small components are strongly altered by quantum mechanical effects which, in many cases, will undermine the classical principles on which most of today's electronic components are based. For these reasons, alternative materials and approaches are currently being explored for novel electronic components, in which the laws of quantum mechanics regulate their functioning in a predictable way. Perhaps in the near future a new generation of computers will rely on fundamental processing units that are made of only a few atoms.

Fortunately, the advent of new methods for the controlled production of nanoscale materials has provided new tools that can be adapted for this purpose. New terms such as nanotubes, nanowires and quantum dots (QDs) are now the common jargon of scientific publications. These objects are among the smallest, man-made units that display physical and chemical properties which make them promising candidates as the fundamental building blocks of novel transistors. The advantages envisaged here are a higher device versatility, a faster switching speed, a lower power dissipation, and the possibility to pack many more transistors on a single chip. Currently, the prototypes of these new single nanotransistors are being fabricated and studied in research laboratories, but are far from commercialization. How millions of such components could be arranged and interconnected in complex architectures, and at low cost, remains a formidable task.

With a completely different objective, the pharmaceutical and biomedical industries have attempted to synthesize large supramolecular assemblies and artificial devices that mimic the complex mechanisms of Nature, or that can potentially be used for more efficient diagnoses and better cures for diseases. Examples in this direction are nanocapsules such as liposomes, embodying drugs that can be selectively released in living organs, or bioconjugate assemblies of biomolecules and magnetic (or fluorescent) nanoparticles that might provide a faster and more selective analysis of biotissues. These prototype systems might one day evolve into more complex nanomachines, with highly sophisticated functional features, capable of carrying out complicated tasks at the cellular level in a living body.

This chapter is not intended as a survey on the present state and future developments of nanoscale science and technology, and the above-mentioned list of examples is far from complete. Nanoscience and nanotechnology will definitely have a strong impact on human-kind in many separate areas. Mention should be made, as the most significant examples, of information technology and the telecommunications industry, and of materials science and engineering, medicine and national security. The aim of this chapter is to highlight the point that any development in nanoscience must necessarily follow an understanding of the physical laws that govern matter at the nanoscale, and how the interplay of the various physical properties of a nanoscopic system translates into a novel behavior, or into a new physical property. In this sense, the chapter will serve as an overview of basic physical rules governing nanoscale materials, with a particular emphasis on QDs, including their various physical realizations and their possible applications. Quantum dots are the ultimate example of a solid, in which all dimensions are shrunk down to a few nanometers. Moreover, semiconductor QDs are, most likely, the most studied nanoscale systems.

The chapter is structured in a comprehensive manner. In Section 2.2, an explanation is provided (with some examples) of why the behavior of nanoscale materials may differ from that of bulk and their atomic counterparts, and how quantum mechanics can help to rationalize this point. A definition is then provided of the "quantum dot." In Section 2.3, a bottom-up approach is followed to provide a simplified picture of a solid as being a very big molecule, where the energy levels of each individual atomic component have merged to form bands. The electronic structure of a QD, as being intermediate between the two extreme cases of single atoms and the bulk, will then be an easier concept to grasp. In Section 2.4, the model of a free electron gas and the concept of quantum confinement is used to explain what happens to a solid when its dimensions shrink, one by one; this will lead to a more accurate definition of the quantum well, quantum wire, and QD. In Section 2.5, the electronic structure of QDs is examined in more detail, although an attempt will be made to keep the level of discussion quite simple. Section 2.6 provides a brief overview of the most popular methods used to fabricate QDs, with various methods leading to different varieties of QDs that can be suited to specific applications. The optical properties of QDs (as discussed in Section 2.7) are unique to this class of materials, and are perhaps the most important reason why research into QDs has exploded during the past decade. The discussion here is focused more on colloidal nanocrystal QDs. Electrical and transport properties are, nonetheless, extremely relevant (as described in Section 2.8), as the addition or subtraction of a charge from a QD may lead to a dramatic modification of its electronic structure, and of the way in which the QD will handle any further addition or subtraction of a charge. This topic will be of fundamental importance for future applications in electronics.

2.2
Nanoscale Materials and Quantum Mechanics

2.2.1
Nanoscale Materials as Intermediate Between Atomic and Bulk Matter

Nanoscale materials frequently demonstrate a behavior which is intermediate between that of a macroscopic solid and that of an atomic or molecular system.

Consider for instance the case of an inorganic crystal composed of very few atoms. Its properties will be different from that of a single atom, but it cannot be imagined that they will be the same as those of a bulk solid. The number of atoms on its surface, for instance, is a significant fraction of the total number of atoms, and therefore will have a large influence on the overall properties of the crystal. It can easily be imagined that this crystal might have a higher chemical reactivity than the corresponding bulk solid, and that it will probably melt at a lower temperature. Consider now the example of a carbon nanotube (CNT), which can be thought of as a sheet of graphite wrapped in a way such that the carbon atoms on one edge of the sheet are covalently bound to the atoms on the opposite edge of the sheet. Unlike its individual components, a CNT is chemically extremely stable because the valences of all its carbon atoms are saturated. Moreover, it might be guessed that CNTs would serve as good conductors because electrons can freely move along these tiny, wire-like structures. Once again, it can be seen that such nanoscopic objects may have properties which do not belong to the realm of their larger (bulk) or smaller (atoms) counterparts. However, there are many additional properties specific to such systems which cannot be easily grasped by a simple reasoning. These properties are related to the sometimes counterintuitive behavior that charge carriers (electrons and holes) can exhibit when they are forced to dwell in such structures. These properties can only be explained by the laws of quantum mechanics.

2.2.2
Quantum Mechanics

A fundamental aspect of quantum mechanics is the particle-wave duality, as introduced by De Broglie, according to which any particle can be associated with a matter wave, the wavelength of which is inversely proportional to the particle's linear momentum. Whenever the size of a physical system becomes comparable to the wavelength of the particles that interact with such a system, the behavior of the particles is best described by the rules of quantum mechanics [7]. All of the information required about the particle is obtained by using its Schrödinger equation, the solutions of which represent the possible physical states in which the system can be found. Fortunately, quantum mechanics is not required to describe the movement of objects in the macroscopic world. The wavelength associated with a macroscopic object is, in fact, much smaller than the object's size, and therefore the trajectory of such an object can be excellently derived with the principles of classical mechanics. The situation changes, for instance, in the case of electrons orbiting around a nucleus, as their associated wavelength is of the same order of magnitude as the electron–nucleus distance.

It is possible to use the concept of particle-wave duality to provide a simple explanation of the behavior of carriers in a semiconductor nanocrystal. In a bulk inorganic semiconductor, conduction band electrons (and valence band holes) are free to move throughout the crystal, and their motion can be described satisfactorily by a linear combination of plane waves, the wavelength of which is generally of the order of nanometers. This means that, whenever the size of a semiconductor solid becomes comparable to these wavelengths, a free carrier confined in this structure will behave as a particle in a potential box [8]. The solutions of the Schrödinger equation are standing waves confined in the potential well, and the energies associated with two distinct wavefunctions are, in general, different and discontinuous. This means that the particle energies cannot take on any arbitrary value, and the system will exhibit a discrete energy level spectrum. Transitions between any two levels are seen as discrete peaks in the optical spectra, for instance; the system is then also referred to as "quantum-confined." If all the dimensions of a semiconductor crystal are shrunk down to a few nanometers, the resultant system is termed a "quantum dot" (QD), and this will form the subject of discussion throughout this chapter. The main point here is that, in order to rationalize (or predict) the physical properties of nanoscale materials – such as their electrical and thermal conductivity, or their absorption and emission spectra – there is first a need to determine their energy level structure.

For quantum-confined systems such as QDs, the calculation of the energy structure is traditionally carried out using two alternative approaches. The first approach was outlined above, whereby a bulk solid is taken and the evolution of its band structure is studied as its dimensions shrink down to a few nanometers (this method is described in more detail in Section 2.4). Alternatively, it is possible to start from the individual electronic states of single isolated atoms (as shown in Section 2.3), and then to study how the energy levels evolve as atoms come closer and begin to interact with each other.

2.3
From Atoms to Molecules and Quantum Dots

From the point of view of a chemist, the basic building blocks of matter are atomic nuclei and electrons. In an atom, electrons orbit around a single nucleus, whereby the number of electrons depends on the element. In the simplest case – the hydrogen atom – one electron orbits around one proton. The electronic states of the hydrogen atom can be calculated analytically [9, 10]. As soon as more than one electron is involved, however, the calculation of the energy levels becomes more complicated since, in addition to the interaction between nucleus and electron, also electron–electron interactions must now also be taken into account. Although the energy states of many-electron atoms can no longer be derived analytically, approximations such as the Hartree–Fock method exist [10]. In this case, each electron can be ascribed to an individual orbit, called the atomic orbital (AO), with an associated discrete energy level. Depending on the angular moment of the orbit, AOs have spherical (s-orbital), club-like (p-orbital) or a more complicated (d-,f-orbitals) shape. The eight valence electrons of a neon atom, for example, occupy one s-orbital and three p-orbitals around the nucleus, with one spin up and one spin down per orbit [10], whereby the energy level of the s-orbital is lower than that of the p-orbitals. Following the rules of quantum mechanics, the energy levels are discrete.

The next-biggest structure obtained from a combination of several atoms is the molecule. The electrons orbit collectively around more than one nucleus, and in a molecule those electrons responsible for covalent bonding between individual atoms can no longer be ascribed to one individual atom, but are "shared." In methane (CH_4), for instance, each of the four sp^3 atomic orbitals of the central carbon atom is linearly combined with the s-orbital of a hydrogen atom to form a bonding (σ) and an anti-bonding (σ^*) orbital, respectively [9]. As these orbitals are "shared" between the atoms, they are called molecular orbitals (MO) (see Figure 2.1). Only the lowest-energy (bonding) orbitals are occupied, and this explains the relative stability of methane [10]. By applying the same principle, it is possible to derive the electronic structure of more complex systems, such as large molecules or atomic clusters. When combining atoms to form a molecule, the scheme starts from the discrete energy levels of the atomic orbitals, such that ultimately discrete levels are obtained for the molecular orbitals [9].

When the size of a polyatomic system becomes progressively larger, the calculation of its electronic structure in terms of combinations of atomic orbitals becomes unfeasible [12–14]. However, simplifications arise if the system under study is a periodic, infinite crystal. The electronic structure of crystalline solids can be in fact described in terms of periodic combinations of atomic orbitals (Bloch-functions) [15, 16]. In this model, a perfect translational symmetry of the crystal structure is assumed, while contributions from the surface of the crystal are neglected by assuming an infinite solid (periodic boundary conditions) (Figure 2.2). The electrons are described as a superposition of plane waves extended throughout the solid. As opposed to the case of atoms and molecules, the energy structure of a solid no longer consists of discrete energy levels, but rather of broad energy bands [15, 16] (as shown

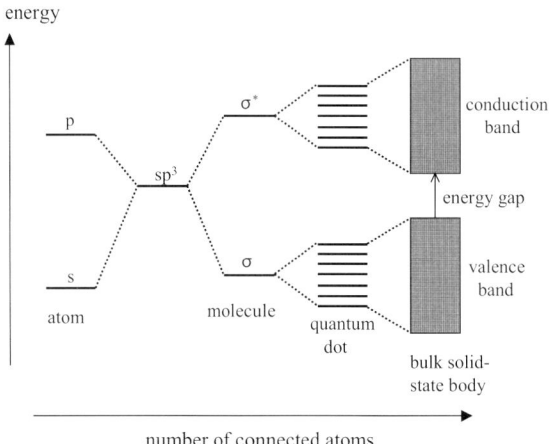

Figure 2.1 Electronic energy levels depending on the number of bound atoms. By binding more and more atoms together, the discrete energy levels of the atomic orbitals merge into energy bands (here shown for a semiconducting material) [11]. Therefore, semiconducting nanocrystals (quantum dots) can be regarded as a hybrid between small molecules and bulk material.

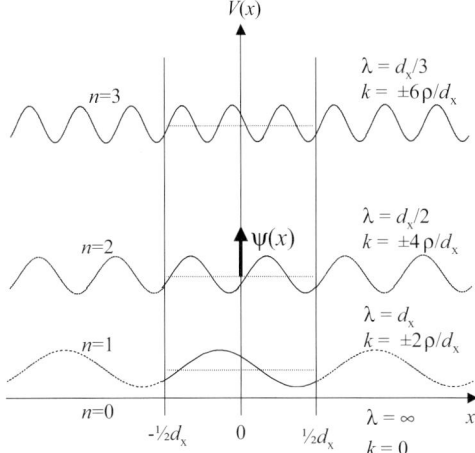

Figure 2.2 Periodic boundary conditions (drawn only for the x-dimension) for a free electron gas in a solid with thickness d. The idea of periodic boundary conditions is to mathematically "simulate" an infinite solid. Infinite extension is similar to an object without any borders; this means that a particle close to the "border" must not be affected by the border, but "behaves" exactly as in the bulk. This can be realized by using a wavefunction ψ(x) that is periodic within the thickness d of the solid. Any electron that leaves the solid from its right boundary would reenter under exactly the same conditions on its left side. For the electron the borders are quasi-nonexistent. The probability density $|\psi(x)|^2$ is the probability that an electron is at the position x in the solid. Different states for the electrons (n = 0, 1, 2, ...) have different wavefunctions. λ is the De Broglie wavelength of the electrons, and k is their corresponding wavenumber. A "real" bulk solid can be approximated by an infinite solid ($d \to \infty$) and its electronic states in k-space are quasi-continuously distributed: $\Delta k = 2\pi/d_x \to 0$.

in Figure 2.1), where every band can be filled only with a limited amount of charge carriers.

In very small crystals of nanometer dimensions, so-called "nanocrystals," the assumptions of translational symmetry and infinite size of the crystal are no longer valid, and thus these systems cannot be described with the same model used for a bulk material. Indeed, it can be imagined that the electronic structure of a nanocrystal should be something intermediate between the discrete levels of an atomic system and the band structure of a bulk solid. This can be evidenced from Figure 2.1: the energy levels of a nanocrystal are discrete, their density is much larger, and their spacing smaller than for the corresponding levels of one atom or a small atomic cluster. Because of their discrete energy levels, such structures are called QDs, although the concept of energy bands and band gap can still be used. The highest occupied atomic levels of the atomic (or ionic) species interact with each other to form the valence band of the nanoparticle. Similarly, the lowest unoccupied levels combine to form the conduction band of the particle. The energy gap between the valence and conduction bands results in the band gap of the nanoparticle. As an example, consider a metallic QD in which the level spacing at the Fermi level is roughly proportional to $\sim E_F/N$, where N is the number of electrons in the QD. Given that E_F is a few eV and that N is close to one per atom, the band gap of a metallic QD becomes observable only at very low temperatures. Conversely, in the case of semiconductor QDs, the band gap is larger and its effects can be observed at room temperature. The size-tunable fluorescence emission of CdSe QDs in the visible region of the spectrum provides a very clear illustration of the presence of a size-dependent band gap.

2.4
Shrinking Bulk Material to a Quantum Dot

In this section, a return will be made to the concept of quantum confinement of carriers in a solid, and the concept used to derive a more detailed description of the electronic band structure in a low-dimensional solid. This description, although more elaborate than that just given above, is indeed more powerful and will catch the general physics of a solid when its dimensions shrink, one by one, down to a few nanometers. Initially, an elementary model of the behavior of electrons in a bulk solid will be considered, and this will then be adapted to the case of confined carriers.

2.4.1
Three-Dimensional Systems (Bulk Material)

The first stage is to consider the case of a three-dimensional (3-D) solid with size d_x, d_y, d_z containing N free electrons (here, "free" means that the electrons are delocalized and thus not bound to individual atoms). The assumption will also be made that the interactions between the electrons, as well as between the electrons and the crystal potential, can be neglected as a first approximation; such a model system is called a "free electron gas" [15, 16]. Astonishingly, this oversimplified model still captures many of the physical aspects of real systems. From more complicated theories, it has been learnt that many of the expressions and conclusions from the free electron model remain valid as a first approximation, even when electron–crystal and electron–electron interactions are taken into account. In many cases it is sufficient to replace the free electron mass m by an "effective" mass m^* which implicitly contains the corrections for the interactions. To keep the story simple, we proceed with the free electron picture. In the free electron model, each electron in the solid moves with a velocity $\vec{v} = (v_x, v_y, v_z)$. The energy of an individual electron is then just its kinetic energy:

$$E = \frac{1}{2} m \vec{v}^2 = \left(v_x^2 + v_y^2 + v_z^2 \right) \tag{2.1}$$

According to Pauli's exclusion principle, each electron must be in a unique quantum state. Since electrons can have two spin orientations ($m_s = +1/2$ and $m_s = -1/2$), only two electrons with opposite spins can have the same velocity $\vec{v}$. This case is analogous to the Bohr model of atoms, in which each orbital can be occupied by two electrons at maximum. In solid-state physics, the wavevector $\vec{k} = (k_x, k_y, k_z)$ of a particle is more frequently used instead of its velocity to describe the particle's state.

Its absolute value $k = |\vec{k}|$ is the wavenumber. The wavevector $\vec{k}$ is directly proportional to the linear momentum $\vec{p}$, and thus also to the velocity $\vec{v}$ of the electron:

$$\vec{p} = m\vec{v} = \frac{h}{2\pi}\vec{k} \tag{2.2}$$

The scaling constant is the Plank constant h, and the wavenumber is related to the wavelength λ associated with the electron through the De Broglie relation [15, 16]:

$$\pm k = \left|\vec{k}\right| \pm \frac{2\pi}{\lambda} \tag{2.3}$$

The wavelengths λ associated with the electrons traveling in a solid are typically of the order of nanometers,[1] much smaller than the dimensions of an ordinary solid.

The calculation of the energy states for a bulk crystal is based on the assumption of periodic boundary conditions, a mathematical trick used to "simulate" an infinite ($d \to \infty$) solid. This assumption implies that the conditions at opposite borders of the solid are identical. In this way, an electron that is close to the border does not really "feel" the border; in other words, the electrons at the borders "behave" exactly as if they were in the bulk. This condition can be realized mathematically by imposing the following condition to the electron wavefunctions: $\psi(x, y, z) = \psi(x + d_x, y, z)$, $\psi(x, y, z) = \psi(x, y + d_y, z)$, and $\psi(x, y, z) = \psi(x, y, z + d_z)$. In other words, the wavefunctions must be periodic with a period equal to the whole extension of the solid [16, 17]. The solution of the stationary Schrödinger equation under such boundary conditions can be factorized into the product of three independent functions $\psi(x, y, z) = \psi(x)\psi(y)\psi(z) = A \exp(ik_x x) \exp(ik_y y) \exp(ik_z z)$. Each function describes a free electron moving along one Cartesian coordinate. In the argument of the functions $k_{x,y,z}$ is equal to $\pm n \Delta k = \pm n\, 2\pi/d_{x,y,z}$ and n is an integer number [15–17]. These solutions are waves that propagate along the negative and the positive direction, for $k_{x,y,z} > 0$ and $k_{x,y,z} < 0$, respectively. An important consequence of the periodic boundary conditions is that all the possible electronic states in the $\vec{k}$ space are equally distributed. There is an easy way of visualizing this distribution in the ideal case of a one-dimensional (1-D) free electron gas: there are two electrons ($m_s = \pm 1/2$) in the state $k_x = 0$ ($v_x = 0$), two electrons in the state $k_x = +\Delta k$ ($v_x = +\Delta v$), two electrons in the state $k_x = -\Delta k$ ($v_x = -\Delta v$), two electrons in the state $k_x = +2\Delta k$ ($v_x = +2\Delta v$), and so on.

For a 3-D bulk material it is possible to follow an analogous scheme. Two electrons ($m_s = \pm 1/2$) can occupy each of the states $(k_x, k_y, k_z) = (\pm n_x \Delta k, \pm n_y \Delta k, \pm n_z \Delta k)$, again with $n_{x,y,z}$ being an integer number. A sketch of this distribution is shown in Figure 2.3. It is possible easily to visualize the occupied states in $\vec{k}$-space because all of these states are included into a sphere, the radius of which is the wavenumber associated with the highest energy electrons. At the ground state, at 0 K, the radius of the sphere is the Fermi wavenumber k_F (Fermi velocity v_F). The Fermi energy $E_F \propto k_F^2$

---

[1] In fact, the wavelength depends on the electron density. The wavelength for electrons in metals is typically around 10 nm, but in semiconductors it may vary between 10 nm and 1 μm.

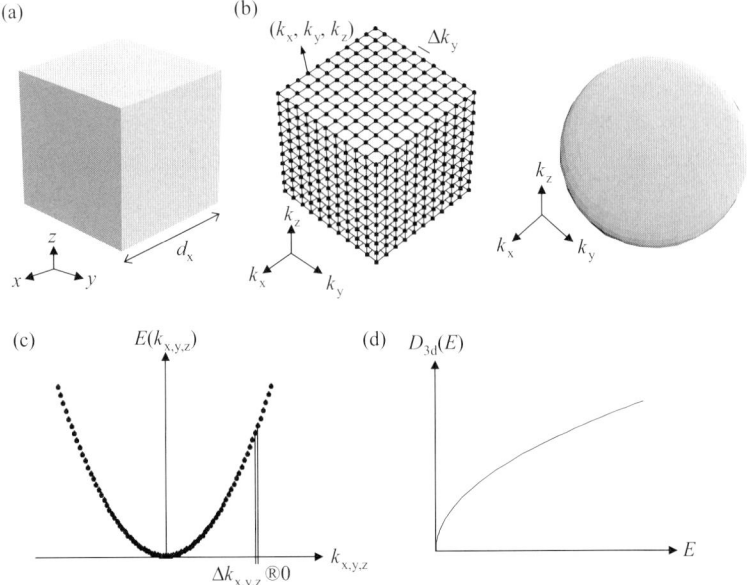

Figure 2.3 Electrons in a three-dimensional (3-D) bulk solid [16]. (a) Such a solid can be modeled as an infinite crystal along all three dimensions, x, y, and z; (b) The assumption of periodic boundary conditions yields standing waves as solutions for the Schrödinger equation for free electrons. The associated wavenumbers (k_x, k_y, k_z) are periodically distributed in the reciprocal k-space [17]. Each of the dots shown in the figure represents a possible electronic state (k_x, k_y, k_z). Each state in k-space can be only occupied by two electrons. In a large solid, the spacing $\Delta k_{x,y,z}$ between individual electron states is very small and therefore the k-space is quasi-continuously filled with states. A sphere with radius k_F includes all states with $k = (k_x^2 + k_y^2 + k_z^2)^{1/2} < k_F$. In the ground state, at 0 K, all states with $k < k_F$ are occupied with two electrons, and the other states are empty. Since the k-space is homogeneously filled with states, the number of states within a certain volume scales with k^3; (c) Dispersion relationship for free electrons in a 3-D solid. The energy of free electrons scales with the square of the wavenumber, and its dependence on k is described by a parabola. For a bulk solid, the allowed states are quasi-continuously distributed and the distance between two adjacent states (here shown as points) in k-space is very small; (d) Density of states D_{3d} for free electrons in a 3-D system. The allowed energies are quasi-continuous, and their density scales with the square root of the energy $E^{1/2}$.

is the energy of the last occupied electronic state. All electronic states with an energy $E \leq E_F$ are occupied, whereas all electronic states with higher energy $E > E_F$ are empty. In a solid, the allowed wave numbers are separated by $\Delta k = \pm n\, 2\pi/d_{x,y,z}$. In a bulk material $d_{x,y,z}$ is large, and so Δk is very small, at which point the sphere of states is filled quasi-continuously [16].

It is now necessary to introduce the useful concept of the density of states $D_{3d}(k)$, which is the number of states per unit interval of wavenumbers. From this definition, $D_{3d}(k)\Delta k$ is the number of electrons in the solid with a wavenumber between k and $k + \Delta k$. If the density of states in a solid is known, then it is possible to calculate, for instance, the total number of electrons having wavenumbers less than a given k_{max}, which is termed $N(k_{max})$. Obviously, $N(k_{max})$ is equal to $\int^{k_{max}} D_{3d}(k)dk$. In the ground state of the solid, all electrons have wavenumbers $k \leq k_F$, where k_F is the Fermi wavenumber. Since in a bulk solid the states are homogeneously distributed in $\vec{k}$-space, it is known that the number of states between k and $k + \Delta k$ is proportional to $k^2 \Delta k$ (Figure 2.3). This can be visualized in the following way. The volume in 3-D$\vec{k}$-space scales with k^3; hence, if only the number of states with a wavenumber between k and $k + \Delta k$ is to be counted, then there is a need to determine the volume of a spherical shell with radius k and thickness Δk. This volume is proportional to the product of the surface of the sphere (which scales as k^2) with the thickness of the shell (which is Δk). $D_{3d}(k)\Delta k$ is thus proportional to $k^2 \Delta k$, and in the limit when Δk approaches zero, it is possible to write:

$$D_{3d}(k) = \frac{dN(k)}{dk} \propto k^2 \qquad (2.4)$$

Instead of knowing the density of states in a given interval of wavenumbers, it is more useful to know the number of electrons that have energies between E and $E + \Delta E$. From Eqs (2.1) and (2.2) it is known that $E(k)$ is proportional to k^2, and thus $k \propto \sqrt{E}$; consequently, $dk/dE \propto 1/\sqrt{E}$. By using Eq. (2.4), it is possible to obtain for the density of states for a 3-D electron gas [17]:

$$D_{3d}(E) = \frac{dN(E)}{dE} = \frac{dN(k)}{dk} \frac{dk}{dE} \propto E \cdot 1/\sqrt{E} \propto \sqrt{E} \qquad (2.5)$$

This can be seen schematically in Figure 2.3. With Eq. (2.5), this simple description of a bulk material can be concluded. The possible states in which an electron can be found are quasi-continuous, with the density of states scaling with the square-root of the energy. Further details regarding the free electron gas model, and more refined descriptions of electrons in solids may be found in any solid-state physics textbook [15].

2.4.2
Two-Dimensional Systems

It is now time to consider a solid that is fully extended along the x and y directions, but which has a thickness along the z-direction (d_z) of only a few nanometers (see Figure 2.5). Free electrons can still move freely in the x–y-plane, but movement in the z-direction is now restricted. Such a system is called a two-dimensional electron gas (2DEG) [18]. As noted in Section 2.2, when one or more dimensions of a solid become smaller than the De Broglie wavelength associated with the free charge carriers, an additional contribution of energy is required to confine the component of the motion of the carriers along this dimension. In addition, the movement of electrons along such a direction becomes quantized; this situation is shown in Figure 2.4, where no electron can leave the solid and electrons that move in the z-direction are trapped in a "box." Mathematically this is described by infinitely high potential wells at the border $z = \pm \frac{1}{2} d_z$.

The solutions for the particle in a box situation can be obtained by solving the 1-D Schrödinger equation for an electron in a potential $V(z)$, which is zero within the box but infinite at the borders. As can be seen in Figure 2.4, the solutions are stationary waves with energies[2] $E_{nz} = \nabla^2 k_z^2/2m = h^2 k_z^2/8\pi^2 m = h^2 n_z^2/8md_z^2$, $n_z = 1, 2, \ldots$ [9, 17]. This is similar to states $k_z = n_z \Delta k_z$ with $\Delta k_z = \pi/d_z$. Again, each of these states can be occupied at maximum by two electrons.

When comparing the states in the k-space for 3-D and 2-D materials (Figure 2.3 and 2.5), for a 2-D solid that is extended in the x–y-plane, only discrete values are allowed for k_z. The thinner the solid in z-direction is, the larger the spacing Δk_z between those allowed states. On the other hand, the distribution of states in the $k_x - k_y$ plane remains quasi-continuous. Therefore, it is possible to describe the possible states in the k-space as planes parallel to the k_x- and k_y-axes, with a separation Δk_z between the planes in the k_z-direction; the individual planes can then be numbered as n_z. Since within one plane the number of states is quasi-continuous, the number of states is proportional to the area of the plane. This means that the number of states is proportional to $k^2 = k_x^2 + k_y^2$. The number of states in a ring with radius k and thickness Δk is therefore proportional to $k \cdot \Delta k$, and integration over all rings yields the total area of the plane in k-space. Here, in contrast to the case of a 3-D solid, the density of states scales linearly with k:

2) The particle-in-a-box approach (Figure 2.4) appears similar to the case of the periodic boundary conditions (Figure 2.2), but there are important differences between the two cases. Periodic boundary conditions "emulate" an infinite solid. A quantum mechanical treatment of this problem yields propagating waves that are periodic within the solid. Such waves can be seen as a superposition of plane waves. For an idealized 1-D solid, with boundaries fixed at $x = \pm d/2$, a combination of plane waves can be for instance $\psi(x) = A \cdot \exp(ikx) + B \cdot \exp(-ikx)$, with $k = n2\pi/d$. Written in another way, the solutions are of the type $\exp(ikx)$, with $k = \pm n2\pi/d$. The solutions for $k = +n2\pi/d$ and $k = -n2\pi/d$ are linearly independent. The waves $\exp(+in2\pi x/d)$ propagate to the right, and the waves $\exp(-in2\pi x/d)$ to the left side of the solid, with neither wave feeling the boundaries. Since $\exp(ikx) = \cos(kx) + i\sin(kx)$ and $\exp(-ikx) = \cos(kx) - i\sin(kx)$, we also can write $\psi(x) = C \cdot \sin(kx) + D \cdot \cos(kx)$ with $k = n2\pi/d$ as solutions. The only constraint here is that the wavefunction must be periodic throughout the solid. The state with wavenumber $k = 0$ is a solution, as $C \cdot \sin(0) + D \cdot \cos(0) = D \neq 0$. Therefore, the state with the lowest kinetic energy is $E \propto k^2 = 0$ for $k = 0$. The individual states in k-space are very close to each other because $\Delta k = 2\pi/d$ tends to 0 when d increases. On the other hand, the particle in a box model describes the case in which the motion of the electrons is confined along one or more directions. Outside the box, the probability of finding an electron is zero. For a 1-D problem, the solutions are standing waves of the type $\psi(x) = A \cdot \sin(kx)$, with $k = n\pi/d$. There is only one solution of this type: the function $\psi(x) = B \cdot \sin(-kx)$ can be written as $\psi(x) = -B \cdot \sin(kx)$, and therefore is still of the type $\psi(x) = A \cdot \sin(kx)$. Because of the boundary conditions $\psi(x = \pm d/2) = 0$, there is no solution of the type $\psi(x) = B \cdot \cos(kx)$. As the standing wave is confined to the box, there is only the solution $k = +n\pi/d > 0$. For a small box, the energy states are far apart from each other in k-space, and the distribution of states and energies is discrete. An important difference with respect to the extended solid is the occurrence of a finite zero-point energy [9]. There is no solution for $k = 0$, since $\psi(0) = A \cdot \sin(0) = 0$. Therefore, the energy of the lowest possible state ($n = 1$) is equal to $E = h^2/8md^2$, i.e. $k = \pi/d$. This energy is called "zero-point energy," and is a purely quantum mechanical effect. It can be understood as the energy that is required to "confine" the electron inside the box. For a large box, the zero-point energy tends to zero, but for small boxes this energy becomes significant, as it scales with the square of the reciprocal of the box size, d^2.

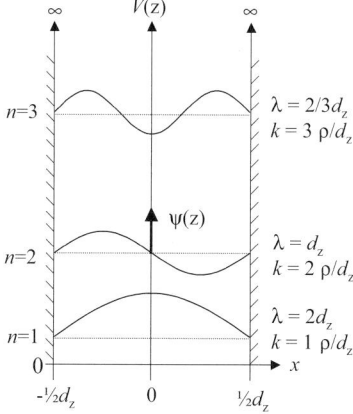

Figure 2.4 Particle in a box model for a free electron moving along in the z axis. The movement of electrons in z-direction is limited to a "box" with thickness d. Since electrons cannot "leave" the solid (the box), their potential energy V(x) is zero within the solid, but is infinite at its borders. The probability density $|\psi(z)|^2$ is the probability that an electron is located at position x in the solid. Different states for the electrons ($n = 1, 2, \ldots$) differ in their wavefunction.

$$D_{2d}(k) = \frac{dN(k)}{dk} \propto k \tag{2.6}$$

In the ground state, all states with $k \leq k_F$ are occupied with two electrons. It is now desirable to know how many states exist for electrons that have energies between E and $E + \Delta E$. The relationship between k and E: $E(k) \propto k^2$ and thus $k \propto \sqrt{E}$ and $dk/dE \propto 1/\sqrt{E}$, is known from Eqs (2.1) and (2.2). Consequently, by using Eq. (2.6), the density of states for a 2-D electron gas can be obtained (see also Figure 2.5) [17]:

$$D_{2d}(E) = \frac{dN(E)}{dE} = \frac{dN(k)}{dk} \frac{dk}{dE} \propto \sqrt{E} \cdot 1/\sqrt{E} \propto 1 \tag{2.7}$$

The density of electronic states in a 2-D solid is therefore remarkably different from the 3-D case. The spacing between the allowed energy levels in the bands increases, because fewer levels are now present. Consequently, as soon as one dimension is reduced to nanometer size, dramatic changes due to quantum confinement occur, as for example the non-negligible zero-point energy. In 2-D materials the energy spectrum remains quasi-continuous, but the density of states now is a step function [17, 19].

The quantum-mechanical behavior of electrons in a 2-D solid is the origin of many important physical effects. With recent progress in nanoscience and technology, the fabrication of 2-D structures has become routine, with such systems usually being formed at interfaces between different materials, or in layered systems in which some of the layers may be only a few nanometers thick. Structures like this can be grown, for example, by the successive deposition of individual layers using molecular

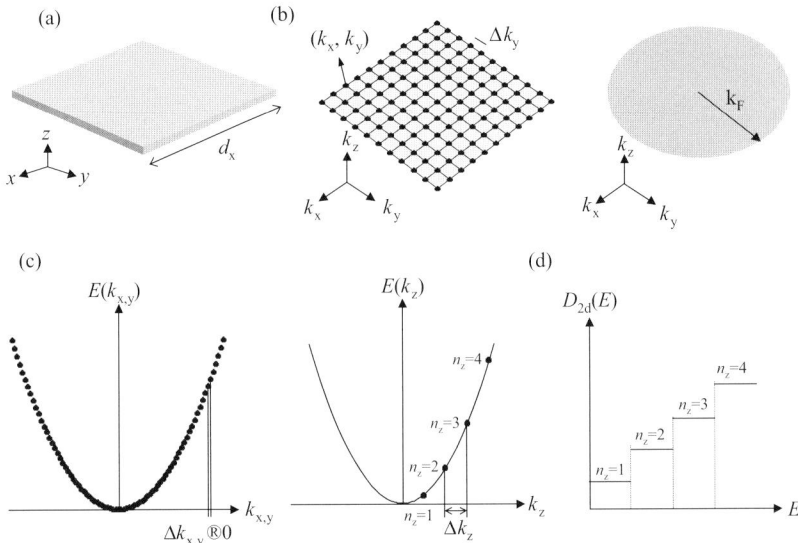

Figure 2.5 Electrons in a two-dimensional (2-D) system. (a) A 2-D solid is (almost) infinitely extended in two dimensions (here x, y), but is very thin along the third dimension (here denoted as z), which is comparable to the De Broglie wavelength of a free electron ($d_z \to \lambda$); (b) Electrons can still move freely along the x- and y- directions. The wavefunctions along such directions can be found again by assuming periodic boundary conditions. The k_x and k_y states are quasi-continuously distributed in k-space. The movement of electrons in the z-direction is restricted and electrons are confined to a "box." Only certain quantized states are allowed along this direction. For a discrete k_z-state, the distribution of states in 3-D k-space can be described as a series of planes parallel to the k_x- and k_y-axes. For each discrete k_z-state, there is a separate plane parallel to the k_x and to the k_y-axes. Here, only one of those planes is shown. The k_x- and k_y-states within one plane are quasi-continuous, since $\Delta k_{x,y} = 2\pi/d_{x,y} \to 0$. The distance between two planes for two separate k_z-states is large, since $\Delta k_z = \pi/d_z \gg 0$. For each k_z-value, the k_x- and k_y states are homogeneously distributed on the k_x–k_y-plane [17]. The number of states within this plane therefore is proportional to the area of a disk around $k_x = k_y = 0$. This means that the number of states for a certain wavenumber scales with k^2. In the ground state, all states with $k \leq k_F$ are occupied with two electrons, while the remaining states are empty; (c) Free electrons have a parabolic dispersion relation ($E(k) \propto k^2$). The energy levels $E(k_x)$ and $E(k_y)$ for the electron motion along the x- and y-directions are quasi-continuous (shown here as circles). The wavefunction $\psi(z)$ at the border of a small "box" must be zero, leading to standing waves inside the box. This constraint causes discrete energy levels $E(k_z)$ for the motion along the z-direction. Electrons can only occupy such discrete states (n_{z1}, n_{z2}, ..., shown here as circles). The position of the energy levels now changes with the thickness of the solid in z-direction, or in other words with the size of the "box"; (d) Density of states for a 2-D electron gas. If electrons are confined in one direction (z) but can move freely in the other two directions (x, y), the density of states for a given k_z-state ($n_z = 1, 2, \ldots$) does not depend on the energy E.

beam epitaxy. In such geometry, the charge carriers (electrons and holes) are able to move freely parallel to the semiconductor layer, but their movement perpendicular to the interface is restricted. The study of these nanostructures led to the discovery of remarkable 2-D quantized effects, such as the Integer and the Fractional Quantum Hall Effect [20–23].

2.4.3
One-Dimensional Systems (Quantum Wires)

If the case is considered where the solid also shrinks along a second (y) dimension, the electrons can move freely only in the x-direction, while their motion along the y- and z-axes is restricted by the borders of the solid (see Figure 2.6). Such a system is called a *quantum wire* or – when the electrons are the charge carriers – a one-dimensional electron system (1DES). The charge carriers and excitations now can move only in one dimension, while occupying quantized states in the other two dimensions.

The states of a 1-D solid now can be obtained by methods analogous to those described for the 3-D and 2-D materials. In the x-direction, the electrons can move freely and again the concept of periodic boundary conditions can be applied. This gives a quasi-continuous distribution of states parallel to the k_x-axis and for the corresponding energy levels. The electrons are confined along the remaining directions, and their states can be derived from the Schrödinger equation for a particle in a box potential; again, this yields discrete k_y and k_z-states. It is now possible to visualize all states as lines parallel to the k_x-axis. The lines are separated by discrete intervals along k_y and k_z, but within one line the distribution of k_x states is quasi-continuous (Figure 2.6). The number of states along one line can be counted by measuring the length of that line, and the number of states is therefore proportional

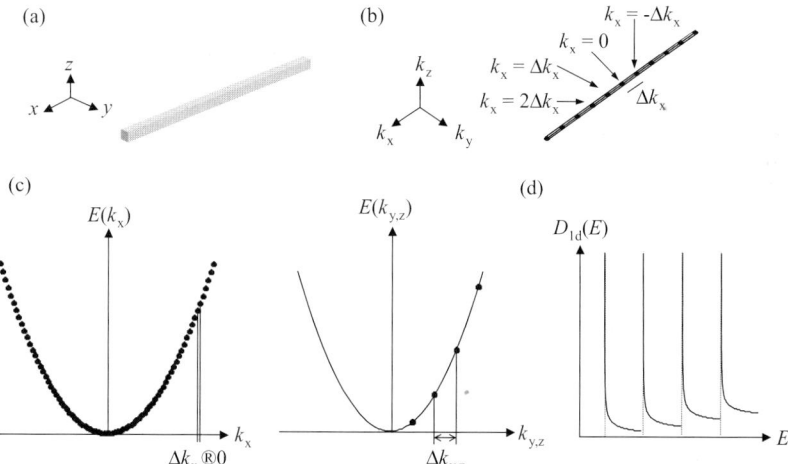

Figure 2.6 (a) A one-dimensional solid; (b) The allowed (k_x, k_y, k_z)-states can be visualized as lines parallel to the k_x-axes in the 3-D k-space. In this figure, only one line is shown as an example. Within each line, the distribution of states is quasi-continuous, since $\Delta k_x \rightarrow 0$. The arrangement of the individual lines is discrete, since only certain discrete k_y- and k_z- states are allowed; (c) This can also be seen in the dispersion relations. Along the k_x-axes the energy band $E(k_x, k_y, k_z)$ is quasi-continuous, but along the k_y- and k_z-axes only certain energies exist; (d) The density of states within one line along the k_x-axes is proportional to $E^{-1/2}$. Each of the hyperbolas shown in the D_{1d}-diagram corresponds to an individual (k_y, k_z)-state.

to $k = k_x$. Hence, the number of states with wavenumbers in the interval between k and $k + \Delta k$ is proportional to Δk:

$$D_{1d}(k) = \frac{dN(k)}{dk} \propto 1 \qquad (2.8)$$

In the ground state, all states with $k \leq k_F$ are occupied with two electrons. From Eqs (2.1) and (2.2) it is possible to determine the relationship between k and E for free electrons: $E(k) \propto k^2$ and thus $k \propto \sqrt{E}$ and $dk/dE \propto 1/\sqrt{E}$. By using Eq. (2.8), the density of states for a 1-D electron gas can be obtained:

$$D_{1d}(E) = \frac{dN(E)}{dE} = \frac{dN(k)}{dk}\frac{dk}{dE} \propto 1 \cdot 1/\sqrt{E} \propto 1/\sqrt{E} \qquad (2.9)$$

The density of states is depicted in Figure 2.6. In 1-D systems the density of states has a $E^{-1/2}$-dependence, and thus exhibits singularities near the band edges [17]. Each of the hyperbolas contains a continuous distribution of k_x states, but only one discrete k_y- and k_z-state.

The quantization of states in two dimensions has important consequences for the transport of charges. Whilst electrons can flow freely along the x-axes, they are limited to discrete states in the y- and z-directions, and therefore are only transported in discrete "conductivity channels." This may be of considerable importance for the microelectronics industry since, if the size of electronic circuits is reduced continually, at a certain point the diameter of wires will become comparable to the De Broglie wavelength of the electrons, and the wire will then exhibit the behavior of a quantum wire. Quantum aspects of 1-D transport were first observed in so-called quantum point contacts, which were defined lithographically in semiconductor heterostructures [24, 25]. More recent examples of such 1-D wires have included short organic semiconducting molecules [26–31], inorganic semiconductor and metallic nanowires [32–37], or break junctions [38–40]. One particular role is played by CNTs [27, 41–47], which have undergone extensive investigations both as model systems for 1-D confinement and for potential applications, such as electron-emitters [48].

2.4.4
Zero-Dimensional Systems (Quantum Dots)

When charge carriers and excitations are confined in all three dimensions the system is called a QD. The division is somewhat arbitrary since, for instance, clusters composed of very few atoms are not necessarily considered as QDs. Although clusters are smaller than the De Broglie wavelength, their properties depend critically on their exact number of atoms. Large clusters have a well-defined lattice and their properties no longer depend critically on their exact number of atoms. With the term QDs, reference will be made to such systems [49, 50].

In a QD, the movement of electrons is confined in all three dimensions, and there are only discrete (k_x, k_y, k_z)-states in the k-space. Each individual state in the k-space can be represented by a point, the final consequence being that only discrete energy

2.5 Energy Levels of a (Semiconductor) Quantum Dot

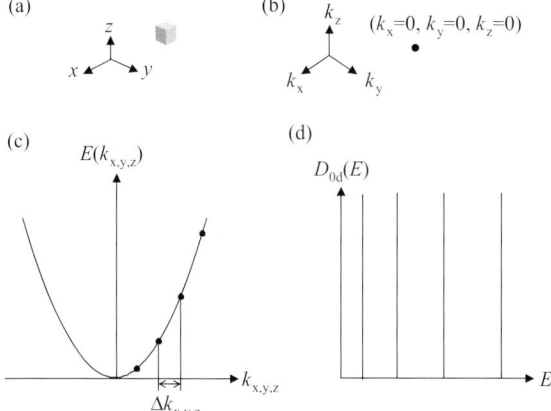

Figure 2.7 A zero-dimensional (1-D) solid. (a) The solid is shrunk in all three dimensions to a thickness that is comparable to the De Broglie wavelength of its charge carriers; (b) Because of such confinement, all states (k_x, k_y, k_z) are discrete points in the 3-D k-space; (c) Only discrete energy levels are allowed; (d) The 1-D density of states $D_{0d}(E)$ contains delta peaks, that correspond to the individual states. Electrons can occupy only states with these discrete energies.

levels are allowed, which appear as delta-peaks in the distribution $D_{0d}(E)$ (Figure 2.7). As can be seen, the energy bands converge to atom-like energy states, with the oscillator strength compressed into a few transitions; this change is most dramatic at the edges of the bands, and influences semiconductors more than metals. In semiconductors, the electronic properties are in fact strongly related to the transitions between the edges of the valence band and the conduction band, respectively. In addition to the discreteness of the energy levels, there is a need to stress again the occurrence of a finite zero-point energy. In a QD – even in the ground state – electrons have energies that are larger than those of bulk electrons at the conduction band edge; these points are discussed in greater detail in Section 2.5.

2.5
Energy Levels of a (Semiconductor) Quantum Dot

As many quantum effects are more pronounced in semiconductors compared to metals, attention will now be focused on the case of a semiconducting material. The changes that occur in the properties of a free electron gas change when the dimensions of the solid are reduced were described in Section 2.4. Although the model of the free electron gas does not include the "nature" of the solid, from a macroscopic point of view it is necessary to distinguish between metals, semiconductors, and insulators [15]. Whilst the model of a free electron gas describes relatively well the case of electrons in the conduction band of metals, the electrons in an insulating material are only poorly described by the free electron model. In order to extend the model of free electrons to semiconducting materials, the concept of a new charge carrier – the hole – was introduced [16]. If an electron from the valence band is excited to the conduction band, the "empty" electronic state in the valence band is called a "hole." Some basic properties of semiconducting materials can be described by the model of free electrons and free holes, where the energy bands for electrons and holes are separated by a band gap [15, 16]. At first approximation, the dispersion relationships for the energy of electrons and holes in a semiconductor are parabolic, but this holds true only for those electrons (holes) occupying levels that lie at the bottom (top) of the conduction (valence) band. Each parabola represents a quasi-continuous set of electron (hole) states along a given direction in k-space. The lowest unoccupied energy band and the highest occupied energy band are separated by an energy gap E_g(bulk) (see Figure 2.8) which, for a bulk semiconductor, can range from a fraction of an electron-volt (eV) up to a few eV.

It might be expected that the energy dispersion relationships are still parabolic in a QD. However, as only discrete energy levels can exist in a QD, each of the original parabolic bands of the bulk case is now fragmented into an ensemble of points. The energy levels of a QD can be estimated using the particle-in-a-box model. As

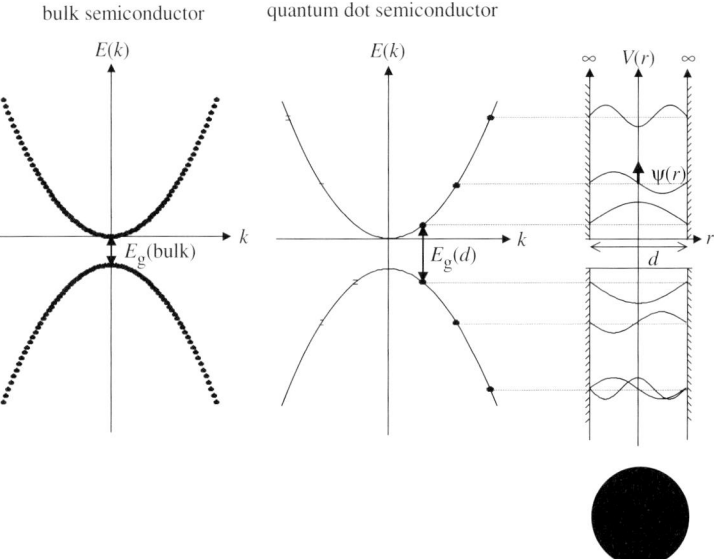

Figure 2.8 Free charge carriers in a solid have a parabolic dispersion relation ($E(k) \propto k^2$). In a semiconductor, the energy bands for free electrons and holes are separated by an energy gap E_g. In a bulk semiconductor, the states are quasi-continuous, and each point in the energy bands represents an individual state. In a quantum dot (QD), the charges are confined to a small volume. This situation can be described as a charge carrier confined in an infinite potential well of width d. Here, the width d of the potential well corresponds to the diameter of the QD. The only allowed states are those whose wavefunctions vanish at the borders of the well [7]. This leads to discrete energy levels [7, 10]. The energy gap between the lowest possible energy level for electrons and holes $E_g(d)$ is larger than that of a bulk material E_g(bulk).

described above (see Figure 2.4), the lowest energy for an electron in a 1-D potential well is:

$$E_{\text{well},1d} = 1/8\, h^2/md^2 \tag{2.10}$$

where d is the width of the well. In a QD, the charge carriers are confined in all three dimensions, and this system can be described as an infinite 3-D potential well. The potential energy is zero everywhere inside the well, but is infinite on its walls; such a well may also be referred to as a "box." The simplest shapes for a 3-D box may be, for instance, a sphere or a cube. If the shape is cubic, then the Schrödinger equation can be solved independently for each of the three translational degrees of freedom, and the overall zero-point energy will simply be the sum of the individual zero point energies for each degree of freedom [9, 51]:

$$E_{\text{well},3d(\text{cube})} = 3\, E_{\text{well},1d} = 3/8\, h^2/md^2 \tag{2.11}$$

If the box is a sphere of diameter d, then the Schrödinger equation can be solved by introducing spherical coordinates and by separating the equation in a radial part, and also in a part that contains the angular momentum [52, 53]. The lowest energy level (with angular momentum = 0) is then:

$$E_{\text{well},3d(\text{sphere})} = 1/2\, h^2/md^2 \tag{2.12}$$

The effect of quantum confinement is again remarkable. More confined charge carriers lead to a larger separation between the individual energy levels, as well as to a greater zero-point energy. If carriers are confined into a sphere of diameter d, then the zero-point energy will be higher than that for charges that are confined to a cube with an edge length equal to d ($E_{\text{well},3d(\text{sphere})} > E_{\text{well},3d(\text{cube})}$); this is simply because such a sphere has a smaller volume ($\pi/6\, d^3$) than the cube (d^3).

An electron–hole pair may be generated in the QD, for instance by a photoinduced process or by charge injection. The minimum energy E_g required to create an electron–hole pair in a QD is composed of several contributions, one of which is the bulk band gap energy, $E_g(\text{bulk})$. Another important contribution is the confinement energy for the carriers, termed call $E_{\text{well}} = E_{\text{well}}(e^-) + E_{\text{well}}(h^+)$. For large particles (bulk: $d \to \infty$), E_{well} tends to zero. It is possible to estimate the overall confinement energy for an electron–hole pair in a spherical QD; this is the zero-point energy of the potential well or, in other words, the energy of the state of a potential box with the lowest energy. This can be written as:

$$E_{\text{well}} = h^2/2m^*d^2 \tag{2.13}$$

where m^* is the reduced mass of the exciton and is given by [54]:

$$1/m^* = 1/m_e + 1/m_h. \tag{2.14}$$

where m_e and m_h are the effective masses for electrons and holes, respectively. In order to calculate the energy required to create an electron–hole pair, another term – the Coulomb interaction (E_{Coul}) – must be considered. This term takes into account the mutual attraction between the electron and the hole, multiplied by a coefficient that describes the screening by the crystal. In contrast to E_{well}, the physical content of this term can be understood within the framework of classical electrodynamics. However, an estimate of such a term is possible only if the wavefunctions for the electron and the hole are known. The strength of the screening coefficient depends on the dielectric constant ε of the semiconductor. An estimate of the coulomb term yields:

$$E_{Coul} = -1.8\, e^2/2\pi\varepsilon\varepsilon_0 d \qquad (2.15)$$

This term may be quite significant, because the average distance between an electron and a hole in a QD dot can be small [55–59]. Thus, it is possible to estimate the size-dependent energy gap of a spherical semiconductor QD, as [54–59]:

$$E_g(dot) = E_g(bulk) + E_{well} + E_{Coul} \qquad (2.16)$$

Then, by inserting Eqs (2.13) and (2.15) into Eq. (2.16), the following is obtained:

$$E_g(d) = E_g(bulk) + h^2/2m^* d^2 - 1.8\, e^2/2\,\pi\varepsilon\varepsilon_0 d \qquad (2.17)$$

where the size-dependence in each term has been emphasized. Equation (2.17) is only a first approximation; rather, many effects – including the crystal anisotropy and spin–orbit coupling – must be considered in a more sophisticated calculation. The basic approximation for the band gap of a QD comprises two size-dependent terms: (i) the confinement energy, which scales as $1/d^2$; and (ii) the Coulomb attraction, which scales as $1/d$. As the confinement energy is always a positive term, the energy of the lowest possible state is always raised with respect to the bulk situation. On the other hand, the Coulomb interaction is always attractive for an electron–hole pair system, and therefore lowers the energy. Because of the $1/d^2$ dependence, the quantum confinement effect becomes the predominant term for very small QD sizes (Figure 2.9).

The size-dependent energy gap may serve as a useful tool when designing materials with well-controlled optical properties, and a much more detailed analysis of this topic is available [62]. Before describing the physical consequences of the size-dependent band gap on the optical and electronic properties, however, a brief overview will be provided of how QDs may be fabricated in practice.

2.6
Varieties of Quantum Dots

The ultimate fabrication technique of QDs should produce significant amounts of sample, with such a high control of QD size, shape and monodispersity that single-particle properties would not be affected by sample inhomogeneity. To date, ensembles of QDs produced by the best available techniques still show a behavior that derives from a distribution of sizes, although this field is evolving very rapidly. In the fabrication of QDs, different techniques may lead to different typologies. Notably,

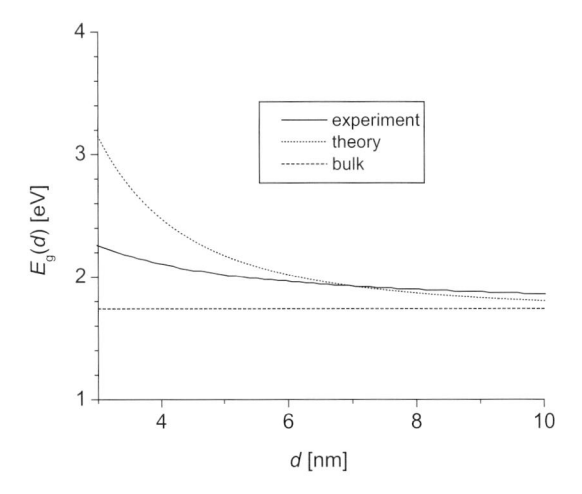

Figure 2.9 Size dependence of the energy gap $E_g(d)$ for colloidal CdSe QDs with diameter d. The bulk value for the energy gap is $E_g(\text{bulk}) = 1.74$ eV [60]. The theoretical curve was obtained using Eq. (2.17) with the following parameters: effective mass of electrons/holes $m_e = 0.13\, m_0$, $m_h = 0.4\, m_0$, $m_0 =$ mass of free electrons ($m = 9.1095 \times 10^{-31}$ kg^{-1}) $\Rightarrow$ $m^* = 0.098\, m$ [60]; dielectric constant $\varepsilon_{CdSe} = 5.8$ [61], permittivity constant $\varepsilon_0 = 8.854 \times 10^{-12}$ C^2 N^{-1} m^{-2}, Planck's constant $h = 6.63 \times 10^{-34}$ J·s, 1 eV $= 1.602 \times 10^{-19}$ J. The experimental values were obtained by the Alivisatos group by recording the absorption spectra of CdSe QDs of different sizes and determining their size using transmission electron microscopy.

confinement may be achieved in several ways, while the QD itself may have a peculiar geometry; for example, it may be embedded into a matrix, grown onto a substrate, or it may be a "free" nanoparticle. Each of these cases is strictly related to the preparative approach chosen.

2.6.1
Lithographically Defined Quantum Dots

Lithographically defined QDs are formed by isolating a small region of a 2-D electron system (see Section 2.4.2) by creating tunneling barriers from its environment. These so-called two-dimensional electron systems (2-DES) and two-dimensional electron gases (2-DEG) are found in metal-oxide-semiconductor field effect transistors (MOSFETs), and also in so-called "semiconductor heterostructures" [17, 18]. The latter are composed of several thin layers of different semiconductor materials grown on top of each other, using the technique of molecular beam epitaxy (MBE). In this case, the layer sequence can be chosen in such a way that all free charge carriers are confined to a thin slice of the crystal, forming essentially a 2-D electron system. A superstructure derived from the periodic repetition of this sequence of layers is termed a "multiple quantum well." One of the most widely investigated systems has been the aluminum gallium arsenide/gallium arsenide (AlGaAs/GaAs) quantum well. AlGaAs has the same lattice constant as GaAs but a wider band gap, the exact

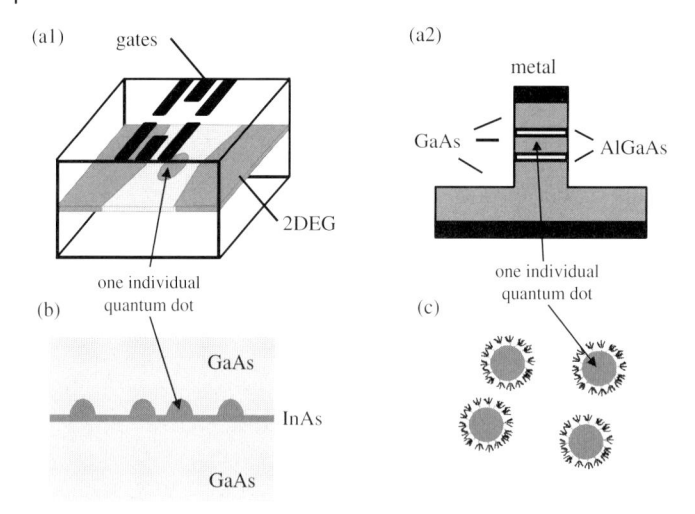

Figure 2.10 Three different types of quantum dot. (a1) A lithographically defined QD in lateral arrangement can be formed by electrostatic depletion of a 2-D electron gas (2DEG, shown in dark gray) via gate electrodes. The 2DEG is formed typically 20–100 nm below the surface of a semiconductor heterostructure (usually GaAs/AlGaAs). The application of negative voltages to metal gates on top of the heterostructure depletes the 2DEG below the gates (shown in light gray), and cuts out a small electron island from the 2DEG. Electrons can still tunnel onto and from the island. Electrical contact to the 2DEG is realized through Ohmic contacts (not shown); (a2) A vertical QD can be formed in a double-barrier heterostructure. A narrow pillar is etched out of a GaAs/AlGaAs/GaAs/AlGaAs/GaAs heterostructure. The AlGaAs layers (light gray) form tunnel barriers that isolate the central GaAs region from the contact region. This central GaAs region behaves now as a QD (shown in dark gray). Electrical contact is made via metal contacts (depicted in black) on top of the pillar and below the heterostructure; (b) Self-assembled QDs: molecular beam epitaxy (MBE) growth of InAs (dark gray) on GaAs (light gray) first leads to the formation of an extended layer of InAs (the wetting layer), and then to the formation of small InAs islands. Single electrons or electron–hole pairs (excitons) can be confined into these InAs QDs, either electrically or optically; (c) Colloidal QDs. These colloidal particles, having a diameter of only a few nanometers, are formed using wet chemistry and can be produced for most of the II-VI, III-V, IV-VI and some type IV semiconductors. The surface of colloidal QDs is coated with a layer of surfactant molecules that prevents aggregation of the particles.

value of which depends on the aluminum content of the layer. As a result, electrons in the GaAs layer are confined to this layer and form a 2-D electron gas.

Currently, QD systems can be generated in either a lateral or a vertical arrangement, as shown in Figure 2.10a. In the lateral geometry, the 2-DEG is locally and electrostatically depleted by applying a negative voltage to the electrodes deposited on top of the crystal. This effect can be understood by applying the following argument. If it is assumed that a negative voltage is applied to the metal electrodes above the 2-DEG then, because of electrostatic repulsions, electrons will be repelled by the electric field of the electrodes and the region of the 2-DEG below the electrodes will be depleted of electrons. As a charge-depleted region behaves like an insulator, the application of an electric field with metal electrodes of an appropriate shape permits the creation of an island of charges that is insulated from the remainder of the 2-DEG and which, if small enough, will behave as a QD. In the vertical geometry, a small pillar of the 2-DEG can be isolated by etching away the surrounding heterostructure. In such an arrangement, the charge carriers will again be confined in all three dimensions.

Most electron transport measurements on QDs performed to date have used the two types of QD described above. The lateral arrangement offers a relatively high degree of freedom for the structure design, which is determined by the choice of electrode geometry. It is also possible to fabricate and study "artificial molecules" [63–69] that are composed of several QDS, linked together. By using the vertical arrangement, structures with very few electrons can be achieved [70] and, indeed, much effort is currently being expended on investigations into many-body phenomena with these QD systems. Relevant examples include studies of the Kondo effect [71–74], as well as the design and control of coherent quantum states with the ultimate goal of quantum information processing (for further information, see Section 2.8).

One remarkable advantage of lithographically defined QDS is that their electrical connection to the "macro-world" is straightforward, with the manufacturing processes used being similar to those employed for chip fabrication. In principle, such structures could be embedded within conventional electronic circuits, although as the geometry of these QDs is determined lithographically it is restricted by the usual size and resolution limits of lithographic techniques. Even the use of electron-beam lithography (EBL) to create QDs does not permit their size to be tailored with nanometer precision. Moreover, as lithographically fabricated QDS are typically larger than 10 nm in size, relatively low lateral confining energies can be achieved.

2.6.2
Epitaxially Self-Assembled Quantum Dots

A major breakthrough in the field of epitaxially grown nanostructures came with the discovery of epitaxial growth regimes that favored the formation of nanometer-sized islands of semiconductor materials on suitable substrates. These islands, which exhibit QD behavior, can be obtained naturally by growing epitaxially a thin layer of a low-band gap material over a higher-band gap material, using either MBE or metalorganic chemical vapor deposition (MOCVD) techniques [50, 75–77]. The respective crystal faces in contact must have a significant lattice mismatch (1–8%), as in the case of InAs on GaAs [78, 79] and Ge on Si [80, 81]. During the growth, a strained film – the "wetting layer" – is initially formed, the maximum thickness of which is related to the difference in lattice constants between the two materials. Beyond this critical thickness, a 2-D $\to$ 3-D transition in the growth regime is observed, with the spontaneous formation of an array of nanometer-sized islands (the Stranski–Krastanov regime), leading to a partial release of the strain. If the growth is not interrupted at this step, then misfit dislocations will form because the energy of formation of these defects becomes smaller than the elastic energy that accumulates in the strained film. The formation of dislocations in highly strained epilayers (when the lattice mismatch is of the order of 10%, or more) before island formation limits the range of possible substrate-island materials. The shape of the islands can also be controlled by varying the growth conditions. Although normally the islands have a truncated pyramidal shape, it is also possible to form ring-shaped QDS [82]. The final step consists of the growth (on top of the islands) of several layers of the substrate material, so that the QDs are completely buried and the interfaces passivated. The relative alignment of the band gaps creates a confining potential for charge carriers that accumulates inside the QDs. In addition, strain fields in the proximity of the island–substrate interface, caused by the lattice mismatch between the two materials, lead to the creation of potentials that modify the band gap of the QDS at the bottom of the island. Notably, holes are more likely to be localized in this region, as they are heavier than electrons.

Since self-assembled QDS may have a diameter as small as a few nanometers, very pronounced quantum size effects may be observed in these systems. Previously, self-assembled QDs have been characterized predominantly by using optical or capacitance spectroscopy in a regime where they only contain a small number of charge carriers. Although measurements on ensembles still suffer from the inhomogeneous broadening of spectroscopic features, in recent years it has become possible to examine only a few or even single self-assembled QDs, by reducing the number of QDs via mesa-etching [83] or with confocal microscopy techniques [84]. Photoluminescence from single self-assembled QDs is a highly efficient process that is characterized by several narrow emission lines that are related to different exciton states in the dots, and is reminiscent of the emission from atoms. As noted above for lithographically defined QDs, many parallels can be drawn between atoms and QDs [83, 85–87], and it is for this reason that QDs have been nicknamed "artificial atoms." Much of the current research effort has been devoted to QD ordering and positioning, as well as to the reduction of QD size distribution. In contrast to lithographically defined QDs, the challenge remains to make electrical contact with self-assembled QDS, such that the most likely applications will be found in the field of optics. A major goal of research into self-assembled QDS is to fabricate nonclassical light sources from single dots; another is to use the dots as light-addressable storage devices (see Section 2.7).

2.6.3
Colloidal Quantum Dots

Colloidal QDS differ remarkably from the QD systems described above, as they are chemically synthesized using wet chemistry, and are free-standing nanoparticles or nanocrystals grown in solution [88]. In this case, colloidal QDs simply represent a subgroup of a broader class of materials that can be synthesized at the nanoscale, using wet chemical methods. In the fabrication of colloidal nanocrystals, the reaction chamber contains a liquid mixture of compounds that control nucleation and growth. In a general synthesis of QDs in solution, each of the atomic species that will form the nanocrystal building blocks is introduced into the reactor as a precursor. (A precursor is a molecule or a complex containing one or more atomic species required for growing the nanocrystal.) Having been introduced into the reaction flask, the precursors will decompose to form new reactive species (the monomers) that will, in turn, cause nucleation and nanocrystal growth. The energy required for precursor decomposition is provided by the liquid in the reactor, either by thermal collisions or by a chemical reaction between the liquid medium and the precursors, or by a combination of these two mechanisms [89].

The key parameter in the controlled growth of colloidal nanocrystals is the presence of one or more molecular species in the reactor, which are broadly termed "surfactants." (A surfactant is a molecule that, under the reaction conditions, is dynamically adsorbed to the surface of the growing QD.) Surfactants must be mobile enough to provide access for the addition of monomer units, but stable enough to prevent nanocrystal aggregation. The choice of surfactant varies from case to case. For example, a molecule that binds too strongly to the surface of the QD would be unsuitable as it would not allow the nanocrystal to grow. In contrast, a weakly coordinating molecule would yield large particles, or aggregates [90]. Examples of suitable surfactants include alkyl thiols, phosphines, phosphine oxides, phosphates, phosphonates, amides or amines, carboxylic acids, and nitrogen-containing aromatics. If the nanocrystal growth is carried out at high temperatures (e.g., 200–400 °C), it is essential that the surfactant molecules are stable under such conditions in order to exert control over the crystal growth.

At low temperatures, or more generally when growth has ceased, the surfactants will be more strongly bound to the surfaces of the nanocrystals, the solubility of which will be increased in a wide range of solvents. This coating allows for a synthetic flexibility, in that it can be exchanged for another coating of organic molecules with different functional groups or polarity. The surfactants may also be temporarily removed, and an epitaxial layer of another material with different electronic, optical, or magnetic properties then grown on the initial nanocrystal [91, 92].

By controlling the mixture of surfactant molecules present during the generation and growth period of the QDS, an excellent control is possible of both their size and shape [89, 93, 94].

As colloidal nanocrystals are dispersed in solution, they are not bound to any solid support, as is the case for the two QD systems described above. Consequently, they can be produced in large quantities in a reaction flask, and later transferred to any desired substrate or object. For example, it is possible to coat the crystal surface with biological molecules (e.g., proteins or oligonucleotides) that are capable of performing tasks of molecular recognition with extremely high efficiency. In particular, ligand molecules may bind with very high specificity to certain receptor molecules, similar to a lock-and-key system. Hence, if a colloidal QD is tagged with ligand molecules, it will bind specifically to all positions where receptor molecules are present. This has led to the creation of small groups of colloidal QDs that are not only mediated by molecular recognition [95–97] but also have the ability to label specific cell compartments with different types of QD [98–100].

Although colloidal QDs are difficult to connect electrically, a number of electron-transport experiments have been conducted in which nanocrystals have been used as

2.7
Optical Properties of Quantum Dots

2.7.1
Absorption and Emission Spectra

In chemistry, size-dependent optical properties of colloidal semiconductor particles have been observed since the early twentieth century (e.g., in CdS colloids [103]), although it was only during the 1980s [57] that this was rationalized in terms of "size quantization." As discussed in Section 2.5, the most striking effect in semiconductor nanoparticles is the widening of the gap E_g between the highest occupied electronic states (the top of the original valence band) and the lowest unoccupied states (the bottom of the original conduction band) [104]. This has a direct effect on the optical properties of QDs as compared to the corresponding bulk material. The minimum energy needed to create an electron–hole pair in a QD (an exciton) is defined by its band gap (E_g), as light with energy lower than E_g cannot be absorbed by the QD. As the band gap depends on the size of a QD, the onset of absorption is also size-dependent [55, 89]. That smaller QDs have an absorption spectrum that is shifted to shorter wavelengths with respect to larger QDs and to the bulk case, is shown in Figure 2.11.

Excitons in semiconductors have a finite lifetime due to a recombination of the photo-excited electron–hole pair. In QDs, the energy released upon exciton annihi-

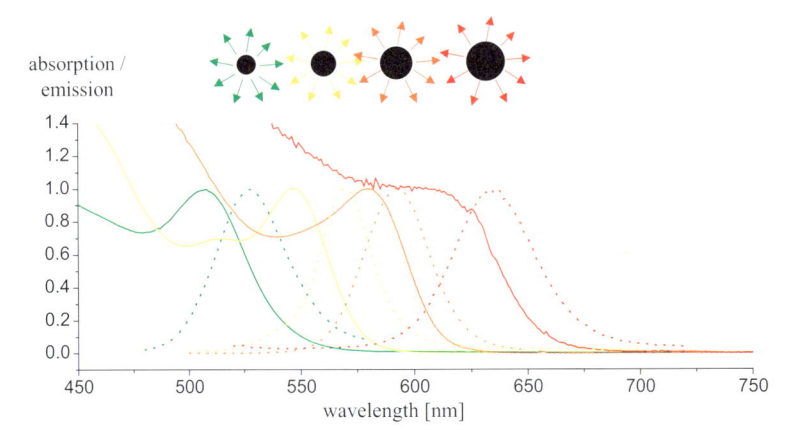

Figure 2.11 Absorption (solid lines) and emission spectra (dotted lines) of colloidal CdSe QDs of different sizes. The absorption peak of green/yellow/orange/red fluorescent nanocrystals of 2.3/4.0/3.8/4.6 nm diameter is at 507/547/580/605 nm, while the fluorescence peaks are at 528/57/592/637 nm.

lation is too large to be dissipated by vibrational modes; instead, it is released in the form of emitted photons. A radiative decay through the emission of photons – in other words, fluorescence – is a highly probable decay channel in QDs [105].

As is the case of organic fluorophores, the range of energies emitted from a colloidal dot sample after excitation is centered at a value that is smaller than that required to excite the sample (and which must be at least as large as its band gap). In other words, the wavelength of fluorescence is longer than that of the absorbed light. The shift between the lowest energy peak in the absorption spectrum of a QD and the corresponding emission peak is termed the "Stokes shift" [9] (see Figure 2.11), the explanation of which requires a more advanced model than is presented in Figure 2.8.

In QDs, the Stokes shift is explained by examining the complex exciton structure [106–109]. Indeed, more complex theoretical models and calculations have shown that the ground state of an exciton in a dot has a total angular momentum equal to zero. In the dipole approximation, the creation of an exciton through the absorption of a photon leads to an exciton state having an angular momentum equal to ± 1. The energy needed for this excitation is the *absorption energy*. In QDs, this excited state then relaxes very rapidly to a state with an angular momentum of 2; such relaxation is nonradiative. In the first order, this state cannot relax to the ground state with angular momentum 0 by emitting a photon, because only transitions that change the angular momentum by ± 1 are allowed; as no photon can be emitted in first order, this state is referred to as a "dark exciton." However, slight perturbations of the crystal lattice and even a weak coupling with phonons will allow this state to relax under the emission of a photon, which results in the decay time of the fluorescence being long and the fluorescence energy being red-shifted with respect to the absorption band edge energy [110]. The dark exciton model has been supported by many experimental data, and in particular by magnetic field-dependent lifetime measurements. The position of the luminescence peak is also dependent on the average QD size, while the peak width is correlated to the nanocrystals size distribution (Figures 2.11 and 2.12). Consequently, the maximum of the emission spectrum and its width can be used to estimate the mean size and the size distribution during nanocrystal growth.

2.7.2
Spectral Diffusion and Blinking

Several crucial differences are apparent between self-assembled and colloidal QDs. For example, stable, ultra-narrow fluorescence peaks have been observed in the emission spectra of single, self-assembled QDs, whereas colloidal semiconductor QDs may have a very narrow size distribution, as observed using transmission electron microscopy (TEM). Nevertheless, in ensemble measurements their emission spectra will still have a full-width at half maximum of several meV [111, 112]. Although this broad range of emission energies was initially ascribed to the residual size distribution, it has now been ascribed to an intrinsic property of colloidal QDs. Whilst the emission peak from a single colloidal QD may be less than 0.1 meV wide, its emission energy shifts randomly over time. Such behavior was referred to as a

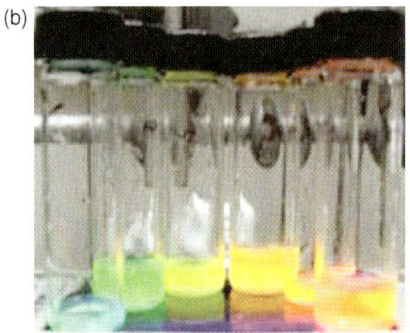

Figure 2.12 Colloidal CdSe QDs of different size dissolved in chloroform. The size of the QDs increases from the left to the right vial. (a) Photograph image of the solutions; (b) Photographic image of the solutions upon UV-illumination from below, showing the different colors of fluorescence.

"spectral jump," or "spectral diffusion" [112]. First observed by Empedocles *et al.* at cryogenic temperatures, spectral diffusion has since also been observed at room temperature [113, 114]. The condition is most likely related to the local environment of the QDs, which create rapidly fluctuating electric fields that may perturb the energy levels of the system. Spectral diffusion may also be observed in single organic fluorophores [115]. Conversely, self-assembled QDs embedded in a matrix do not exhibit spectral jumps, because their local environment does not change with time.

Although, in self-assembled QDs, multi-exciton states can be observed and studied at high pumping power, they have never been observed in colloidal QDs. The absence of multi-excitons in single colloidal QDs is believed to correlate with the fluorescence intermittence observed in these dots [116]. The fluorescence emission from a single QD exhibits a dramatic on/off behavior that is referred to as "blinking," and represents another spectroscopic feature to distinguish between colloidal and self-assembled QDs [117, 118]. This blinking behavior is similar to that of single organic fluorophores [119], and in CdSe QDs it can be readily observed using standard epifluorescence microscopy. In a nanocrystal, the "off" periods of blinking may vary from milliseconds to several minutes; the mechanism responsible is believed to involve an Auger photoionization of the colloidal QD. In this model, if two electron–hole pairs are present simultaneously inside the QDs, then the energy released by the annihilation of one electron–hole pair is transferred to the other pair. The resultant excess of energy may cause one of the carriers to be ejected into the environment of the QD, leaving it charged. If an electron–hole pair is created during this time, the energy released by its recombination is transferred to the third remaining carrier via a nonradiative process – which is why ionized QDs do not emit. If the ejected carrier returns inside the QD, or if the QD is neutralized, then the radiative emission is restored. The probability of Auger processes occurring in nanocrystals is greater than in the bulk due to a breakdown of translational symmetry. Such probability is also related to the spatial overlap of the carrier wavefunctions, and for this reason it is higher in colloidal than in self-assembled QDs, with the former usually being much smaller than the latter. The probability of Auger processes is increased further in the case of finite and defective barriers, which offer a wide range of electronic states where the excited carriers can be localized. Again, this is the case with colloidal QDs but not with self-assembled QDs, which are buried in an inorganic, defect-free, thick matrix. A more detailed description of the Auger effect in colloidal QDs may be found elsewhere [112, 120, 121].

2.7.3
Metal Nanoparticles

Colloidal metal nanoparticles may have optical absorption spectra with an absorption peak that resembles that of colloidal semiconductor particles (cf. Figure 2.11 and Figure 2.13). However, this absorption does not derive from transitions between quantized energy states. Instead, in metal particles the collective modes of motion of the electron gas may be excited, and these are referred to as "surface plasmons" [122, 123]. The peak in the absorption spectrum is the resonance frequency for the generation of surface plasmons. The size-dependence of the plasmon frequency is

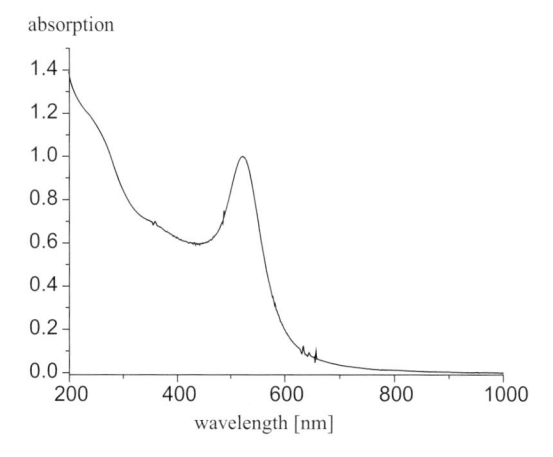

Figure 2.13 Absorption spectrum of colloidal gold nanocrystals of 10 nm diameter.

negligible; for instance, the absorption maximum for colloidal gold nanocrystals remains essentially unshifted for nanoparticles in the size range between 5 and 30 nm. This contrasts with the dramatic size-related effect in the absorption of colloidal semiconductor nanocrystals, when the diameter is changed by only fractions of a nanometer (see Figure 2.11).

2.7.4
Overview of Some Selected Applications

Many applications of the quantum-mechanical aspects of QDs can be found in optics. As in the more general case of atoms or molecules, QDs may be excited either optically or electrically although, regardless of the nature of the excitation, they may emit photons as they relax from an excited state to a ground state. Based on these properties, QDs may be used as lasing media, as single photon sources, as optically addressable charge storage devices, and/or as fluorescent labels.

Self-assembled QDs incorporated into the active layer of a quantum well laser have caused significant improvements in the laser's operating characteristics, due to the zero-dimensional density of the states of the QDs. In QD lasers, the threshold current density is reduced, the temperature stability is improved, and the differential gain is increased. A first demonstration of the lasing operation of a QD laser structure was provided back in 1994 [124, 125], since which time the lasing characteristics have been greatly improved due to a better control of the growth of self-organized QD layers. In fact, QD lasers are today rapidly approaching commercial status [126]. Both, optical gain and stimulated emission have also been observed from CdSe and CdS colloidal nanocrystal QDs [127]. On the basis of these results, it is conceivable that future optical devices might also be built using a self-assembly of colloidal QDs.

In the past, QDs have been used not only as conventional laser sources but also as "nonclassical" light sources. Photons emitted from thermal light sources have characteristic statistical correlations; in terms of arrival time, they tend to "bunch" together (super-poissonian counting statistics). For applications in quantum information processing, however, it would be desirable to emit only single photons at a time (sub-poissonian counting statistics, or anti-bunching). During the past few years, it has become possible to demonstrate the first prototypes of such single-photon sources, based on single QDs [128–132].

The use of self-assembled QDs has also been discussed as the basis of an all-optical storage device in which excitons are optically generated and the electrons and holes are stored separately in coupled QD pairs [133]. By applying an electric field, the electron and hole could be forced to recombine so as to generate a photon which would provide an optical read-out.

Colloidal QDs have also been used in the development of light-emitting diodes [134, 135], by their incorporation into a thin film of conducting polymer. Moreover, colloidal QDs have also been used in the fabrication of photovoltaic devices [136, 137].

Chemically synthesized QDs fluoresce in the visible range with a wavelength that is tunable by the size of the colloids. The possibility to control the onset of absorption and color of fluorescence by tailoring the size of colloidal QDs makes them objects of interest for the labeling of biological structures [98, 99], in the form of a new class of fluorescent marker. Such tenability, combined with their extremely reduced photobleaching, makes colloidal QDs an interesting alternative to conventional fluorescent molecules [88]. Possible biological applications of fluorescent colloidal QDs are discussed elsewhere in this book (see Chapter 6).

2.8
Some (Electrical) Transport Properties of Quantum Dots

Electron transport through ultrasmall structures such as QDs is governed by charge and energy quantization effects. Charge quantization comes into play for structures with an extremely small capacitance. The capacitance of a nanostructure – which, roughly speaking, is proportional to its typical linear dimension – may become so small that the energy required to charge the structure with one additional charge carrier (electron, hole, copper pair) exceeds the thermal energy available. In this case, charge transport through the structure is blocked – an effect which appropriately has been termed "Coulomb blockade" (CB). As shown below, this effect can be exploited to manipulate single electrons within nanostructures. Likewise, due to the small size of the structures, energy quantization may become considerable. In contrast to bulk structures, charge carriers within a QD are only allowed to occupy discrete energy levels, similar to electrons within an atom. In a manner which resembles the scattering effects known from atomic and nuclear physics, the occurrence of these discrete energy levels can lead to a modification of the charge transport characteristics of ultrasmall devices. In metallic nanostructures, Coulomb blockade can also occur without "quantum aspects," as the energy level spacing in these structures is usually too small to be observable. Yet, energy quantization is in fact what makes the QD "quantum," and differentiates it from most other metallic nanostructures; indeed, in extremely small metallic nanoparticles and clusters the quantum effects may be considerable. At this point, a brief introduction is provided of the history and fundamentals of Coulomb blockade. Single-electron tunneling transport through QDs is also discussed, and an overview provided of the possible applications of these effects.

2.8.1
Coulomb Blockade: Basic Theory and Historical Sketch

The Coulomb blockade effect was originally observed in experiments on small metallic or superconducting particles [138–141], in which nanometer-sized metallic grains were embedded within a metal oxide–metal tunnel junction. Electronic measurements performed on these systems at low temperatures ($T \sim 1$ K) revealed an anomalous behavior of the resistance (or differential capacitance) at zero bias. It was realized that this behavior was caused by the extremely small capacitance of the metallic particles. In a simple model, a spherical particle of diameter d embedded in a dielectric with permittivity ε has a capacitance

$$C = 2\pi\varepsilon\varepsilon_0 d \tag{2.18}$$

and therefore at low enough temperatures the charging energy[3]

$$E_C = \frac{e^2}{2C} = \frac{e^2}{4\pi\varepsilon\varepsilon_0 d} \tag{2.19}$$

required for the addition of a single electron to the particle may well exceed the thermal energy $k_B T$ (where k_B is the Boltzmann constant). As an example, a metallic particle of radius 10 nm embedded in a dielectric with $\varepsilon = 4$ has a charging energy of $E_C = 18$ meV, which corresponds to a temperature of about 200 K. Below this temperature (as a rule of thumb, $E_C > 10\, k_B T$ should be satisfied), it is impossible to add or remove a single charge to the particle at zero bias, and therefore electronic transport is blocked. This explains the highly increased resistance of the tunnel junctions at low temperatures observed in the early experiments. However, the Coulomb blockade can be lifted if enough energy is supplied by applying a bias over the structure. It is found that for $|V_{bias}| > e/2C$ the conductance starts to rise from its suppressed value. The voltage interval $[-e/2C, +e/2C]$ over which conduction is suppressed is often referred to as the "Coulomb gap." Apart from the small size of the conductive island, the other essential requirement for Coulomb blockade to be observed is that the metallic island or particle be isolated from the contacts via tunnel barriers. This ensures that the charge on the island is "sufficiently well quantized." If the island were coupled more strongly to the contacts, charge could "leak out" and thus destroy the effect. To be more precise, the energy uncertainty of the electrons due to the finite lifetime τ of its energy state (given roughly by the RC time of the island, where R is resistance) should be less than the charging energy, that is:

$$\Delta E \approx \frac{\hbar}{\tau} = \frac{\hbar}{RC} < \frac{e^2}{C} \tag{2.20}$$

Solving for R yields as a condition for the resistance $R > \frac{h}{e^2}$ which just means that R should be on the order of the resistance quantum.[4]

Many years after the initial experiments, the Coulomb blockade effect was observed for the first time in a lithographically defined metallic structure [142]. In contrast to the earlier experiments, the effect could now be observed for a single metallic island rather than averaged over a large population of nanoparticles. Shortly afterwards, Coulomb blockade was also observed in a semiconductor microstructure [143]. The use of lithographic techniques facilitates the realization of well-defined three-terminal geometries. In such a geometry, in addition to the source and

---

3) Although the charging energy $E_C = e^2/4\pi\varepsilon\varepsilon_0 d$ (Eq. 2.19) closely resembles the Coulomb energy $E_{Coul} = -1.8 \cdot e^2/2\pi\varepsilon\varepsilon_0 d$ (Eq. 2.15), both terms describe different situations. The charging energy describes the energy required to add an additional electron to an already negatively charged system (a sphere with capacity C), and therefore it is repulsive ($E > 0$). The Coulomb energy describes the attraction ($E < 0$) between an electron and a hole within a sphere.

4) $\hbar/e^2$ is the "resistance quantum"; it is equal to the resistance of a single nondegenerate conduction mode of a 1-D conducting channel connecting two large reservoirs.

drain contacts, a gate electrode is placed near the electron island which allows the electrostatic energy of the island to be tuned; in this way, the number of electrons on the island can be changed, one by one. A variation of the gate voltage will therefore lead to almost periodic conductance oscillations which indicate the charging of the island with single electrons (see discussion below).

In most cases, for a metallic nanostructure Coulomb blockade is a purely classical (electrostatic) phenomenon, in the sense that the energy level spacing ΔE in the structure is much smaller than the thermal energy $k_B T$ which is due to the metal's high density of states at the Fermi level. This is strikingly different for semiconductor nanostructures, for which the level spacing can become comparable to, or even larger than, the charging energy. Hence, in semiconductor nanostructures the energy required to add or remove single electrons is strongly modified by the quantum-mechanical energy levels. For this reason, the term QDs is normally used for such ultrasmall semiconductor structures. Following the initial experiments mentioned above, a huge body of data has been produced on the transport properties of metallic islands and semiconductor QDS in the Coulomb blockade regime.

2.8.2
Single-Electron Tunneling

A number of excellent reviews on the theoretical and experimental aspects of Coulomb blockade and single-electron tunneling have been produced during the past decade. Among these, an early review on single-electron tunneling in metallic islands was provided by Averin and Likharev [144], while an updated treatment by Schön emphasized other aspects [145]. The theoretical aspects of Coulomb blockade in semiconductor nanostructures have also been surveyed [146]. An excellent review of transport experiments on semiconductor QDs [147] is recommended, together with other reviews [148, 149]. At this point, a brief introduction is provided to a simple model of a single-electron transistor which captures many of the relevant aspects. (For more detailed discussions, the reader is referred to the more specialized data and references listed in these reviews.)

A schematic representation of a single-electron transistor is depicted in Figure 2.14. This consists of a small conducting island that is connected to its surroundings via two tunneling barriers and a gate electrode. The capacitances of the barriers and the gate are denoted C_S, C_D and C_G, the drain-source and gate voltages $V = V_D - V_S$ and V_G. The total capacitance of the island is given by $C = C_S + C_D + C_G$. The electrostatic energy of the island in this model is given by:

$$E(N, Q_G) = \frac{(Ne - Q_G)^2}{2C} \tag{2.21}$$

where the integer N is the number of electrons on the island, e^- is the electronic charge, and the gate charge $Q_G = C_D V_D + C_G V_G + C_S V_S$ (in the circuit diagram, V_S has been set at 0, the reference potential). Terms independent of N have been omitted in this expression. $E(N, Q_G)$ defines a set of parabolas with minima at $Ne = Q_G$. Whereas, the gate charge Q_G can be varied continuously by the external voltage

2 Quantum Dots

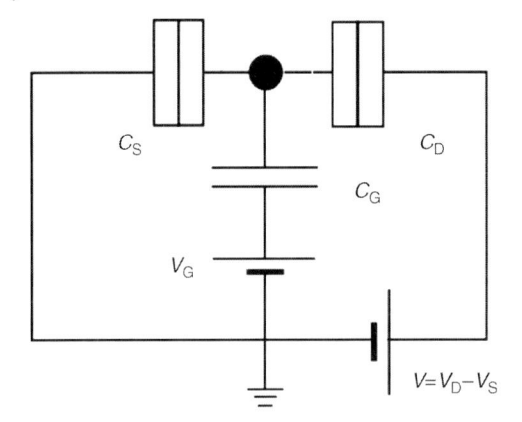

Figure 2.14 A schematic circuit diagram for a single-electron transistor. The electron island, indicated by the black dot, is connected to source and drain contacts via tunneling barriers having capacitances C_S and C_D. Additionally, the electrostatic energy of the island can be tuned with a capacitively coupled gate (capacitance C_G). In this circuit, the source contact has been set to ground, and the gate voltage is applied with respect to the ground potential.

sources, in the Coulomb blockade regime the charge on the island can only vary in integer numbers. For different gate charges (gate voltages) the island may therefore be occupied by a different number of electrons; in other words, the gate voltage can be used to tune the number of electrons on the island. The charge can fluctuate, however, if the energies for two successive occupation numbers are degenerate, that is, if $E(N+1,Q_G) = E(N,Q_G)$. At these points, the Coulomb blockade is lifted and charges can be added to or removed from the dot; that is, the conductance of the dot becomes finite. Solving for the gate charge leads to the condition for charge fluctuation[5] $Q_G^{(N)} = (N+\tfrac{1}{2})e$. The distance between two adjacent degeneracy points is therefore $Q_G^{(N+1)} - Q_G(N) = e$. By considering the special case $V_{DS} = 0$ (zero bias), then $Q_G = C_G V_G$ and $\Delta V_G = V_G^{(N+1)} - V_G^{(N)} = e/C_G$. This shows that the gate voltages at which a finite conductance can be measured are equally spaced – leading to the periodic conductance oscillations of a single-electron transistor in the Coulomb blockade regime (Figure 2.15).

In a thermodynamic description of the system, the charge on the island can fluctuate if the probability to find N electrons on the island equals that to find $N+1$

---

5) By using Eq. (2.19), the condition for charge fluctuation $E(N+1,Q_G) = E(N,Q_G)$ becomes:

$$((N+1)e - Q_G)^2/2C = (Ne - Q_G)^2/2C$$
$$\Leftrightarrow (N+1)^2 e^2 - 2(N+1)eQ_G + Q_G^2 = N^2 e^2 - 2NeQ_G + Q_G^2$$
$$\Leftrightarrow N^2 e^2 + 2Ne^2 + e^2 - 2NeQ_G - 2eQ_G + Q_G^2 = N^2 e^2 - 2NeQ_G + Q_G^2$$
$$\Leftrightarrow 2Ne^2 + e^2 - 2eQ_G = 0$$
$$\Leftrightarrow 2eQ_G = (2N+1)e^2$$
$$\Leftrightarrow Q_G = (N+1/2)e$$

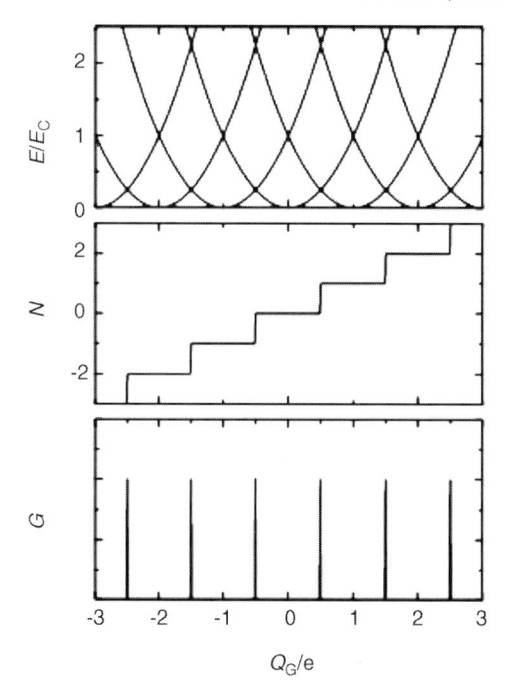

Figure 2.15 Schematic depiction of electrostatic energy E/E_C, occupation number N and conductance G of the single-electron transistor shown in Figure 2.14, as a function of the gate charge Q_G. The occupation of the single-electron transistor changes when the energy for N electrons equals that of $N + 1$. At those points where the charge can fluctuate, the conductance becomes finite and the Coulomb blockade is lifted. In real systems, the line shape of the conductance peaks is determined by temperature and/or lifetime broadening.

electrons, or $P(N) = P(N + 1)$, where the equilibrium probability P is given by:

$$P(N) = \frac{1}{Z}\exp(-(E(N)-\mu N)/k_B T) \tag{2.22}$$

where Z is the partition function (Zustandssumme) [150] and μ is the chemical potential of the contacts. The condition for fluctuation thus becomes $E(N + 1) = E(N) = \mu$. The left-hand side of this equation is called the addition energy for the $(N + 1)$-th electron. If a finite bias V is applied over the system, the electrochemical potentials of the two reservoirs differ by eV. In this case, a "conductance window" opens and the Coulomb blockade is lifted for

$$\mu - e\chi V < E(N+1) - E(N) < \mu + e(1-\chi)V \tag{2.23}$$

where $\chi (0 \leq \chi \leq 1)$ is the fractional voltage drop over the drain barrier.

Current–voltage (*I–V*) curves can be calculated from a master equation for the occupation probability of the Coulomb blockade island. For the single-electron transistor with two tunneling barriers and one island, the master equation is [144, 145]:

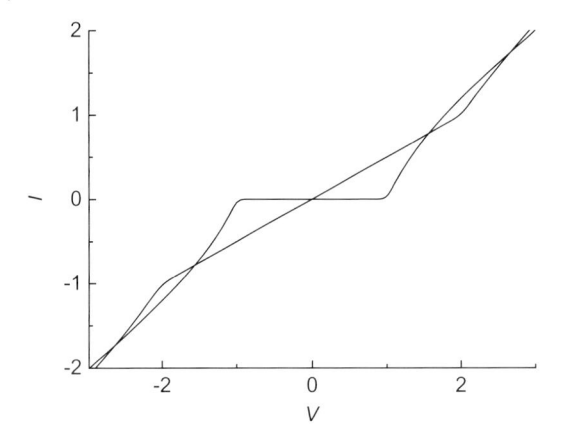

Figure 2.16 Calculated current–voltage (I–V) curves for a single-electron transistor for two different values of the gate charge. The source and drain tunnel resistances and capacitances are assumed equal, the charging energy is 100 k_BT. For gate charge $Q_G = 0$, the single-electron transistor is in the Coulomb blockade regime, and the I–V curve exhibits the "Coulomb gap." For a gate charge $Q_G = 0.5$, the Coulomb blockade is lifted (cf. Figure 2.15), and the I–V curve is linear around $V = 0$. The calculation was performed using the program SETTRANS by A. N. Korotkov, available for noncommercial use at: http://hana.physics.sunysb.edu/set/software/index.html

$$\frac{d}{dt}p(N,t) = [\Gamma_{L \to I}(N-1) + \Gamma_{R \to I}(N-1)]p(N-1,t) + [\Gamma_{I \to L}(N+1)$$
$$+ \Gamma_{I \to R}(N+1)]p(N+1,t) - [\Gamma_{L \to I}(N) + \Gamma_{I \to R}(N)$$
$$+ \Gamma_{R \to I}(N) + \Gamma_{I \to R}(N)]p(N,t) \qquad (2.24)$$

where $p(N,t)$ is the nonequilibrium probability to find N electrons on the island at time t. The $\Gamma_{X \to Y}(N)$ are tunneling rates for transitions from X to Y, where X and Y can be L (left contact), R (right contact), or I (island). An example for a typical I–V curve obtained from the master equation is shown in Figure 2.16. In the Coulomb blockade regime (when the gate charge is $Q_G = Ne$), the current in the Coulomb gap around $V = 0$ is zero. For $Q_G = (N + 1/2)e$, the Coulomb blockade is lifted and the I–V curve is linear with finite slope at $V = 0$.

For a QD, it is also necessary to account for the energy quantization of the electrons on the island. In the simplest model, which sometimes is referred to as the constant interaction (CI) model, it is assumed that the electrons successively occupy single-particle energy levels ε_i (see Figure 2.4[6]) and the mutual electrostatic repulsion of the electrons is accounted for by a classical capacitance. Thus, the energy for N electrons on the QD is

---

[6] In order to avoid confusion, here are denoted the discrete energy levels of electrons in a potential well with ε_i. These energy levels are identical to those shown in Figure 2.4. For a 3-D potential box, we obtain $\varepsilon_i = i^2 h^2/2md^2$, $i = 1, 2, 3,\ldots$ [cf. Eq. (2.12)]. The lowest possible energy ($i = 1$) is $h^2/2md^2$.

$$E(N) \approx \frac{(Ne)^2}{2C} + \sum_1^N \varepsilon_i. \qquad (2.25)$$

The addition energy therefore becomes:

$$E(N+1) - E(N) = \frac{(2N+1)e^2}{2C} + \varepsilon_{N+1} \qquad (2.26)$$

and the distance between two adjacent conductance resonances follows as:

$$\Delta\mu_N = E(N+1) - 2E(N) + E(N-1) = \frac{e^2}{C} + \Delta\varepsilon_N \qquad (2.27)$$

where $\Delta\varepsilon_N = \varepsilon_{N+1} - \varepsilon_N$ is the N-th level spacing. In the CI model, the distance between adjacent conductance maxima is given by a classical charging term and the spacing between single particle energy levels. This shows how the quantum nature of QDs directly enters its electrical properties.

In systems where the CI model can be applied, the spectrum of the QD can, in principle, be deduced from transport or capacitance measurements – experimental techniques which therefore have been termed transport or capacitance spectroscopy. The CI model is quite reasonable for systems with very few electrons, but in many situations many-body effects lead to strong deviations from this simple behavior. The relative importance of energy quantization versus charging energy can be judged from the magnitude of the terms e^2/C and $\Delta\varepsilon$ in the expression above. Typically, the capacitance C scales with the diameter d of a system like $C \approx 2\pi\varepsilon\varepsilon_0 d$ [cf. Eq. (2.18)], and thus $E_C \propto e^2/\varepsilon\varepsilon_0 d$ [cf. Eq. (2.19)], whereas the quantization energies are on the order of $\Delta\varepsilon \approx h^2/2m^*d$ [cf. Eq. (2.13)]. Most trivially, for smaller systems the quantum effects become stronger.

2.8.3
Tunneling Transport: The Line Shape of Conductance Peaks

A slightly different approach to treat electron transport through QDs considers conduction as a transmission problem through an electron island. In this formulation, single-electron tunneling through QDs bears similarities to the well-known description of *resonant tunneling transport*. The latter occurs when a system with discrete energy levels is connected to two reservoirs via tunneling barriers. In the case of QDs, the discreteness of the energies naturally comes in either due to Coulomb blockade or to quantum confinement, or both. Such a system is depicted schematically in Figure 2.17, where the tunneling rates through the left and right tunnel barriers are denoted by Γ_L and Γ_R, and the quantum level in the dot has energy ε_1. It can be shown that the transmission at energy E through such a system is given approximately by:

$$T(E) = \frac{\Gamma_L \Gamma_R}{(E-\varepsilon_1)^2 + \left(\frac{\Gamma_L + \Gamma_R}{2}\right)^2} \qquad (2.28)$$

2 Quantum Dots

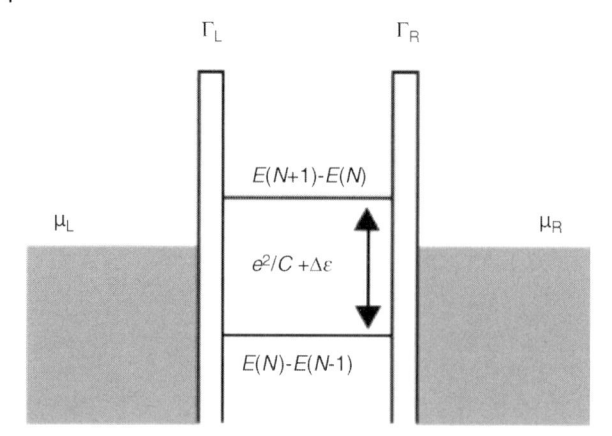

Figure 2.17 Electron transport through a QD, viewed as double barrier tunneling. The "resonant energies" of the QD are given by the addition energies, as discussed in the text. In the constant interaction model, the difference between adjacent resonances is approximated by a charging term and the single particle energy level spacing. The chemical potentials of the reservoirs are denoted μ_L and μ_R, and the tunneling rates through the barriers are Γ_L and Γ_R.

where the tunneling rates have been assumed energy independent. The current is then given by:

$$I = \frac{2e}{h} \int [f(E) - f(E - eV_{ds})] T(E) dE \quad (2.29)$$

where $f(E)$ denotes the Fermi function and the factor of two accounts for spin degeneracy. From this expression, the conductance of the system can be determined and hence the shape of the conductance oscillations in the Coulomb blockade regime. Depending on the relative size of the coupling energies $h\Gamma_L, h\Gamma_R$ and the thermal energy $k_B T$, the line shape is dominated by lifetime broadening of the energy state on the island or by temperature broadening. In the low-bias regime ($eV_{DS} \approx 0$), the Fermi function in the above expression can be expanded which yields for the conductance

$$G = -\frac{2e^2}{h} \int f'(E) T(E) dE \quad (2.30)$$

Thus, if temperature broadening dominates, the conductance oscillations have the shape of the derivative of the Fermi function ($\propto \cos h^{-2}(E/2k_B T)$). However, if lifetime broadening dominates, the conductance peaks are of a Lorentzian shape.

2.8.4
Some Applications

The peculiar properties of QDs described above may lead to their use in a variety of device applications. The most important of these features are the discrete energy spectrum, the smallness of the capacitance, and, in relation, the possibility to manipulate single charges. The survey of device applications provided in the following sections is by no means complete; the intention is simply to demonstrate the depth of investigations conducted in this field, and the many directions pursued.

The possibility to manipulate single charges in single-electron tunneling devices led to early proposals for metrological applications of the Coulomb blockade effect. For this, devices referred to as "electron turnstiles" can be used to count single electrons, forming the basis of accurate current standards based on the definition of the second. An electron turnstile consists of a number of electron islands connected together via tunneling barriers, with the potentials of the islands being controlled individually by the use of gates. If the gate voltages are changed in a particular operation cycle, the potentials of the islands can be changed in such a manner that single electrons are "pumped" through the system. If the operation cycle is be repeated at a frequency f, the resulting current ($I = ef$) is linked directly to the definition of the second via frequency [151–153]. The same principles can be used to construct capacitance standards [154].

Single-electron transistors have also been proposed as a basis for logic circuits, with attempts to implement conventional CMOS (complementary metal-oxide-semiconductor) logic by using single-electron transistors as replacements for n-MOS and p-MOS transistors (depending on whether the single-electron transistor is in the Coulomb blockade region, or whether the blockade is lifted) [155, 156]. For other concepts, the logical states are essentially represented by single electrons, for example, in QD cellular automata [157]. Whilst all of these proposals face strong conceptual and technical challenges, to date it seems unlikely that a single-electron approach can offer a competitive alternative to traditional CMOS-based logic. Among problems encountered have been the required smallness of the QDs for reliable room temperature operation, the sensitivity to background charges, the need for strategies for reproducible placement and interconnection of QDs to build very-large-scale integration (VLSI) structures, and the lack of gain of single-electron transistors [158].

Due to their low capacitance, nanometer-sized metallic islands or QDs are extremely sensitive to neighboring charges. Whereas, in some situations this is unfavorable – for example, if fluctuating background charges disturb the operation of a single-electron transistor – this effect can also be exploited for device applications. One promising application of the charge sensitivity of QDs is the fabrication of *single-electron memories*. In one approach, the floating gate memory technique (which is already used in CMOS memories) is adapted to utilize single-electron effects. For this, a single-electron transistor is used as an electrometer to sense the charging/discharging of a floating gate [159]. The charge state of the floating node can thus be used as memory, the state of which is read by the single-electron transistor.

Some of the most promising applications of single-electron devices lie in the area of sensorics, the most obvious being that of charge sensing and electrometry. In this respect, single-electron transistors are the most sensitive charge detectors developed to date, with an equivalent charge noise of the order of 10^{-3} to 10^{-5} $e/\sqrt{Hz}$ at 10 Hz operation frequency [160, 161]. A charge noise of approximately $2 \cdot 10^{-5}$ $e/\sqrt{Hz}$ may be achieved at a higher operation frequency of 4 kHz [162], and even $1.2 \cdot 10^{-5}$ $e/\sqrt{Hz}$ at

1.1 MHz (with an "RF-single-electron transistor") [163]. Single-electron transistor-based electrometers were also used to monitor other physical systems; typical examples included the characterization of quantum Hall edge states [164] and the capacitance spectroscopy of vertical QDs [165]. Single-electron transistor electrometers have also been used as charge sensors for scanning probe microscopy (SPM) [166]. One interesting application not based explicitly on charge sensitivity, was to use QDs to detect single photons in the far-infrared range [167]. More generally, single-electron transistors are expected to prove useful in many low-noise analog applications, notably in the amplification of quantum signals [168].

Most of the above-mentioned device proposals do not rely on the quantum nature of QDs, but simply make use of the Coulomb blockade effect, based on the ultrasmall capacitance of small electron islands. Although the majority of applications of energy quantization within QDS are naturally found in optics (see Section 2.7), the possible use of QD energy levels for quantum information processing has recently attracted considerable attention in the field of electronics. In order to build a quantum computer, techniques must first be developed to produce and manipulate "qubits" (quantum bits), the information unit employed in quantum information processing and quantum computing. Qubits can be realized in any physical system with two quantum states that can be identified as logical "0" or "1" (in Dirac notation $|0\rangle$ and $|1\rangle$). The difference between classical and quantum logical states is that a quantum system can also be in a superposition of $|0\rangle$ and $|1\rangle$ [169]. Certain classes of computational problems – for example, the factorization of large numbers – are expected to be solved more efficiently on quantum computers. Possible realizations of qubits are the spin states of particles (spin up–spin down), the polarization of photons (vertical–horizontal), or other physical systems which can be approximately described as two-level quantum systems. The implementation of QDs in quantum computers makes use of the quasi-molecular states formed in two coupled QDs [170]. The $|0\rangle$ and $|1\rangle$ states correspond to an electron localized on either of the two dots, and superpositions can be formed by strongly coupling the two dots together. The first experimental realizations of such artificial molecular states in QDs have already been reported [171, 172]. Compared to other proposals for quantum computers, QD-based schemes have the important advantage of being compatible with conventional solid-state electronics. One major problem, however, is posed by the relatively short decoherence times which may corrupt the desired evolution of the system states.

References

1 Lane, N. (2001) *J. Nanopart. Res.*, **3**, 95–103.
2 Service, R.F. (2000) *Science*, **290**, 1526–1527.
3 Kingon, A.I., Maria, J.-P., and Streiffer, S.K. (2000) *Nature*, **406**, 1032–1038.
4 Lloyd, S. (2000) *Nature*, **406**, 1047–1054.
5 Ito, T. and Okazaki, S. (2000) *Nature*, **406**, 1027–1031.
6 Peercy, P.S. (2000) *Nature*, **406**, 1023–1026.
7 Cohen-Tannoudji, C., Diu, B., and Laloe, F. (1997) *Quantum Mechanics*, 1st edn, John Wiley & Sons, New York.
8 Yoffe, A.D. (2001) *Adv. Phys.*, **50**, 1–208.

9 Atkins, P.W. (1986) *Physical Chemistry*, 3rd edn, Oxford University Press, Oxford.

10 Karplus, M. and Porter, R.N. (1970) *Atoms and Molecules*, 1st edn, W. A. Benjamin, Inc., New York.

11 Alivisatos, A.P. (1997) *Endeavour*, **21**, 56–60.

12 Harrison, W.A. (1989) *Electronic Structure and the Properties of Solids: The Physics of the Chemical Bond*, Dover Publications, Dover.

13 Kobayashi, A., Sankey, O.F., Volz, S.M., and Dow, J.D. (1983) *Phys. Rev. B*, **28**, 935–945.

14 Burdett, J.K. (1984) *Prog. Solid State Chem.*, **15**, 173–255.

15 Kittel, C. (1989) *Einführung in die Festkörperphysik*, 8th edn, R. Oldenbourg Verlag, München, Wien.

16 Ashcroft, N.W. and Mermin, N.D. (1976) *Solid State Physics*, Saunders College, Philadelphia.

17 Davies, J.H. (1998) *The Physics of Low-Dimensional Semiconductors*, Cambridge University Press, Cambridge.

18 Ando, T., Fowler, A.B., and Stern, F. (1982) *Rev. Mod. Phys.*, **54**, 437–672.

19 Moriarty, P. (2001) *Rep. Prog. Phys.*, **64**, 297–381.

20 Zhitenev, N.B., Fulton, T.A., Yacoby, A., Hess, H.F., Pfeiffer, L.N., and West, K.W. (2000) *Nature*, **404**, 473–476.

21 Suen, Y.W., Engel, L.W., Santos, M.B., Shayegan, M., and Tsui, D.C. (1992) *Phys. Rev. Lett.*, **68**, 1379–1382.

22 Stormer, H.L. (1998) *Solid State Commun.*, **107**, 617–622.

23 Stormer, H.L., Du, R.R., Kang, W., Tsui, D.C., Pfeiffer, L.N., Baldwin, K.W., and West, K.W. (1994) *Semicond. Sci. Technol.*, **9**, 1853–1858.

24 Wharam, D.A., Thornton, T.J., Newbury, R., Pepper, M., Ahmed, H., Frost, J.E.F., Hasko, D.G., Peacock, D.C., Ritchie, D.A., and Jones, G.A.C. (1988) *J. Phys. C*, **21**, L209.

25 van Wees, B.J., Houten, H.v., Beenakker, C.W.J., Williams, J.G., Kouwenhoven, L.P., Marel, D.v.d., and Foxon, C.T. (1988) *Phys. Rev. Lett.*, **60**, 848.

26 Bumm, L.A., Arnold, J.J., Cygan, M.T., Dunbar, T.D., Burgin, T.P., Jones, L. II, Allara, D.L., Tour, J.M., and Weiss, P.S. (1996) *Science*, **271**, 1705–1707.

27 Anantram, M.P., Datta, S., and Xue, Y.Q. (2000) *Phys. Rev. B*, **61**, 14219–14224.

28 Cobden, D.H. (2001) *Nature*, **409**, 32–33.

29 Freemantle, M. (2001) *Chem. Eng. News*, **March 5**, 38.

30 Cui, X.D., Primak, A., Zarate, X., Tomfohr, J., Sankey, O.F., Moore, A.L., Moore, T.A., Gust, D., Harris, G., and Lindsay, S.M. (2001) *Science*, **294**, 571–574.

31 Reed, M.A. (2001) *MRS Bull.*, 113–120.

32 Hu, J.T., Odom, T.W., and Lieber, C.M. (1999) *Acc. Chem. Res.*, **32**, 435–445.

33 Cui, Y., Duan, X., Hu, J., and Lieber, C.M. (2000) *J. Phys. Chem. B*, **104**, 5213–5216.

34 Rodrigues, V., Fuhrer, T., and Ugarte, D. (2000) *Phys. Rev. Lett.*, **85**, 4124–4127.

35 Rao, C.N.R., Kulkarni, G.U., Govindaraj, A., Satishkumar, B.C., and Thomas, P. (2000) *J. Pure Appl. Chem.*, **72**, 21–33.

36 Häkkinen, H., Barnett, R.N., Scherbakov, A.G., and Landman, U. (2000) *J. Phys. Chem. B*, **104**, 9063–9066.

37 Cui, Y. and Lieber, C.M. (2001) *Science*, **291**, 851–853.

38 Reed, M.A., Zhou, C., Muller, C.J., Burgin, T.P., and Tour, J.M. (1997) *Science*, **278**, 252–254.

39 van den Brom, H.E., Yanson, A.I., and Ruitenbeek, J.M. (1998) *Physica B*, **252**, 69–75.

40 Xe, H.X., Li, C.Z., and Tao, N. (2001) *J. Appl. Phys. Lett.*, **78**, 811–813.

41 Tans, S.J., Devoret, M.H., Dai, H., Thess, A., Smalley, R.E., Geerligs, L.J., and Dekker, C. (1997) *Nature*, **386**, 474–477.

42 Saito, S. (1997) *Science*, **278**, 77–78.

43 McEuen, P.L., Bockrath, M., Cobden, D.H., and Lu, J.G. (1999) *Microelectron. Eng.*, **47**, 417–420.

44 Yao, Z., Postma, H.W.C., Balents, L., and Dekker, C. (1999) *Nature*, **402**, 273–276.

45 Odom, T.W., Huang, J.-L., Kim, P., and Lieber, C.M. (2000) *J. Phys. Chem. B*, **104**, 2794–2809.

46 McEuen, P.L. (2000) *Phys. World*, 31–36.

47 Jacoby, M. (2001) *Chem. Eng. News*, **April 30**, 13.

48 de Heer, W.A. et al. (1995) *Science*, **270**, 1179.
49 Bastard, G. and Brum, J.A. (1986) *IEEE J. Quantum Electron.*, **22**, 1625–1644.
50 Petroff, P.M., Lorke, A., and Imamoglu, A. (2001) *Phys. Today*, 46–52.
51 Landau, L.D. and Lifschitz, E.M. (1979) *Quantenmechnik*, 9th edn, vol. **3** Akademie-Verlag, Berlin.
52 Schwabl, F. (1990) *Quantenmechanik*, 2nd edn, Springer Verlag, Berlin.
53 Messiah, A. (1976) *Quantenmechanik (Band 1)*, Walter de Gruyter, Berlin.
54 Trindade, T., O'Brien, P., and Pickett, N.L. (2001) *Chem. Mater.*, **13**, 3843–3858.
55 Bawendi, M.G., Steigerwald, M.L., and Brus, L.E. (1990) *Annu. Rev. Phys. Chem.*, **41**, 477–496.
56 Brus, L.E. (1983) *J. Chem. Phys.*, **79**, 5566–5571.
57 Brus, L.E. (1984) *J. Chem. Phys.*, **80**, 4403–4409.
58 Brus, L. (1986) *J. Phys. Chem.*, **90**, 2555–2560.
59 Steigerwald, M.L. and Brus, L.E. (1990) *Acc. Chem. Res.*, **23**, 183–188.
60 Trindade, T., O'Brien, P., and Pickett, N.L. (2001) Nanocrystalline semiconductors: synthesis, properties, and perspectives. *Chem. Mater.*, **13** (11), 3843–3858.
61 Gorska, M. and Nazarewicz, W. (1974) Application of the random-element isodisplacement model to long-wavelength optical phonons in CdSexTe1−x mixed crystals. *Phys. Status Solidi B*, **65**, 193–202.
62 Efros, A.L. and Rosen, M. (2000) *Annu. Rev. Mater. Sci.*, **30**, 475–521.
63 Fuhrer, A., Lüscher, S., Ihn, T., Heinzel, T., Ensslin, K., Wegscheider, W., and Bichler, M. (2001) *Nature*, **413**, 822–825.
64 Blick, R.H., Weide, D.W.v.d., Haug, R.J., and Eberl, K. (1998) *Phys. Rev. Lett.*, **81**, 689–692.
65 Kemerink, M. and Molenkamp, L.W. (1994) *Appl. Phys. Lett.*, **65**, 1012.
66 Waugh, F.R., Berry, M.J., Mar, D.J., Westervelt, R.M., Campman, K.L., and Gossard, A.C. (1995) *Phys. Rev. Lett.*, **75**, 705.
67 Hofmann, F. and Wharam, D.A. (1995) *Adv. Solid State Phys.*, **35**, 197.
68 Blick, R.H., Haug, R.J., Weis, J., Pfannkuche, D., Klitzing, K.v., and Eberl, K. (1996) *Phys. Rev. B*, **53**, 7899–7902.
69 Bayer, M., Gutbrod, T., Reithmaier, J.P., Forchel, A., Reinecke, T.L., Knipp, P.A., Dremin, A.A., and Kulakovskii, V.D. (1998) *Phys. Rev. Lett.*, **81**, 2582–2585.
70 Tarucha, S., Austing, D.G., Honda, T., Hage, R.J.v.d., and Kouwenhoven, L.P. (1996) *Phys. Rev. Lett.*, **77**, 3613.
71 Goldhaber-Gordon, D., Shtrikman, H., Mahalu, D., Abusch-Magder, D., Meirav, U., and Kastner, M.A. (1998) *Nature*, **391**, 156.
72 Cronenwett, S.M., Oosterkamp, T.H., and Kouwenhoven, L.P. (1998) *Science*, **281**, 540.
73 Simmel, F., Blick, R.H., Kotthaus, J.P., Wegscheider, W., and Bichler, M. (1999) *Phys. Rev. Lett.*, **83**, 804.
74 Schmid, J., Weis, J., Eberl, K., and Klitzing, K.v. (2000) *Phys. Rev. Lett.*, **84**, 5824.
75 Cho, A.Y. (1999) *J. Cryst. Growth*, **202**, 1–7.
76 Fafard, S., Leon, R., Leonard, D., Merz, J.L., and Petroff, P.M. (1994) *Superlattices Microstruct.*, **16**, 303–309.
77 Petroff, P.M. and DenBaars, S.P. (1994) *Superlattices Microstruct.*, **15**, 15–21.
78 Leon, R., Petroff, P.M., Leonard, D., and Fafard, S. (1995) *Science*, **267**, 1966–1968.
79 Luyken, R.J., Lorke, A., Govorov, A.O., Kotthaus, J.P., Medeiros-Ribeiro, G., and Petroff, P.M. (1999) *Appl. Phys. Lett.*, **74**, 2486–2488.
80 Liu, Z.F., Shen, Z.Y., Zhu, T., Hou, S.F., Ying, L.Z., Shi, Z.J., and Gu, Z.N. (2000) *Langmuir*, **16**, 3569–3573.
81 Paul, D. (1999) *J. Adv. Mater.*, **11**, 191–204.
82 Garcia, J.M., Medeiros-Ribeiro, G., Schmidt, K., Ngo, T., Feng, J.L., Lorke, A., Kotthaus, J., and Petroff, P.M. (1997) *Appl. Phys. Lett.*, **71**, 2014–2016.
83 Bayer, M., Stern, O., Hawrylak, P., Fafard, S., and Forchel, A. (2000) *Nature*, **405**, 923–926.
84 Warburton, R.J., Schäflein, C., Haft, D., Bickel, F., Lorke, A., Karrai, K.,

Garcia, J.M., Schoenfeld, W., and Petroff, P.M. (2000) *Nature*, **405**, 926–929.

85 Tarucha, S. (1998) *MRS Bull.*, **23**, 49–53.

86 Lorke, A. and Luyken, R. (1998) *J. Phys. B: Condens. Matter*, **256–258**, 424–430.

87 Gammon, D. (2000) *Nature*, **405**, 899–900.

88 Alivisatos, A.P. (1996) *Science*, **271**, 933–937.

89 Murray, C.B., Norris, D.J., and Bawendi, M.G. (1993) *J. Am. Chem. Soc.*, **115**, 8706–8715.

90 Peng, X., Wickham, J., and Alivisatos, A.P. (1998) *J. Am. Chem. Soc.*, **120**, 5343–5344.

91 Dabbousi, B.O., Rodriguez-Viejo, J., Mikulec, F.V., Heine, J.R., Mattoussi, H., Ober, R., Jensen, K.F., and Bawendi, M.G. (1997) *J. Phys. Chem. B*, **101**, 9463–9475.

92 Peng, X., Schlamp, M.C., Kadavanich, A.V., and Alivisatos, A.P. (1997) *J. Am. Chem. Soc.*, **119**, 7019–7029.

93 Kudera, S., Carbone, L. et al. (2006) *Phys. Status Solidi C*, **203** (6), 1329–1336.

94 Cozzoli, P.D., Pellegrino, T. et al. (2006) *Chem. Soc. Rev.*, **35** (11), 1195–1208.

95 Mirkin, C.A., Letsinger, R.L., Mucic, R.C., and Storhoff, J. (1996) *Nature*, **382**, 607–609.

96 Krüger, C., Agarwal, S. et al. (2008) *J. Am. Chem. Soc.*, **130** (9), 2710–2711.

97 Zanchet, D., Micheel, C.M., Parak, W.J., Gerion, D., Williams, S.C., and Alivisatos, A.P. (2002) *J. Phys. Chem. B*, **106**, 11758–11763.

98 Charvet, N., Reiss, P. et al. (2004) *J. Mater. Chem.*, **14** (17), 2638–2642.

99 Parak, W.J., Pellegrino, T. et al. (2005) *Nanotechnology*, **16**, R5–R25.

100 Wu, X. and Bruchez, M.P. (2004) *Methods Cell Biol.*, **75**, 171–183.

101 Klein, D.L., McEuen, P.L., Bowen-Katari, J.E., Roth, R., and Alivisatos, A.P. (1996) *Appl. Phys. Lett.*, **68**, 2574–2576.

102 Klein, D.L., Roth, R., Lim, A.K.L., Alivisatos, A.P., and McEuen, P.L. (1997) *Nature*, **389**, 699–701.

103 Jaeckel, G. (1926) *Z. Tech. Phys.*, **6**, 301–304.

104 Gaponenko, S.V. (1998) *Optical Properties of Semiconductor Nanocrystals*, Cambridge University Press, Cambridge.

105 Klimov, V.I., McBranch, D.W., Leatherdale, C.A., and Bawendi, M.G. (1999) *Phys. Rev. B - Condens. Matter*, **60**, 13740–13749.

106 Nirmal, M., Norris, D.J., Kuno, M., Bawendi, M.G., Efros, A.L., and Rosen, M. (1995) *Phys. Rev.*, **75**, 3728–3731.

107 Efros, A.L., Rosen, M., Kuno, M., Nirmal, M., Norris, D.J., and Bawendi, M. (1996) *Phys. Rev. B*, **54**, 4843–4856.

108 Reboredo, F.A., Franceschetti, A., and Zunger, A. (2000) *Phys. Rev. B*, **61**, 13073–13087.

109 Johnston-Halperin, E., Awschalom, D.D., Crooker, S.A., Efros, A.L., Rosen, M., Peng, X., and Alivisatos, A.P. (2001) *Phys. Rev. B*, **63**, 205309–205323.

110 Dahan, M., Laurence, T., Pinaud, F., Chemla, D.S., Alivisatos, A.P., Sauer, M., and Weiss, S. (2001) *Opt. Lett.*, **26**, 825–827.

111 Empedocles, S.A., Norris, D.J., and Bawendi, M.G. (1996) *Phys. Rev. Lett.*, **77**, 3873–3876.

112 Empedocles, S. and Bawendi, M. (1999) *Acc. Chem. Res.*, **32**, 389–396.

113 Empedocles, S.A. and Bawendi, M.G. (1999) *J. Phys. Chem. B*, **103**, 1826–1830.

114 Neuhauser, R.G., Shimizu, K.T., Woo, W.K., Empedocles, S.A., and Bawendi, M.G. (2000) *Phys. Rev. Lett.*, **85**, 3301–3304.

115 Basche, T. (1998) *J. Lumin.*, **76–7**, 263–269.

116 Nirmal, M., Dabbousi, B.O., Bawendi, M.G., Macklin, J.J., Trautman, J.K., Harris, T.D., and Brus, L.E. (1996) *Nature*, **383**, 802–804.

117 Efros, A.L. and Rosen, M. (1997) *Phys. Rev. Lett.*, **78**, 1110–1113.

118 Kuno, M., Fromm, D.P., Hamann, H.F., Gallagher, A., and Nesbitt, D.J. (2000) *J. Chem. Phys.*, **112**, 3117–3120.

119 Dickson, R.M., Cubitt, A.B., Tsien, R.Y., and Moerner, W.E. (1997) *Nature*, **388**, 355–358.

120 Nirmal, M. and Brus, L. (1999) *Acc. Chem. Res.*, **32**, 407–414.

121 Empedocles, S.A., Neuhauser, R., Shimizu, K., and Bawendi, M.G. (1999) *Adv. Mater.*, **11**, 1243–1256.

122 Hodak, J.H., Henglein, A., and Hartland, G.V. (2000) *Pure Appl. Chem.*, **72**, 189–197.

123 Mulvaney, P. (1996) *Langmuir*, **12**, 788–800.

124 Ledentsov, N.N., Ustinov, V.M., Egorov, A.Y., Zhukov, A.E., Maksimov, M.V., Tabatadze, I.G., and Kop'ev, P.S. (1994) *Semiconductors*, **28**, 832.

125 Kirstaedter, N., Ledentsov, N.N., Grundmann, M., Bimberg, D., Ustinov, V.M., Ruvimov, S.S., Maximov, M.V., Kop'ev, P.S., Alferov, Z.I., Richter, U., Werner, P., Gösele, U., and Heydenreich, J. (1994) *Electron. Lett*, **30** 1416.

126 Bimberg, D., Grundmann, M., Heinrichsdorff, F., Ledentsov, N.N., Ustinov, V.M., Zhukov, A.E., Kovsh, A.R., Maximov, M.V., Shernyakov, Y.M., Volovik, B.V., Tsatsul'nikov, A.F., Kop'ev, P.S., and Alferov, Z.I. (2000) *Thin Solid Films*, **367**, 235.

127 Klimov, V.I., Mikhailovsky, A.A., Xu, S., Malko, A., Hollingsworth, J.A., Leatherdale, C.A., Eisler, H.-J., and Bawendi, M.G. (2000) *Science*, **290**, 314–317.

128 Michler, P., Imamoglu, A., Mason, M.D., Carson, P.J., Strouse, G.F., and Buratto, S.K. (2000) *Nature*, **406**, 968–970.

129 Michler, P., Kiraz, A., Becher, C., Schoenfeld, W.V., Petroff, P.M., Zhang, L., Hu, E., and Imamoglu, A. (2000) *Science*, **290**, 2282.

130 Michler, P., Imamoglu, A., Kiraz, A., Becher, C., Mason, M.D., Carson, P.J., Strouse, G.F., Buratto, S.K., Schoenfeld, W.V., and Petroff, P.M. (2002) *Phys. Status Solidi B - Basic Res.*, **229**, 399–405.

131 Zwiller, V., Blom, H., Jonsson, P., Panev, N., Jeppesen, S., Tsegaye, T., Goobar, E., Pistol, M.-E., Samuelson, L., and Björk, G. (2001) *Appl. Phys. Lett.*, **78**, 2476.

132 Santori, C., Fattal, D., Vuckovic, J., Solomon, G.S., and Yamamoto, Y. (2002) *Nature*, **419**, 594.

133 Lundstrom, T., Schoenfeld, W., Lee, H., and Petroff, P.M. (1999) *Science*, **286**, 2312.

134 Colvin, V.L., Schlamp, M.C., and Alivisatos, A.P. (1994) *Nature*, **370**, 354–357.

135 Rizzo, A., Li, Y.Q. et al. (2007) *Appl. Phys. Lett.*, **90** (5), Article no. 051106.

136 Huynh, W.U., Dittmer, J.J., and Alivisatos, A.P. (2002) *Science*, **295**, 2425–2427.

137 Huynh, W.U., Dittmer, J.J. et al. (2003) *Phys. Rev. B*, **67**, 115326-1–115326-12.

138 Gorter, C. (1951) *J. Phys.*, **17**, 777.

139 Giaever, I. and Zeller, H.R. (1968) *Phys. Rev. Lett.*, **20**, 1504.

140 Zeller, H.R. and Giaever, I. (1969) *Phys. Rev.*, **181**, 791.

141 Lambe, J. and Jaklevic, R.C. (1969) *Phys. Rev. Lett.*, **22**, 1371.

142 Fulton, T.A. and Dolan, G. (1987) *Phys. Rev. Lett.*, **59**, 109–112.

143 Scott-Thomas, J.H.F., Field, S.B., Kastner, M.A., Smith, H.I., and Antoniadis, D.A. (1989) *Phys. Rev. Lett.*, **62**, 583.

144 Averin, D.V. and Likharev, K.K. (1991) *Mesoscopic Phenomena in Solids* (eds B.L. Altshuler, P.A. Lee, and R.A. Webb), Elsevier, Amsterdam.

145 Schön, G. (1998) *Quantum Transport and Dissipation* (eds T. Dittrich, P. Hänggi, G.-L. Ingold, B. Kramer, G. Schön, and W. Zwerger), Wiley-VCH, Weinheim.

146 van Houten, H., Beenakker, C.W.J., and Staring, A.A.M. (1991) *Single Charge Tunneling*, (NATO ASI Series B) (eds H. Grabert and M.H. Devoret), Plenum, New York.

147 Kouwenhoven, L.P., Marcus, C.M., McEuen, P.L., Tarucha, S., Westervelt, R.M., and Wingreen, N.S. (1997) *Mesoscopic Electron Transport, (NATO ASI Series E)* (eds L.L. Sohn and L.P. Kouwenhoven), Kluwer, Dordrecht.

148 Kastner, M.A. (1993) *Phys. Today*, 24–31.

149 Ashoori, R.C. (1996) *Nature*, **379**, 413–419.

150 Reichl, L.E. (1998) *A Modern Course in Statistical Physics*, 2nd edn, John Wiley & Sons, New York.

151 Geerligs, L.J., Anderegg, V.F., Holweg, P.A.M., Mooij, J.E., Pothier, H.,

Esteve, D., Urbina, C., and Devoret, M.H. (1990) *Phys. Rev. Lett.*, **64**, 2691.

152 Keller, M.W., Martinis, J.M., Zimmerman, N.M., and Steinbach, A.H. (1996) *Appl. Phys. Lett.*, **69**, 1804.

153 Keller, M.W., Martinis, J.M., and Kautz, R.L. (1998) *Phys. Rev. Lett.*, **80**, 4313.

154 Keller, M.W., Eichenberger, A.L., Martinis, J.M., and Zimmerman, N.M. (1999) *Science*, **285**, 1706.

155 Tucker, J.R. (1992) *J. Appl. Phys.*, **72**, 4399.

156 Chen, R.H., Korotkov, A.N., and Likharev, K.K. (1996) *Appl. Phys. Lett.*, **68**, 1954.

157 Amlani, I., Orlov, A.O., Toth, G., Bernstein, G.H., Lent, C.S., and Snider, G.L. (1999) *Science*, **284**, 289.

158 Likharev, K.K. (1999) *Proc. IEEE*, **87**, 606.

159 Chen, C.D., Nakamura, Y., and Tsai, J.S. (1997) *Appl. Phys. Lett.*, **71**, 2038.

160 Visscher, E.H., Lindeman, J., Verbrugh, S.M., Hadley, P., and Mooij, J.E. (1996) *Appl. Phys. Lett.*, **68**, 2014.

161 Krupenin, V.A., Presnov, D.E., Scherer, H., Zorin, A.B., and Niemeyer, J. (1998) *J. Appl. Phys.*, **84**, 3212.

162 Starmark, B., Henning, T., Claeson, T., Delsing, P., and Korotkov, A.N. (1999) *J. Appl. Phys.*, **86**, 2132.

163 Schoelkopf, R.P., Wahlgren, P., Kozhevnikov, A.A., Delsing, P., and Prober, D.E. (1998) *Science*, **280**, 1238–1242.

164 Wei, Y.Y., Weis, J., Klitzing, K.v., and Eberl, K. (1998) *Phys. Rev. Lett.*, **77**, 1674.

165 Koltonyuk, M., Berman, D., Zhitenev, N.B., Ashoori, R.C., Pfeiffer, L.N., and West, K.W. (1999) *Appl. Phys. Lett.*, **74**, 555.

166 Yoo, M.J., Fulton, T.A., Hess, H.F., Willett, R.L., Dunkleberger, L.N., Chichester, R.J., Pfeiffer, L.N., and West, K.W. (1997) *Science*, **276**, 579–582.

167 Komiyama, S., Astafiev, O., Antonov, V., Kutsuwa, T., and Hirai, H. (2000) *Nature*, **403**, 405.

168 Devoret, M.H. and Schoelkopf, R. (2000) *Nature*, **406**, 1039–1046.

169 Nielsen, M.A. and Chuang, I.L. (2000) *Quantum Computation and Quantum Information*, Cambridge University Press, Cambridge.

170 Loss, D. and Di Vincenzo, D.P. (1998) *Phys. Rev. A*, **57**, 120.

171 Bayer, M., Hawrylak, P., Hinzer, K., Fafard, S., Korkusinski, M., Wasilewski, Z.R., Stern, O., and Forchel, A. (2001) *Science*, **291**, 451–453.

172 Holleitner, A.W., Blick, R.H., Hüttel, A.K., Eberl, K., and Kotthaus, J.P. (2002) *Science*, **297**, 70.

3
Syntheses and Characterizations

3.1
Zintl Ions

3.1.1
Homoatomic and Intermetalloid Tetrel Clusters – Synthesis, Characterization, and Reactivity

Sandra Scharfe and Thomas F. Fässler

3.1.1.1 Introduction

The elements of the carbon group (tetrels) have the outstanding property that they undergo a transition from one of the best insulating materials (diamond) through the most commonly used intrinsic semiconductor (silicon) to typical metals (lead). Even the various allotropes of the individual elements mirror these dramatic changes, such as insulating diamond versus metallic graphite or semi-metallic α-Sn versus metallic β-Sn. Another related property of these elements is the formation of larger homoatomic aggregates, illustrated either by their different modifications (such as carbon fullerenes or nanotubes) or in terms of homoatomic cluster anions, as will be outlined below. These ligand-free homoatomic polyhedral anions (Zintl ions) show a rich variety in structures, and are also fascinating because of their simplicity. The investigation of their versatile chemical reactivity revealed their potential as starting materials for the formation of nanostructured materials such as hexaporous germanium, or the synthesis of well-defined clusters which consist exclusively of two or more types of (semi)metals (intermetalloid clusters).

This chapter reviews the synthesis, structures and physical properties of homoatomic Zintl ions and recent developments in the formation of mixed clusters consisting of the tetrel elements Ge, Sn, and Pb and various d-block elements. A brief summary is provided of the chemistry of homoatomic clusters (this has been reviewed earlier [1, 2]), with the main focus being oriented towards the latest results in the field of intermetalloid clusters [3, 4]

Nanoparticles: From Theory to Application. Edited by Günter Schmid
Copyright © 2010 WILEY-VCH Verlag GmbH & Co. KGaA, Weinheim
ISBN: 978-3-527-32589-4

3.1.1.2 Homoatomic Clusters of Tetrel Elements

3.1.1.2.1 Discrete Clusters in Neat Solids and from Solutions The chemistry of homoatomic main group element clusters can be traced back to the observation made by A. Joannis in 1891, who described the dissolution of a lead bar in sodammonium [5]; this started the fascinating field of the chemistry of polyanions of main group elements. Joannis' report delineates the first observation of the green solution of a polyanion – many years later determined as $[Pb_9]^{4-}$ – from the reaction of Na and Pb in liquid ammonia. The successive studies of Krauss, Smyth, and Zintl established the existence of highly charged particles of several Group 14, 15, and 16 elements [6]. The existence of such polyanions was finally confirmed by the structure determination of an $[Sn_9]^{4-}$ ion, in 1976 [7].

In the meantime, homoatomic ligand-free Zintl clusters of tetrel atoms are known for compositions $[E_4]^{4-}$, $[E_5]^{2-}$, and $[E_9]^{n-}$ with E = Si, Ge, Sn, Pb and n = 2–4. A unique example of a cluster with ten vertices is $[Pb_{10}]^{2-}$ with a bicapped quadratic anti-prismatic structure (Figure 3.1). The smallest clusters $[E_4]^{4-}$ were first prepared in solid-state reactions of alkali metals (A = Na, K, Rb, Cs) with elemental tetrels (E = Si, Ge, Sn, Pb) in a 1 : 1 ratio [8].

In a salt-like description of these compounds, AE, the alkali metal A transfers its valence electron to the tetrel E, and – in agreement with the Zintl–Klemm–Busmann concept [9] – the tetrel elements form P_4-analogous tetrahedral clusters $[E_4]^{4-}$. These highly charged clusters $[E_4]^{4-}$ are retained in a liquid ammonia solution of RbPb, as shown by Korber *et al.* via the crystallization of the ammoniate $Rb_4Pb_4(NH_3)_2$ [10]. The isotypic compounds $A_4Sn_4(NH_3)_2$ were obtained for A = Rb and Cs from reactions of the element Sn, as well as from Cs with $Sn(C_6H_5)_4$ in liquid ammonia [10].

The structures of D_{3h}-symmetric $[Sn_5]^{2-}$ and $[Pb_5]^{2-}$ clusters were elucidated some time ago, in 1977 [11]. They form in ethylenediamine solutions of A–E alloys with different A:E ratios in the presence of [2.2.2]crypt, and without any additional oxidizing agents. A similar route led to the discovery of $[Ge_5]^{2-}$ about 20 years later [12]. This anion was gained from reactions in liquid ammonia, and also characterized as an $[A^+([2.2.2]crypt)]$ salt with A = K or Rb [13]. For the synthesis of the corresponding $[Si_5]^{2-}$, the binary compound $K_{12}Si_{17}$ was extracted

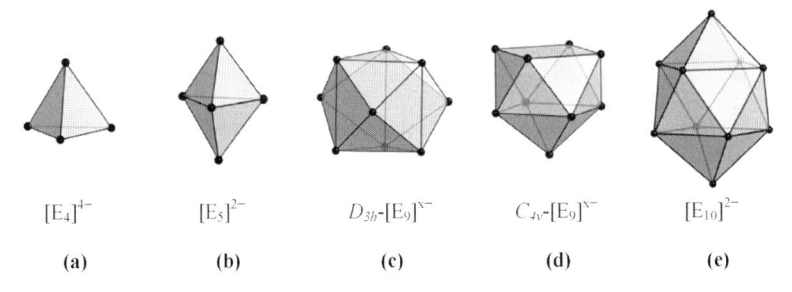

(a)	(b)	(c)	(d)	(e)
$[E_4]^{4-}$	$[E_5]^{2-}$	D_{3h}-$[E_9]^{x-}$	C_{4v}-$[E_9]^{x-}$	$[E_{10}]^{2-}$

Figure 3.1 Structures of known homoatomic tetrel clusters.

with liquid ammonia in the presence of [2.2.2]crypt and treated with triphenylphosphine [14].

Intermetallic compounds of the general composition $A_{12}E_{17}$ contain both tetrahedral clusters $[E_4]^{4-}$ and $[E_9]^{4-}$ polyanions (A = Na, K, Rb, Cs and E = Si, Ge, Sn) [15, 16]. In these compounds, two $[E_4]^{4-}$ units and one anionic cluster $[E_9]^{4-}$ are counterbalanced by 12 positively charged alkali metal cations.

However, nine-atom clusters are most prominent in the intermetallic compounds A_4E_9 (A = Na, K, Rb, Cs; E = Si, Sn, Pb). Cs_4Ge_9 was the first representative of which the crystal structure was presented by Sevov et al. in 1997 [17]. At this time, $[E_9]^{4-}$ clusters were conventionally extracted from intermetallic alloys of the nominal composition A_4E_9 with polar aprotic solvents such as ethylenediamine (en), liquid ammonia, and dimethylformamide (dmf). Intensive investigations have since followed, both in the solid-state and in solution, and the crystal structures of K_4Ge_9, K_4Sn_9 and of A_4Pb_9 with A = K, Rb, Cs have been elucidated [16, 18]. So far, a compound containing exclusively $[Si_9]^{4-}$ clusters and alkali metal cations has not been obtained, and all experiments aimed at obtaining nine-atom silicon clusters in solution started from $A_{12}Si_{17}$ Zintl phases (A = K, Rb, K/Rb) [19–21].

Nonetheless, in most cases the potassium salts K_4E_9 (E = Ge, Sn, Pb) are selected to study the chemistry of Ge_9, Sn_9 and Pb_9 clusters [2, 4], whereas the first crystal structure determination was presented for $Na_4Sn_9(en)_7$ by Diehl, Kummer and Strähle [7].

A major step forward in order to obtain crystalline material suitable for the structure determination of soluble Zintl ions by X-ray crystallography was achieved by using alkali metal ion-sequestering agents such as [2.2.2]crypt [22] as employed by Corbett [23], and [18]crown-6 [24] used by Fässler [25], thereby enhancing the solubility of the intermetallic phases. Sequestered cations also have a major influence on the ion packing in the solid, and the emerging products display an exciting structural diversity.

The structures of $[E_9]^{4-}$ clusters in alkali metal-crypt or -crown-ether salts [2] were consecutively determined for E = Ge, Sn and Pb, and the series was recently completed by Korber, who presented the crystal structure of [Rb([18]crown-6)] $Rb_3Si_9(NH_3)_4$, crystals, obtained by dissolving the ternary phase $K_6Rb_6Si_{17}$ in liquid ammonia in the presence of [18]crown-6. The structure of the $[Si_9]^{4-}$ anion deviated slightly from the expected monocapped square antiprism, and was coordinated by seven Rb^+ cations [21].

Nine-atom clusters with a threefold negative charge $[E_9]^{3-}$ were isolated from solution for all heavier Group 14 elements (E = Si to Pb), and obtained exclusively as $[K^+([2.2.2]crypt)]$ salts [14, 26–28]. These compounds contain either one or two symmetrically independent E_9 clusters per asymmetric unit, beside [2.2.2]crypt-sequestered potassium cations and various amounts of solvent molecules. In the solid state, all compounds show anisotropic electron paramagnetic resonance (EPR) signals, with increasing line widths from Ge to Pb. In compounds with two E_9 clusters per asymmetric unit, one cluster always is well ordered, whereas the second cluster shows disorder for some of the tetrel atoms. Detailed magnetic studies on [K ([2.2.2]crypt)]$_6E_9E_9(en)_{1.5}(tol)_{0.5}$ (E = Sn, Pb) revealed that only 50% of the clusters were paramagnetic [26]. Consequently, a threefold negative charge for the ordered cluster and a superposition of $[E_9]^{2-}$ and $[E_9]^{4-}$ clusters – and thus disproportionation of $[E_9]^{3-}$ for the disordered one cluster – have been assumed.

In the mixed-valent compound $[K^+([2.2.2]crypt)]_6[Ge_9]^{2-}[Ge_9]^{4-}(en)_{2.5}$, a cluster with the shape of a strongly distorted tricapped trigonal prism was interpreted as $[Ge_9]^{2-}$ [29]. The charge distribution, however, has not been verified by magnetic measurements. Aside from this result, we and other groups repeatedly obtained crystals that crystallized in the hexagonal crystal system and contained strongly disordered $[Ge_9]^{2-}$ clusters that were counterbalanced by two [2.2.2]crypt-sequestered K cations per cluster unit. In spite of numerous attempts, the crystal structure could not be solved to date, because of heavy disorder. However, the structure of a compound with the same cell parameters and disordered cluster units, but different color, was reported to contain $[Ge_{10}]^{2-}$ anions [30]. $[E_9]^{3-}$ clusters have not yet been observed in neat solids.

Until now, a nine-atom cluster with an unambiguous charge of −2 has been isolated and structurally characterized only in $[K([18]crown-6)]_2Si_9(py)$ [19]. This compound was obtained by the dissolution of $K_{12}Si_{17}$ in liquid ammonia in the presence of Ph_3GeCl, whose role during the synthesis remains unexplained. Subsequently, the reaction product was crystallized from pyridine.

Finally, the *closo* cluster $[Pb_{10}]^{2-}$ was obtained after the controlled oxidation of an ethylenediamine solution of K_4Pb_9 with $(PPh_3)Au(I)Cl$ [31]. This ten-vertices cluster represents the largest empty tetrel atom cluster.

Homoatomic clusters with six, seven, eight, or more than ten vertices have not yet been isolated from solution, although both gas-phase experiments and theoretical investigations have predicted a remarkable stability for the stannaspherene $[Sn_{12}]^{2-}$ and the plumbaspherene $[Pb_{12}]^{2-}$ [32]. Octahedral $[E_6]^{2-}$ clusters have only been isolated as transition metal-stabilized species $\{[EM(CO)_5]_6\}^{2-}$ (E = Ge and Sn; and M = Cr, Mo or W). These display an almost-perfect E_6 octahedron, with each E atom coordinated to an $M(CO)_5$ fragment [33].

3.1.1.2.2 **Cluster Shapes and Ion Packing** The structures of tetrahedral $[E_4]^{4-}$, trigonal bipyramidal $[E_5]^{2-}$, and bicapped quadratic–antiprismatic $[E_{10}]^{2-}$ clusters strictly follow Wade's rules and form $12 = 4 \times 2 + 4$) skeletal electron (ske) *nido*-, $12 = 5 \times 2 + 2$) ske *closo*-, and $22 = 10 \times 2 + 2$) ske *closo*-clusters, respectively [34]. Each Group 14 atom has four valence electrons, two of which remain as a "lone pair" at each vertex equivalent to the covalent *exo*-BH bonds in boranes and are not involved in the framework bonding.

The crystal structures of the ammoniates containing $[E_4]^{4-}$ clusters show remarkable similarities to those of the corresponding intermetallic 1:1 phases with respect to cluster shape and coordination spheres [10]. Thus, strong contacts between the alkali metal cations and the tetrel atoms abound in all compounds.

A deltahedral *closo*-cluster with nine atoms ($n = 9$) corresponds to a tricapped trigonal prism and D_{3h} point group symmetry (Figure 3.1c) if $2n + 2 = 20$ ske are available for the cluster skeleton bonding. Hence, a cluster with a twofold negative charge, $[E_9]^{2-}$, is appropriate for a *closo*-cluster, and $[E_9]^{4-}$ is on par with a *nido*-cluster due to $2n + 4 = 22$ ske, forming a C_{4v}-symmetric monocapped square antiprism

(Figure 3.1d). The shape of a 21 ske cluster $[E_9]^{3-}$ should integrate between the *closo*- and the *nido*-form in a C_{2v}-symmetric structure, which derives from a tricapped trigonal prism with one or two elongated prism heights.

Numerous crystal structures that contain $[E_9]^{4-}$ clusters obtained from liquid ammonia and ethylenediamine solutions for E = Ge, Sn, and Pb exist [2]. A careful examination of the atomic displacement parameters of the cluster atoms has shown that the static structure usually obtained by X-ray structure determination does not unambiguously allow an assignment to one of the boundary structures. Indeed, the energy barrier for the intermolecular conversion between the D_{3h}-*closo* and the C_{4v}-*nido* structure is very low [35]. Nuclear magnetic resonance (NMR) experiments in solution proofed that $[Sn_9]^{4-}$ and $[Pb_9]^{4-}$ clusters exhibit fluxional behavior on the NMR time scale [36]. Likewise, $[E_9]^{n-}$ ($n = 2-4$) clusters frequently display distortions from the expected ideal cluster shapes in the solid-state structures. As a consequence of the small energy barriers between the various conformers, ion packing also highly influences the cluster shape in the solid. In binary solids, an A_4E_9 multi-faceted coordination of the alkali metal cations to the $[E_9]^{4-}$ clusters occurs. These contacts are feasible towards the cluster vertices, edges and triangular faces, although η^4-coordination to the open square face of a *nido*-cluster was also found.

In various $[E_9]^{4-}$ compounds obtained from solution, the A^+ ions are either surrounded by crypt molecules and by less encapsulating crown ethers, though the compounds can also contain unsequestered cations. Compared to the solid-state structures of the binary intermetallic compounds A_4E_9, the number of $A-E_9$ contacts gradually decreases with increasing content of solvent and sequestering molecules, resulting in lower-dimensional networks – that is, layers and chains (Figure 3.2).

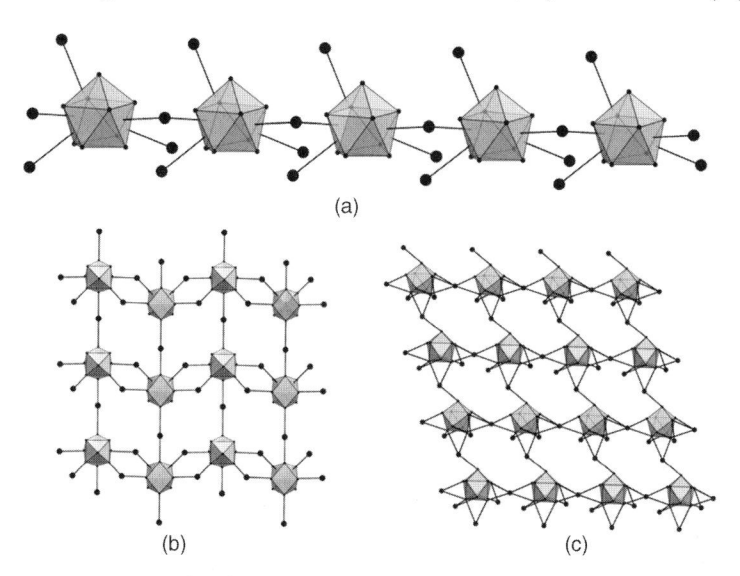

Figure 3.2 Examples of structures with ion packings resembling one- and two-dimensional intermetallic motifs of (a) a linear chain $^1_\infty[K_4Sn_9]$ in $[K([18]crown-6)]_3KE_9$ [25], (b) a 2-D-layer $^2_\infty[(K_3Se_9)^-]$ in $[K([2.2]crypt)][K_3Ge_9](en)_2$ [39] and (c) a 2-D-layer $^2_\infty[K_4Sn_9]$ in $[K([18]crown-6)]_2K_2Sn_9(en)_{1.5}$ [42].

In solvate crystals obtained from liquid ammonia and ethylenediamine germanium, *nido*-$[Ge_9]^{4-}$ clusters were found together with "naked" K or Rb and Rb or Cs cations, respectively [37, 38], in which all cations were coordinated to the cluster anions in numerous contacts, resulting in a three-dimensional (3-D) network very similar to that in the binary intermetallic phases. The solvent molecules in the lattice are attached to the alkali metal atoms, but the compounds $A_4Ge_9(en)$ (A = Rb, Cs) lose their ethylenediamine easily upon heating, thereby leaving A_4Ge_9 [38].

A two-dimensional (2-D) network of $[Ge_9]^{4-}$ clusters was found in [K([2.2]crypt)][K$_3$Ge$_9$](en)$_2$, where anionic $^2_\infty[(K_3Se_9)^-]$ layers were separated by layers of [K$^+$([2.2]crypt)] cations that were coordinated to the Ge$_9$ *nido*-cluster in a η^4-capping position (Figure 3.2) [39].

A solvate compound with the composition $Na_4Sn_9(en)_7$ was identified, though the results of its solid-state structure determination remain unsatisfactory [7]. Apparently, the $[Sn_9]^{4-}$ cluster has contacts to two sodium atoms via triangular faces, and adopts the shape of a distorted tricapped trigonal prism which is, in fact, contradictory to Wade's rules. Further investigations yielded the isotypic compounds [K([2.2.2]crypt)]$_3$KE$_9$ [28, 40] and a series of different compounds with the general composition [K([18]crown-6)]$_{4-x}$K$_x$E$_9$(sol) ($x = 0$–2 for E = Sn; $x = 0$, 2 for E = Pb) [25, 41–43] which all display multiple cation–anion contacts in the solid state. Although most of the clusters have approximate C_{4v} symmetry, $[Sn_9]^{4-}$ also appears as a distorted tricapped trigonal prism in [K([18]crown-6)]$_4$Sn$_9$ and [K([18]crown-6)]$_3$KSn$_9$(en) (Figure 3.3a) [25]. In the latter case, infinite inorganic chains of $^1_\infty[K_4Sn_9]$ lance through the crystal (Figure 3.2a), and are surrounded by crown ether molecules. Similar chains are found in [K([2.2.2]crypt)]$_3$KE$_9$ although, due their spherical coordination by [2.2.2]crypt, the encapsulated cations exhibit no contacts to the almost C_{4v}-symmetric anions, which results in a $^1_\infty[(K_3Se_9)^{3-}]$ chain [28, 40].

In [K([18]crown-6)]$_2$K$_2$E$_9$(en)$_{1.5}$, the slabs of $^2_\infty[K_4Sn_9]$ are separated by [18]crown-6 [42, 43] molecules, as also observed in the isotypic compound [Rb([18]crown-6)]$_2$Rb$_2$Sn$_9$(en)$_{1.5}$ [44]. Two and three [K([18]crown-6)] units are coordinated to the E$_9$ cluster in [K([18]crown-6)]$_4$Sn$_9$ and [K([18]crown-6)]$_4$Pb$_9$(en)(tol), leading to isolated structure motifs that are furthest away from those found in the binary intermetallics (Figure 3.3) [25, 41].

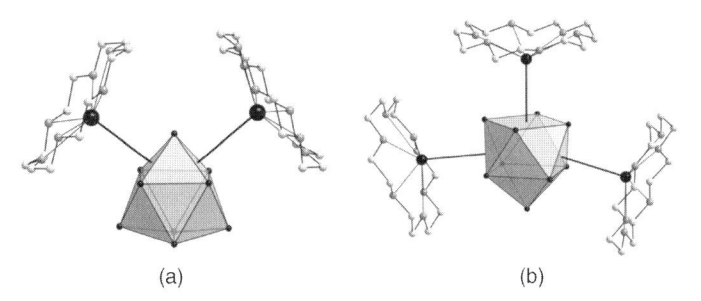

Figure 3.3 Structure detail of (a) [K([18]crown-6)]$_3$K[Sn$_9$] [25] and (b) [K([18]crown-6)]$_4$[Pb$_9$] [41] with two and three connected [K([18]crown-6)] units, respectively.

In the crystal structure of [Na([2.2.2]crypt)]$_4$Sn$_9$ all cations are completely surrounded by cryptand molecules, and no interactions occur between the anionic *nido*-clusters and Na$^+$ [23].

[E$_9$]$^{4-}$ clusters without direct contacts to alkali metal cations also exist in the compounds [Li(NH$_3$)$_4$]$_4$E$_9$(NH$_3$) with E = Sn, Pb, where the coordination sphere of the Li ions is completed by four ammonia molecules [45]. This situation resembles that of all structures containing paramagnetic [E$_9$]$^{3-}$ clusters, in which no cation–anion interactions were ever found. All of the [E$_9$]$^{3-}$ units adopt the shape of a distorted tricapped trigonal prism with approximate C_{2v} symmetry, except for [Sn$_9$]$^{3-}$ which, surprisingly, retains almost perfect D_{3h} symmetry.

Similarly, the *closo*-[Si$_9$]$^{2-}$ cluster noticeably deviates from an ideal D_{3h} cluster shape expected according to its 20 ske [19]. The [Si$_9$]$^{2-}$ unit interacts with two K cations via two deltahedral cluster faces, and the shape of the cluster is best described as a distorted tricapped trigonal prism with one elongated prism height.

3.1.1.2.3 Linked E$_9$ Clusters

The formation of stable radicals [E$_9^\bullet$]$^{3-}$ also opens the possibility of cluster dimerization and – through further oxidation – the formation of polymeric chains with *exo*-bonds between the clusters. Hitherto, such oxidative coupling reactions have been reported exclusively for Ge$_9$ clusters (Figure 3.4). In

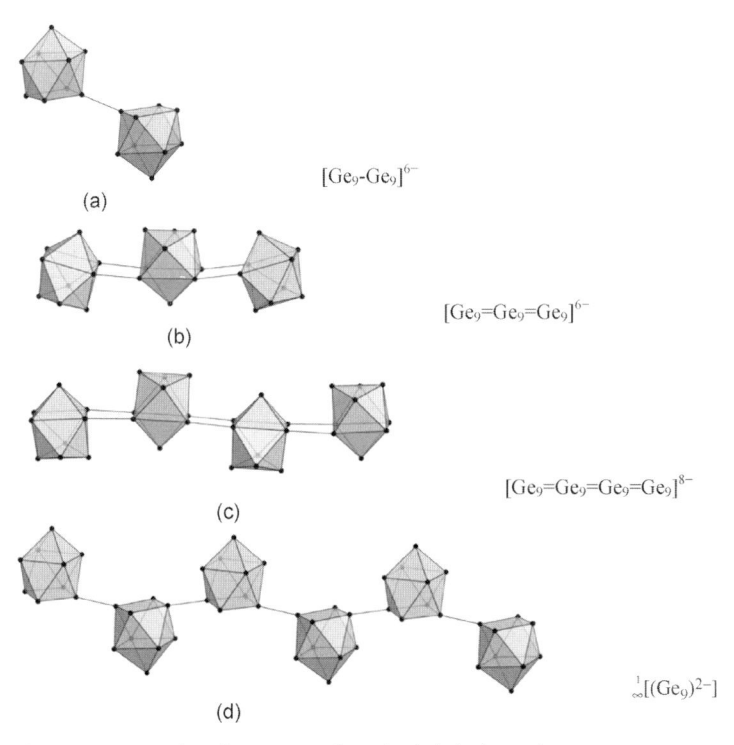

Figure 3.4 Examples of structures of covalently linked Ge$_9$ clusters.

dimeric cluster anions, two *nido*-shaped Ge$_9$ units are connected via an *exo*-bond between two atoms of the open rectangular cluster faces [37, 46–48]. The clusters deviate from C_{4v} symmetry and, in general, the diagonal of the open face that points towards the *exo*-bonded atom is significantly shorter than the second diagonal. In all cases several alkali metal cations coordinate to the dimeric cluster units [Ge$_9$–Ge$_9$]$^{6-}$, which leads to low-dimensional overall structures such as chains [46, 47] or layers [37, 46, 48] in the solid state. Apparently, the complete encapsulation of the alkali metal atoms by the sequestering agent seems to be the crucial factor for the stabilization of free cluster radicals in the crystals. Further oxidation to formal [Ge$_9$]$^{2-}$ then leads to the one-dimensional (1-D) cluster polymer $^1_\infty[(Ge_9)^{2-}]$ [39, 49, 50]. The connection always proceeds via two opposite vertex atoms of the open faces of the *nido*-clusters. In the structure of [K([18]crown-6)]$_2\{^1_\infty[Ge_9]\}$(en) – which was the first example of its kind studied using X-ray crystallography – one of the two [K$^+$([18]crown-6)] units is seen to cap a triangular face of the cluster, while the other unit serves as spacer between the polymeric chains, which thus are separated by more than 14 Å [49].

Other experiments with weak oxidation agents led to the discovery of a linear trimeric and a linear tetrameric anion, where three or four Ge$_9$ units are doubly linked with each other via two parallel *exo*-bonds. The individual clusters in [Ge$_9$=Ge$_9$=Ge$_9$]$^{6-}$ [51] and [Ge$_9$=Ge$_9$=Ge$_9$=Ge$_9$]$^{8-}$ [52, 53] adopt the shape of a tricapped trigonal prism, and these units are collinearly interconnected via two of their three prism heights that are consequently elongated. The intercluster bonds are longer than the *exo*-bonds observed in [Ge$_9$–Ge$_9$]$^{6-}$, and thus are supposed to have a bond order less than one, which is supported by the overall cluster charge of −6 and −8, respectively. Thus, the intercluster bonds within the trimer and tetramer should not be regarded as 2-electron–2-center *exo*-bonds, but rather as part of a delocalized electron system that comprises the whole anion [52], with the cluster charge being equally distributed among the clusters units. It is worth mentioning at this point that most of the dimeric, oligomeric and polymeric cluster compounds have been obtained without the addition of oxidation agents.

3.1.1.3 Intermetalloid Clusters of Tetrel Elements

3.1.1.3.1 Complexes of Zintl Ions Soluble homoatomic Zintl ions [E$_9$]$^{4-}$ launch a unique possibility for the feasible formation of heteroatomic clusters. The starting material is easily accessible and available in good quantities. For potassium-containing representatives, the elements are fused at high temperatures in metal ampoules in the ratio K:E = 4:9. Originally, the phases were dissolved in liquid ammonia, but the resulting K$_4$E$_9$ phases are also soluble in ethylenediamine, as shown by Diehl and Kummer [54]. In order to increase the synthetic potential, other solvents have been employed, for example, dmf which was successfully debuted for the synthesis of an [Ge$_9$–Ge$_9$]$^{6-}$ anion [48]. Another simple and efficient one-pot route for the synthesis of compounds containing the elements Sn and Pb proceeds via their reduction with Na or K in liquid [18]crown-6 at 40 °C [25].

The study of the reactions of Zintl ions with transition metal compounds started in 1988 when Eichhorn and Haushalter discovered the anion [Sn$_9$Cr(CO)$_3$]$^{4-}$ as the first

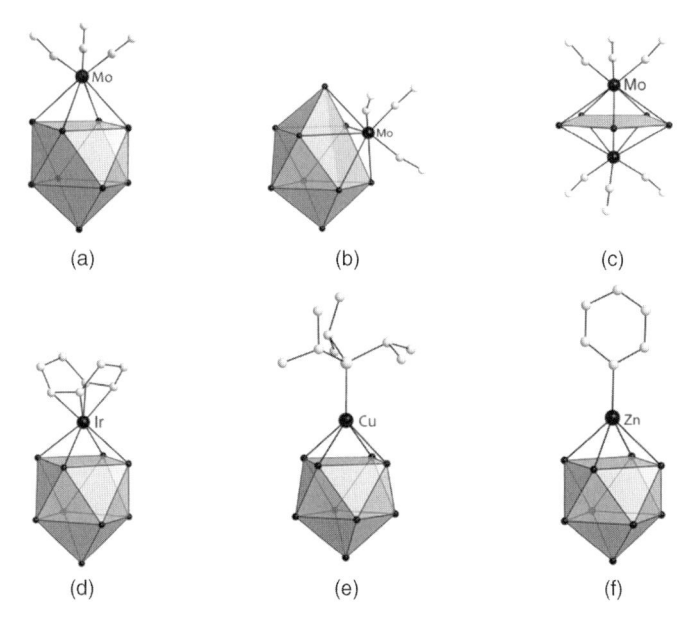

Figure 3.5 Coordination compounds with Zintl ions. (a, b) Isomers of [E$_9$-M(CO)$_3$]$^{4-}$ (E = Sn, Pb; M = Cr, Mo, W) [55–57]; (c) [(CO)$_3$Mo(Pb$_5$)Mo(CO)$_3$]$^{4-}$ [89]; (d) [Sn$_9$-Ir(cod)]$^{3-}$ [58]; (e) [Ge$_9$-Cu(P^{i}Pr$_3$)]$^{3-}$ [59]; (f) [E$_9$-Zn(C$_6$H$_5$)]$^{3-}$ (E = Si to Pb) [20].

nonborane transition metal–main group element deltahedral cluster [56]. Subsequently, the synthesis of other complexes [E$_9$M(CO)$_3$]$^{4-}$ (E = Sn, Pb; M = Cr, Mo, W) of this series was attained (Figure 3.5a) [55, 57]. The *nido*-E$_9$ cluster predominantly coordinates with its four atoms of the open square to the transition metals in a η^4-fashion, but isomerization of the resulting *closo*-clusters also gives access to η^5-E$_9$ complexes (Figure 3.5b) [55]. The η^5-Pb$_5$ coordination is rather similar to that found in [(CO)$_3$Mo(Pb$_5$)Mo(CO)$_3$]$^{4-}$ [89] (Figure 3.5c), with a planar cyclo-Pb$_9$ ligand. The latter anion was obtained as K$_2$[K([2.2]-crypt)]$_2$[Pb$_5${Mo(CO)$_3$}$_2$](en)$_3$ using [2.2]-crypt (diazo[18]crown-6) instead of the more frequently used [2.2.2]-crypt, in moderate and reproducible yields, from the reaction of a [Pb$_9$]$^{4-}$ solution with [MesMo(CO)$_3$]. Transition metal complexes, for example, (CO)$_3$Cr(η^6-1,3,5-C$_6$Me$_3$H$_3$) [56], react with Group 14 Zintl ions in ligand-exchange reactions, and in the products the E$_9$ clusters act as six-electron donors. Recently, other complexes involving d-block elements have been isolated: [Sn$_9$–Ir(cod)]$^{3-}$ [58] (Figure 3.5d), [Ge$_9$–Cu(P^{i}Pr$_3$)]$^{3-}$ [59] (Figure 3.5e), [E$_9$–Zn(C$_6$H$_5$)]$^{3-}$ (E = Si–Pb) [20], (Figure 3.5f) and [Sn$_9$–Cd(Sn^{n}Bu$_3$)]$^{3-}$ [60].

According to the isolobal concept, M(CO)$_3$ fragments with M = Cr, Mo, and W, as well as Ir(cod)$^+$, CuPR$_3$$^+$ and MR$^+$ (M = Zn, Cd and R = C$_6$H$_5$ and Sn(alkyl)$_3$) units act as zero-electron building blocks. Therefore – based on the Wade–Mingos rules – the cluster expansion corresponds to the transformation of a nine-atom *nido*-cluster to a ten-atom *closo*-cluster with 9 × 2 + 4 = 22 ske and 10 × 2 + 2 = 22 ske, respectively, as the number of delocalized skeleton electrons does not change through the coordination of the transition metal fragment.

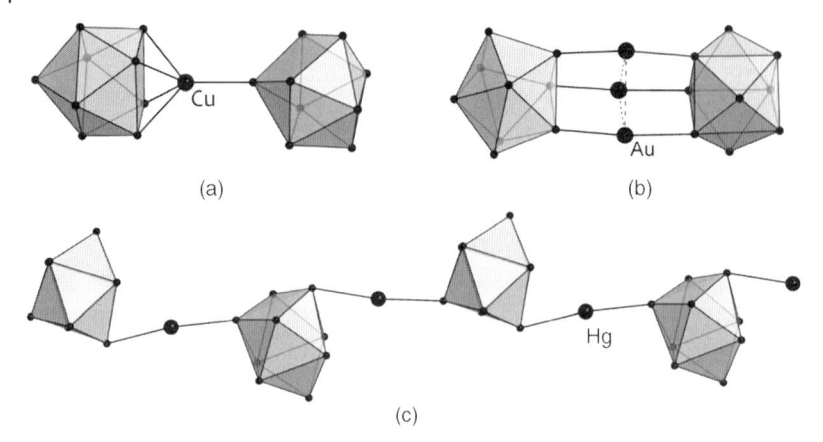

Figure 3.6 Zintl ions with ligand-free d-block metal atoms. (a) $[(\eta^4\text{-Ge}_9)\text{Cu}(\eta^1\text{-Ge}_9)]^{7-}$ [59]; (b) $[\text{Ge}_9\text{Au}_3\text{Ge}_9]^{5-}$ [65]; (c) $^1_\infty[(\text{Hg Ge}_9)^{2-}]$ [61].

3.1.1.3.2 Ligand-Free Heteroatomic Cluster: Intermetalloids The isolation of a compound containing a $^1_\infty[(\text{Hg Ge}_9)^{2-}]$ [61] chain (Figure 3.6c) as one of the first ligand-free complexes of a Zintl anion with a d-block element (Figure 3.6c) initiated a further round in Zintl ion chemistry. The formation of heterometallic ligand-free anions now allows an approach of intermetallic phases from a "bottom-up" synthesis. In analogy to Schnöckel's term of metalloid clusters [62], the expression "intermetalloid clusters" was introduced [3] for cluster compounds, that consist exclusively of atoms of at least two different (semi)metallic elements and which might even exhibit topological similarities to structural motifs of related intermetallic compounds.

Whereas, the anionic polymer in Figure 3.6c was obtained by the direct reaction of a solution containing $[\text{Ge}_9]^{4-}$ ions and elemental Hg [61, 63], an oligomer cutout with almost identical connectivity of the atoms, $[\text{Hg}_3(\text{Ge}_9)_4]^{10-}$, was generated recently by the reaction of the Zintl ion with $\text{Hg}(\text{C}_6\text{H}_5)_2$ [64]. These complexes are best described by assuming covalent Ge–Hg contacts between Ge$_9$ clusters. Each *exo*-bond of the $[\text{Ge}_9]^{4-}$ *nido*-cluster reduces the charge by one; hence, with two *exo*-bonds per cluster, the charge of the *nido*-clusters arises to -2.

The anion $[(\eta^4\text{-Ge}_9)\text{Cu}(\eta^1\text{-Ge}_9)]^{7-}$ (Figure 3.6a) is closely related to $[(\eta^4\text{-Ge}_9)\text{Cu-P}(i\text{-Pr}_3)]^{3-}$ (Figure 3.5d), but P(i-Pr)$_3$ is substituted by a second $[\text{Ge}_9]^{4-}$ ligand [59]. Therefore, this anion represents a rare case where the homoatomic cluster acts as a two-electron σ-donor ligand. A similar bonding situation has been observed before only in the octahedral complexes $[\text{E}(\text{Cr(CO)}_5)_6]^{2-}$ (E = Ge and Sn) [33].

The replacement of Cu by Au using (PPh$_3$)Au(I)Cl as reactand leads to the related cluster $[\text{Ge}_9\text{Au}_3\text{Ge}_9]^{5-}$, as shown in Figure 3.6b [65]. In $[\text{Au}_3\text{Ge}_{18}]^{5-}$, three Au atoms are arranged in a triangle that bridges two clusters, with the triangular gold unit being almost coplanar to the two deltahedral faces of the neighboring $[\text{Ge}_9]^{4-}$ clusters; rather short Au–Au contacts within the triangle hint at aurophilic interactions [90]. The positive nature of the charge of the Au atoms was confirmed using density functional theory (DFT) calculations.

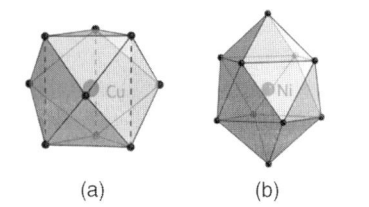

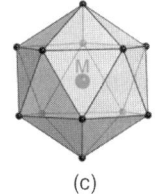

Figure 3.7 Representative Zintl ions with one endohedral metal atom. (a) [Cu@E$_9$]$^{3-}$ (E = Sn, Pb) [68]; (b) [Ni@Pb$_{10}$]$^{2-}$ [69]; (c) [M@Pb$_{12}$]$^{2-}$ (M = Ni, Pd, Pt) [71, 72] and [Ir@Sn$_{12}$]$^{3-}$ [58].

Stripping off the ligands in complexes shown in Figure 3.5 can lead to filled nine-atom clusters. The smallest transition metal-filled Zintl clusters were found in [Ni@Ge$_9$]$^{3-}$ [66, 67] and [Cu@E$_9$]$^{3-}$ for E = Sn and Pb [68] (Figure 3.7). [Ni@Ge$_9$]$^{3-}$ was obtained from reactions of K$_4$Ge$_9$ with Ni(cod)$_2$. [Cu@E$_9$]$^{3-}$ resulted from reactions of Cu–Mes with the respective Zintl phases K$_4$Sn$_9$ and K$_4$Pb$_9$ in dmf. In these compounds, the E$_9$ cluster skeleton is retained, and a transition metal is incorporated into it. The clusters can be considered as paramagnetic [Ni^0@(Ge$_9$)$^{3-}$] and diamagnetic [Cu$^+$@(E$_9$)$^{4-}$]. The valence state of nickel in [Ni@Ge$_9$]$^{3-}$ was confirmed using EPR measurements [66, 67], while for the copper compounds ^{119}Sn-, ^{207}Pb- and ^{63}Cu–NMR investigations approved diamagnetism by sharp NMR signals [68]. The charge of Cu(I) was also confirmed by DFT calculations.

The structure refinement of [Ni@Ge$_9$]$^{3-}$ suffers from an intense disorder [66, 67], in that the cluster apparently adopts the shape of a strongly distorted tricapped trigonal prism with unequally elongated prism heights in the range of 3.089–3.560 Å. In [Cu@E$_9$]$^{3-}$, the cluster atoms also build a tricapped trigonal prism with elongated heights. Maximizing the strength of nine Cu–E interactions leads to an almost perfect spherical shape of the nine-atom cluster, and not to the typical *nido*-type structure of 22 skeleton electron [E$_9$]$^{4-}$ clusters. The prism heights in this anion differ from each other by maximal 5% (E = Sn) or 6% (E = Pb), but they are extended by 17% and 15%, respectively, compared to empty clusters with a similar skeleton (as shown in Figure 3.1d). As a result of this elongation, the Cu–E distances vary in a very narrow range of ±0.05 Å for both compounds, with slightly longer contacts to the capping atoms leading to an almost perfect sphere of E atoms around the Cu atom. NMR experiments using [Cu@E$_9$]$^{3-}$ solutions again illustrated the structural flexibility of the E$_9$ clusters [68].

The endohedral ten-vertices cluster [Ni@Pb$_{10}$]$^{2-}$ (Figure 3.7b) was obtained from the reaction of Ni(cod)$_2$ with K$_4$Pb$_9$ in ethylenediamine [69]. Herein, the Ni^0 atom occupies the center of a bicapped square antiprism, as predicted from DFT calculations [70]. ^{207}Pb-NMR studies again revealed a fast atom exchange of the cluster atoms in solution, as only one (broad) signal at −996 ppm was detected [69].

The process of formation of the ten-vertices clusters remains unexplained to date. Apparently, an oxidation with the cod ligand serving as oxidizing agent takes places, as the resulting cyclo-octene could be detected using gas chromatography-mass spectrometry (GC-MS) measurements [71]. However, as noted above, oxidation reactions have also been observed without the addition of further reactants. The icosahedral cluster $[Ni@Pb_{12}]^{2-}$ was found as a minor byproduct from the reaction of $Ni(cod)_2$ with K_4Pb_9 (Figure 3.7c) [71]. For the heavier homologs Pd and Pt, the isostructural clusters $[Pd@Pb_{12}]^{2-}$ and $[Pt@Pb_{12}]^{2-}$ were obtained from the reaction of K_4Pb_9 with $Pd(PPh_3)_4$ and $Pt(PPh_3)_4$, respectively [71, 72]. Again, the cluster atoms are oxidized during the reaction which, in this case, is ascribed to PPh_3. The cluster shape follows Wade's rules that predict a *closo*-structure for a cluster with twelve vertices and $2n + 2 = 26$ electrons. Although the corresponding Pd and Pt analogs of $[Ni@Pb_{10}]^{2-}$ have not yet been studied in the solid state or in solution, they are known to occur in the gas phase and were identified using laser desorption–ionization–time-of-flight (LDI-TOF) mass spectroscopic measurements on dmf solutions of crystalline samples of $[K(2.2.2crypt)]_2[M@Pb_{12}]$ (solv.) [71]. These studies also demonstrated the presence of the empty Zintl clusters $[Pb_{10}]^{2-}$ and $[Pb_{12}]^{2-}$. All of these gas-phase species must arise during the fragmentation process of $[M@Pb_{12}]^{2-}$.

A comparison of the solid-state structures of $[M@Pb_{12}]^{2-}$ confirms that the smaller the transition metal is, the stronger distorted is the lead skeleton [71]. While the interstitial Pt atom is surrounded by an almost perfect icosahedron with virtually equal Pb–Pb distances, the Pb–Pb contacts for the interstitial Pd and Ni atoms vary over a larger range with increasing variances. The cluster distortion manifests in the distribution of the M–Pb distances that increases from Pt to Ni. The arising anisotropy certainly indicates the instability of the cluster, whereas for M = Ni the ten-vertices cluster is clearly favored. In solution, one ^{207}Pb-NMR resonance was found for $[M@Pb_{12}]^{2-}$ (Ni: 1167 ppm, Pd: 1520 ppm, Pt: 1780 ppm); this either reflects the high symmetry of the cluster, or it can be ascribed to a fluxional behavior [71].

An analogous germanium compound displaying a metal atom-centered Ge_{12} cluster has not been reported to date.

Recently, the successful synthesis of an endohedral twelve-atom tin cluster was conducted, $[Ir@Sn_{12}]^{3-}$ [58] which, apparently, is formed in a stepwise reaction which starts from an ethylenediamine solution of K_4Sn_9 and $[IrCl(cod)]_2$ (Scheme 3.1). Initially, the Ir(cod)-capped Sn_9 cluster is formed (Figure 3.5d), but if the ethylenediamine solution of this ten-vertices cluster is heated to 80 °C it is transformed into $[Ir@Sn_{12}]^{3-}$ under loss of the cod ligand. The reaction is accelerated by the addition of PPh_3 or dppe. The phosphines act as oxidation agents, as it has also been shown previously [71]. The cluster can be written as $[Ir^{1-}@(Sn_{12})^{2-}]$, and the cluster skeleton fulfills Wade's rules; notably, the twelve tin atoms form an almost perfect icosahedron, with Sn–Sn distances in a slightly larger range than are found in $[Pt@Pb_{12}]^{2-}$.

$$2\,Sn_9^{4-} + [IrCl(C_8H_{12})]_2 \longrightarrow 2\,[Sn_9Ir(C_8H_{12})]^{3-} + cod + 2\,Cl^-$$

$$2\,[Sn_9Ir(C_8H_{12})]^{3-} \xrightarrow{oxidation} [Ir@Sn_{12}]^{3-} + 2\,cod + \text{"Ir/Sn}_x\text{"}$$

Scheme 3.1 Step-by-step synthesis of the endohedral cluster $[Ir@Sn_{12}]^{3-}$ [58].

Hints for the formation of larger assemblies comprising 17 or 18 tetrel atoms are given by $[(\eta^4\text{-}Ge_9)Cu(\eta^1\text{-}Ge_9)]^{7-}$ (as shown in Figure 3.6a), and by a class of E_9 clusters which are filled and simultaneously capped by a transition metal, such as the anions $[(Ni@Sn_9)Ni\text{-}CO]^{3-}$ and $[(Pt@Sn_9)PtPPh_3]^{3-}$ [73] (see Figure 3.8a and b, respectively) as well as by modified Ge_9 clusters $[(Ni@Ge_9)Ni\text{-}R]$ (R = CO, C–C–Ph, en) [67]. Assuming that a ligand exchange in $[Ge_9\text{-}Cu(P^iPr_3)]^{3-}$ (as discussed above) would also take place in $[(Pt@Sn_9)PtPPh_3]^{3-}$, the formation of a cluster species that include two transition metals and 18 tetrel atoms (see e.g., Figure 3.8f) could be achieved. However, more complex redox reactions must be assumed. The educts that led to the discovery of $[M@Pb_{12}]^{2-}$ gave clusters of different sizes when reacting with Sn_9 clusters. Three new clusters of higher nuclearity, $[Ni_2@Sn_{17}]^{4-}$ [74], $[Pt_2@Sn_{17}]^{4-}$ [75], and $[Pd_2@Sn_{18}]^{4-}$ [76, 77], were synthesized in ethylenediamine solution from $Ni(cod)_2$, $Pt(PPh_3)_4$, and $Pd(PPh_3)_4$, respectively (Figure 3.8c, d, and f). The cluster formation requires the partial oxidation of the Sn_9 clusters, along with the reduction of cyclo-octadiene involving also the solvent ethylenediamine. In $[Ni_2@Sn_{17}]^{4-}$, two $[Ni@Sn_9]^{2-}$ subunits are connected via one apex-like, central Sn atom that participates in both Sn_9 clusters [74]. The clusters deviate significantly from deltahedral structures. With a more open shape, $[Ni_2@Sn_{17}]^{4-}$ displays D_{2d} point group symmetry, and the central Sn atom is surrounded by a pseudo-cube of eight Sn atoms. The high coordination number of the central Sn atom causes remarkably long Sn–Sn contacts that are more common in solid-state compounds such as $BaSn_5$, in which Sn atoms even reach a twelve-fold coordination by other tin atoms [78]. The dynamic behavior of this cluster was examined by temperature-dependent ^{119}Sn-NMR experiments in dmf solutions [74]. At a temperature of −64 °C, four resonances appeared in the ^{119}Sn-NMR spectrum (−1713 ppm, −1049 ppm, −1010 ppm, +228 ppm) that were assigned to the different cluster atoms via peak intensity analysis, while the central Sn atom generated the most deshielded signal. At 60 °C, only one signal was observed at −1167 ppm, which indicated fast atom-to-atom exchange reactions. In contrast, $[Pt_2@Sn_{17}]^{4-}$ (Figure 3.8d) generated only one ^{119}Sn-NMR resonance over the whole temperature range (−742.3 ppm), and apparently was fluxional in dmf solution even at −60 °C [75]. The solid-state structure of $[Pt_2@Sn_{17}]^{4-}$ differs significantly from that of the isoelectronic $[Ni_2@Sn_{17}]^{4-}$. The Sn_{17} cluster cage encloses two Pt atoms at a distance of 4.244 Å, which does not indicate the presence of bonding contacts between the two Pt atoms [75].

A similar ellipsoid structure was found for the cluster $[Pd_2@Sn_{18}]^{4-}$ [76, 77] that is accessorily isoelectronic to the earlier-discovered anion $[Pd_2@Ge_{18}]^{4-}$ [79]. In both clusters, two Pd^0 atoms are encapsulated by 18 tetrel atoms that form a prolate deltahedral cluster; these units are the largest endohedral tetrel atom clusters with a completely closed cluster shell yet discovered. The interstitial Pd atoms with d^{10} electron configuration approximately occupy the ellipse focuses. Even though the Pd–Pd distance of 2.831 Å is shorter in the smaller Ge cage [79] than in the Sn cage (3.384 Å) [76], no bonding Pd–Pd contacts were considered in both cases [79]. The $[Pd_2@Sn_{18}]^{4-}$ cluster was shown to be fluxional at room

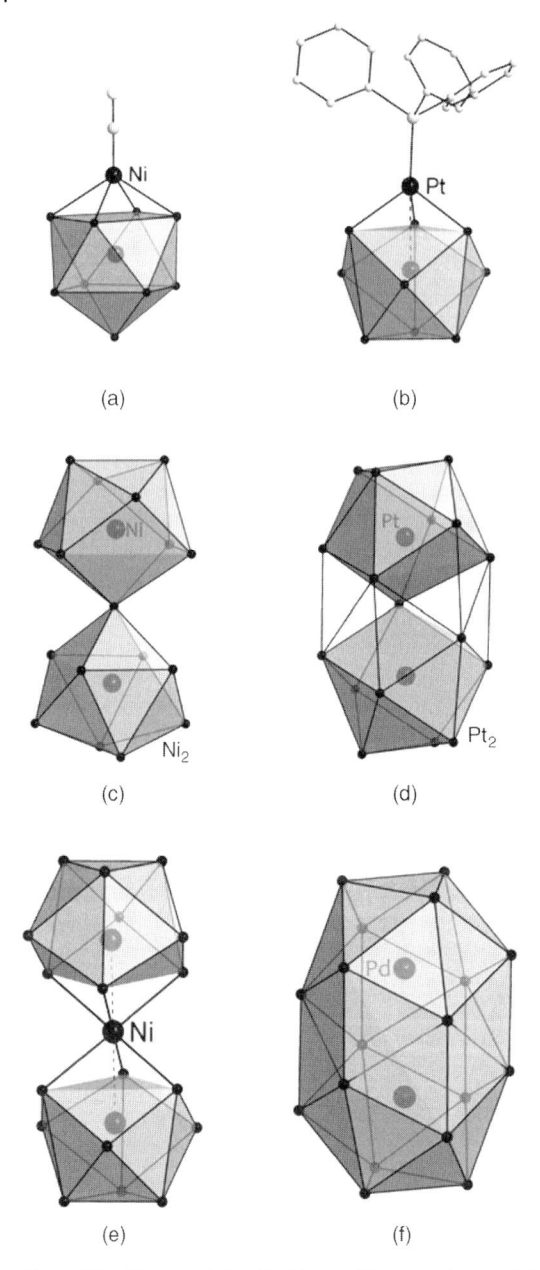

Figure 3.8 Representative Zintl ions with one or two endohedral metal atoms. For structure assignment, see the text.

temperature in ethylenediamine solution, with only one ^{119}Sn-NMR resonance at −751.3 ppm [76].

The intermetalloid cluster [Ni$_3$@Ge$_{18}$]$^{4-}$ displayed a somewhat similar structure, where a linear trimer of Ni atoms was found to bind to two separated, widely opened Ge$_9$ clusters [66]. These two Ge$_9$ units featured the same shape as that of the two E$_9$ subunits found in [Pd$_2$@E$_{18}$]$^{4-}$. Although the Ge atoms of the two cluster units are not connected, the relative orientation of the atoms of each open face is already staggered, as it is required for an 18-vertices cage [79].

For all endohedral clusters with more than nine skeleton atoms (and even for those), the mechanism of the cluster formation remains unclear [4]. However, observations made during the synthesis of several empty transition metal E$_9$ clusters clearly flag a possible reaction path. The above-mentioned anions [Ge$_9$Cu–P(i-Pr$_3$)]$^{3-}$ and [Ge$_9$Cu–Ge$_9$]$^{7-}$ are formed in reactions of K$_4$Ge$_9$ with (i-Pr$_3$)P–CuCl in liquid ammonia, but at different temperatures [59]. At −70 °C, the Cu–P bond remains intact, and [Ge$_9$Cu–P(i-Pr$_3$)]$^{3-}$ crystallizes from the solution. After storing this solution at −40 °C, [Ge$_9$Cu–Ge$_9$]$^{7-}$ forms – that is, the phosphine ligand is substituted by a σ-donating [Ge$_9$]$^{4-}$ unit. Two consequences are clear from this result. First, the ligand at the transition metal can be stripped off after the metal atom is coordinated. Second, two reaction paths are possible: (i) the transition metal atom can subsequently slip into the E$_9$ cluster if the cage size is suitable, as occurs for [Cu@E$_9$]$^{3-}$ and [Ni@Ge$_9$]$^{3-}$; or (ii) the empty coordination site of the d-block metal can add a second cluster under formation of [Ge$_9$Cu–Ge$_9$]$^{4-,7-}$. Yet, there remain some open questions: for example, it is not clear why a Ni atom builds a ten-vertices cluster during the reaction with the Pb$_9$ unit, whereas nine much smaller Ge atoms are able to enclose the same transition metal suitably. Apparently, the numerous structures of endohedral clusters obtained from solution chemistry and reported in the literature are constructed in a completely different way, as it is assumed from the results of gas-phase experiments where the tetrel atoms assemble step-by-step around the "doping" element [80].

3.1.1.4 Beyond Deltahedral Clusters

The bottom-up synthesis of well-defined (semi)metal and intermetalloid clusters that can reach the nanometer scale represents a major challenge. Deltahedral structures reach a limit in size unless more capable templates are found. However, Zintl ions themselves can serve as building units for the formation of larger structures, including classical and nonclassical tetrel–tetrel bonds. Two further examples justify the plan to use such tetrel clusters as precursors for nanoscale clusters:

1) Recently, the family of intermetalloid clusters was expanded to include electron-poor transition elements such as Co, in which a nondeltahedral structure which is much more similar to structures usually observed in intermetallic compounds was found. The anion [Co@Ge$_{10}$]$^{3-}$ (Figure 3.9a) was formed during the reaction of K$_4$Ge$_9$ with Co(C$_8$H$_{12}$)(C$_8$H$_{13}$), and represents a completely new, unprecedented cluster type [81]. The germanium atoms build a D_{5h}-symmetric pentagonal prism which contains no triangular

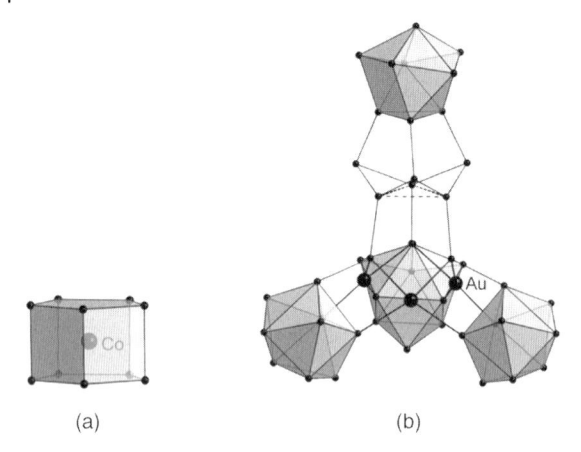

Figure 3.9 (a) The nondeltahedral endohedrally filled anion [Co@Ge$_{10}$]$^{3-}$ [81]; (b) The largest structurally characterized cluster of heavier tetrel elements [Au$_3$Ge$_{45}$]$^{9-}$ [86].

faces at all. This cluster shape indicates that neither the electron count nor the bonding situation follow conventional rules. According to Wade's rules, this cluster is expected to have a D_{4d}-symmetric bicapped square antiprismatic structure. However, DFT calculations have revealed that for [Co@Ge$_{10}$]$^{3-}$ the potential surface for this cluster is rather flat with an energy difference of 13.3 kcal mol^{-1} between the stationary points. That is, the cluster unit can easily switch from the favored D_{5h}- to the energetically unpropitious D_{4d}-symmetric atom arrangement [81]. For comparison, the same calculations for [Ni@Ge$_{10}$]$^{2-}$ revealed an energetic minimum for a D_{4d}-symmetric arrangement, while the pentagonal prism corresponded to a ground state at 5.33 kcal mol^{-1} [81, 82]. Thus, a global minimum cannot be determined indisputably for these clusters, whereas the empty [Ge$_{10}$]$^{2-}$ units clearly prefer a bicapped square antiprism [83]; this was also found from calculations for the clusters in [Cu@Ge$_{10}$]$^{-}$ and [Zn@Ge$_{10}$] [82]. Natural bond orbital (NBO) analyses of [Co@Ge$_{10}$]$^{3-}$ indicated a natural charge of −1.05 for the incorporated Co atom, and consequently a d^{10} electron configuration [81]. The interatomic Co–Ge and Ge–Ge distances in [Co@Ge$_{10}$]$^{3-}$ resemble a bonding situation encountered in the binary intermetallic compound CoGe$_2$ [84]. In the latter case, the Co atom is located within a distorted cubic cavity of eight Ge atoms. The reaction of Fe (2,6-Mes$_2$C$_6$H$_3$)$_2$ with K$_4$Ge$_9$ leads to the formation of the analogous cluster [Fe@Ge$_{10}$]$^{3-}$ [85]. Due to the odd number of electrons, the anion would be expected to be paramagnetic, although no magnetic data have been reported to date.

2) A change of the conditions of the reaction of K$_4$Ge$_9$ and Ph$_3$PAu(I)Cl in ethylenediamine solutions leads to the formation of a byproduct which contains the fascinating large cluster [Au$_3$Ge$_{45}$]$^{9-}$ [86]. In this ligand-free intermetalloid cluster, five Ge$_9$ units are covalently connected, four of which are deltahedral clusters that are interconnected by the fifth. Within the overall structure, a large

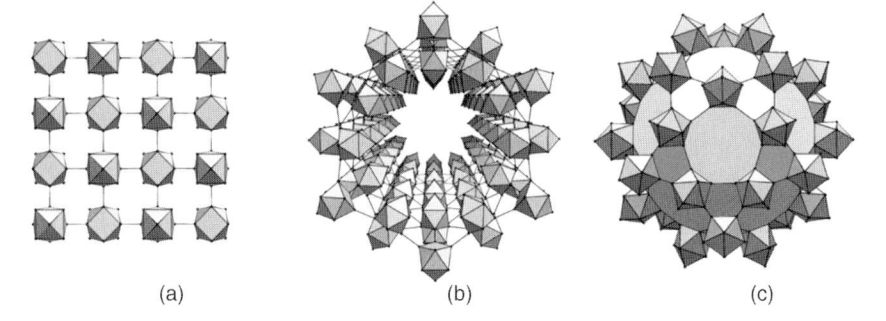

Figure 3.10 Proposed Ge allotropes based on Ge$_9$ building units. (a) Two-dimensional connection of nido-Ge$_9$ clusters forming a $\{_\infty^2[Ge_9]_n\}$ sheet; [88] (b) $\{_\infty^1[Ge_9]_n\}$ nanotube composed of Ge$_9$ clusters. The unit cell of the tube is composed of six unit cells of the sheet shown in Figure 3.10a; (c) (Ge$_9$)$_{30}$ cage derived from a truncated icosadodecahedron [87].

variety of bonding schemes appear which are reminiscent of the bonding situations encountered in boranes. Multicenter bonds arise within the deltahedral clusters, and covalent two-center, two-electron bonds link them together. Long Ge–Ge contacts (shown as dashed lines in Figure 3.9b) form a Ge triangle between the five-coordinated Ge atoms, and correspond to three-center, two-electron bonds. The five-membered Ge faces resemble the structural motif of 3-D solids such as alkali and alkaline earth metal germanides with clathrate structures. In contrast to the linear arrangement in the structure of oligomeric $[(Ge_9)_4]^{8-}$ (Figure 3.4c) [52, 53], the Ge–Au bonds in $[Au_3Ge_{45}]^{9-}$ confine the cluster to a more or less spherical shape. Interestingly $[Au_3Ge_{45}]^{9-}$ exhibits no Au–Au contacts, despite the presence of a large negative cluster charge [86].

Following this concept of cluster linking, a new Ge modification consisting exclusively of Ge$_9$ clusters has already been proposed (Figure 3.10a) [88]. Subsequent quantum-chemical calculations have shown that even nanotubes composed of nine-atom clusters (Figure 3.10b), as well as sizable clusters consisting of 30 Ge$_9$ building blocks, should be rather stable (Figure 3.10c) [87].

Acknowledgments

The authors like to thank the Deutsche Forschungsgemeinschaft (FA 198/6-1) for financial support, and Dr A. Schier for revising the manuscript.

Reference

1 Corbett, J.D. (1985) *Chem. Rev.*, **85**, 383.
2 Fässler, T.F. (2001) *Coord. Chem. Rev.*, **215**, 347.
3 (a) Fässler, T.F. and Hoffmann, S.D. (2004) *Angew. Chem.*, **116**, 6400;
(b) Fässler, T.F. and Hoffmann, S.D. (2004) *Angew. Chem., Int. Ed.*, **43**, 6242.

4 Sevov, S.C. and Goicoechea, J.M. (2006) *Organometallics*, **25**, 5678.
5 Joannis, A. and Hebd, C.R. (1891) *Seances Acad. Sci.*, **113**, 795.
6 (a) Kraus, C.A. (1924) *Trans. Am. Electrochem. Soc.*, **45**, 175; (b) Smyth, F.H. (1917) *J. Am. Chem. Soc.*, **39**, 1299; (c) Zintl, E. and Harder, A. (1931) *Z. Phys. Chem. A*, **154**, 47.
7 Diehl, L., Khodadadeh, K., Kummer, D., and Strähle, J. (1976) *Chem. Ber.*, **109**, 3404.
8 (a) Müller, W. and Volk, K. (1977) *Z. Naturforsch.*, **32b**, 709; (b) Witte, J. and von Schnering, H.G. (1964) *Z. Anorg. Allg. Chem.*, **327**, 260; (c) Baitinger, M., Grin, Y., Kniep, R., and von Schnering, H.G. (1999) *Z. Kristallogr. NCS*, **214**, 457; (d) Baitinger, M., Peters, K., Somer, M., Carrillo-Cabrera, W., Grin, Y., Kniep, R., and von Schnering, H.G. (1999) *Z. Kristallogr. NCS*, **214**, 455; (e) Grin, Y., Baitinger, M., Kniep, R., and von Schnering, H.G. (1999) *Z. Kristallogr. NCS*, **214**, 453; (f) Hewaidy, I.F., Busmann, E., and Klemm, W. (1964) *Z. Anorg. Allg. Chem.*, **328**, 283; (g) Marsh, R.E. and Shoemaker, D.P. (1953) *Acta Crystallogr.*, **6**, 197; Hoch, C. and Röhr, C. (2002) *Z. Anorg. Allg. Chem.*, **628**, 1541.
9 (a) Klemm, W. (1959) *Proc. Chem. Soc. London*, 329; (b) Zintl, E. (1939) *Angew. Chem.*, **52**, 1.
10 Wiesler, K., Brandl, K., Fleischmann, A., and Korber, N. (2009) *Z. Anorg. Allg. Chem.*, **635**, 508.
11 Edwards, P.A. and Corbett, J.D. (1977) *Inorg. Chem.*, **16**, 903.
12 Campbell, J. and Schrobilgen, G.J. (1997) *Inorg. Chem.*, **36**, 4078.
13 Suchentrunk, C. and Korber, N. (2006) *New J. Chem.*, **30**, 1737.
14 Goicoechea, J.M. and Sevov, S.C. (2004) *J. Am. Chem. Soc.*, **126**, 6860.
15 (a) von Schnering, H.G., Baitinger, M., Bolle, U., Carrillo-Cabrera, W., Curda, J., Grin, Y., Heinemann, F., Llanos, J., Peters, K., Schmeding, A., and Somer, M. (1997) *Z. Anorg. Allg. Chem.*, **623**, 1037; (b) Quéneau, V., Todorov, E., and Sevov, S.C. (1998) *J. Am. Chem. Soc.*, **120**, 3263.
16 Hoch, C., Wendorff, M., and Röhr, C. (2003) *J. Alloys Compd.*, **361**, 206.
17 Quéneau, V. and Sevov, S.C. (1997) *Angew. Chem.*, **109**, 1818.
18 (a) Hoch, C., Wendorff, M., and Röhr, C. (2002) *Acta Crystallogr. C*, **58**, I45; (b) Ponou, S. and Fässler, T.F. (2007) *Z. Anorg. Allg. Chem.*, **633**, 393; (c) Quéneau, V. and Sevov, S.C. (1998) *Inorg. Chem.*, **37**, 1358; (d) Todorov, E. and Sevov, S.C. (1998) *Inorg. Chem.*, **37**, 3889.
19 Goicoechea, J.M. and Sevov, S.C. (2005) *Inorg. Chem.*, **44**, 2654.
20 Goicoechea, J.M. and Sevov, S.C. (2006) *Organometallics*, **25**, 4530.
21 Joseph, S., Suchentrunk, C., Kraus, F., and Korber, N. (2009) *Eur. J. Inorg. Chem.*, 4641.
22 [2.2.2]crypt = 4,7,13,16,21,24-hexaoxa-1,10-diazabicyclo- [8.8.8]hexacosane].
23 Corbett, J.D. and Edwards, P.A. (1977) *J. Am. Chem. Soc.*, **99**, 3313.
24 [18]crown-6 = 1,7,10,13,16-hexaoxacyclooctadecane.
25 (a) Fässler, T.F. and Hoffmann, R. (1999) *Angew. Chem.*, **111**, 526; (b) Fässler, T.F. and Hoffmann, R. (1999) *Angew. Chem., Int. Ed.*, **38**, 543.
26 (a) Fässler, T.F. and Schütz, U. (1999) *Inorg. Chem.*, **38**, 1866; (b) Fässler, T.F., Hunziker, M., Spahr, M., and Lueken, H. (2000) *Z. Anorg. Allg. Chem.*, **626**, 692.
27 (a) Angilella, V. and Belin, C. (1991) *J. Chem. Soc. Faraday Trans.*, **87**, 203; (b) Critchlow, S.C. and Corbett, J.D. (1983) *J. Am. Chem. Soc.*, **105**, 5715; (c) Fässler, T.F. and Hunziker, M. (1994) *Inorg. Chem.*, **33**, 5380; (d) Fässler, T.F. and Hunziker, M. (1996) *Z. Anorg. Allg. Chem.*, **622**, 837; (e) Yong, L., Hoffmann, S.D., and Fässler, T.F. (2005) *Z. Kristallogr. NCS*, **220**, 49.
28 Campbell, J., Dixon, D.A., Mercier, H.P.A., and Schrobilgen, G.J. (1995) *Inorg. Chem.*, **34**, 5798.
29 Belin, C.H.E., Corbett, J.D., and Cisar, A. (1977) *J. Am. Chem. Soc.*, **99**, 7163.
30 Belin, C., Mercier, H., and Angilella, V. (1991) *New J. Chem.*, **15**, 931.
31 (a) Spiekermann, A., Hoffmann, S.D., and Fässler, T.F. (2006) *Angew. Chem.*, **118**, 3538; (b) Spiekermann, A., Hoffmann,

S.D., and Fässler, T.F. (2006) *Angew. Chem., Int. Ed.*, **45**, 3459.

32. (a) Chen, Z., Neukermans, S., Wang, X., Janssens, E., Zhou, Z., Silverans, R.E., King, R.B., Schleyer, P.v.R., and Lievens, P. (2006) *J. Am. Chem. Soc.*, **128**, 12829; (b) Cui, L.F., Huang, X., Wang, L.M., Li, J., and Wang, L.S. (2006) *J. Phys. Chem. A*, **110**, 10169; (c) Cui, L.F., Huang, X., Wang, L.M., Zubarev, D.Y., Boldyrev, A.I., Li, J., and Wang, L.S. (2006) *J. Am. Chem. Soc.*, **128**, 8390.

33. (a) Kirchner, P., Huttner, G., Heinze, K., and Renner, G. (1998) *Angew. Chem.*, **110**, 1754; (b) Kirchner, P., Huttner, G., Heinze, K., and Renner, G. (1998) *Angew. Chem., Int. Ed.*, **37**, 1664; (c) Schiemenz, B. and Huttner, G. (1993) *Angew. Chem.*, **105**, 295; (d) Schiemenz, B. and Huttner, G. (1993) *Angew. Chem., Int. Ed.*, **32**, 297.

34. (a) Wade, K. (1972) *Nucl. Chem. Lett.*, **8**, 559; (b) Wade, K. (1976) *Adv. Inorg. Chem. Radiochem.*, **18**, 1; (c) Corbett, J.D. (1997) *Structure and Bonding*, **87**, 158; (d) Fässler, T.F. (1999) *Metal Clusters in Chemistry* (eds P. Braunstein, L.A. Oro, and P.R. Raithby), Wiley-VCH Verlag, Weinheim, p. 1612.

35. Rosdahl, J., Fässler, T.F., and Kloo, L. (2005) *Eur. J. Inorg. Chem.*, 2888.

36. Rudolph, R.W., Wilson, W.L., Parker, F., Taylor, R.C., and Young, D.C. (1978) *J. Am. Chem. Soc.*, **100**, 4629.

37. Suchentrunk, C., Daniels, J., Somer, M., Carrilo-Cabrera, W., and Korber, N. (2005) *Z. Naturforschung*, **60b**, 277.

38. (a) Carrillo-Cabrera, W., Aydemir, U., Somer, M., Kircali, A., Fässler, T.F., and Hoffmann, S.D. (2007) *Z. Anorg. Allg. Chem.*, **633**, 1575; (b) Somer, M., Carrillo-Cabrera, W., Peters, E.M., Peters, K., and von Schnering, H.G. (1998) *Z. Anorg. Allg. Chem.*, **624**, 1915.

39. Downie, C., Mao, J.-G., and Guloy, A.M. (2001) *Inorg. Chem.*, **40**, 4721.

40. Burns, R. and Corbett, J.D. (1985) *Inorg. Chem.*, **24**, 1489.

41. Fässler, T.F. and Hoffmann, R. (1999) *J. Chem. Soc., Dalton Trans.*, 3339.

42. Hauptmann, R. and Fässler, T.F. (2003) *Z. Kristallogr. NCS*, **218**, 458.

43. Yong, L., Hoffmann, S.D., and Fässler, T.F. (2006) *Inorg. Chim. Acta*, **359**, 4774.

44. Hauptmann, R. and Fässler, T.F. (2002) *Z. Anorg. Allg. Chem.*, **628**, 1500.

45. Korber, N. and Fleischmann, A. (2001) *J. Chem. Soc., Dalton Trans.*, 383.

46. Hauptmann, R. and Fässler, T.F. (2003) *Z. Anorg. Allg. Chem.*, **629**, 2266.

47. (a) Hauptmann, R. and Fässler, T.F. (2003) *Z. Kristallogr. NCS*, **218**, 461; (b) Xu, L. and Sevov, S.C. (1999) *J. Am. Chem. Soc.*, **121**, 9245.

48. Nienhaus, A., Hoffmann, S.D., and Fässler, T.F. (2006) *Z. Anorg. Allg. Chem.*, **632**, 1752.

49. (a) Downie, C., Tang, Z., and Guloy, A.M. (2000) *Angew. Chem.*, **112**, 346; (b) Downie, C., Tang, Z., and Guloy, A.M. (2000) *Angew. Chem., Int. Ed.*, **39**, 338.

50. Ugrinov, A. and Sevov, S.C. (2005) *Comp. Rend. Chimie*, **8**, 1878.

51. (a) Ugrinov, A. and Sevov, S.C. (2002) *J. Am. Chem. Soc.*, **124**, 10990; (b) Yong, L., Hoffmann, S.D., and Fässler, T.F. (2005) *Z. Anorg. Allg. Chem.*, **631**, 1149.

52. Ugrinov, A. and Sevov, S.C. (2003) *Inorg. Chem.*, **42**, 5789.

53. Yong, L., Hoffmann, S.D., and Fässler, T.F. (2004) *Z. Anorg. Allg. Chem.*, **630**, 1977.

54. (a) Kummer, D. and Diehl, L. (1970) *Angew. Chem.*, **82**, 881; (b) Kummer, D. and Diehl, L. (1970) *Angew. Chem., Int. Ed.*, **9**, 895.

55. (a) Yong, L., Hoffmann, S.D., and Fässler, T.F. (2005) *Eur. J. Inorg. Chem.*, **18**, 3363; (b) Kesanli, B., Fettinger, J., and Eichhorn, B. (2001) *Chem. Eur. J.*, **7**, 5277.

56. Eichhorn, B.W. and Haushalter, R.C. (1988) *J. Am. Chem. Soc.*, **110**, 8704.

57. (a) Eichhorn, B.W. and Haushalter, R.C. (1990) *J. Chem. Soc., Chem. Commun.*, 937; (b) Campbell, J., Mercier, H.P.A., Franke, H., Santry, D.P., Dixon, D.A. and Schrobilgen, G.J. (2002) *Inorg. Chem.*, **41**, 86.

58. Wang, J.-Q., Stegmaier, S., Wahl, B., and Fässler, T.F. (2010) *Chem. Eur. J.*, **16**, 1793.

59. Scharfe, S. and Fässler, T.F. (2010) *Eur. J. Inorrg. Chem.*, 1207.

60. Zhou, B., Denning, M.S., Chapman, T.A.D., and Goicoechea, J.M. (2009) *J. Am. Chem. Soc*, **48**, 2899.

61. (a) Nienhaus, A., Hauptmann, R., and Fässler, T.F. (2002) *Angew. Chem.*, **114**, 3352; (b) Nienhaus, A., Hauptmann, R.,

and Fässler, T.F. (2002) *Angew. Chem., Int. Ed.*, **41**, 3213.

62. (a) Schnepf, A. and Schnöckel, H. (2002) *Angew. Chem.*, **114**, 3682; (b) Schnepf, A. and Schnöckel, H. (2002) *Angew. Chem., Int. Ed.*, **41**, 3533.
63. Boeddinghaus, M.B., Hoffmann, S.D., and Fässler, T.F. (2007) *Z. Anorg. Allg. Chem.*, **633**, 2338.
64. Denning, M.S. and Goicoechea, J.M. (2008) *Dalton Trans.*, 5882.
65. (a) Spiekermann, A., Hoffmann, S.D., Kraus, F., and Fässler, T.F. (2007) *Angew. Chem.*, **119**, 1663; (b) Spiekermann, A., Hoffmann, S.D., Kraus, F., and Fässler, T.F. (2007) *Angew. Chem., Int. Ed.*, **46**, 1638.
66. (a) Goicoechea, J.M. and Sevov, S.C. (2005) *Angew. Chem.*, **117**, 4094;(b) Goicoechea, J.M. and Sevov, S.C. (2005) *Angew. Chem., Int. Ed.*, **44**, 4026.
67. Goicoechea, J.M. and Sevov, S.C. (2006) *J. Am. Chem. Soc.*, **128**, 4155.
68. Scharfe, S., Fässler, T.F., Stegmaier, S., Hoffmann, S.D., and Ruhland, K. (2008) *Chem. Eur. J.*, **14**, 4479.
69. Esenturk, E.N., Fettinger, J., and Eichhorn, B.W. (2005) *J. Chem. Soc. Chem. Commun.*, 247.
70. Schrodt, C. and Ahlrichs, W.F.R. (2002) *Z. Anorg. Allg. Chem.*, **628**, 2478.
71. Esenturk, E.N., Fettinger, J., and Eichhorn, B.W. (2006) *J. Am. Chem. Soc.*, **128**, 9178.
72. (a) Esenturk, E.N., Fettinger, J., Lam, Y.-F., and Eichhorn, B. (2004) *Angew. Chem.*, **116**, 2184; (b) Esenturk, E.N., Fettinger, J., Lam, Y.-F., and Eichhorn, B. (2004) *Angew. Chem., Int. Ed.*, **43**, 2132.
73. Kesanli, B., Fettinger, J., Gardner, D.R., and Eichhorn, B. (2002) *J. Am. Chem. Soc.*, **124**, 4779.
74. Esenturk, E.N., Fettinger, J.C., and Eichhorn, B.W. (2006) *J. Am. Chem. Soc.*, **128**, 12.
75. Kesanli, B., Halsig, J.E., Zavalij, P., Fettinger, J.C., Lam, Y.F., and Eichhorn, B.W. (2007) *J. Am. Chem. Soc.*, **129**, 4567.
76. Kocak, F.S., Zavalij, P., Lam, Y.-F., and Eichhorn, B.W. (2008) *Inorg. Chem.*, **47**, 3515.
77. Sun, Z.M., Xiao, H., Li, J., and Wang, L.S. (2007) *J. Am. Chem. Soc.*, **129**, 9560.
78. Fässler, T.F., Hoffmann, S., and Kronseder, C. (2001) *Z. Anorg. Allg. Chem.*, **620**, 2486.
79. Goicoechea, J.M. and Sevov, S.C. (2005) *J. Am. Chem. Soc.*, **127**, 7676.
80. Zhang, X., Li, G., and Gao, Z. (2001) *Rapid Commun. Mass Spectrom*, **15**, 1573.
81. (a) Wang, J.-Q., Stegmaier, S., and Fässler, T.F. (2009) *Angew. Chem.*, **121**, 2032; (b) Wang, J.-Q., Stegmaier, S., and Fässler, T.F. (2009) *Angew. Chem., Int. Ed.*, **48**, 1998.
82. King, R.B., Silaghi-Dumistrescu, I., and Uta, M.-M. (2006) *Chem. Eur. J.*, **14**, 4542.
83. King, R.B., Silaghi-Dumistrescu, I., and Uta, M.-M. (2006) *Inorg. Chem.*, **45**, 4974.
84. Schubert, K. and Pfisterer, H. (1950) *Z. Metallkd.*, **41**, 433.
85. Zhou, B., Denning, M.S., Kays, D.L., and Goicoechea, J.M. (2009) *J. Am. Chem. Soc.*, **131**, 2802.
86. (a) Spiekermann, A., Hoffmann, S.D., Fässler, T.F., Krossing, I., and Preiss, U. (2007) *Angew. Chem.*, **119**, 5404; (b) Spiekermann, A., Hoffmann, S.D., Fässler, T.F., Krossing, I., and Preiss, U. (2007) *Angew. Chem., Int. Ed.*, **46**, 5310.
87. Karttunen, A.J., Fässler, T.F., Linnolahti, M., and Pakkanen, T.A. (2010) *Chem. Phys. Chem.*, in press.
88. (a) Fässler, T.F. (2001) *Angew. Chem.*, **113**, 4289; (b) Fässler, T.F. (2001) *Angew. Chem., Int. Ed.*, **40**, 4164.
89. Yong, L., Hoffmann, S.D., Fässler, T.F., Riedel, S., and Kaupp, M. (2005) *Angew. Chem., Int. Ed.*, **44**, 2092.
90. Schmidbaur, H.(ed.) (1999) *Gold – Progress in Chemistry, Biochemistry and Technology*, John Wiley & Sons, Chichester.

3.2
Semiconductor Nanoparticles

3.2.1
Synthesis and Characterization of II–VI Nanoparticles

Alexander Eychmüller

During the past two decades, the "synthesis" or "preparation" of II–VI semiconductor nanoparticles has experienced an enormous development, to the point where the published material related to the topic has become virtually unmanageable. Nevertheless, the aim of this chapter is to provide a chronological overview of some of the major historical lines in this area, starting with the earliest studies with CdS nanocrystals prepared in aqueous solution. At several points in the story – mostly when successful preparations are first described – the chapter branches into evolving sub-fields, leading to Sections 3.2.1.2 and 3.2.1.3. The remainder of the review then relates to matters distinct from these preparational approaches. More complicated nanoheterostructures, in which two compounds are involved in the build-up of spherically layered particles, are detailed in Sections 3.2.1.4 and 3.2.1.5.

Naturally, in all chapters, the view of the author is indicated with regards to the importance of the studies under review for the evolution of the subject as a whole. Besides this necessarily subjective viewpoint, it is not impossible that other important work might have been overlooked. Thus, even if a particular chapter provides little more than a list of references, questions such as "What has been done in this field" may be answered, but never those such as "What has *not* been done yet!"

3.2.1.1 Historical Review

Commonly, even the shortest history of the preparation of II–VI semiconductor nanoparticles starts with the studies of A. Henglein during the early 1980s [1]. Like others [2–7], Henglein's report dealt with surface chemistry, photodegradation, and catalytic processes in colloidal semiconductor particles. Yet, it was this very report that revealed the first absorption spectrum of a colloidal solution of size-quantized CdS nanocrystals. The community, however, obviously has lost sight of a report by Katzschmann, Kranold and Rehfeld [8], in which the optical absorption of II–VI semiconductor particles embedded in glasses was examined in conjunction with X-ray small-angle scattering (SAS) measurements. These experiments revealed the presence of small spherical particles with radii in the range of 1–5 nm, giving rise to a variation of the optical constants. Particles above 5 nm were found to have approximately the optical constants of the macroscopic crystal, while for smaller particles the gap change was found to follow the law $E_g = A\text{-}r^2 + B$, where E_g is the bandgap of the particles and A and B are constants (e.g., for CdSe A = 1.70 eV (nm)^2 and B = 1.73 eV). The authors' explanation that, "A lattice contraction is considered to be the cause for this optical anomaly," was not correct, however.

Henglein's CdS nanoparticles were prepared from Cd(ClO$_4$)$_2$ and Na$_2$S on the surface of commercial silica particles. The absorption onset is shifted considerably to higher energies with respect to the bulk bandgap of CdS (515 nm). In addition, the sol emits light upon excitation at 390 nm, which was also a matter of investigation in this report. The first correct interpretation of the observed blue shift of the absorption as a quantum mechanical effect stems was made by Brus [9] when, in the framework of the effective mass approximation, the shift in kinetic energy of the charge carriers due to their spatial restriction to the volume of the nanometer-sized semiconductors was calculated. (It should be noted that comparable experimental [10] and theoretical studies [11] were carried out almost simultaneously on the I–VII compound CuCl in the Soviet Union.)

In 1993, another milestone in the preparation of II–VI semiconductor nanocrystals was the study of Murray, Norris and Bawendi [12], whose synthesis was based on the pyrolysis of organometallic reagents such as dimethyl cadmium and trioctylphosphine selenide, after injection into a hot coordinating solvent. This approach provided a temporally discrete nucleation and permitted a controlled growth of the nanocrystals. The evolution of the absorption spectra of a series of CdSe crystallites, ranging in size from about 1.2 to 11.5 nm, is shown in Figure 3.11.

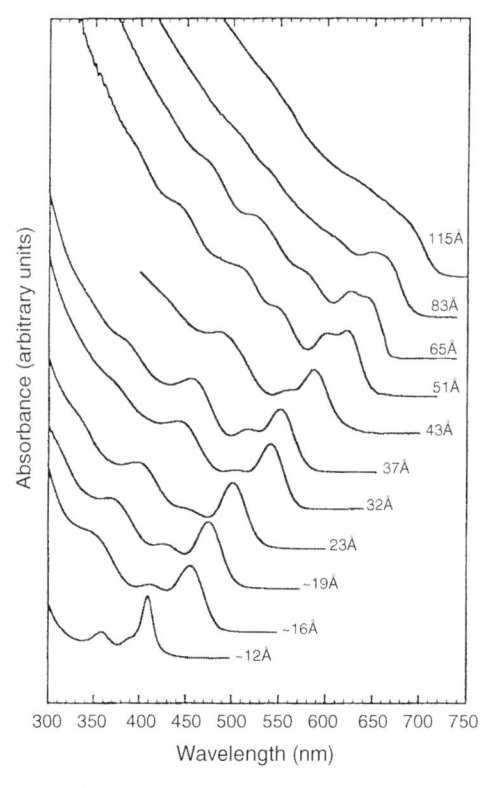

Figure 3.11 1993 state-of-the-art absorption spectra of CdSe colloidal solutions ("Hot-injection synthesis"). From Ref. [12].

This series spanned a range of sizes, from more or less molecular species to fragments of the bulk lattice containing tens of thousands of atoms. The remarkable quality of these particles can be seen clearly in several features of the absorption spectra, such as the steepness of the absorption onset, the narrow absorption bands, and the occurrence of higher energy transitions. Subsequently, major progress was made in the preparation of such systems within about a decade of these initial findings.

In order to illustrate the occurrences of this time, at this point it is worth detailing both the preparation and characterization studies conducted in various different laboratories, worldwide. For example, Rossetti et al. described the quantum size effects in redox potentials and the electronic spectra of CdS nanocrystals [13]. On this case, resonance Raman spectra were recorded using freshly prepared samples and from larger, "aged" particles, while the size distributions were studied using transmission electron microscopy (TEM); the crystallinity and crystal structures of the samples were also determined. In a series of reports, Henglein and coworkers investigated the dissolution of CdS particles [14], enlarged the field by preparing ZnS and ZnS–CdS co-colloids (including via TEM characterization) [15, 16], provided insight into photophysical properties such as time-resolved fluorescence decays [17], and introduced the term "magic agglomeration numbers" [18]. Likewise, Ramsden and Grätzel described the synthesis of CdS particles formed in aqueous solution from $Cd(NO_3)_2$ and H_2S or Na_2S [19]. A combination of luminescence data and TEM imaging led to the particles obtained being assigned to the cubic phase. In 1985, A. J. Nozik entered the arena with a report on phosphate-stabilized CdS and small PbS colloids [20] whilst, after yet another theoretical treatise [21], Brus again produced a form of review in 1986 [22] which summarized the then current status in the field of semiconductor clusters, from both experimental and theoretical viewpoints. Also in 1986, attempts at fractional separation (using exclusion chromatography) of an as-prepared sol of hexametaphosphate (HMP; more correctly, Grahams salt)-stabilized CdS particles were described, together with stopped-flow experiments on extremely small clusters [23]. The binding of simple amines to the surface of CdS particles of various sizes, prepared not only via conventional routes but also in $AOT–H_2O$–heptane reverse micelles, led to a dramatic enhancement of the luminescence quantum yield [24]. This effect was interpreted as a modification of the mid-gap states that might play an important role in the nonradiative recombination processes of electron and hole in the materials under investigation. Subsequently, a series of novel preparative approaches was presented, which included the preparation of CdS particles in dihexadecyl phosphate (DHP) surfactant vesicles [25], and also the synthesis of colloidal CdSe [26]. Indeed, the latter approach has today become the most important model system for investigating the properties of nanocrystalline matter. As determined using pulse radiolysis experiments, an excess electron on the surface of CdS colloids causes a bleaching of the optical absorption at wavelengths close to the onset of absorption [27]. A year later, the results of a study were reported (originally from the Henglein group, but frequently cited since then [28]) which referred to CdS sols with particle sizes ranging from 4 to 6 nm (as measured via

TEM-based particle size histograms) that had been prepared by first precipitating Cd^{2+} ions with a stoichiometric quantity of injected H_2S. This was followed by an "activation" of the colloid by the addition of NaOH and excess Cd^{2+} ions, and resulted in a strong emission band close to the energy of the band edge. The influence of the "starting pH" on the color of the emitted light following this activation process, as well as on the particle size, is shown in Figure 3.12. This activation is interpreted in terms of the precipitation of cadmium hydroxide onto the surface of the CdS particles. By using this procedure, the photostability of the sols would be greatly increased, so as to approach that of organic dyes such as Rhodamine 6G.

The reported fluorescence quantum yields (>50% in this report and 80% in reports by others [24]) seem overoptimistic. From a modern viewpoint, the "activation" should rather be interpreted as being the result of the creation of a shell of the large bandgap material $Cd(OH)_2$ on the surface of the CdS particles (cf. also Section 3.2.1.4). This results in the formation of an electronic barrier at the surface of the particles, preventing the charge carriers from escaping from the particles' core. This preparation operates as the basis for core–shell structures and multilayered nanotopologies (as described in greater detail in Sections 3.2.1.4 and 3.2.1.5).

An enormous step forward [especially in terms of the characterization involving, among others, infrared (IR) and NMR spectroscopy, high-resolution transmission electron microscopy (HR-TEM) and X-ray diffraction (XRD)] was achieved during studies conducted from 1988 onwards by the group of Steigerwald and Brus at

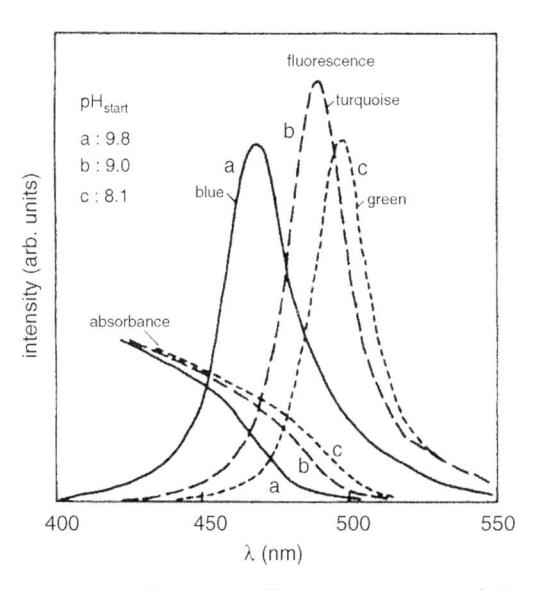

Figure 3.12 Absorption and luminescence spectra of CdS sols ("polyphosphate preparation") with varying starting pH. From Ref. [28].

AT&T [29]. In this case, organometallic compounds were used to synthesize nanometer-sized clusters of CdSe in inverse micellar solutions. Following chemical modification of the surface of these clusters, the cluster molecules could be removed and isolated from the reaction mixture; a detailed powder XRD investigation was subsequently conducted. Later during that year, several reports describing microheterogeneous systems were made which involved CdS, $Zn_xCd_{1-x}S$, and CdS coated with ZnS, all of which were formed in and stabilized by (again) DHP vesicles [30]. Additionally, layers of CdS particles (5 nm in size) were prepared via the Langmuir–Blodgett technique [31]. These and many more results of studies involving small particle research, both on metal and on semiconductor particles, have been summarized in reviews by Henglein [32, 33] and by Steigerwald and Brus [34]. Furthermore, considerable progress has been made in the radiolytic production of CdS particles [35], in the preparation of CdTe and ZnTe [36], and in the application of scanning tunneling microscopy (STS) to quantum-sized CdS [37]. The use of thiophenol as a stabilizing agent for CdS clusters has been reported by Herron et al. [38]. In this case, the competitive reaction chemistry between CdS core cluster growth and surface capping by thiophenolate is assumed to be responsible for the evolution of clusters, the cores of which consist essentially of cubic CdS, and of which the reactive surface has been passivated by covalently attached thiophenyl groups. The synthesis of porous, quantum-size CdS membranes has been achieved in three subsequent steps: (i) an initial colloid preparation following standard procedures [28]; (ii) concentration of the suspension and dialysis; and (iii) a controlled evaporation of the residual solvent [39]. In a study of photoreductive reactions on small ZnS particles, an elaborate HR-TEM analysis was conducted which enabled a determination of the actual particle size by simply counting the lattice planes [40]. The postpreparative size separation of colloidal CdS by gel electrophoresis was also demonstrated [41], and an ion-dilution technique utilized in order to generate ultrasmall particles of CdS and CdSe in Nafion membranes [42]. Following a series of initial investigations into the recombination processes of electron–hole pairs in clusters [17, 43–45], two reports were made on this topic, both of which proposed a mechanism of delayed emission for the radiative "band-to-band" recombination [46, 47]. At about the same time, several reports described the processes of quenching and "antiquenching" (i.e., an enhancement of the luminescence of nanosized CdS), all of which contributed in a qualitative manner to a deeper understanding of the optical properties of suspended nanocrystals [48–53]. CdS nanoparticles were synthesized in several nonaqueous solvents, and their catalytic efficiency in the photoinitiation of the polymerization of various vinylic monomers examined in detail [54].

Early in 1993, a special issue of *The Israel Journal of Chemistry* which was devoted to quantum-size particles, contained several reports that were concerned predominantly with the synthesis and characterization of II–VI-semiconductor nanoparticles [55–61]. Later that year, two (now frequently cited) reports provided a summary of the achievements in the field [62, 63]. Notably, the characterization of wet-chemical-synthesized CdTe and HgTe, as well as of ternary HgCdTe compounds (including XRD, TEM and photoluminescence), formed the focus of a series of studies by Müllenborn et al. [64]. In this case, the nature of the stabilizer was shown to play an important role in the efficiency of luminescence quenching, as demonstrated by the comparison of inverse micelle–HMP and thiophenol-capped CdS [65]. Thiol capping was also the clue to synthesizing differently sized CdTe particles prepared in aqueous solution [66], whereby the bandgaps of 2.0 nm, 2.5 nm, 3.5 nm, and 4.0 nm crystallites were estimated from pulse radiolysis experiments. Interestingly, their dependence on the size of the particles was shown to better agree with pseudopotential [67] and tight-binding models [68, 69] than with the effective mass approximation [21]. One important result of the studies conducted by Rajh *et al.* was the observation that the extinction coefficient per particle was independent of the size, as had been predicted theoretically [70–72]. In going full circle, this brings to attention the studies performed by Murray *et al.* [12], of which yet another aspect should now be mentioned, namely the extremely thorough characterization of the particles prepared. Among other findings, HR-TEM images revealed the presence of stacking faults in certain crystallographic directions; they also showed that the particles were prolate, and that the crystallites would arrange themselves freely into secondary structures. These investigations, which probed only a limited number of particles, were supplemented by a carefully conducted XRD which included structural simulations. These studies formed the basis of the "organometallic preparative route" (also named the "TOP–TOPO method," "hot injection synthesis," or "synthesis in organics"). It should be noted that some of these names refer to alterations of the original preparative route, while others do not (though no attempt will be made to judge the validity of this point). A deeper discussion – including an attempt to distinguish between "hot-injection" and "heating-up" methods – has been provided by Reiss [73]. Some of the more recent accomplishments of these synthetic approaches are outlined in Section 3.2.1.3, while details of investigations into core–shell nanoparticles are provided in Section 3.2.1.4.

Thiol stabilization also proved to offer a very successful route towards the creation of monosized nanoparticles of various compounds; however, as it has also attracted widespread interest it will be outlined separately in Chapter 3.2.1.2. In this chapter, from this point onwards, attention will be focused on recent developments that differ from the "hot injection synthesis" and from the thiol technique.

A narrowing of the size distribution of a quite polydisperse CdS colloid, by virtue of size-selective photocorrosion, was demonstrated by Matsumoto *et al.* [74]. In order to replace the hazardous metal alkyls in the organometallic synthesis, a single-source approach to the synthesis of II–VI nanocrystallites was presented by Trindade and O'Brian [75]. In this case, the CdSe particles were formed from methyldiethyldiselenocarbamato cadmium (II), a compound synthesized beforehand from dimethyl cadmium and bisdiethyldiselenocarbamato cadmium (II). The vesicles were again used as the reaction compartments [76], while the nanophase reactors were generated by binary liquids at the surface of colloidal silica particles, which also served as reaction compartments for small CdS and ZnS particles [77]. The creation of composites from CdSe nanoclusters and phosphine-functionalized block copolymers was the aim of Fogg *et al.* [78]. CdS particles formed from Na_2S and $CdSO_4$ have been studied extensively using ^{113}Cd-and ^1H-NMR [79]. Based on the results of a careful analysis of the temperature-dependent NMR spectra, three different Cd^{2+} species (located in the core, at the surface, and in solution) could be distinguished, while a considerable quantity of water was found to be present in the dry samples. Several procedures have been reported by which the surface could be protected (e.g., with SiO_2 [80]), or an encapsulation achieved [81, 82]. However, there appears to be a complex interplay between the nature of the solvent, the anions of the cadmium salt, the ratio of salt to thiourea, and the temperature, which controls the morphology and sizes of the nanocrystalline CdS products. As shown with XRD and TEM, spheres of various sizes, rods, and even nanowires of up to 900 nm in length were generated using a solvothermal process [83]. The luminescence properties of CdS particles – prepared either in the presence of a poly(aminoamine) dendrimer or polyphosphate – were reinvestigated, with special emphasis placed on the polarization of the emitted light [84] and on the luminescence lifetimes [85]. Further synthetic efforts yielded HgE (E = S, Se, Te) nanoparticles [86], PbS sols with various capping agents [such as poly(vinyl alcohol; PVA), poly(vinylpyrrolidone) (PVP), gelatin, DNA, polystyrene, and poly(methylmethacrylate) (PMMA) [87]], and ZnS nanoparticles in a silica matrix [88]. Likewise, small CdTe particles were shown to be accessible by conducting their preparation in reverse micelles [89], while exciton interactions in CdS nanocrystal aggregates in reverse micelle have been examined [90] and "peanut-like" nanostructures of CdS and ZnS were obtained by reactions at a static organic–aqueous biphase boundary [91]. In 2001, the group of Lahav reported details of the organization of CdS and PbS nanoparticles in hybrid organic–inorganic Langmuir–Blodgett films, and characterization of the layers using XRD and TEM techniques [92]. The zeolite MCM-41 was shown to be suitable as a host for CdS nanocrystals, which were formed via an ion-exchange reaction (as shown by Zhang and coworkers [93]). In the past, mesoporous materials have played a clear role in hosting nanosized semiconductor particles in many reports (e.g., [94–98]). An example of this was a synthetic approach to cadmium chalcogenide nanoparticles in the mesopores of SBA-15 silica (as the host matrix), whereby the use of cadmium organochalcogenolates allowed nanoparticles of all three cadmium chalcogenides to be prepared, following the same experimental protocol [95]. A completely new approach to the synthesis of Cd-chalcogenide nanoparticles was demonstrated by Peng and Peng, who used CdO as the cadmium precursor and strong ligands (such as hexylphosphonic acid or tetradecylphosphonic acid [99]). In this way, very fine ("high-quality") nanocrystals could be prepared, and a limited degree of shape control (i.e., spheres or rods) was also possible. Four types of binding site at the surface of CdS particles, prepared by precipitation from a solution of cadmium carboxylate in dimethyl sulfoxide (DMSO), were identified by means of IR and NMR spectroscopy – namely oxidized sulfur, adsorbed acetate, oxygen-bound DMSO, and hydroxyl ions [100]. Micelles of various compositions have been involved in the preparation of CdS (block-copolymer (compound) micelles [101, 102]) and triangular CdS nanocrystals ($Cd(AOT)_2$–iso-octane–water reverse micelles [103]). The concept of the encapsulation of nanoparticles in micelles has recently been investigated by Fan [104], who described a rapid, interfacially driven microemulsion process in which the flexible surface chemistry of the nanocrystal (NC) micelles caused them to be water-soluble and to provide biocompatibility. The protection of CdS particles by capping with a shell of silica was reported by Farmer and Patten [105]. Furthermore, organic dendrons have been used to stabilize CdSe quantum dots (QDs) [106]. The optical properties of CdS QDs in Nafion was the focus of a study conducted by Nandakumar et al. [107], namely the two-photon absorption and the photon-number squeezing with CdSe particles in PMMA [108]. Molecular species such as $(Li)_4$-$[Cd_{10}Se_4(SPh)_{16}]$ as fragments of the bulk structure may serve as templates for the growth of nanocrystalline CdSe in hexadecylamine as solvent [109]. This approach may also be applied to the synthesis of ZnSe and CdSe–ZnS core–shell nanocrystals. The use of dendrimers as stabilizing entities has also been investigated recently [110, 111], and has led to advanced highly sensitive methods for detecting oligonucleotide targets using QD-functionalized nanotubes containing a cascade energy band-gap architecture [112].

The details of many such efforts have been included in review-like articles [62, 63, 113–119], with empiricism, chemical imagination, and intuition largely forming the basis for the preparation of nanosized objects in solution. In the present author's opinion, this will continue for a number of years despite attempts to successfully predict superior preparative routes by means of combinatorial and computer-aided scientific approaches [120].

3.2.1.2 Thiol-Stabilized Nanoparticles

The use of thiols as capping agents in the synthesis of II–VI semiconductor nanocrystals has been referred to several times above. Following preliminary studies [55, 121, 122] based on earlier investigations conducted by Dance et al. [123–125] and others [126, 127], this approach came to full fruition for the first time in a report by Herron et al. [128], in which the synthesis and optical properties of a cluster with a 1.5 nm CdS core having the total formula $[Cd_{32}S_{14}(SPh)_{36}]$ 4 DMF (SPh = thiophenol, DMF = N,N-dimethylformamide), was described. The structure of this material, as part of a cubic lattice determined with single-crystal XRD, is discussed in greater detail in Section 4.1.1. In 1994, a report was made on the synthesis, characterization and optical properties of thiol- and polyphosphate-stabilized CdS particles which ranged in size from 1.3 nm to about 10 nm, as determined using small- and wide-angle XRD and TEM imaging [129]. The report, which also provided chemical and thermogravimetric analysis data, explained (also for CdS small particles) the independence of the extinction coefficient per particle on the particle size, identified the temperature dependence of the first electronic transition for the different-sized particles, and also described the observation of a "reversible absorbance shift." The latter was detected as the difference between the transition energy recorded in solution and that of a film formed from the particles on a quartz slide.

Initially, this was taken as an indication of a cluster–cluster interaction that was operative in the films, but a year later Nosaka et al. described the formation process of "ultrasmall particles" from cadmium complexes of 2-mercaptoethanol [130]. In this very thorough study, the stopped-flow technique was used to gain insight into the very first steps of particle evolution, while NMR yielded information as to the presence of several different cluster species in solution. Among these, $[Cd_{17}S4(SR)_{24}]^{2+}$ (RS = 2-mercaptoethanol) exhibited the same tetrahedral core structure as the crystallized $[Cd_{17}S_4(SR)_{26}]$ cluster of Vossmeyer et al. [131].

The same preparative approach – that is, the dissolution of a metal salt in water in the presence of a stabilizing thiol – yielded nanocrystals of CdSe [132], CdTe [133–136], HgTe [137], ZnSe [138], CdHgTe [139–141], ZnCdSe [142], and CdSeTe [143] (for recent reviews on this subject, see also Refs [144, 145]). The thiols used included 1-thioglycerol, 2-mercaptoethanol, 1-mercapto-2-propanol, 1,2-dimercapto-3-propanol, thioglycolic acid, thiolactic acid, and cysteamine [146]. In order to achieve specific functionalities, this "classical" list of thiol-stabilizers may be easily extended to include for example, dihydrolipoic acid (DHLA) [147], l-glutathione (GSH) [148], and 2-mercaptopropionic acid (2-MPA) [149]. Subsequent to the first steps in the preparation, heating of the reaction solution may be applied in order to initiate particle growth. Thus, this preparative approach relies on the separation of nucleation and growth in a very similar manner to the organometallic TOP–TOPO (tri-octyl phosphine–tri-octyl phosphine oxide) preparation (which is outlined in more detail in Section 3.2.1.3 [12]). Yet another similarity between the two strategies is the postpreparative treatment that involves the application of a size-selective precipitation to the crude cluster solutions [150]. The sizes of the evolving nanocrystals depend mainly on the concentration ratios of the chemicals involved, and on the duration of the heat treatment. Some typical room-temperature phospholuminescence (PL) spectra (normalized) of colloidal solutions of ZnSe, CdTe, HgTe together with alloyed nanocrystals are shown in Figure 3.13. In this case,

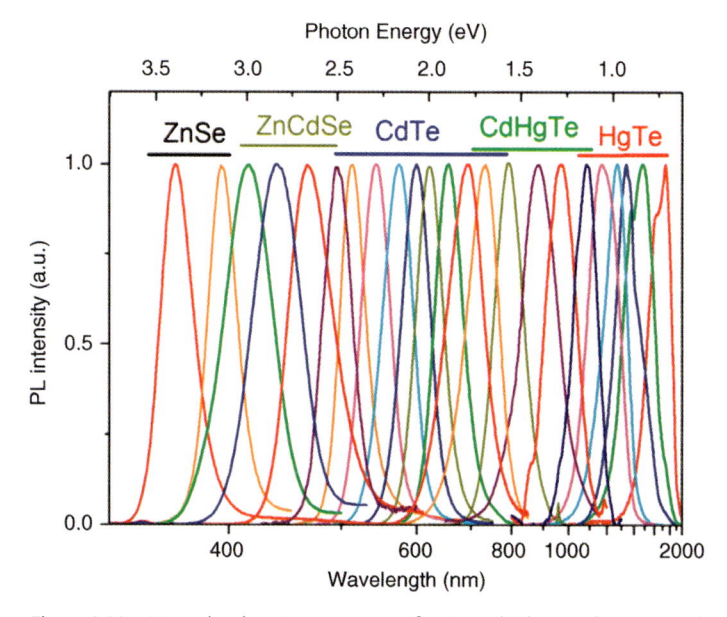

Figure 3.13 Normalized emission spectra of various thiol-capped nanocrystals.

the PL bands are located typically close to the absorption thresholds (so-called band-edge or "excitonic" PL), and are sufficiently narrow (full-width at half maximum as low as 35 nm being increased up to 55–60 nm for fractions of larger NCs). The position of the PL maximum of the smallest (~2 nm) luminescing CdTe NCs is located at 510 nm (green emission), whereas the largest (~8 nm) CdTe nanocrystals obtained emit in the far red with a PL maximum at 800 nm. The whole spectral range between these two wavelengths is covered by the intermediate sizes of CdTe NCs. The PL quantum yield of as-synthesized CdTe nanocrystals depends on the nature of the stabilizing agent [144, 145]. The PL quantum yield lies typically below 10% for thioglycerol-, dithioglycerol- and mercaptoethanol-stabilized NCs, while values of 30–35% for 2-(dimethylamino)ethanethiol-stabilized nanocrystals have been attained. Recent improvements in synthetic protocols have demonstrated means by which 40–60% PL quantum yields [151] may be received for as-prepared CdTe NCs in the presence of thioglycolic acid (TGA) or MPA, if the Cd: stabilizer ratio present at the start of the synthesis was in the range of 1–1.5. A detailed investigation of this phenomenon shows that the formation of relatively low-solubility uncharged Cd–TGA complexes under these conditions is favorable for the further controlled growth of higher quality NCs [152]. In most cases, the PL quantum yield can be sufficiently improved by post-preparative treatments of the NCs, such as size-selective precipitation and photochemical etching [145].

The findings of standard characterization studies on nanosized HgTe particles have been described [137]; an example is shown in Figure 3.14, which demonstrates

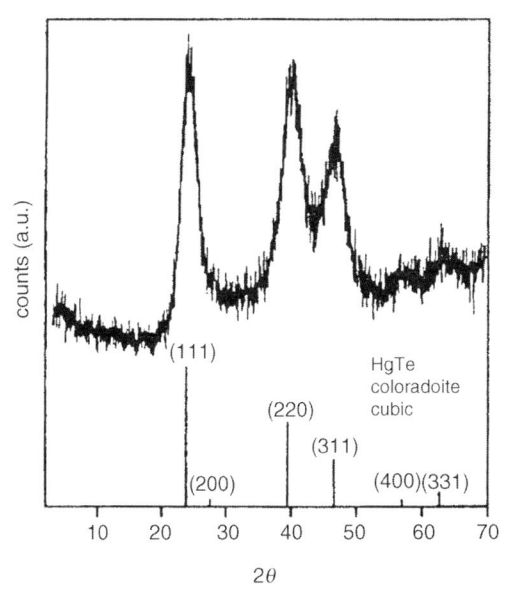

Figure 3.14 Powder X-ray diffractogram of HgTe nanocrystals ("thiol synthesis") with corresponding bulk values given as a line spectrum. From Ref. [37].

the powder X-ray diffraction (P-XRD) pattern of a fraction of HgTe nanoparticles. The crystalline structure derived from the positions of the wide-angle diffraction peaks is clearly the cubic (coloradoite) HgTe phase, while the wide-angle diffraction peaks are broadened because of the small particle size. The mean cluster sizes estimated from the full-width at half-maximum intensity of the (111) reflection, according to the Scherrer equation, gives a value of about 3.5 nm. The weakly resolved small-angle XRD peak indicates a rather broad size distribution of the HgTe nanocrystals in the aqueous colloidal solution.

The synthetic procedure for ZnSe NCs [138] was very similar to that of CdTe NCs. The stabilizer of choice was TGA, and the synthesis of ZnSe NCs was carried out at pH 6.5, giving rise to stable colloidal solutions with a moderate luminescence [153]. As a beneficial post-preparative treatment, photochemical etching was applied and studied in detail; the result was that, under optimal conditions, a PL quantum yield of 25–30% could be achieved. Not only the optical spectra but also P-XRD and HR-TEM analyses showed that the improvement in PL quantum yield was accompanied by growth of the NCs. The XRD reflexes also shifted to values which were characteristic of ZnSe–ZnS alloys (the sulfur appears as a product of photodecomposition of the TGA in solution [154]). The formation of such a shell of a larger band gap material provided an additional stabilization of the core particle, which led in turn to better and more robust optical properties.

The photophysical properties of these thiol-stabilized particles, such as PL [155] and optically detected magnetic resonance (ODMR) [156], their dispersion in polymer films in order to fabricate light-emitting devices [157–159], their possible use in telecommunication devices [160], further structural properties unraveled by EXAFS (extended X-ray absorption fine structure) [161–165], and their incorporation into photonic crystals [166–168] and silica spheres [169], have been described and in part reviewed in a number of survey reports [144, 145, 170–173].

As an example of the capabilities of EXAFS spectroscopy, the mean Cd–S distances as a function of the diameters of a series of CdS nanocrystals are depicted in Figure 3.15 [163]. These data are gained from a thorough temperature-dependent study of the size dependence of various structural and dynamic properties of CdS nanoparticles ranging in size from 1.3 to 12.0 nm. The properties studied include the static and the dynamic mean-square relative displacement, the asymmetry of the interatomic Cd–S pair potential, with conclusions drawn as to the crystal structure of the nanoparticles, the Debye temperatures, and the Cd–S bond lengths. As seen from Figure 3.15, the thiol-stabilized particles (samples 1–7) show an expansion of the mean Cd–S distance, whereas the phosphate-stabilized particles (samples 8–10) are slightly contracted with respect to the CdS bulk values. The difference between the two series of nanocrystals relates to the type of interaction between the ligands and the Cd atoms at the surfaces. The thiols are covalently bound, and through this they have an influence on the structure of the particles as a whole (most probably including steric requirements of the ligands at the surfaces), whereas the phosphates are only loosely bound, which allows the particles to contract slightly, in the same way that a liquid droplet tries to minimize its surface.

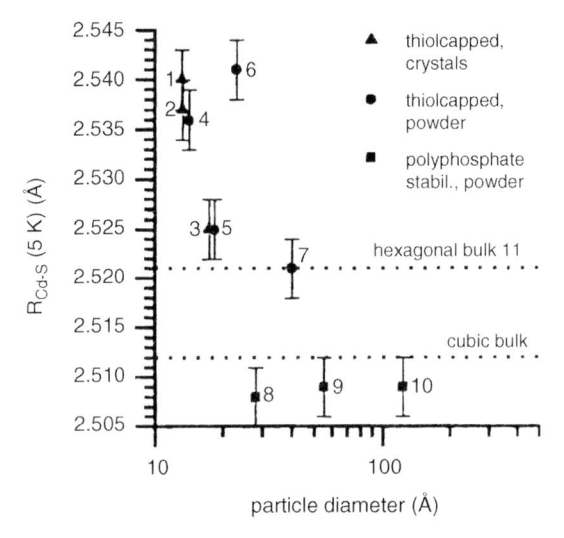

Figure 3.15 Mean Cd–S distances of thiol- and phosphate-stabilized CdS nanocrystals determined by extended X-ray absorption fine structure (EXAFS) spectroscopy at 5 K as a function of the particle diameter. From Ref. [163].

A large percentage of the atoms of particles in the nanometer size regime are located at their surface, which makes investigations into the surface structure highly desirable. The surface structure of thiophenol-stabilized CdS nanocrystals has been the subject of two reviews (in 1997 and 1998) [174, 175], in which both research groups used NMR techniques in order to study the functionalization, structure, and dynamics of the terminating ligands. The PbS [176] and ZnS [177, 178] nanocrystals may also be prepared with thiols as capping agents, showing again the versatility of this approach. The interaction of γ-radiation with Cd^{2+}- and thiol-containing sols yields small CdS particles, with the dose of the radiation determining the size of the evolving particles [179]. The reverse process – that is, the size-selective dissolution of preformed CdS nanocrystals by photoetching – was demonstrated recently by Torimoto et al. [180]. In an impressive study, Aldana, Wang and Peng provided a detailed insight into the photochemical degradation of thiol-stabilized CdSe particles [181]. At least three steps were involved in the process: (i) a photocatalytic oxidation of the thiol ligands on the surface; (ii) photo-oxidation of the nanocrystals; and (iii) precipitation of the nanocrystals. A bichromophoric nanoparticle system consisting of thioglycolic acid-stabilized CdTe particles and layer-by-layer (LbL) assembled films of an anionic polyelectrolyte was presented by Westenhoff and Kotov [182]. The first investigation into the electronic structure of a nanoparticle dimer was reported by Nosaka and Tanaka [183]. The semiempirical molecular orbital calculations first investigated the influence of phenyl-capping of polynuclear cadmium complexes, and subsequently yielded a negligible influence on the excited state by dimer formation. The stability of thiol-stabilized semiconductor particles was investigated in another report (see Ref. [33]). Various analytical methods such as analytical ultracentrifugation, UV-visible absorption spectroscopy, NMR spectroscopy, and powder XRD have also been applied. Reduced particle sizes are found in solution at low particle concentrations. Most probably, the assumed covalently bound thiols desorb from the surface of the particles, leaving behind vulnerable unstabilized particles that might even undergo a continuous decay over time. Additionally, breaking of the intra-ligand S—C bond is observed, resulting in a cadmium sulfide particle synthesis introducing sulfur only by the ligands, without any additional sulfide ions.

Finally, it may be noted that the nanocrystals prepared in aqueous solutions may be transferred into different nonpolar organic solvents [184]. This is accomplished by a partial ligand exchange, with acetone playing an important role in the efficient phase transfer. As the nanocrystals retain their luminescence properties after being transferred, this procedure provides a new source of easily processable luminescent materials for possible applications. The use of aqueous solutions and the avoidance of high temperatures and unstable precursors allow this synthetic approach to be easily up-scaled. Moreover, as the reaction yields (which approach gram quantities of dried products) are limited by the laboratory facilities, they may be further increased by using industrial-scale equipment.

3.2.1.3 The "Hot-Injection" Synthesis

The term "Hot-injection" synthesis, which refers to an organometallic approach to the synthesis of CdE (E = S, Se, Te) nanocrystals, was first introduced in 1993 by Murray et al. [12]. The preparative route is based on the pyrolysis of organometallic reagents [e.g., dimethylcadmium and bis(trimethylsilyl)selenium] by injection into a hot coordinating solvents [e.g., tri-n-octylphosphine oxide (TOPO) and tri-n-octylphosphine (TOP)]. This provides temporally a discrete nucleation and permits the controlled growth of nanocrystals; a subsequent size-selective precipitation leads to isolated samples with very narrow size distributions. The widespread success of this procedure is due largely to its versatility, its reproducibility and, last but not least, to the high quality of the particles in terms of their crystallinity and uniformity. The versatility is seen, for example, in a successful transformation of the preparation principles beyond the II–VI materials to the classes of the III–V and IV–VI semiconductor nanocrystals [185–216]. It is probably not an exaggeration to identify the first reports on the organometallic synthesis also as the trigger for elaborate theoretical studies on the size-dependent properties of semiconductor clusters, such as the electronic structure (see, e.g., Refs [217–233]). The years immediately after the "invention" of the hot-injection synthesis were characterized by investigations into the optical properties of CdSe particles of a wide range of sizes, and the surface properties of the evolving nanocrystals. Details of many of these exquisite studies were presented in an excellent review by the founders of the field [234], including the outstanding investigations into Förster energy transfer which takes place in close-packed assemblies of CdSe QD solids [235, 236]. Their photostability and high luminescence quantum yield led to the particles becoming ideal candidates for single-particle emission studies [237–245].

Some of the early studies on the "high-tech" applications of semiconductor nanoparticles have been based on the above-mentioned preparative route, namely the construction of a light-emitting diode (LED) produced from CdSe nanocrystals and a semiconducting polymer [246, 247]; in fact, even a single-electron transistor has been created by using CdSe nanocrystals [248].

The fate of the photogenerated charge carriers has been the subject of several studies of particles derived via the organometallic procedure (e.g., see Refs [249, 250] and references therein). The spin spectroscopy of "dark excitons" in CdSe particles of various sizes was conducted as a function of temperature and magnetic field strength [251]. The influence on the sizes, size distributions, and luminescence quantum yields of various amines on addition to the reaction mixtures was studied by Talapin and coworkers [252–254]. Alternative routes towards high-quality CdSe nanocrystals were identified by Qu, Peng and Peng, by varying the solvents–ligands and the precursors such as CdO and $Cd(Ac)_2$ instead of $Cd(CH_3)_2$ [99, 255]. Nowadays, ZnE (E = S, Se, Te) materials can also be prepared with high optical quality (ZnS [256]; ZnSe [257]; ZnTe [258]) via similar high-temperature syntheses such as those employed for the CdE materials. Synthetic progress has also been achieved by signifying that the solvent in the reaction does not necessarily need to be coordinating by itself, but that high-boiling noncoordinating solvents (e.g., octadecene) [259] may also allow size and shape control of II-VI semiconductor materials when coordinating agents such as carboxylic acids, phosphonic acids or amines are present in the reaction mixture in small amounts.

More recently, the dangerous organometallic precursors (especially dimethylcadmium) have been replaced by less dangerous cadmium compounds, such as CdO [99] or Cd-acetate [260], both of which can be dissolved in TOPO or ODE in the presence of carboxylic or phosphonic acids. This makes the high-temperature synthesis of nanocrystals in organic solvents much easier (and therefore accessible), without losing any of the excellent optical properties of the particles.

Organometallically synthesized CdSe (and InAs) nanocrystals have been employed as experimental examples in a very interesting study concerning the kinetics of nanocrystal growth [238]. It has been shown that diffusion-limited growth can lead to a narrowing of size distributions with time and that, under certain conditions, the kinetics of nanocrystal growth may also be influenced by the variation of the surface energy with size. In a theoretical study, the proposals of Alivisatos [261] were reconsidered, providing an even more detailed insight into the evolution of ensembles of nanoparticles in colloidal solutions [262]. An outline is provided of how the growth of nanoparticles in a diffusion-controlled regime may result in better final size distributions than those found for nanoparticles grown in a reaction-controlled regime. In addition, possible ways of controlling the particle size distributions in colloidal solutions have been discussed; these include, for example, an increase in the surface tension at the solvent–nanoparticle interface, and a decrease in the diffusion or mass transfer coefficient. A related experimental study was conducted by Qu and Peng in 2002 [263] in which the photoluminescence quantum yield (PL-QY) of organometallically prepared CdSe nanocrystals was studied during their growth. A "bright point" was observed which denoted the fact that, under a given set of initial conditions of the colloidal solution, the PL-QY was increased to a maximum value before decreasing again. Very similar results were reported in an even more thorough investigation by Talapin *et al.* [264]. In this case, results concerning the dynamic distribution of growth rates within ensembles of CdSe and InAs nanoparticles prepared organometallically, and CdTe prepared in aqueous solutions, were presented which indicated that the observed phenomenon of fraction-dependent PL-QY was of a more general character. By relating these results to the above-mentioned theoretical studies, it is proposed that the nanocrystal fractions with the best surface quality (in terms of a minimum of defects) corresponded to particles with the smallest net growth rate within the ensemble, because annealing of surface defects was most effective at this "zero growth rate."

In two studies, the absorption cross-section (or extinction coefficient) of CdSe nanocrystals has been under investigation [265, 266]. Initially, Striolo *et al.* [265] reported (among other results) that the oscillator strength of the first exciton transition per CdSe unit was more or less independent of the particle size, in accord with Schmelz *et al.* [267]. In contrast, Bawendi and coworkers clearly showed that as the QD radius decreased, the oscillator strength per CdSe unit of the lowest-energy transition increased (as had been reported by others for CdS [129] and CdTe [133], prepared in different ways), whereas the spectra converged for higher energies. For InAs QDs of mean radii ranging from 1.6 to 3.45 nm, the per-QD particle oscillator strength of the first-exciton transition was found to be constant for all sizes studied [268]. For small PbSe particles it was reported that, at high photon energies, ε scaled with the nanocrystal volume, irrespective of the particle size, and that from this size-independent absorption coefficients could be calculated. At the band gap, however, ε was size-dependent, while the resulting absorption coefficient increased quadratically with decreasing PbSe size. [269]. A theoretical study conducted by Sun and Goldys [270] aimed at providing an understanding of the experimental findings (see Ref. [271] and the references above). The experimental results for CdSe, CdS, and CdTe which, depending on the QD size, crossed over from a strong to an intermediate confinement regime with a radius greater than the hole Bohr radius, and for PbS nanocrystals which fell into the strong confinement regime were well reproduced. Due to model limitations, however, departures of the theoretical curves from the experimental data were more pronounced at lower energies than for larger nanoparticles, which were more weakly confined.

3.2.1.4 Core–Shell Nanocrystals

Coating a given ensemble of nanoparticles with another material yields "core–shell" nanocrystals (see above; see also Refs [272, 273]). In these structures, the cores may be any type of colloidal particle, such as metals, insulators, and all classes of semiconductors. Likewise, the shells may consist of type of material, including organics. At this point, attention is focused on semiconductor cores and inorganic shells, mainly of semiconductor materials.

In the majority of cases, the electronic structure of the core–shell particles is as follows. The core material, having a certain bandgap, is capped by a material with a larger bandgap, such that the conduction band energy of the capping material is higher (more negative in an electrochemical sense) than that of the core material, while the valence band energy of the capping material is lower (more positive on the electrochemical energy scale) than that of the core material. This energetic situation is frequently referred to as a type-I structure. The main consequence of this capping is that the exciton photogenerated in the core is prevented from spreading over the entire particle, and in this way it is forced to recombine while being spatially confined to the core. In most cases, this is accompanied by enhanced luminescence.

Early reports on the formation of coated nanoparticles include the aqueous polyphosphate-stabilized $CdS-Cd(OH)_2$ system [28]. Here, the preparative procedure was to add to a given sol of CdS particles a large excess of Cd ions, while increasing the pH to achieve strongly basic conditions. The obtained increase in luminescence quantum yield in comparison with the uncoated CdS particles was explained by a saturation of free valences at the particle surface, and this was taken as an indication of the formation of the proposed structure. The report was speculative in a sense that a structural characterization was not performed. Yet, this was not very surprising, since obtaining clear-cut evidence for shell formation of a few monolayers of material on nano-objects represents a characterizational challenge that has been accepted more or less since then in a wealth of reports on systems such as Ag_2S on CdS, ZnS on CdS, Ag_2S on AgI, CdSe on ZnS, CdSe on ZnSe, HgS on CdS, PbS on CdS, CdS on HgS, ZnSe on CdSe, ZnS on CdSe, CdS on CdSe, and *vice versa*. As many of these endeavors have already been reviewed (see Refs [33, 62, 63, 113, 114, 172, 272, 273]), they will not be mentioned here in detail. A very fine example of the most careful characterization of the overgrowth of CdSe nanocrystals with ZnS was reported by Dabbousi *et al.* [274], where for shell growth on preformed CdSe nanocrystals, diethylzinc and hexamethyldisilathiane were used as Zn and S precursors. The dissolution of equimolar amounts of the precursors in suitable solvents was followed by a dropwise addition to vigorously stirred CdSe colloidal solutions held at temperatures between 140 and 220 °C, depending on the core sizes. Besides optical spectroscopy, the authors performed wavelength dispersive X-ray spectroscopy, X-ray photoelectron spectroscopy (XPS), small- and wide-angle X-ray scattering, and (high-resolution) electron microscopy. Figure 3.16, which serves as an example of the thorough characterization conducted in these studies, shows the X-ray scattering in the small-angle region of bare 4.2 nm CdSe dots (curve a) and the increase of the dot diameters with the increase in the ZnS shell thickness from 0.65 monolayers (curve b) to 5.3 monolayers (curve e). The gradual shift of the ringing pattern to smaller angles indicated that ZnS was in fact growing onto the CdSe dots, instead of nucleating separate ZnS nanocrystals. This point was also deduced from the size histograms gained from electron microscopy studies of the same series of particles. In Figure 3.17, the uniform growth of the particles is clearly demonstrated, together with a broadening of the size distribution. The aspect ratio histograms (Figure 3.17b) of the slightly elongated particles are also monitored, and used to generate accurate fits to the small-angle X-ray data shown above. In an effort to relate the structural to the optical properties, growth of the ZnS shell on the CdSe core is described as "locally epitaxial." The ZnS shell tends to create defects at coverages of more than 1.3 monolayers, accompanied by a decrease in luminescence quantum yield.

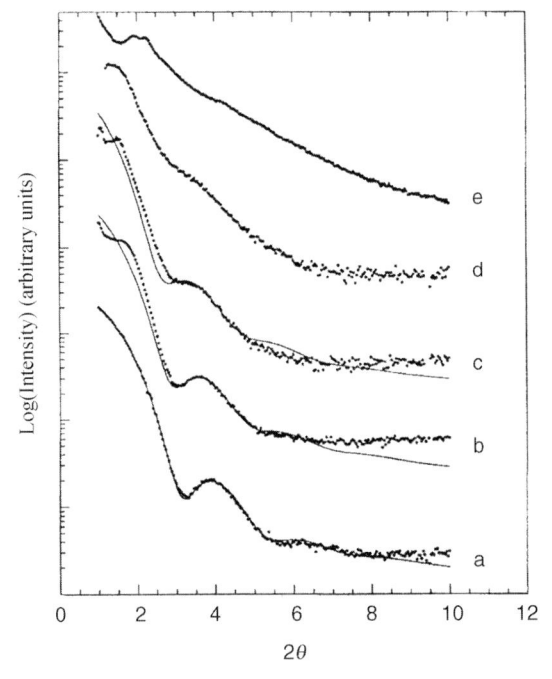

Figure 3.16 Small-angle X-ray scattering data from bare 4.2 nm CdSe particles (a) overgrown by 0.65 (b) through 5.3 (e) monolayers of ZnS. From Ref. [274].

Based on the results of these studies, progress has been made with similar particles, for example, in understanding the photo- and electrodarkening effects and in the assembly of luminescing semiconductor nanocrystals with biological objects in order to create fluorescent probes for sensing, imaging, or other diagnostic applications. To name a few, in other reports of the synthesis of CdSe–ZnS core–shell particles, high-resolution spectral hole burning has been performed, or effects such as photo-oxidation and photobleaching have been investigated on the single particle level. Likewise, the emission properties of single ZnS- or CdS-overcoated CdSe particles have been investigated by several groups. Other reports have included structures in which either the core or the shell is a II–VI semiconductor material, such as CdS capped by sulfur or polysulfide, InAs capped by CdSe, ZnSe and ZnS, or InP capped by $ZnCdSe_2$ or ZnS. Other synthetic approaches have included the formation of core–shell-type particles from CdTe and CdS, ZnSe on CdSe, and CdS on HgTe [276]. In the latter report, a synthetic route to strongly IR-emitting nanocrystals was outlined, whereby the capping enhanced the robustness of the particles rather than the (already high) luminescence quantum yield. The body of these studies has been summarized elsewhere (see Refs [171, 272, 273, 275]).

Early reports of ternary core–shell–shell (CSS) structures stem from Reiss et al. and Talapin et al. [277, 278]. In the latter case, CdSe nanocrystals stabilized with TOP and TOPO were coated with a shell of CdS or ZnSe. The second (i.e., outermost) shell was composed of ZnS in both cases. According to the authors, the main purpose of the

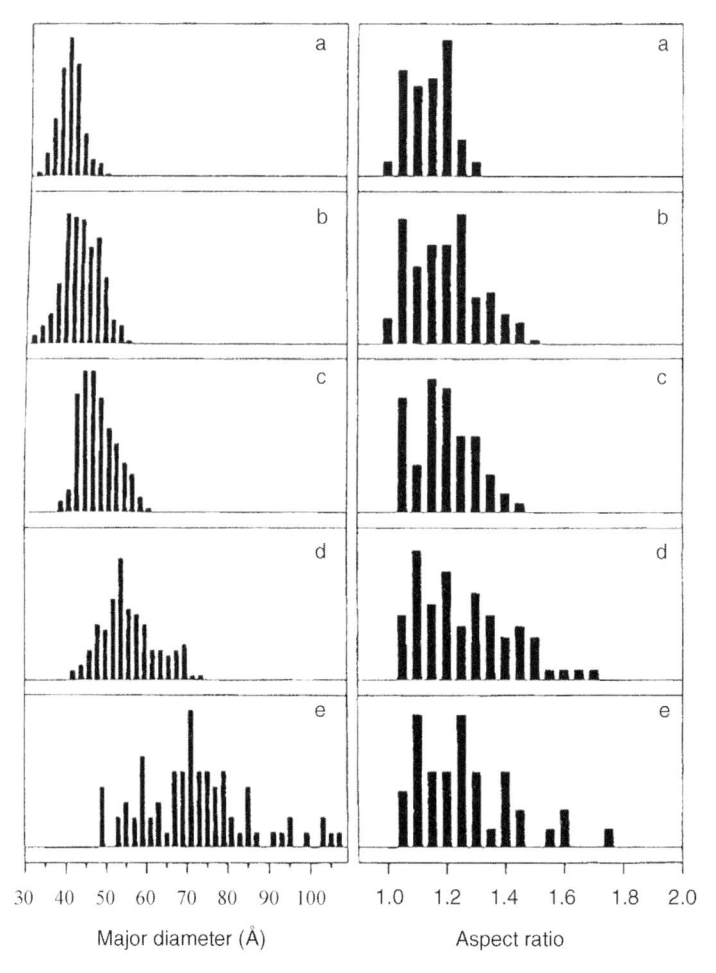

Figure 3.17 Left column: Size histograms of a series of CdSe nanocrystals overgrown with 0.65 (b) to 5.3 (e) monolayers of ZnS (a) and corresponding aspect ratios (right column). From Ref. [274].

outermost ZnS shell was to avoid charge carrier penetration towards the surface of the particles. The problem of the large lattice mismatch between CdSe and ZnS was overcome by using ZnSe or CdS as an intermediate shell, which effectively acted as a "lattice adapter." Another benefit of the CSS structures was their greatly enhanced photostability, which was assigned to an improved stability against photo-oxidation of the CSS structures in comparison with the core–shell particles. Recently, Jun *et al.* presented a simplified one-step synthesis for CdSe–CdS–ZnS CSS nanocrystals [279]. Another example of CSS structures with a lattice-adapting layer was reported by Mews *et al.* in 2005 [280]; in this case, the authors presented the possibility of including a 50:50 alloyed layer of $Zn_{0.5}Cd_{0.5}S$ in the structure, which resulted in CdSe–CdS–$Zn_{0.5}Cd_{0.5}$S–ZnS CSS particles. The coating steps were performed using

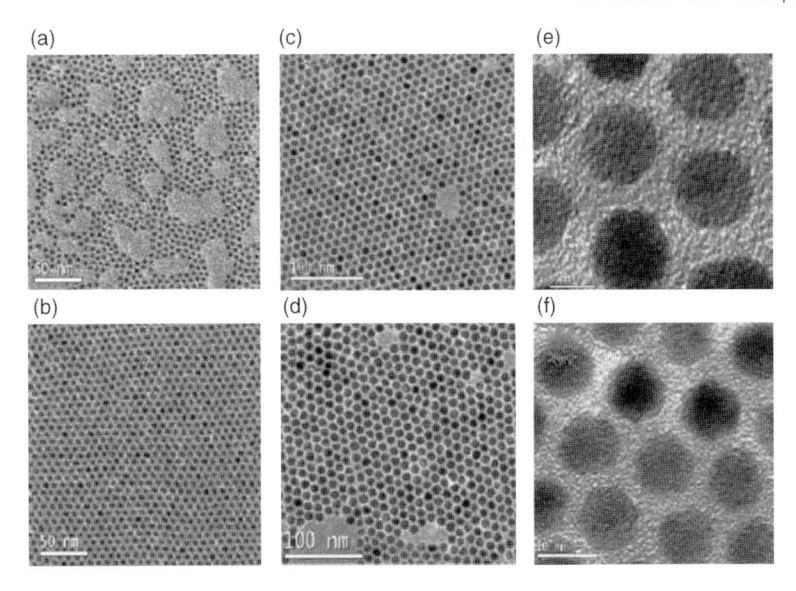

Figure 3.18 TEM images of the plain CdSe cores and core–shell nanocrystals obtained under typical reaction conditions. (a) TEM images of CdSe cores (before injection of Cd^{2+} solution); (b) (a) plus two monolayers of CdS; (c–e) (b) plus 3.5 monolayers of $Zn_{0.5}Cd_{0.5}S$; (d–f) (c) plus two monolayers of ZnS. From Ref. [280].

the SILAR (successive ion layer addition and reaction) technique, as introduced by the Peng group [281] which, again, utilizes high-boiling but noncoordinating solvents. Figure 3.18 shows impressively the quality of the particles in terms of uniformity, crystallinity, and homogeneity. Another interesting structure developed by Peng and coworkers was the CdSe–ZnS–CdSe CSS system [282], where the ZnS was the high band-gap material and the two CdSe units were electronically separated, giving rise to two distinct emission signals emanating from one nanocrystal, where none of the signals was trap-related. The two distinct emissions that occurred with relatively thick intermediate ZnS layers were interpreted as a complete electronic decoupling of the outer CdSe quantum well and the CdSe core. Most recently, this concept has yielded white light emission from semiconductor QDs [283].

While type-I nanoheterostructures are superior compared to homogeneous QDs in terms of fluorescence quantum yield and photostability, the type-II heterostructures also exhibit several interesting properties (as pointed out above). The staggered band alignment gives rise to a spatially indirect transition which occurs at lower energies than both band gap energies of the two materials used. Thus, type-II core–shell QDs can be emitters at a wavelength that cannot be achieved with either one of the two materials alone. In addition, the emission lifetime of type-II heterostructures is strongly increased. The potential applications of type-II core–shell QDs are based on the lasing of the nanoparticles [284]. A comparative study of CdTe–CdS core–shell nanocrystals (type-I situation) and CdTe–CdSe core–shell nanocrystals (type-II situation) has been presented recently [285]. In addition, these authors presented results obtained with ternary heterostructures containing CdTe, CdS, and CdSe. The relative positions of the conduction and valence band edges of the relevant semiconductors were such that the hole would be expected to be located in the CdTe-core of the particles in all cases discussed. The electron, however, would be delocalized over the whole nanocrystal in the case of CdTe–CdS nanocrystals, but would be located in the CdSe shell in the case of CdTe–CdSe and CdTe–CdS–CdSe nanocrystals. Thus, any overlap of the electron and hole wave functions would be expected to vary significantly in the different heterostructures. All structures showed large quantum efficiencies, whereas in all cases the overgrowth of the CdTe core with at least one layer of another II-VI material was found to lead to a reduction in the nonradiative rate. In the case of the CdSe-containing systems, in which a type-II structure was formed, this was confirmed by a red shift of the absorption and emission spectra, and also by a prolonged radiative lifetime. In contrast, the CdSe free nanoparticles showed a type-I behavior; that is, only a small red shift and an unchanged radiative lifetime. By sandwiching a CdS shell between the CdTe core and the CdSe shell, the charge carrier separation was shown to be tunable, and with this also the luminescence lifetimes of the nanoheterostructures. This would be of major importance in future applications, such as bio-labeling and lasing of semiconductor nanocrystals.

3.2.1.5 Quantum Dot Quantum Wells

As an extension of the studies involving the precipitation of HgS on polyphosphate-stabilized CdS and CdS on HgS, a wet chemical synthesis has been developed which yields a nanoheterostructure consisting of three domains, viz., the system CdS–HgS–CdS. The basic preparative idea involves the substitution reaction

$$CdS + nHg^{2+} \rightarrow CdS + HgS + nCd^{2+} \tag{3.1}$$

taking place at the surface of the preformed CdS particles. This reaction proceeds because of the at least 22 orders of magnitude smaller solubility product of HgS compared to that of CdS. The art here consists of managing the reaction to take place only at the surface, and this is largely achieved by applying a strict control to the experimental conditions with regards to pH, concentrations, duration of the addition, stirring, and so on. When a monolayer of HgS has been built on the CdS cores, this structure is capped by more CdS via the addition of H_2S and further Cd ions. The first study described the use of structures with a CdS core of about 5 nm surrounded by a shell of one to three monolayers of HgS, which again was coated by up to five monolayers of CdS [286, 287]. Subsequently, β-HgS with a bulk band gap of 0.5 eV formed a quantum well inside the cubic CdS (bulk band gap 2.5 eV), which caused the authors to describe these new structures as "quantum dot quantum wells" (QDQWs), although perhaps an even better term would be "quantum well in a quantum dot." Colloquially, the term "nano-onions" was established in the meantime. The first four reports on these particles, published in 1993 and 1994, described the synthesis, characterization, transient bleaching after laser excitation [288], and a widely applied theory based on the "effective mass approximation" [289]. As with the other findings, the bleaching experiments on nanoheterostructures caused them to appear as homogeneous electronic systems, and the results could not be explained by a superposition of the properties of the segments. Together, these efforts have been summarized in several reviews (see Refs [114, 290, 291]). It should be noted that Ref. [290] also incorporates the results of Mews and Alivisatos [292, 293], including an impressive HR-TEM investigation (see also Ref. [116]) with accompanying simulations presenting information concerning the shape and crystallinity of the particles. Hole burning and fluorescence line narrowing experiments have provided evidence on the degree of charge carrier localization in the HgS layer. Theoretical reports (see Refs [294, 295]) have referred to these experiments and deduced that a "dark exciton" also represents the lowest excitation state in these nanoheterostructures. A recent tight-binding study conducted by Perez-Conde and Bhattacharjee also detected bright and dark states, together with the strong localization of electron and hole states within the HgS well [296]. The tight-binding approach was also utilized by Bryant and Jaskolski to provide a good description for nanosystems with monolayer variations in composition [297]. It should be noted that trap states cannot be described solely as states in the quantum well; instead, depending on the shell thickness and band offset, trap states can have a large interface character.

Two reports by El-Sayed and coworkers [298, 299] published in 1996 and 1998, proved experimentally that the electronic structure was more complicated and could be observed using advanced spectroscopic techniques. These authors carried out experiments which were analogous to the transient bleaching at CdS and CdS–HgS–CdS particles, but with femtosecond time resolution. By this means, it was possible to observe the kinetics of the charge carrier capture in traps and in the HgS quantum well, respectively. Clearly, the spectral diffusion of the bleaching (of the "optical hole," as referred to by the authors) takes place over an intermediate state, to which the authors assigned an energetically relatively high "dark exciton" state, as predicted by theory [295]. In this state, the electron is situated in the HgS well, whereas the hole is in the CdS cladding layer of the particle and the "well" has a rather barrier-like effect on the hole. In a subsequent study, similar experiments were carried out on particles with a two monolayer-thick well of HgS, with the results agreeing well with theoretical predictions regarding the exciton energies [300]. The dynamics of the localization of the exciton in CdS–HgS QDQWs has also been the subject of an investigation by Yeh et al. [301]. By comparing the results from nanosecond hole-burning experiments with those having femtosecond time resolution, Yeh et al. concluded that the primary optical interaction would excite electrons from a delocalized QDQW ground state, and not from a localized HgS well state. Only after longer times (of the order of 400 fs) would the exciton finally be localized in the HgS well. Optically detected magnetic resonance (ODMR) was applied to QDQW samples with varying layer thicknesses of HgS, as well as of the outer CdS, by Lifshitz and coworkers [302]. Because of the technique's sufficient longevity, it proved possible to study (in particular) the trap luminescence which, when compared to the near band edge fluorescence, was red-shifted by about 400–700 meV. Consequently, changes in luminescence that were induced by magnetic resonance in the excited state were detected. Based on measurements carried out that depended on the intensity of the luminescence excitation, and on the frequency of the coupled microwaves, with subsequent simulations of the spectra, it was concluded that there were at least two different recombination channels via traps in the particles. These traps would be localized in the CdS–HgS interfaces, and might possibly be twin grain boundaries and edge dislocations. Subsequently, in a photoluminescence experiment, an enhancement of the band edge emission was observed upon excitation with two differently colored beams, instead of with one beam as in an ordinary experiment [303]. This result was explained as being the consequence of a partial quenching of the defect emission, and suggested a mutual interaction of the two luminescence events.

A strong reduction in the spectral and intensity fluctuations was observed when QDQWs of CdS and HgS were compared with pure CdS particles in single-particle emission studies [304]. This, again, was explained by charge carrier localization in the HgS region of the nanocrystals.

A QDQW with a large lattice mismatch, namely ZnS–CdS–ZnS, was prepared by the group of El-Sayed [305], where established techniques were combined to synthesize small seeds of CdS and ZnS, with nanoepitaxy later being applied to grow ZnS–CdS nanoheterostructures. The evolving absorption spectra of the particles were presented, together with a theoretical treatment.

Yet another QDQW – namely that of composition CdS–CdSe–CdS – was recently synthesized from TOPSe and cadmium oleate in octadecene [306]. The absorption, emission, and Raman scattering spectra of colloidal CdS–CdSe–CdS QDQWs with different thicknesses of the CdSe well have been investigated and discussed.

Recently, the Georgia Tech group of El-Sayed reported the initial results of a QDQW system containing two wells of HgS, separated by a wall of CdS. The two-well system, when characterized by absorption and emission spectroscopy, clearly revealed the formation of a two-well system rather than a single double-layer structure [307].

During the same year, Dorfs described a similar effort, namely the preparation and characterization of a series of QDQWs containing two wells [308]. The preparative route first followed established procedures which yielded single-well particles consisting of a CdS core of diameter 4.7 nm, surrounded by one or two monolayers of HgS capped by two, three, or four layers of CdS. Following the same concept as that for the single-well particles – that is, a repetition of the replacement of the surface CdS by HgS – the addition of a second monolayer of HgS and capping of the whole structure by three monolayers of CdS yielded a whole new family of particles. With regards to nomenclature, the authors suggested for these nanoheterostructures the following (cf. Figure 3.19): CdS–HgS is represented by "abcd," where, starting with the CdS core, "a" is the number of HgS monolayers, "b" is the number of monolayers of CdS as the barrier between the wells, "c" is the number of monolayers of the second HgS well, and "d" is the number of cladding layers.

In Figure 3.20 are summarized the absorption spectra of a total of eight different double-well QDQWs prepared according to the above-outlined procedures. All eight samples had in common the core size 4.7 nm and the outer cladding of three monolayers of CdS. However, they differed in the three inner compartments: the

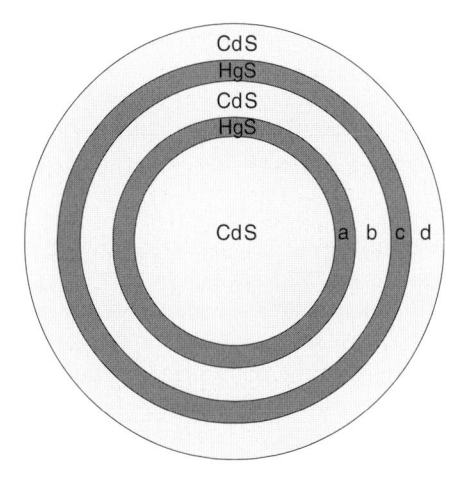

Figure 3.19 Schematic drawing of the double-well nanoheterostructures described in the text. The wells (a and c) consist of the small bandgap material β-HgS, the core, the barrier between the wells (b) and the cladding layer (d) are made from CdS From Ref. [208].

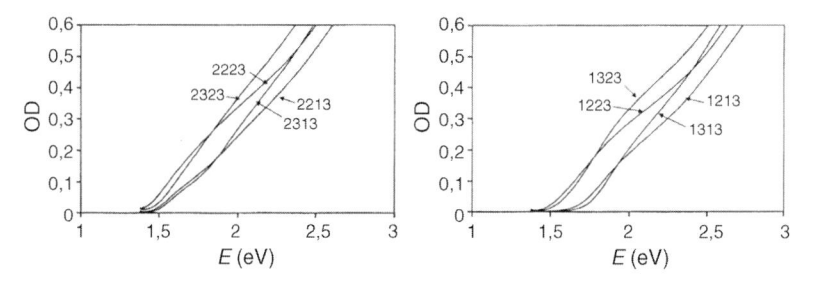

Figure 3.20 Absorption spectra of eight different double-well nanoheterostructures. In CdS nanocrystals, two wells of HgS are embedded, well thickness and well separations are varied (for details of nomenclature, see the text). From Ref. [308].

inner well consisted of either one or two monolayers of HgS, the barrier between the wells were either two or three monolayers wide, and the outer well again consisted of one or two monolayers of HgS. Thus, the structures CdS–HgS 1213, 1223, 1313, 1323, 2213, 2223, 2313, and 2323 were generated. The most remarkable findings gained from the absorption spectra and the accompanying TEM study were as follows:

1) Judging from the particle size histograms, there was no indication of separately formed CdS or HgS. Exclusively homogeneous particle growth was observed.
2) The greater proportion of HgS present in the structures, the further into the red spectral region was the position of the first electronic transition.
3) When comparing the four pairs of samples with the same well composition, the transition was located further into the red when the wells were separated by only

two monolayers of CdS instead of three; this was explained by an enhanced electronic coupling on decreasing the well separation.

4) The onset of absorption of the particles exhibiting an asymmetric inner potential (e.g., CdS–HgS 2313 and 1323) took place at the same energy.
5) The high-energy absorption depended on the total amount of material present in solution.

Some of these findings have been treated from a theoretical basis already, and have provided an insight into the very interesting electronic properties and charge separation features [309]. In an investigation using XPS with tunable synchrotron radiation, a member of this family of particles was compared to single-well CdS–HgS–CdS QDQWs [310]. As might be expected, the photoemission spectra of the CdS–HgS–CdS–HgS–CdS double-quantum-well nanocrystals appeared quite similar to those of the single-well species. In addition to a bulk species, high-resolution spectra also revealed a surface environment for Cd, while different S species could not be resolved. To some extent, the XPS experiments allowed the characterization and verification of the onion-like structure of the QDQW nanocrystals, as shown also by the extension of a previously developed model for core–shell nanocrystals [311].

Reference

1 Henglein, A. (1982) *Ber. Bunsenges. Phys. Chem.*, **86**, 301.
2 Kalyanasundaram, K., Borgarello, E., Duonghong, D., and Grätzel, M. (1981) *Angew. Chem., Int. Ed. Engl.*, **20**, 987.
3 Darwent, J.R. and Porter, G. (1981) *J. Chem. Soc., Chem. Commun.*, 145.
4 Rossetti, R. and Brus, L. (1982) *J. Phys. Chem.*, **86**, 4470.
5 Duonghong, D., Ramsden, J., and Grätzel, M. (1982) *J. Am. Chem. Soc.*, **104**, 2977.
6 Kuczynski, J. and Thomas, J.K. (1982) *Chem. Phys. Lett.*, **88**, 445.
7 Fox, M.A., Lindig, B., and Chem, C.C. (1982) *J. Am. Chem. Soc.*, **104**, 5828.
8 Katzschmann, R., Kranold, R., and Rehfeld, A. (1977) *Phys. Status Solidi A*, **43**, K161.
9 Brus, L.E. (1983) *J. Chem. Phys.*, **79**, 5566.
10 Ekimov, A.I. and Onushchenko, A.A. (1982) *Sov. Phys. - Semiconductors*, **16**, 775.
11 Efros, A.L. and Efros, A.L. (1982) *Sov. Phys. - Semiconductors*, **16**, 772.
12 Murray, C.B., Norris, D.J., and Bawendi, M.G. (1993) *J. Am. Chem. Soc.*, **115**, 8706.
13 Rossetti, R., Nakahara, S., and Brus, L.E. (1983) *J. Chem. Phys.*, **79**, 1086.
14 Gutierrez, M. and Henglein, A. (1983) *Ber. Bunsenges. Phys. Chem.*, **87**, 474.
15 Henglein, A. and Gutierrez, M. (1983) *Ber. Bunsenges. Phys. Chem.*, **87**, 852.
16 Henglein, A., Gutierrez, M., and Fischer, C. (1984) *Ber. Bunsenges. Phys. Chem.*, **88**, 170.
17 Weller, H., Koch, U., Gutierrez, M., and Henglein, A. (1984) *Ber. Bunsenges. Phys. Chem.*, **88**, 649.
18 Fojtik, A., Weller, H., Koch, U., and Henglein, A. (1984) *Ber. Bunsenges. Phys. Chem.*, **88**, 969.
19 Ramsden, J.J. and Grätzel, M. (1984) *J. Chem. Soc., Faraday Trans.*, **1**, 80, 919.
20 Nozik, A.J., Williams, F., Nenadovic, M.T., Rajh, T., and Micic, O.I. (1985) *J. Phys. Chem.*, **89**, 397.
21 Brus, L.E. (1984) *J. Chem. Phys.*, **80**, 4403.
22 Brus, L.E. (1986) *J. Phys. Chem.*, **90**, 2555.
23 Fischer, C.-H., Weller, H., Fojtik, A., Lume-Pereira, C., Janata, E., and

Henglein, A. (1986) *Ber. Bunsenges. Phys. Chem.*, **90**, 46.

24 Dannhauser, T., O'Neil, M., Johansson, K., Whitten, D., and McLendon, G. (1986) *J. Phys. Chem.*, **90**, 6074.

25 Tricot, Y.-M. and Fendler, J.H. (1986) *J. Phys. Chem.*, **90**, 3369.

26 Nemeljokvic, J.M., Nenadovic, M.T., Micic, O.I., and Nozik, A.J. (1986) *J. Phys. Chem.*, **90**, 12.

27 Henglein, A., Kumar, A., Janata, E., and Weller, H. (1986) *Chem. Phys. Letters*, **132**, 133.

28 Spanhel, L., Haase, M., Weller, H., and Henglein, A. (1987) *J. Am. Chem. Soc.*, **109**, 5649.

29 Steigerwald, M.L., Alivisatos, A.P., Gibson, J.M., Harris, T.D., Kortan, R., Muller, A.J., Thayer, A.M., Duncan, T.M., Douglass, D.C., and Brus, L.E. (1988) *J. Am. Chem. Soc.*, **110**, 3046.

30 Youn, H.-C., Baral, S., and Fendler, J.H. (1988) *J. Phys. Chem.*, **92**, 6320.

31 Smotkin, E.S., Lee, C., Bard, A.J., Campion, A., Fox, M.A., Mallouk, T.E., Webber, S.E., and White, J.M. (1988) *Chem. Phys. Lett.*, **152**, 265.

32 Henglein, A. (1988) *Top. Curr. Chem.*, **143**, 113.

33 Henglein, A. (1989) *Chem. Rev.*, **89**, 1861.

34 Steigerwald, M.L. and Brus, L.E. (1990) *Acc. Chem. Res.*, **23**, 183.

35 Hayes, D., Micic, O.I., Nenadovic, M.T., Swayambunathan, V., and Meisel, D. (1989) *J. Phys. Chem.*, **93**, 4603.

36 Resch, U., Weller, H., and Henglein, A. (1989) *Langmuir*, **5**, 1015.

37 Zen, J.-M., Fan, F.-R.F., Chen, G., and Bard, A.J. (1989) *Langmuir*, **5**, 1355.

38 Herron, N., Wang, Y., and Eckert, H. (1990) *J. Am. Chem. Soc.*, **112**, 1322.

39 Spanhel, L. and Anderson, M.A. (1990) *J. Am. Chem. Soc.*, **112**, 2278.

40 Yanagida, S., Yoshiya, M., Shiragami, T., Pac, C.J., Mori, H., and Fujita, H. (1990) *J. Phys. Chem.*, **94**, 3104.

41 Eychmüller, A., Katsikas, L., and Weller, H. (1990) *Langmuir*, **6**, 1605.

42 Smotkin, E.S., Brown, R.M., Rabenberg, L.K., Salomon, K., Bard, A.J., Campion, A., Fox, M.A., Mallouk, T.E., Webber, S.E., and White, J.M. (1990) *J. Phys. Chem.*, **94**, 7543.

43 Chestnoy, N., Harris, T.D., Hull, R., and Brus, L.E. (1986) *J. Phys. Chem.*, **90**, 3393.

44 Wang, Y. and Herron, N. (1987) *J. Phys. Chem.*, **91**, 257.

45 Wang, Y. and Herron, N. (1988) *J. Phys. Chem.*, **92**, 4988.

46 O'Neil, M., Marohn, J., and McLendon, G. (1990) *J. Phys. Chem.*, **94**, 4356.

47 Eychmüller, A., Hässelbarth, A., Katsikas, L., and Weller, H. (1991) *Ber. Bunsenges. Phys. Chem.*, **95**, 79.

48 Chandler, R.R. and Coffer, J.L. (1991) *J. Phys. Chem.*, **95**, 4.

49 Hiramoto, M., Hashimoto, K., and Sakata, T. (1991) *Chem. Phys. Lett.*, **182**, 139.

50 Uchihara, T., Nakamura, T., and Kinjo, A. (1992) *J. Photochem. Photobiol. A: Chem.*, **66**, 379.

51 Resch, U., Eychmüller, A., Haase, M., and Weller, H. (1992) *Langmuir*, **8**, 2215.

52 Kumar, A. and Kumar, S. (1992) *J. Photochem. Photobiol. A: Chem.*, **69**, 91.

53 Bigham, S.R. and Coffer, J. (1992) *J. Phys. Chem.*, **96**, 10581.

54 Hoffman, A.J., Mills, G., Yee, H., and Hoffmann, M.R. (1992) *J. Phys. Chem.*, **96**, 5546.

55 Wang, Y., Harmer, M., and Herron, N. (1993) *Isr. J. Chem.*, **33**, 31.

56 Fendler, J.H. (1993) *Isr. J. Chem.*, **33**, 41.

57 Kamat, P.V., Vanwijngaarden, M.D., and Hotchandani, S. (1993) *Isr. J. Chem.*, **33**, 47.

58 Luangdilok, C. and Meisel, D. (1993) *Isr. J. Chem.*, **33**, 53.

59 Micic, O.I., Rajh, T., Nedeljkovic, J.M., and Comor, M.I. (1993) *Isr. J. Chem.*, **33**, 59.

60 Johansson, K., Cowdery, R., Oneil, M., Rehm, J., McLendon, G., Marchetti, A., and Whitten, D.G. (1993) *Isr. J. Chem.*, **33**, 67.

61 Weller, H., Eychmüller, A., Vogel, R., Katsikas, L., Hässelbarth, A., and Giersig, M. (1993) *Isr. J. Chem.*, **33**, 107.

62 Weller, H. (1993) *Angewandte Chemie*, **32**, 41.

63 Weller, H. (1993) *Adv. Mater.*, **5**, 88.

64 Müllenborn, M., Jarvis, R.F., Yacobi, B.G., Kaner, R.B., Coleman, C.C., and

Haegel, N.M. (1993) *Appl. Phys. A*, **56**, 317.
65 Chandler, R.R. and Coffer, J.L. (1993) *J. Phys. Chem.*, **97**, 9767.
66 Rajh, T., Micic, O.I., and Nozik, A.J. (1993) *J. Phys. Chem.*, **97**, 11999.
67 Rama Krishna, M.V. and Friesner, R.A. (1991) *J. Chem. Phys.*, **95**, 8309.
68 Lippens, P.E. and Lannoo, M. (1989) *Phys. Rev. B*, **39**, 10935.
69 Lippens, P.E. and Lannoo, M. (1990) *Phys. Rev. B*, **41**, 6079.
70 Schmitt-Rink, S., Miller, D.A.B., and Chemla, D.S. (1987) *Phys. Rev. B*, **35**, 8113.
71 Hanamura, E. (1988) *Phys. Rev. B*, **37**, 1273.
72 Kayanuma, Y. (1988) *Phys. Rev. B*, **38**, 9797.
73 Reiss, P. (2008) *Semiconductor Nanocrystal Quantum Dots* (ed. A. Rogach), Springer, Vienna, New York, p. 35.
74 Matsumoto, H., Sakata, T., Mori, H., and Yoneyama, H. (1996) *J. Phys. Chem.*, **100**, 13781.
75 Trindade, T. and O'Brian, P. (1996) *Adv. Mater.*, **8**, 161.
76 Korgel, B.A. and Monbouquette, H.G. (1996) *J. Phys. Chem.*, **100**, 346.
77 Dekany, I., Nagy, L., Turi, L., Kiraly, Z., Kotov, N.A., and Fendler, J.H. (1996) *Langmuir*, **12**, 3709.
78 Fogg, D.E., Radzilowski, L.H., Blanski, R., Schrock, R.R., and Thomas, E.L. (1997) *Macromolecules*, **30**, 417.
79 Ladizhansky, V., Hodes, G., and Vega, S. (1998) *J. Phys. Chem.*, **102**, 8505.
80 Correa-Duarte, M.A., Giersig, M., and Liz-Marzan, L.M. (1998) *Chem. Phys. Lett.*, **286**, 497.
81 Kurth, D.G., Lehmann, P., and Lesser, C. (2000) *Chem. Commun.*, 949.
82 Lemon, B.I. and Crooks, R.M. (2000) *J. Am. Chem. Soc.*, **122**, 12886.
83 Yu, S.-H., Yang, J., Han, Z.-H., Zhou, Y., Yang, R.-Y., Qian, Y.-T., and Zhang, Y.-H. (1999) *J. Mater. Chem.*, **9**, 1283.
84 Lakowicz, J.R., Gryczinski, I., Gryczinski, Z., and Murphy, C.J. (1999) *J. Phys. Chem. B*, **103**, 7613.
85 Wu, F., Zhang, J.Z., Kho, R., and Mehra, R.K. (2000) *Chem. Phys. Lett.*, **330**, 237.
86 Li, Y., Ding, Y., Liao, H., and Qian, Y. (1999) *J. Phys. Chem. Sol.*, **60**, 965.
87 Patel, A.A., Wu, F., Zhang, J.Z., Torres-Martinez, C.L., Mehra, R.K., Yang, Y., and Risbud, S.H. (2000) *J. Phys. Chem.*, **104**, 11598.
88 Hebalkar, N., Lobo, A., Sainkar, S.R., Pradhan, S.D., Vogel, W., Urban, J., and Kulkarni, S.K. (2001) *J. Mater. Sci.*, **36**, 4377.
89 Ingert, D., Feltin, N., Levry, L., Gouzerh, P., and Pileni, M.P. (1999) *Adv. Mater.*, **11**, 220.
90 Cao, L., Miao, Y.M., Zhang, Z.B., Xie, S.S. Yang, G.Z., and Zou B.S., (2005) *J. Chem. Phys.*, 123, 024702.
91 Xie, Y., Huang, J., Li, B., Liu, Y., and Qian, Y. (2000) *Adv. Mater.*, **12**, 1523.
92 Konopny, L., Berfeld, M., Popovitz-Biro, R., Weissbuch, I., Leiserowitz, L., and Lahav, M. (2001) *Adv. Mater.*, **13**, 580.
93 Zhang, Z., Dai, S., Fan, X., Blom, D.A., Pennycook, S.J., and Wei, Y. (2001) *J. Phys. Chem. B*, **105**, 6755.
94 Besson, S., Gacoin, T., Ricolleau, C., Jacquiod, C., and Boilot, J.P. (2002) *Nano Lett.*, **2**, 409.
95 Yosef, M., Schaper, A.K., Fröba, M., and Schlecht, S. (2005) *Inorg. Chem.*, **44**, 5890.
96 Brieler, F.J., Grundmann, P., Fröba, M., Chen, L.M., Klar, P.J., Heimbrodt, W., von Nidda, H.A.K., Kurz, T., and Loidl, A. (2005) *Eur. J. Inorg. Chem.*, 3597.
97 Turner, E.A., Huang, Y.N., and Corrigan, J.F. (2005) *Eur. J. Inorg. Chem.*, 4465.
98 Dimos, K., Koutselas, I.B., and Karakassides, M.A. (2006) *J. Phys. Chem. B*, **110**, 22339.
99 Peng, Z.A. and Peng, X. (2001) *J. Am. Chem. Soc.*, **123**, 183.
100 Elbaum, R., Vega, S., and Hodes, G. (2001) *Chem. Mater.*, **13**, 2272.
101 Zhao, H., Douglas, E.P., Harrison, B.S., and Schanze, K.S. (2001) *Langmuir*, **17**, 8428.
102 Zhao, H. and Douglas, E.P. (2002) *Chem. Mater.*, **14**, 1418.
103 Pinna, N., Weiss, K., Sack-Kongehl, H., Vogel, W., Urban, J., and Pileni, M.P. (2001) *Langmuir*, **17**, 7982.
104 Fan, H.Y. (2008) *Chem. Commun.*, 1383.
105 Farmer, S.C. and Patten, T.E. (2001) *Chem. Mater.*, **13**, 3920.

106 Wang, Y.A., Li, J.J., Chen, H., and Peng, X. (2002) *J. Am. Chem. Soc.*, **124**, 2293.
107 Nandakumar, P., Vijayan, C., and Murti, Y.V.G.S. (2002) *J. Appl. Phys.*, **91**, 1509.
108 Ispasoiu, R.G., Jin, Y., Lee, J., Papadimitrakopoulos, F., and Goodson, T. III (2002) *Nano Lett.*, **2**, 127.
109 Cumberland, S.L., Hanif, K.M., Javier, A., Khitrov, G.A., Strouse, G.F., Woessner, S.M., and Yun, C.S. (2002) *Chem. Mater.*, **14**, 1576.
110 Pan, M.F., Gao, F. He, R., Cui, D.X., and Zhang, Y.F. (2006) *J. Colloid Interface Sci.*, **297**, 151.
111 Ling, J. and Cong, R.M. (2008) *Acta Chim. Sinica*, **66**, 2070.
112 Feng, C.L., Yu, Y.M., Zhong, X.H., Steinhart, M., Majoral,, J.P., and Knoll, W. (2009) Smart Optics, in *Advances in Science and Technology*, vol. **55** (eds P. Vincenziniand G. Righini), TransTech Publications, p. 84.
113 Henglein, A. (1995) *Ber. Bunsenges. Phys. Chem.*, **99**, 903.
114 Weller, H. and Eychmüller, A. (1995) *Adv. Photochem.*, **20**, 165.
115 Alivisatos, A.P. (1996) *Science*, **271**, 933.
116 Alivisatos, A.P. (1996) *J. Phys. Chem.*, **100**, 13226.
117 Weller, H. and Eychmüller, A. (1996) *Semiconductor Nanoclusters*, vol. **103** (eds P.V. Kamatand D. Meisel), Elsevier Science B. V., Amsterdam, p. 5.
118 Alivisatos, P.A., Barbara, P.F., Castleman, A.W., Chang, J., Dixon, D.A., Klein, M.L., McLendon, G.L., Miller, J.S., Ratner, M.A., Rossky, P.J., Stupp, S.I., and Thompson, M.E. (1998) *Adv. Mater.*, **10**, 1297.
119 Trindade, T., O'Brian, P., and Pickett, N.L. (2001) *J. Mater. Chem.*, **13**, 3843.
120 Whitling, J.M., Spreitzer, G., and Wright, D.W. (2000) *Adv. Mater.*, **12**, 1377.
121 Farneth, W.E., Herron, N., and Wang, Y. (1992) *Chem. Mater.*, **4**, 916.
122 Herron, N., Suna, A., and Wang, Y. (1992) *J. Chem. Soc., Dalton Trans.*, 2329.
123 Dance, I.G., Choy, A., and Scudder, M.L. (1984) *J. Am. Chem. Soc.*, **106**, 6285.
124 Lee, G.S.H., Craig, D.C., Ma, I., Scudder, M.L., Bailey, T.D., and Dance, I.G. (1988) *J. Am. Chem. Soc.*, **110**, 4863.
125 Lee, G.S.H., Fisher, K.J., Craig, D.C., Scudder, M.L., and Dance, I.G. (1990) *J. Am. Chem. Soc.*, **112**, 6435.
126 Strickler, P. (1969) *J. Chem. Soc., Chem. Commun.*, 655.
127 Schwarzenbach, G., Gautschi, K., Peter, J., and Tunaboylu, K. (1972) *Trans. R. Inst. Technol. Stockholm.*, **271**, 295.
128 Herron, N., Calabrese, J.C., Farneth, W.E., and Wang, Y. (1993) *Science*, **259**, 1426.
129 Vossmeyer, T., Katsikas, L., Giersig, M., Popovic, I.G., Diesner, K., Chemseddine, A., Eychmüller, A., and Weller, H. (1994) *J. Phys. Chem.*, **98**, 7665.
130 Nosaka, Y., Shigeno, H., and Ikeuchi, T. (1995) *J. Phys. Chem.*, **99**, 8317.
131 Vossmeyer, T., Reck, G., Katsikas, L., Haupt, E.T.K., Schulz, B., and Weller, H. (1995) *Inorg. Chem.*, **34**, 4926.
132 Rogach, A.L., Kornowski, A., Gao, M., Eychmüller, A., and Weller, H. (1999) *J. Phys. Chem. B*, **103**, 3065.
133 Rajh, T., Micic, O.I., and Nozik, A.J. (1993) *J. Phys. Chem.*, **97**, 11999.
134 Rogach, A.L., Katsikas, L., Kornowski, A., Su, D.S., Eychmüller, A., and Weller, H. (1996) *Ber. Bunsenges. Phys. Chem.*, **100**, 1772.
135 Rogach, A.L., Katsikas, L., Kornowski, A., Su, D., Eychmüller, A., and Weller, H. (1997) *Ber. Bunsenges. Phys. Chem.*, **101**, 1668.
136 Gao, M., Kirstein, S., Möhwald, H., Rogach, A.L., Kornowski, A., Eychmüller, A., and Weller, H. (1998) *J. Phys. Chem. B*, **102**, 8360.
137 Rogach, A.L., Kershaw, S.V., Burt, M., Harrison, M., Kornowski, A., Eychmüller, A., and Weller, H. (1999) *Adv. Mater.*, **11**, 552.
138 Shavel, A., Gaponik, N., and Eychmüller, A. (2004) *J. Phys. Chem. B*, **108**, 5905.
139 Kershaw, S.V., Burt, M., Harrison, M., Rogach, A.L., Weller, H., and Eychmüller, A. (1999) *Appl. Phys. Lett.*, **75**, 1694.
140 Harrison, M.T., Kershaw, S.V., Burt, M.G., Eychmüller, A., Weller, H., and Rogach, A.L. (2000) *Mater. Sci. Eng. B*, **69**, 355.
141 Rogach, A.L., Harrison, M.T., Kershaw, S.V., Kornowski, A., Burt, M.G.,

Eychmüller, A., and Weller, H. (2001) *Phys. Status Solidi B*, **224**, 153.

142 Lesnyak, V., Plotnikov, A., Gaponik, N., and Eychmüller, A. (2008) *J. Mater. Chem.*, **18**, 5142.

143 Piven, N., Susha, A., Döblinger, M., and Rogach, A.L. (2008) *J. Phys. Chem. C*, **112**, 15253.

144 Gaponik, N., Talapin, D.V., Rogach, A.L., Hoppe, K., Shevchenko, E.V., Kornowski, A., Eychmüller, A., and Weller, H. (2002) *J. Phys. Chem. B*, **106**, 7177.

145 Rogach, A.L., Franzl, T., Klar, T.A., Feldmann, J., Gaponik, N., Lesnyak, V., Shavel, A., Eychmüller, A., Rakovich, Y.P., and Donegan, J.F. (2007) *J. Phys. Chem. C*, **111**, 14628.

146 Hoppe, K., Geidel, E., Weller, H., and Eychmüller, A. (2002) *Phys. Chem. Chem. Phys.*, **4**, 1704.

147 Fang, Z., Liu, L., Xu, L., Yin, X., and Zhong, X. (2008) *Nanotechnology*, **19**, 235603.

148 Fang, Z., Li, Y., Zhang, H., Zhong, X., and Zhu, L. (2009) *J. Phys. Chem. C*, **113**, 14145.

149 Acar, H.Y., Kas, R., Yurtsever, E., Ozen, C., and Lieberwirth, I. (2009) *J. Phys. Chem. C*, **113**, 10005.

150 Chemseddine, A. and Weller, H. (1993) *Ber. Bunsenges. Phys. Chem.*, **97**, 636.

151 Grabolle, M., Spieles, M., Lesnyak, V., Gaponik, N., Eychmüller, A., and Resch-Genger, U. (2009) *Anal. Chem.*, **81**, 6285.

152 Shavel, A., Gaponik, N., and Eychmüller, A. (2006) *J. Phys. Chem. B*, **110**, 19233.

153 Murase, N., Gao, M.Y., Gaponik, N., Yazawa, T., and Feldmann, J. (2001) *Int. J. Mod. Phys. B*, **15**, 3881.

154 Döllefeld, H., Hoppe, K., Kolny, J., Schilling, K., Weller, H., and Eychmüller, A. (2002) *Phys. Chem. Chem. Phys.*, **4**, 4747.

155 Kapitonov, A.M., Stupak, A.P., Gaponenko, S.V., Petrov, E.P., Rogach, A.L., and Eychmüller, A. (1999) *J. Phys. Chem. B*, **103**, 10109.

156 Glozman, A., Lifshitz, E., Hoppe, K., Rogach, A.L., Weller, H., and Eychmüller, A. (2001) *Isr. J. Chem.*, **41**, 39.

157 Gao, M., Lesser, C., Kirstein, S., Möhwald, H., Rogach, A.L., and Weller, H. (2000) *J. Appl. Phys.*, **87**, 2297.

158 Gaponik, N.P., Talapin, D.V., Rogach, A.L., and Eychmüller, A. (2000) *J. Mater. Chem.*, **10**, 2163.

159 Talapin, D.V., Poznyak, S.K., Gaponik, N.P., Rogach, A.L., and Eychmüller, A. (2002) *Physica E.*, **14**, 237.

160 Harrison, M.T., Kershaw, S.V., Burt, M.G., Rogach, A.L., Kornowski, A., Eychmüller, A., and Weller, H. (2000) *Pure Appl. Chem.*, **72**, 295.

161 Rockenberger, J., Tröger, L., Kornowski, A., Eychmüller, A., Feldhaus, J., and Weller, H. (1997) *J. Phys. IV France*, **7**, 1213.

162 Rockenberger, J., Tröger, L., Kornowski, A., Voßmeyer, T., Eychmüller, A., Feldhaus, J., and Weller, H. (1997) *Ber. Bunsenges. Phys. Chem.*, **101**, 1613.

163 Rockenberger, J., Tröger, L., Kornowski, A., Vossmeyer, T., Eychmüller, A., Feldhaus, J., and Weller, H. (1997) *J. Phys. Chem. B*, **101**, 2691.

164 Rockenberger, J., Tröger, L., Rogach, A.L., Tischer, M., Grundmann, M., Weller, H., and Eychmüller, A. (1998) *Ber. Bunsenges. Phys. Chem.*, **102**, 1561.

165 Rockenberger, J., Tröger, L., Rogach, A.L., Tischer, M., Grundmann, M., Eychmüller, A., and Weller, H. (1998) *J. Chem. Phys.*, **108**, 7807.

166 Gaponenko, S.V., Kapitonov, A.M., Bogomolov, V.N., Prokofiev, A.V., Eychmüller, A., and Rogach, A.L. (1998) *JETP Lett.*, **68**, 142.

167 Gaponenko, S.V. (1998) *Optical Properties of Semiconductor Nanocrystals*, Cambridge University Press, Cambridge.

168 Gaponenko, S.V., Bogomolov, V.N., Petrov, E.P., Kapitonov, A.M., Eychmüller, A., Rogach, A.L., Kalosha, I.I., Gindele, F., and Woggon, U. (2000) *J. Lumin.*, **87–89**, 152.

169 Rogach, A.L., Nagesha, D., Ostrander, J.W., Giersig, M., and Kotov, N.A. (2000) *Chem. Mater.*, **12**, 2676.

170 Eychmüller, A. and Rogach, A.L. (2000) *Pure Appl. Chem.*, **72**, 179.

171 Eychmüller, A. (2000) *J. Phys. Chem. B*, **104**, 6514.

172 Rogach, A.L. (2000) *Mater. Sci. Eng. B*, **69–70**, 435.

173 Rogach, A.L. (2000) Optical Properties of Colloidally Synthesized II-VI

Semiconductor Nanocrystals, in *Optical Properties of Semiconductor Nanostructures* (ed. M.L. Sadowski), Kluwer Academic Publishers, p. 379.

174 Veinot, J.G.C., Ginzburg, M., and Pietro, W.J. (1997) *Chem. Mater.*, **9**, 2117.

175 Sachleben, J.R., Colvin, V., Emsley, L., Wooten, E.W., and Alivisatos, A.P. (1998) *J. Phys. Chem. B*, **102**, 10117.

176 Chen, S., Truax, L.A., and Sommers, J.M. (2000) *Chem. Mater.*, **12**, 3864.

177 Kumbhojkar, N., Nikesh, V.V., Kshirsagar, A., and Mahamuni, S. (2000) *J. Appl. Phys.*, **88**, 6260.

178 Nanda, J., Sapra, S., Sarma, D.D., Chandrasekharan, N., and Hodes, G. (2000) *Chem. Mater.*, **12**, 1018.

179 Mostafavi, M., Liu, Y., Pernot, P., and Belloni, J. (2000) *Radiat. Phys. Chem.*, **59**, 49.

180 Torimoto, T., Kontani, H., Shibutani, Y., Kuwabata, S., Sakata, T., Mori, H., and Yoneyama, H. (2001) *J. Phys. Chem. B*, **105**, 6838.

181 Aldana, J., Wang, Y.A., and Peng, X. (2001) *J. Am. Chem. Soc.*, **123**, 8844.

182 Westenhoff, S. and Kotov, N.A. (2002) *J. Am. Chem. Soc.*, **124**, 2448.

183 Nosaka, Y. and Tanaka, H. (2002) *J. Phys. Chem. B*, **106**, 3389.

184 Gaponik, N., Talapin, D.V., Rogach, A.L., Eychmüller, A., and Weller, H. (2002) *Nano Lett.*, **2**, 803.

185 Micic, O.I., Curtis, C.J., Jones, K.M., Sprague, J.R., and Nozik, A.J. (1994) *J. Phys. Chem.*, **98**, 4966.

186 Micic, O.I., Sprague, J.R., Curtis, C.J., Jones, K.M., Machol, J.L., Nozik, A.J., Giessen, H., Fluegel, B., Mohs, G., and Peyghambarian, N. (1995) *J. Phys. Chem.*, **99**, 7754.

187 Micic, O.I., Sprague, J., Lu, Z.H., and Nozik, A.J. (1996) *Appl. Phys. Lett.*, **68**, 3150.

188 Guzelian, A.A., Banin, U., Kadavanich, A.V., Peng, X., and Alivisatos, A.P. (1996) *Appl. Phys. Lett.*, **69**, 1432.

189 Guzelian, A.A., Katari, J.E.B., Kadavanich, A.V., Banin, U., Hamad, K., Juban, E., Alivisatos, A.P., Wolters, R.H., Arnold, C.C., and Heath, J.R. (1996) *J. Phys. Chem.*, **100**, 7212.

190 Micic, O.I., Cheong, H.M., Fu, H., Zunger, A., Sprague, J.R., Mascarenhas, A., and Nozik, A.J. (1997) *J. Phys. Chem. B*, **101**, 4904.

191 Banin, U., Lee, C.J., Guzelian, A.A., Kadavanich, A.V., Alivisatos, A.P., Jaskolski, W., Bryant, G.W., Efros, A.L., and Rosen, M. (1998) *J. Chem. Phys.*, **109**, 2306.

192 Cao, Y.W. and Banin, U. (2000) *J. Am. Chem. Soc.*, **122**, 9692.

193 Tessler, N., Medvedev, V., Kazes, M., Kan, S.H., and Banin, U. (2002) *Science*, **295**, 1506.

194 Kim, S.W., Zimmer, J.P., Ohnishi, S., Tracy, J.B., Frangioni, J.V., and Bawendi, M.G. (2005) *J. Am. Chem. Soc.*, **127**, 10526.

195 Aharoni, A., Mokari, T., Popov, I., and Banin, U. (2006) *J. Am. Chem. Soc.*, **128**, 257.

196 Zimmer, J.P., Kim, S.-W., Ohnishi, S., Tanaka, E., Frangioni, J.V., and Bawendi, M.B. (2006) *J. Am. Chem. Soc.*, **128**, 2526.

197 Murray, C.B., Sun, S., Gaschler, W., Doyle, H., Betley, T.A., and Kagan, C.R. (2001) *IBM J. Res. Dev.*, **45**, 47.

198 Sashchiuk, A., Langof, L., Chaim, R., and Lifshitz, E. (2002) *J. Cryst. Growth*, **240**, 431.

199 Wehrenberg, B.L., Wang, C., and Guyot-Sionnest, P. (2002) *J. Phys. Chem. B*, **106**, 10634.

200 Du, H., Chen, C., Krishnan, R., Krauss, T.D., Harbold, J.M., Wise, F.W., Thomas, M.G., and Silcox, J. (2002) *Nano Lett.*, **2**, 1321.

201 Hines, M.A. and Scholes, G.D. (2003) *Adv. Mater.*, **15**, 1844.

202 Schaller, R.D., Petruska, M.A., and Klimov, V.I. (2003) *J. Phys. Chem. B*, **107**, 13765.

203 Bakueva, L., Musikhin, S., Hines, M.A., Chang, T.-W.F., Tzolov, M., Scholes, G.D., and Sargent, E.H. (2003) *Appl. Phys. Lett.*, **82**, 2895.

204 Steckel, J.S., Coe-Sullivan, S., Bulovic, V., and Bawendi, M.G. (2003) *Adv. Mater.*, **15**, 1862.

205 Yu, W.W., Falkner, J.C., Shih, B.S., and Colvin, V.L. (2004) *Chem. Mater.*, **17**, 3318.

206 Pietryga, J.M., Schaller, R.D., Werder, D., Stewart, M.H., Klimov, V.I., and Hollingsworth, J.A. (2004) *J. Am. Chem. Soc.*, **126**, 11752.

207 Finlayson, C.E., Amezcua, A., Sazio, P.J.A., Walker, P.S., Grossel, M.C., Curry, R.J., Smith, D.C., and Baumberg, J.J. (2005) *J. Mod. Opt.*, **7**, 955.

208 McDonald, S.A., Konstantatos, G., Zhang, S.G., Cyr, P.W., Klem, E.J.D., Levina, L., and Sargent, E.H. (2005) *Nat. Mater.*, **2**, 138.

209 Solomeshch, O., Kigel, A., Saschiuk, A., Medvedev, V., Aharoni, A., Razin, A., Eichen, Y., Banin, U., Lifshitz, E., and Tessler, N. (2005) *J. Appl. Phys.*, **98**, 074310.

210 Levina, L., Sukhovatkin, W., Musikhin, S., Cauchi, S., Nisman, R., Bazett-Jones, D.P., and Sargent, E.H. (2005) *Adv. Mater.*, **17**, 1854.

211 Konstantatos, G., Huang, C.J., Levina, L., Lu, Z.H., and Sargent, E.H. (2005) *Adv. Funct. Mater.*, **15**, 1865.

212 Xu, J., Cui, D.H., Zhu, T., Paradee, G., Liang, Z.Q., Wang, Q., Xu, S.Y., and Wang, A.Y. (2006) *Nanotechnology*, 21, 5428.

213 Finlayson, C.E., Sazio, P.J.A., Sanchez-Martin, R., Bradley, M., Kelf, T.A., and Baumberg, J.J. (2006) *Semicond. Sci. Technol.*, **3**, L21.

214 Tan, T.T., Selvan, S.T., Zhao, L., Gao, S.J., and Ying, J.Y. (2007) *Chem. Mater.*, **19**, 3112.

215 Pietryga, J.M., Werder, D.J., Williams, D.J., Casson, J.L., Schaller, R.D., Klimov, V.I., and Hollingsworth, J.A. (2008) *J. Am. Chem. Soc.*, **130**, 4879.

216 Bourdakos, K.N., Dissanayake, D.M.N.M., Lutz, T., Silva, S.R.P., and Curry, R.J. (2008) *Appl. Phys. Lett.*, **92**, 153311.

217 Hill, N.A. and Whaley, K.B. (1993) *J. Chem. Phys.*, **99**, 3707.

218 Zorman, B., Ramakrishna, M.V., and Friesner, R.A. (1995) *J. Phys. Chem.*, **99**, 7649.

219 Wang, L.W. and Zunger, A. (1996) *Phys. Rev. B*, **53**, 9579.

220 Efros, A.L., Rosen, M., Kuno, M., Nirmal, M., Norris, D.J., and Bawendi, M.G. (1996) *Phys. Rev. B*, **54**, 4843.

221 Chamarro, M., Gourdon, C., Lavallard, P., Lublinskaya, O., and Ekimov, A.I. (1996) *Phys. Rev. B*, **53**, 1336.

222 Eichkorn, K. and Ahlrichs, R. (1998) *Chem. Phys. Lett.*, **288**, 235.

223 Wei, S.H., Zhang, S.B., and Zunger, A. (2000) *J. Appl. Phys.*, **87**, 1304.

224 Perez-Conde, J. and Bhattacharjee, A.K. (2001) *Phys. Rev. B*, **63**, 245318.

225 Diener, J., Kovalev, D., Heckler, H., Polisski, G.O., Kunzner, N., Koch, F., Efros, A.L., and Rosen, M. (2001) *Opt. Mater.*, **17**, 135.

226 Pokatilov, E.P., Klimin, S.N., Fomin, V.M., Devreese, J.T., and Wise, F.W. (2002) *Phys. Rev. B.*, **75**, 075316.

227 He, L.X., Bester, G., and Zunger, A. (2005) *Phys. Rev. Lett.*, **94**, 086801.

228 Erwin, S.C., Zu, L.J., Haftel, M.I., Efros, A.L., Kennedy, T.A., and Norris, D.J. (2005) *Nature*, **436**, 91.

229 An, J.M., Franceschetti, A., and Zunger, A. (2007) *Phys. Rev. B*, **76**, 045401.

230 Frenzel, J., Joswig, J.O., and Seifert, G. (2007) *J. Phys. Chem. C*, **111**, 10761.

231 Crawford, N.R.M., Schrodt, C., Rothenberger, A., Shi, W.F., and Ahlrichs, R. (2008) *Chem. Eur. J.*, **14**, 319.

232 Luo, J.W., Franceschetti, A., and Zunger, A. (2009) *Nano Lett.*, **9**, 2648.

233 Wang, X.Y., Ren, X.F., Kahen, K., Hahn, M.A., Rajeswaran, M., Maccagnano-Zacher, S., Silcox, J., Cragg, G.E., Efros, A.L., and Krauss, T.D. (2009) *Nature*, **459**, 686.

234 Murray, C.B., Kagan, C.R., and Bawendi, M.G. (2000) *Annu. Rev. Mater. Sci.*, **30**, 545.

235 Kagan, C.R., Murray, C.B., Nirmal, M., and Bawendi, M.G. (1996) *Phys. Rev. Lett.*, **76**, 3043.

236 Kagan, C.R., Murray, C.B., and Bawendi, M.G. (1996) *Phys. Rev. B*, **54**, 8633.

237 Blanton, S.A., Hines, M.A., and Guyot-Sionnest, P. (1996) *Appl. Phys. Lett.*, **69**, 3905.

238 Nirmal, M., Dabbousi, B.O., Bawendi, M.G., Macklin, J.J., Trautman, J.K., Harris, T.D., and Brus, L.E. (1996) *Nature*, **383**, 802.

239 Empedocles, S.A. and Bawendi, M.G. (1997) *Science*, **278**, 2114.

240 Banin, U., Bruchez, M., Alivisatos, A.P., Ha, T., Weiss, S., and Chemla, D.S. (1999) *J. Chem. Phys.*, **110**, 1195.

241 Koberling, F., Mews, A., and Basché, T. (2001) *Adv. Mater.*, **13**, 672.

242 Schlegel, G., Bohnenberger, J., Potapova, I., and Mews, A. (2002) *Phys. Rev. Lett.*, **88**, 137401.

243 Koberling, F., Kolb, U., Potapova, I., Phillip, G., Basché, T., and Mews, A. (2003) *J. Phys. Chem. B*, **107**, 7463.

244 Chung, I., Witkoskie, J.B., Cao, J.S., and Bawendi, M.G. (2006) *Phys. Rev. E*, **73**, 011106.

245 Huang, H., Dorn, A., Nair, G., Bulovic, V., and Bawendi, M.G. (2007) *Nano Lett.*, **7**, 3781.

246 Colvin, V.L., Schlamp, M.C., and Alivisatos, A.P. (1994) *Nature*, **370**, 354.

247 Dabbousi, B.O., Bawendi, M.G., Onitsuka, O., and Rubner, M.F. (1995) *Appl. Phys. Lett.*, **66**, 1316.

248 Klein, D.L., Roth, R., Lim, A.K.L., Alivisatos, A.P., and McEuen, P.L. (1997) *Nature*, **389**, 699.

249 Klimov, V.I. (2000) *J. Phys. Chem. B*, **104**, 6112.

250 Donega, C.D., Liljeroth, P., and Vanmaekelbergh, D. (2005) *Small*, **1**, 1152.

251 Johnston-Halperin, E., Awschalom, D.D., Crooker, S.A., Efros, A.L., Rosen, M., Peng, X., and Alivisatos, A.P. (2001) *Phys. Rev. B*, **63**, 205309.

252 Talapin, D.V., Haubold, S., Rogach, A.L., Kornowski, A., Haase, M., and Weller, H. (2001) *J. Phys. Chem. B*, **105**, 2260.

253 Talapin, D.V., Rogach, A.L., Kornowski, A., Haase, M., and Weller, H. (2001) *Nano Lett.*, **1**, 207.

254 Talapin, D.V., Rogach, A.L., Mekis, I., Haubold, S., Kornowski, A., Haase, M., and Weller, H. (2002) *Colloids Surf. A*, **202**, 145.

255 Qu, L., Peng, A., and Peng, X. (2001) *Nano Lett.*, **1**, 333.

256 Li, Y.C., Li, X.H., Yang, C.H., and Li, Y.F. (2004) *J. Phys. Chem. B*, **108**, 16002.

257 Hines, M.A. and Guyot-Sionnest, P. (1998) *J. Phys. Chem. B*, **102**, 3655.

258 Xie, R.G., Zhong, X.H., and Basché, T. (2005) *Adv. Mater.*, **17**, 2741.

259 Yu, W.W. and Peng, X.G. (2002) *Angew. Chem., Int. Ed. Engl.*, **41**, 2368.

260 Mekis, I., Talapin, D.V., Kornowski, A., Haase, M., and Weller, H. (2003) *J. Phys. Chem. B*, **107**, 7454.

261 Peng, X., Wickham, J., and Alivisatos, A.P. (1998) *J. Am. Chem. Soc.*, **120**, 5343.

262 Talapin, D.V., Rogach, A.L., Haase, M., and Weller, H. (2001) *J. Phys. Chem. B*, **105**, 12278.

263 Qu, L. and Peng, X. (2002) *J. Am. Chem. Soc.*, **124**, 2049.

264 Talapin, D.V., Rogach, A.L., Shevchenko, E.V., Kornowski, A., Haase, M., and Weller, H. (2002) *J. Am. Chem. Soc.*, **124**, 5782.

265 Striolo, A., Ward, J., Prausnitz, J.M., Parak, W.J., Zanchet, D., Gerion, D., Milliron, D., and Alivisatos, A.P. (2002) *J. Phys. Chem. B*, **106**, 5500.

266 Leatherdale, C.A., Woo, W., Mikulec, F.V., and Bawendi, M.G. (2002) *J. Phys. Chem. B*, **106**, 7619.

267 Schmelz, O., Mews, A., Basche, T., Herrmann, A., and Müllen, K. (2001) *Langmuir*, **17**, 2861.

268 Yu, P.R., Beard, M.C., Ellingson, R.J., Ferrere, S., Curtis, C., Drexler, J., Luiszer, F., and Nozik, A.J. (2005) *J. Phys. Chem. B*, **109**, 7084.

269 Moreels, I., Lambert, K., De Muynck, D., Vanhaecke, F., Poelman, D., Martins, J.C., Allan, G., and Hens, Z. (2007) *Chem. Mater.*, **19**, 6101.

270 Sun, J.J. and Goldys, E.M. (2008) *J. Phys. Chem. C*, **112**, 9261.

271 Yu, W.W., Qu, L., Guo, W., and Peng, X. (2003) *Chem. Mater.*, **15**, 2854.

272 Dorfs, D. and Eychmüller, A. (2006) *Z. Phys. Chem.*, **220**, 1539.

273 Reiss, P., Protiere, M., and Li, L. (2009) *Small*, **5**, 154.

274 Dabbousi, B.O., Rodriguez-Viejo, J., Mikulec, F.V., Heine, J.R., Mattoussi, H., Ober, R., Jensen, K.F., and Bawendi, M.B. (1997) *J. Phys. Chem. B*, **101**, 9436.

275 Empedocles, S.A., Neuhauser, R., Shimizu, K., and Bawendi, M.G. (1999) *Adv. Mater.*, **11**, 1243.

276 Harrison, M.T., Kershaw, S.V., Rogach, A.L., Kornowski, A., Eychmüller, A., and Weller, H. (2000) *Adv. Mater.*, **12**, 123.

277 Reiss, P., Carayon, S., Bleuse, J., and Pron, A. (2003) *Synth. Met.*, **139**, 649.
278 Talapin, D.V., Mekis, I., Götzinger, S., Kornowski, A., Benson, O., and Weller, H. (2004) *J. Phys. Chem. B*, **108**, 18826.
279 Jun, S., Jang, E., and Lim, J.E. (2006) *Nanotechnology*, **17**, 3892.
280 Xie, R., Kolb, U., Li, J., Basche, T., and Mews, A. (2005) *J. Am. Chem. Soc.*, **127**, 7480.
281 Li, J.J., Wang, Y.A., Guo, W., Keay, J.C., Mishima, T.D., Johnson, M.B., and Peng, X. (2003) *J. Am. Chem. Soc.*, **125**, 12567.
282 Battaglia, D., Blackman, B., and Peng, X.G. (2005) *J. Am. Chem. Soc.*, **127**, 10889.
283 Sapra, S., Mayilo, S., Klar, T.A., Rogach, A.L., and Feldmann, J. (2007) *Adv. Mater.*, **19**, 569.
284 Klimov, V.I., Ivanov, S.A., Nanda, J., Achermann, M., Bezel, I., McGuire, J.A., and Piryatinski, A. (2007) *Nature*, **447**, 441.
285 Dorfs, D., Franzl, T., Osovsky, R., Brumer, M., Lifshitz, E., Klar, T.A., and Eychmüller, A. (2008) *Small*, **4**, 1148.
286 Eychmüller, A., Mews, A., and Weller, H. (1993) *Chem. Phys. Lett.*, **208**, 59.
287 Mews, A., Eychmüller, A., Giersig, M., Schooß, D., and Weller, H. (1994) *J. Phys. Chem.*, **98**, 934.
288 Eychmüller, A., Voßmeyer, T., Mews, A., and Weller, H. (1994) *J. Lumin.*, **58**, 223.
289 Schooss, D., Mews, A., Eychmüller, A., and Weller, H. (1994) *Phys. Rev. B*, **49**, 17072.
290 Weller, H. and Eychmüller, A. (1995) *Adv. Photochem.*, **20**, 165.
291 Mews, A. and Eychmüller, A. (1998) *Ber. Bunsenges. Phys. Chem.*, **102**, 1343.
292 Mews, A., Kadavanich, A.V., Banin, U., and Alivisatos, A.P. (1996) *Phys. Rev. B*, **53**, 13242.
293 Mews, A., Banin, U., Kadavanich, A.V., and Alivisatos, A.P. (1997) *Ber. Bunsenges. Phys. Chem.*, **101**, 1621.
294 Bryant, G.W. (1995) *Phys. Rev. B*, **52**, R16997.
295 Jaskolski, W. and Bryant, G.W. (1998) *Phys. Rev. B*, **57**, R4237.
296 Perez-Conde, J. and Bhattacharjee, A.K. (2002) *Phys. Status Solidi B*, **229**, 485.
297 Bryant, G. and Jaskolski, W. (2001) *Phys. Status Solidi B*, **224**, 751.
298 Kamalov, V.F., Little, R., Logunov, S.L., and El-Sayed, M.A. (1996) *J. Phys. Chem.*, **100**, 6381.
299 Little, R.B., Burda, C., Link, S., Logunov, S., and El-Sayed, M.A. (1998) *J. Phys. Chem. A*, **102**, 6581.
300 Braun, M., Burda, C., Mohamed, M., and El-Sayed, M. (2001) *Phys. Rev. B*, **64**, 035317.
301 Yeh, A.T., Cerullo, G., Banin, U., Mews, A., Alivisatos, A.P., and Shank, C.V. (1999) *Phys. Rev. B*, **59**, 4973.
302 Lifshitz, E., Porteanu, H., Glozman, A., Weller, H., Pflughoefft, M., and Eychmüller, A. (1999) *J. Phys. Chem. B*, **103**, 6875.
303 Porteanu, H., Lifshitz, E., Pflughoefft, M., Eychmüller, A., and Weller, H. (2001) *Phys. Status Solidi B*, **226**, 219.
304 Koberling, F., Mews, A., and Basché, T. (1999) *Phys. Rev. B*, **60**, 1921.
305 Little, R.B., El-Sayed, M.A., Bryant, G., and Burke, S. (2001) *J. Chem. Phys.*, **114**, 1813.
306 Le, B.H., Nguyen, X.N., Pham, L.N., Vu, D.C., Pham, T.L., and Nguyen, T.T.T. (2009) Physics and Engineering of New Materials, in *Proceedings in Physics*, **127** (eds D.T. Cat, A. Pucci, and K. Wandelt,), Springer, p. 79.
307 Braun, M., Burda, C., and El-Sayed, M.A. (2001) *J. Phys. Chem. A*, **105**, 5548.
308 Dorfs, D. and Eychmüller, A. (2001) *Nano Lett.*, **1**, 663.
309 Chang, K. and Xia, J.-B. (1998) *Phys. Rev. B*, **57**, 9780.
310 Borchert, H., Dorfs, D., McGinley, C., Adam, S., Möller, T., Weller, H., and Eychmüller, A. (2003) *J. Phys. Chem. B*, **107**, 7481.
311 Borchert, H., Haubold, S., Haase, M., Weller, H., McGinley, C., Riedler, M., and Möller, T. (2002) *Nano Lett.*, **2**, 151.

3.2.2
Synthesis and Characterization of III-V Semiconductor Nanoparticles

Uri Banin

3.2.2.1 Introduction

The preparation of nanocrystals of group III–V semiconductors has presented a significant and important challenge to research groups in the field for over a decade [1–7]. Unlike silicon, many of the group III–V semiconductors are direct gap semiconductors, and this has led to their widespread implementation as the central building blocks of present-day optoelectronic devices such as diode lasers that are ubiquitous in telecommunication applications. They have also provided an obvious technological motivation for the development of wet chemical synthesis approaches used to prepare nanocrystals of these materials. Furthermore, the opportunity to develop such routes for the preparation of III–V semiconductor nanocrystals presents a challenge for synthetic chemistry. While group II–VI semiconductors can be created directly in aqueous media by using ionic precursors, no equivalent synthetic procedures has yet been developed for the III–V semiconductor compounds. However, this results from the more covalent bonding that is typical of III–V compounds, and to the high energetic barriers involved in the formation of semiconductors from suitable precursors.

The availability of high-quality group III–V semiconductor nanocrystals has contributed significantly to fundamental studies of the size-dependent electronic, optical, electrical and structural properties of nanocrystals [8–10]. An interesting case is that of InAs, a narrow band semiconductor (bulk band gap $E_g = 0.42$ eV). Under the conditions of quantum confinement, the band gap of InAs nanocrystals can be tuned over a broad range in the near-IR, where the selection of traditional organic dyes is poor and their stability is low. This points out potential applications of such nanocrystals in optoelectronic devices relevant for telecommunications [11], and as biological fluorescent markers in the near-IR spectral region [12].

As group III–V semiconductor QDs (e.g., InAs) can also be prepared via strain-induced growth mechanisms using molecular beam epitaxy (MBE) [13, 14], there is an opportunity for a direct comparison of the properties of two QD systems: colloidal-grown nanocrystals versus MBE-grown dots [15].

Early attempts to prepare group III–V semiconductor nanocrystals included the synthesis of GaAs nanocrystals in 1990 by Olshavsky *et al.* [16], using a dehalosilylation reaction of $GaCl_3$ and $As(Si(CH_3)_3)_3$ as developed by Wells, [17], where quinoline was used as both the solvent and as the surface passivating ligand. Although these reactions produced nanocrystals with sizes of about 3–4 nm diameter, the optical absorption of the particles showed some interferences due to suspected Ga–quinoline complexes or quinoline oligomeric species [18]. To date, most studies on the preparation of III–V nanocrystalline materials have been based on similar dehalosilylation reactions with variations on the reactants, reaction conditions, and solvents.

An additional route for the preparation of GaAs was reported by Nozik et al. [19], who used Ga(acac)$_3$ and As(Si(CH$_3$)$_3$)$_3$ with triglyme as the solvent to produce GaAs nanocrystals. However, the absorption spectra did not show a well-defined feature that shifted with particle diameter, which indicated a poor size distribution or other problems associated with nanocrystal quality. Wells et al. [20, 21] have also prepared nanocrystalline GaAs and GaP using the metathetical reaction of GaCl$_3$ and (Na–K)$_3$E (E = P, As) in glymes. Whilst these reactions produced nanocrystalline material, there was no separation of sizes, and the optical spectra did not exhibit the characteristic well-defined absorption feature.

InP nanocrystals were prepared by Nozik et al. using the dehalosilylation reaction of an uncharacterized InCl$_3$–oxalate complex and P(Si(CH$_3$)$_3$)$_3$ with trioctylphosphine oxide (TOPO) serving as the solvent and surface-passivating group [4]. This synthesis procedure provided high-quality nanocrystals, and a size-dependent feature was observed in the absorption spectra. Nozik also reported the synthesis of nanocrystals of the indirect gap semiconductor GaP and the ternary material GaInP$_2$, using analogous reactions. Additionally, Guzelian et al. reported the successful synthesis of size-controlled InP nanocrystals [3] where InCl$_3$, dissolved in TOP, was reacted with P(Si(CH$_3$)$_3$)$_3$ in TOPO. The preparation of InAs nanocrystals provided an example where a very high level of control on particle size and monodispersity has been demonstrated [5]. Thus, the decision was taken to focus on this particular system, in order to obtain a more detailed view of the reaction procedure. Recently, Nozik and coworkers also reported the synthesis of colloidal GaN nanocrystals, by using a polymeric gallium imide precursor, {Ga(NH)$_{3-2}$}$_n$, that was reacted in trioctylamine at 360 °C to produce GaN nanocrystals [22].

Following improvements in the synthesis of InAs and InP nanocrystals, various research groups in the field have implemented a variety of strategies to improve the fluorescence and stability of the functional particles. The optical and electronic properties of the particles can be further controlled by fabricating advanced core–shell nanocrystal structures, where a shell of a higher band gap semiconductor is grown on the core, yielding superior surface passivation as demonstrated first for group II–VI semiconductor nanocrystals [23–26], and more recently for group III–V nanocrystals. The synthesis of core–shell nanocrystals with InAs cores was recently developed and will be outlined here [24–26]. The preparation of InP–ZnS core–shells with improved luminescence properties was reported by Weller and coworkers [27], while Nozik and colleagues described the synthesis of lattice-matched shells of ZnCdSe$_2$ on InP cores [28]. Another approach to improve the luminescence of InP nanocrystals was reported by Nozik and coworkers, by etching with HF [29].

This section first provides a general discussion of the synthetic strategy employed for group III–V semiconductor nanocrystals, followed by a description of the synthesis of InAs nanocrystals as a prototypical example. The characterization of the particles produced, using a variety of methods, is briefly reviewed, after which details are provided of the synthesis of core–shell nanocrystals with III–V semiconductor cores, focusing particularly on InAs cores.

3.2.2.2 Synthetic Strategy

The first step in the synthesis of a nanocrystal is a nucleation of the particles, as a successful nucleation will indicate whether the reaction in question is appropriate for the formation of nanocrystals. For a monodisperse sample, the nucleation event must be well separated in time from the growth step; this generally means that nucleation must occur on a short time scale, on the order of a fraction of a second. In contrast, if nucleation is spread out in time, such that the nucleation and growth overlap, then different nucleation sites will undergo growth of varying durations, resulting in a broad size distribution.

One effective way to separate nucleation from growth is to inject suitable precursors into a solvent at high temperature, such that nucleation occurs immediately on contact with the hot solvent. This method has been perfected for the preparation of CdSe nanocrystals [30, 31], and has also been implemented for InP and InAs nanocrystals [3–5]. One difficulty in fully realizing the benefits of the injection method in the case of group III–V semiconductor nanocrystals lies in the need for prolonged reaction times in order to overcome the high reaction barriers, along with a need to anneal at high temperature in order to achieve a better crystallinity of the nanocrystals.

The next stage in nanocrystal synthesis is growth of the particles. As with nucleation, intermediate species and the growing particle must stay in solution throughout the process. Thus, the growth is completed when the reagents are consumed, although a further period of high-temperature annealing may be required to obtain good crystallinity. Under prolonged high-temperature annealing, the possibility of Ostwald ripening arises as an undesired mechanism that will broaden the size distribution [32]. Growth can also be effectively quenched by cooling the reaction mixture. A typical behavior during the growth stage was demonstrated by Alivisatos and coworkers, who investigated the growth of both CdSe and InAs nanocrystals, and introduced the important concept of "size focusing" [31]. This concept is based on the existence of a critical size that depends on the reaction conditions (concentration of reactants, temperature, solvent, etc.). Above the critical size, the growth rate is decreased with increased size, and this leads to a narrowing–focusing of the size distribution. In contrast, below the critical size defocusing will result because the growth rate of smaller particles is slower than that of the bigger particles. In order to achieve a narrow distribution the conditions must be optimized so that, during growth, the reaction remains within size-focusing conditions.

An additional critical issue to produce high-quality nanocrystals is passivation of the surface. During the synthesis, surface passivation helps to control the growth of the nanocrystal, prevents agglomeration and fusing of particles, and also provides for the solubility of the nanocrystals in common solvents. Solvation is important not only for the nanocrystals, but also for the reactants and intermediates. Surface passivation is achieved by using an organic molecule to coordinate or bond to the nanocrystal surface and, most commonly, this is the solvent used for the reaction. This strategy benefits from having the capping molecule in huge excess, and from including a common solvation medium for the reactants, intermediates, and nanocrystals.

Capping groups typically contain an electron-rich donating group such as phosphine oxides, phosphines, amines, or thiols, and behave as a Lewis base which coordinates to the electron-poor Lewis acid-like metal of the semiconductor, such as cadmium or indium. This coordination passivates the dangling orbitals at the nanocrystal surface, preventing further growth or agglomeration. The other end of the ligand imparts solubility to the nanocrystal by providing the particle with a hydrophilic or hydrophobic surface. For example, the alkyl groups of TOPO or TOP result in nanocrystals that are soluble in relatively nonpolar solvents, whereas the use of ligands such as mercaptobenzoic acid renders the particles soluble in polar solvents (e.g., methanol) [33].

Whilst surface passivation plays an important role during the synthesis, it can also be used to further derivatize the surface of the particles, providing an additional chemical control to attach nanocrystals to a surface [34], to electrodes [35], to other nanocrystals, and to biomolecules [36]. Due to the relatively weak coordination bonds of most surface-capping groups, these molecules can be easily exchanged after the initial synthesis. Many of the properties of nanocrystals (notably their photoluminescence) are sensitive to the surface passivation, and manipulation of the surface-capping groups can be used to tune these properties [37, 38]. The core–shell approach discussed below has proved to be a powerful method for enhancing not only the fluorescence quantum yield but also the stability of particles against oxidation and photodegradation.

The improvement of size distribution after synthesis represents a widely used strategy for achieving monodisperse samples. Size-selective precipitation is commonly applied to obtain fractions of nanocrystals with improved size distribution from a given reaction [5, 30]. The method is based on the size-dependent solubility of the nanocrystals, that can be used to first destabilize the dispersion of large particles. This is achieved by adding controlled amounts of a nonsolvent to the solution of particles, which in turn leads to the precipitation of large nanocrystals that can be then separated either by filtration or by centrifugation. These steps can be carried out consecutively so as to obtain a sequence of nanocrystal fractions with increasingly smaller diameters. A similar procedure may, in principle, be applied to each fraction once again to further narrow the size distribution.

While the precise chemical mechanisms of nanocrystal syntheses have not been investigated, several general comments can be made regarding the effect of the reaction mechanism on the nanocrystal product. In order to accomplish the ideal nucleation event discussed above, the reaction must be both rapid and complete – characteristics which point towards simple mechanisms, using reagents with low barriers to reaction. For example, at one extreme is the aqueous or micelle synthesis of CdS or CdSe, where cadmium salts such as $CdCl_4$ serve as the cadmium source and H_2S or H_2Se serves as the group VI source. In these types of reaction, the metal ion is available for reaction with a very simple group VI source; moreover, these reactions are rapid and can even occur at room temperature. On the downside, the reactions are so facile that it is difficult to control and isolate the nucleation.

The chemistry of group III–V materials has proven to be more difficult with regards to designing a system with the attributes of both rapid and complete, yet controlled, nucleation. In contrast to cadmium, the group III metals are too reactive for use as ionic sources as they would react too strongly with the solvent medium. Attempts to develop reactions involving group III alkyls analogous to $(CH_3)_2Cd$ (a reagent commonly used in the synthesis of CdSe nanocrystals) also appear to be too reactive as they tend to proceed uncontrollably to the bulk phase, regardless of any capping group.

A general strategy to reduce the reactivity of these materials would be to attach ligands to the group III and group V centers, and then to react these molecules, presumably in a more controlled fashion. Over the past few years, much interest has arisen in novel routes to bulk group III–V materials [39–42], particularly GaAs, due to the difficult and hazardous nature of traditional reactions involving $Ga(CH_3)_3$ and AsH_3. Substituted gallium and arsenic compounds were developed in the hope of reducing reactant toxicity, eliminating gaseous reagents, and lowering the reaction temperatures. As mentioned, Wells and coworkers [17] developed a dehalosilylation reaction using $GaCl_3 + As(Si(CH_3)_3)_3$ that produced GaAs with the elimination of CH_3SiCl, while Barron et al. [43] used an analogous reaction to produce bulk InP. These reactions are driven by formation of the CH_3SiCl bond, in combination with the good lability of the $Si(CH_3)_3$ ligand. Such reactions are used widely in the synthesis of group III–V semiconductor nanocrystals.

3.2.2.3 InAs and InP Nanocrystals

3.2.2.3.1 Synthesis of InAs and InP Nanocrystals
As the syntheses of InP and InAs are analogous, at this point InAs will be treated in detail. All steps of the reaction and manipulation of the reagents are carried out under argon, or in a dry box. The set-up to perform the synthesis is shown schematically in Figure 3.21. The nanocrystals

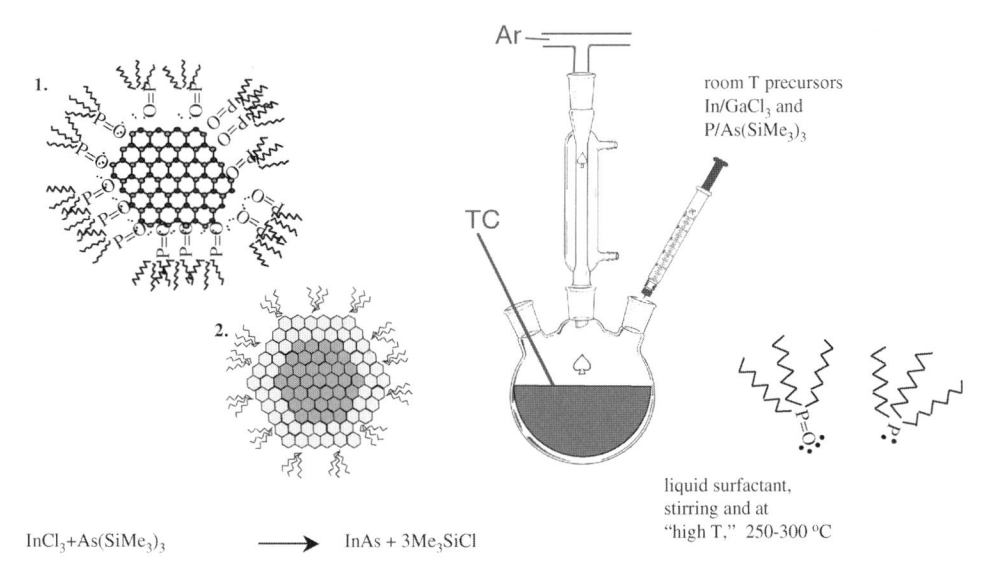

Figure 3.21 Schematic reaction set-up and reaction scheme for III–V semiconductor nanocrystals, either for cores coated by organic ligands (1), or for core–shell nanocrystals (2).

produced are crystalline and highly soluble in a variety of organic solvents, including hexanes, toluene, and pyridine. Particles with diameters ranging from ~2 to 10 nm, and with narrow size distributions can be prepared. The surface of the nanocrystals can be further derivatized with a variety of ligands, including amines, thiols, and phosphines.

The reaction can be written as:

$$InCl_3 + As(Si(CH_3)_3)_3 \rightarrow InAs + 3Si(CH_3)_3Cl \tag{3.2}$$

Whilst details of this synthesis are reported elsewhere (see Refs [5, 7, 31]), at this point it is pertinent to discuss a typical preparation. For this, 3 g of TOP was heated in a three-necked flask on a Schlenk line under an Ar atmosphere to a temperature of 300 °C, with vigorous stirring. A 1 ml aliquot of stock solution (see below) was rapidly injected and the solution cooled to 260 °C for further growth; the growth was monitored by recording the absorption spectra of aliquots removed from the reaction solution. Additional injections were used to grow larger-diameter cores and, on reaching the desired size, the reaction mixture was allowed to cool to room temperature before being transferred into the glove box. Anhydrous toluene was added to the reaction solution, and the nanocrystals were precipitated by adding anhydrous methanol. The size distribution of the nanocrystals in a typical reaction was on the order of ±15%; however, this was improved using a size-selective precipitation, with toluene and methanol as the solvent and nonsolvent, respectively.

Stock solution of InAs Core Under an Ar atmosphere, 9 g of $InCl_3$ was dissolved in 30 ml of TOP at 260 °C, with stirring. Then solution was then cooled and taken into a glove box. The stock solution was prepared by mixing a desired amount of $(TMS)_3As$ and $InCl_3$-TOP solution with the As : In molar ratios at 1 : 2 to 1 : 1.5.

In the stock solution, $InCl_3$ forms a complex with TOP. Although this complex has not been isolated, several examples of $InCl_3$–phosphine oxide complexes have been discussed [44], and an interaction between the Lewis acid-like $InCl_3$ and the strongly donating phosphine or phosphine oxide ligand would be expected. This complex is necessary for success of the reaction, as it helps to maintain the solubility of the reaction intermediates and also establishes the interaction that ultimately passivates the surface of the nanocrystals.

At this point, it is valuable to examine this nanocrystal synthesis using the methodology of nucleation, growth, and termination described earlier. Whilst details of the reaction's initial stages have not yet been elucidated, the studies of Barron [43] and Wells [45] can provide insight into the structure of these initial reaction intermediates in the case of InP. Both research groups have used the $InCl_3 + P(Si(CH_3)_3)_3$ reaction to synthesize bulk InP at temperatures above 400 °C, while investigating intermediate production at lower temperatures. Thus, a stepwise elimination of the three equivalents of $(CH_3)_3SiCl$ was reported, starting with the formation of $[Cl_2InP(SiCH_3)_2]_x$ at room temperature or below. The exact intermediate product in the reaction discussed here most likely has a slightly different structure, due to the action of the coordinating solvent and evidenced by the increased solubility of the product, although the important element is the formation of the In–P bond. Wells further established this In–P interaction by providing a crystal structure for the related 1 : 1 adduct $I_3In \cdot P(Si(CH_3)_3)_3$. Barron provided evidence of the elimination of additional CH_3SiCl at temperatures as low as 150 °C, along with the appearance of peaks in the XPS spectrum characteristic of indium in an InP environment. However, in order to eliminate the final equivalent of CH_3SiCl and to obtain a crystalline material, higher temperatures are required with the reactions producing InP or InAs nanocrystals at a minimum of about 240 °C.

Termination of the particle is accomplished by the coordinating solvent medium. Both, TOPO and TOP are effective in passivating the particle surface, with TOP in particular recognized as a strong donor ligand with both high polarity and polarizability contributing to its donor strength [44, 46, 47]. Based on the analogy with CdSe and standard donor–acceptor analysis, TOP would: (i) coordinate to acceptor indium surface sites, thus providing a passivating shell to terminate growth; (ii) prevent agglomeration among particles; and (iii) through its alkyl groups, provide for excellent solubility in organic solvents such as toluene and hexanes.

In order to study size-dependent phenomena, a variety of different sizes – each with a narrow size distribution – must be isolated. By taking advantage of the differential solubility of various sizes of nanocrystals, it is possible to obtain narrow size distributions by using size-selective precipitation techniques. The solubility of a specific nanocrystal is determined by several factors, most notably the size of the particle, its shape, and the nature of the particle surface. Due to the alkyl groups of the TOP cap, the particles are soluble in relatively nonpolar solvents. However, if a more polar solvent is added to a solution of nanocrystals, the ability of the combined solvent system to solvate the nanocrystals will be reduced. The mechanism of agglomeration is dominated by attractive van der Waals forces among particles [48–51]; however, at some composition of solvents the solvation ability of the system will no longer overcome these attractive forces, and the particles will begin to agglomerate and precipitate. The key to size-selective precipitation is that different sizes, shapes, and surface coverages will have different solubilities, and thus may be separated. Assuming that the dominant difference in the nanocrystals is that of particle size, then the ensemble of nanocrystals will be separated by size, while the degree of resolution among sizes will be determined by the smaller differences in the sample, such as shape or surface structure. The largest particles will precipitate first, due to their stronger attraction, with the primary solvent system employed being toluene–methanol (the nanocrystals are soluble in toluene, but insoluble in highly polar methanol). Thus, starting with the initial reaction mixture diluted in toluene, the incremental addition of methanol will result in a size-selective precipitation of the nanocrystals and the isolation of individual distributions.

3.2.2.3.2 Structural and Basic Optical Characterization of InAs and InP Nanocrystals

The evaluation of nanocrystal structure encompasses several important areas, including the identity and crystallinity of the core, the morphology of the particles,

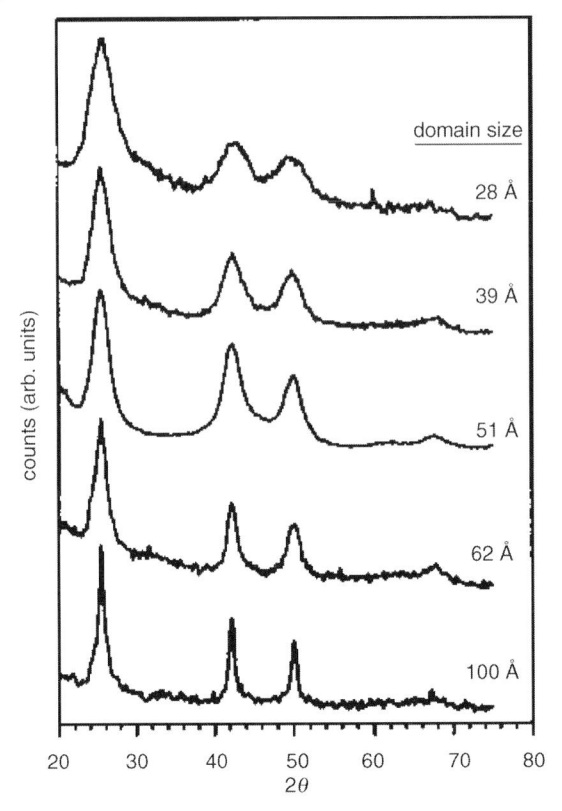

Figure 3.22 Powder X-ray diffraction spectra of InAs nanocrystals with different sizes. The domain size was calculated from the width of the reflections, as detailed in the text. Reproduced from Ref. [188].

and the composition of the nanocrystal surface. The crystallinity of the core can be studied using both powder XRD and electron diffraction, while the morphology of the particles, including their shape and size distributions, may be monitored using TEM. X-ray photoelectron spectroscopy (XPS) can also be used to examine the nanocrystal composition, with particular attention being paid to the surface-capping groups (as discussed in Section NaN, concerning the core–shell nanocrystals). At this point, the InAs and InP nanocrystals will be treated together, with representative data shown from one or both systems.

The powder XRD analysis of a series of InAs nanocrystal sizes is shown in Figure 3.22, where the peak positions are seen to index well with the bulk InAs lattice reflections. Information on the domain size of the sample can be obtained from the width of the lattice reflections, using the Debye–Scherrer formula for spherical particles [52]:

$$D = \frac{1.2\lambda}{[(\Delta 2\theta)(\cos\theta)]} \tag{3.3}$$

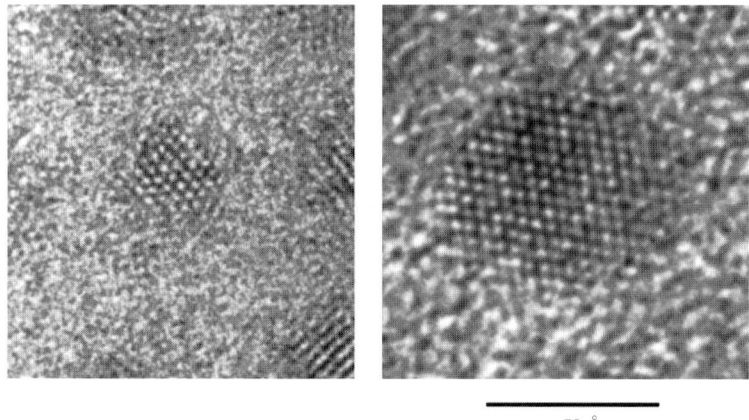

Figure 3.23 HR-TEM images of two InAs nanocrystals, 2.2 and 5 nm in diameter. Reproduced from Ref. [188].

In this equation, λ is 1.5418 Å for CuKα radiation, $\Delta 2\theta$ is the measured linewidth in radians, θ is 1–2 of the measured diffraction angle, and D is the estimated particle size (in Angstroms). An average domain size for a given sample was obtained by averaging D for the (111), (220), and (311) peaks, and this is shown in the figure.

Currently, TEM is crucial for investigating the structure of the nanocrystals, as it allows a direct evaluation of both the size and shape of the particles. A high-resolution TEM image of InAs nanocrystals, 2 and 6 nm in diameter, is shown in Figure 3.23. Here, cross fringes are observed, indicating a high degree of crystallinity, with the lattice spacings indexing to those of InAs; the nanocrystals are near-spherical in shape.

The use of TEM is also essential when calculating the size and size distribution of a nanocrystal sample and, for the wide variety of size-dependent properties exhibited by the nanocrystals, this is clearly a crucial measurement. Size distributions can be obtained by the direct measurement of nanocrystals images for a large number of particles, typically a few hundred. The results of these measurements for InP show size distributions with standard deviations of about 20% in diameter, whereas for InAs the distributions are substantially narrower than 10%.

The quality of the size distribution is most clearly borne out in Figure 3.24, which shows ordered, self-assembled solids of InAs nanocrystals, with diameter of 5.8 nm, dissolved in toluene, and deposited onto the TEM grid. The real space images for three different superlattice (SL) faces detected on the grid are exhibited in Figure 3.24a–c, while Figure 3.24d–f represent the optical diffraction patterns measured for the corresponding TEM negatives. In this case, ordered domains, extending over regions with micron length scales, were observed. Both, the real space images and the diffraction patterns – which in effect correspond to a 2-D Fourier transform of the image – allow an assignment of the three different faces to different views along three zone axes of the close-packed face-centered cubic (fcc) structure.

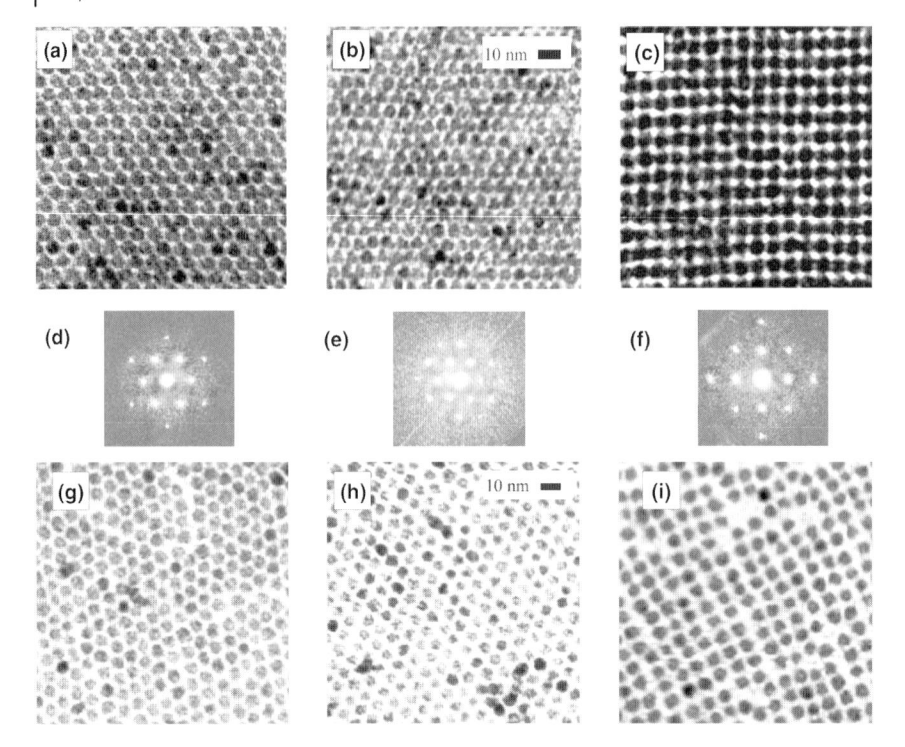

Figure 3.24 TEM images of arrays and superlattices prepared from InAs nanocrystals, 5.8 nm in diameter. Frames (a–c) show ordered superlattices prepared from InAs nanocrystals, while frames (d–f) show the optical diffraction pattern from the micrographs, respectively. Frames (g–i) show TEM images of the monolayers. See text for details.

Figure 3.24a, and its corresponding diffraction pattern Figure 3.24d, represent the superlattice when viewed along the $(111)_{SL}$ zone axis, Figure 3.24b (diffraction pattern Figure 3.24e) represent the $(110)_{SL}$ zone axis, and Figure 3.24c (diffraction pattern Figure 3.24f) represent the $(100)_{SL}$ zone axis. Figure 3.24g–i show images of ordered monolayers with three different packings corresponding to three different superlattice faces resolved in Figure 3.24a–c, respectively.

Quantum confinement effects of nanocrystals are evidenced most clearly in the optical properties of the system, as the electronic energy levels of the clusters become a function of size (a detailed account of this aspect is provided in Chapter 5). The basic optical characterization of semiconductor nanocrystals provides important information on particle size – from the position of the band gap energy, and the size distribution – from the sharpness of peaks in absorption and luminescence. Figure 3.25 shows the room-temperature absorption spectra for a series of InAs nanocrystal sizes, along with the photoluminescence spectra. The quantum confinement effects are clearly evident from the size-dependent nature of the spectra, with the band gap in all samples being shifted substantially from the bulk InAs gap of 0.42 eV. In all samples, the absorption onset is characterized by a distinct feature at

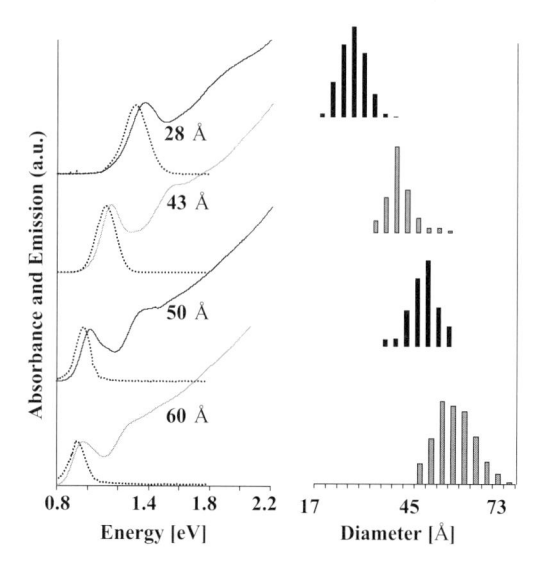

Figure 3.25 The right-hand panel shows the distribution of diameters for four InAs nanocrystal samples obtained from HR-TEM measurements. The left panel displays the room-temperature absorption (solid lines), and emission spectra (dashed lines) for the corresponding samples. The mean diameter for each sample is noted. Reproduced from Ref. [15].

the absorption edge corresponding to the first excitonic transition. Additional features are resolved at a higher energy, corresponding to higher excited states. All states shift with nanocrystal size. The width of the features is primarily due to inhomogeneous broadening resulting from size variations of the particles within a specific sample. The size distribution, as monitored using TEM, is shown in the right-hand frame of Figure 3.25. The absorption of InP nanocrystals showed similar effects, with an absorption onset shifting to the blue with decreasing cluster diameter (as discussed in Ref. [4]).

3.2.2.4 Group III–V Core–Shell Nanocrystals: Synthesis and Characterization

The emission color from semiconductor nanocrystal QDs is tunable by the size as a result of the quantum-confinement effect [53, 54]. Harnessing this emission for real-world applications, such as biological fluorescence marking [55, 56] and optoelectronic devices [11, 57–60], represents an important challenge which imposes stringent requirements of a high fluorescence quantum yield (QY), and of stability against photodegradation. These characteristics are difficult to achieve in semiconductor nanocrystals coated by organic ligands, due to an imperfect surface passivation. In addition, the organic ligands are labile for exchange reactions because of their weak bonding to the nanocrystal surface atoms [37]. As discussed above, a proven strategy for increasing both the fluorescence QY and the stability is to grow a shell of a higher band gap semiconductor on the core nanocrystal [23–27, 61]. In such composite core–shell structures, the shell type and shell thickness provide further control for tailoring the optical, electronic, electrical, and chemical properties of semiconductor nanocrystals. As several reports of Group III–V core–shell nanocrystals have been already described, attention will be focused here on the present authors' detailed study of core–shell nanocrystals with InAs cores [6, 7]. The band gap energy of these core–shell nanocrystals is tunable in the NIR spectral range, covering the wavelengths that are important for telecommunication applications. Recently, they were also successfully incorporated with semiconducting polymers to form an LED structure that provides efficient electroluminescence at 1.3 μm. Such core–shell nanocrystals may be further developed to serve as efficient fluorescent labels for biological applications in the NIR range.

The core–shell approach is conceptually closely related to the approach used in 2-D quantum wells (QWs) [62]. In a QW, a thin nanometric dimensional layer of a low band gap semiconductor is sandwiched between thick layers of a high band gap semiconductor, thus forming a square potential well for the electron and hole wavefunctions. The determining factor for the growth of QWs is the lattice mismatch between the two semiconductors, while the electronic properties of the QW are determined primarily by the energetic offsets between the conduction and valence band edges of the two semiconductors. Thus, the influence of the lattice mismatch and band offsets on the growth and electronic properties of core–shell nanocrystals with InAs cores was investigated. Figure 3.26 shows, schematically, the values of both

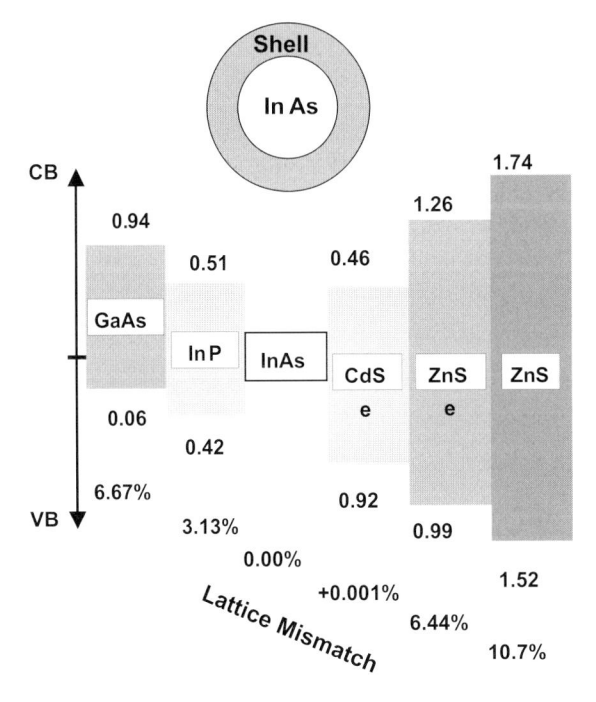

Figure 3.26 Summary of the band offsets (in eV) and lattice mismatch (in %) between the core InAs and the III–V semiconductor shells (left side), and II–VI semiconductor shells (right side) grown in these studies. CB = conduction band; VB = valence band. Reproduced from Ref. [7].

parameters for InAs and a variety of shell materials [63]. The lattice mismatch ranges from nearly zero for CdSe shells, to as high as 10.7% for ZnS shells; the band offsets also cover a broad range of values. It should be noted that the shell materials include both group II–VI and III–V semiconductors, a selection that provides substantial tunability for the properties of the composite core–shell nanocrystals.

3.2.2.4.1 **Synthesis of Core–Shell Nanocrystals with InAs Cores** Preparation of the InAs core–shell nanocrystals is carried out in a two-step process. In the first step, the InAs cores are prepared using the injection method with TOP as solvent. This allowed hundreds of milligrams of nanocrystals per synthesis to be obtained (as detailed above), with a size-selective precipitation being used to improve the size distribution of cores to $\sigma \sim 10\%$. In the second step, the shells of the various materials were grown on the prepared cores. This two-step approach followed methods developed for the synthesis of group II–VI core–shell nanocrystals.

Details of the synthesis of core–shell nanocrystals are available elsewhere [6, 7]. Briefly, TOP-capped InAs cores (5–20 mg) were dissolved in 3–6 g of TOP (or a mixture with TOPO) in a three-necked flask. Under an Ar flow on a Schlenk line, the nanocrystal solution was heated to 260 °C, and the shell precursor solution introduced into the hot solution by drop-wise addition. The growth of core–shells was monitored using UV-visible spectroscopy of aliquots taken from the reaction flask. Following growth of the desired shell thickness the reaction mixture was cooled to room temperature, after which the core–shell nanocrystals passivated by TOP were obtained by precipitation, using a mixture of methanol and toluene.

An attempt to grow the group III–V semiconductor shells, InP and GaAs, at a lower temperature of 160 °C in TOP proved to be unsuccessful, as the low temperature reduced the rate of nucleation of nanocrystals of the shell material, with the latter process competing with shell growth. The alloying process of the core and shell materials was also less likely at low temperatures and, indeed, growth could not be achieved at 160 °C. Rather, these shells could be grown only at higher temperatures ($T > 240$ °C). A minimal temperature is required to overcome the reaction barrier for the precursors, similar to the conditions required for growth of III–V semiconductor nanocrystal cores. Above this limit, a controlled growth of InP shells of varying thickness was achieved.

Although thin GaAs shells could also be grown, unlike InP shells the growth was limited to a thickness of less than two monolayers. This difference may be related to a strong bonding of Ga to TOP, consistent with the observation that it is difficult to synthesize GaAs nanocrystals in this way [64]. An additional difference between the GaAs and InP shells can be seen in the solubility of the core–shell nanocrystals of each type; typically, the InAs–InP core–shells are readily soluble in organic solvents, following precipitation of the nanocrystals from the growth solution. Special conditions were required to obtain a good solubility of the InAs–GaAs core–shells [1].

In contrast to the group III–V semiconductor shells, growth of the II–VI semiconductor shells (CdSe, ZnSe, and ZnS) was observed at low temperatures above 150 °C. For the CdSe shells, by limiting the nucleation such that a controlled shell growth could be achieved, TOP could be used instead of TOPO as the growth medium. This finding was consistent with the stronger bonding of Cd to TOP compared to TOPO.

In the case of InAs–ZnSe and InAs–ZnS core–shells grown in TOP, their solubility following separation from the growth solution was poor, and the TOP ligands could easily be removed by using methanol. Additionally, when growing thick ZnS shells, a substantial nucleation of the ZnS nanocrystals was observed by XRD measurements. In order to overcome these difficulties, a mixture of TOP–TOPO was used as the growth solution instead of TOP; this allowed a minimization of nucleation such that soluble nanocrystals were obtained. The growth rate of ZnSe and ZnS shells was substantially slower for the TOP–TOPO mixture compared to TOP alone. These features were consistent with the stronger bonding of Zn to TOPO than to TOP, as had been reported for the synthesis of ZnSe nanocrystals [65]. Cd is a softer Lewis acid than Zn, and binds more strongly to TOP, which is a softer Lewis base than TOPO; in contrast, Zn (the harder Lewis acid) binds more strongly to TOPO (the harder Lewis base). These characteristics follow the general hard–soft concept for interaction strength between Lewis acids and bases [66].

3.2.2.4.2 Optical Characterization of the Core–Shell Nanocrystals

The most direct and immediate probes for shell growth are the absorption and fluorescence spectra. Figure 3.27 shows the sequence of absorption spectra measured for aliquots taken from the reaction solution during the growth of InP shells on InAs cores with an initial radius of 1.3 nm. In this case, the first absorption peak shifts to the red upon shell growth. The red shift occurs because the conduction band offset between InAs and InP is smaller than the confinement energy of the electron; thus, as the shell grows the electron wavefunction extends to a larger box and its confinement energy is lowered. As the electron effective mass, m_e, in InAs

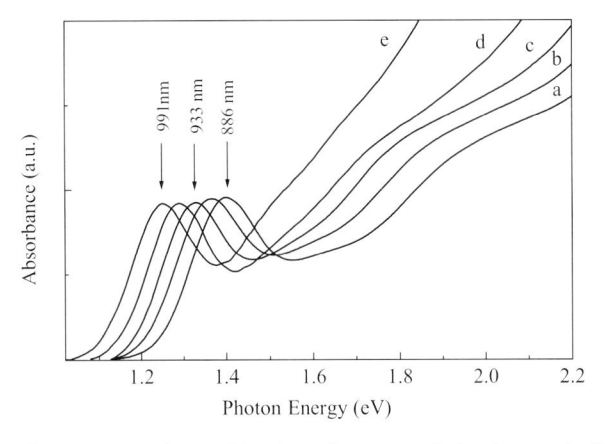

Figure 3.27 Evolution of the absorption spectra during the growth of InP shells on InAs cores with an initial radius of 1.3 nm (a). The InP shell thickness in number of monolayers is: (b) 0.5, (c) 1.1, (d) 1.7, (e) 2.5. Reproduced from Ref. [7].

is extremely small ($m_e = 0.024 m_e$, where m_e is the mass of the free electron [67]), it is highly delocalized and a large potential step is required for it to be confined to the core. The spectral features remain sharp during the reaction, indicating that the growth is controlled and that the size distribution is maintained. However, the fluorescence of the InAs–InP core–shells is quenched substantially when compared to the original cores.

For CdSe, which has a conduction band offset similar to that of InP, shell growth also leads to a red shift of the absorption onset. Figure 3.28a and b shows the evolution of the absorption and emission spectra during the growth of InAs–CdSe core–shells with two different initial core radii, namely 1.2 nm (Figure 3.28a), and 2.5 nm (Figure 3.28b). As in the case of InP shells, this red shift resulted from the lower confinement energy of the electron, the wavefunction of which extended to the shell region. In this case, the sharpness of the spectral features was partially washed out during the growth. In contrast to InAs–InP core–shells, the band gap fluorescence QY for InAs–CdSe core–shells was substantially enhanced, up to a maximum value of

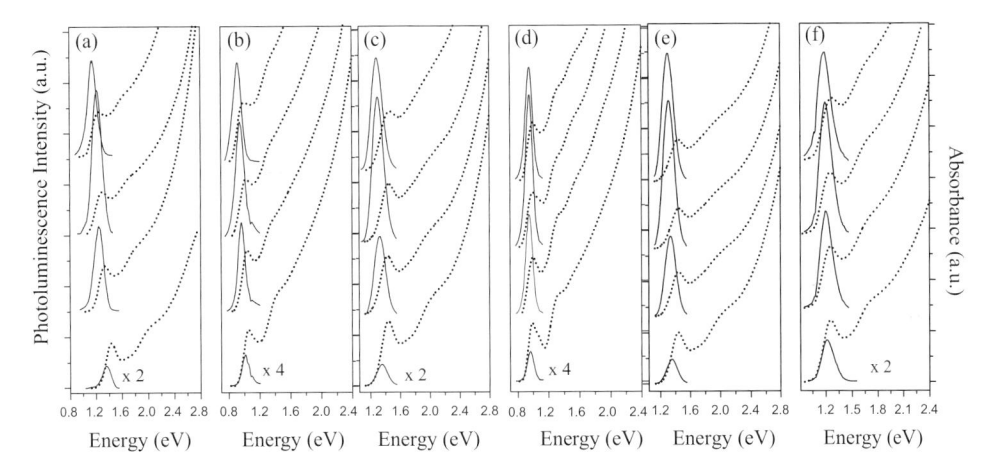

Figure 3.28 Evolution of absorption (dashed lines), and photoluminescence (solid line) for growth of core–shells. The PL spectra are given on a relative scale for comparison of the enhancement of QY with shell growth. Reproduced from Ref. [7]. (a) InAs–CdSe with initial core radius of 1.2 nm. The shell thickness (in number of monolayers) and QY for the traces from bottom to top are respectively: 0, 1.2%; 0.6, 13%; 1.2, 21%; 1.8, 18%; (b) InAs–CdSe with initial core radius of 2.5 nm. The shell thickness (in number of monolayers) and QY for the traces from bottom to top are respectively: 0, 0.9%; 0.7, 11%; 1.2, 17%; 1.6, 14%; (c) InAs–ZnSe with initial core radius of 1.2 nm. The shell thickness (in number of monolayers) and QY for the traces from bottom to top are respectively: 0, 1.2%; 0.6, 9%; 1.5, 18%; 2.5, 14%; (d) InAs–ZnSe with initial core radius of 2.8 nm. The shell thickness (in number of monolayers) and QY for the traces from bottom to top are respectively: 0, 0.9%; 0.7, 13%; 1.3, 20%; 2.2, 15%; (e) InAs–ZnS with initial core radius of 1.2 nm. The shell thickness (in number of monolayers) and QY for the traces from bottom to top are respectively: 0, 1.2%; 0.7, 4%; 1.3, 8%; 1.8, 7%; (f) InAs–ZnS with initial core radius of 1.7 nm. The shell thickness (in number of monolayers) and QY for the traces from bottom to top are respectively: 0, 1.1%; 0.6, 5%; 1.3, 7.1%; 2.2, 6.3%.

21%, and almost 20-fold larger than the QY of the cores. These values of QY competed favorably with the QY values for organic NIR laser dyes [12]. This pronounced difference in the emission of InP versus CdSe shells was assigned to the different quality of the outer surface of the nanocrystals. In both cases, the similar and relatively small conduction band offset led to a substantial probability of the presence of an electron wavefunction at the nanocrystal surface. As a result, for both cases the emission was sensitive to the outer surface, and susceptible to trapping in unpassivated surface sites. Indeed, this problem has been reported for InP nanocrystals that have been prepared in TOP–TOPO and which have a very low fluorescence QY that could be enhanced by surface treatments such as oxidation and etching [22]. CdSe–TOP–TOPO-coated nanocrystals, on the other hand, display band gap emission with a room-temperature QY of a small percentage.

ZnSe and ZnS are shell materials with much larger band offsets relative to InAs. The evolution of the absorption and emission spectra for ZnSe and ZnS shell growth on cores with various radii, is shown in Figure 3.28c–f, respectively. For both core–shells, the absorption onset is almost the same as in the original core, a fact which can be explained by the large conduction and valence band offsets of ZnSe and ZnS relative to InAs, which create a substantial potential barrier for the electron and hole wavefunctions at the core–shell interface. In this case, unlike the CdSe and InP shells, the electron and hole wavefunctions are both confined primarily to the core region, such that the band gap does not shift from the value of the core. As shown in Figure 3.28c–f, the fluorescence QY is enhanced for both shells: for ZnSe, the maximum QY values were 20%, while in ZnS a maximum QY of 8% was achieved for a shell thickness of 1.2 to 1.8 monolayers. With further shell growth, the QY was seen to decrease, a reduction that may be assigned to trapping of the charge carriers at the core–shell interface. The lattice mismatch between InAs and both ZnSe and ZnS was large. In thin shells, the strain could be sustained and epitaxial growth of the shell on the core could occur. With further shell growth, however, defects may form at the core–shell interface that could trap the carriers and lead to a gradual reduction of QY in the thicker shells. A similar observation was reported for the group II–VI semiconductor core–shells, CdSe–ZnS and CdSe–CdS [25, 26].

The selection of core–shells extends the control afforded for the electronic properties of the composite nanocrystals. As an example of flexible control, two types of core–shells can be demonstrated, InAs–ZnSe and InAs–CdSe, both of which emit strongly at 1.3 μm, a wavelength with significant technological importance in fiber-optic communications. Figure 3.28b shows the absorption and emission spectra for a CdSe shell overgrown on a core with a radius of 2.5 nm, where the core band gap emission is at 1220 nm; in this case, with growth of the shell the emission shifts to the red, accompanied by a substantial enhancement of the QY, to a value of 17% achieved at 1306 nm. However, for ZnSe shells the band gap hardly shifts. Figure 3.28d shows that, by using a much larger core, with radius of 2.8 nm, a high QY of 20% at 1298 nm could be achieved by growing the ZnSe shell. Such core–shell nanocrystals have been used recently as the optically active chromophores in efficient LEDs covering the NIR spectral range [11]. In this case, the LED was based on a nanocrystal–semiconductor nanocrystal composite.

3.2.2.4.3 Chemical and Structural Characterization

X-Ray Photoelectron Spectroscopy (XPS) The technique of XPS can be used effectively to examine the chemical composition of the core–shells [68]. Figure 3.29 shows the XPS survey spectra for InAs cores, 1.7 nm in radius, and for InAs–CdSe core–shells prepared with the same core and with a shell thickness of three monolayers. The indium and arsenic peaks are clearly resolved for the core (lower spectrum). Additional peaks belonging to Cd and Se (the shell materials) can be identified in the core–shells (top spectrum). The ratio of the XPS peak heights between core and core–shell for the core atoms, is energy-dependent. The relative intensity of the peaks at high binding energies (e.g., In$_{MNN}$), which have a small kinetic energy and thus a small escape depth, is quenched more upon shell growth, compared to the peaks at small binding energy and large escape depth (e.g., As$_{LMN}$).

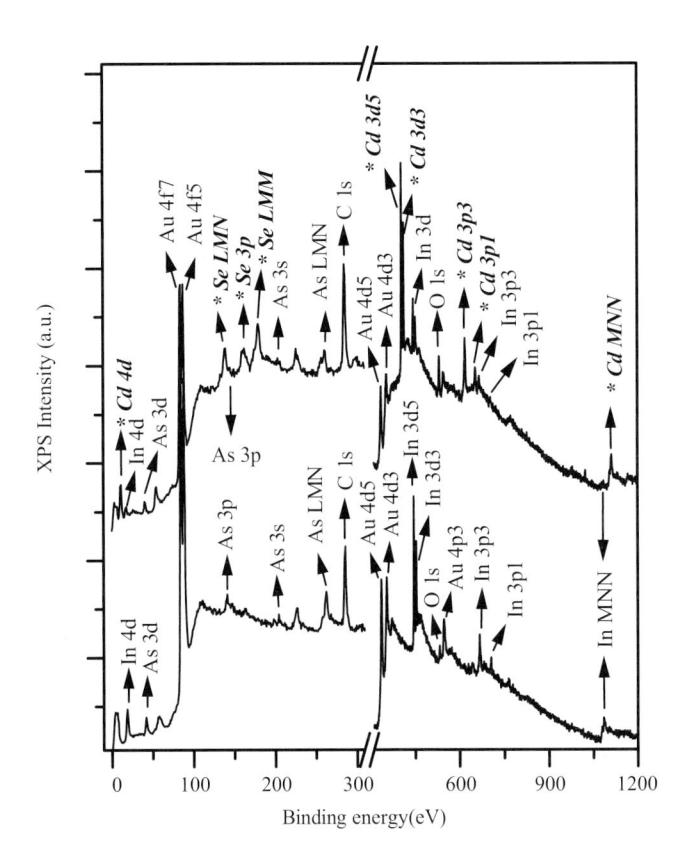

Figure 3.29 XPS survey spectrum for InAs cores with radius of 1.7 nm (lower trace), and for InAs–CdSe core–shells with shell thickness of three monolayers (top trace). The assignment of the peaks is indicated. The new peaks in the XPS spectrum for the core–shells associated with Cd and Se are emphasized in bold italic type. Reproduced from Ref. [7].

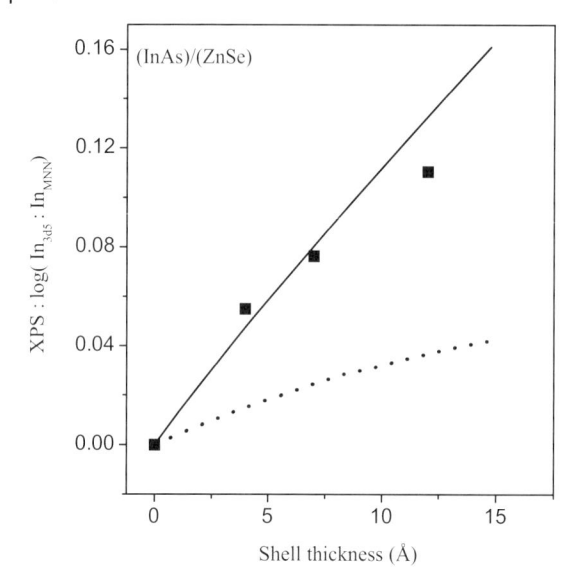

Figure 3.30 Summary of high-resolution XPS data for InAs–ZnSe. The log of the ratio of the intensities of the In_{3d5} to the In_{MNN} is shown, versus the shell thickness. The ratio is normalized to the ratio in the cores. Squares = experimental data; solid line = calculated ratio for core–shell structure; dotted line = calculated ratio for alloy formation. Reproduced from Ref. [7].

High-resolution XPS provides quantitative evidence for shell growth based on the finite escape depth, λ, of photoelectrons from the core atoms [26]. The typical escape depths are on the order of the shell thickness, and the photoelectron signal from core atoms should decrease accordingly in the core–shell structure [68].

The application of XPS to study shell growth is demonstrated in Figure 3.30, which shows the XPS data for InAs–ZnSe core–shells with an initial core radius of 1.7 nm. For this, high-resolution XPS measurements were performed on a sequence of core–shell samples with varying thickness. Figure 3.30 shows the experimental results (squares) for the log of the ratio of the In_{3d5} to the In_{MNN} Auger peak, normalized by the ratio in the core; the ratio was increased upon ZnSe shell growth. The number of atoms for the two peaks was identical, and the increase in the ratios was due to the difference in escape depths, which was lower for the Auger peak. As a result, shell growth led to larger reduction in the relative intensity of the Auger peak. To check these effects, the expected ratio for a spherical geometry was simulated assuming either a core–shell configuration (solid line) or alloy formation (dashed line). Clearly, only the calculated ratio for the core–shell structure was in agreement with experimental data.

Transmission Electron Microscopy The HR-TEM images of core and core–shell nanocrystals are shown in Figure 3.31, where Figure 3.31b shows an InAs core with radius of 1.7 nm, and Figure 3.31a and c show, respectively, the InAs–InP and

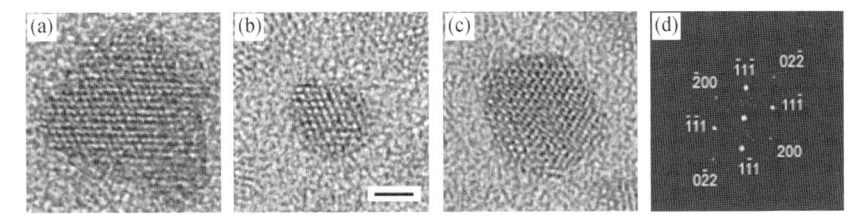

Figure 3.31 HR-TEM images of (a) InAs–InP core–shell (core radius 1.7 nm, shell thickness 2.5 nm), (b) InAs core (core radius 1.7 nm), and (c) InAs–CdSe core–shell (core radius 1.7 nm, shell thickness 1.5 nm). The scale bar is 2 nm. The nanocrystals are viewed along the [011] zone axis; (d) Fourier transform of image (c). The pattern corresponds to the diffraction pattern from the [011] zone of the cubic crystal structure. Reproduced from Ref. [7].

InAs–CdSe core–shell nanocrystals with thick shells grown on similar cores. Here, the crystalline interior is viewed along the [011] zone axis of the cubic lattice. The cubic lattice is also resolved for the InAs–CdSe core–shell, as clearly revealed by the Fourier transform of the image presented in Figure 3.31d. Although, in general, CdSe nanocrystals grown under such conditions form the wurtzite structure, here they adopt the cubic structure on the InAs core. For all three cases, the fringes are visible across the entire nanocrystals, in accordance with epitaxial shell growth in these particles.

X-Ray Diffraction The powder XRD patterns for the InAs core, 1.7 nm in radius, and for InAs–InP core–shells with increasing thickness, are presented in Figure 3.32. The InAs XRD pattern consists of the characteristic peaks of cubic InAs, which are broadened because of the finite crystalline domain. With InP shell growth, the diffraction peaks shift to larger angles, consistent with the smaller lattice constant for InP compared with InAs. The shift is most clearly resolved for the high-angle peaks and, in addition, the diffraction peaks are narrowed; this is demonstrated for the (111) peak shown in the inset of Figure 3.32. Such narrowing indicates that the crystalline domain is larger for the core–shells, thus providing direct evidence for the epitaxial growth mode of the shell. The relatively simple diffraction pattern for cubic InAs allows for a clear observation of the narrowing as the shell is grown. In contrast, in the case of core–shells with CdSe cores, the narrowing is masked by the complex diffraction pattern of the wurtzite structure [15].

The XRD patterns for a series of core–shells with different shell materials and varying thickness are displayed in Figure 3.33 (filled circles), where the general pattern of the cubic lattice is maintained for all materials. Although the diffraction peaks narrow with shell growth in the case of CdSe, ZnSe and GaAs shells (for the reasons discussed above), this is not the case for ZnS, most likely due to the very large lattice mismatch of ZnS and InAs (10.7%). In this case, the large strain may lead to cracking at the InAs–ZnS interface at an early growth stage. Moreover, ZnS nanocrystals grown under these conditions form the wurtzite structure preferen-

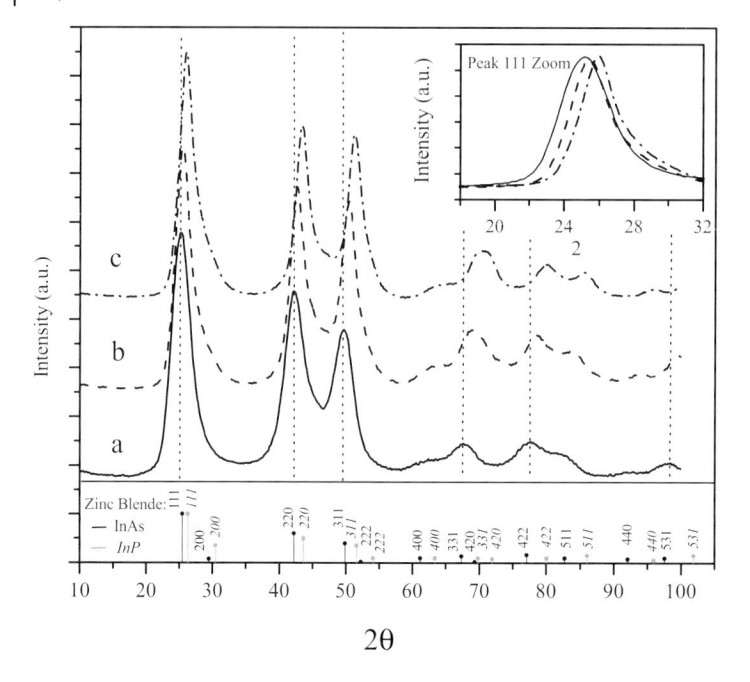

Figure 3.32 X-ray diffraction patterns for InAs cores with radius of 1.7 nm (trace a, solid line), and InAs–InP core–shells with shell thickness of 2.4 monolayers (trace b, dashed line), and 6.2 monolayers (trace c, dot-dashed line). The inset shows a zoom of the 111 peak. The diffraction peak positions of bulk cubic InAs (dark) and *InP* (light, italic) are indicated. Reproduced from Ref. [7].

tially, and this may create additional defects on shell growth. Both effects may explain the relatively small enhancement of the phospholuminescence QY for the ZnS shells.

To further prove the interpretation of the XRD data, and to obtain more quantitative information concerning the core–shell structures, the powder diffraction patterns were simulated [7, 69]. Nanocrystals were built by stacking planes along the (111) axis of the cubic lattice, and the sum of the specified core radius, r_c, and shell thickness, r_s, was used to carve out the nanocrystal, assuming a spherical shape.

The experimental data (filled circles), and the simulation results (thin lines) for the core and a series of core–shells are displayed in Figure 3.33, where simulation of the cores was seen to fit the experimental data very well. The fit was obtained using the simulated XRD pattern for an equally weighted combination of two types of core nanocrystal structures with the same radii, differing in the (111) plane stacking sequence. Both nanocrystal structures had three stacking faults, namely that the experimental diffraction intensity between the (220) and the (311) peaks did not go to zero, while in the simulated pattern for nanocrystals without stacking faults the value was close to zero. Thus, three stacking faults along the (111) direction were required to quantitatively reproduce the experimental

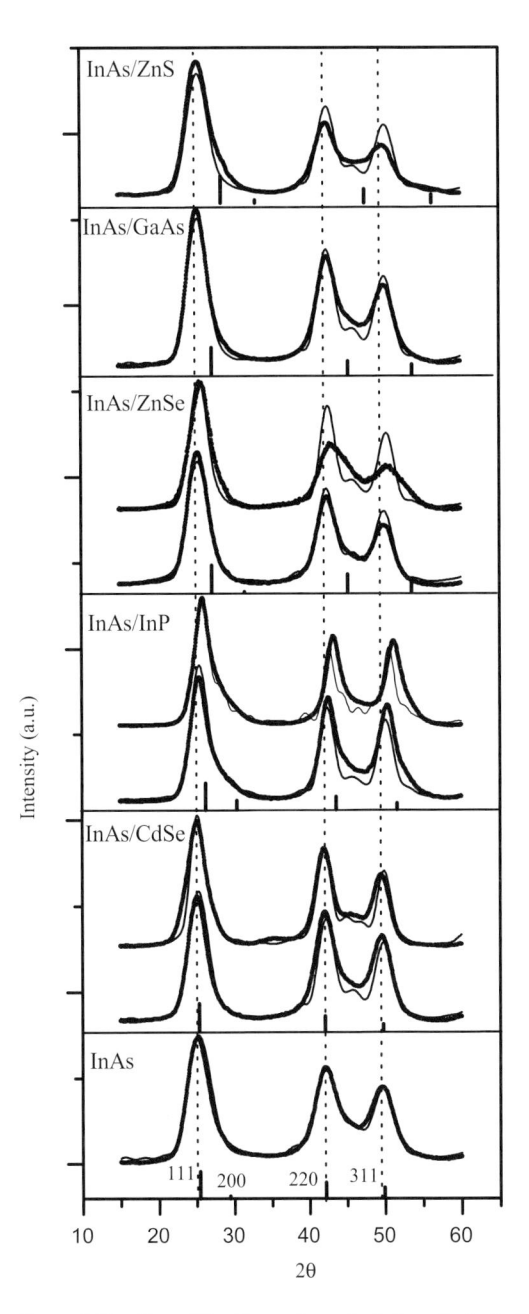

Figure 3.33 X-ray diffraction patterns for InAs cores and various core shells. The experimental curves (filled circles), are compared with the simulated curves (thin solid lines). See text and Table 1 for details of the simulated structures. The markers at the bottom of each frame indicate the diffraction peak positions and the relative intensities for the InAs core material (lower frame), and the various shell materials (other traces). The vertical dashed lines indicate the positions of the InAs core nanocrystal diffraction peaks. Reproduced from Ref. [7].

pattern; however, the introduction of surface disorders had a minimal effect on the simulated patterns.

For InAs–CdSe core–shells, the lattice mismatch was zero, and the experimental peak positions did not shift with shell growth (this was well reproduced in the simulation). An additional stacking fault was added in the shell region for the thicker shell of three monolayers, so as to better reproduce the experimental pattern.

For the other core–shells there was a lattice mismatch which ranged from 3.13% for InP to 10.7% for ZnS. A gradual shift of the diffraction peaks with shell growth towards larger angles was observed for these core–shells (Figures 3.32 and 3.33), which meant that the lattice spacing had been modified in the shell region. In order to achieve an epitaxial growth mode, atoms in both the core and shell regions at the core–shell interface must have identical lattice spacings, and to simulate the smooth switching of the lattice spacing from the core to the shell a Fermi-like switching function was used. This provided a physical model for understanding the epitaxial growth mode, while reproducing the observed change in the peak positions on shell growth.

The simulated patterns reproduced the two main effects observed upon shell growth in InAs–InP, InAs–ZnSe, and InAs–GaAs core–shells, namely the shifting of all diffraction peaks to larger angles, and the narrowing of the peaks (Figure 3.33), and provided further evidence for the epitaxial shell growth mode. In the case of InAs–ZnS, no narrowing was observed in the experimental diffraction pattern; this indicated that, in the InAs–ZnS core–shells, the interface region was not fully epitaxial (as explained above). Similar to the InAs–CdSe core–shells, in the case of the thick InP shells (6.2 monolayers) it was necessary to add stacking faults in the shell region to better reproduce the experimental pattern. On average, a stacking fault was seen to exist for every four to five layers in InAs core–shell nanocrystals, and within these stacking faults the bonds of atoms remained fully saturated and charge carrier traps did not necessarily form. Thus, it is likely that these planar stacking faults do not cause any substantial reduction in the fluorescence QY.

3.2.2.4.4 Model Calculations for the Band Gap

The core–shell band offsets provide control for modifying the electronic and optical properties of these composite nanocrystals. To examine the effect of the band offsets of various shells on the band gap of the composite nanocrystals, calculations using a particle in a spherical box model were performed [16, 70]. Briefly, in this model the electron and hole wavefunctions are treated separately, after which the coulomb interaction is added within a first-order perturbation theory [71]. Three radial potential regions should be considered in the core–shell nanocrystals, namely the core, the shell, and the surrounding organic layer. Continuity is required for the radial part of the wavefunctions for both electron and hole at the interfaces. In addition, the probability current, $\frac{1}{m_i^*}\frac{d}{dr}R_i(k_i r)$, where m_i^* is the effective mass in region i, R_i is the radial part of the lowest energy $1S_{e-h}$ electron or hole wavefunction, and k_i is the wave vector in region i, must be continuous. The effective masses, and dielectric constants of the bulk semiconductors were used in these calculations [67], while the band offsets were taken from Ref. [63]. First, for InAs cores of 1.7 nm radius, a barrier height of 4.5 eV was used for the surrounding organic layer for both carriers. The confinement energy for the electron is sensitive to the barrier height because of the small electron effective mass, whereas the heavier hole is much less sensitive. The experimental value obtained was in reasonable agreement with the calculated band gap, although there was a minor deviation that was understandable in this simplistic model [72]. The calculated versus measured relative change of the band gap upon growth of the different shells, is plotted for various core–shells in Figure 3.34. The error in the band gap shifts should be small compared to that in the absolute band gap energies, and the agreement for the band gap shifts was good. For the ZnS shells (the dot-dashed line), hardly any shift was observed, both in the experiment and in the calculations, and this was consistent with the large band offsets between InAs and ZnS. For both the InP (dashed line) and CdSe (solid line) shells, the experimental and calculated band gaps shifted to the red upon shell growth, due mainly to a reduction in the electron confinement energy in the core–shells. The lowest $1S_e$ level of the electron was above the core–shell barrier for both InP and CdSe shells, which led to a high probability of the presence of electrons in the shells.

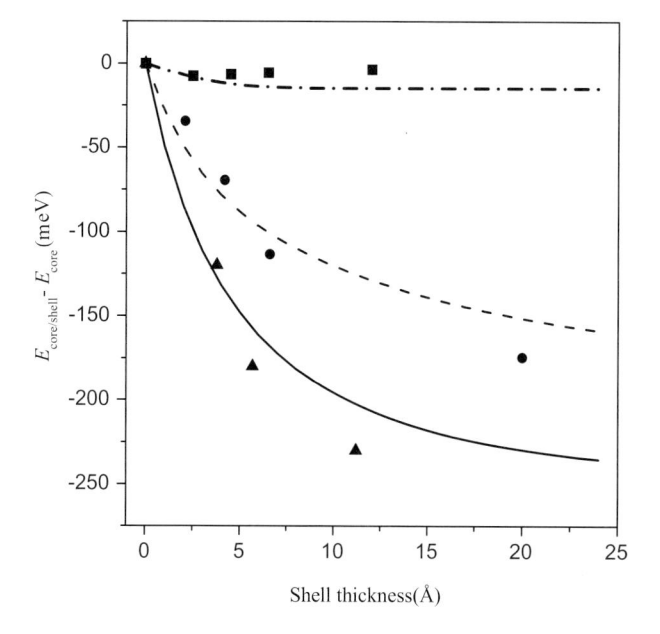

Figure 3.34 Experimental and calculated shifts of the band gap energy in various core–shells versus the shell thickness. InAs–ZnS core–shells: experimental data = squares, calculated shift of gap = dot- dashed line. InAs–InP core–shells: experimental data = circles, calculated shift of gap = dashed line. InAs–CdSe core–shells: experimental data = triangles, calculated shift of gap = solid line. Reproduced from Ref. [7].

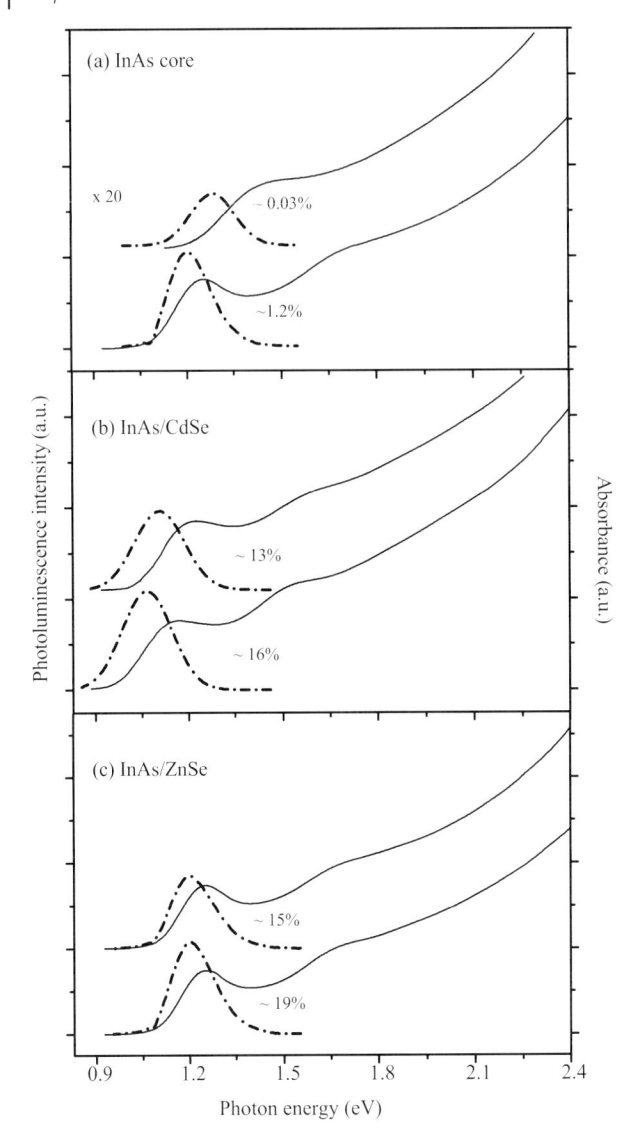

Figure 3.35 Comparison of stability of (a) cores, (b) InAs–CdSe core–shells, and (c) InAs–ZnSe core–shells. The absorption spectra (solid lines) and PL spectra (dashed lines) are shown on the same energy scale for fresh (bottom traces in each frame), and for nanocrystals kept in an oxygen saturated solution for 10 months (top traces in each frame). The photoluminescence QY is also indicated in each case. Reproduced from Ref. [7].

3.2.2.4.5 **Stability of Core–Shell Nanocrystals** Core–shell passivation with a shell of a semiconductor material that has large band offsets compared to the core, provides also increased protection and stability compared to organically passivated core nanocrystals. The photostability of the core–shells were compared to

IR140, a typical NIR organic laser dye [73]. Solutions of nanocrystals and of the dye saturated with oxygen, were irradiated at 810 nm with a Ti-sapphire laser for 30 min, using an intensity of 250 mW such that the optical density of the irradiated solutions was 0.2 at 810 nm. Under these conditions, each nanocrystal and dye molecule absorbed a total of approximately 0.5×10^6 photons. For the dye, following irradiation, the main absorption peak at 830 nm vanished, which indicated that the dye had completely degraded. For InAs cores (radius 1.7 nm), the absorption maximum shifted to the blue by 10 nm, the optical density was slightly decreased, and the phospholuminescence intensity was decreased by a factor of 2.1. For InAs–ZnSe core–shells (core radius 1.7 nm, shell thickness approximately two monolayers), the absorption did not change upon irradiation, and the phospholuminescence intensity was decreased by a factor of 1.7, though to a value which was still eightfold stronger than for the fresh core. Finally, for InAs–CdSe core–shells (core radius 1.7 nm, shell thickness ~1.8 monolayers), the absorption shifted to the blue by 5 nm, and the phospholuminescence intensity was reduced by only 10%. The results of these experiments demonstrated the improved photostability of the core–shells compared to a typical NIR laser dye.

These core–shells also displayed a greatly increased stability against oxidation compared to the cores alone. The stability of the bare (TOP-coated) InAs cores, and of the two core–shell samples similar to those examined above, were studied. For this, the absorption and emission spectra of fresh samples, and of samples maintained for ten months in a solution saturated with oxygen under daylight conditions, were compared (as summarized in Figure 3.35). Under these conditions, the InAs cores exhibited a considerable blue shift of the absorption, accompanied by a washing out of the spectral features, and the emission was quenched by a factor of 40 compared to the fresh cores (Figure 3.35a). Taken together, these phenomena indicated that a substantial oxidation of the bare cores had occurred. Yet, under similar conditions the core–shells showed very different behaviors: for InAs–CdSe core–shells the absorption shifted slightly to the blue while the QY was decreased from 16% to 13% (Figure 3.35b). By comparison, for the InAs–ZnSe core–shells the absorption was hardly shifted and the emission QY was reduced from 19% to 15% (Figure 3.35c).

References

1 Alivisatos, A.P. (1996) *Science*, **271**, 933.
2 Nozik, A.J. and Micic, O.I. (1998) *MRS Bull.*, **23**, 24.
3 Guzelian, A.A., Katari, J.E.B., Kadavanich, A.V., Banin, U., Hamad, K., Juban, E., Alivisatos, A.P., Wolters, R.H., Arnold, C.C., and Heath, J.R. (1996) *J. Phys. Chem.*, **100**, 7212.
4 Micic, O.I., Curtis, C.J., Jones, K.M., Sprague, J.R., and Nozik, A.J. (1994) *J. Phys. Chem.*, **98**, 4966.
5 Guzelian, A.A., Banin, U., Kadanich, A.V., Peng, X., and Alivisatos, A.P. (1996) *Appl. Phys. Lett.*, **69**, 1462.
6 Cao, Y.W. and Banin, U. (1999) *Angew. Chem., Int. Ed. Engl.*, **38**, 3692.

7 Cao, Y.W. and Banin, U. (2000) *J. Am. Chem. Soc.*, **122**, 9692.
8 Banin, U., Cao, Y.W., Katz, D., and Millo, O. (1999) *Nature*, **400**, 542.
9 Micic, O.I., Cheong, H.M., Fu, H., Zunger, A., Sprague, J.R., Mascarenhas, A., and Nozik, A.J., (1997) *J. Phys. Chem. B*, **101**, 4904.
10 Banin, U., Lee, J.C., Guzelian, A.A., Kadavanich, A.V., Alivisatos, A.P., Jaskolski, W., Bryant, G.W., Efros, Al.L., and Rosen, M. (1998) *J. Chem. Phys.*, **109**, 2306.
11 Tessler, N., Medvedev, V., Kazes, M., Kan, S., and Banin, U. (2002) *Science*, **295**, 1506.
12 Daehne, S., Resch-Genger, U., and Wolfbeis, O.S.(eds) (1997) *Near-Infrared Dyes for High Technology Applications*, Kluwer Academic Publishers, Dordrecht.
13 Grundmann, M., Christen, J., Ledentsov, N.N. *et al.* (1995) *Phys. Rev. Lett.*, **74**, 4043.
14 Leon, R., Petroff, P.M., Leonard, D., and Fafard, S. (1995) *Science*, **267**, 1966.
15 Banin, U., Lee, J.C., Guzelian, A.A., Kadavanich, A.V., and Alivisatos, A.P. (1997) *Superlattices Microstruct.*, **22**, 559–568.
16 Olshavsky, M.A., Goldstein, A.N., and Alivisatos, A.P. (1990) *J. Am. Chem. Soc.*, **112**, 9438.
17 Wells, R.L., Pitt, C.G., McPhail, A.T., Purdy, A.P., Shafieezad, S., and Hallock, R.B. (1989) *Chem. Mater.*, **1**, 4.
18 Uchida, H., Curtis, C.J., and Nozik, A.J. (1991) *J. Phys. Chem.*, **95**, 5382.
19 Uchida, H., Curtis, C.J., Kamat, P.V., Jones, K.M., and Nozik, A.J. (1992) *J. Phys. Chem.*, **96**, 1156.
20 Kher, S.S. and Wells, R.L. (1994) *Chem. Mater.*, **6**, 2056–2062.
21 Kher, S.S. and Wells, R.L. (1996) *Nanostruct. Mater.*, **7**, 591.
22 Micic, O.I., Ahrenkiel, S.P., Bertram, D., and Nozik, A.J. (1999) *Appl. Phys. Lett.*, **75**, 478.
23 Mews, A., Eychmüller, A., Giersig, M., Schoos, D., and Weller, H. (1994) *J. Phys. Chem.*, **98**, 934.
24 Hines, M.A. and Guyot-Sionnest, P.J. (1996) *J. Phys. Chem.*, **100**, 468.
25 Peng, X., Schlamp, M.C., Kadavanich, A.V., and Alivisatos, A.P. (1997) *J. Am. Chem. Soc.*, **119**, 7019.
26 Dabbousi, B.O., Rodriguez-Viejo, J., Mikulec, F.V., Heine, J.R., Mattoussi, H., Ober, R., Jensen, K.F., and Bawendi, M.G. (1997) *J. Phys. Chem. B*, **101**, 9463.
27 Talapin, D.V., Rogach, A.L., Kornowski, A., Haase, M., and Weller, H. (2001) *Nano Lett.*, **1**, 207.
28 Micic, O.I., Smith, B.B., and Nozik, A.J. (2000) *J. Phys. Chem. B*, **104**, 12149.
29 Micic, O.I., Sprague, J., Lu, Z.H., and Nozik, A.J. (1996) *Appl. Phys. Lett.*, **68**, 3150.
30 Murray, C.B., Norris, D.J., and Bawendi, M.G. (1993) *J. Am. Chem. Soc.*, **115**, 8706.
31 Peng, X., Wickham, J., and Alivisatos, A.P. (1998) *J. Am. Chem. Soc.*, **120**, 5343.
32 Smith, A.L. (1983) *Particle Growth in Suspensions*, Academic Press, London.
33 Chan, W.C.W. and Nie, S. (1998) *Science*, **281**, 2016.
34 Katari, J.E.B., Colvin, V.L., and Ailivisatos, A.P. (1994) *J. Phys. Chem.*, **98**, 4109.
35 (a) Klein, D.L., Roth, R., Lim, A.K.L., Alivisatos, A.P., and McEuen, P.L. (1997) *Nature*, **389**, 699; (b) Klein, D.L. *et al.* (1996) *Appl. Phys. Lett.*, **68**, 2574.
36 Loweth, C.J., Caldwell, W.B., Peng, X.G., Alivisatos, A.P., and Schultz, P.G. (1999) *Angew. Chem., Int. Ed. Engl.*, **38**, 1812.
37 Kuno, M., Lee, J.K., Dabbousi, B.O., Mikulec, F.V., and Bawendi, M.G. (1997) *J. Chem. Phys.*, **106**, 9869.
38 Jacobsohn-Eilon, M., Mokari, T., and Banin, U. (2001) *J. Phys. Chem. B*, **105**, 12726–12731.
39 Cowley, A.H. and Jones, R.A. (1989) *Angew. Chem., Int. Ed. Engl.*, **28**, 1208.
40 Wells, R.L. (1992) *Coord. Chem. Rev.*, **112**, 273.
41 Cowley, A.H., Harris, P.R., Jones, R.A., and Nunn, C.M. (1991) *Organometallics*, **10**, 652.
42 Miller, J.E., Kidd, K.B., Cowley, A.H., Jones, R.A., Ekerdt, J.G., Gysling, H.J., Wernberg, A.A., and Blanton, T.N. (1990) *Chem. Mater.*, **2**, 589.
43 Healy, M.D., Laibinis, P.E., Stupik, P.D., and Barron, A.R. (1989) *J. Chem. Soc., Chem. Commun.*, 359.
44 Robinson, W.T., Wilkins, C.J., and Zeing, Z. (1990) *J. Chem. Soc. Dalton Trans.*, 219.

45 Wells, R.L., Aubuchon, S.R., Sher, S.S., Lube, M.S., and White, P.S. (1995) *Chem. Mater.*, **7**, 793.
46 Jin, S., McKee, V., Nieuwenhuyzen, M., Robinson, W.T., and Wilkens, C.J. (1993) *J. Chem. Soc., Dalton Trans.*, 3111.
47 Inoue, K., Yoshizuka, K., and Yamaguchi, S. (1994) *J. Chem. Eng. Japan*, **27**, 737.
48 Sato, T. and Ruch, R. (1980) *Stabilization of Colloidal Dispersions by Polymer Adsorption*, Marcel Dekker, New York, pp. 46–51.
49 Kimura, K. (1994) *J. Phys. Chem.*, **98**, 11997.
50 Hamaker, H.C. (1937) *Recl. Trav. Chim.*, **56**, 3.
51 Verwey, E.J. and Overbeek, J.Th.G. (1948) *Theory of Stability of Lyophobic Colloid*, Elsevier, Amsterdam, p. 160.
52 Guinier, A. (1963) *X-Ray Diffraction*, Freeman, San Francisco.
53 Nirmal, M. and Brus, L. (1999) *Acc. Chem. Res.*, **32**, 407.
54 Weller, H. (1993) *Angew. Chem., Int. Ed. Engl.*, **32**, 41.
55 Bruchez, M., Moronne, M., Gin, P., Weiss, S., and Alivisatos, A.P. (1998) *Science*, **281**, 2013.
56 Mitchell, G.P., Mirkin, C.A., and Letsinger, R.L. (1999) *J. Am. Chem. Soc.*, **121**, 8122.
57 Colvin, V.L., Schlamp, M.C., and Alivisatos, A.P. (1994) *Nature*, **370**, 354.
58 Dabboussi, B.O., Bawendi, M.G., Onitsuka, O., and Rubner, M.F. (1995) *Appl. Phys. Lett.*, **66**, 1316.
59 Schlamp, M.C., Peng, X.G., and Alivisatos, A.P. (1997) *J. Appl. Phys.*, **82**, 5837.
60 Mattoussi, H., Radzilowski, L.H., Dabbousi, B.O., Thomas, E.L., Bawendi, M.G., and Rubner, M.F. (1998) *J. Appl. Phys.*, **83**, 7965.
61 Harrison, M.T., Kershaw, S.V., Rogach, A.L., Kornowski, A., Eychmüller, A., and Weller, H. (2000) *Adv. Mater.*, **12**, 123.
62 Allan, G., Bastard, G., Bocara, N., Lanoo, M., and Voos, M.(eds) (1986) *Heterojunctions and Semiconductor Superlattices*, Springer, Berlin, Heidelberg.
63 Wei, S. and Zunger, A. (1998) *Appl. Phys. Lett.*, **72**, 2011.
64 Guzelian, A.A. (1996) Synthesis and Characterization of III-V Semiconductor Nanocrystals, Ph.D. Thesis, University of California at Berkeley.
65 Hines, M.A. and Guyot-Sionnest, P. (1998) *J Phys. Chem. B*, **102**, 3655.
66 Lagowski, J.J. (1973) *Modern Inorganic Chemistry*, Marcel Dekker, New York, pp. 320–322.
67 Madelung, O., Schulz, M., and Heiss, H.(eds) (1982) *Landolt-Bornstein: Numerical Data and Functional Relationships in Science and Technology*, New Series, vol. **17**, Springer-Verlag, New York.
68 Hoener, C.F., Allan, K.A., Bard, A.J., Campion, A., Fox, M.A., Mallouk, T.E., Webber, S.E., and White, J.M. (1992) *J. Phys. Chem.*, **96**, 3812.
69 Wickham, J.N., Herhold, A.B., and Alivisatos, A.P. (2000) *Phys. Rev. Lett.*, **84**, 923.
70 Schoos, D., Mews, A., Eychmüller, A., and Weller, H. (1994) *Phys. Rev. B*, **49**, 17072.
71 Brus, L.E. (1984) *J. Chem. Phys.*, **80**, 4403.
72 Williamson, A.J. and Zunger, A. (1999) *Phys. Rev. B*, **59**, 15819.
73 Leduc, M. and Weisbuch, C. (1978) *Opt. Commun.*, **26**, 78.

3.2.3
Synthesis and Characterization of Ib–VI Nanoclusters

Stefanie Dehnen, Andreas Eichhöfer, John F. Corrigan, Olaf Fuhr, and Dieter Fenske

3.2.3.1 Introduction

The synthesis of chalcogen-bridged molecular clusters of the coinage metal elements, copper, silver, and gold, represents an area of increasing activity in recent chemical and materials science research. This can be attributed in large part to two facts. First, binary coinage metal chalcogenides feature relatively high ionic [1] and even higher electric [2] conductivity in the solid state, leading to properties that are positioned between those of semi-conducting and metallic phases. Second, an enormous general interest exists in the size dependency of the chemical, physical, and structural properties of substances when going from small molecules to bulk materials [3]. Therefore, numerous research groups are concentrating on the syntheses and properties of colloidal nanoparticles with narrow size distributions [4]. An alternative approach is the formation and isolation of crystalline cluster compounds that are suitable for single-crystal X-ray diffraction (XRD) studies. In both cases, the molecules must be kinetically protected from further condensation reactions so as to form the thermodynamically favored binary phases. Therefore, a ligand sphere coordinating to either surface metal centers or chalcogen atoms is necessary, and this in most cases consists uniquely of tertiary phosphine molecules or phosphines in combination with other organic groups. Herein, attention is focused on ligand-stabilized, chalcogen-bridged metal nanoclusters.

Both, experimental [5–33] and quantum chemical [34–37] investigations have been undertaken during the past decade in order to study the transition from the obviously nonmetal Cu_2E monomer (E = S, Se, Te) to bulk Cu_2E [38–40]. From the synthetic point of view, the merit of the endeavors was the perception of a close relationship of the chosen reaction conditions with the formation of the desired products, as well as the use of the acquired knowledge for the synthesis of different cluster types. Furthermore, the great number of isolated and structurally characterized cluster compounds and comprehensive computational studies on this subject have allowed the development and explanation of size-dependent structure principles observed in the molecules. Lastly, various calculations and measurements have been carried out in order to obtain an insight into the physical properties of the cluster materials [10(b)–12].

3.2.3.2 Chalcogen-Bridged Copper Clusters

3.2.3.2.1 Synthesis Routes
The compounds are – with some exceptions – highly air- and moisture-sensitive; consequently, all reaction steps must be carried out with the rigorous exclusion of air and moisture in a high-purity, inert atmosphere by employing standard Schlenkline techniques. All solvents are dried and freshly distilled prior to use. Most chalcogen-bridged copper compounds have been synthesized according to the following general reaction pathway, and possess one of the formulas shown in Scheme 3.2.

From Scheme 3.2, it is possible to recognize the importance of the CuX to phosphine ratio, the counterion X of the copper salt, and especially the nature of the phosphine used. In the first step, the latter complexes to the copper center to result in a coordination complex, the reactivity of which can be subtly modified by the choice of X and PR_2R'. Furthermore, the driving force of the reactions is the formation of thermodynamically stable $SiMe_3X$, and is thus determined by the choice of X via the bond strength of the Si−X bond [41]. Finally, the steric demand of the PR_2R' molecules coating the cluster surface – as measured by Tolman's cone angle

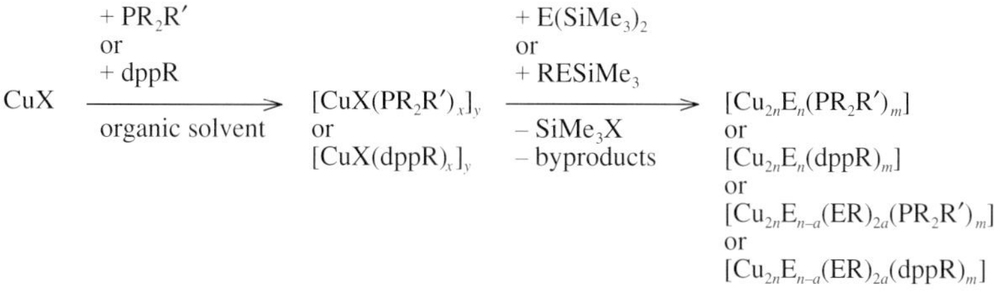

3.2 Semiconductor Nanoparticles

$$\text{CuX} \xrightarrow[\text{organic solvent}]{\substack{+ \text{PR}_2\text{R}' \\ \text{or} \\ + \text{dppR}}} \substack{[\text{CuX}(\text{PR}_2\text{R}')_x]_y \\ \text{or} \\ [\text{CuX}(\text{dppR})_x]_y} \xrightarrow[\substack{- \text{SiMe}_3\text{X} \\ - \text{byproducts}}]{\substack{+ \text{E}(\text{SiMe}_3)_2 \\ \text{or} \\ + \text{RESiMe}_3}} \substack{[\text{Cu}_{2n}\text{E}_n(\text{PR}_2\text{R}')_m] \\ \text{or} \\ [\text{Cu}_{2n}\text{E}_n(\text{dppR})_m] \\ \text{or} \\ [\text{Cu}_{2n}\text{E}_{n-a}(\text{ER})_{2a}(\text{PR}_2\text{R}')_m] \\ \text{or} \\ [\text{Cu}_{2n}\text{E}_{n-a}(\text{ER})_{2a}(\text{dppR})_m]}$$

X = Cl, OAc (Ac = OCCH$_3$)
R, R' = organic group
dppR = bis-(diphenylphosphino)-alkane, -alkene, alkyne or -aryl
E = S, Se, Te
main byproducts are Cu$_2$E, E=PR$_2$R', (Me$_3$Si)$_2$, (RE)$_2$

Scheme 3.2 General synthesis route for the formation of phosphine-ligated, chalcogen-bridged copper clusters.

Θ [42] – is, though not linearly, closely related to the final size and shape of the copper chalcogenide core.

In addition to these variables, other reaction conditions, such as the chosen solvent and the reaction temperature or temperature program, strongly influence the course of the synthesis. It has been shown that, under similar reaction conditions, the use of derivatives RE(SiMe$_3$) (R = organic group) leads to the formation of completely different structures compared to reactions employing Se(SiMe$_3$)$_2$. The reason for this is the different reactivities of the compared substances, as well as a partly incomplete Se–C bond cleavage. Thus, some RE$^-$ fragments are bonded at the periphery of the cluster framework and also act as surface ligands in conjunction with the phosphines.

The copper atoms in the vast majority of the clusters can be assigned a formal charge of +1, while the chalcogen ligands are formally viewed as E^{2-} or RE$^-$ groups. Some of the selenium-bridged species, however – and nearly all copper telluride clusters – form nonstoichiometric compounds that display mixed valence metal centers in the formal oxidation states 0 and +I or +I and +II. These observations correlate with those made for the binary phases Cu$_2$S, Cu$_{2-x}$Se, and Cu$_{2-x}$Te [38–40].

Even though most ligand-covered copper chalcogenide clusters were prepared using the above-described methods, other approaches for the generation of ligated or naked [Cu$_{2n}$E$_n$] cluster compounds have been reported previously. In addition to syntheses in solvent environments [5–31] that led to the formation of, for example, [PhMe$_3$As]$_4$[Cu$_8${S$_2$C$_2$(CN)$_2$}$_6$] [5], (Me$_4$N)$_2$[Cu$_5$(SPh)$_7$] [6], (Ph$_4$P)$_4$[Cu$_{12}$S$_8$] [7], or [Cu$_6${Se(2,4,6-iPr$_3$C$_6$H$_2$)}$_6$] [8], ligand-free particles have been generated and characterized by means of laser ablation techniques, in combination with mass spectroscopy.

3.2.3.2.2 Sulfur-Bridged Copper Clusters The reactions yielding the phosphine-coated sulfur-bridged copper clusters that have been isolated and structurally

Scheme 3.3

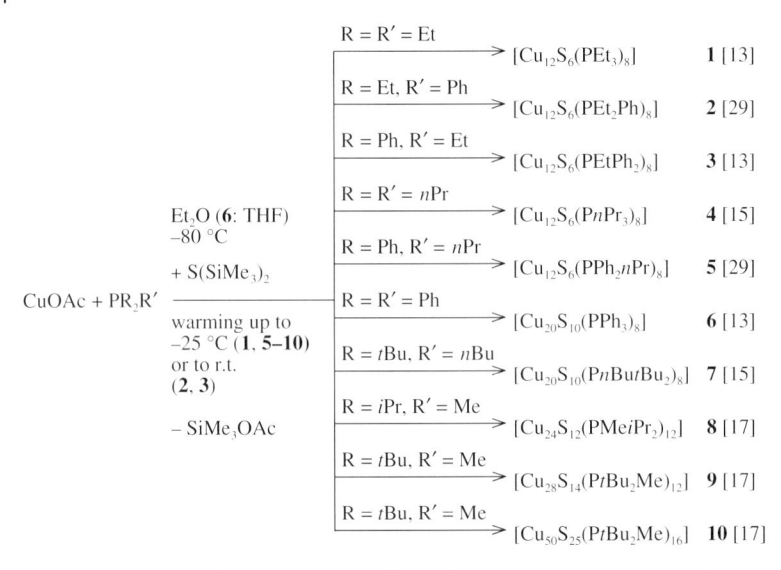

CuOAc + PR₂R′
Et₂O (**6**: THF) −80 °C
+ S(SiMe₃)₂
warming up to −25 °C (**1, 5–10**) or to r.t. (**2, 3**)
− SiMe₃OAc

R, R′	Product	#	Ref
R = R′ = Et	$[Cu_{12}S_6(PEt_3)_8]$	**1**	[13]
R = Et, R′ = Ph	$[Cu_{12}S_6(PEt_2Ph)_8]$	**2**	[29]
R = Ph, R′ = Et	$[Cu_{12}S_6(PEtPh_2)_8]$	**3**	[13]
R = R′ = nPr	$[Cu_{12}S_6(PnPr_3)_8]$	**4**	[15]
R = Ph, R′ = nPr	$[Cu_{12}S_6(PPh_2nPr)_8]$	**5**	[29]
R = R′ = Ph	$[Cu_{20}S_{10}(PPh_3)_8]$	**6**	[13]
R = tBu, R′ = nBu	$[Cu_{20}S_{10}(PnBu tBu_2)_8]$	**7**	[15]
R = iPr, R′ = Me	$[Cu_{24}S_{12}(PMe iPr_2)_{12}]$	**8**	[17]
R = tBu, R′ = Me	$[Cu_{28}S_{14}(PtBu_2Me)_{12}]$	**9**	[17]
R = tBu, R′ = Me	$[Cu_{50}S_{25}(PtBu_2Me)_{16}]$	**10**	[17]

Scheme 3.3 Survey of the synthesis of sulfur-bridged copper clusters protected by terminal phosphine ligands.

characterized to date are summarized in Scheme 3.3, in order of increasing cluster size.

Sulfur-bridged clusters could not be obtained by reactions starting from copper halides. The reason for this is the lower reactivity of a primary copper acetate–phosphine complex when compared to analogous halide complexes due to chelating acetate ligands that leave in a double-step substitution of sulfur ligands for oxygen [43]. Reactants of higher reactivity lead directly to the formation and precipitation of Cu_2S, even at low temperatures. The kinetic restrictions at the beginning of the synthesis combine with those at the end of the cluster growth. The phosphine molecules must display a certain minimum steric demand, as it has not yet proved possible to generate sulfur-bridged copper clusters stabilized with PMe_3 ligands using this strategy.

Half of the number of known copper sulfide clusters adhere to the general formula $[Cu_{12}S_6(PR_2R′)_8]$, which represents the smallest known copper chalcogenide $[Cu_{2\cdot nEn}]$ cluster core (**1–5**) [13, 15, 18]. The respective molecular structure corresponds to one of two possible isomers that can both be derived from two highly symmetric polyhedra of metal or chalcogen atoms. The highly symmetric conformation is the reason for the strong preference for "Cu_{12}" species, where an S_6 octahedron is penetrated by a Cu_{12} cubo-octahedron. The latter is distorted, however, because of the coordination of the metal centers to phosphine ligands, as there is only enough space for eight instead of 12 phosphine molecules at the cluster surface within reasonable Cu–P bond distances. Different distributions of the terminal ligands to the cubo-octahedron atoms lead to the formation of isomers I or II, respectively. Both cluster structures as well as a computed hypothetical "naked," undistorted $[Cu_{12}S_6]$ [35] core are shown in Figure 3.36.

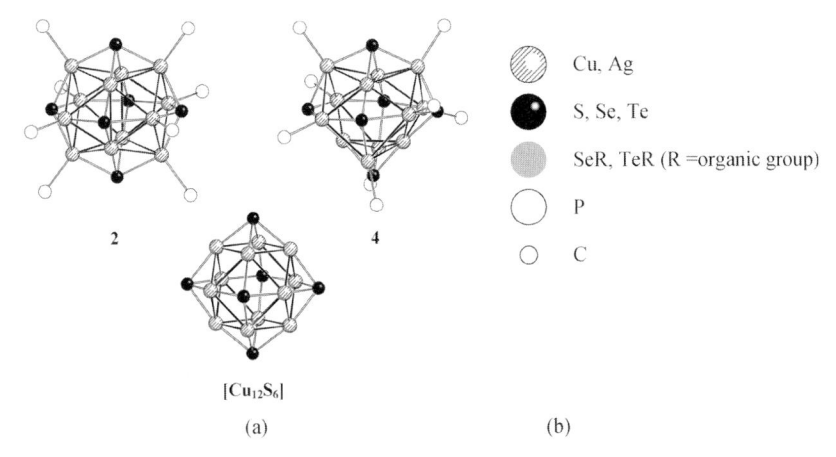

Figure 3.36 Molecular structures of [Cu$_{12}$S$_6$(PEt$_2$Ph)$_8$] (**2**) and [Cu$_{12}$S$_6$(PnPr$_3$)$_8$] (**4**) (without organic groups) and of a calculated "naked" [Cu$_{12}$S$_6$] core. Cu: shaded spheres; S: black spheres; P: white spheres.

In the "Cu$_{12}$" clusters, two coordination modes for copper atoms can be observed. The copper atoms that do not bind to phosphine ligands are nearly linearly coordinated by sulfur neighbors, and positioned either in the equatorial Cu$_4$ ring of the cuboctahedron (I) or two each in the top and bottom of the Cu$_4$ planes (II). Copper atoms that are attached to PR$_2$R' groups are surrounded in a nearly trigonal planar manner by two sulfur neighbors and one phosphorus atom. The phosphine molecules formally "pull" the respective metal atoms out of the based Cu$_{12}$ polyhedron, whereas the "naked" copper atoms are slightly pushed towards the cluster center. All sulfur ligands act as μ_4 bridges between copper atoms. In molecules of type I, the eight phosphorus atoms are arranged around the cluster surface in a cubic manner, whereas a P8 deltadodecahedron is found in the other isomer. Both arrangements are counted among the most stable polyhedra of eight points on a spherical surface.

Quantum chemical investigations with the program system TURBOMOLE [44, 45], employing MP2 [46] and density functional theory (DFT) [47, 48] methods, were carried out assigning the highest possible symmetry in order to explain the obvious experimentally observed preference for type I clusters in comparison to those of type II [36]. For model ligands PH$_3$ (MP2 and DFT, highest possible symmetry: $D4h$ for type I or $S4$ for type II), both structures that represent local minima on the energy hyperface were isoenergetic within the limits of the methods used. On calculating the clusters [Cu$_{12}$S$_6$(PR$_3$)$_8$] with R = Et or R = Pr according to both structural isomers (DFT, highest possible symmetry: $C4h$ or $S4$), that is, one real (**1**, **4**) and one hypothetical species each, there is no direct correlation of a certain type of substituent at the phosphorus atoms with purely thermodynamic stabilization. Thus, further kinetic effects such as the solubility and mobility of the hydrocarbyl groups at the experimental temperatures (g $\gg$ K) were found to play key roles in the discrimination of isomers. The same cone angle values of PEt$_3$ and PnPr$_3$ also serve to underline this conclusion.

3 Syntheses and Characterizations

A second pair of isomers is formed in the presence of sterically more demanding phosphines, PPh$_3$ or PnButBu$_2$, that contain 20 copper atoms in the cluster framework [13, 15]. The positions of eight Cu–P groups again determine the shape of the entire cluster, the undistorted base of which can be viewed as a formal condensation product of the two [Cu$_{12}$S$_6$] units given in Figure 3.36.

A continuation of cluster growth also takes place with PMeiPr$_2$. Being too bulky for an arrangement of eight molecules around a [Cu$_{12}$S$_6$] core and not generating a suitable ligand sphere with six terminal ligands, the phosphine groups force a formal dimerization of "Cu$_{12}$" clusters to the [Cu$_{24}$S$_{12}$] core in **8** [17]. Figure 3.37 shows the formal dimerization of two [Cu$_{12}$S$_6$(PR$_2$R')$_8$] clusters of type II, which results in the molecular structure of **8** after the loss of two phosphine groups.

In contrast to the formal condensation, $2 \times$ "Cu$_{12}$" $\to$ "Cu$_{20}$" $-$ 4 Cu, that leads to the sharing of a Cu$_4$ rectangular face in **6**, a conjunction can be observed via two Cu$_3$

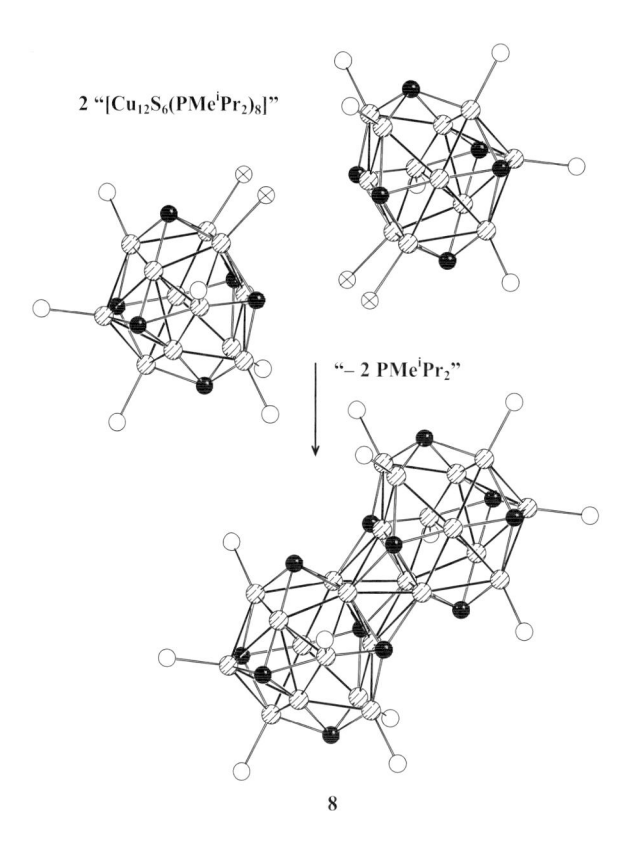

2 "[Cu$_{12}$S$_6$(PMe^{i}Pr$_2$)$_8$]"

"– 2 PMe^{i}Pr$_2$"

8

Figure 3.37 Formal dimerization–condensation reaction of two hypothetical clusters "[Cu$_{12}$S$_6$(MeiPr$_2$)$_8$]," resulting in the molecular structure of [Cu$_{24}$S$_{12}$(PMeiPr$_2$)$_{12}$] (**8**) (without organic groups). Cu: shaded spheres; S: black spheres; P: white spheres. Phosphine ligands that are formally lost during the dimerization–condensation are marked by a cross.

triangular faces of the two based cubo-octahedra. Thus, a Cu_6 octahedron is formed around the inversion center of the cluster. These six copper atoms show a novel coordination pattern: they are surrounded in a trigonal planar manner exclusively by sulfur neighbors. The four sulfur atoms in the cluster center are found to act as μ_5 bridges, whereas all sulfur ligands in the previously described clusters and all other additional sulfur ligands in **8** are μ_4 bridging. Naturally, the μ_5 bridges exhibit longer average Cu–S bond distances when compared to μ_4 bridging sulfur ligands by binding to four copper neighbors within normal distances (2.159(3)–2.378(3) Å) and possessing one significantly longer Cu–S bond each to copper atoms of the opposite asymmetric unit (2.485(3), 2.509(3) Å).

In the molecular structure of **8**, the start of the transition is observed, from the so-called "small" copper sulfide clusters to the "middle-sized" species. Both cluster sizes possess spherical Cu–S–P frameworks with copper centers near the Sn polyhedral edges, but the "middle-sized" examples show some new structural features: coordination numbers other than four for sulfur atoms, and copper centers that bind to more than two sulfur neighbors.

The same observations are made for clusters **9** and **10**, and to a greater extent for the latter [17]. Unlike the clusters described above, clusters **9** and **10** are topologically independent from smaller complexes. They are, however, themselves structurally related and are indeed produced side-by-side from the same reaction. The molecular structures of **9** and **10** are given in Figure 3.38, emphasizing their structural relationship. In Figure 3.39, the based S_{14} or S_{25} polyhedra are contrasted.

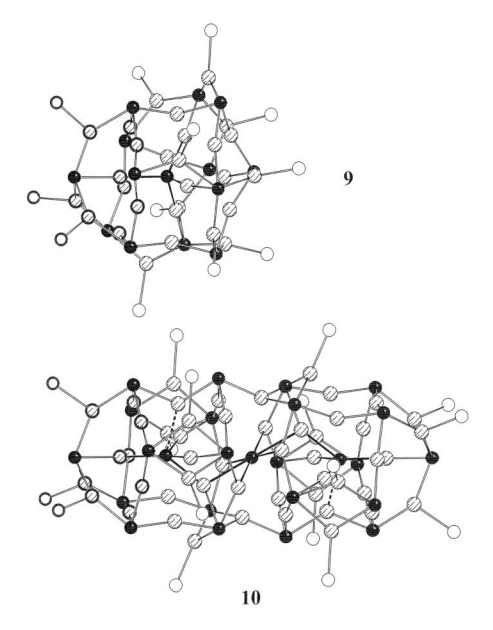

Figure 3.38 Molecular structures of $[Cu_{28}S_{14}(PtBu_2Me)_{12}]$ (**9**) and $[Cu_{50}S_{25}(PtBu_2Me)_{16}]$ (**10**) (without organic groups). Cu: shaded spheres; S: black spheres; P: white spheres. The atoms of topologically equal zones are highlighted in bold.

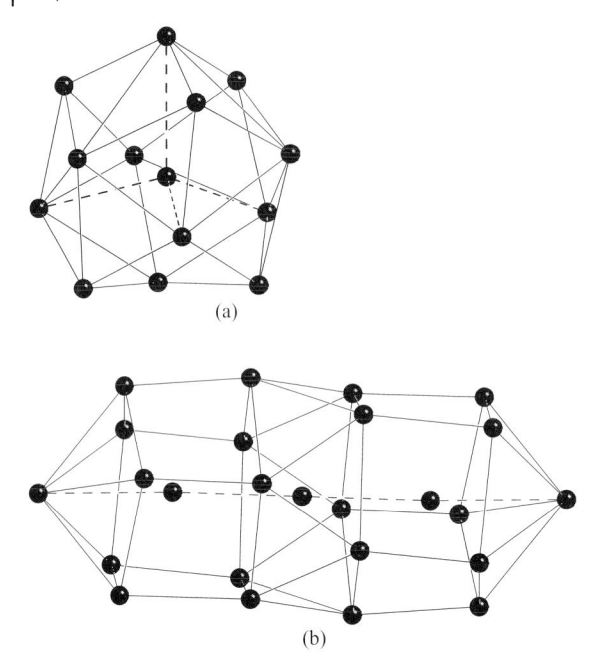

Figure 3.39 S_{14} (a) and S_{25} (b) polyhedra in **9** and **10**, respectively. The drawn S—S contacts do not represent binding interactions, but serve only to explain the geometries of the polyhedra.

Besides various bridging modes of the sulfur atoms (μ_3–μ_5 in **9**, μ_4–μ_6 in **10**) and the higher numbers of three sulfur ligands around several copper atoms in **10**, both clusters show another new structural feature characterizing "middle-sized" Cu–S clusters. Only for these two largest examples of copper sulfide clusters does one observe sulfur atoms inside an outer Cu–S shell, generating inner Cu–S units. In **9**, one sulfur ligand centers a distorted S_{13} deltahedron with four copper atoms bonded to the central sulfur atom to yield an inner [SCu$_4$] unit. The remaining Cu centers are positioned either slightly below the edges of the S_{13} polyhedron if they show no Cu–P bonds, or slightly above if bonded to phosphine ligands. Near C_3 symmetry of the Cu–S core is distorted by an "irregular" position of a single copper atom that also promotes the μ_3 or μ_5 bridging modes of two sulfur ligands. The spherical cluster core is enclosed by a distorted icosahedron of the 12 phosphorus atoms. The sulfur substructure of **10** represents a cylinder-shaped polyhedron of near D5d symmetry with a length of 17.14 Å. Underlining the supposition that **10** is formed from **9** through rough dimerization–condensation, three inner sulfur atoms are present. Two of them – formally arising from the central sulfur atom in **9** – are also μ_4 bridging, resulting in an [SCu$_4$] fragment. One of the corresponding copper centers in each of these units, however, belongs to the cluster surface rather than remaining an inner metal atom because of structural differences between **9** and **10** when approaching the center in **10**. Here can be observed the most unusual feature of any of the copper sulfide clusters reported: the central sulfur ligand binds to six copper atoms in a slightly corrugated chair-type manner. This arrangement, which is unknown for all other copper sulfide compounds, displays large Cu–S distances in the range 2.633(4)–2.673(5) Å, which are about 0.3–0.5 Å longer than usual Cu–S bonds. Considering the next ten copper atoms at distances of within 2.896(5)–3.814(5) Å around the sulfur center, a spherical [SCu$_{16}$] arrangement results in the cluster hub. The structural observations give evidence for an interstitial S^{2-} anion rather than a more covalently bonded sulfur ligand.

By a comprehensive *ab initio* study of clusters [Cu$_{2n}$S$_n$(PH$_3$)$_m$](n = 1–4, 6, 10; m = 0, 2, 4, 6, 8) with an MP2 level (program system TURBOMOLE), stabilization energies for the sequential attachment of one Cu$_2$S unit to a given "naked" cluster (m = 0, Figure 3.40), as well as the course of the phosphine-binding energies for clusters containing up to 12 copper atoms (m > 0, Figure 3.41) were developed [35]. For "naked" clusters, the stabilization energies increase continuously with the cluster size, which proves the investigated molecules to be thermodynamically unstable species *in vacuo*, and thus must be stabilized by ligands. Moreover, the increment of the stabilization energy per monomer unit decreases with increasing cluster size as expected, since it should theoretically converge with infinite cluster size – that is, solid Cu$_2$S.

As a third result, it can be gathered from the calculations that the Cu–P binding energy per Cu–P bond decreases rapidly from a value of about 140 kJ mol^{-1} for [Cu$_2$S(PH$_3$)$_2$] – with a short relaxation when the Cu:P ratio is decreased (1 for n = 1–3, >1 for n = 4, 6) – to a value of about 56 kJ mol^{-1} for the "Cu$_{12}$" species. Even if a higher Cu–P binding energy is assigned to the "real" tertiary phosphines with organic side chains, it is evident that the phosphine ligands are bonded very weakly to the copper sulfide core. Therefore, an additional stabilization by solvents or a crystal lattice is required.

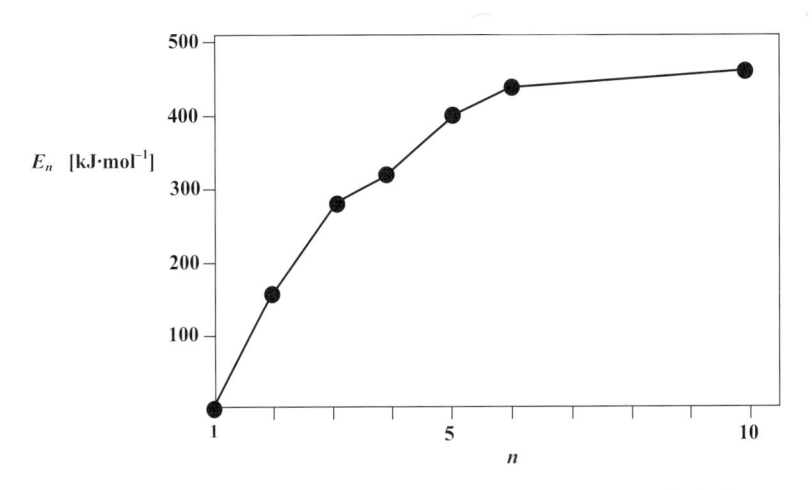

Figure 3.40 Course of the stabilization energies per monomer unit E_n in "naked" clusters [Cu$_{2n}$S$_n$] (n = 1–4, 6, 10). The given values consider the total energies of the most stable structural isomers computed on MP2 level.

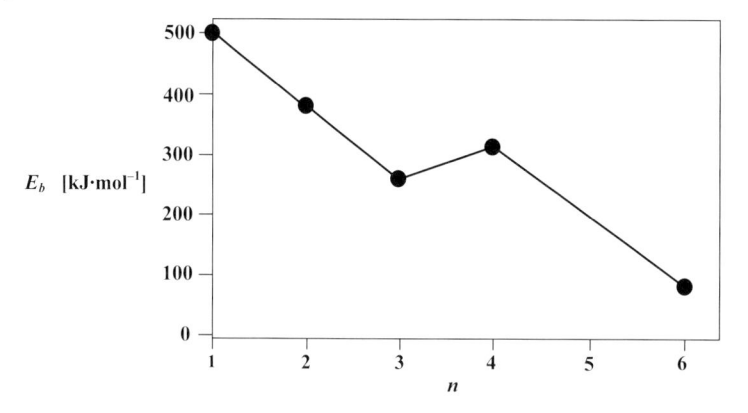

Figure 3.41 Course of the Cu–P binding energies per Cu–P bond E_b in phosphine-ligated clusters [Cu$_{2n}$S$_n$(PH$_3$)m] ($n = 1$–4, 6; $m = 2, 4, 6, 8$). The given values consider the total energies of the most stable structural isomers computed on MP2 level with respect to the total energy of PH$_3$ calculated by the same method.

3.2.3.2.3 Selenium-Bridged Copper Clusters

The reactions yielding all PR$_2$R′-coated selenium-bridged copper clusters that have been isolated and structurally characterized to date, in order of increasing cluster size, are summarized in Scheme 3.4.

In contrast to the sulfur-bridged compounds, the synthesis of selenium-bridged clusters can start from either copper acetate or copper chloride. This can be ascribed to the lower reactivity of Se(SiMe$_3$)$_2$ when compared to the sulfur analog [49, 50]. Thus, the formation of Cu$_2$Se does not occur as rapidly as the formation of Cu$_2$S, and the cluster growth and ligand protection can often be observed even at room temperatures. However, by reacting copper acetate and by choosing low temperatures, it is possible to obtain smaller clusters that can be characterized as being intermediates during the course of the formation of larger molecules. In addition to a great number of "stoichiometric" compounds that formally contain Cu$^+$ and Se^{2+}, some clusters were characterized that exhibited a Cu:Se ratio of less than two, indicating mixed valence metal centers. All selenium ligands feature the formal charge 2, as no Se–Se binding interactions can be found, such that Cu^I and Cu^{II} can be assigned within the same cluster core in **14, 17, 27**, and **33**. The phosphine ligands again play a key role in the determination of the observed cluster size and shape; however, the number of known compositions is approximately fourfold that of the copper sulfide system, and the cluster growth proceeds to a much higher extent. Again, the lower tendency to precipitate the binary phase can be held responsible for this observation.

Another route into selenium-bridged copper polynuclear complexes or clusters is their synthesis using RSeSiMe$_3$ (R = organic group) in the presence of either tertiary phosphines or chelating ligands such as 1,2-bipyridyl, diphenylphosphino acid or bidentate phosphines dppR (R = organic spacer). The compounds thus prepared are listed in Scheme 3.5, and ordered according to their core size.

3.2 Semiconductor Nanoparticles | 137

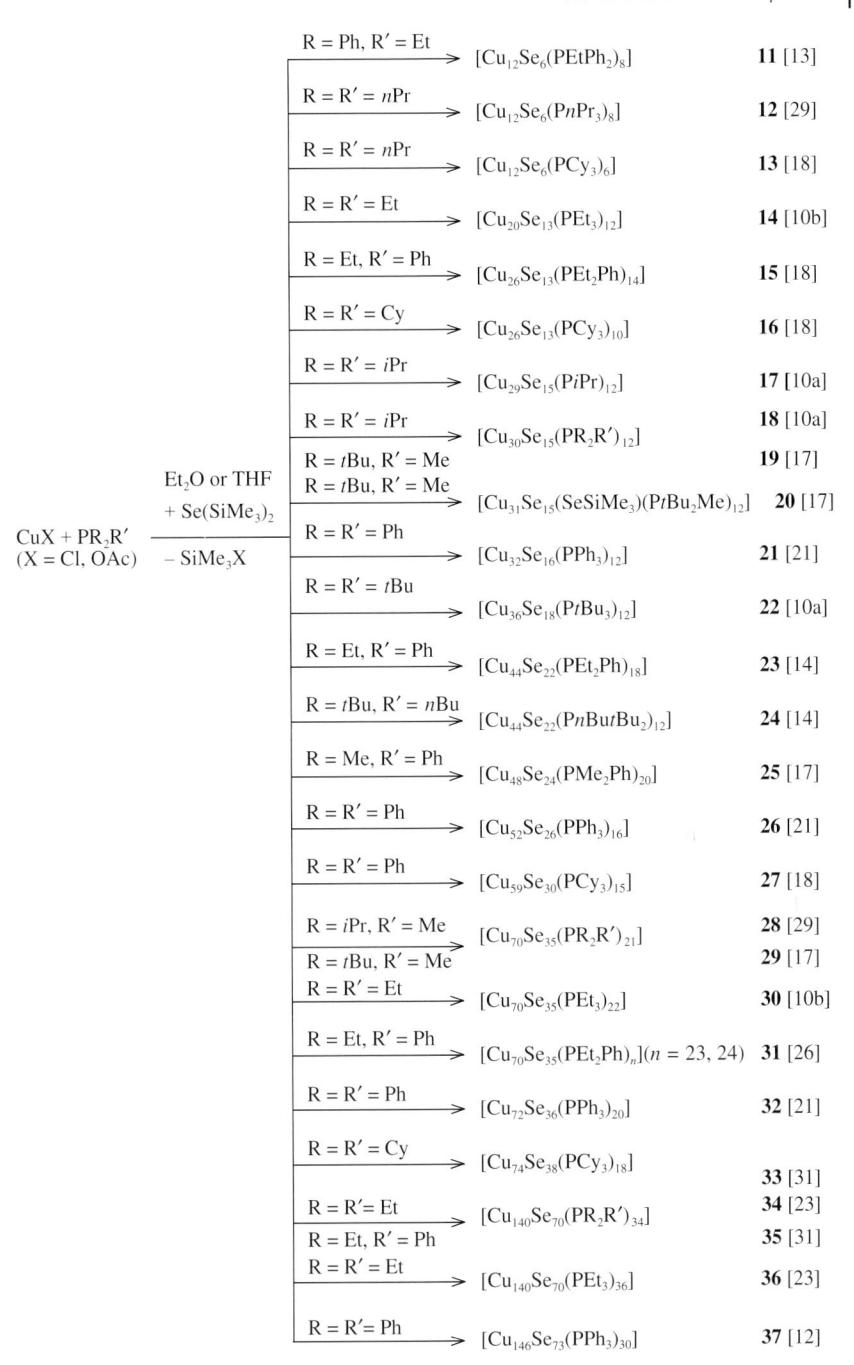

R = Ph, R′ = Et → [Cu$_{12}$Se$_6$(PEtPh$_2$)$_8$]	**11**	[13]
R = R′ = nPr → [Cu$_{12}$Se$_6$(PnPr$_3$)$_8$]	**12**	[29]
R = R′ = nPr → [Cu$_{12}$Se$_6$(PCy$_3$)$_6$]	**13**	[18]
R = R′ = Et → [Cu$_{20}$Se$_{13}$(PEt$_3$)$_{12}$]	**14**	[10b]
R = Et, R′ = Ph → [Cu$_{26}$Se$_{13}$(PEt$_2$Ph)$_{14}$]	**15**	[18]
R = R′ = Cy → [Cu$_{26}$Se$_{13}$(PCy$_3$)$_{10}$]	**16**	[18]
R = R′ = iPr → [Cu$_{29}$Se$_{15}$(PiPr)$_{12}$]	**17**	[10a]
R = R′ = iPr → [Cu$_{30}$Se$_{15}$(PR$_2$R′)$_{12}$]	**18**	[10a]
R = tBu, R′ = Me	**19**	[17]
R = tBu, R′ = Me → [Cu$_{31}$Se$_{15}$(SeSiMe$_3$)(PtBu$_2$Me)$_{12}$]	**20**	[17]

CuX + PR$_2$R′ (X = Cl, OAc) — Et$_2$O or THF + Se(SiMe$_3$)$_2$ − SiMe$_3$X →

R = R′ = Ph → [Cu$_{32}$Se$_{16}$(PPh$_3$)$_{12}$]	**21**	[21]
R = R′ = tBu → [Cu$_{36}$Se$_{18}$(PtBu$_3$)$_{12}$]	**22**	[10a]
R = Et, R′ = Ph → [Cu$_{44}$Se$_{22}$(PEt$_2$Ph)$_{18}$]	**23**	[14]
R = tBu, R′ = nBu → [Cu$_{44}$Se$_{22}$(PnButBu$_2$)$_{12}$]	**24**	[14]
R = Me, R′ = Ph → [Cu$_{48}$Se$_{24}$(PMe$_2$Ph)$_{20}$]	**25**	[17]
R = R′ = Ph → [Cu$_{52}$Se$_{26}$(PPh$_3$)$_{16}$]	**26**	[21]
R = R′ = Ph → [Cu$_{59}$Se$_{30}$(PCy$_3$)$_{15}$]	**27**	[18]
R = iPr, R′ = Me → [Cu$_{70}$Se$_{35}$(PR$_2$R′)$_{21}$]	**28**	[29]
R = tBu, R′ = Me	**29**	[17]
R = R′ = Et → [Cu$_{70}$Se$_{35}$(PEt$_3$)$_{22}$]	**30**	[10b]
R = Et, R′ = Ph → [Cu$_{70}$Se$_{35}$(PEt$_2$Ph)$_n$] (n = 23, 24)	**31**	[26]
R = R′ = Ph → [Cu$_{72}$Se$_{36}$(PPh$_3$)$_{20}$]	**32**	[21]
R = R′ = Cy → [Cu$_{74}$Se$_{38}$(PCy$_3$)$_{18}$]	**33**	[31]
R = R′ = Et → [Cu$_{140}$Se$_{70}$(PR$_2$R′)$_{34}$]	**34**	[23]
R = Et, R′ = Ph	**35**	[31]
R = R′ = Et → [Cu$_{140}$Se$_{70}$(PEt$_3$)$_{36}$]	**36**	[23]
R = R′ = Ph → [Cu$_{146}$Se$_{73}$(PPh$_3$)$_{30}$]	**37**	[12]

Scheme 3.4 Survey of the synthesis of selenium-bridged copper clusters protected by terminal phosphine ligands.

3 Syntheses and Characterizations

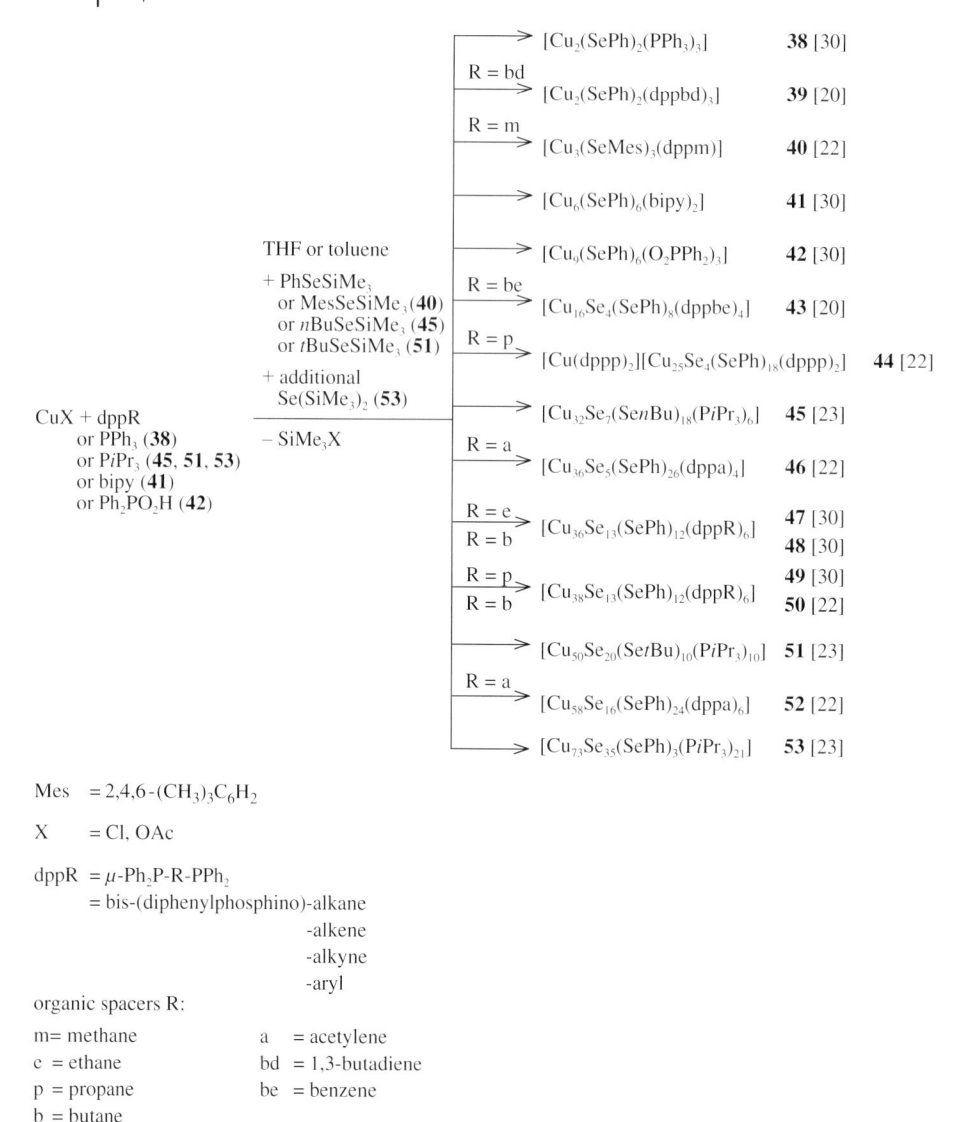

Scheme 3.5 Reaction scheme:

CuX + dppR
  or PPh$_3$ (**38**)
  or PiPr$_3$ (**45, 51, 53**)
  or bipy (**41**)
  or Ph$_2$PO$_2$H (**42**)

THF or toluene
 + PhSeSiMe$_3$
  or MesSeSiMe$_3$ (**40**)
  or nBuSeSiMe$_3$ (**45**)
  or tBuSeSiMe$_3$ (**51**)
 + additional Se(SiMe$_3$)$_2$ (**53**)
 − SiMe$_3$X

Products:

- R = bd → [Cu$_2$(SePh)$_2$(PPh$_3$)$_3$]   **38** [30]
- R = bd → [Cu$_2$(SePh)$_2$(dppbd)$_3$]   **39** [20]
- R = m → [Cu$_3$(SeMes)$_3$(dppm)]   **40** [22]
- → [Cu$_6$(SePh)$_6$(bipy)$_2$]   **41** [30]
- → [Cu$_9$(SePh)$_6$(O$_2$PPh$_2$)$_3$]   **42** [30]
- R = be → [Cu$_{16}$Se$_4$(SePh)$_8$(dppbe)$_4$]   **43** [20]
- R = p → [Cu(dppp)$_2$][Cu$_{25}$Se$_4$(SePh)$_{18}$(dppp)$_2$]   **44** [22]
- → [Cu$_{32}$Se$_7$(SenBu)$_{18}$(PiPr$_3$)$_6$]   **45** [23]
- R = a → [Cu$_{36}$Se$_5$(SePh)$_{26}$(dppa)$_4$]   **46** [22]
- R = e → [Cu$_{36}$Se$_{13}$(SePh)$_{12}$(dppR)$_6$]   **47** [30]
- R = b →   **48** [30]
- R = p → [Cu$_{38}$Se$_{13}$(SePh)$_{12}$(dppR)$_6$]   **49** [30]
- R = b →   **50** [22]
- → [Cu$_{50}$Se$_{20}$(SetBu)$_{10}$(PiPr$_3$)$_{10}$]   **51** [23]
- R = a → [Cu$_{58}$Se$_{16}$(SePh)$_{24}$(dppa)$_6$]   **52** [22]
- → [Cu$_{73}$Se$_{35}$(SePh)$_3$(PiPr$_3$)$_{21}$]   **53** [23]

Mes = 2,4,6-(CH$_3$)$_3$C$_6$H$_2$

X = Cl, OAc

dppR = μ-Ph$_2$P-R-PPh$_2$
     = bis-(diphenylphosphino)-alkane
                              -alkene
                                -alkyne
                                -aryl

organic spacers R:
- m = methane
- e = ethane
- p = propane
- b = butane
- a = acetylene
- bd = 1,3-butadiene
- be = benzene

Scheme 3.5 Survey of the synthesis of selenido–selenolato-bridged copper clusters protected by monodentate or chelating bidentate ligands.

Again, clusters containing exclusively Cu^I, as well as compounds with mixed-valence metal centers, can be found. In contrast to the compounds presented in Schemes 3.3 and 3.4, one example of a selenido–selenolato-bridged cluster exists apparently containing Cu^I and Cu^0. Apart from this peculiarity, the cluster size and conformation of the Se^{2-}–SeR-bridged compounds is now not only influenced by the nature and steric demand of the phosphine used, but it is also dependent on the spatial properties of the organic substituent at the selenium reactant. This determines both the Se^{2-}:SeR^- ratio and the ensuing arrangement of Se^{2-} ligands positioned inside the cluster core and SeR^- ligands at the periphery of the molecules.

The delicate, nonlinear structural influence of the phosphine ligands allows for the formation of different clusters of often significantly varying size and shape in the presence of the same phosphine. The isolation of various cluster compounds is then enabled only by the judicious choice of reaction conditions. In contrast, the use of different PR_2R' ligands can lead to clusters of the same $[Cu_{2n}Se_n]$ cores that possibly display differing numbers of terminal ligands and/or structural isomerism. These observations have already been illustrated and explained for the sulfur-bridged copper clusters, but they are more pronounced for the Cu–Se species, and a much higher structural variety is found. As for the copper sulfide compounds, it is possible to discriminate between "small" (11–13) and "middle-sized" (14–27) clusters with respect to their shape and the coordination properties of the atoms concerned. However a novel group of "large" clusters (28–37) exists for the Cu–Se system. Whereas, the transition from the first to the second type of molecular size occurs almost seamlessly, a clear break occurs when the number of copper atoms exceeds 59.

The only composition and structure that is identical for both the copper sulfide and selenide clusters is that for $[Cu_{12}E_6(PR_2R')_8]$ (E = S, Se; PR_2R' = $PEtPh_2$, $PnPr_3$; respective cone angles: 141, 132). Compounds **11** [13] and **12** [29] are topologically identical to their sulfur analogs. The use of a much larger phosphine ligand PCy_3 does not lead to a condensation or dimerization of cluster cores as observed for **6–8** with PR_2R' groups of medium or large size around. Unlike the copper sulfide system, in the presence of PPh_3, $PnButBu_2$, or $PMeiPr_2$, a reduction in the number of terminal ligands to six is acceptable in the case of **13** [18], and the resulting ligand shell is suitable for the cluster core. The clearly larger cone angle of PCy_3 when compared to PPh_3, $PnButBu_2$, or $PMeiPr_2$ (170° versus 145, 146, and 165) allows a distorted P_6 octahedron to sufficiently protect the copper selenide framework, although a third variation of a distortion of the $[Cu_{12}E_6]$ core is observed.

With the clusters discussed, strong preference is perceived for highly symmetric substructures of all components involved in the heavy-atom framework. Capped or uncapped, sometimes condensed octahedral, prismatic, cubic, dodecahedral, cubo-octahedral, or icosahedral arrangements of either metal, chalcogen or phosphorus atoms penetrate or enclose each other in all structures. Among the discussed compounds, the "Cu_{12}" molecules possess the most symmetric substructures, although selenium ligands evidently prefer higher bridging modes. Therefore, the formation of "small" clusters with a strong μ_4 bridging restriction is disadvantaged and thus rarer than the formation of larger clusters. For the Cu–Se clusters, three additional "middle-sized" compounds are interesting with respect to their symmetry properties, as well as being an additional illustration of the phosphines' structural influence. The cluster frameworks of **14** [10b], and the Cu–Se isomers **15** [18] and **16** [18] can, again, be described as an alternating packing of copper, selenium, and phosphorus polyhedra, respectively. Regular polyhedra are, however, only observed in the structures of **14** and **15**, whereas the $[Cu_{26}Se_{13}]$ isomeric core of **16**,

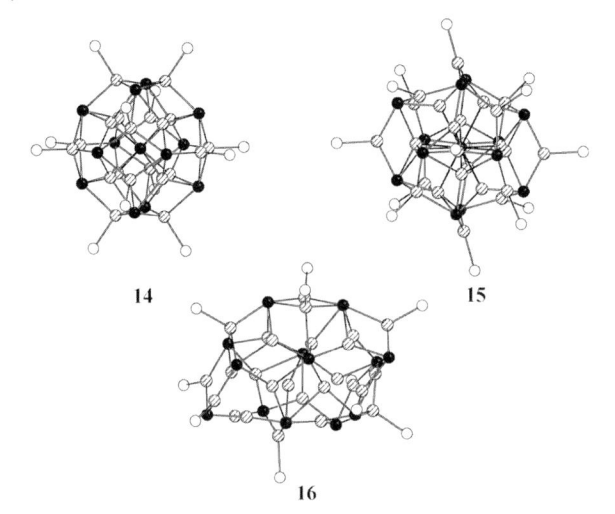

Figure 3.42 Molecular structures of [$Cu_{20}Se_{13}(PEt_3)_{12}$] (**14**), [$Cu_{26}Se_{13}(PEt_2Ph)_{14}$] (**15**), and [$Cu_{26}Se_{13}(PCy_3)_{10}$] (**16**) (without organic groups). Cu: shaded spheres; Se: black spheres; P: white spheres.

being surrounded by only ten PCy_3 (versus 14 PEt_2Ph groups), does not follow a known polyhedral pattern. The molecular structures of **14–16** are shown in Figure 3.42.

Clusters **14** and **15** are topologically related, as both contain a centered icosahedron of selenium atoms, although the penetrating polyhedra of copper atoms are arranged in different manners according to the different number of copper atoms and Cu–PR_2R' fragments (**14**: R = R' = Et, **15**: R = Et, R' = Ph) in both clusters (Figure 3.43).

In **14**, a Cu_8 cube surrounds the inner selenium atom below the Se_3 faces of the Se_{12} icosahedron. An outer Cu_{12} icosahedron follows, which is positioned above and perpendicular to the Se_{12} shell. The eight copper centers inside the cluster thus form tetrahedral [$CuSe_4$] groups, while the 12 outer copper centers display a [Se_3CuP] environment by binding to the equally oriented P_{12} icosahedron of the PEt_3 ligands. In **15**, the inner selenium atom has 12 copper next-nearest neighbors that form an icosahedron, each edge being positioned below an Se–Se edge of the enclosing Se_{12} polyhedron and perpendicular to it, resulting in an coordinative occupation of one copper atom per Se_3 face. Slightly above the Se_{12} icosahedral surface, an arrangement of 14 Cu–PEt_2Ph units can be described as a sixfold capped cube. The 12 inner copper atoms and the eight metal centers of the capped cube that are all placed near Se_3 faces are approximately tetrahedrally coordinated by either four selenium ligands or three selenium atoms and one phosphorus ligand – characteristic of clusters being larger than "small" ones. The six remaining Cu–PCy_3 fragments are situated above six Se–Se edges, and therefore represent three-coordinated [Se_2CuP] units, common for both "small" and "middle-sized" clusters. When comparing the Cu: Se ratio for **14** and **15**, it is easy to see the fundamental difference that may be responsible for the

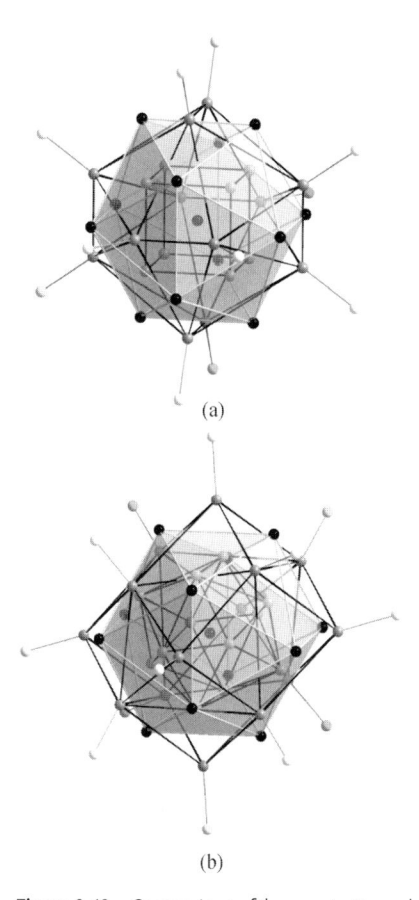

Figure 3.43 Comparison of the penetrative polyhedra in **14** (a) and **15** (b) that form the heavy atom cluster cores. Cu: gray spheres; S: black spheres; P: white spheres. The drawn lines only represent polyhedral edges and do not specify binding interactions.

structural differences. Cluster **14** represents a mixed-valence compound featuring formally six Cu^{II} and 14 Cu^{I} centers to balance the 26$^-$ charge of the Se$_{13}$ substructure. The intermetallic distances in **14** however (Cu(1)aCu(1): 2.821(4), Cu(1)aCu(2): 2.678(4) Å) do not suggest any significant interatomic interactions, despite the partial deficiency of electrons. The only structural hint of different valence situations in **14** or **15** is that, apart from equal average Cu–P bond lengths (2.247 Å in **14** and **15**) and similar average intermetallic distances Cu–Cu (2.707 Å in **14**, 2.682 Å in **15**), significantly shorter average Cu–Se bonds are measured in the mixed-valence compound **14** (2.365 Å) versus **15** (2.457 Å).

The molecular structure of **16** differs from those of **14** and **15**, since a very irregular Se$_{13}$ deltahedron forms the selenium substructure. Twelve of the copper centers and ten Cu–PCy$_3$ fragments are bonded to the Se$_{13}$ polyhedron in such a way that they are again three- or four-coordinate. Unlike the situation in **14** or **15**, it is also possible to observe four copper atoms that show a near-linear coordination to two selenium neighbors. This pattern is typical of "small" Cu–E clusters, and again underlines the partial propagation of these properties in the "middle-sized" compounds.

Similar to the "Cu$_{12}$"a"Cu$_{20}$", "Cu$_{12}$"a"Cu$_{24}$" and "Cu$_{28}$"a"Cu$_{50}$" relationships discussed for copper sulfide clusters, the molecular structures of **14** and **15** have proved to be closely related, even though the numbers and ratios of heavy atoms are very different. In contrast, a similar number of copper and selenium atoms can (but need not necessarily) lead to a topological relationship. Compounds **17–21** [10a, 17, 21] contain 29, 30, 31 and 32 copper atoms, respectively, in addition to 15 or 16 selenium ligands. Nevertheless, three completely different molecular frameworks are observed. The first structural type is adopted by **17** [10a], **18** [10a], and **19** [17]; as an example, the molecular structure of **17** is shown in Figure 3.44.

Although the number of copper atoms is reduced by one in **17** in comparison to **18** or **19**, indicating a mixed valence situation, all clusters show identical Cu–Se–P cores. The additional copper atom present in **18** and **19** is positioned in the cluster center. A reaction employing the same reactants and temperatures as for the synthesis of **19**, but using a 40% higher PtBu$_2$Me: CuOAc ratio, yields compound **20** [17]. Analogous atomic arrangements or corresponding polyhedra are not found in **19** and **20**, even though the cluster core of **20** (Figure 3.45) can equally be described as consisting of five parallel planes of copper and selenium atoms that are surrounded by 12 phosphine ligands.

The SeSiMe$_3$ group in **20**, the occurrence of which is unique for copper chalcogenide clusters, not only influences the molecular structure of **20** but also clarifies the obvious reason for the structural differences for this series. Cluster **20** can be viewed as a "frozen" intermediate on the way to the formation of a larger cluster such as the "Cu$_{70}$" cluster **29** [17] (see below), as a leaving group is still present. The latter acts uniquely as a protecting unit with the phosphine ligands, and therefore further cluster growth is slowed to such an extent that the compound

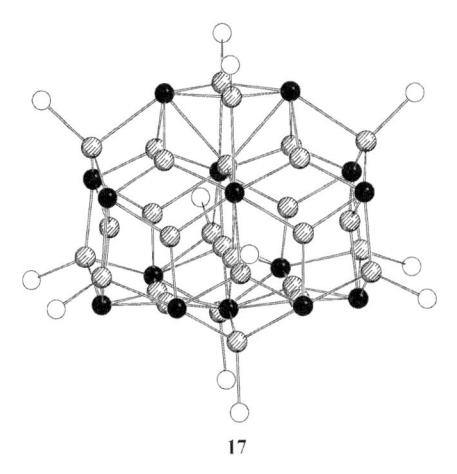

Figure 3.44 Molecular structure of [Cu$_{29}$Se$_{15}$(P*i*Pr)$_{12}$] (**17**) (without organic groups). Cu: shaded spheres; Se: black spheres; P: white spheres.

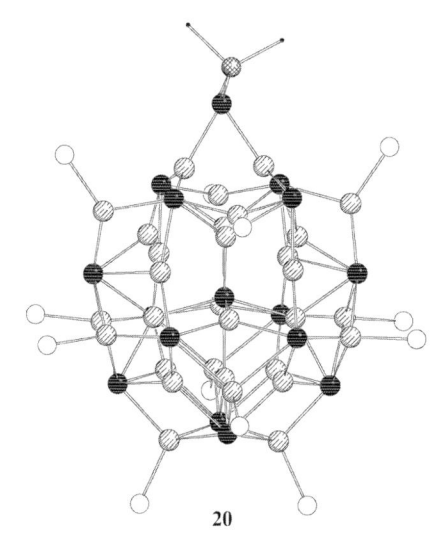

Figure 3.45 Molecular structure of [Cu$_{31}$Se$_{15}$(SeSiMe$_3$)(PtBu$_2$Me)$_{12}$] (**20**) (without PtBu$_2$Me groups). Cu: shaded spheres, Se: black spheres; P: white spheres; Si: criss-cross pattern; C: small black spheres.

crystallizes preferentially. Its formation is thus naturally independent of the cluster growth yielding **19**. A third structural type for similar cluster sizes is found in **21** [21], in which the selenium atoms form a flattened polyhedron consisting of nonbonded Se$_3$ triangles. The copper atoms show quasi-linear, distorted trigonal planar or tetrahedral coordination geometries. These coordination patterns and μ_5 and μ_6 bridging modes of the selenium ligands are common for "middle-sized" Cu–E clusters. Complexes **17** and **21** represent exceptions for their cluster size in another way, as they do not contain copper or selenium atoms inside a spherically tolerable Cu–Se shell. All other "middle-sized" clusters with 20 copper atoms and upwards (beginning with **14**) have either central copper atoms (**18**, **19**) or central selenium ligands (**14–16**, **20**). The larger molecules **22–27** [10a, 14, 17, 18, 21a] also enclose selenium atoms within a Se$_n$ polyhedron ($n = 18$, 22, 24, 26, or 30). The number of inner selenium ligands consequently increases with increasing cluster size from one central atom (**22**) [10a] to two inner ones (**23–25**) [14, 17] to three selenium centers inside the cluster core (**26, 27**) [18, 21]. The respective underlying selenium substructures are summarized and contrasted in Figure 3.46.

Although these six compounds are all classified as being spherical "middle-sized" clusters, and all have central units within a Cu–Se surface, it is possible to distinguish two different shapes. Clusters **22** and **26** (Figure 3.47) adopt a triangular topology, whereas **23–25** and **27** form ellipsoid-type cluster surfaces (Figures 3.48 and 3.49).

The molecular structures of both **22** and **26** display idealized C$_3$ symmetry that is not realized in the crystal (**22**: C$_1$ symmetry, **26**: C$_2$ symmetry). Along the idealized threefold axis, a selenium ligand is found at the top and at the bottom of the

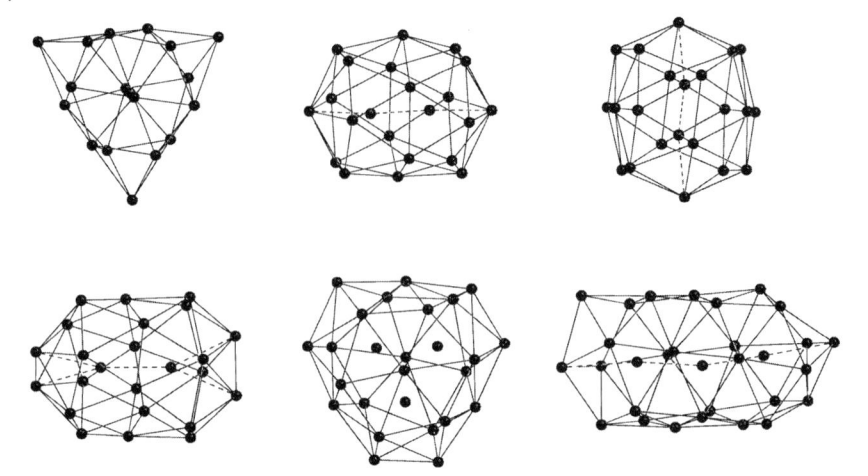

Figure 3.46 Selenium substructures in **22–27** (from top left to bottom right). The drawn Se–Se contacts do not represent bonding interactions, but rather illustrate the geometries of the polyhedra.

flattened trigonal prism, as roughly formed by the cluster atoms. In **22**, a third selenium ligand is positioned between the latter two in the cluster center, whereas in **26**, three inner selenium atoms are grouped around the center of the Cu–Se core and two copper atoms are arranged along the pseudo C_3 axis between the two abovementioned selenium ligands. Beginning with the central unit, the cluster is formed by the usual alternating bonds from copper to selenium atoms. For **22**, [Se$_2$Cu], [Se$_3$Cu], or [Se$_2$CuP] coordination environments around copper centers and μ_3–μ_6 bridging selenium ligands result, with Cu–Se distances of between 2.347(4) and 2.875(5) Å. The larger cluster core of **26** features consequently higher coordination

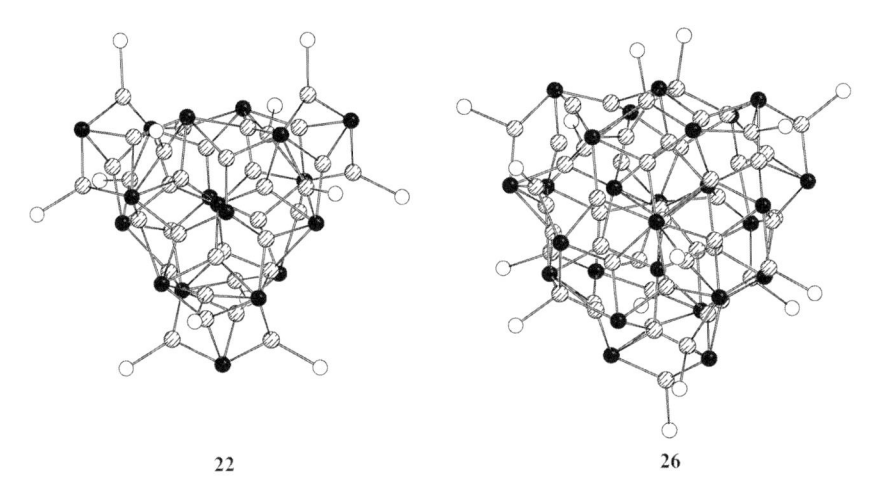

22   **26**

Figure 3.47 Molecular structures of [Cu$_{36}$Se$_{18}$(PtBu$_3$)$_{12}$] (**22**) and [Cu$_{52}$Se$_{26}$(PPh$_3$)$_{16}$] (**26**) (without organic groups). Cu: shaded spheres; Se: black spheres; P: white spheres.

numbers that show near-tetrahedral [Se$_3$CuP] arrangements, in addition to two- and three-coordinated copper atoms as in **22**, and the selenium atoms act as μ_4–μ_8 bridges within a wide Cu–Se range of distances ranging from 2.19 to 2.91 Å. Twelve PtBu$_3$ ligands enclose the copper selenide core of **22**, while 16 PPh$_3$ ligands are bonded to the Cu–Se core of **26**.

The structural isomers **23** and **24** are obtained via completely different reaction conditions, including the use of phosphine ligands, different molar ratios of CuOAc: PR2R′, and the exposure of the reaction mixtures to different final reaction temperatures.

The spatial demands of the phosphine ligands allowed ligand shells of either 18 PEt$_2$Ph or 12 PnButBu$_2$ groups to sufficiently stabilize a [Cu$_{44}$Se$_{22}$] core. In **23**, the phosphorus atoms form a regular deltahedron of alternating, eclipsed rings P$_3$,P$_6$,P$_6$, and P$_3$ along the long axis of the cluster. The 12 phosphorus atoms at the periphery of the cluster core in **24** define a highly distorted icosahedrons (Figure 3.48). As a result of the different number, and thus arrangement, of PR2R′ ligands that expectedly affect the core geometry by "pulling" the attached copper atoms somewhat out of the Cu–Se core, some differences are observed on examining the copper selenide frameworks in detail. This is clearly seen when the respective Se$_{22}$ substructures are compared with each other (Figure 3.46). Apart from different distortions of the described peripheral Se$_{20}$ polyhedron (total lengths or widths of the Se$_{22}$ substructures according to the orientations in Figure 3.49: 11.43 × 8.64 Å in **23**, 8.34 × 10.60 Å in **24**), the relative positions of the inner Se$_2$ "dumbbells" are perpendicular to each other. Thus, a fragment of 12 copper and two selenium atoms can be assigned in both cluster cores that are positioned completely differently, which is initiated by the significantly different situations on the cluster surface mentioned above. The isomerism observed with the "Cu$_{44}$" clusters again shows that certain [Cu$_{2n}$E$_n$]

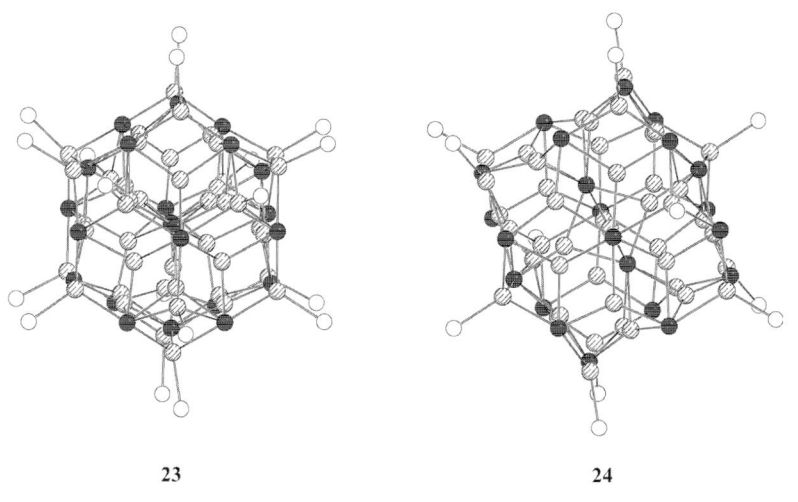

23   **24**

Figure 3.48 Molecular structures of [Cu$_{44}$Se$_{22}$(PEt$_2$Ph)$_{18}$] (**23**) and [Cu$_{44}$Se$_{22}$(PnButBu$_2$)$_{12}$] (**24**) (without organic groups). Cu: shaded spheres; Se: black spheres; P: white spheres.

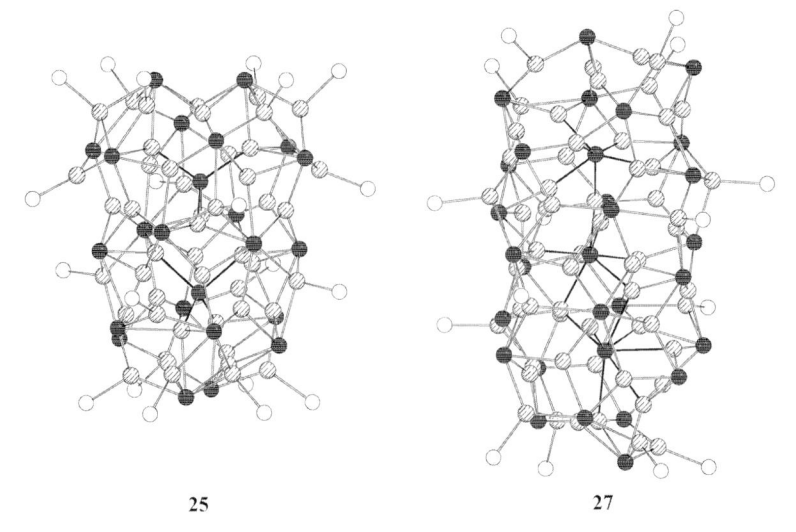

Figure 3.49 Molecular structures of [Cu$_{48}$Se$_{24}$(PMe$_2$Ph)$_{20}$] (**25**) and [Cu$_{59}$Se$_{30}$(PCy$_3$)$_{15}$] (**27**) (without organic groups). Cu: shaded spheres; Se: black spheres; P: white spheres.

compositions represent "isles of stability" – within given limits of structural isomers – that are experimentally obtained whenever a suitable ligand shell can be realized by means of the selected phosphine molecules.

Two further ellipsoid-like arrangements are found for clusters **25** and **27** (Figure 3.49). Apparently, it is not possible to create a phosphine shell for surrounding a [Cu$_{44}$Se$_{22}$] cluster by a suitable number and arrangement of PMe$_2$Ph groups (cone angle: 127). Instead, the smallest compound to be stable and insoluble enough for isolation is the slightly larger cluster **25**, which is however structurally related to the somewhat smaller species **23**. The ellipsoid-like Se$_{24}$ substructure of **25** (Figure 3.46) can be derived from the Se$_{22}$ arrangement in **23** via substitution of two selenium ligands capping the outer Se$_6$ rings for the single Se center in the smaller polyhedron. Again, all coordination features typical of "middle-sized" Cu–Se clusters are found, with four copper atoms that show a near-tetrahedral environment of four selenium ligands. Unlike all smaller clusters that contain selenium atoms inside the cluster core, **25** displays two tetrahedral [SeCu$_4$] units – linked by a trans-edge junction via a compressed Cu$_4$ tetrahedron – with a uniquely small bridging number of four at the two inner selenium ligands (cf. bridging modes of the inner selenium ligands in **14–16**, **20**, **22–24**: μ_5–μ_8, μ_{12}). Together with **14** and **17**, **27** is the third copper selenide cluster that displays a number of metal centers that is too small for an overall $+$I oxidation state at the copper atoms. A single Cu^{2+} center cannot however be assigned, and the "lack" of one electron is likely to be delocalized over the cluster framework. As in **23–25**, the molecule is oblong in shape and represents the largest spherical copper selenide cluster known. It concludes the series of "middle-sized" Cu–Se clusters. The selenium atoms form a highly irregular Se$_{27}$ polyhedron around three inner chalcogen centers

(Figure 3.46) that can alternatively be viewed as a rough arrangement of selenium ligands in three layers. The molecular structure of **27** includes all possible coordination environments around copper atoms, [Se$_2$Cu], [Se$_3$Cu], [Se$_4$Cu], [Se$_2$CuP], and [Se$_3$CuP]. Selenium ligands are observed to act as μ_4, μ_5, μ_6, or μ_7 bridges.

Summarizing the structural properties of the "small" Cu–Se clusters **11–13** and the "middle-sized" clusters **14–27**, it is possible to perceive the spherical shape and the absence of any topological relationship with that of the bulk material Cu$_2$Se as a central feature. Only the tendency toward higher coordination numbers three and four at the copper centers with increasing cluster size approaches the situation in the binary phase. The principal difference between "small" and "middle-sized" clusters is the occurrence of inner atoms that appears firstly with a copper selenide core of 33 atoms (**14**). Typical of the coordination patterns in the "middle-sized" molecules is a mixture of higher and lower coordination numbers or bridging modes, the lower being observed in "small" clusters without variation.

Although only 11 additional copper atoms are observed in compounds **28–31** [17, 10b, 29, 26], these represent the smallest examples of a new type of copper selenide clusters. The so-called "large" Cu–Se clusters are characterized by significantly different cluster shapes, structural principles and relationships to the bulk material when compared to the smaller species. Clusters **28–37** display A–B–A-type packed layers of selenium atoms that form large triangular frameworks. A weak image of the molecular structures of **22** or **26** arises; however, a closer examination clearly enables the latter series to be distinguished from the layer-type structures. Clusters **28–37** are all based on three such selenium layers, of which the middle one possesses the most chalcogen atoms. Figure 3.50 shows the selenium networks of the "large" clusters **28** [29] (as an example of the Cu$_{70}$Se$_{35}$ cores), **32** [21] (Cu$_{72}$Se$_{36}$ core), **33** [31] (Cu$_{74}$Se$_{38}$ core), **34** [23] (as an example of the Cu$_{140}$Se$_{70}$ cores), and **37** [12] (Cu$_{146}$Se$_{73}$ core). It is worth mentioning that this structural pattern seems to be exceptionally advantaged for large copper selenide clusters, as it is observed equally for a size range from "Cu$_{70}$" to "Cu$_{146}$," that is, even with a doubling of the molecular mass.

Clusters **28–31** contain Se$_{10}$, Se$_{15}$ and Se$_{10}$ layers that represent perfect triangles. The next largest selenium substructure should be formed by an arrangement of Se$_{15}$, Se$_{21}$, and Se$_{15}$ triangular layers, each displaying one more row of selenium atoms. A corresponding "[Cu$_{102}$Se$_{51}$]" cluster core is as yet unknown; rather, the formation of smaller clusters in between this range that show selenium layers with the middle one deviating from a perfect triangular shape, is observed. In **32**, one corner of the middle layer is modified with respect to the corresponding corner in the "Cu$_{70}$" clusters in such a way that two selenium atoms replace one. All three corners are similarly changed, to result in the Se$_{10}$Se$_{18}$Se$_{10}$ selenium framework of **33**, and this compound is another cluster featuring fewer copper centers than Cu$_{2n}$ with respect to Se$_n$. The third "idealized" cluster composition is actually realized with **34**, **35** and **36**. The underlying selenium layers contain 21, 28, and 21 atoms. By replacing all corner selenium centers of the middle triangle by two chalcogen atoms, the Se$_{73}$ substructure that is observed in **37** is obtained. For this third possible size, a successive

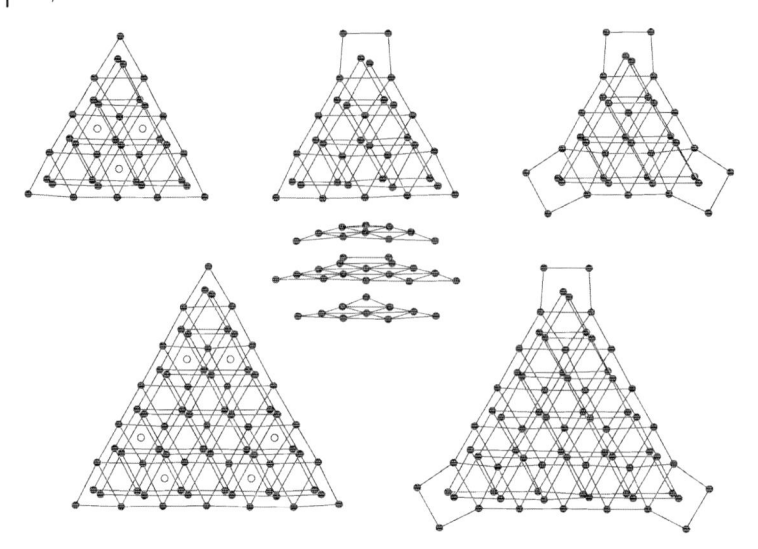

Figure 3.50 Selenium substructures in **28**, **32**, **33**, **34**, and **36** (from top left to bottom right) which contain 35, 36, 38, 70, or 73 selenium atoms. The drawn Se–Se contacts do not represent binding interactions, but only serve to visualize the A–B–A packing of selenium atoms in three layers, which is additionally demonstrated by the supplementary view of the Se_{36} substructure of **32** (top mid). The open circles represent the positions of disordered copper atoms (see text and following figures).

substitution of triangle corners in the middle layer could not as yet be experimentally proven, whereas for the first, smaller size (35 selenium atoms), a modification of one or three corners was observed.

The reason for the missing "Cu_{102}" cluster might be related to the primary steps of the cluster growth mechanism. Selenium networks consisting of three triangle hexagonal layers based on $0.5[n(n + 1)]$ selenium atoms ($n =$ number of straight rows in the triangle network) in A–B–A order offer three possible situations for the center of the triangle: this results in either finds a hole (representing the unoccupied C position), one selenium atom in the middle layer (located on a B position), or two selenium atoms belonging to the two outer layers (located on A positions). Since only such "large" Cu–Se cluster that follow the last two possibilities were observed, the presence of a central Cu_xSe_y axis (SeA–Cu–Cu–SeA in **28–33**, Cu–SeB–Cu in **34–37**) seems to be necessary for the formation of a stable cluster core. Assuming that the cluster growth begins in the center of the molecule, the presupposition of such a $[Cu_2Se]$ or a $[Cu_2Se_2]$ fragment being situated at this place appears feasible for a nonspherical cluster. Therefore, only those molecules that contain either one (middle; cf. **34–37**) or two (outer; cf. **28–33**) selenium layers with a central selenium atom instead of a central triangular hole are formed and experimentally observed. Indeed, neither Se_{15} nor Se_{21} layers that would be exclusively underlying a "Cu_{102}" molecule contain central selenium atoms and should therefore be only found together with other triangles.

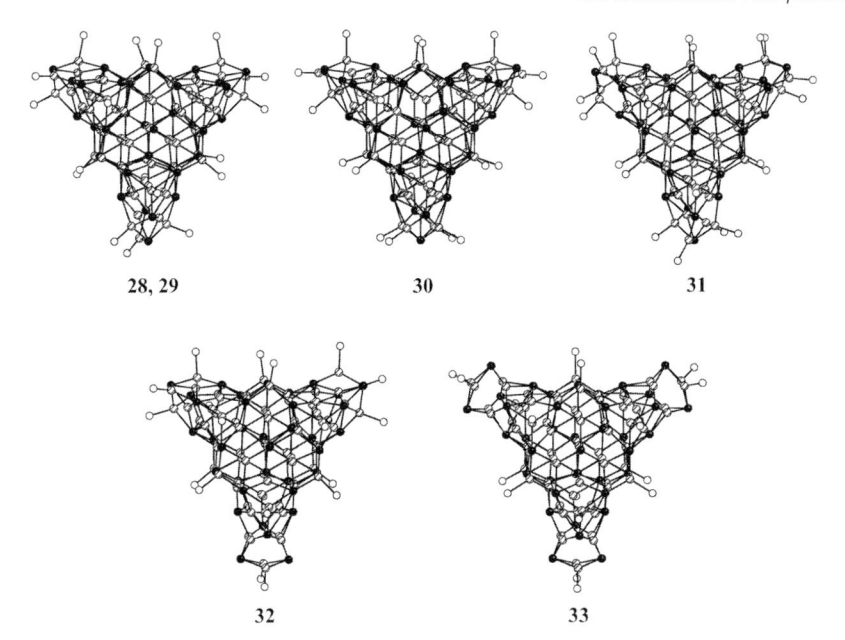

Figure 3.51 Molecular structures of [Cu$_{70}$Se$_{35}$(PR$_2$R′)$_{21}$] (R = iPr, tBu; R′ = Me) (**28, 29**), [Cu$_{70}$Se$_{35}$(PEt$_3$)$_{22}$] (**30**), [Cu$_{70}$Se$_{35}$(PEt$_2$Ph)$_n$] (n = 23, 24) (**31**), [Cu$_{72}$Se$_{36}$(PPh$_3$)$_{20}$] (**32**), and [Cu$_{74}$Se$_{38}$(PCy$_3$)$_{18}$] (**33**) (without organic groups). Cu: shaded spheres; Se: black spheres; P: white spheres. Two copper atoms that are positioned around the central Se–Cu–Cu–Se axis are statistically distributed with a 0.67 occupation over three positions in **28, 29**, or **31**.

In Figure 3.51, the cores of the clusters containing 70 (**28–31**), 72 (**32**) and 74 (**33**) copper atoms are compared.

The "Cu$_{70}$" core is quite tolerant with regard to its ligand shell. An increase in the spatial demand of the PR$_2$R′ molecules is not strictly coupled with a decreased number of terminal ligands. However, there is still a rough correlation, since 21 groups of the bulkier phosphines PMeiPr$_2$ (cone angle: 146) and PtBu$_2$Me (cone angle: 161°) are arranged around the Cu–Se core, whereas 22 and 23 terminal ligands are observed for the smaller phosphines PEt$_3$ or PEt$_2$Ph (cone angles: 132 or 136° respectively). The clusters **28, 29** [17], and **31** [26] differ marginally with respect to their molecular structures. In **28** and **29**, the only observable difference is a slightly less complete ligand shell achieved by the less demanding PMeiPr$_2$ groups. Cluster **31** displays a disorder problem with the PEt$_2$Ph groups, such that it cannot be determined with certainty whether the total number of phosphine ligands sums to 23 or 24. Accordingly, for the formation of the cluster, another two (or three) additional phosphine molecules – versus **28** or **29** – are formally arranged in the triangle corners in such a way that two (or three) corners possess four instead of three Cu–PEt$_2$Ph groups and one corner (or none) retains three of them. An intermediate number of phosphine ligands, 22, is observed with a more distinct structural modification, but it is again the corner region that is affected. In contrast to all of the corner arrangements of copper and selenium atoms in the other "Cu$_{70}$" clusters that cannot be described by means of regular patterns, in **30** [10b] only two corners follow this pattern. The third shows a highly ordered [Cu$_9$Se$_3$(PEt$_3$)$_4$] configuration featuring a centered cube of copper atoms that is capped at three faces by µ$_4$-selenium ligands and carries four phosphine groups at the bottom atoms. This peculiarity of one cluster corner in **30** is one of two reasons for the missing pseudo-threefold molecular symmetry. A C$_3$ symmetry of the [Cu$_{70}$Se$_{35}$] core of **28**, **29** or **31** is additionally approached – apart from one "wrong" Cu–PR2R' position in **28** or **29** – by the statistical disorder of two copper atoms. The latter are distributed over three positions – with an accordant 0.67 occupancy – around the Cu–Se–Se–Cu pseudo-threefold axis running through the center of the molecule. The mentioned disorder situation of PEt$_2$Ph ligands in **31** also leads to a pseudo-C$_3$ axis as far as 24 phosphine groups are assigned to the ligand shell. In contrast, a close look to the cluster center in **30** shows an open copper atom position. The two other metal atom sites are fully occupied, leading to a lower – but crystallographically realized – C$_2$ symmetry.

The molecular structure of **32** can be derived from that of **28** or **29** in a simple manner if, first, all three positions of the copper atoms around the Cu–Se–Se–Cu axis mentioned above are fully occupied, and second, if one [Se$_2$CuPPh$_3$] unit is formally substituted for one corner selenium atom such that a modification of the central selenium layer occurs. The number of phosphine groups that enclose the respective corner is additionally reduced to two, as two of the three copper atoms that bind PR$_2$R' ligands in the analogous corners in **28** or **29** remain "naked" in **32**. Such (formal) alteration of the cluster corner, and full occupation of the Cu$_3$ group around the cluster center, cause the formation of the mixed-valence cluster framework in **33**. With pseudo-C$_{3h}$ symmetry, the point group symmetry of the molecule does not exceed C$_2$ in the crystal because of slight distortions and the positions of the organic groups bonded to phosphorus atoms.

The largest copper selenide clusters isolated to date possess 140 or 146 copper atoms with 70 or 73 selenium ligands, respectively. The molecular structures of **34**, **36** [23], and **37** are shown in Figure 3.52.

The different number of PR$_2$R' groups observed in **34** or **35** on the one hand and **36** on the other hand, affects the corresponding molecular structures in a manner similar to that for the "Cu$_{70}$" species. On re-examining the corners, **34**, which contains 34 terminal PEt$_3$ ligands, reveals an enlarged analog of **30**. As in the smaller compound, two corners are arranged irregularly, whereas the third corner shows the regular, centered cubic fragment. By increasing the number of PEt$_3$ groups to 36 in **36**, all three corners show an ordered arrangement of copper, selenium, and phosphorus atoms. This structural unit, which is only found for the PEt$_3$-ligated "large" Cu–Se clusters, underlines the structural relationship of **30**, **34**, and **36** and demonstrates another steric influence of certain types of PR2R' molecules. All other phosphine ligands that surround "large" copper selenide clusters feature larger cone angles, which might be the reason for a forced modification to a more staggered arrangement if four corner Cu–PR$_2$R' units were present. Consequently, in **35**, all three corners show arrangements similar to that in the smaller analog **31**. Disordered copper atoms that increase the molecular

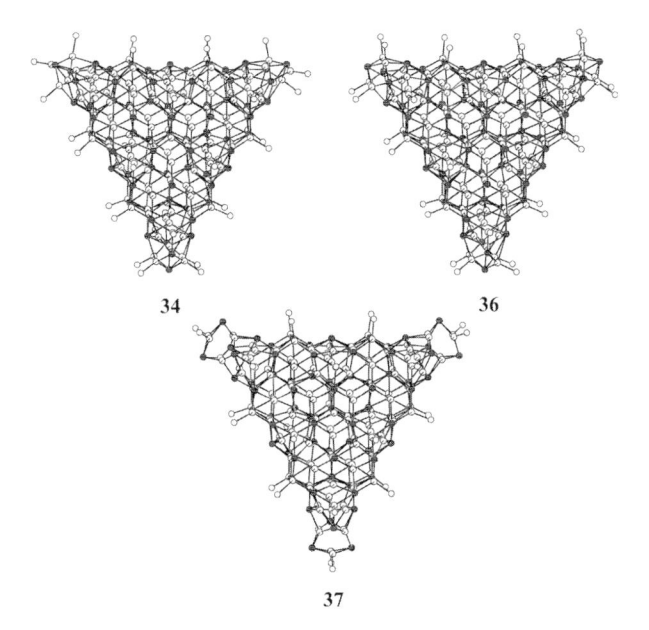

Figure 3.52 Molecular structures of [$Cu_{140}Se_{70}(PEt_3)_{34}$] (**34**), [$Cu_{140}Se_{70}(PEt_3)_{36}$] (**36**), and [$Cu_{146}Se_{73}(PPh_3)_{30}$] (**37**) (without organic groups). Cu: shaded spheres; Se: black spheres; P: white spheres. In **34** and **36**, the two central copper atoms are each half-occupied to result in one atom in the sum, being statistically distributed over two sites. Six further copper atoms are assigned a 0.67 occupation resulting in four instead of six metal atoms in the sum.

pseudo-symmetry are again present in the "Cu_{140}" clusters. Additionally, there is a twofold disorder of one central copper atom on two positions to result in an actual "$Cu_{0.5}$"–Se–"$Cu_{0.5}$" axis. Such metal atoms are again affected by a two-thirds occupancy factor, and lie in the plane spanned by the middle selenium layer. In addition, the six next-but-one metal atoms viewed from the cluster center are disordered statistically to amount to four atoms in the sum.

Three formal replacements [$Se_2Cu(PR_2R'_2)$] of three corner selenium atoms removal of six PR_2R' molecules in **36**, and full occupation of all split positions present in the "Cu_{140}" clusters yield the composition and D3d symmetric molecular structure of **37**. Thus, the structural transitions of the "Cu_{70}" compounds to slightly larger species, and the transition occurring with the largest known copper selenide clusters, are related.

All of the "large" Cu–Se clusters show an A–B–A packing of selenium hexagonal layers, the metal centers occupying either trigonal or tetrahedral holes or being situated outside the selenium layers in a likewise trigonal or tetrahedral coordination environment. Thus, the layer-type clusters clearly show a closer relationship to solid-state structures than do the spherical ones. For the $Cu_{2x}Se$ phases, however, only cubic and not hexagonal modifications are known. Interestingly, for the Cu_2S (chalcosine) high-temperature phase [38], a hexagonal lattice with mainly trigonally surrounded copper atoms in addition to tetrahedrally and linearly coordinated atoms does exist. In contrast, the cubic $Cu_{2x}Se$ crystal structures [39] can be derived from an anti-fluorite lattice [51] by positioning the selenium atoms on defined lattice places. In this case, six to eight copper atoms per unit cell – according to the phase composition – are localized in trigonal, tetrahedral or octahedral holes. In α-Cu_2Se, the metal centers are placed in an ordered manner within the selenium lattice, whereas in β-Cu_2Se they are statistically distributed. Even though the reported structural properties of copper selenide binary phases are not uniform, there is a clear agreement regarding the dominating coordination number of four for the copper atoms and the absence of any linearly coordinated metal centers. Because of the hexagonal modification and the preferred coordination number of three for the copper atoms, the layer-type Cu–Se clusters approach the bulk material situation only in principle.

For the copper selenide clusters presented herein, it can be perceived that it is not possible to assign a special range of cluster sizes to one given phosphine ligand, being followed by another PR_2R' type to protect the next larger clusters, and so on. Rather, it can be recognized that even very different types of phosphines – perhaps varying in size – are in the position to provoke the formation and crystallization of similar, isomeric, or even identical Cu–Se cores if the ligand shells comply with certain requirements of regular geometry, and thus protective properties. In contrast, an examination of Scheme 3.4 rationalizes the formation of several clusters that may vary significantly in size, in the presence of the same phosphine, where different reaction conditions permit various suitable ligand shells to be realized. In order to illustrate the dependence of the size of the cluster core on the use of different copper salts, stoichiometries, solvents, and reaction temperatures, the cluster formation reactions in the presence of PEt_2Ph as phosphine ligand are shown in Scheme 3.6, with a thorough description of the respective conditions.

As for the copper sulfide clusters, comprehensive theoretical investigations within the MP2 approximation (program system TURBOMOLE) were carried out for the $[Cu_{2n}Se_n]$ clusters [34]. Again, stabilization energies per monomeric unit for a "naked" cluster growth from $n=1$ to $n=6$ and $n=10$ were calculated, as were the phosphine-binding energies for the mono-, di-, tri-, tetra-, and hexameric species.

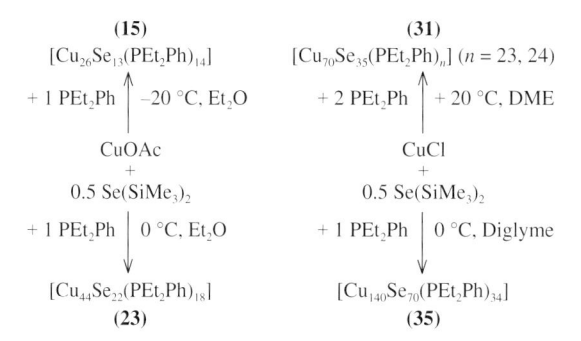

(15) $[Cu_{26}Se_{13}(PEt_2Ph)_{14}]$
↑ + 1 PEt_2Ph | −20 °C, Et_2O

CuOAc
+
0.5 $Se(SiMe_3)_2$

+ 1 PEt_2Ph | 0 °C, Et_2O ↓

$[Cu_{44}Se_{22}(PEt_2Ph)_{18}]$
(23)

(31) $[Cu_{70}Se_{35}(PEt_2Ph)_n]$ ($n = 23, 24$)
↑ + 2 PEt_2Ph | ↑ + 20 °C, DME

CuCl
+
0.5 $Se(SiMe_3)_2$

+ 1 PEt_2Ph | 0 °C, Diglyme ↓

$[Cu_{140}Se_{70}(PEt_2Ph)_{34}]$
(35)

Scheme 3.6 Synthesis of various copper selenide clusters in the presence of PEt_2Ph under different reaction conditions.

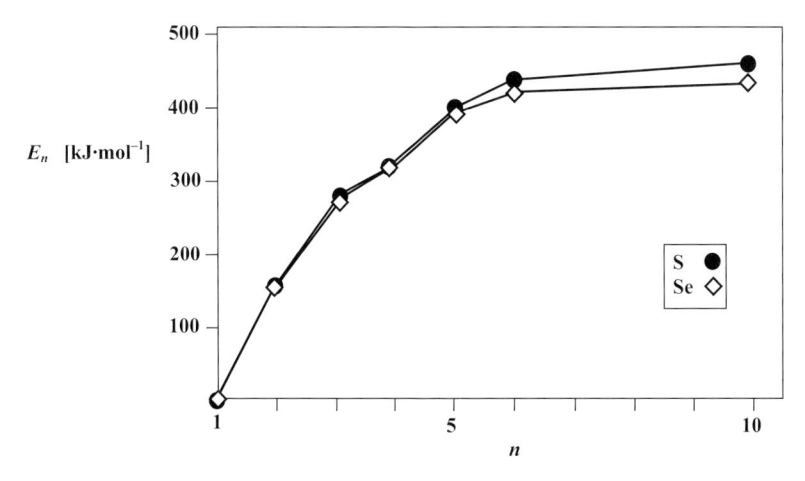

Figure 3.53 Course of the stabilization energies per monomer unit E_n in "naked" clusters $[Cu_{2n}E_n]$ (E = S, Se; n = 1–4, 6, 10). The given values consider the total energies of the most stable structural isomers computed on MP2 level.

The results are given in Figures 3.53 and 3.54, respectively. The corresponding graphs for the Cu–S clusters (see Figures 3.40 and 3.41) are reproduced once more for comparison.

The course of both energies is equivalent for the quantum chemically investigated sulfur- or selenium-bridged copper clusters. Thus, the statement describing the kinetic stability of the species and the very low Cu–P binding energy is the same as that for the Cu–S compounds. Additionally, a comparison of the energetic properties of sulfur- versus selenium-bridged copper clusters produces two results. First, with

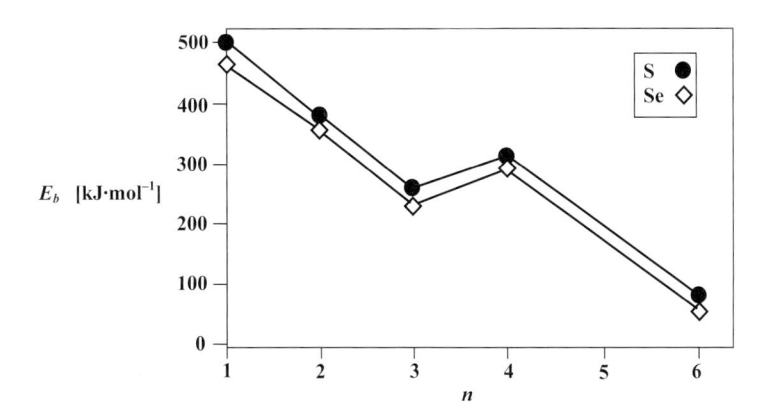

Figure 3.54 Course of the Cu–P binding energies per Cu–P bond E_n in phosphine-ligated clusters $[Cu_{2n}E_n(PH_3)_m]$ (E = S, Se; n = 1–4, 6; m = 2, 4, 6, 8). The given values consider the total energies of the most stable structural isomers computed on MP2 level with respect to the total energy of PH_3 calculated by the same method.

near-identical courses of the stabilization energy (maximum differences: G5 kJ mol^{-1}, <2%) up to $n=4$, there is a stabilization gain for the Cu–S species from $n=6$ onwards. This correlates with a larger absolute value for the formation enthalpy of solid Cu$_2$S (ΔH_f: 83.4 kJ mol^{-1}) [49] when compared to solid Cu$_2$Se (ΔH_f: 59.3 kJ mol^{-1}) [50]. The stabilization energies per monomeric unit for the calculated, rather small, clusters however do not exceed differences of 5%. Thus, it is comprehensible why the laser ablation spectra of copper chalcogenides does not vary much for the different chalcogen types. Second, the phosphine molecules are generally bonded by 4–7 kJ mol^{-1} (4–9%) higher values of the Cu–P binding energy for the sulfur-bridged compounds. This correlates with the discrepancy between the experiments reported by Dance et al. [32], which showed a similar behavior for different chalcogens – in agreement with the calculations of "naked" species – and the experimental observations of significant differences in the syntheses and structures of phosphine-clad sulfur or selenium-bridged copper clusters, respectively. The larger Cu–P binding energies computed for Cu–S clusters, when compared to the Cu–P binding energies of their Cu–Se analogs, help to explain the different synthetic product spectra. Assuming that the activation energy E_a for Cu–P bond cleavage increases by a value comparable to the Cu–P binding energy itself when going from the Cu–Se to the Cu–S systems, a decrease can be calculated in the respective reaction rate constant k by means of the Arrhenius equation [Eq. (3.4)] [52].

$$k = A \cdot e^{-E_a/(R \cdot T)} \tag{3.4}$$

For a typical cluster formation temperature T (e.g., 250 K), an increase in the activation energy E_a of about 5 kJ mol^1 causes a reduction in the rate constant k by one order of magnitude. Thus, the cluster growth that is initiated by the descent of a primary terminal phosphine ligand shell will occur less intensely for sulfur-bridged clusters. Consequently, unlike the selenium-bridged copper clusters, the Cu–S species show a much lower tendency to exceed the aggregation beyond "Cu$_{12}$."

The use of alkylated or arylated derivatives RSeSiMe$_3$ (R = organic group) provokes syntheses according to another mechanism that leads to the formation of selenido-selenolato-bridged copper clusters. The given synthesis route (Schemes 3.2 and 3.5) in no case produces selenolato-free species. This differs from similar reactions with tellurolato reactants RTeSiMe$_3$ (see below). The experimentally isolated and characterized compounds that are listed in Scheme 3.5 display significantly different structures with regard to the purely E^{2-}-bridged molecules (E = S, Se). This applies even for reactions in the presence of tertiary phosphines PPh$_3$ or PiPr$_3$ that were likewise used for the syntheses of copper selenide clusters. The expanded ligand properties of SeR, which can act as both a bridging and a terminal ligand, leads to the crystallization of low nuclearity complexes with two, three, six, or nine metal centers respectively (**38–42**) [20, 22, 30] if Se–C bond cleavage is suppressed. Larger cluster compounds can be isolated whenever Se^{2-} particles are produced during the reaction to make inner selenium bridges available. However, these molecules containing selenide and selenolato ligands show structural patterns that are also common for Cu–Se clusters as far as the

3.2 Semiconductor Nanoparticles | 155

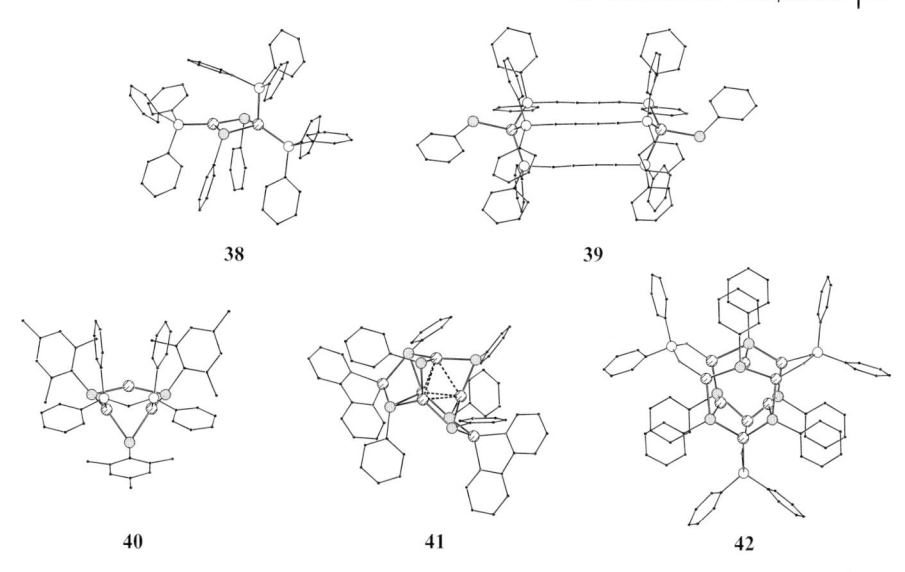

Figure 3.55 Molecular structures of [Cu$_2$(SePh)$_2$(PPh$_3$)$_3$] (**38**), [Cu$_2$(SePh)$_2$(dppbd)$_3$] (**39**), [Cu$_3$(SeMes)$_3$(dppm)] (**40**), [Cu$_6$(SePh)$_6$(bipy)$_2$] (**41**), and [Cu$_9$(SePh)$_6$(O$_2$PPh$_2$)$_3$] (**42**) (without hydrogen atoms). Cu: shaded spheres; Se: gray spheres; P: white spheres; C: small black spheres.

inner core is concerned. In Figure 3.55, the compounds illustrated (**38–42**) contain only SeR$^-$ and no Se^{2-} ligands.

Binuclear complexes **38** [30] and **39** [20] formally show analogous compositions, but differ structurally according to the restrictions of the different types of ligating component present. These restrictions have structural relevance for all Se^{2-}–SeR-bridged copper clusters as follows. First, bidentate phosphines dppR (R = organic spacer) bridge two copper centers at the periphery of the polynuclear complex or cluster molecule, and therefore act simultaneously as protecting groups. Second, selenolato ligands SeR$^-$ usually behave like dppR molecules, although they are able to bridge two or three metal centers. Additionally, they can be terminally bonded, as found in **39** when the Cu–Cu distance (9.01 Å) is too large for a μ$_2$-SeR bridge. Third, tertiary phosphines PR$_2$R' act exclusively as terminal ligands, and thus the copper atoms in **38** are held by SePh$^-$ fragments at a distance of 2.846(1) Å. The molecular structure of **40** [22] also follows these patterns, again with the final purpose of optimal protection of the heavy-atom framework by organic groups. Compounds **41** [30] and **42** [30] are the only examples that could be isolated by employing the bidentate amine ligand bipyridyl (**41**) or diphenylphosphino acid Ph$_2$PO$_2$H(**42**). Cluster **41** seems to represent a "snapshot" molecule of cluster growth, since two [Cu(bipy)]$^+$ fragments are unsymmetrically attached to a quite regular cluster center. The latter is formed from a distorted tetrahedron Cu$_4$ (Cu–Cu: 2.650(4)–2.824(4) Å), the six edges of which are μ$_2$-bridged by SePh$^-$ ligands.

Four of the phenyl selenolato groups, however, act as μ$_3$ bridges, as each pair binds to the additional Cu(bipy)$^+$ units. Much higher molecular and crystallographic symmetry is found for **42**. Two six-membered Cu$_3$(SePh)$_3$ rings are conjunct by three "naked" copper centers, leading to a μ_3 bridging function of the SePh$^-$ groups. Additionally, the rings are held by three Ph$_2$PO$_2^-$ ligands that bind to the six copper atoms. Twelve phenyl groups envelop the uncharged cluster, and therefore enable the termination of atomic aggregation and crystallization in tetrahydrofuran (THF).

As for the copper chalcogenide clusters **1–37**, the nature of the phosphine ligands plays a key role in the size and structures of the observed product molecules – even though they are no longer the only protective groups at the cluster peripheries. Success in the synthesis of copper selenide–selenolate clusters was achieved by the use of bidentate phosphines. The variation of organic spacers between the two ligating [Ph$_2$P] fragments replaces the former variation of organic substituents R of PR$_2$R' with the selenolato-free clusters. Again, one given dppR ligand does not necessarily restrict the synthesis to the formation of one single cluster product; rather, the outer reaction conditions select between several potential compounds that can all be conveniently protected by the phosphine employed. Thus, with ethane, propane, butane, acetylene, or benzene spacers, clusters are obtained with 16, 25, 36, 38, or 58 copper centers (**43** [20], **44** [22], **46–50** [22, 30], **52** [22]). The only tertiary phosphine PR$_3$ that was successfully used for the formation of higher nuclearity Se^2–SePh-bridged copper clusters was PiPr$_3$. The synthesis of molecules featuring 32, 50, or even 73 metal centers (**45**, **51**, **53**) [23] was observed by variation of the reaction conditions with this phosphine. Figure 3.56 shows the molecular structures

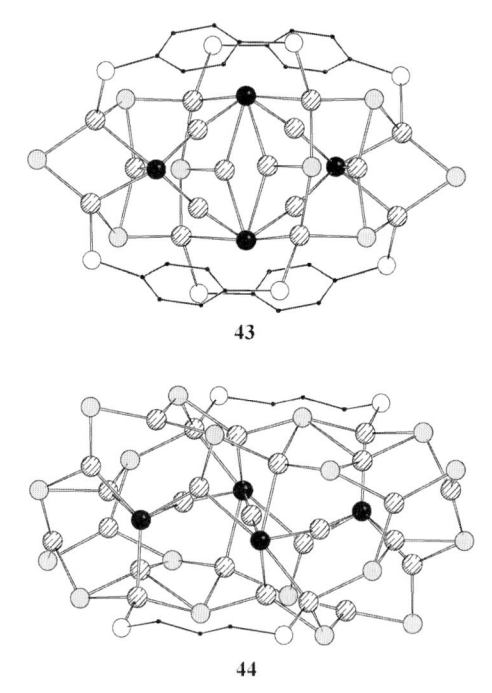

Figure 3.56 Molecular structures of [Cu$_{16}$Se$_4$(SePh)$_8$(dppbe)$_4$] (**43**) and [Cu(dppp)$_2$][Cu$_{25}$Se$_4$(SePh)$_{18}$(dppp)$_2$] (**44**) (without terminal phenyl groups). Cu: shaded spheres; Se: black spheres; SeR: gray spheres; P: white spheres; C: small black spheres.

of **43** and the anion in **44**, the only ionic copper selenide cluster compound isolated to date via this synthetic method.

The clusters **43** and **44** are both oblong in shape. The selenium substructure of **43** can, however, be viewed as a beginning of hexagonal packing of selenium atoms. The middle "layer" consists of six atoms arranged in zigzag fashion, the two enclosing "layers" each containing three selenium centers occupying the quasi-A position with respect to the quasi-B position of the atoms in between. Four of the chalcogen (Se^{2-}) bridges act as μ_4 or μ_6 bridging ligands. The $SePh^-$ ligands are bonded to two or three copper neighbors each, and at the same time contribute to the protection of the Cu–Se core by means of their organic substituents. Six copper atoms are surrounded linearly by two selenium ligands: two of these show an almost trigonal planar [Se_3Cu] coordination, whereas the remaining eight metal centers occupy the centers of slightly distorted [Se_3P] tetrahedra. In contrast to **43**, the chalcogen framework of **44** rather represents an ellipsoid-type deltahedron of four Se^{2-} ligands and 18 selenium atoms of the $SePh^-$ groups. Again, the bridging grade of the chalcogen centers is higher for the "naked" ligands (μ_5 or μ_7) than for the $SePh^-$ fragments (μ_2–μ_4 with regards to the Cu–Se bonds). There are 18 copper atoms positioned at the cluster periphery, and these belong to almost planar [Se_3Cu] or distorted tetrahedral [Se_3CuP] units. The observation that no linear [Se_2Cu] arrangements are present in **44** underlines the trend to higher coordination numbers at the metal center with increasing molecular size, as reported for the copper selenide clusters without selenolato ligands. The cluster corpus is large enough to accommodate some inner copper atoms that are either two-, three-, or – rarely – five-coordinated by selenium neighbors. Comparing the compositions of **43** and the anion in **44**, it can be perceived that there is an increasing preference for ligation by $SePh^-$ ligands rather than by phosphine donors. This may be explained by the ambivalent character of the $SePh^-$ groups, which first form fairly stable Cu–Se cluster bonds, and second, carry protective organic substituents. Thus, the ratio of "terminally ligating selenium atoms" to phosphorus atoms increases from 1 : 1 in **43** to 4.5 : 1 in **44**.

The molecular structures of the cluster compounds **45** and **46** are given in Figure 3.57, while Figure 3.58 shows the packing of selenium atoms in **45** and the selenium substructure in **46**.

As for the comparison of **43** and **44**, it is possible to discuss a smaller cluster, the structure of which is based on a regular packing of selenium atoms in layers (**45**) and a larger, rather spherically shaped compound, that features a polyhedral selenium framework (**46**). The selenium layers in the Se*n*Bu-bridged species **45** are packed in an A–B–C fashion similar to the Cu_2Se solid-state structure, which is so far unparalleled for all known copper chalcogenide clusters. This similarity between the atomic arrangement of a cluster molecule and the nonmolecular Cu_2Se suggests that a fragment of a bulk structure may indeed be stabilized to nanosize by coating the surface with suitable groups. This tendency is also observed in three silver selenide–selenolate clusters, [$Ag_{112}Se_{32}(SenBu)_{48}(PtBu_3)_{12}$], [$Ag_{114}Se_{34}(SenBu)_{46}(PtBu_3)_{14}$], and [$Ag_{172}Se_{40}(SenBu)_{92}(dppp)_4$] (see Section 3.1.3.3) [53]. Cluster **45** is the only copper selenide–selenolate compound that contains Se*n*Bu$^-$ groups like these silver clusters. It is not yet proven, but it is at least possible, that the packing of

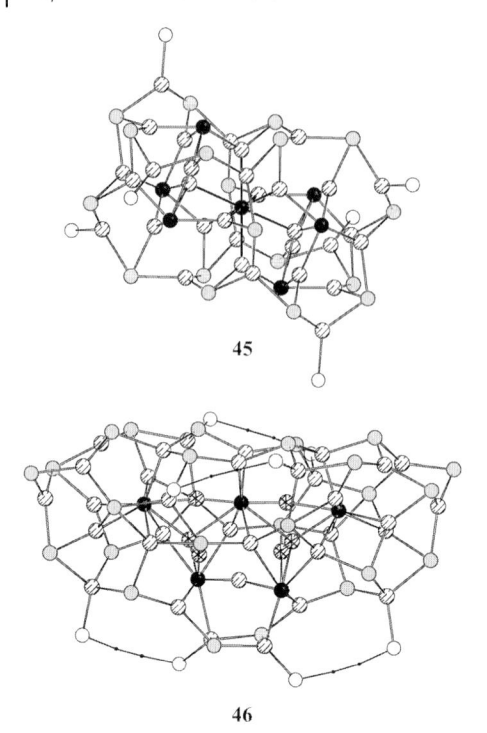

Figure 3.57 Molecular structures of [Cu$_{32}$Se$_7$(SenBu)$_{18}$(PiPr$_3$)$_6$] (**45**) and [Cu$_{36}$Se$_5$(SePh)$_{26}$(dppa)$_4$] (**46**) (without n-butyl, i-propyl or phenyl groups). Cu: shaded spheres; Se: black spheres; SeR: gray spheres; P: white spheres; C: small black spheres. Three copper atoms (marked with a cross) in the cluster center of **46** are distributed over six sites with an occupation of 0.5.

the selenium atoms may be influenced by the nature of the organic group attached to the selenolato ligand. The chalcogen layers in **45** are centered by a large planar rhombus of nine selenium atoms; five of these are "naked" Se^{2-} ligands in the cluster network, while the four vertex atoms are SenBu$^-$ groups. Above and below this middle layer can be found two truncated rhombuses consisting of only eight selenium atoms, with the "central" atoms not being bonded to organic substituents. However, the size dependency of structural properties seems to be different from that of the selenolato-free copper clusters: layer-type structures are observed even for smaller molecules such as **43** or **45**, in addition to larger, spherical structures. A reversion to lower average coordination numbers in larger clusters may also appear, as observed in **45**. Of the 32 copper atoms in **45**, twelve have a trigonal planar coordination geometry and two show an ideal tetrahedral environment; the remaining 18 metal atoms are shifted away from the centers of tetrahedral holes. Hence, each of these displays only three normal Cu–Se bonds, with the fourth Cu–Se distance being relatively long (3.18–3.35 Å). The tendency to occupy tetrahedral holes – even without perfect realization – again correlates with the structural properties of

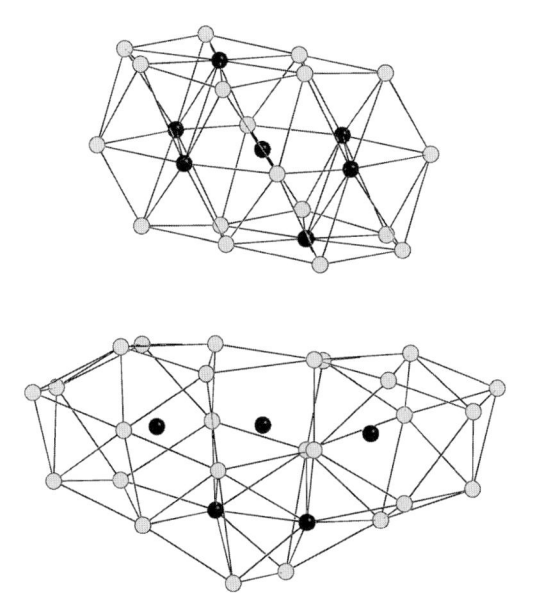

Figure 3.58 Selenium substructures in **45** (top) and **46** (bottom). Se: black spheres; SeR: gray spheres. The drawn Se–Se contacts do not represent bonding interactions, but only serve to visualize the A–B–C packing of selenium atoms in three layers in **45** or explain the geometry of the polyhedron in **46**.

bulk Cu_2Se. In **46**, five inner Se^{2-} ligands are enclosed by a shell of 26 $SePh^-$ fragments. The bridging grades of the Se^{2-} ligands amount to μ_7 or μ_8, whereas the selenium atoms of the $SePh^-$ groups act only as μ_2–μ_4 bridges between next copper atoms. The copper centers feature coordination numbers of two, three, or four, within $[Se_2Cu]$, $[Se_3Cu]$, $[Se_4Cu]$, or $[Se_3CuP]$ units, respectively. Depending on their position inside the cluster or at the surface, the selenium neighbors of the metal atoms are either exclusively Se^{2-} or a combination of Se^{2-} and $SePh^-$ ligands. As for **43** and **44**, the number of terminal selenolato groups is a multiple of the number of ligating phosphorus atoms (SeR: P ratios are 3 : 1 in **45**, and 3.25 : 1 in **46**).

Four closely related Cu–Se–SePh clusters are formed in the presence of either dppe, dppp, or dppb (**47–50**). The syntheses of isomeric, mixed valence compounds **47** and **48** containing 36 copper centers were carried out in THF; the isomeric "Cu_{38}" clusters **49** and **50** that formally display exclusively Cu^+ crystallize from toluene. As well as being performed in different solvents, the reactions producing the dppp-ligated compounds **48** or **50** differed in the CuCl: dppp ratios (3.1 : 1 for **48**, 5 : 1 for **50**). The molecular structures of **47** and **49** are given as examples in Figure 3.59.

Cluster **47** crystallizes in the trigonal space group R_3, whereas **48–50** crystallize in triclinic space (P1). Despite the different crystallographic symmetry, all four compounds display very similar structural properties. A closer look at the atomic arrangements helps to put the preference of this prototypic cluster framework down to highly symmetric substructures. A central selenium atom positioned on the

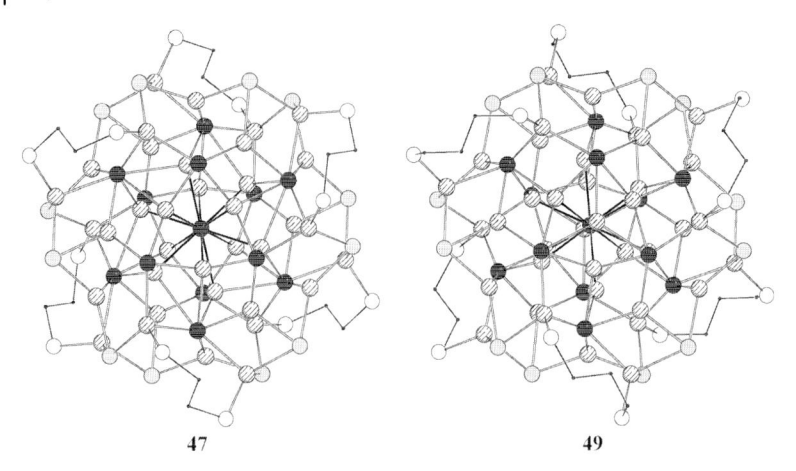

Figure 3.59 Molecular structures of $[Cu_{36}Se_{13}(SePh)_{12}(dppe)_6]$ (**47**) and $[Cu_{38}Se_{13}(SePh)_{12}(dppp)_6]$ (**49**) (without phenyl groups). Cu: shaded spheres; Se: black spheres; SeR: gray spheres; P: white spheres; C: small black spheres.

inversion centers of the clusters is surrounded by a distorted icosahedron of 12 copper atoms. The latter is itself enclosed by another distorted Se_{12} icosahedron. An alternate packing of Cu_{12} and Se_{12} icosahedra has already been discussed for the Cu–Se clusters **14** and **15** (see Figures 3.42 and 3.43). In these compounds, one polyhedron is positioned with its edges perpendicular to the edges of the polyhedron below. In contrast to this pattern, the Se_{12} and Cu_{12} icosahedra in **47–50** are arranged such that one of the idealized fivefold axes of one polyhedron parallels one of the idealized threefold axes of the other, and *vice versa*. The central $[Se_{13}Cu_{12}]$ unit of **47** is shown as an example in Figure 3.60.

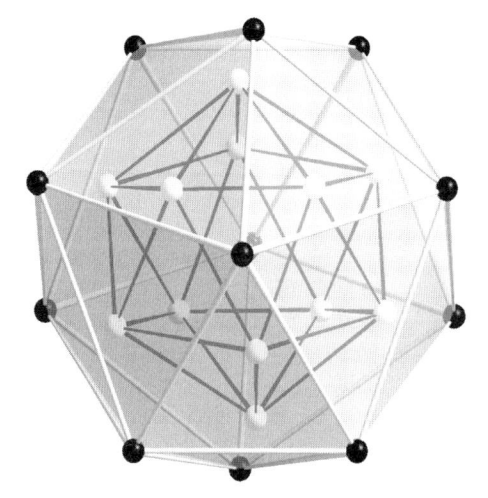

Figure 3.60 Central $[Se_{13}Cu_{12}]$ unit in **47**. Cu: white spheres; Se: black spheres. The drawn lines only represent polyhedral edges and do not specify bonding interactions.

In this way, six edges and six faces of the Cu$_{12}$ icosahedron are bridged by the selenide ligands that form the Se$_{12}$ polyhedron. The cluster core expands out to the periphery by coordination of the Se$_{12}$ substructure to another 24 copper atoms, 12 of which are bonded to the six dppR ligands. These copper atoms are bridged by 12 μ$_3$-SePh$^-$ groups that are positioned at the cluster surface with the phosphine molecules. Except for the μ$_{12}$ bridging central selenium atom, the selenide ligands act as μ$_6$ (each six in **47** and **48**, all in **49** and **50**) or μ$_5$ bridges (each six in **47** and **48**) between copper centers. The different numbers of copper neighbors per selenium atom in **47** and **48** on the one hand and **49** and **50** on the other hand is caused by two copper atoms being formally "missing" in the molecular structures of **47** or **48**, resulting in the mixed-valence situation. The metal centers are distorted trigonal planar or tetrahedrally surrounded. All Cu–Se distances in the mixed-valence clusters lie within the range of the Cu–Se bond lengths in all four clusters (2.301(4)–2.888(3) Å), and thus give no evidence of any localization of the Cu^{2+} centers. However, the range of the Cu–Cu contacts in **47** or **48** (2.511(3)–2.943(4) Å) is somewhat narrower than that found in **49** or **50** (2.305(4)–3.040(3) Å). The spherical, somewhat oblate molecules show maximum diameters of 23.9–25.7 Å.

The use of different selenolato sources and different optimized reaction conditions leads to three various copper selenide–selenolate clusters with terminal PiPr$_3$ ligands, **45** (Figure 3.57) **51** (Figure 3.61), and **53** (Figure 3.64). The latter is synthesized in an uncommon manner by the use of both PhSeSiMe$_3$ and Se(SiMe$_3$)$_2$ in the same reaction, and is discussed below. The syntheses of the PiPr$_3$-clad clusters, specifying the reaction conditions in detail, are summarized in Scheme 3.7, while Figure 3.61 shows the molecular structure of compound **51**.

On examining Figure 3.61, it can be surmised that a "normal" cluster growth occurred in the presence of a tertiary phosphine – indeed under Se–C bond cleavage conditions – to result in the formation of a copper selenide cluster which is represented by the [Cu$_{40}$Se$_{20}$] central unit of **51**. Only the sidewise attachment of

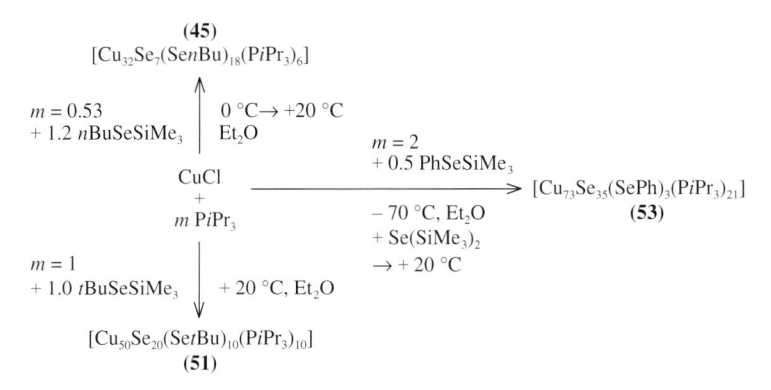

(45)
[Cu$_{32}$Se$_7$(SenBu)$_{18}$(PiPr$_3$)$_6$]

$m = 0.53$
$+ 1.2\ n$BuSeSiMe$_3$

0 °C → +20 °C
Et$_2$O

CuCl
+
m PiPr$_3$

$m = 2$
$+ 0.5$ PhSeSiMe$_3$

-70 °C, Et$_2$O
$+$ Se(SiMe$_3$)$_2$
$\rightarrow +20$ °C

[Cu$_{73}$Se$_{35}$(SePh)$_3$(PiPr$_3$)$_{21}$]
(53)

$m = 1$
$+ 1.0\ t$BuSeSiMe$_3$

$+ 20$ °C, Et$_2$O

[Cu$_{50}$Se$_{20}$(SetBu)$_{10}$(PiPr$_3$)$_{10}$]
(51)

Scheme 3.7 Synthesis of various selenido–selenolato-bridged copper clusters in the presence of PiPr$_3$ under different reaction conditions.

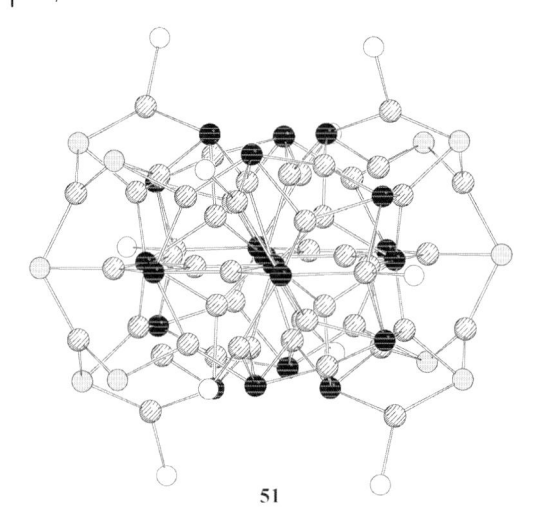

Figure 3.61 Molecular structure of [Cu$_{50}$Se$_{20}$(SetBu)$_{10}$(PiPr$_3$)$_{10}$] (**51**) (without organic groups). Cu: shaded spheres; Se: black spheres; SeR: gray spheres; P: white spheres.

two symmetry-equivalent [Cu$_5$(SetBu)$_5$] fragments reflects the presence of the selenolato reactant. This is in contrast to the related PiPr$_3$-ligated structure of **45**, which shows only seven Se^{2-} ligands with 18 SetBu$^-$ groups. A definite correlation between the significantly different course of the cluster formation and the respective reaction conditions cannot be evaluated based on the low statistic, although it is very likely that the Cu: P ratio of the reactants and the temperature play an important role. The copper atoms in **51** bind to Se^{2-}, SePh, and PiPr$_3$ ligands in different ratios to achieve coordination numbers of two (with a bend coordination geometry indicating a weak third interaction), three, and four. The selenium atoms that carry t-butyl groups act as μ_3 bridges between copper atoms, whereas the Se^{2-} ligands bridge between five, six, or seven copper neighbors. The molecules are regularly protected by PiPr$_3$ ligands and the organic substituents of the SetBu$^-$ groups.

The molecular structure of another dppR-ligated copper cluster (R = a = acetylene) with both selenido and selenolato ligands, **52**, again displays high symmetry, and here the molecular symmetry is underlined by very high crystallographic symmetry. The compound crystallizes in the cubic space group F$\bar{4}$3c, and is thus the second copper chalcogenide cluster known besides **14** (space group Fm3) with a cubic crystal lattice. The molecular structure of **52** is illustrated in Figure 3.62.

As observed for **47–50**, the cluster core is oblate in shape although, in contrast to **47–50**, it is not based on icosahedral fragments. Instead, a central tetrahedron Cu$_4$ can be perceived which shares its corners with four further Cu$_4$ units that themselves share the trans face with each octahedral Cu$_6$. The remaining 30 copper atoms are situated between these polyhedral, or are connected at the outside of the polyhedral arrangement. The copper substructure of **52** is shown in Figure 3.63.

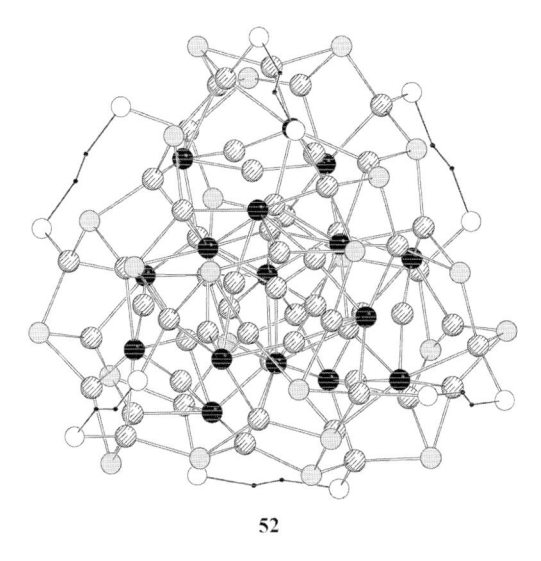

Figure 3.62 Molecular structure of [Cu$_{58}$Se$_{16}$(SePh)$_{24}$(dppa)$_6$] (**52**) (without phenyl groups). Cu: shaded spheres; Se: black spheres; SeR: gray spheres; P: white spheres; C: small black spheres.

Even though the Cu$_{58}$ substructure must be assigned the formal charge 56$^+$ (assuming Se^{2-} and SePh$^-$ ligands as well as electroneutrality in the sum), which leads to an extraordinary mixed valence Cu$^+$–Cu^0 compound, there is no suggestions of any exceptional Cu–Cu binding interactions. The Cu–Cu distances lie within the usual range of 2.540(3) and 3.012(4) Å. Thus, the cluster

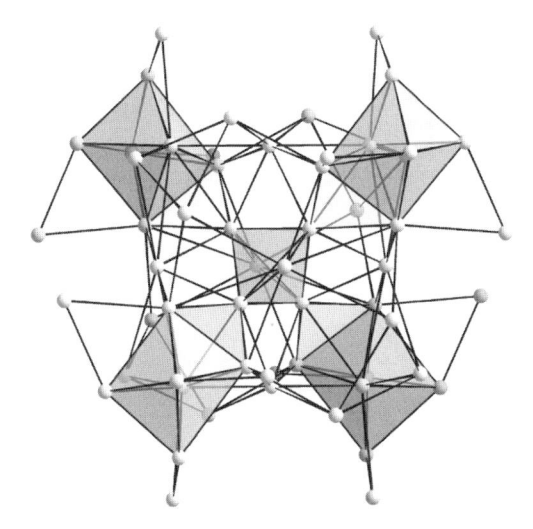

Figure 3.63 Copper substructure in **52**. The connected tetrahedra Cu$_4$ and octahedra Cu$_6$ are highlighted by gray shading. Cu–Cu distances are drawn up to 3.012 Å.

core is habitually held together by bridging selenido or selenolato ligands, where the former act as μ_5 or μ_6 bridges, and the latter are μ_3 bridges between next neighboring copper atoms. Despite the large cluster size, near-linear as well as trigonal planar or tetrahedrally coordinated metal centers, are observed. Therefore, it is again proved that the size-dependence of the structural principles discussed for chalcogenolato-free copper sulfide or copper selenide clusters are widely invalid in the Cu−Se−SeR cluster systems. Six dppa ligands and 24 phenyl groups of the SePh$^-$ fragments ligate the cluster molecule. Hence, **52** continues the series of compounds that show a significant preference for protection by SeR$^-$ groups rather than phosphine ligands. The clusters **43** and **47–51** represent exceptions to this series, as their SeR:P ratios are all 1:1.

The largest copper cluster with both Se^{2-} and SeR$^-$ ligands in the framework is **53**, the molecular structure of which is shown in Figure 3.64. The cluster displays a [Cu$_{70}$Se$_{35}$] core that is identical to that observed in **28** or **29** (see Figures 3.15 and 3.16). Only three additional Cu−SePh fragments positioned at the corners of the triangular molecule distinguish **53** from the slightly smaller clusters. The fact that a typical copper selenide cluster structure with only a few selenolato groups present can be observed is surprising, as PhSeSiMe$_3$ was the first selenium source to be added during the synthesis. This indicates again the distinct preference for the "Cu$_{70}$" molecular structure that is formed whenever the reaction conditions (see Scheme 3.7) and the solubility situation enables cluster growth to proceed until crystallization of this large cluster compound can occur.

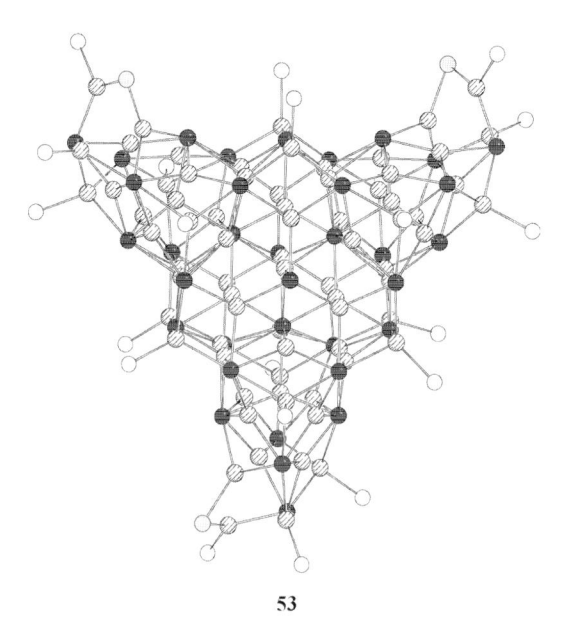

53

Figure 3.64 Molecular structure of [Cu$_{73}$Se$_{35}$(SePh)$_3$(P*i*Pr$_3$)$_{21}$] (**53**) (without organic groups). Cu: shaded spheres; Se: black spheres; SeR: gray spheres; P: white spheres.

3.2.3.2.4 Tellurium-Bridged Copper Clusters

In this section, attention is focused on describing the syntheses and structures of ligand-stabilized copper telluride cluster molecules. Solid-state copper tellurides, as well as ternary alkali metal copper telluride compounds, which consist of n-dimensionally linked copper telluride clusters, are beyond the scope of this review. Although their syntheses have recently undergone an acceleration, only a limited number yet exist, including KCuTe [54], NaCuTe [55], KCu$_3$Te$_2$ [56], NaCu$_3$Te$_2$ [57], K$_2$Cu$_5$Te$_5$ [58], and K$_4$Cu$_8$Te$_{11}$ [59], all of which have been prepared by means of melt reactions. The recently synthesized compound K$_3$Cu$_{11}$Te$_{16}$ [60] is the first example of a ternary alkali metal copper telluride crystallized from supercritical 1,2-diaminoethane.

According to the general reaction pathway shown in Schemes 3.8 and 3.9, the syntheses of ligand-stabilized copper telluride clusters have mainly been achieved in one of three different ways [11, 16, 19, 20, 24, 25, 31, 61].

The reaction of copper(I) acetate with Te(SiMe$_3$)$_2$ at low temperatures (-50 to $-20\,^\circ$C) yields smaller cluster compounds with stoichiometric composition. However, at higher reaction temperatures (up to $25\,^\circ$C), larger clusters can be isolated that are mostly of the mixed-valence type. Similar mixed-valence compounds have also been isolated from reactions of copper(I) chloride with Te(SiMe$_3$)$_2$ at lower temperatures (down to $-70\,^\circ$C). By treating copper(I) chloride or copper(I) acetate with tellurolate compounds RTeSiMe$_3$ (R = tBu, nBu, Ph), it is possible to obtain clusters that contain both RTe$^-$ and Te^{2-} ligands as a result of the facile cleavage of the Te–C bond. In some cases, clusters featuring Te–Te bridges were isolated. Examples of the synthesis of larger copper telluride clusters, starting from cluster precursors, include the light-induced formation of [Cu$_{50}$Te$_{17}$(TePh)$_{20}$(PEtPh$_2$)$_8$][PEtPh$_3$]$_4$ (**72**) from [Cu$_6$(TePh)$_6$(PEtPh$_2$)$_5$] (**64**) in benzene [19], and the co-condensation of [Cu$_{12}$Te$_3$(TePh)$_6$(PEt$_3$)$_6$] (**67**) with [Cu$_5$(TePh)$_6$(PEt)$_3$][PEt$_3$Ph] (**63**) to form [Cu$_{29}$(TePh)$_{12}$Te$_9$(PEt$_3$)$_8$][PEt$_3$Ph] (**71**), as outlined in Scheme 3.9 [61].

All of the Cu–Te cluster compounds that have been synthesized and structurally characterized to date, using single-crystal XRD analysis, are summarized in Table 3.1. The species can be subdivided into four main groups, three of which consist of stoichiometric compositions [Cu$_{2n}$Te$_{nx}$(TeR)$_{2x}$(PR$_2$R')$_m$] that differ in the nature of their tellurium ligands, while the fourth group represents the mixed-valence clusters.

The structures of the tellurium-bridged copper clusters (Figures 3.65–3.69; also see below) differ from those of the copper sulfide or selenide clusters, with the exception of some smaller clusters. This can be more easily understood against the background of the larger atomic and ionic radii of the tellurium atoms compared to the lighter chalcogen atoms, which allow higher coordination numbers and therefore new binding geometries that lead to different structures. In addition, there is a greater tendency to form mixed-valence clusters. This is in line with the existence of many stable binary compounds with compositions Cu$_{2x}$Te [40], and with the situation found in the ternary alkali copper telluride compound K$_2$Cu$_5$Te$_5$ [58].

In the case of copper telluride clusters, the turning point in size has not yet been reached where the whole core structures display bulk structure characteristics, as seen for the copper selenide species. Nevertheless, the tellurium frameworks in the largest cluster compounds, **85–87** display hexagonal structure properties. Powder

CuOAc + 0.5 Te(SiMe$_3$)$_2$ $\xrightarrow[-\text{SiMe}_3\text{OAc}]{+\text{PR}_3 \text{ or } \text{PR}_2\text{R}'}$

[Cu$_4$Te(Te$_2$)$_2$(PCyPh$_2$)$_4$]	(59)
[Cu$_6$Te$_3$(PCyPh$_2$)$_5$]	(54)
[Cu$_8$Te$_4$(PPh$_3$)$_7$]	(55)
[Cu$_{12}$Te$_6$(PPh$_3$)$_8$]	(56)
[Cu$_{16}$Te$_8$(PPhnPr$_2$)$_{10}$]	(57)
[Cu$_{16}$Te$_9$(PPh$_3$)$_8$]	(73)
[Cu$_{23}$Te$_{13}$(PPh$_3$)$_{10}$]	(76)
[Cu$_{24}$Te$_{12}$(PiPr$_3$)$_{12}$]	(58)
[Cu$_{26}$Te$_{12}$(PEt$_2$Ph)$_{12}$]	(79)
[Cu$_{44}$Te$_{23}$(PPh$_3$)$_{15}$]	(82)
[Cu$_{44}$Te$_{23}$(PPhnPr$_2$)$_{15}$]	(83)
[Cu$_{44}$Te$_{23}$(PCyPh$_2$)$_{15}$]	(84)
[Cu$_{58}$Te$_{32}$(PCy$_3$)$_{16}$]	(85)

CuCl + 0.5 Te(SiMe$_3$)$_2$ $\xrightarrow[-\text{SiMe}_3\text{Cl}]{+\text{PR}_3 \text{ or } \text{PR}_2\text{R}'}$

[Cu$_4$(Te$_2$)$_2$(PiPr$_3$)$_4$]	(60)
[Cu$_{16}$Te$_9$(PEt$_3$)$_8$]	(74)
[Cu$_{16}$Te$_9$(PEt$_2$Ph)$_8$]	(75)
[Cu$_{23}$Te$_{13}$(PiPr$_3$)$_{10}$]	(77)
[Cu$_{26}$Te$_{10}$(Te$_2$)$_3$(PtBu$_3$)$_{10}$]	(61)
[Cu$_{28}$Te$_{13}$(Te$_2$)$_2$(PEt$_2$Ph)$_{12}$]	(62)
[Cu$_{29}$Te$_{16}$(PiPr$_3$)$_{12}$]	(81)

CuOAc + nBuTe(SiMe$_3$)$_2$ $\xrightarrow[-\text{SiMe}_3\text{OAc}]{+\text{PR}_3 \text{ or } \text{PR}_2\text{R}'}$ [Cu$_{11}$Te(TenBu)$_9$(PPh$_3$)$_5$] (65)

CuCl + nBuTe(SiMe$_3$)$_2$ $\xrightarrow[-\text{SiMe}_3\text{Cl}]{+\text{PR}_3 \text{ or } \text{PR}_2\text{R}'}$

[Cu$_{18}$Te$_6$(TenBu)$_6$(PnPr$_3$)$_8$]	(69)
[Cu$_{23}$Te$_{13}$(PEt$_3$)$_{12}$]	(78)
[Cu$_{58}$Te$_{32}$(PPh$_3$)$_{16}$]	(86)

CuCl + tBuTe(SiMe$_3$)$_2$ $\xrightarrow[-\text{SiMe}_3\text{Cl}]{+\text{PR}_3 \text{ or } \text{PR}_2\text{R}'}$

[Cu$_{18}$Te$_6$(TetBu)$_6$(PEtPh$_2$)$_7$]	(68)
[Cu$_{19}$Te$_6$(TetBu)$_7$(PEt$_3$)$_8$]	(70)
[Cu$_{27}$Te$_{15}$(PMeiPr$_2$)$_{12}$]	(80)
[Cu$_{58}$Te$_{32}$(PnButBu$_2$)$_{14}$]	(87)

CuCl + PhTe(SiMe$_3$)$_2$ $\xrightarrow[-\text{SiMe}_3\text{Cl}]{+\text{PR}_3 \text{ or } \text{PR}_2\text{R}'}$

[Cu$_6$(TePh)$_6$(PEtPh$_2$)$_5$]	(64)
[Cu$_{12}$Te$_3$(TePh)$_6$(PPh$_3$)$_6$]	(66)
[Cu$_{12}$Te$_3$(TePh)$_6$(PEt$_3$)$_6$]	(67)
[Cu$_5$(TePh)$_6$(PEt$_3$)$_3$][PEt$_3$Ph]	(63)

Scheme 3.8 Survey of the reaction pathways for the synthesis of ligand-stabilized tellurido or tellurido–tellurolato-bridged copper clusters.

[Cu$_{12}$Te$_3$(TePh)$_6$(PEt$_3$)$_6$] (67)

[Cu$_5$(TePh)$_6$(PEt$_3$)$_3$][PEt$_3$Ph] (63) $\xrightarrow[-1,5\text{ TePh}_2]{\text{THF / RT}}$ [Cu$_{29}$(TePh)$_{12}$Te$_9$(PEt$_3$)$_8$][PEt$_3$Ph] (71)

[Cu$_6$(TePh)$_6$(PEtPh$_2$)$_5$] (64) $\xrightarrow[-\text{TePh}_2]{h\nu \text{ / C}_6\text{H}_6}$ [Cu$_{50}$Te$_{17}$(TePh)$_{20}$(PEtPh$_2$)$_8$][PEtPh$_3$]$_4$ (72)

Scheme 3.9 Reaction pathways for the formation of larger telluride–tellurolato-bridged copper clusters, starting from cluster precursors.

Table 3.1 Classification of copper telluride clusters into stoichiometric or mixed valence, according to the types of tellurium ligand.

Stoichiometric			Mixed-valence
Only with Te^{2-} ligands	With additional Te-Te bridges	With additional TeR' ligands	
[Cu$_6$Te$_3$(PCyPh$_2$)$_5$] (54)[31]	[Cu$_4$Te(Te$_2$)(PCyPh$_2$)$_4$] (59)[31]	[Cu$_5$(TePh)$_6$(PEt$_3$)$_3$][PEt$_3$Ph] (63)[69]	[Cu$_{16}$Te$_9$(PPh$_3$)$_8$] (73)[24]
			[Cu$_{16}$Te$_9$(PEt$_3$)$_8$] (74)[11]
			[Cu$_{16}$Te$_9$(PEt$_2$Ph)$_8$] (75)[11]
[Cu$_8$Te$_4$(PPh$_3$)$_7$] (55)[24]	[Cu$_4$(Te$_2$)$_2$(PiPr$_3$)$_4$] (60)[11]	[Cu$_6$(TePh)$_6$(PEtPh$_2$)$_5$] (64)[19]	[Cu$_{23}$Te$_{13}$(PPh$_3$)$_{10}$] (76)[24]
			[Cu$_{23}$Te$_{13}$(PiPr$_3$)$_{10}$] (77)[11]
			[Cu$_{23}$Te$_{13}$(PEt$_3$)$_{12}$] (78)[16]
[Cu$_{12}$Te$_6$(PPh$_3$)$_8$] (56)[24]	[Cu$_{26}$Te$_{10}$(Te$_2$)$_3$(PtBu$_3$)$_{10}$] (61)[11]	[Cu$_{11}$Te(TenBu)$_9$(PPh$_3$)$_5$] (65)[16]	[Cu$_{26}$Te$_{12}$(PEt$_2$Ph)$_{12}$] (79)[24]
[Cu$_{16}$Te$_8$(PPhnPr$_2$)$_{10}$] (57)[24]	[Cu$_{28}$Te$_{13}$(Te$_2$)$_2$(PEt$_2$Ph)$_{12}$] (62)[11]	[Cu$_{12}$Te$_3$(TePh)$_6$(PPh$_3$)$_6$] (66)[19]	[Cu$_{27}$Te$_{15}$(PMeiPr$_2$)$_{12}$] (80)[25]
		[Cu$_{12}$Te$_3$(TePh)$_6$(PEt$_3$)$_6$] (67)[69]	
[Cu$_{24}$Te$_{12}$(PiPr$_3$)$_{12}$] (58)[24]		[Cu$_{18}$Te$_6$(TetBu)$_6$(PEtPh$_2$)$_7$] (68)[25]	[Cu$_{29}$Te$_{16}$(PiPr$_3$)$_{12}$] (81)[11]
		[Cu$_{18}$Te$_6$(TenBu)$_6$(PnPr$_3$)$_8$] (69)[16]	
		[Cu$_{19}$Te$_6$(TetBu)$_7$(PEt$_3$)$_8$] (70)[25]	
			[Cu$_{44}$Te$_{23}$(PPh$_3$)$_{15}$] (82)[24]
			[Cu$_{44}$Te$_{23}$(PPhnPr$_2$)$_{15}$] (83)[24]
			[Cu$_{44}$Te$_{23}$(PCyPh$_2$)$_{15}$] (84)[24]
		[Cu$_{29}$(TePh)$_{12}$Te$_9$(PEt$_3$)$_8$][PEt$_3$Ph] (71)[69]	[Cu$_{58}$Te$_{32}$(PCy$_3$)$_{16}$] (85)[31]
			[Cu$_{58}$Te$_{32}$(PPh$_3$)$_{16}$] (86)[16]
		[Cu$_{50}$Te$_{17}$(TePh)$_{20}$(PEtPh$_2$)$_8$][PEtPh$_3$]$_4$(72)[19]	[Cu$_{58}$Te$_{32}$(PnBut$n$Bu$_2$)$_{14}$] (87)[25]

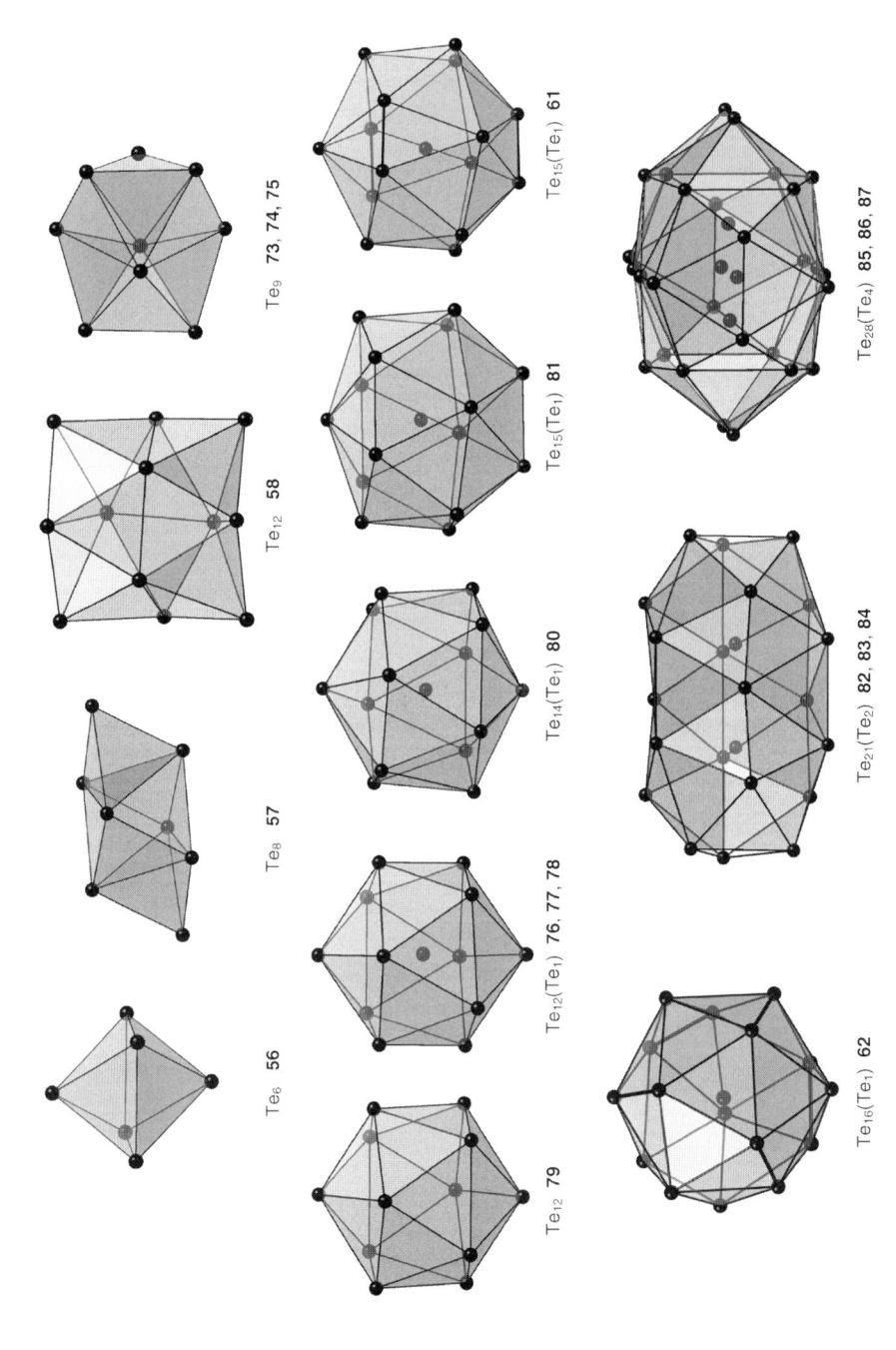

Figure 3.65 Tellurium polyhedra in copper telluride clusters. Connections to the central tellurium atoms are omitted for clarity. Direct Te—Te bonds in **61** and **62** are drawn as thicker lines.

diffraction patterns of Cu_2Te could be also indexed on the basis of a hexagonal cell of tellurium atoms. However, the structural details for bulk Cu_2Te have not yet been completely established [40b].

In all the copper tellurium cluster molecules, the tellurium atoms form polyhedra with triangular faces. In the larger molecules, some of the tellurium atoms are located inside the polyhedral, whereas for the smaller cluster molecules the tellurium polyhedra can usually be derived from classical polyhedra (Figure 3.65). Those of the larger molecules often show unusual cage structures. The distances between the tellurium atoms are usually nonbonding, except for the Te—Te units in **59–62**.

The copper atoms reside in the holes and on the surface of the polyhedral, and are coordinated by either two, three, or four tellurium atoms, with some being additionally coordinated by one further phosphine ligand.

The Cu—Te distance ranges that are regarded as bonding contacts range from 2.403 to 3.337 Å for all the clusters under discussion. Even though the coordination numbers (CN) for Te^{2-} ligands range from 4 to 12, while those for TeR^- are usually

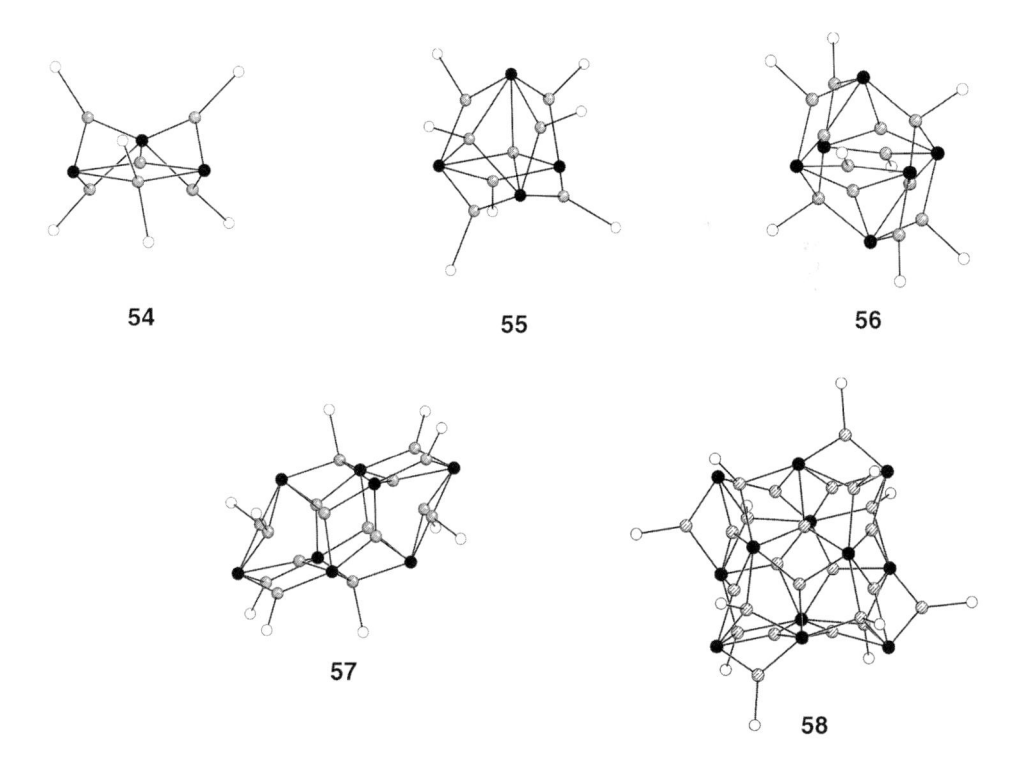

54

55

56

57

58

Figure 3.66 Molecular structures of stoichiometric copper telluride clusters $[Cu_6Te_3(PCyPh_2)_5]$ (**54**), $[Cu_8Te_4(PPh_3)_7]$ (**55**), $[Cu_{12}Te_6(PPh_3)_8]$ (**56**), $[Cu_{16}Te_8(PPhnPr_2)_{10}]$ (**57**), and $[Cu_{24}Te_{12}(PiPr_3)_{12}]$ (**58**). C and H atoms are omitted for clarity. Cu: shaded spheres; Te: black spheres; P: white spheres.

smaller (CN = 2, 3, 4), there is no significant difference between the average Cu–Te distances for the Te^{2-} and TeR^- ligands.

Those tellurium atoms that are located in the center of the cluster molecules tend to bond to a large number of copper atoms. In some compounds, the copper atoms themselves form polyhedra, described as Frank–Kasper polyhedra [62]. In these cases, the clusters consist of interpenetrating Frank–Kasper polyhedra, which are [34b, 63] well known from a number of intermetallic phases (e.g., Laves phases).

However, the observed Cu–Cu distances provide no indication of any strong metal–metal (i.e., d^{10}–d^{10}) interactions, and theoretical investigations have pointed to the same conclusions [63]. The shortest Cu–Cu contacts were observed in the molecular structures of **62** (2.383 Å), and **78** (2.382 Å), but these involved copper atoms which exhibited elongated thermal ellipsoids. Aside from these, the shortest Cu–Cu distances were interestingly observed for the largest clusters, namely **85** (2.426 Å), **86** (2.448 Å), and **87** (2.433 Å).

In the following section, all cluster structures will be briefly described and discussed according to the division into four groups indicated in Table 3.1. The molecular structures of the stoichiometric copper telluride clusters **54–58** are shown in Figure 3.66.

In **54**, three tellurium atoms form a triangle, in the center of which a copper atom is coordinated in a slightly distorted triangular fashion. Four [CuPPh$_2$Cy] units are doubly bridging two of the edges, while the third edge is bridged by a copper atom which is bonded to two phosphine ligands.

The tellurium atoms in **55** form a tetrahedron, the six edges of which are bridged by [CuPPh$_3$] groups. An additional μ_3-[CuPPh$_3$] fragment is bonded to a Te$_3$ face, and one naked copper atom is located in the center of the Te$_4$ tetrahedron.

Cluster **56** features a Te$_6$ octahedron, in which six [CuPPh$_3$] groups act as m2-bridges above six of the octahedral edges. Two further [CuPPh$_3$] groups are coordinated by three tellurium atoms, giving a tetrahedral environment around the copper atoms. The other four copper atoms are each coordinated by three tellurium atoms. Thus, **56** represents a novel isomer of the [Cu$_{12}$E$_6$(PR$_3$)$_8$] clusters (see above) which was not observed for E = S, Se.

An octahedron of tellurium atoms is also found in **57**; however, two further tellurium atoms additionally cap two opposite triangular faces. Eight [CuPPh$_n$Pr$_2$] groups are bonded along edges of this polyhedron. Of the other eight copper centers, two are tetrahedrally coordinated by one phosphine and a Te$_3$ face, while the six remaining metal atoms are bonded in a distorted trigonal planar mode by three tellurium atoms and are located inside the cluster cavity.

The Te$_{12}$ substructure of **58** cannot be described as a classical deltahedron, but rather shows an arrangement of two highly distorted face-sharing tetragonal antiprisms. Four [CuPiPr$_3$] groups bridge one Te–Te non-bonding edge each, and eight [CuPiPr$_3$] units act as μ_3-bridges above Te$_3$ faces. The other 12 copper atoms are each surrounded by three tellurium atoms, and are slightly shifted away from the Te$_3$ planes toward the center of the molecule.

Some reactions of copper(I) chloride or copper(I) acetate lead to the formation of copper telluride clusters that contain Te–Te bridges (**59–62**; see Figure 3.67).

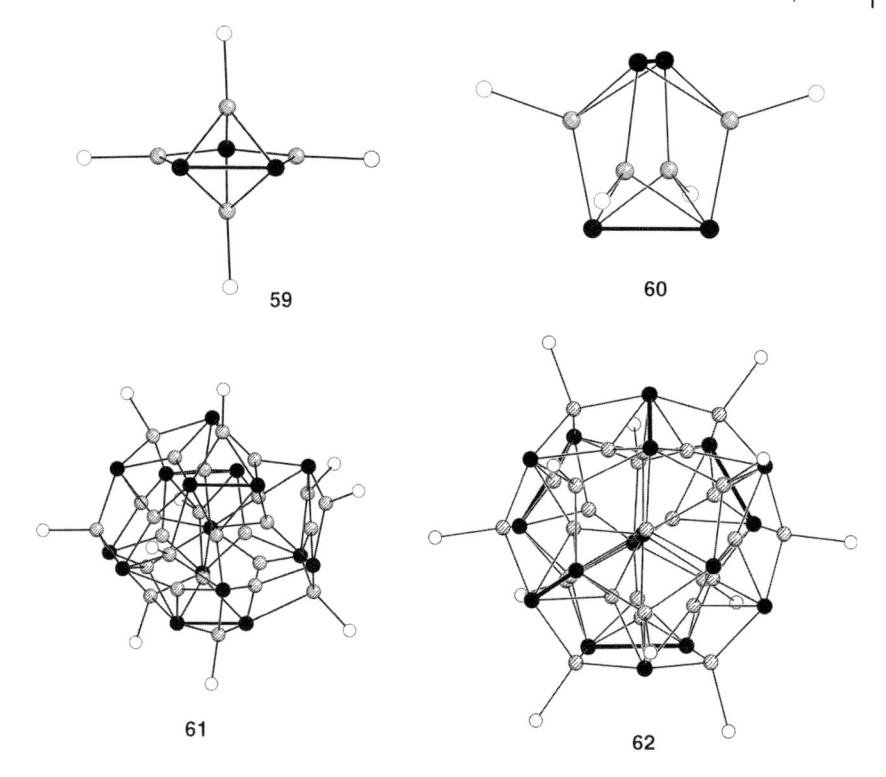

Figure 3.67 Molecular structures of stoichiometric copper telluride clusters [Cu$_4$Te(Te$_2$)$_2$(PCyPh$_2$)$_4$] (**59**), [Cu$_4$(Te$_2$)$_2$(PiPr$_3$)$_4$] (**60**), [Cu$_{26}$Te$_{10}$(Te$_2$)$_3$(PtBu$_3$)$_{10}$] (**61**), and [Cu$_{28}$Te$_{13}$(Te$_2$)$_2$(PEt$_2$Ph)$_{12}$] (**62**) containing Te–Te bridges. C and H atoms are omitted for clarity. Cu: shaded spheres; Te: black spheres; P: white spheres.

In **59**, three tellurium atoms form an approximate isosceles triangle with a Te–Te bond along the short edge (Te–Te: 2.791 Å). The two longer edges of the triangle are bridged by one [CuPPh$_2$Cy] unit each, while the two other copper atoms are located above and below the triangular face, and additionally coordinated by phosphine ligands.

The four copper atoms in **60** form a "butterfly" structure bridged by two ditelluride ligands (Te–Te: 2.800–2.812 Å). Every copper atom is additionally coordinated by a phosphine ligand.

Clusters **61** and **62** are spherical molecules based on tellurium polyhedra, parts of which are strongly distorted because of the presence of Te–Te bonds. Both compounds display Frank–Kasper polyhedra containing either 15 (**61**) or 16 (**62**) tellurium atoms centered by one additional tellurium ligand. The tellurium polyhedron of **61** is similar to that observed in **20**, but differs by the presence of three Te–Te bonds (2.936–3.011 Å). The copper atoms are bonded to the tellurium atoms in a different arrangement to that in **20** with eight [CuPtBu$_3$] units being situated above the Te$_3$ faces of the polyhedron, while two such units are edge-bridging. Eleven further

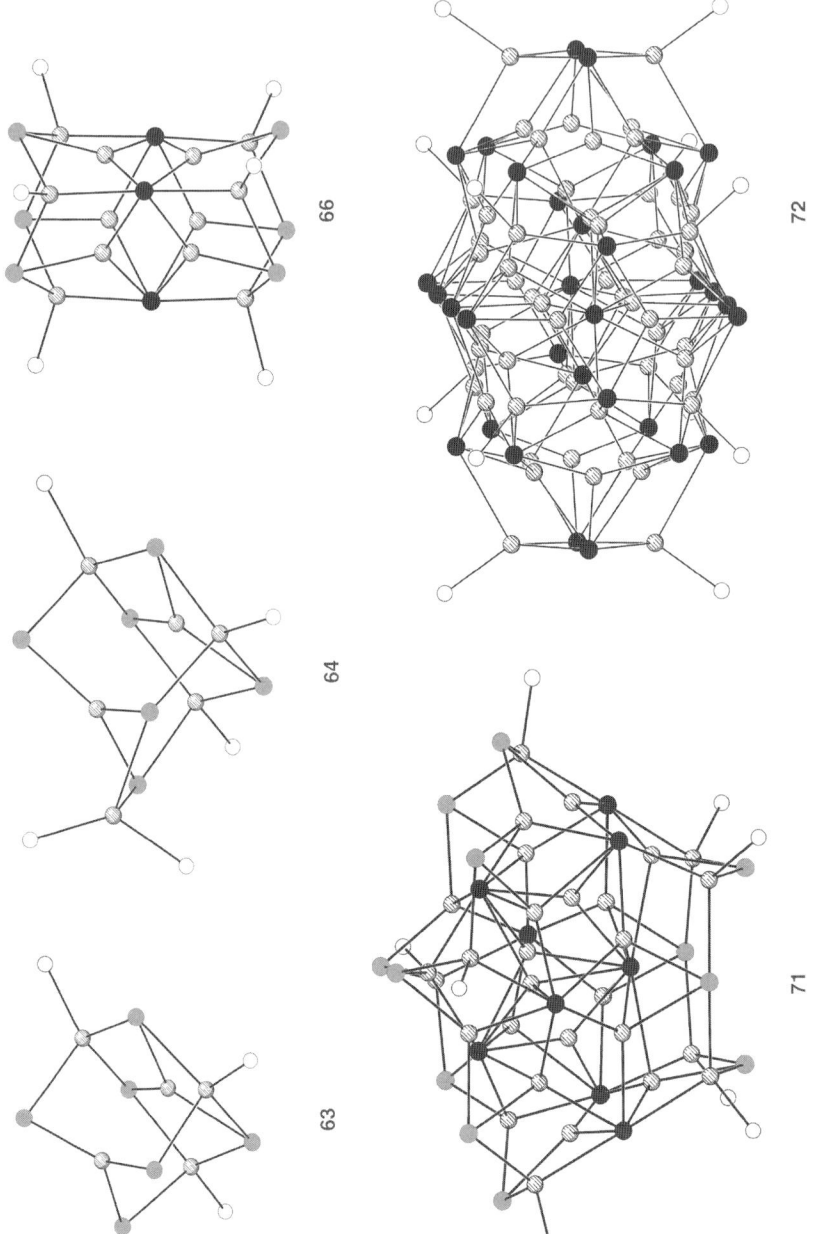

Figure 3.68 Molecular structures of stoichiometric copper telluride–tellurolate clusters [Cu$_5$(TePh)$_6$(PEt$_3$)$_3$][PEt$_3$Ph] (**63**), [Cu$_6$(TePh)$_6$(PEtPh$_2$)$_5$] (**64**), [Cu$_{12}$Te$_3$(TePh)$_6$(PPh$_3$)$_6$] (**66**), [Cu$_{29}$(TePh)$_{12}$Te$_9$(PEt$_3$)$_8$][PEt$_3$Ph] (**71**), and [Cu$_{50}$Te$_{17}$(TePh)$_{20}$(PEtPh$_2$)$_8$][PEt$_3$Ph]$_4$ (**72**) containing TeR' ligands. C and H atoms are omitted for clarity. Cu: shaded spheres; Te: black spheres; TeR: gray spheres; P: white spheres.

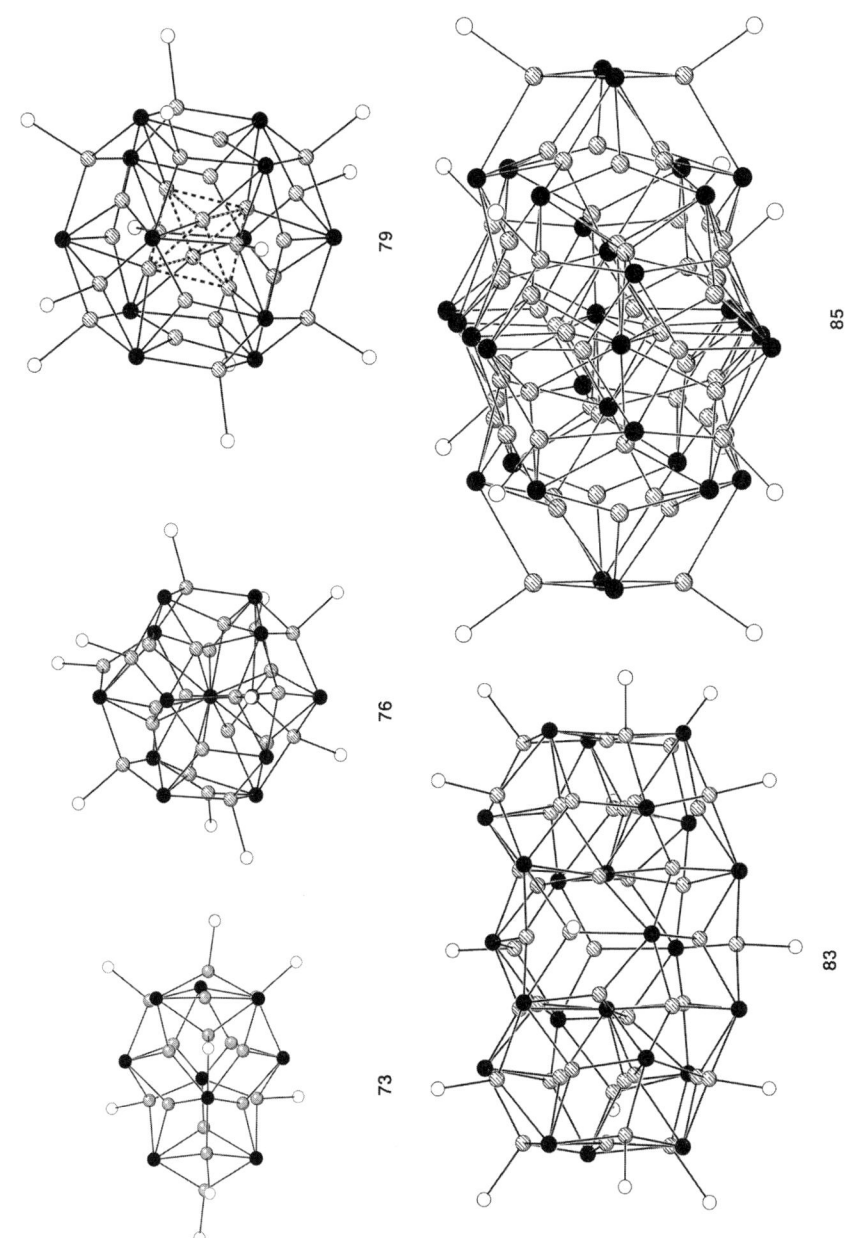

Figure 3.69 Molecular structures of mixed-valence copper telluride clusters [Cu$_{16}$Te$_9$(PPh$_3$)$_8$] (**73**), [Cu$_{26}$Te$_{12}$(PEt$_2$Ph)$_{12}$] (**79**), [Cu$_{44}$Te$_{23}$(PPhnPr$_2$)$_{15}$] (**83**), and [Cu$_{58}$Te$_{32}$(PCy$_3$)$_{16}$] (**85**). C and H atoms are omitted for clarity. Cu: shaded spheres; Te: black spheres; P: white spheres.

copper atoms are coordinated by three tellurium atoms, eight of which are localized below Te_3 faces, while the other three bind between the central tellurium atom and the tellurium atoms of the polyhedron. The remaining copper atoms are localized in the center of the molecule and are each tetrahedrally surrounded by four tellurium atoms.

In **62**, two of the copper centers display relatively high thermal parameters, from which it can be concluded that their positions are only partly occupied. A deficiency of electrons at the tellurium atoms resulting from the partial occupation of copper sites could either be formally delocalized in the valence band or localized as a ditelluride group, following a structural deformation similar to a Peierls deformation [64]. Considering the tellurium polyhedra, the molecular structure of **62** shows two short Te–Te bonds (2.848 and 2.876 Å) and four Te–Te distances in the range 3.150–3.351 Å. The latter tellurium atoms have thermal ellipsoids which are elongated in the direction of the formal Te–Te bond. As the Te–Te bonds and copper centers seem to be disordered to a certain extent in **62**, the given formula represents the idealized composition of a stoichiometric cluster compound with three Te–Te bonds. The situation in **62** can therefore be viewed as being somewhat similar to that in CuS, a well-known, mixed-valence compound that can be described as $(Cu^+)_2Cu^{2+}(S_2)^{2-}S^{2-}$ or as $(Cu^+)_3(S_2)^{2-}S^{2-}$ [65].

In a different approach to the synthesis of copper telluride cluster compounds, alkyl-and aryl-(trimethylsilyl)tellurium compounds $RTeSiMe_3$ (R = organic group) were used. Because of the facile cleavage of the C–Te bond, clusters that feature either exclusively TeR^- ligands, a mixture of TeR^- and Te^{2-} ligands, or solely Te^{2-} ligands, are obtained. The tellurium atoms of the tellurolato ligands bridge two or three copper atoms at most, whereas "naked" Te^{2-} ligands can coordinate to up to seven copper atoms. The polyhedra of tellurium atoms usually possess unusual structures, even in the smaller clusters. In the larger clusters, the characteristics of layer-type arrangements of tellurium atoms can be observed. As in the cluster compounds discussed above, the copper atoms are coordinated in linear, trigonal, or tetrahedral geometries by tellurium atoms, while the metal centers coordinated by two or three chalcogens may also be ligated by a phosphine ligand. The molecular structures of the copper telluride–tellurolate clusters **63**, **64**, **66**, **71**, and **72** are shown in Figure 3.68.

The ionic cluster compound **63** contains six $TePh^-$ ligands, the tellurium atoms of which define an octahedron. In contrast to **64** (see below), all tellurolato ligands act as m_2-bridges. There are two distinct coordination geometries around the copper, and two of the metal centers are located inside the cluster below Te_3 faces. The others are coordinated by three tellurium atoms above Te_3 faces and are also each ligated by one phosphine ligand each.

The structure of **64** is also based on a nonbonded octahedral array of tellurolato ligands. With the exception of one tellurium atom, which forms two Cu–Te bonds, all of these atoms are bridging three copper sites. Four of the copper atoms adopt tetrahedral geometries, either with three tellurium atoms and one phosphorus or through two Cu–Te and two Cu–P bonds. The other two copper atoms lie in opposite deltahedral Te_3 faces of the Te_6 octahedron. Thus, **64** is related to the recently described homoleptic hexanuclear copper selenolate complex [Cu$_6${Se(2,4,6iPr$_3$C$_6$H$_2$)}$_6$] [8].

Clusters **66** and **67** are identical except for the organic groups R of the PR$_3$ ligands, as both contain two tellurolato layers of three TePh$^-$ groups each, and one central tellurido layer of three Te^{2-} ligands. The resulting Te$_9$ polyhedron can alternatively be viewed as two face-sharing Te$_6$ octahedra. Six of the 12 copper atoms are coordinated in a distorted trigonal fashion by three tellurium atoms, resulting in shorter contacts to Te^{2-} ligands than to the TePh. The other six copper atoms are tetrahedrally surrounded by three tellurium atoms and one phosphine ligand. Compounds **66** and **67** may be viewed as condensation products of two (TePh)$_6$ frameworks, similar to those found in **64**.

The synthesis of the ionic cluster compound **71** is the only example to date of a co-condensation reaction of two smaller clusters, namely **67** and **63**, to form a larger cluster. However, the structure of **71** cannot be considered as being built up from recognizable structural fragments of the precursor compound. The tellurium atoms form an unusual polyhedron consisting of 12 TePh$^-$ ligands and nine Te^{2-} ligands, where two of the tellurolato ligands act as a μ_4-bridges and the other ten act as μ_3-bridging atoms between the copper centers. The telluride ligands each coordinate six, seven, eight, or ten copper atoms, with Cu–Te distances between 2.58 and 3.14 Å. Eight of the copper atoms which are bonded to phosphine ligands cap Te$_3$ faces at the cluster surface, while four further copper atoms without phosphine ligands are coordinated to Te$_3$ faces from the inside of the cluster core. The remaining 17 copper atoms are surrounded in a tetrahedral manner in the center of the molecule. Interestingly, a structurally related silver tellurium compound, [Ag$_{30}$(TePh)$_{12}$Te$_9$(PEt$_3$)$_{12}$] [71], can be prepared by the direct reaction of AgCl with Te(SiMe$_3$)$_2$–PhTeSiMe$_3$, which is in contrast to the synthesis of **71**. This cluster represents the only example of a copper–telluride–tellurolate cluster yet isolated with a distinct structural relationship with its silver analog.

Upon standing in daylight, solutions of **64** in benzene gradually darken to a brown color and, over the course of several days, brown platelets of the ionic, mixed tellurido–tellurolato cluster **72** crystallize from these solutions. On the other hand, when solutions of **64** in benzene are protected from light, they show no sign of darkening, even after several weeks at room temperature. The detection of a ^{125}Te{^{1}H}-NMR-signal at d = 688 ppm reveals the formation of TePh$_2$ in the solution, providing good evidence for the mechanism of formation of the naked tellurido ligands. The most striking feature of the cluster core is the manner in which the tellurium ligands are arranged: the molecule consists of "top" and "bottom" layers of μ_3-TePh$^-$ ligands with a central tellurido (Te^{2-}) layer. Sandwiched between these are the 50 copper atoms, along with the remaining seven tellurido ligands. The 20 phenyltellurolato ligands are each bonded to three copper atoms and, in conjunction with the eight coordinated PEt$_2$Ph ligands, serve to effectively stabilize the inner copper telluride core. The telluride ligands of the middle Te$_{10}$ layer act as μ_6- and μ_7-bridges between the copper sites. The seven remaining Te^{2-} centers function as μ_6-, μ_7-, or μ_8-ligands. Eight of the copper atoms are bonded to Te$_3$ faces at the surface of the cluster molecule and to one phosphine ligand. The 42 copper atoms in the center of the molecule are either trigonally (**18**) or tetrahedrally (**24**) coordinated by the tellurium neighbors.

Counting the electrons for **73–85** on the assumption of Te^{2-} and Cu^+ leads to an overall electron excess for **79** and an electron deficiency for the other compounds; these compounds can therefore be thought of as being of mixed-valence type. The molecular structures of **73**, **76**, **79**, **83**, and **85** are depicted in Figure 3.69.

The smallest mixed-valence cluster molecules, **73–75**, feature 16 copper atoms arranged around a Te_9 polyhedron. However, the tellurium substructure cannot be derived from the skeleton of **57** by simply adding one tellurium atom, as might have been assumed. Two of the copper atoms that carry a phosphine ligand bridge Te–Te nonbonding edges, while the other six [$CuPR_2R'$] units bridge three tellurium atoms, such that the copper atoms are tetrahedrally coordinated. The remaining eight copper atoms are coordinated by three tellurium atoms in a distorted trigonal planar manner. Assuming that the tellurium atoms have a formal charge of 2, it is possible to formally assign 14 Cu^{1+} centers and two Cu^{2+} centers. However, the co-existence of Cu^{2+} with Te^{2-} is unlikely for thermodynamic reasons, because of an electron-transfer process from the reducing Te^{2-} ligand to the oxidizing Cu^{2+} center. An assignment of 16 Cu^+, seven Te^{2-} and two Te^{1-} is not supported by structural means. The compounds should, therefore, rather be described in terms of a "pure" Cu^+ substructure, ligated by tellurium centers with an average charge of 1.78. This electron deficiency can be delocalized in the valence band of the compound, as has recently been shown by calculations on the electron-rich compound **79** (see below) [66].

Thirteen tellurium atoms form a distorted, centered icosahedron in the closely related compounds **76–78**. Ten of the 20 Te_3 faces are bridged by copper–phosphine units, whereas ten more copper atoms are situated below the remaining faces, either trigonally coordinated or tetrahedrally surrounded, because of an additional Cu–Te bond to the central tellurium atom. Compound **78** requires 12 phosphine ligands for a sufficient shielding of the cluster core, but this can be put down to the smaller steric demand of the phosphine PEt_3 compared to PPh_3. As a result, in **78**, twelve of the copper atoms are coordinated above Te_3 faces, and eight below. The three remaining copper atoms in all three compounds are positioned inside the icosahedrons; each of these has three tellurium neighbors, two of which belong to the Te_{12} shell, while one is the central tellurium ligand. The formal counting of charges leads to an electron excess of three and, by assigning a charge of +1 to every copper atom (as discussed above), an average charge of 1.77 is obtained for the 13 tellurium ligands.

An icosahedron of tellurium atoms is also found in **79** although, in contrast to those observed in **76–78**, it does not contain a central atom. The arrangement of the copper atoms above the Te_3 faces is similar to that found in **78**. This might have been anticipated, given the Tolman angle for both phosphines. Twelve [$CuPEt_2Ph$] units are bonded above Te_3 faces, while eight copper atoms cap the other Te_3 faces from the interior. In contrast to the situation in **76–78**, in **79** six rather than three copper atoms are located in the center of the cluster molecule, forming an octahedron with intermetallic distances between 2.592 and 2.607 Å. These distances lie within the range of Cu–Cu distances measured in the other copper telluride clusters under investigation. Therefore, it cannot be determined from these distances whether d^{10}–d^{10} interactions are present within the Cu_6 octahedron. However, it is remarkable that two of the copper atoms must be formally uncharged Cu^0 atoms lying alongside 24 Cu^+ centers, if a formal charge of 2^- is assigned to all of the tellurium ligands. Calculations of the electronic band structure reveal that a localization of the oxidation states does not occur, and that the excess electron density cannot be assigned to d^{10}–d^{10} interactions [66]. The "additional" electron pair, which arises if exclusively Cu^+ and Te^{2-} centers are assigned, is better described as an MO embedded in the valence band of tellurium 5p orbitals and copper 4s orbitals.

The largest copper telluride cluster molecules synthesized to date are also mixed-valence compounds. Compounds **82–87** display ellipsoidal tellurium deltahedra with an increasing number of inner tellurium atoms, ranging from two inner tellurium atoms in **82–84** to four such tellurium atoms in the centers of the molecular structures of **85–87**. As usual, the copper atoms that are coordinated by phosphine ligands are coordinated above the Te_3 triangles at the cluster surface. The other copper atoms are either coordinated by three or, especially in the cluster center, by four tellurium ligands. The structures of **85–87** already suggest the formation of distorted layers of tellurium atoms showing incipient characteristics of bulk Cu_2Te [40]. However, the turning point has not yet been reached for copper telluride cluster compounds. In view of the reported tendency for ever bigger molecules to be necessary in order that bulk structure properties will be observed on going from Cu–S to Cu–Se clusters, the cluster size that might begin to show bulk Cu_2Te structural characteristics should in fact be considerably larger than a "Cu_{58}" species, and probably even larger than a "Cu_{70}" analog of the turning point to "large" clusters found for the copper selenide system.

3.2.3.3 Chalcogen-Bridged Silver Clusters

The development of the chemistry of Cu_2S, Cu_2Se and Cu_2Te complexes has yet to be mirrored for the heavier metals Ag and Au. Monodentate phosphine ligands have generally been shown to be incapable of providing a kinetically stabilizing sphere when Ag(I) and Au(I) salts are reacted with $E(SiMe_3)_2$, and these reactions lead instead to the formation of amorphous binary solids.

It is clear from the development of the chemistry of the copper chalcogenide cluster complexes described above that: (i) the incorporation of additional surface ligands in the form of SeR and TeR functionalities; and/or (ii) the use of bidentate phosphine ligands, offer additional structural flexibility. Additionally, they may be used to stabilize AgSe, AgTe, and AuSe cluster cores, which themselves are synthesized with bis(silylated) reagents. The use of monosilylated chalcogen reagents (as the hydrocarbon moiety is relatively inert and unreactive toward metal salts) could therefore aid in inhibiting the formation of the bulk, amorphous solids. Furthermore, the reaction chemistry of metal–selenolate and metal–tellurolate complexes can also provide the source of E^{2-} ligands required for the formation of larger polymetallic complexes [19, 61]. Thus, the use of monosilylated reagents RESiMe$_3$ (R = aryl, alkyl) has led to the development of nanometer scale Ag–Se–SeR and Ag–Te–TeR complexes. The carbon chain on the chalcogen center is readily modified, and, as discussed below, this plays an important role in governing the structures formed. Hence, the general synthesis route is very similar to that shown in Scheme 3.2. A survey on the accordant reactions and the resulting structures, as well as the structural principles of the products, is provided in the following sections.

3.2.3.3.1 Sulfur-Bridged Silver Clusters

The clusters with the highest number of metal atoms synthesized and structurally characterized to date can be obtained by the combination of the elements silver and sulfur. In contrast to the synthesis of copper chalcogenide clusters described in Section 3.1.3.2, reactions of silver halides with S(SiMe$_3$)$_2$ and stabilizing phosphanes always lead to the precipitation of silver sulfide. This problem can be overcome by the use of silver carboxylates (AgO$_2$CR). In the presence of tertiary phosphanes, these carboxylates react with silylated sulfur compounds to yield ligand-protected sulfur-bridged silver clusters (Scheme 3.10). The driving force in these reactions is the formation of silylesters.

Another possible means of obtaining multinuclear sulfur-bridged silver clusters would be to react silver thiolates (AgSR) with carbon disulfide (Scheme 3.11).

The largest clusters synthesized to date have been obtained by the reaction of silver thiolates with S(SiMe$_3$)$_2$ in the presence of bidentate phosphanes. Depending on the reactants, solvents and molar ratios used, clusters containing several hundreds of silver atoms can be synthesized (Scheme 3.12). The formation of these metal-rich compounds can only be observed in the presence of bidentate phosphanes. Although

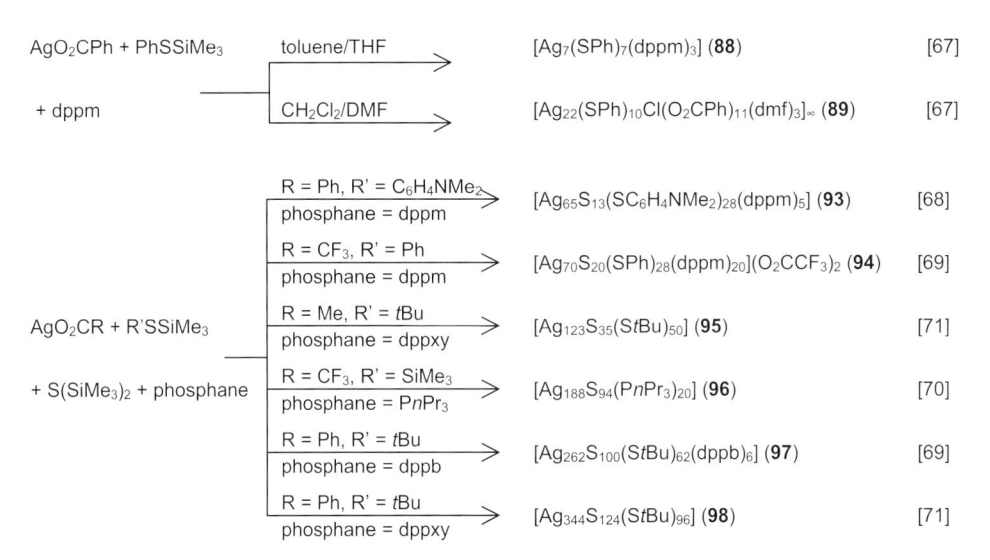

Reactants	Conditions	Product	Ref.
AgO$_2$CPh + PhSSiMe$_3$ + dppm	toluene/THF	[Ag$_7$(SPh)$_7$(dppm)$_3$] (**88**)	[67]
	CH$_2$Cl$_2$/DMF	[Ag$_{22}$(SPh)$_{10}$Cl(O$_2$CPh)$_{11}$(dmf)$_3$]$_\infty$ (**89**)	[67]
AgO$_2$CR + R'SSiMe$_3$ + S(SiMe$_3$)$_2$ + phosphane	R = Ph, R' = C$_6$H$_4$NMe$_2$, phosphane = dppm	[Ag$_{65}$S$_{13}$(SC$_6$H$_4$NMe$_2$)$_{28}$(dppm)$_5$] (**93**)	[68]
	R = CF$_3$, R' = Ph, phosphane = dppm	[Ag$_{70}$S$_{20}$(SPh)$_{28}$(dppm)$_{20}$](O$_2$CCF$_3$)$_2$ (**94**)	[69]
	R = Me, R' = tBu, phosphane = dppxy	[Ag$_{123}$S$_{35}$(StBu)$_{50}$] (**95**)	[71]
	R = CF$_3$, R' = SiMe$_3$, phosphane = PnPr$_3$	[Ag$_{188}$S$_{94}$(PnPr$_3$)$_{20}$] (**96**)	[70]
	R = Ph, R' = tBu, phosphane = dppb	[Ag$_{262}$S$_{100}$(StBu)$_{62}$(dppb)$_6$] (**97**)	[69]
	R = Ph, R' = tBu, phosphane = dppxy	[Ag$_{344}$S$_{124}$(StBu)$_{96}$] (**98**)	[71]

Scheme 3.10 Synthesis of various sulfur-bridged silver clusters by reactions of silver carboxylates with silylated sulfur compounds in the presence of tertiary phosphanes.

AgSC$_6$H$_2$iPr$_3$ + CS$_2$ $\longrightarrow$ [Ag$_{10}$(SC$_6$H$_2$iPr$_3$)$_{10}$] (**90**) [72]

AgSPh + CS$_2$ + PPh$_3$ $\longrightarrow$ [Ag$_{14}$S(SPh)$_{12}$(PPh$_3$)$_8$] (**91**) [73]

(HNEt$_3$)$_{2n}$[Ag$_{10}$(SC$_6$H$_4$tBu)$_{12}$]$_n$ + CS$_2$ $\longrightarrow$ (HNEt$_3$)$_4$[Ag$_{50}$S$_7$(SC$_6$H$_4$tBu)$_{40}$] (**92**) [74]

Scheme 3.11 Synthesis of various sulfur-bridged silver clusters by reaction of silver thiolates with carbon disulfide.

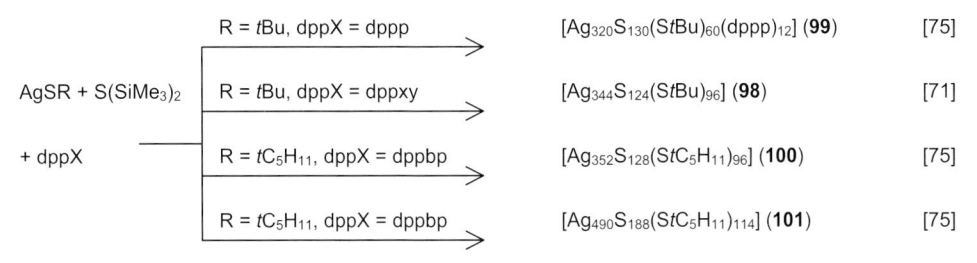

AgSR + S(SiMe$_3$)$_2$ + dppX

R = tBu, dppX = dppp	[Ag$_{320}$S$_{130}$(StBu)$_{60}$(dppp)$_{12}$] (**99**)	[75]
R = tBu, dppX = dppxy	[Ag$_{344}$S$_{124}$(StBu)$_{96}$] (**98**)	[71]
R = tC$_5$H$_{11}$, dppX = dppbp	[Ag$_{352}$S$_{128}$(StC$_5$H$_{11}$)$_{96}$] (**100**)	[75]
R = tC$_5$H$_{11}$, dppX = dppbp	[Ag$_{490}$S$_{188}$(StC$_5$H$_{11}$)$_{114}$] (**101**)	[75]

Scheme 3.12 Synthesis of various metal-rich, sulfur-bridged silver clusters by reactions of silver thiolates S(SiMe$_3$)$_2$ in the presence of bidentate phosphanes.

they are not built into the ligand shell, these phosphanes seem to be essential for the process of cluster formation.

In [Ag$_7$(SPh)$_7$(dppm)$_3$] (**88**), the seven silver atoms adopt a topology analogous to the P$_7^{3-}$ anion, in which the edges and edges and faces are μ_2-, μ_3- and μ_4-bridged by SPh$^-$ ligands, respectively. The core in [Ag$_{10}$(SC$_6$H$_2$iPr$_3$)$_{10}$] (**90**) is a highly folded 20-membered ring of alternating silver and sulfur atoms in which all silver atoms are linearly coordinated by two silver atoms; both cluster cores are shown in Figure 3.70.

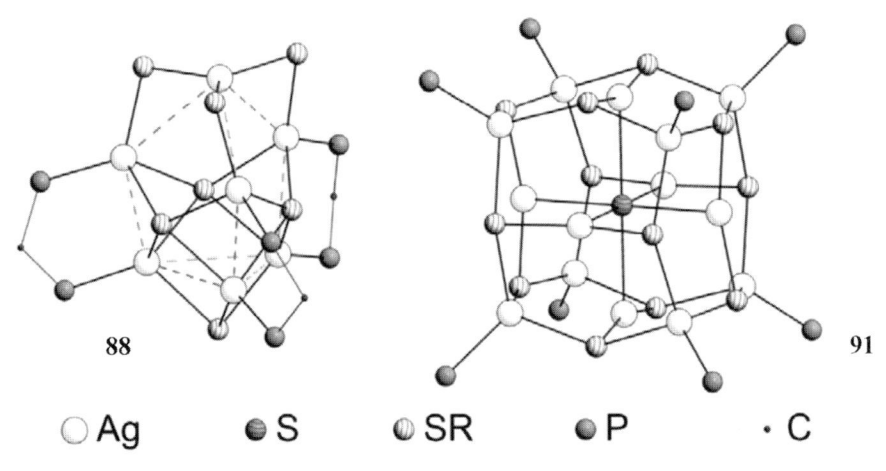

○ Ag ● S ◐ SR ● P · C

Figure 3.70 Molecular structures of [Ag$_7$(SPh)$_7$(dppm)$_3$] (**88**) (left) and [Ag$_{14}$S(SPh)$_{12}$(PPh$_3$)$_8$] (**91**) (right) without Ph groups.

○ Ag   ○ SR   ○ Cl   ○ O   ○ N   • C

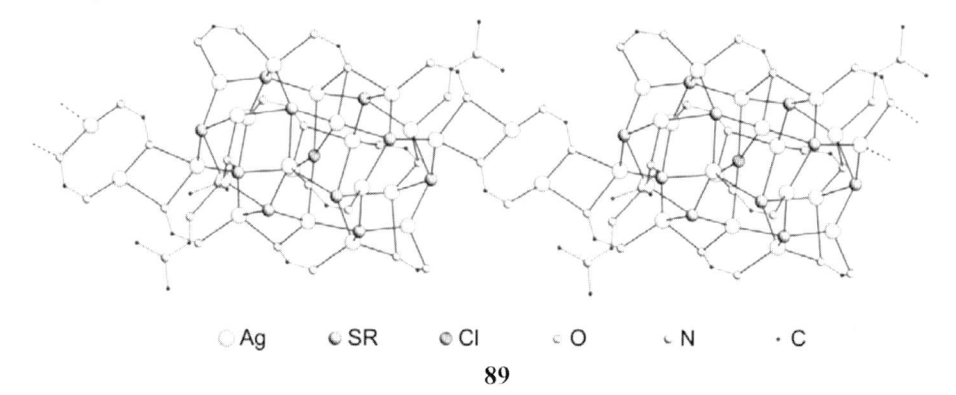

89

Figure 3.71 Section of the polymeric compound $[Ag_{22}(SPh)_{10}Cl(O_2CPh)_{11}(dmf)_3]_\infty$ (**89**).

The compound $[Ag_{22}(SPh)_{10}Cl(O_2CPh)_{11}(dmf)_3]_\infty$ (**89**; Figure 3.71) consists of a polymeric chain in which $[Ag_{20}(SPh)_{10}Cl]$ units are connected via $[Ag_2(O_2CPh)_4]$ fragments. $[Ag_{14}S(SPh)_{12}(PPh_3)_8]$ (**91**) is composed of an Ag_6 octahedron, an Ag_8 cube, and a S_{12} icosahedron centered around a μ_6-S^{2-} ion. The inner ion is weakly bonded to the six silver atoms of the octahedron.

The larger the clusters become, the more complex structures are found. The cluster anion $[Ag_{50}S_7(SC_6H_4tBu)_{40}]^{4-}$ in (**92**) shows a discus-like structure with a pseudo-fivefold axis. In the inner part there is an Ag_{15} pentagonal prismatic unit which is surrounded by another Ag_{20} decagonal prism, of which the five Ag_4 rectangle fragments cap the five Ag_6 sidefaces of the inner pentagonal prism. The whole structure is finally surrounded by another 15 silver ions. The structure of this S_{13} core in $[Ag_{65}S_{13}(SC_6H_4NMe_2)_{28}(dppm)_5]$ (**93**) can be described as a small part of a double hexagonal closest packing, with three sulfur atoms representing an A layer, six sulfur atoms a B layer, three sulfur atoms another A layer, and the last atom a C layer. This part of the cluster is surrounded by a shell of 39 sulfur atoms, each of which is bonded to a dimethylanilino group. However, these sulfur atoms are not arranged in close-packed layers. For charge balance, 65 silver ions are incorporated into the sulfur substructure. In the X-ray structure analysis, 60 of these ions were unambiguously located; the five remaining silver ions were disordered over 10 positions. A view of both clusters is provided in Figure 3.72.

In the case of selenium-bridged copper clusters, it was found that there is a structural transition from molecular spherical structures towards cutouts of the bulk structure when the clusters reach 70 copper atoms. An analogous structural transformation has not yet been found in the case of sulfur-bridged silver clusters. For example, the dication $[Ag_{70}S_{20}(SPh)_{28}(dppm)_{20}]^{2+}$ in **94** shows a shell-like sulfur substructure that consists of an inner S_8 and an outer S_{40} polyhedron (Figure 3.73).

Up to this composition, it is always possible to find a charge balancing between Ag^+ cations on the one hand, and S^{2-} and SR^- anions on the other hand. Assuming that silver is always in the oxidation state +1, the cluster $[Ag_{123}S_{35}(StBu)_{50}]$ (**95**; Figure 3.74), when synthesized from silver acetate and a mixture of tBuSSiMe$_3$ and S

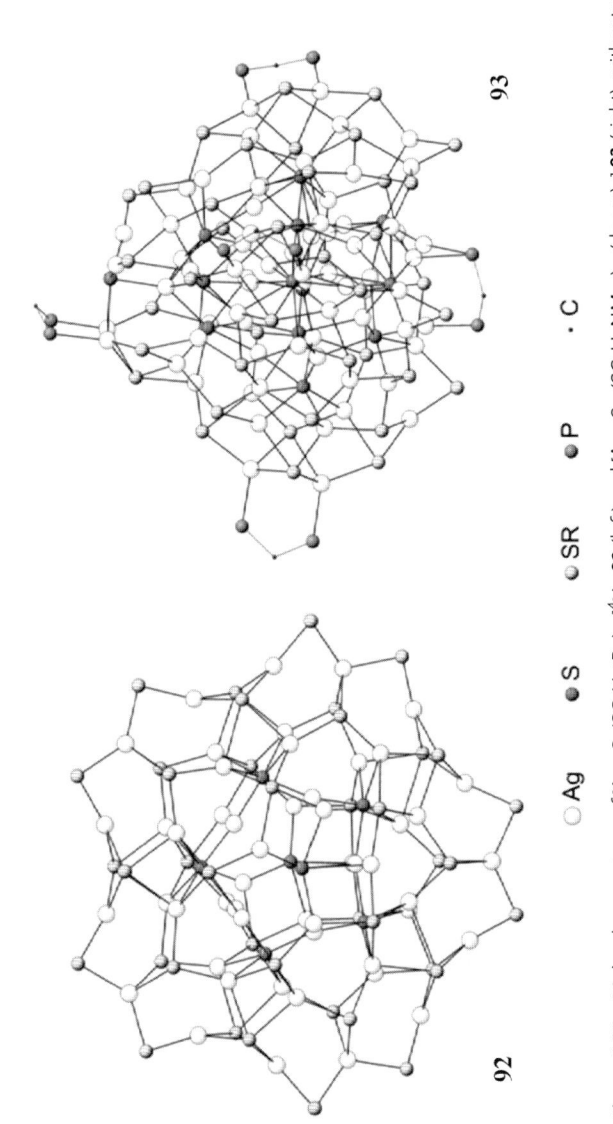

Figure 3.72 Molecular structures of $[Ag_{50}S_7(SC_6H_4tBu)_{40}]^{4-}$ in **92** (left) and $[Ag_{65}S_{13}(SC_6H_4NMe_2)_{28}(dppm)_5]$ **93** (right), without organic groups at the RS ligands.

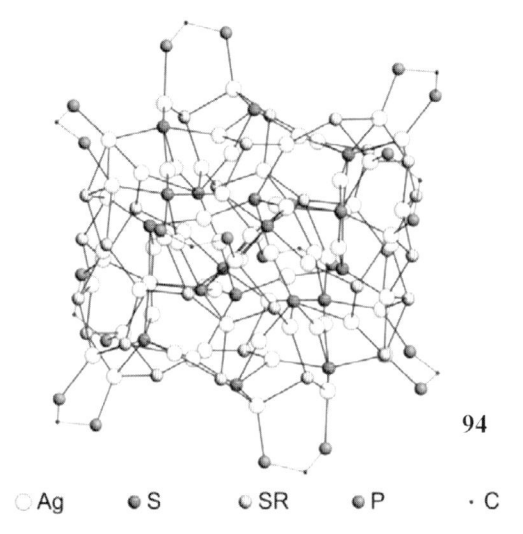

Figure 3.73 The molecular structure of $[Ag_{70}S_{20}(SPh)_{28}(dppm)_{20}]^{2+}$ in **94**.

$(SiMe_3)_2$, should be a triply charged cation. However, there is no evidence of any counterions, and it is assumed therefore that there is a nonstoichiometric silver-to-sulfur ratio. This situation is not surprising, because such an effect is also observed in binary Ag_2S.

A nonstoichiometric composition between silver and is not necessarily caused by the cluster size, however. For example, the spherical cluster $[Ag_{188}S_{94}(PnPr_3)_{20}]$ (**96**; Figure 3.74), which can be synthesized from silver trifluoroacetate with $S(SiMe_3)_2$ in

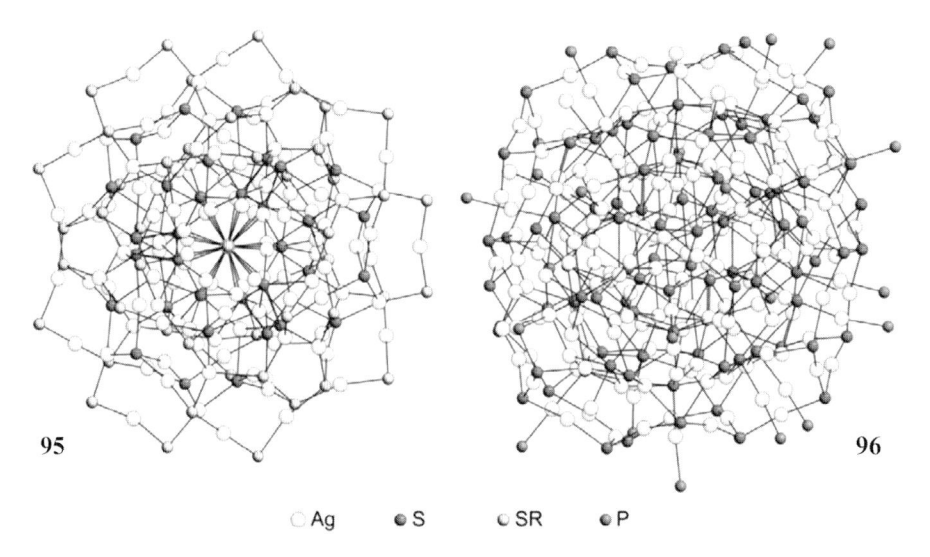

Figure 3.74 Molecular structures of $[Ag_{123}S_{35}(StBu)_{50}]$ (**95**) (left) and $[Ag_{188}S_{94}(PnPr_3)_{20}]$ (**96**) (right).

the presence of PnPr$_3$, is strictly stoichiometric. Nonetheless, it can be said that the metal-rich compounds synthesized to date have shown a increasing tendency to have non stoichiometric compositions with increasing cluster size.

The use of silver benzoate as a metal source and a mixture of S(SiMe$_3$)$_2$ and tBuSSiMe$_3$ in the presence of the bidentate ligand 1,4-bis(diphenylphosphino)butane (dppb), yields the ellipsoid-shaped [Ag$_{262}$S$_{100}$(StBu)$_{62}$(dppp)$_6$] (**97**). A similar reaction using 1,4-bis(diphenylphosphanylmethyl)benzene (dppxy) as a ligand leads to the formation of [Ag$_{344}$S$_{124}$(StBu)$_{96}$] (**98**), where no phosphine ligands are present in the ligand shell; nevertheless, the presence of the phosphine is essential for the cluster formation. Recently, a two-step route was established for the synthesis of giant silver sulfide clusters. For this, the first step involved the isolation of silver thiolates (e.g., AgStBu), the reaction of which with S(SiMe$_3$) in the presence of phosphines led to the clusters [Ag$_{320}$S$_{130}$(StBu)$_{60}$(dppp)$_{12}$] (**99**), [Ag$_{352}$S$_{128}$(StC$_5$H$_{11}$)$_{96}$] (**100**), and [Ag$_{490}$S$_{188}$(StC$_5$H$_{11}$)$_{114}$] (**101**). The first two clusters in this series still had an ellipsoid shape, whereas the largest cluster could be described as a double-ellipsoid. In contrast to the silver selenide clusters with increasing cluster size, there was no structural transition towards motifs that were related to solid-state phases of Ag$_2$S. The cluster cores of the giant clusters **97**, **98**, **99** and **100** are shown in Figure 3.75, while the structure of the cluster core of **101** is illustrated in Figure 3.76.

The composition of these compounds can also be proved by using matrix-assisted laser desorption/ionization-time-of-flight (MALDI-TOF) mass spectrometry (Figure 3.77), whereas signals for higher masses indicated an aggregation of clusters in the gas phase. By thermal decomposition, it is possible to displace the organic groups to produce either nanosized mixed crystals of Ag–Ag$_2$S or silver nanoparticles, where the size of the particles depends on the composition of the initial clusters.

The structural determinations of these large clusters with nuclearities greater than ∼120 metal atoms proved problematic. For such clusters, the intensities of the reflections dropped off rather sharply above $2\theta \approx 40°$ (for MoKα radiation), and the structure refinement usually resulted in unsatisfactorily high R factors, with a high residual electron density close to heavy atoms within the cluster molecule. More satisfying R factors can only be obtained if this electron density can be modeled as disordered atoms which, on the other hand, complicates efforts to provide precise estimates of the molecular formulae. These effects may result from a range of factors:

- There is no perfect translational order in the lattice.
- With the silver-chalcogenide clusters there is a tendency towards nonstoichiometry, as seen for the binary phases [76]. This behavior could be a consequence of the rather similar electronegativities of silver and the chalcogenides. There is no clear distinction between Ag$^+$ and E^{2-} (E = S, Se, Te), and the clusters behave rather like alloys [77].
- The surface tension of the spherical molecules generates a Laplace pressure within the molecule, which can result in a disorder or even a phase transition.
- Interactions between defects within the clusters can themselves lead to an increase in the defect concentration [78].

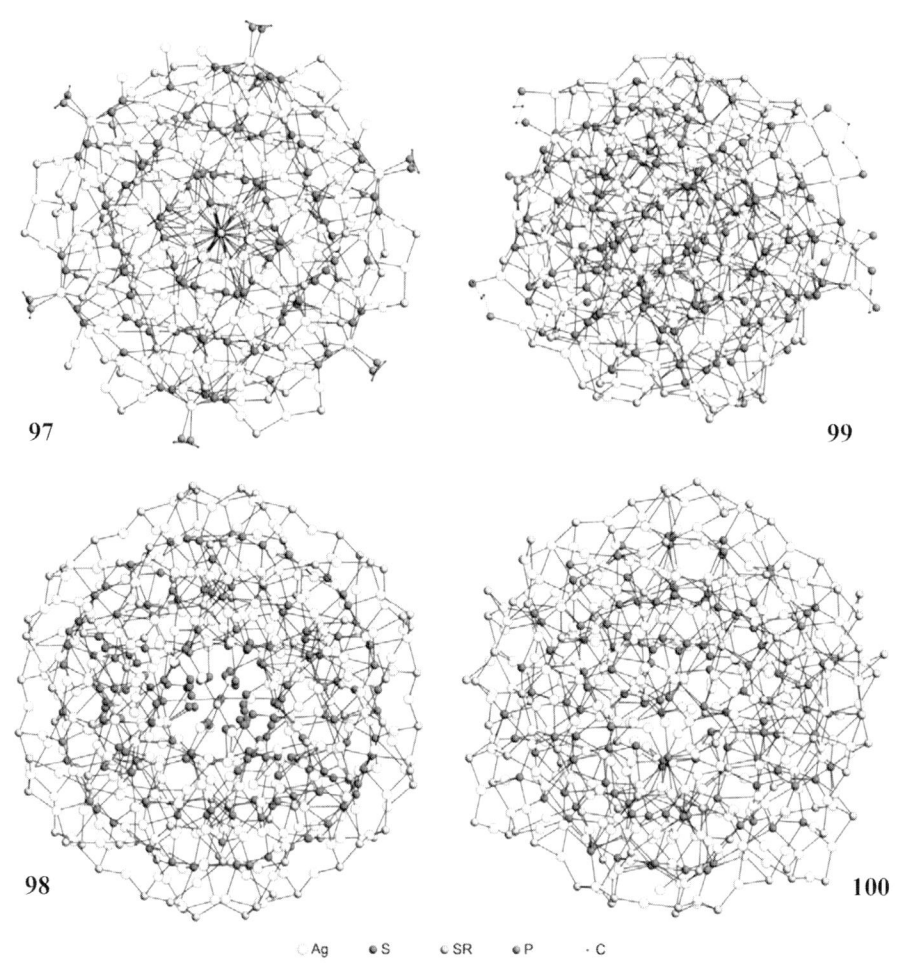

Figure 3.75 Molecular structures of [Ag$_{262}$S$_{100}$(StBu)$_{62}$(dppp)$_6$] (**97**) (top left), [Ag$_{344}$S$_{124}$(StBu)$_{96}$] (**98**) (bottom left), [Ag$_{320}$S$_{130}$(StBu)$_{60}$(dppp)$_{12}$] (**99**) (top right), and [Ag$_{352}$S$_{128}$(StC$_5$H$_{11}$)$_{96}$] (**100**) (bottom right).

It is, therefore, no longer possible (in the strictest sense) to obtain precise molecular formulae for these large clusters from their structure determinations. Rather, it must be considered in terms of "idealized" formulae, resulting from the assumptions that nonbonded Ag$\cdots$Ag distances are larger than 280 pm, and that nonbonding S$\cdots$S distances are at least 410 pm. It is conceivable that different spatial arrangements of Ag$^+$ and chalcogenide ions can be accommodated below the ordered surface of the cluster. This possibility is supported by the observation that datasets measured from crystals obtained from different synthetic reactions can often not be refined with the same set of coordinates for the silver, chalcogen, and carbon atoms, even though they have the same unit cells. Furthermore, a broad

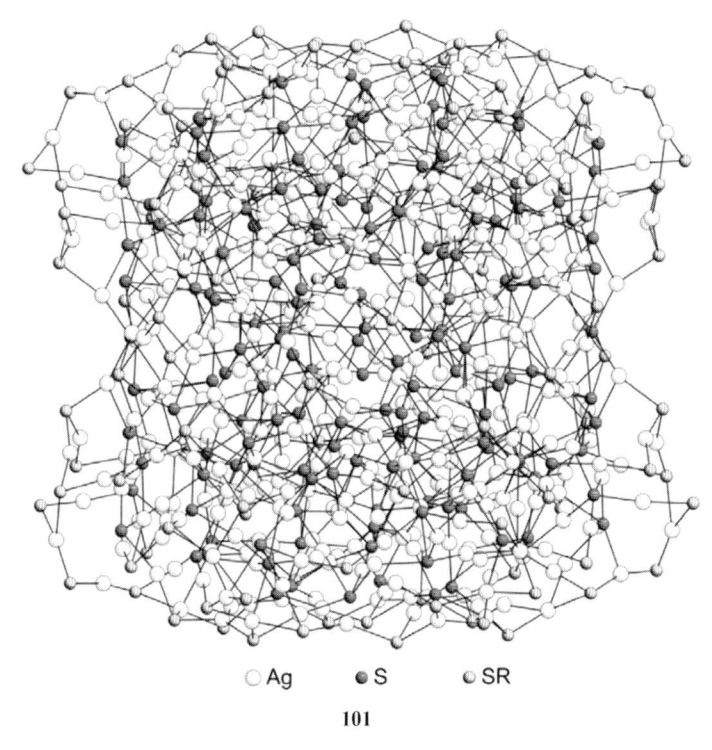

Figure 3.76 The molecular structure of [Ag$_{490}$S$_{188}$(StC$_5$H$_{11}$)$_{114}$] (**101**) without organic groups.

distribution of molecular ions in the mass spectrum is often observed. Today, the structural determination of such large cluster molecules is pushing the currently available techniques to their limits; indeed, it must be accepted that, at present, such techniques may not always be capable of providing satisfactory answers.

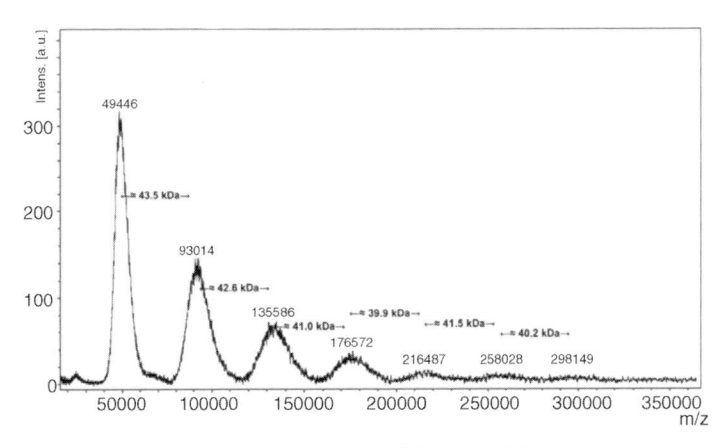

Figure 3.77 MALDI-TOF mass spectrum of [Ag$_{344}$S$_{124}$(StBu)$_{96}$] (**98**).

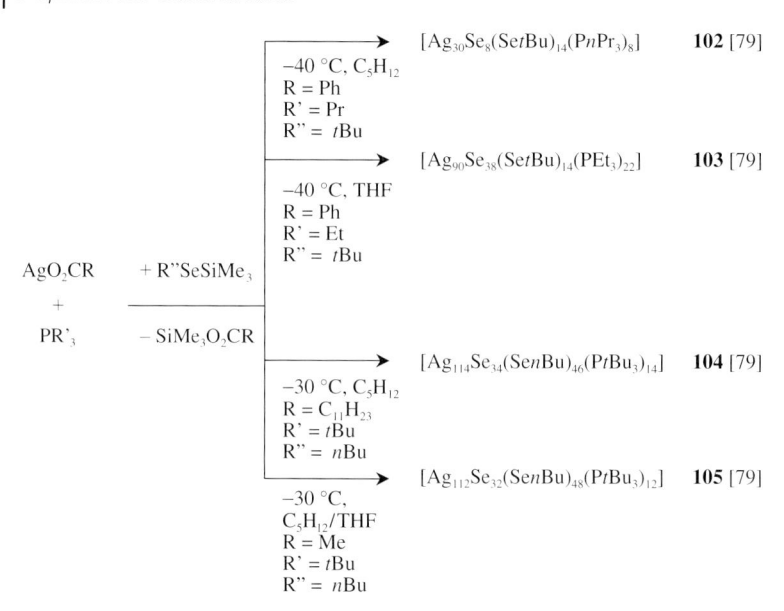

Scheme 3.13

```
AgO₂CR     + R"SeSiMe₃
   +
  PR'₃     − SiMe₃O₂CR
```

→ −40 °C, C₅H₁₂; R = Ph, R' = Pr, R" = tBu → [Ag₃₀Se₈(SetBu)₁₄(PnPr₃)₈]   **102** [79]

→ −40 °C, THF; R = Ph, R' = Et, R" = tBu → [Ag₉₀Se₃₈(SetBu)₁₄(PEt₃)₂₂]   **103** [79]

→ −30 °C, C₅H₁₂; R = C₁₁H₂₃, R' = tBu, R" = nBu → [Ag₁₁₄Se₃₄(SenBu)₄₆(PtBu₃)₁₄]   **104** [79]

→ −30 °C, C₅H₁₂/THF; R = Me, R' = tBu, R" = nBu → [Ag₁₁₂Se₃₂(SenBu)₄₈(PtBu₃)₁₂]   **105** [79]

Scheme 3.13 The synthesis of selenido–selenolato-bridged silver clusters with monodentate PR₃ ligands.

3.2.3.3.2 Selenium-Bridged Silver Clusters The reaction chemistry of Ag(I) and RESiMe₃ complexes in the presence of tertiary phosphine ligands is summarized in Scheme 3.13. Of note is the requirement for acetate salts of silver for the successful isolation of polynuclear complexes, as AgCl invariably leads to Ag₂Se formation. This contrasts with the reaction chemistry of Cu(I) described above, and also with the chemistry of silver–tellurium complexes (see below). Of equal importance is the facile formation of selenide (Se^{2-}) ligands from RSe fragments, as this allows for the formation of extremely large Ag–Se frameworks. Although this reactivity limits the number of low-nuclearity complexes that can be isolated, and thus limits the amount of information that can be obtained regarding condensation pathways, it offers the distinct advantage of the formation of the largest structurally characterized metal–chalcogenide complexes reported to date.

The reaction of silver benzoate with tBuSeSiMe₃ and PPr₃ at 40 °C in pentane as solvent affords the light-sensitive, spherical cluster [Ag₃₀Se₈(SetBu)₁₄(PPr₃)₈] **102** in good (60%) yield [79]. The structure of **102** (Figure 3.78) shows the surface groups (PR3 and tBu) stabilizing the AgSe core. The SetBu ligands bridge two and three silver centers, while the Se^{2-} ligands adopt coordination numbers five and six.

There are ten silver centers that are two-coordinate, bonded to two SetBu ligands, two Se^{2-} ligands, or one SetBu ligand and one Se^{2-} ligand. The remaining 20 silver atoms, including the eight that are bonded to PPr₃ ligands, exhibit distorted trigonal planar or tetrahedral coordination geometries. As is the case for the stoichiometric copper–chalcogenide complexes described above, the shortest nonbonding contacts

102

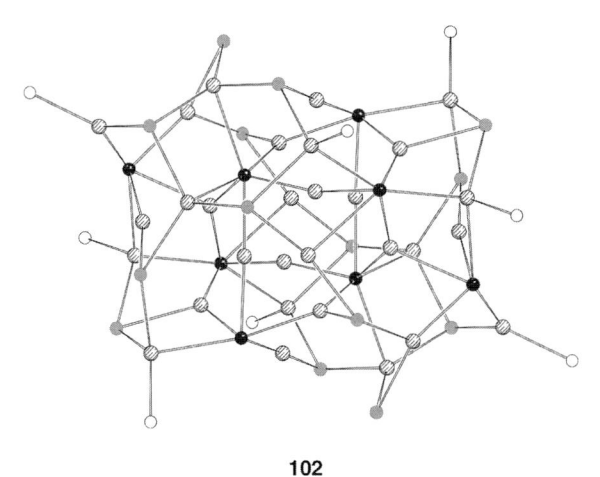

Figure 3.78 The molecular structure of [Ag$_{30}$Se$_8$(SetBu)$_{14}$(PnPr$_3$)$_8$] (**102**) (without organic groups). Ag: shaded spheres; Se: black spheres; SeR: gray spheres; P: white spheres.

(2.904–2.991 Å) between the metal centers are consistent with the +1 oxidation states assigned to all Ag atoms.

A similar reaction for the formation of **102** with PEt$_3$ in THF leads to the larger cluster [Ag$_{90}$Se$_{38}$(SetBu)$_{14}$(PEt$_3$)$_{22}$] (**103**) in 60% yield (Figure 3.79) [79]. The similar basicities and Tolman cone angles [42] for PPr$_3$ and PEt$_3$ suggest that the use of a different phosphine ligand for **103** does not account for the different product formation. A more likely reason is the combination of a different solvent (THF

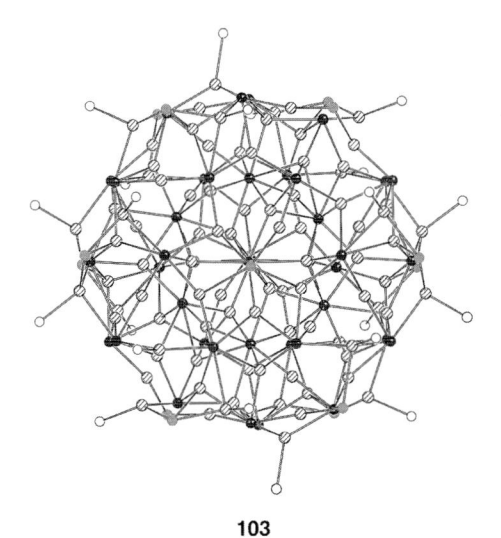

103

Figure 3.79 The core atoms in [Ag$_{90}$Se$_{38}$(SetBu)$_{14}$(PEt$_3$)$_{22}$] (**103**). Ag: shaded spheres, Se: black spheres, SeR: gray spheres, P: white spheres.

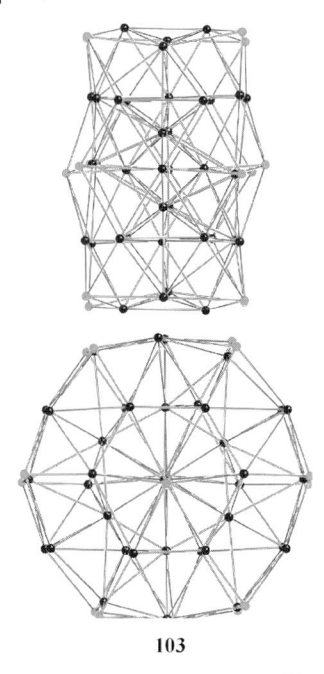

Figure 3.80 Two projections of the Se$_{52}$ selenium substructure in **103**. Se: black spheres, SeR: gray spheres.

versus pentane) with higher reaction–crystallization temperatures (20 °C for **103**; 60 °C for **102**). Higher reaction temperatures for these clusters leads to the formation of bulk Ag$_2$Se and other unidentified amorphous materials. The 38 Se^{2-} ligands in the core of **103** result in a larger, pseudo-spherical framework with core dimensions of $1.1 \times 1.7 \times 1.9$ nμ^3. The 90 Ag centers are ligated to the 38 Se^{2-} and 14 SetBu groups, with 22 also bonded to the surface PEt$_3$ ligands. The metal atoms display linear, trigonal planar, and tetrahedral coordination. The SetBu$^-$ groups ligate in μ_2- and μ_3-fashion, and the Se^{2-} centers also exhibit varying coordination modes, with those in the outer shell bridging four and five silver atoms and the core ligands forming five to eight bonds to the metals. The arrangement of the selenium centers in **103** is best described in two ways (Figure 3.80). The projection at the top of Figure 3.80 illustrates the approximately D6h symmetry found in the nonbonded Se$_{52}$ polyhedron, forming two "belts" of selenium ligands. The outer belt consists of 12 SetBu$^-$ and 18 Se^{2-}, whereas the inner belt is composed exclusively of 18 Se^{2-}. A perpendicular view of the Se framework emphasizes the development of a layered arrangement of these ligands, which dominates the overall structural motif observed in the larger Ag–Se cluster complexes, as described below.

The $2.3 \times 2.3 \times 2.5$ nm^3 AgSe core in [Ag$_{114}$Se$_{34}$(SenBu)$_{46}$(PtBu$_3$)$_{14}$] **104** (Figure 3.81) is reproducibly formed from the reaction of silver laurate (C$_{11}$H$_{23}$CO$_2$Ag), PtBu$_3$ and nBuSeSiMe$_3$ in pentane at 30 °C [79]. The increased size of **104** compared to **102** and **103** is accompanied by a rather marked change in the

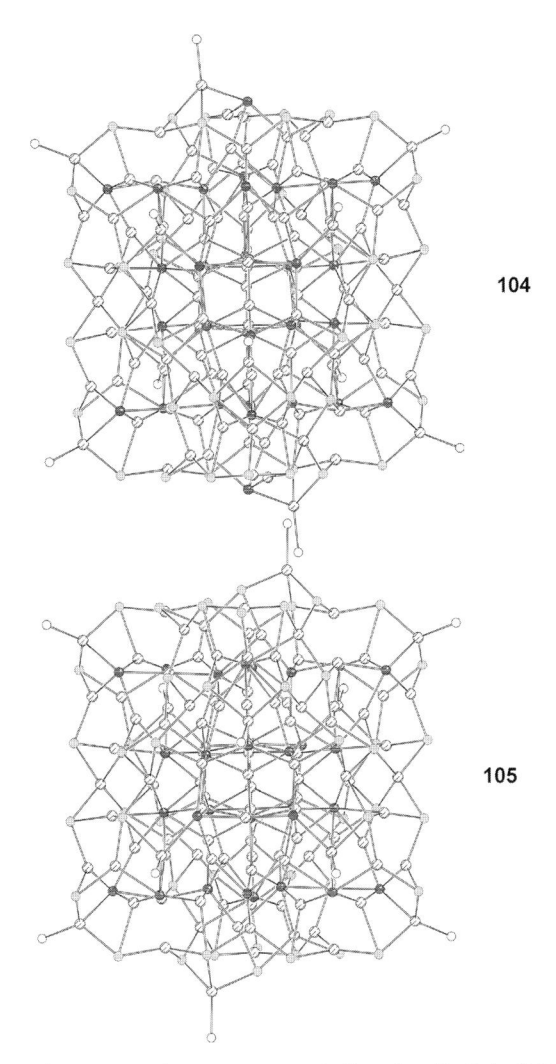

Figure 3.81 The core structures in [Ag$_{114}$Se$_{34}$(SenBu)$_{46}$(PtBu$_3$)$_{14}$] (**104**) (top) and [Ag$_{112}$Se$_{32}$(SenBu)$_{48}$(PtBu$_3$)$_{12}$] (**105**) (bottom). Ag: shaded spheres, Se: black spheres, SeR: gray spheres, P: white spheres.

gross structural features exhibited. Specifically, the arrangement of Se atoms in **104** (Figure 3.82) corresponds to a distorted cubic-centered structure that is also observed in solid Ag$_2$Se. The Ag atoms have trigonal and tetrahedral coordination, and are distributed between the layers of Se, which is also consistent with the structural features observed in the bulk solid. This "transition" – that is, the size-dependent structural changes – is also observed in the PR$_3$-stabilized Cu$_2$Se clusters, where a clear relationship between molecule and bulk is observed for structures containing 70 or more Cu atoms, as described in Sections 3.2.3.2.3. The effect of the different

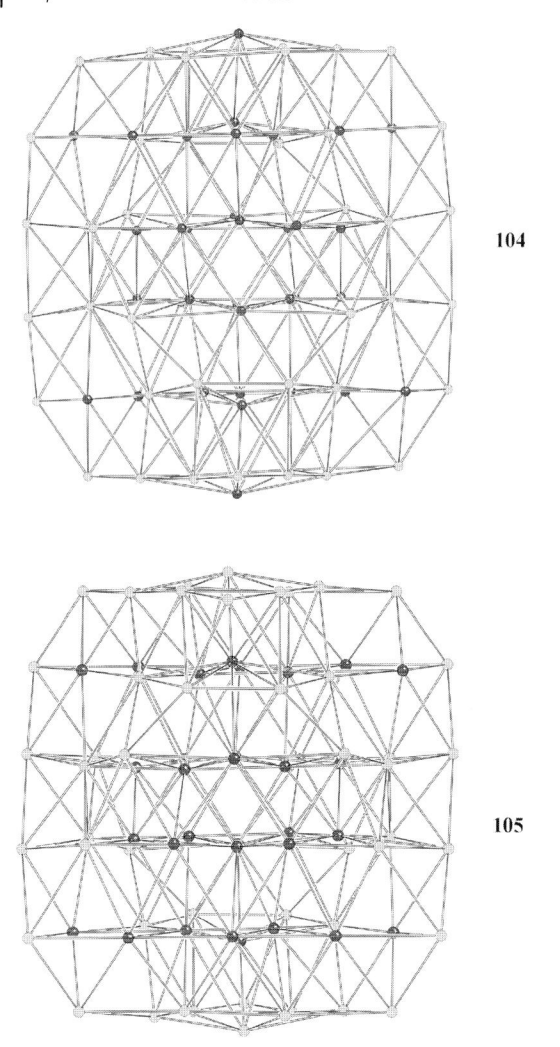

Figure 3.82 The selenium frameworks in **104** (top) and **105** (bottom). Se: black spheres; SeR: gray spheres.

substituents on the selenolate ligands (nBu versus tBu) cannot be discounted, however. The nBuSe$^-$ ligands in **104** bridge three and four Ag(I) centers, and the selenide ligands again are within the bonding distance range to five, six, seven, and eight metals.

The cluster [Ag$_{112}$Se$_{32}$(SenBu)$_{48}$(PtBu$_3$)$_{12}$] **105** (Figure 3.81) is closely related to that of **90**, and is formed from the reaction of AgOAc, PtBu$_3$ and nBuSeSiMe$_3$ in pentane–THF mixtures at 30 °C [79]. The number of selenium atoms in **104** and **105** is identical; however, the number of surface selenolate groups is greater in the latter (48 versus 46), with a concomitant reduction in the number of PtBu$_3$ ligands

3.2 Semiconductor Nanoparticles

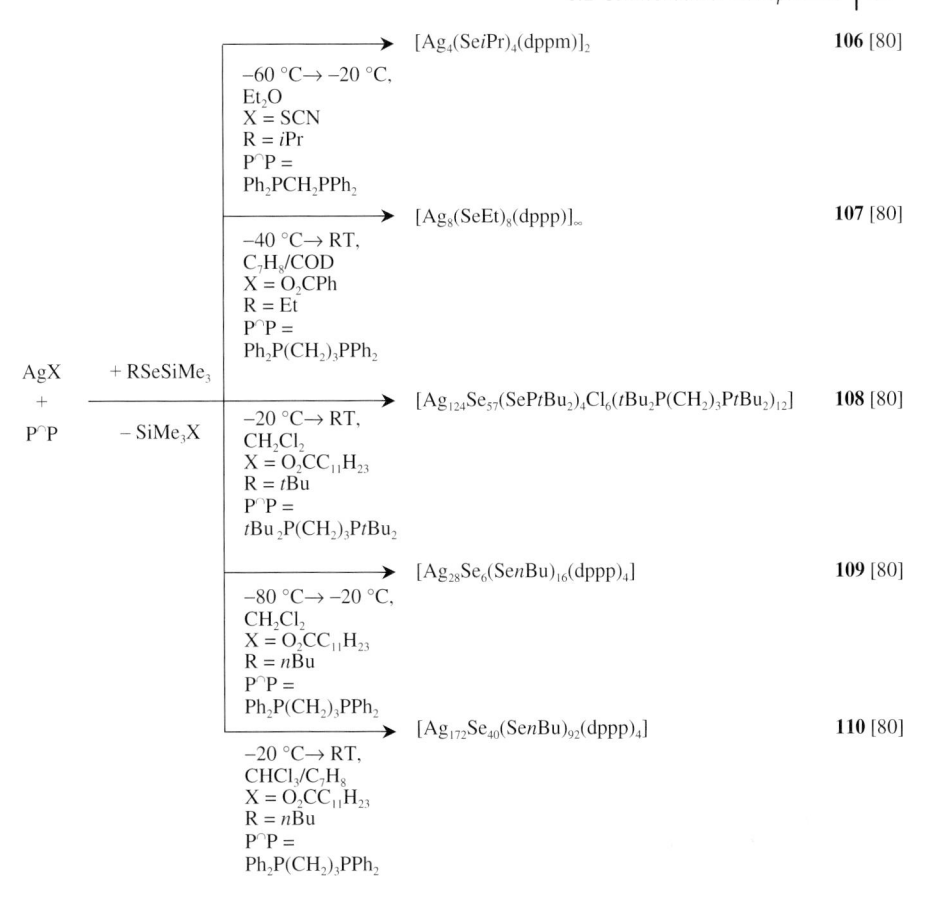

```
                                    → [Ag₄(SeiPr)₄(dppm)]₂                    106 [80]
                    -60 °C → -20 °C,
                    Et₂O
                    X = SCN
                    R = iPr
                    P⌒P =
                    Ph₂PCH₂PPh₂
                                    → [Ag₈(SeEt)₈(dppp)]∞                     107 [80]
                    -40 °C → RT,
                    C₇H₈/COD
                    X = O₂CPh
                    R = Et
                    P⌒P =
                    Ph₂P(CH₂)₃PPh₂
AgX    + RSeSiMe₃                   → [Ag₁₂₄Se₅₇(SePtBu₂)₄Cl₆(tBu₂P(CH₂)₃PtBu₂)₁₂]   108 [80]
 +                  -20 °C → RT,
P⌒P    - SiMe₃X     CH₂Cl₂
                    X = O₂CC₁₁H₂₃
                    R = tBu
                    P⌒P =
                    tBu₂P(CH₂)₃PtBu₂
                                    → [Ag₂₈Se₆(SenBu)₁₆(dppp)₄]               109 [80]
                    -80 °C → -20 °C,
                    CH₂Cl₂
                    X = O₂CC₁₁H₂₃
                    R = nBu
                    P⌒P =
                    Ph₂P(CH₂)₃PPh₂
                                    → [Ag₁₇₂Se₄₀(SenBu)₉₂(dppp)₄]            110 [80]
                    -20 °C → RT,
                    CHCl₃/C₇H₈
                    X = O₂CC₁₁H₂₃
                    R = nBu
                    P⌒P =
                    Ph₂P(CH₂)₃PPh₂
```

Scheme 3.14 The synthesis of selenido–selenolato-bridged silver clusters with bidentate phosphine ligands.

(12 versus 14). Two fewer (112) Ag centers in **105** relative to **104** is consistent with the lower Se^{2-}: $nBuSe^-$ ratio in the former (32 : 48 and 34 : 46, respectively).

Bidentate phosphine ligands have been successfully employed for the synthesis of Cu–E–ER complexes (E = Se, Te) from silylated reagents, as described above. Their utility in stabilizing Ag–SeR and Ag–Se–SeR frameworks using analogous reaction strategies is demonstrated and summarized in Scheme 3.14.

The use of bidentate phosphine ligands in conjunction with low-temperature conditions is apparently sufficient to enable the isolation of low-nuclearity Ag–SeR complexes from Ag(I) salts and $RESiMe_3$, which hitherto was impossible when working with PR_3 ancillary ligands. For example, the cluster $[Ag_4(SeiPr)_4$ $(Ph_2PCH_2PPh_2)_2]$ (**106**) is isolated in 45% yield from silver thiocyanate (or silver benzoate), bis(diphenylphosphino)methane and $iPrSeSiMe_3$ at low temperatures [80]. Cluster **106** (Figure 3.83) consists of an Ag_4 tetrahedron with each nonbonded Ag–Ag edge bridged by a μ_2-$iPrSe^-$ ligand. The trigonal planar

106

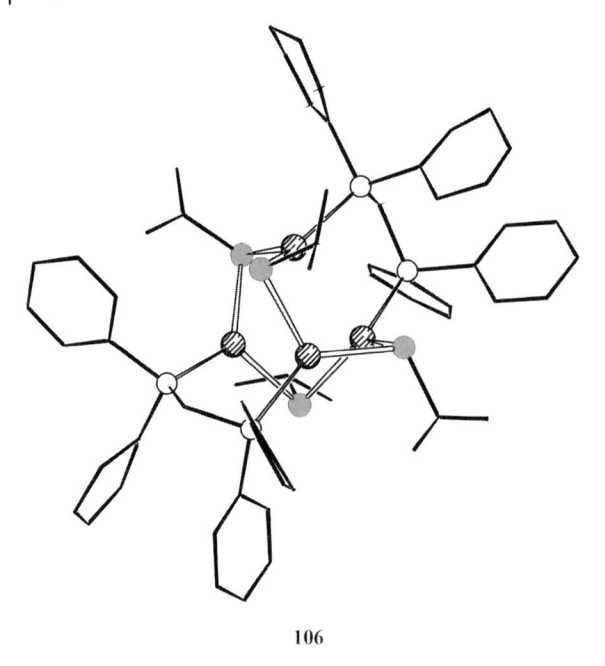

Figure 3.83 The molecular structure of [Ag$_4$(Se*i*Pr)$_4$(Ph$_2$PCH$_2$PPh$_2$)$_2$] (**106**) (without H atoms). Ag: shaded spheres; SeR: gray spheres; P: white spheres; C: wire representation.

coordination geometry about the Ag(I) centers is completed via coordination to a phosphine ligand.

The μ_3-bridging capacity of alkylselenolate ligands and the flexible coordination ability of bis(diphenylphosphino)propane ligands lead to an infinite repeat of [Ag$_8$(SeEt)$_8$(Ph$_2$PCH$_2$CH$_2$CH$_2$PPh$_2$)] units (Figure 3.84) in **107** [80]. This polymeric structure consists of a strand of distorted Se$_6$ octahedral arrangements of ethylselenolate ligands, each of the EtSe$^-$ ligands functioning as μ_3-bridges to the Ag(I) centers. There is no discrepancy between "intra" and "inter" unit bonding distances, which range from 2.509(4) to 2.847(3) Å. In contrast to Ag$_4$ cluster **106**, the silver atoms in the Ag$_8$ cluster **107** display different coordination geometries: a linearly coordinated Ag is bonded to two EtSe ligands above an edge, while a trigonal planar Ag center resides in a deltahedral face of the Se$_6$ octahedron; two additional silver

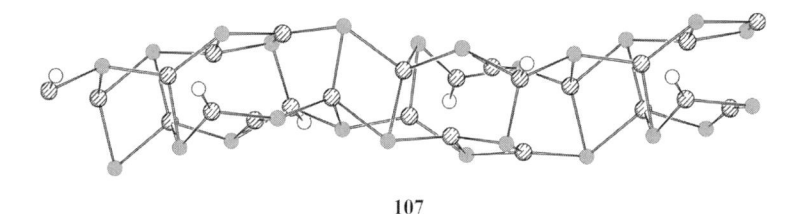

107

Figure 3.84 A projection of the infinite repeat in [Ag$_8$(SeEt)$_8$(Ph$_2$P(CH$_2$)$_3$PPh$_2$)]$_\infty$ (**107**) (without organic groups). Ag: shaded spheres; SeR: gray spheres; P: white spheres.

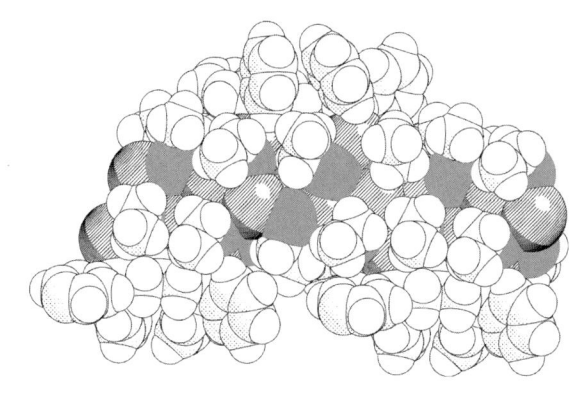

Figure 3.85 A space-filling model of polymeric **107**. Ag: shaded spheres; Se: black spheres; SeR: gray spheres; P: white spheres; C: light gray spheres; H: small white spheres.

atoms show trigonal planar geometry "between" the octahedral subunits; and the remaining silver atoms are three- and four-coordinate, occupying positions above and within the Se_6 octahedra. Although the ethyl groups bonded to the selenium sites have poor shielding ability, the AgSe strand is effectively insulated and stabilized by the dppp ligands (Figure 3.85).

Much larger Ag–Se frameworks are formed using bidentate ligands if the reaction conditions and the selenium reagent are modified. The cluster $[Ag_{124}Se_{57}(SePtBu_2)_4Cl_6(tBu_2P(CH_2)_3PtBu_2)_{12}]$ (**108**; Figure 3.86) is isolated as black crystals

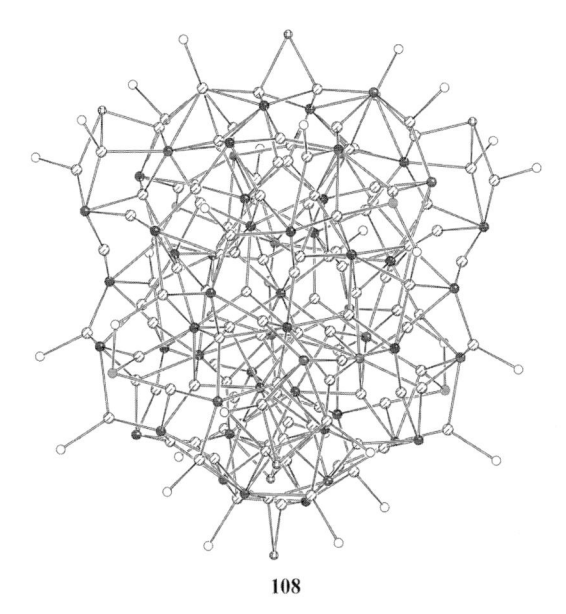

108

Figure 3.86 A projection of the molecular structure of $[Ag_{124}Se_{57}(SePtBu_2)_4Cl_6(tBu_2P(CH_2)_3\text{-}PtBu_2)_{12}]$ (**108**) (without organic groups). Ag: shaded spheres; Se: black spheres; SeR: gray spheres; P: white spheres.

from the reaction of silver laurate, 1,3-bis(di-*tert*-butylphosphino)propane, and *tert*-butylselenotrimethylsilane in methylene chloride at 20 °C [80]. Cluster **108**, with a spherical ~2 nm AgSe core, is a very unusual member of the series of high-nuclearity silver selenide clusters, bears no structural relationship to the clusters **104** and **105** described above, and displays no structural characteristics of the bulk phases of Ag_2Se.

The important structural features of **108** are the Frank–Kasper polyhedra surrounding the central selenide ligand. A nonbonded (Se...Se = 4.93–5.36 Å) Se_{16} polyhedron surrounds the central Se^{2-}, and is itself enveloped by 44 selenium centers, four of which are from the surface $SePtBu^{2-}$ ligands (Figure 3.87). This defined Se_{44} polyhedron (Se...Se = 3.80–5.99 Å) possesses 84 Δ_3 faces that can be described as a growth of the inner Se_{16} deltahedron. The Se^{2-} ligands in **108** act as μ_3, μ_4, μ_5, μ_6, μ_7 or μ_8 bridges to the silver atoms, with the Ag–Se distances ranging from 2.491(3) to 3.101(3) Å. The four selenodi-*tert*-butylphosphido ligands, SePtBu$_2$, are μ_3-η^1(P): η^2(Se)-bonded to 12 Ag(I) centers. The formation of these ligands from the reaction conditions must result from the decomposition of the *t*BuSe-SiMe$_3$ and *t*Bu$_2$P(CH$_2$)$_3$P*t*Bu$_2$ ligands. For their part, the six surface chlorides, which arise from the reaction solvent methylene chloride, are μ_3- and m_2-bridging to the silver centers.

The combination of 57 Se^2, six Cl^- and four $SePtBu^{2-}$ ligands again allows for +1 oxidation states to be assigned to the silver atoms. The 124 Ag(I) centers are distributed as follows: 12 of the 28 tetrahedral sites of the central Se_{16} deltahedron surrounding the central Se are occupied by silver atoms in distorted geometries; 80 silver atoms located between the Se_{16} and Se_{44} polyhedra exhibit linear, trigonal planar and tetrahedral coordination; and the remaining 32 silver atoms are located at the cluster surface with coordination numbers three and four.

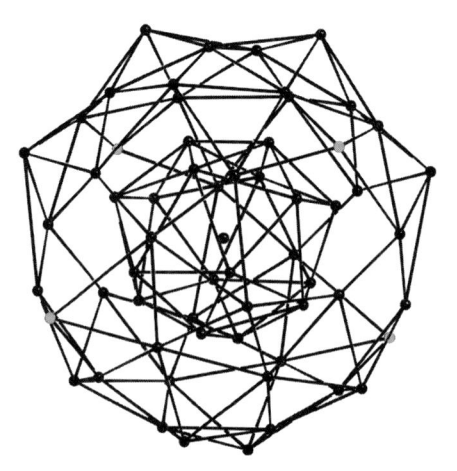

Figure 3.87 The selenium substructure in **108**. The central Se atom is surrounded by the Se_{16} Frank–Kasper polyhedron, which is itself surrounded by the Se_{44} polyhedron in a second coordination sphere.

3.2 Semiconductor Nanoparticles

The effects of solvent and reaction–crystallization conditions on product formation are clearly evident in the syntheses of [Ag$_{28}$Se$_6$(SenBu)$_{16}$(dppp)$_4$] (**109**) and [Ag$_{172}$Se$_{40}$(SenBu)$_{92}$(dppp)$_4$] (**110**). The former is obtained in 45% yield from 1 : 0.25 : 1 mixtures of AgO$_2$CC$_{11}$H$_{23}$: Ph$_2$P(CH$_2$)$_3$PPh$_2$: nBuSeSiMe$_3$ in CH$_2$Cl$_2$ at $-80\,^\circ$C, slowly rising to $-20\,^\circ$C [80]. In contrast, the same combination of reagents in CHCl$_3$–toluene at $20\,^\circ$C generates cluster **110** in 20% yield when the reaction solutions are allowed to warm to room temperature [79].

The red crystals of **109** formed from the reaction mixtures are very sensitive and decompose at temperatures above $10\,^\circ$C. An X-ray analysis of the highly solvated crystals is possible, however, and the structure of **109** (Figure 3.88) displays the features expected for a "medium-sized" Ag–Se complex that is not structurally dissimilar to [Ag$_{30}$Se$_8$(SetBu)$_{14}$(PPr$_3$)$_8$] (**102**). Each of the nBuSe$^-$ ligands is μ_3-bonded on the surface of the cluster, and the six Se$^-$ atoms are arranged to form a nonbonded octahedron at the center of the cluster. For their part, the Ag(I) atoms display the (distorted) linear, trigonal planar and tetrahedral coordination geometries expected in such a medium-sized cluster.

The cluster $2.4 \times 2.4 \times 2.0\,\text{nm}^3$ AgSe core in [Ag$_{172}$Se$_{40}$(SenBu)$_{92}$(dppp)$_4$] (**110**) represents the largest structurally characterized metal–selenide complex reported to date (Figure 3.89) [79]. With increasing cluster size, the distribution of the silver centers appears to become chaotic, and the refinement of the X-ray data for **110** is correspondingly problematic. In this respect, **110** reflects the structure of solid Ag$_2$Se, where the Ag$^+$ cations are also disordered. On the cluster surface, 92 nBuSe ligands are μ_3- and μ_4-bonded to the silver atoms (Se–Ag = 2.498(3)–3.064(3) Å). Noteworthy is the fact that there are only eight phosphine–silver bonds (four dppp ligands) on the surface, as the butyl chains of the selenolate ligands provide the majority of the stabilization of the silver–selenide core. The 40 Se^{2-} ligands act as μ_5, μ_6, μ_7 and μ_8 bridges (Ag–Se = 2.44–3.30 Å), and are arranged with the Se centers of the nBuSe$^-$ groups to form layered arrangements of selenium.

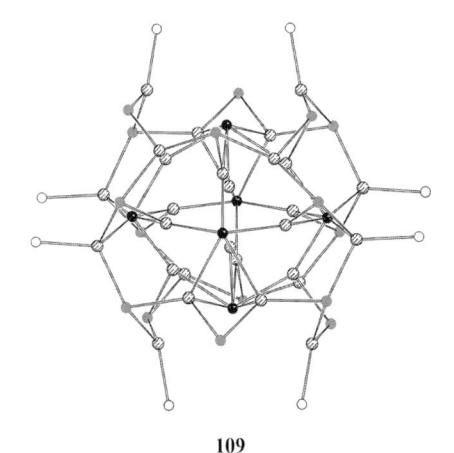

Figure 3.88 A projection of the core atoms in [Ag$_{28}$Se$_6$(SenBu)$_{16}$(dppp)$_4$] (**109**) (without organic groups). Ag: shaded spheres; Se: black spheres; SeR: gray spheres; P: white spheres.

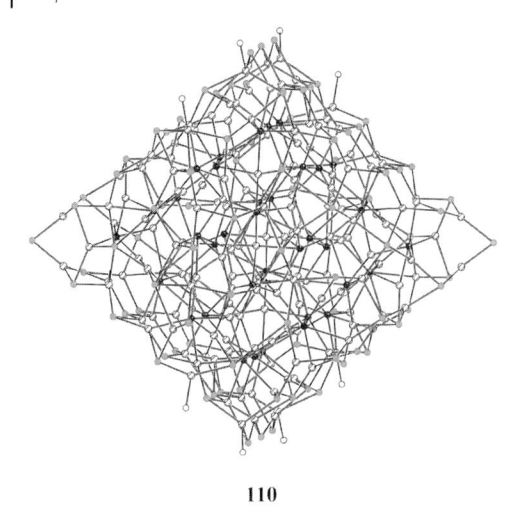

Figure 3.89 The Ag, Se, and P atoms in the 2.4 × 2.4 × 2.0 nm^3 AgSe core in [Ag$_{172}$Se$_{40}$(SenBu)$_{92}$(dppp)$_4$] (**110**).

3.2.3.3.3 Tellurium-Bridged Silver Clusters Silver chloride is dissolved readily in ether or hydrocarbon solvents using excess trialkylphosphine ligands. At room temperature, the addition of nBuTeSiMe$_3$ to these solutions proceeds with the precipitation of insoluble, brick-red, amorphous materials or the formation of Ag$_2$Te. Low-temperature reactions again provide a route to access crystalline products suitable for X-ray analysis (Scheme 3.15) [81]. The addition of nBuTeSiMe$_3$ to pentane solutions of a 1 : 3 mixture of AgCl : PEt$_3$ at 40 °C yields a gold-colored solution from which bright yellow crystals of [Ag$_6$(TenBu)$_6$(PEt$_3$)$_4$] (**111**; Figure 3.90) grow within several days in modest yield (30%). The six tellurolate ligands define a nonbonded octahedron, with four of the nBuTe$^-$ ligands μ$_3$-bonded and two nBuTe$^-$ ligands bridging two Ag atoms. Two Ag(I) centers are tetrahedral, bonded to two PEt$_3$ and two nBuTe$^-$ ligands, while the four remaining silver atoms are trigonal planar and bonded to the butyltellurolate ligands only, residing in the

AgCl + x PEt$_3$	$+ n$BuTeSiMe$_3$ / $-$ SiMe$_3$Cl → $-40\,°\text{C} \to -30\,°\text{C}$, C$_5H_{12}$, $x = 3$ → [Ag$_6$(TenBu)$_6$(PEt$_3$)$_6$]	**111** [81]
	$-40\,°\text{C} \to -30\,°\text{C}$, C$_5H_{12}$, $x = 1$ → [Ag$_{32}$Te$_7$(TenBu)$_{18}$(PEt$_3$)$_6$]	**112** [81]
	$-40\,°\text{C} \to -30\,°\text{C}$, C$_5H_{12}$, $x = 2$ → [Ag$_{48}$Te$_{12}$(TenBu)$_{24}$(PEt$_3$)$_{14}$]	**113** [81]

Scheme 3.15 A summary of the reaction chemistry of Et$_3$P–AgCl and nBuTeSiMe$_3$.

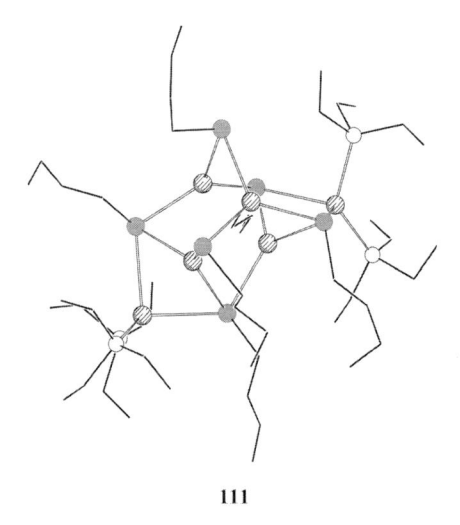

Figure 3.90 The molecular structure of [Ag$_6$(TenBu)$_6$(PEt$_3$)$_4$] (**111**) (without H atoms). Ag: shaded spheres; TeR: gray spheres; P: white spheres; C: wire representation.

deltahedral faces of the Te$_6$ polyhedron. The importance of th phosphine: metal ratio in the synthesis of metal–chalcogenide complexes with silylated reagents is again highlighted with the synthesis of [Ag$_{32}$(TenBu)$_{18}$Te$_7$(PEt$_3$)$_6$] (**112a**): if the ratio of phosphine to silver is only 1 : 1, there is no evidence of **111**, and cluster **112a** is formed in low yield. The projection illustrates the layered nature of cluster **112a**, which resides on a crystallographic inversion center (Figure 3.91). The "top" layer of **112a** consists of three μ_3-nTe(Bu)$^-$ groups, whereas the second tellurium-rich layer contains three μ_3-tellurolate and three μ_5-telluride (Te^{2-}) ligands. The latter form a near-planar hexagonal Te$_3$Ag$_3$ ring (maximum deviation = 0.10 Å), which lies above (0.61 Å) a tetrahedral Ag center. The central tellurium layer contains six μ_3-nTe (Bu)$^-$ ligands and one, central Te^{2-} atom, which is within the range of bonding distances to eight silver atoms. The arrangement of Te is repeated in the bottom half of the structure. The result is that 24 Ag atoms all form three bonds to neighboring Te atoms, and six Ag atoms achieve trigonal planar coordination by bonding to two tellurolate ligands and one phosphine ligand. The butyltellurolate ligands in **112** symmetrically bridge three metal centers. The combination of 18 tellurolate (**18**) and seven telluride ligands (**14**) again suggests a +1 oxidation state for the 32 silver atoms. The yield of cluster **112a** is low (~5%), although the synthesis is reproducible. There appears to be some inherent stability associated with the Ag–Te framework, as analogous [Ag$_{32}$(TenBu)$_{18}$Te$_7$(PR$_3$)$_6$] clusters can be isolated using other tertiary phosphine ligands, and [Ag$_{32}$(TenBu)$_{18}$Te$_7$(PEt$_2$Ph)$_6$] (**112b**), [Ag$_{32}$(TenBu)$_{18}$Te$_7$(PnPr$_3$)$_6$] (**112c**) and [Ag$_{32}$(TenBu)$_{18}$Te$_7$(PiPr$_3$)$_6$] (**112d**) are also similarly prepared. Unlike **112a–112c**, for which yields are low, the yield of cluster **112d** (R = iPr) is relatively good (40%), as reaction solutions (of AgCl, PiPr$_3$ and nBuTeSiMe$_3$) can be stirred at room temperature prior to crystallization without the formation of insoluble, solid materials. Presumably, this facilitates

112a

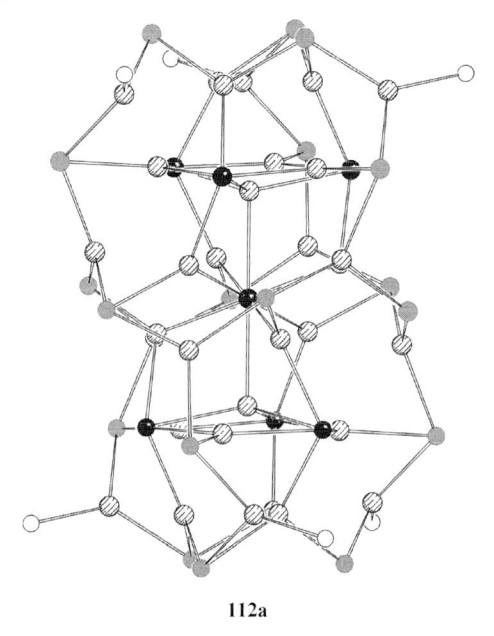

Figure 3.91 The molecular structure of [Ag$_{32}$(TenBu)$_{18}$Te$_7$(PEt$_3$)$_6$] (**112a**) (without organic groups). Ag: shaded spheres; Te: black spheres; TeR: gray spheres; P: white spheres.

the formation of additional Te^{2-} ions via the tellurium–carbon bond cleavage of butyltellurolate moieties. The range of cone angles of the phosphines (132–160°) for which identical cluster frameworks are observed illustrates the prominent role that the alkyl chains on the Te centers play in stabilizing the cluster core, consistent with the limited number (six) of phosphines on the surface of the cluster. Attempts to improve the yields of clusters **112a–d** by using a mixture of nBuTeSiMe$_3$ and Te(SiMe$_3$)$_2$ have not yielded any crystalline materials.

By modifying the ratio of triethylphosphine to silver chloride, such that AgCl: PEt$_3$: nBuTeSiMe$_3$ was 1 : 2 : 1, the cluster [Ag$_{48}$(TenBu)$_{24}$Te$_{12}$(PEt$_3$)$_{14}$] (**113**; Figure 3.92) was isolated as dark-red crystals after several days at low temperatures. The cluster consisted of two spherical Ag$_{24}$Te$_{16}$ subunits linked by four µ$_3$-TenBu$^-$ ligands about a crystallographic inversion center. In contrast to **112**, the 14 Ag–PEt$_3$ centers in **113** are all tetrahedral. The remaining silver atoms within the cluster core have coordination number three (trigonal planar) or four (tetrahedral), while longer Ag...Te contacts also suggest intermediate bonding descriptions. The 24 nBuTe$^-$ ligands and 12 Te^{2-} atoms were in agreement with a +1 oxidation state to be assigned to the Ag sites, and the Ag(I)...Ag(I) contacts ranged between 2.802(2) and 3.222(2) Å.

Although the synthesis of butyltellurolate–silver complexes requires low temperatures in order to avoid the precipitation of amorphous solids, and relatively few structural types of these compounds have been isolated (see above), the chemistry of phenyltellurolate–silver complexes from PhTeSiMe$_3$ and AgCl is rich and varied with respect to the polynuclear frameworks formed. These aryltellurolate reagents can also

113

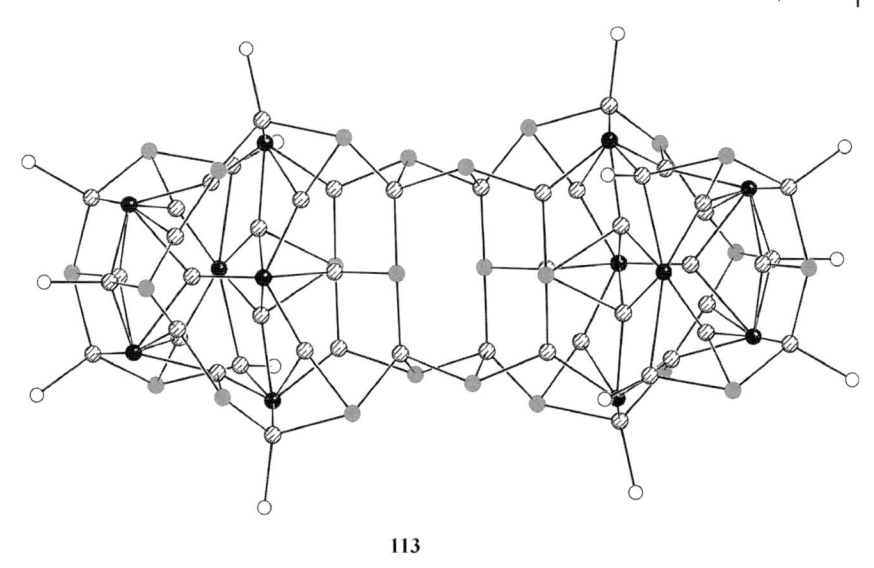

Figure 3.92 The molecular structure of [Ag$_{48}$(TenBu)$_{24}$Te$_{12}$(PEt$_3$)$_{14}$] (**113**) (without organic groups). Ag: shaded spheres; Te: black spheres; TeR: gray spheres; P: white spheres.

be classified as being "gentler," in that reactions between the phosphine-solubilized metal salt and the silylated chalcogens can often be performed at room temperature so as to yield medium-sized polynuclear complexes.

The reaction of a suspension of AgCl : 2PMe$_3$ with PhTeSiMe$_3$ yields two products, [Ag$_{10}$(μ$_3$-TePh)$_{10}$(PMe$_3$)$_2$]$_\infty$ (**114**) and [Ag$_{14}$Te(TePh)$_{12}$(PMe$_3$)$_8$] (**115**) and in 40% and 20% yield, respectively (Scheme 3.16) [82].

Complex **114** consists of a polymeric, one-directional arrangement, in which the PhTe$^-$ ligands serve as both intra- and intermolecular bridges. A crystallographic inversion center relates the two halves of the Ag$_{10}$ cluster frame (Figure 3.93), which is repeated to form the polymeric chain. All PhTe$^-$ ligands are bonded to three Ag(I)

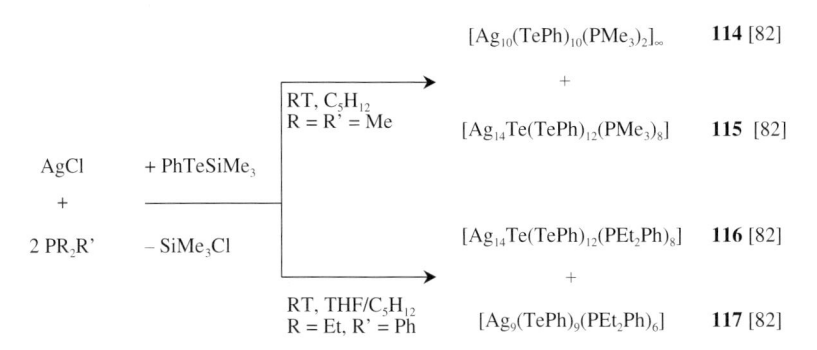

AgCl + PhTeSiMe$_3$
+
2 PR$_2$R'   − SiMe$_3$Cl

RT, C$_5$H$_{12}$
R = R' = Me → [Ag$_{10}$(TePh)$_{10}$(PMe$_3$)$_2$]$_\infty$ **114** [82]
+
[Ag$_{14}$Te(TePh)$_{12}$(PMe$_3$)$_8$]   **115** [82]

[Ag$_{14}$Te(TePh)$_{12}$(PEt$_2$Ph)$_8$]   **116** [82]
+
RT, THF/C$_5$H$_{12}$
R = Et, R' = Ph → [Ag$_9$(TePh)$_9$(PEt$_2$Ph)$_6$]   **117** [82]

Scheme 3.16 The synthesis of telluride–tellurolato-bridged silver clusters with monodentate phosphine ligands.

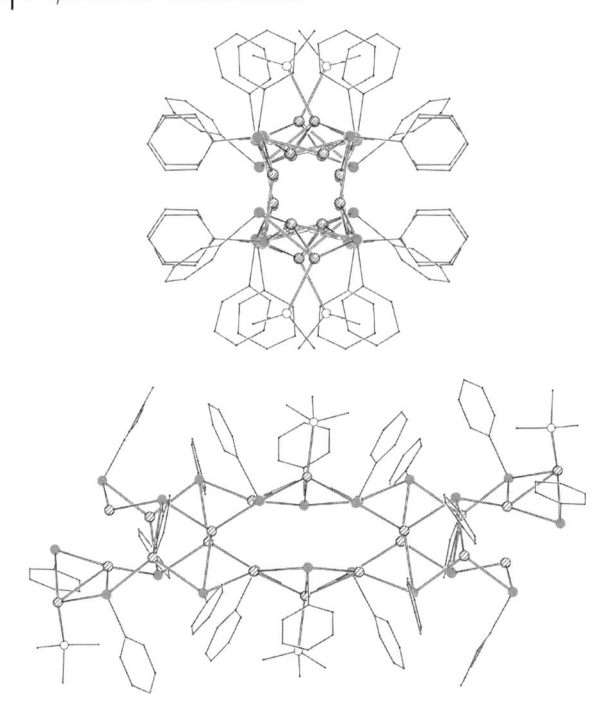

Figure 3.93 Two different views of the infinite repeat observed in [Ag$_{10}$(μ_3-TePh)$_{10}$(PMe$_3$)$_2$]$_\infty$ (**114**) (H atoms omitted). Ag: shaded spheres; TeR: gray spheres; P: white spheres; C: wire representation.

atoms, and the silver centers are three-coordinate and trigonal planar (sum of the angles = 350.3–357.2) with the exception of those bonded to the PMe$_3$ ligand, which adopt a tetrahedral coordination geometry. There is no significant difference in length between the inter- and intracluster Te–Ag contacts. The structure of **114** consists of a rather shell-like make-up: the inner core consists of silver atoms surrounded by a Te–Ag layer which is itself surrounded by PMe$_3$ and Ph ligands (Figure 3.93). The structure consists of a repeating clustered arrangement of silver and tellurium, forming an AgTe "wire."

The second product formed in the reaction of AgCl: 2PMe$_3$ with PhTeSiMe$_3$ is the molecular complex [Ag$_{14}$Te(TePh)$_{12}$(PMe$_3$)$_8$] (**115**) [82]. In **115** (Figure 3.94), the 14 silver centers are bonded to μ_3-, three μ_4- and two m-TePh$^-$ ligands with a central μ_6-Te^{2-}. The Ag–Te distances associated with the two m-TePh ligands are the shortest (on average) within the cluster. The nonbonded Te$_{13}$ framework can be described as a capped, centered icosahedron in which one μ_5 vertex is missing. The eight silver centers bonded to phosphine ligands exhibit tetrahedral coordination geometry, the remaining sites being either trigonal planar or tetrahedral.

The highly distorted and asymmetric nature of the cluster **115** can be contrasted with the framework observed in [Ag$_{14}$Te(TePh)$_{12}$(PEt$_2$Ph)$_8$] (**116**; Figure 3.95). Although the ratio of Ag: TeR: Te: PR$_3$ is the same in **115** and **116**, there are marked

115

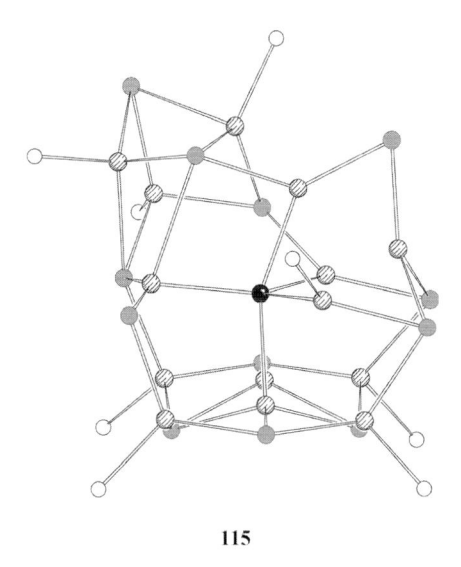

Figure 3.94 The Ag–Te–P framework in [Ag$_{14}$Te(TePh)$_{12}$(PMe$_3$)$_8$] (**115**) (without organic groups). Ag: shaded spheres; Te: black spheres; TeR: gray spheres; P: white spheres.

differences in the structures of the two clusters [82]. The eight PEt$_2$Ph ligands in **116** are located at the corners of a cube, while the 12 TePh$^-$ ligands are evenly distributed, forming three (distorted) layers that are each bonded to three silver atoms. The central telluride is within the range of bonding distances (2.768(1)–2.825(1) Å) to six silver atoms. The Te$_{13}$ framework can be described as a highly distorted centered icosahedron. Of the 20 deltahedral faces of the icosahedron, eight are capped by the

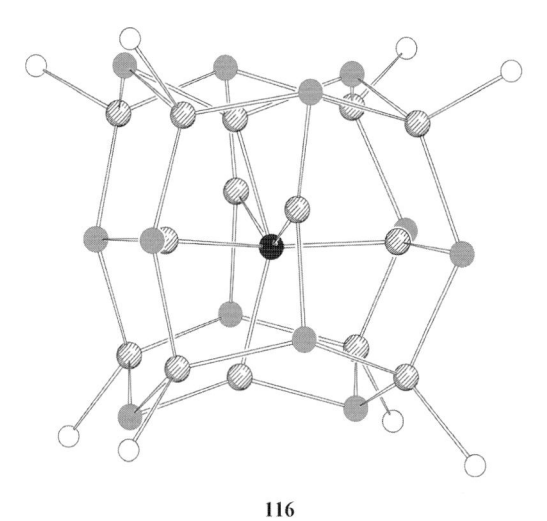

116

Figure 3.95 The Ag–Te–P framework in [Ag$_{14}$Te(TePh)$_{12}$(PEt$_2$Ph)$_8$] (**116**) (without organic groups). Ag: shaded spheres; Te: black spheres; TeR: gray spheres; P: white spheres.

tetrahedral Ag(PEt$_2$Ph) sites. The Ag–Te distances at these four-coordinate silver sites are longer than the six three-coordinate Ag(I), which lie within the Te$_{13}$ frame: the latter have Ag–Te contacts ranging from 2.660(1) to 2.825(1) Å, whereas for the former set the Ag–Te contacts are longer, ranging from 2.838(1) to 3.026(1) Å. The coordination about the three-coordinate silver centers is, however, highly asymmetric, as invariably one of the Te–Ag–Te angles lies close to 90°.

The second, and major, product (60%) from the reaction of AgCl: PEt$_2$Ph with Te (Ph)SiMe$_3$ is the cluster [Ag$_9$(TePh)$_9$(PEt$_2$Ph)$_6$] (**117a**; Scheme 3.16) [82]. The Ag–Te framework (Figure 3.96) sits on a threefold rotation axis that bisects three three-coordinate Ag(I) atoms in the center of the rather columnar cluster that are bonded to three TePh$^-$ ligands in trigonal planar fashion. The other six silver atoms are each bonded to one PEt$_2$Ph and three TePh$^-$ ligands, and are tetrahedral. All TePh$^-$ groups bridge three Ag(I) atoms. The coordination environment about the three central TePh$^-$ ligands is unusual in the Ag–TePh clusters described, as it consists of a "T-shaped" coordination about tellurium. The central layer consists of a planar Ag(TePh)$_3$ arrangement, with the three phenyl rings lying in this plane. The extended packing arrangement of **117a** results in the central Ag(TePh)$_3$ layers being aligned, in a mixed aromatic-Ag–Te infinite repeat.

With the exception of [Ag$_{32}$(TenBu)$_{18}$Te$_7$(PiPr$_3$)$_6$] (**112d**), the systematic access of higher nuclearity, nanoscale AgTe clusters in good yield using monosilylated reagents R–ArTeSiMe$_3$ is not as effective as the corresponding technique for Cu–Te or Ag–Se complexes. However, a combination of silylated tellurium sources (i.e., ArTeSiMe$_3$ and Te(SiMe$_3$)$_2$) can be used to access metal–telluride clusters in good yield, if the substituent on the tellurolate ligand is an aryl group. Thus, although the reaction of AgCl: PR$_3$ solutions and Te(SiMe$_3$)$_2$ leads to the formation of bulk Ag$_2$Te, AgCl dissolved in ethereal solvents with triethylphosphine reacts with a combination

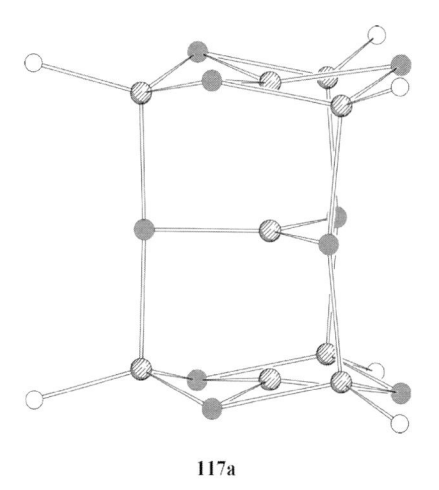

117a

Figure 3.96 The Ag–Te–P framework in [Ag$_9$(TePh)$_9$(PEt$_2$Ph)$_6$] (**117a**) (without organic groups). Ag: shaded spheres; TeR: gray spheres; P: white spheres.

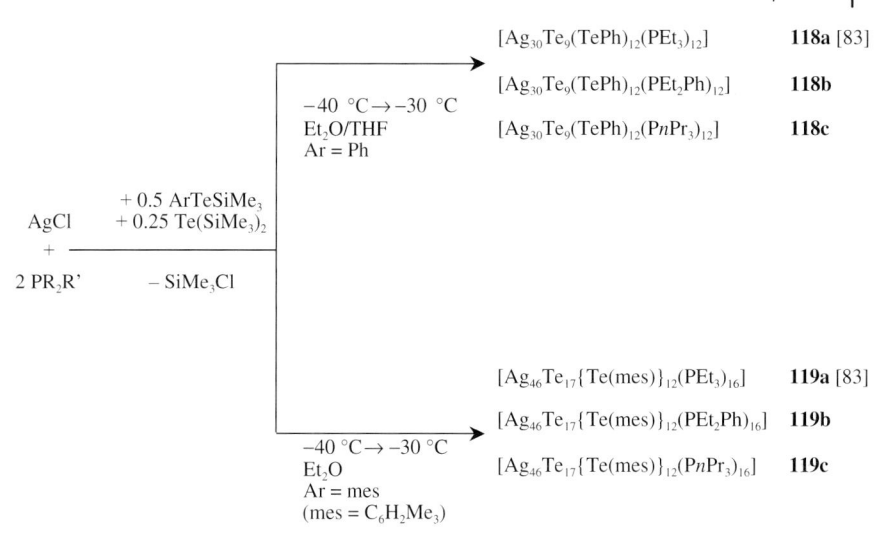

3.2 Semiconductor Nanoparticles | 203

$$\text{AgCl} + 2\,\text{PR}_2\text{R'} \xrightarrow[\text{- SiMe}_3\text{Cl}]{+\,0.5\ \text{ArTeSiMe}_3\ +\,0.25\ \text{Te(SiMe}_3)_2}$$

−40 °C → −30 °C
Et$_2$O/THF
Ar = Ph

[Ag$_{30}$Te$_9$(TePh)$_{12}$(PEt$_3$)$_{12}$]	118a [83]
[Ag$_{30}$Te$_9$(TePh)$_{12}$(PEt$_2$Ph)$_{12}$]	118b
[Ag$_{30}$Te$_9$(TePh)$_{12}$(PnPr$_3$)$_{12}$]	118c

−40 °C → −30 °C
Et$_2$O
Ar = mes
(mes = C$_6$H$_2$Me$_3$)

[Ag$_{46}$Te$_{17}${Te(mes)}$_{12}$(PEt$_3$)$_{16}$]	119a [83]
[Ag$_{46}$Te$_{17}${Te(mes)}$_{12}$(PEt$_2$Ph)$_{16}$]	119b
[Ag$_{46}$Te$_{17}${Te(mes)}$_{12}$(PnPr$_3$)$_{16}$]	119c

Scheme 3.17 The synthesis of higher nuclearity telluride–tellurolato-bridged silver clusters from AgCl, PR$_2$R', Te(SIMe$_3$)$_2$, and ArTeSiMe$_3$.

of Te(SiMe$_3$)$_2$ and ArTeSiMe$_3$ (Ar = Ph, mes; mes (mesityl) = C$_6$H$_2$Me$_3$) to yield the cluster complexes [Ag$_{30}$(TePh)$_{12}$Te$_9$(PEt$_3$)$_{12}$] (**118a**) and [Ag$_{46}$(Temes)$_{12}$Te1$_7$(PEt$_3$)$_{16}$] (**119a**), respectively (Scheme 3.17) [83].

When an AgCl: PEt$_3$ (1:2) solution is treated with 0.5 equivalents of PhTeSiMe$_3$ and 0.25 equivalents of Te(SiMe$_3$)$_2$ at 40 °C, red crystals of **118a** form within a few days at 30 °C in fair yield (20%). The ∼1.3 × 1.1 × 0.8 nm^3 cluster core is comprised of 12 tellurolate (PhTe$^-$), nine telluride (Te^{2-}), and 30 silver atoms (Figure 3.97).

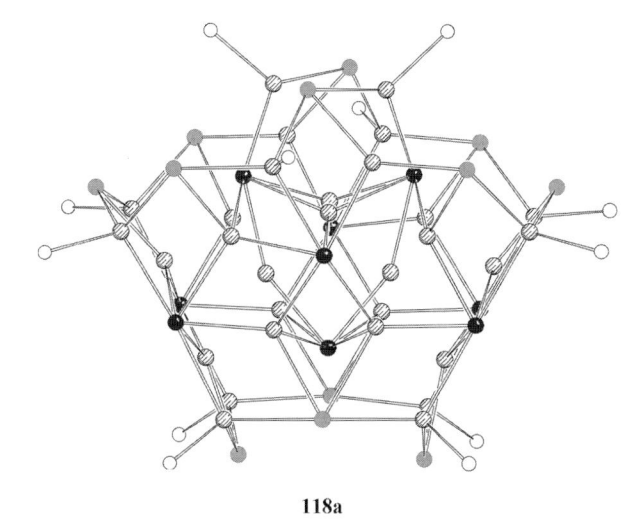

118a

Figure 3.97 A projection of the core atoms in [Ag$_{30}$(TePh)$_{12}$Te$_9$(PEt$_3$)$_{12}$] (**118a**). Ag: shaded spheres; Te: black spheres; TeR: gray spheres; P: white spheres.

Similar complexes can be prepared using the different phosphine ligands [PEt$_2$Ph (**118b**) and PnPr$_3$ (**118c**)].

The order of addition of the silylated reagents does not affect the yield of product. The phenyltellurolate ligands in **118a** act as both μ_3- and μ_4-bridges between the Ag (I) centers. Each of the nine Te^{2-} atoms forms six bonding contacts to the Ag centers (Ag−Te = 2.703(3)–3.185(3) Å). Of the 12 phosphine-bonded silver sites, all but two – which are trigonal planar – exhibit tetrahedral coordination. The coordination environments of the silver atoms bonded only to tellurium are varied: two silver atoms form two bonds, each [<(Te−Ag−Te) = 153.1(1)–153.22(9)°], while eight silver atoms are three-coordinate and six are tetrahedral. This varied coordination geometry about silver in the cluster contrasts with the trigonal planar geometry of the majority of the silver centers in **112** and **113**, but is consistent with the more close-packed nature of the AgTe framework in **118** and the higher coordination modes of the interstitial Te ligands. Similar varied coordination geometries have been observed for the copper–chalcogenide clusters described in the preceding sections: whereas, for copper–sulfide clusters containing up to 28 Cu centers, near-linear S−Cu−S and trigonal planar S$_2$−Cu−P arrangements are observed, the copper centers in larger selenium clusters exhibit tetrahedral and trigonal planar coordination. Cluster **118a** is structurally related to the copper–tellurium cluster [Cu$_{29}$(TePh)$_{12}$Te$_9$(PEt$_3$)$_8$][PEt$_3$Ph] (**71**), which is formed from the co-condensation reaction of the two smaller clusters [Cu$_{12}$Te$_3$(TePh)$_6$(PEt$_3$)$_6$] (**63**) and [Cu$_5$(TePh)$_6$-(PEt$_3$)$_3$][PEt$_3$Ph] (**67**) [61]. This is the only example identified to date where there is a distinct structural relationship between a copper–telluride–tellurolate nanocluster and a silver analog. There are four additional PEt$_3$ ligands on the surface of **118a** compared to **63**, and the phosphine-bonded silver centers lie well "outside" of the tellurium polyhedron, whereas the copper atoms "missing" the phosphine in **63** are drawn into the cluster framework. The near-linear coordination geometries of the two two-coordinate Ag centers in **118a** are replaced with trigonal planar copper sites in **63**, and this is accompanied by an increase in the coordination numbers of two telluride ligands.

The substituted aryl ligand 2,4,6-Me$_3$C$_6$H$_2$Te$^-$ (mesTe) is also suitable for stabilizing polymetallic AgTe frameworks. Thus, in an analogous reaction to that for **118**, AgCl: PR$_3$ (PR$_3$ = PEt$_3$,PnPr$_3$, PEt$_2$Ph), solutions in ether solvents are reacted with a combination of Te(mes)SiMe$_3$ and Te(SiMe$_3$)$_2$. The clusters [Ag$_{46}$(Temes)$_{12}$Te$_{17}$(PR$_3$)$_{16}$] (**119**) are formed as dark red-black crystals in excellent yield (80–85%) [83]. Cluster **119a** (PR$_3$ = PEt$_3$) contains a distribution of telluride ligands (Figure 3.98) in the central portion of the cluster, and tellurolate groups and PEt$_3$ are on the cluster surface. About the central Te^{2-} are 12 additional telluride ligands arranged to form a distorted [nonbonded Te...Te = 4.523(2)–5.288(2) Å] centered icosahedron. The central atom telluride is within the range of bonding distances to eight silver metal atoms [2.814(2)–3.198(2) Å], which occupy positions within the Te$_{13}$-centered icosahedron, and an additional six Ag atoms reside within its deltahedral faces. This central Ag$_{34}$Te$_{13}$ unit is surrounded by 12 μ_3-Te(mes)$^-$, four Te^{2-}, and 12 Ag centers. The coordination number about the 46 silver atoms in **119a** varies from two (linear) to four (tetrahedral). The more condensed nature of the

3.2 Semiconductor Nanoparticles | 205

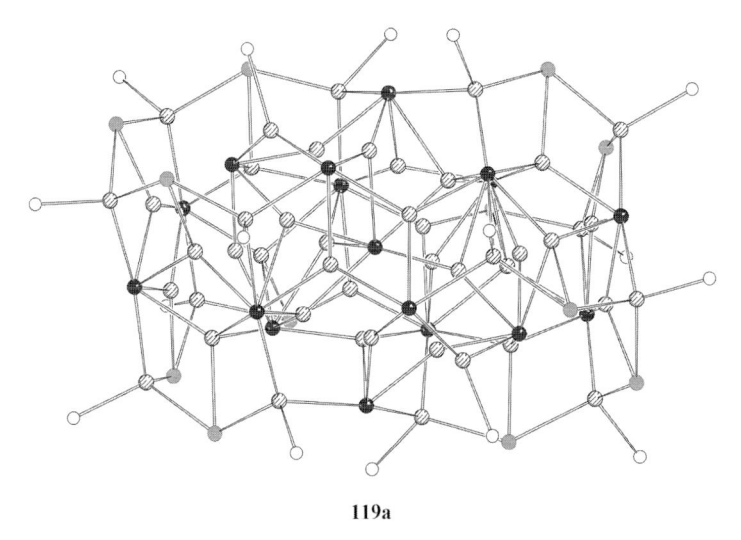

119a

Figure 3.98 The molecular structure of [Ag$_{46}$(Temes)$_{12}$Te$_{17}$(PEt$_3$)$_{16}$] (**119a**) (without organic groups). Ag: shaded spheres; Te: black spheres; TeR: gray spheres; P: white spheres.

framework in **119a** is reflected in its color of dark red-black as compared to the more lightly colored (red) **118a**.

Bidentate phosphine ligands have also been shown to be effective in stabilizing AgTe polynuclear frameworks, and, as in the case of AgSe complexes, they can serve as both inter- and intracluster bridges. Unlike the entry into PR$_3$–Ag–Te–TeR complexes, where AgCl is used as a source of Ag(I), the successful preparation of {[Ag$_5$(TePh)$_6$(Ph$_2$PCH$_2$CH$_2$PPh$_2$)(Ph$_2$PCH$_2$CH$_2$PPh$_3$)]}$_y$ (**120**), [Ag$_{18}$Te(TePh)$_{15}$(Ph$_2$P(CH$_2$)$_3$PPh$_2$)$_3$Cl] (**121**), and [Ag$_{38}$Te$_{13}$(TetBu)$_{12}$(Ph$_2$PCH$_2$CH$_2$PPh$_2$)$_3$] (**122**) requires carboxylate salts of Ag(I) (Scheme 3.18) [84].

A suspension of silver laurate and a half-equivalent of dppe in toluene can be solubilized via the addition of one equivalent of tri-n-propylphosphine. The addition of PhTeSiMe$_3$ at room temperature to this solution results in the crystallization of {[Ag$_5$(TePh)$_6$(Ph$_2$PCH$_2$CH$_2$PPh$_2$)(Ph$_2$PCH$_2$CH$_2$PPh$_3$)]}$_y$ (**120**) in moderate yield (28%) after several days [84]. The structure of **120** (Figure 3.99) consists of [Ag$_5$(TePh)$_6$] units linked by a dppe ligand. The arrangement of silver centers and tellurolate ligands is similar to that observed in the copper–tellurium complex [Cu$_5$(TePh)$_6$(PEt$_3$)$_3$][PEt$_3$Ph]. There are three μ_3-TePh groups (Ag–Te = 2.786–2.993 Å) around the "base" of the cluster, and three μ_2-TePh bridges (Ag–Te = 2.662–2.837 Å), with the Te centers arranged to form a distorted nonbonded Te$_6$ octahedron. There are four silver atoms in the base of the cluster, and a lone, trigonal planar silver atom bonded to the three μ_2-TePh bridges. The lone Ag atom centered in the base of the cluster has noticeably shorter Ag–Ag contacts to the other three basal Ag sites (2.815(7)–2.870(6) Å) than do the other silver atoms in **106** (>3.1 Å). The three additional Ag(I) atoms are tetrahedral, bonded to one PPh$_2$(CH$_2$)-group. The dppe ligand serves as an intercluster bridge between two [Ag$_5$(TePh)$_6$]$^-$ cluster units, to form the

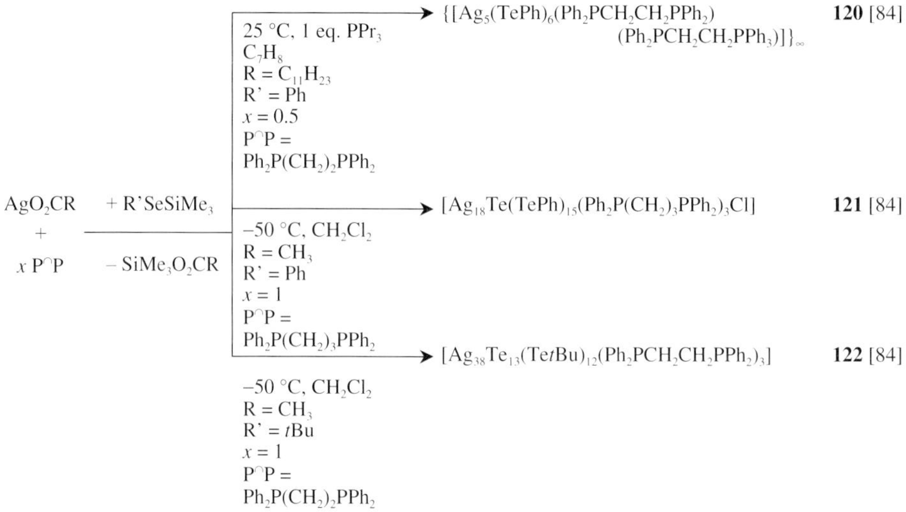

Scheme 3.18 The synthesis of telluride–tellurolato-bridged silver clusters with bidentate phosphine ligands.

infinite repeat observed in **106**. The pendant [Ph$_2$PCH$_2$CH$_2$PPh$_3$]$^+$ cation balances the negative charge on the cluster framework. Although the origin of this cation is difficult to rationalize, the generation of phosphonium centers during photochemically induced condensation reactions of Cu–TePh complexes is well established [19, 61].

[Ag$_{18}$Te(TePh)$_{15}$(Ph$_2$P(CH$_2$)$_3$PPh$_2$)$_3$Cl] (**121**) is synthesized by the 1:1:1 reaction of AgOAc, Ph$_2$P(CH$_2$)$_3$PPh$_2$ and PhTeSiMe$_3$ in CH$_2$Cl$_2$, the highly symmetric cluster residing on a crystallographic C$_3$ axis (Figure 3.100) [84]. Cluster **121** possesses a pseudo close-packed arrangement of Te centers, with a central telluride ligand present from Te–C$_{phenyl}$ bond cleavage of a TePh fragment. The five silver atoms

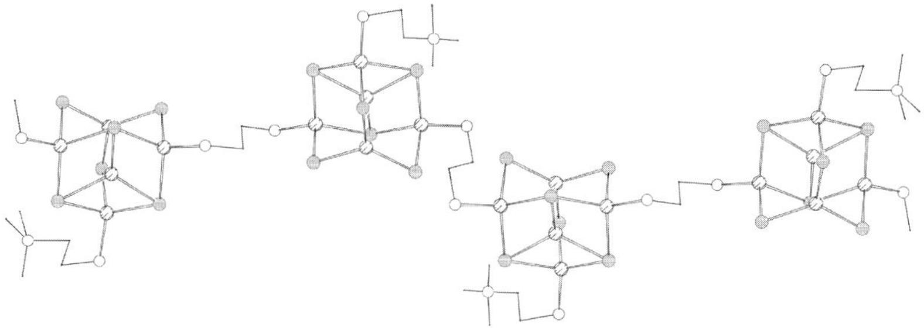

Figure 3.99 A projection of the polymeric repeat in {[Ag$_5$(TePh)$_6$(Ph$_2$P(CH$_2$)$_2$PPh$_2$)(Ph$_2$P(CH$_2$)$_2$PPH$_3$)]}$_\infty$ (**120**). Ag: shaded spheres; TeR: gray spheres; P: white spheres; C: wire representation. For clarity, only the *ipso* carbon atoms of the –PPh$_3^+$ centers and the carbon atoms of the methylene units on phosphorous are illustrated.

121

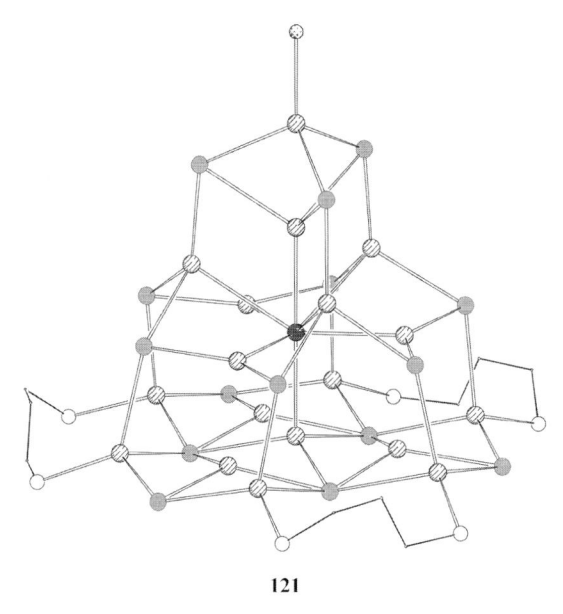

Figure 3.100 The core atoms in [Ag$_{18}$Te(TePh)$_{15}$(Ph$_2$P(CH$_2$)$_3$PPh$_2$)$_3$Cl] (**121**). Ag: shaded spheres; Te: black spheres; TeR: gray spheres; P: white spheres.

within the range of bonding distances to this central Te^{2-} are arranged to form near-perfect trigonal bipyramidal coordination geometry about the tellurium. The PhTe$^-$ ligands bridge either three or four Ag atoms, and, consistent with the rather open nature of the Ag–Te framework, the majority (nine) of the 11 interstitial silver atoms are three-coordinate, the other two adopting tetrahedral coordination. All phosphine-bonded Ag(I) sites and the lone Ag–Cl fragment are also bonded to three PhTe$^-$ ligands. The dppp ligands are located exclusively around the "base" of the pyramidal AgTe framework, with the chloride occupying the apical position on the cluster.

The largest structurally characterized Ag–Te–TeR cluster stabilized with bidentate phosphine ligands, [Ag$_{38}$Te$_{13}$(TetBu)$_{12}$(Ph$_2$PCH$_2$CH$_2$PPh$_2$)$_3$] (**122**) is obtained without the use of Te(SiMe$_3$)$_2$ as a source of telluride ligands [84]. Cluster **122** forms as small, black crystals in ~50% yield from the reaction of AgOAc, dppe and tBuTeSiMe$_3$ in a 4:4:3 ratio at low temperatures (Scheme 3.18). Gradual warming of the reaction solutions to room temperature leads to the formation of **122** over several weeks, and the stability of these solutions at higher temperature (versus the precipitation of amorphous solids) accounts for the formation of such a "telluride-rich" cluster.

The structure of **122** (Figure 3.101) contains a spherical Ag–Te framework. The 13 Te^{2-} ligands are spatially arranged in the cluster core to form a centered icosahedron, as is observed in [Ag$_{46}$(Temes)$_{12}$Te$_{17}$(PR$_3$)$_{16}$] (**119**) [83]. The central telluride ligand of this deltahedral sub-unit forms eight Ag–Te bonding interactions (3.049–3.199 Å), and the polyhedron is in fact surrounded by 12 Ag(I) atoms, which themselves are

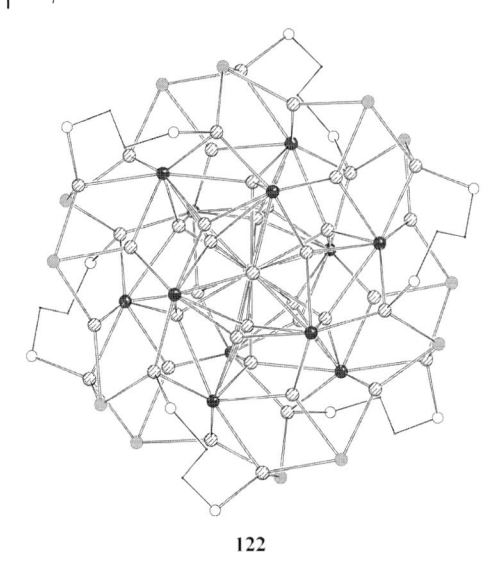

Figure 3.101 The core atoms in [Ag$_{38}$Te$_{13}$(TetBu)$_{12}$(Ph$_2$P(CH$_2$)$_2$PPh$_2$)$_3$] (**122**). Ag: shaded spheres; Te: black spheres; TeR: gray spheres; P: white spheres.

arranged to form an Ag$_{12}$ (nonbonded) icosahedron. The outer Te$_{12}$ and inner Ag$_{12}$ "shells" are then surrounded by the 12 μ$_3$-tBuTe$^-$ surface ligands and the additional 26 Ag(I) atoms. The 12 phosphine-bonded silver atoms have tetrahedral coordination, as they are each also bonded to three Te centers. Sixteen of the remaining Ag(I) atoms are trigonal planar, and ten Ag(I) atoms are tetrahedrally coordinated to four Te ligands, in asymmetric fashion.

3.2.3.4 Selenium-Bridged Gold Clusters

The growing interest in gold–chalcogenide polynuclear chemistry can be attributed partly to the demonstration that these systems exhibit rich photoluminescence behavior, and partly to the interesting structural frameworks that are generated as a result of weak Au(I)–Au(I) interactions [85].

The use of bidentate phosphine ligands in stabilizing metal–chalcogenide cores is again demonstrated by the synthesis of [Au$_{10}$Se$_4$(Ph$_2$PCH$_2$PPh$_2$)$_4$]Br$_2$ (**123**) and [Au$_{18}$Se$_8$(Ph$_2$PCH$_2$CH$_2$PPh$_2$)$_6$]Cl$_2$ (**124a**, Scheme 3.19) from the reaction of Se(SiMe$_3$)$_2$, [(AuBr)$_2$(dppm)] and [(AuCl)$_2$(dppe)] at ambient temperature, respectively. The complex dication **123** is illustrated in Figure 3.102, and features four μ$_3$-selenide ligands and four bidentate DPPM ligands bonded to ten Au(I) centers [86]. The μ$_3$-Se^{2-} ligands are symmetrically bonded to three gold atoms (Se–Au = 2.423(1)–2.461(1) Å), with two ligands on either side of the Au10 "plane." The arrangement of the ten gold atoms has eight Au(I) atoms forming an outer, corrugated octahedron that surrounds two central Au(I) atoms. The molecule can also be regarded as consisting of two [Au$_5$Se$_2$]$^+$ units held together by two dppm ligands, together with a contribution from the aurophilic interaction of the central Au(I)–Au(I) atoms [3.127(1) Å]. These two gold atoms are bonded to two selenide ligands, with a noticeable deviation

3.2 Semiconductor Nanoparticles | 209

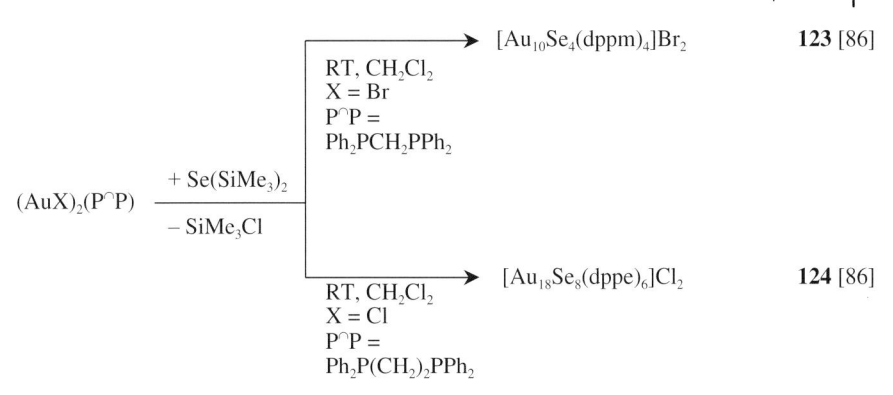

$(AuX)_2(P^\frown P)$ $\xrightarrow[- SiMe_3Cl]{+ Se(SiMe_3)_2}$

- RT, CH$_2$Cl$_2$; X = Br; P$^\frown$P = Ph$_2$PCH$_2$PPh$_2$ → [Au$_{10}$Se$_4$(dppm)$_4$]Br$_2$ **123** [86]
- RT, CH$_2$Cl$_2$; X = Cl; P$^\frown$P = Ph$_2$P(CH$_2$)$_2$PPh$_2$ → [Au$_{18}$Se$_8$(dppe)$_6$]Cl$_2$ **124** [86]

Scheme 3.19 The synthesis of selenido-bridged gold clusters with bidentate phosphine ligands.

from linear geometry [164.7(1)–168.6(1)] due to the metal–metal interaction. The eight additional gold sites are each bonded to one phosphorus and one selenium ligand, again with moderate deviations from linear geometries [<P–Au–Se = 168.5(1)–177.2(2)] and Au(I)–Au(I) distances of 3.031(1)–3.364(1) Å, similar to those observed in the structurally related complex [Au$_{12}$(dppm)$_6$S$_4$]$^{4+}$.

The larger gold-selenide cluster [Au$_{18}$Se$_8$(Ph$_2$PCH$_2$CH$_2$PPh$_2$)$_6$]Cl$_2$ (**124a**; Figure 3.103) is prepared in 90% yield from [(AuCl)$_2$(dppe)] and Se(SiMe$_3$)$_2$. The structure of the complex dication in **124** consists of two [Au$_9$(μ$_3$-Se)$_4$(dppe)$_3$]$^+$ fragments formed from three Au$_4$Se$_3$dppe "rings." Each of these forms a catenated structure with the corresponding ring in the second Au$_9$ fragment, such that there are three pairs of interlocked rings, resulting in a central Au$_6$Se$_2$ heterocubane unit. The six gold atoms in this heterocubane are two-coordinate, each linearly bonded to two selenide ligands [<Se–Au–Se = 173.7(1)–174.9(1)]. The Au(I)...Au(I) distances in

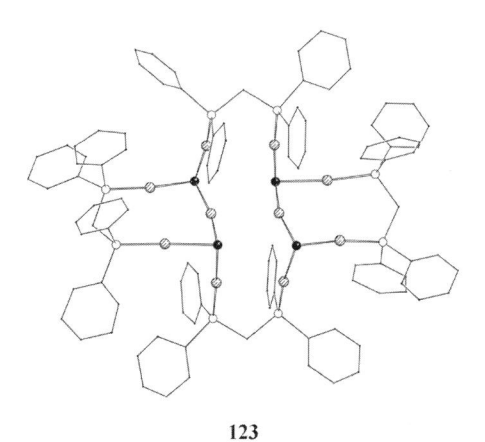

123

Figure 3.102 The molecular structure of the cationic framework in [Au$_{10}$Se$_4$(Ph$_2$PCH$_2$PPh$_2$)$_4$]Br$_2$ (**123**) (without H atoms). Au: shaded spheres; Se: black spheres; P: white spheres; C: wire representation.

124

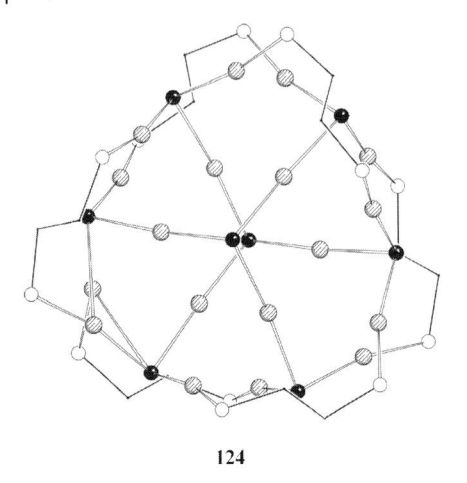

Figure 3.103 A projection of the molecular structure of $[Au_{18}Se_8(Ph_2P(CH_2)_2PPh_2)_4]^{2+}$ (**124**) (H atoms and phenyl rings omitted for clarity). Au: shaded spheres; Se: black spheres; P: white spheres; C: wire representation.

the Au_6Se_2 unit are in the range 2.940(2)–3.002(2) Å, and in the rest of the cluster are in the range 2.948(2)–3.327(3) Å. The remaining 12 Au(I) centers display the same, expected coordination geometries, as they are bonded to one phosphine and one selenide ligand [<P−Au−Se = 66.8(1)–173.1(1)]. The μ_3-Se ligands symmetrically bridge the Au(I) centers (2.430(2)–2.481(3) Å), with six of the selenide ligands distributed around the periphery of the AuSe frame. It is noteworthy that reactions of [(AuCl)$_2$(dppe)] with RESiMe$_3$ (R = Et, iPr, tBu, nBu) invariably lead to the formation of $[Au_{18}Se_8(dppe)_6]Cl_2$ (**124a**) via Se−C bond cleavage, with no evidence to date of any intermediate, Au−SeR, containing complexes that can be isolated. However, the chloride counterions may easily be substituted via the reaction of **124a** with KPF$_6$ and NaBPh$_4$ to yield $[Au_{18}Se_8(dppe)_6](PF_6)_2$ (**124b**) and $[Au_{18}Se_8(dppe)_6](BPh_4)_2$ (**124c**), respectively.

High-resolution electrospray ionization mass spectrometric investigations of **124b** have displayed the expected isotopic distribution centered around m–z 3283.89, corresponding to the dication **124**, with no evidence of fragmentation. Controlled, collision-induced fragmentation reactions of **124** with argon have indicated that the preferred dissociation path is into two Au−Se fragments, $[Au_{13}Se_6(dppe)_3]^+$ and $[Au_5Se_2(dppe)_2]^+$, with the concomitant loss of dppe. The results of these experiments suggest similar Au−Se and Au−P binding energies in **124**.

Reference

1 Fedorin, V.A. (1993) *Semiconductors*, **27**, 196.
2 (a) Mansour, B.A., Demian, S.E., and Zayed, H.A. (1992) *J. Mater. Sci.*, **3**, 249; (b) Vučić, Z., Milat, O., Horvatić, V., and Ogorelek, Z. (1981) *Phys. Rev.*, **B24**, 5398.
3 (a) Nimtz, G., Marquard, P., and Gleiter, H. (1988) *J. Cryst. Growth*, **86**, 66;

(b) Koutecky, J., and Fantucci, P. (1986) *Chem. Rev.*, **86**, 539; (c) Morse, M.D. (1986) *Chem. Rev.*, **86**, 1049; (d) Kappes, M.M. (1988) *Chem. Rev.*, **88**, 369; (e) Weller, H. (1993) *Angew. Chem.*, **105**, 43.

4. (a) Weller, H. (1998) *Angew. Chem.*, **110**, 1748; (b) Alivisatos, A.P. (1996) *Science*, **271**, 933; (c) Weller, H. (1996) *Angew. Chem.*, **108**, 1159; (d) Schmid, G. (ed.) (1995) *Clusters and Colloids*, VCH, Weinheim.

5. (a) Fackler, J.P. Jr and Coucouvanis, D. (1966) *J. Am. Chem. Soc.*, **88**, 3913; (b) McCandlish, L.E., Bissell, E.C., Coucouvanis, D., Fackler, J.P., and Knox, K. (1968) *J. Am. Chem. Soc.*, **90**, 7357.

6. Dance, I.G. (1978) *Aust. J. Chem.*, **31**, 617.

7. (a) Betz, P., Krebs, B., and Henkel, G. (1984) *Angew. Chem.*, **96**, 293; (b) Betz, P., Krebs, B., and Henkel, G. (1984) *Angew. Chem., Int. Ed. Engl.*, **23**, 311.

8. Ohlmann, D., Pritzkow, H., Grützmacher, H., Anthamatten, M., and Glaser, P. (1995) *J. Chem. Soc. Chem. Commun.*, 1011.

9. Recent review: Dehnen, S., Eichhöfer, A., and Fenske, D. (2002) *Eur. J. Inorg. Chem.*, 279–317.

10. (a) Fenske, D., Krautscheid, H., and Balter, S. (1990) *Angew. Chem.*, **102**, 799; (b) Fenske, D., and Krautscheid, H. (1990) *Angew. Chem.*, **102**, 1513.

11. (a) Fenske, D. and Steck, J.C. (1993) *Angew. Chem.*, **105**, 254; (b) Fenske, D. and Steck, J.C. (1993) *Angew. Chem., Int. Ed. Engl.*, **32**, 238.

12. Krautscheid, H., Fenske, D., Baum, G., and Semmelmann, M. (1993) *Angew. Chem.*, **105**, 1364.

13. Dehnen, S., Schäfer, A., Fenske, D., and Ahlrichs, R. (1994) *Angew. Chem.*, **106**, 786.

14. Dehnen, S. and Fenske, D. (1994) *Angew. Chem.*, **106**, 2369–2372.

15. Dehnen, S., Fenske, D., and Deveson, A.C. (1996) *J. Cluster Sci.*, **7** (3), 351–369.

16. Corrigan, J.F., Balter, S., and Fenske, D. (1996) *J. Chem. Soc., Dalton Trans.*, 729.

17. Dehnen, S. and Fenske, D. (1996) *Chem. Eur. J.*, **2**, 1407–1416.

18. Deveson, A., Dehnen, S., and Fenske, D. (1997) *J. Chem. Soc., Dalton Trans.*

19. (a) Corrigan, J.F., and Fenske, D. (1997) *Angew. Chem.*, **109**, 2070; (b) Corrigan, J.F., and Fenske, D. (1997) *Angew. Chem., Int. Ed. Engl.*, **36**, 1981.

20. Semmelmann, M., Fenske, D., and Corrigan, J.F. (1998) *J. Chem. Soc., Dalton Trans.*, 2541.

21. Eichhöfer, A. and Fenske, D. (1998) *J. Chem. Soc., Dalton Trans.*, 2969.

22. Bettenhausen, M., Eichhöfer, A., Fenske, D., and Semmelmann, M. (1999) *Z. Anorg. Allg. Chem.*, **625**, 593.

23. Zhu, N. and Fenske, D. (1999) *J. Chem. Soc., Dalton Trans.*, 1067.

24. Eichhöfer, A., Corrigan, J.F., Fenske, D., and Tröster, E. (2000) *Z. Anorg. Allg. Chem.*, **626**, 338.

25. Zhu, N. and Fenske, D. (2000) *J. Cluster Sci.*, **11**, 135.

26. Krautscheid, H. (1991) Ph.D. Thesis, University of Karlsruhe.

27. Steck, J.C. (1992) Ph.D. Thesis, University of Karlsruhe.

28. Balter, S. (1994) Ph.D. Thesis, University of Karlsruhe.

29. Dehnen, S. (1996) Ph.D. Thesis, University of Karlsruhe.

30. Semmelmann, M. (1997) Ph.D. Thesis, University of Karlsruhe.

31. Eichhöfer, A., and Fenske, D.,unpublished results.

32. El Nakat, J.H., Dance, I.G., Fisher, K.J., and Willet, G.D. (1991) *Inorg. Chem.*, **30**, 2957.

33. (a) van der Putten, D., Olevano, D., Zanoni, R., Krautscheid, H., and Fenske, D. (1995) *J. Electron. Spectrosc. Relat. Phenom.*, **76**, 207; (b) Enderle, A. (1993) Diploma Thesis, University of Karlsruhe; (c) Stöhr, U. (1993) Diploma Thesis, University of Karlsruhe.

34. (a) Schäfer, A., Huber, C., Gauss, J., and Ahlrichs, R. (1993) *Theor. Chim. Acta*, **87**, 29; (b) Schäfer, A., and Ahlrichs, R. (1994) *J. Am. Chem. Soc.*, **116**, 10686.

35. Dehnen, S., Schäfer, A., Ahlrichs, R., and Fenske, D. (1996) *Chem. Eur. J.*, **2**, 429.

36. Eichkorn, K., Dehnen, S., and Ahlrichs, R. (1998) *Chem. Phys. Lett.*, **284**, 287.

37. Schäfer, A. (1994) Ph.D. Thesis, University of Karlsruhe.

38. (a) Wells, A.F. (1984) *Structural Inorganic Chemistry*, 5th edn, Clarendon Press,

Oxford; (b) Evans, H.T. (1979) *Z. Kristallogr.*, **150**, 299.

39 (a) Heyding, R.D., and Murray, R.M. (1976) *Can. J. Chem.*, **54**, 841; (b) Rahlfs, P. (1936) *Z. Phys. Chem.*, **B31**, 157; (c) Borchert, W. (1945) *Z. Kristallogr.*, **106**, 5; (d) Stevels, A.L.N. (1969) *Philips Res. Rep. Suppl.*, **9**, 124; (e) Stevels, A.L.N., and Jellinek, F. (1971) *Recl. Trav. Chim.*, **90**, 273; (f) Tonjec, A., Ogorelec, Z., and Mestnik, B. (1975) *J. Appl. Crystallogr.*, **8**, 375.

40 (a) Blachnik, R., Lasocka, M., and Walbrecht, U. (1983) *J. Solid State Chem.*, **48**, 431; (b) Asadov, Y.G., Rustamova, L.V., Gasimov, G.B., Jafarov, K.M., and Babajev, A.G. (1992) *Phase Transitions*, **38**, 247.

41 Aylward, G.H. and Findlay, T.J.V. (1986) *Datensammlung Chemie*, VCH, Weinheim, p. 125.

42 Tolman, C.A. (1977) *Chem. Rev.*, **77**, 313.

43 (a) Wilkinson, G. (ed.) (1987) *Comprehensive Coordination Chemistry*, vol. **5**, Pergamon Press, Oxford, p. 583; (b) Drew, M.G.B., Othman, A.H.B., Edwards, D.A., and Richards, R. (1975) *Acta Crystallogr. Sect. B*, **B31**, 2695.

44 (a) Ahlrichs, R., Bär, M., Häser, M., Horn, H., and Kölmel, C. (1995) *Chem. Phys. Lett.*, **242**, 652; (b) Treutler, O., and Ahlrichs, R. (1995) *J. Chem. Phys.*, **102**, 346; (c) Schäfer, A., Horn, H., and Ahlrichs, R. (1992) *J. Chem. Phys.*, **97**, 2571.

45 (a) Eichkorn, K., Treutler, O., Öhm, H., Häser, M., and Ahlrichs, R. (1995) *Chem. Phys. Lett.*, **242**, 652; (b) Treutler, O., and Ahlrichs, R. (1995) *J. Chem. Phys.*, **102**, 346;(c) Eichkorn, K., Weigend, F., Treutler, O., and Ahlrichs, R. (1997) *Theor. Chim. Acta*, **97**, 119.

46 Møller, C. and Plesset, M.S. (1934) *Phys. Rev.*, **46**, 618.

47 Parr, R.G. and Yang, W. (1988) *Density Functional Theory of Atoms and Molecules*, Oxford University Press, New York.

48 Ziegler, T. (1991) *Chem. Rev.*, **91**, 651.

49 Lax, E. (1967) *Taschenbuch für Chemiker und Physiker*, vol. I, Springer, Berlin, Heidelberg, p. 324.

50 Lax, E. (1967) *Taschenbuch für Chemiker und Physiker*, vol. I, Springer, Berlin, Heidelberg, p. 326.

51 Hollemann, A.F. and Wiberg, E. (1985) *Lehrbuch der Anorganischen Chemie*, Walter de Gruyter, Berlin, p. 148.

52 Wedler, G. (1987) *Lehrbuch der Physikalischen Chemie*, VCH, Weinheim, p. 168.

53 (a) Fenske, D., Zhu, N., and Langetepe, T. (1998) *Angew. Chem.*, **110**, 2784; (b) Fenske, D., Zhu, N., and Langetepe, T. (1998) *Angew. Chem., Int. Ed.*, **37**, 2640.

54 Park, Y. (1992) Ph.D. Thesis, Michigan State University.

55 Savelsberg, G. and Schäfer, H. (1978) *Z. Naturforsch.*, **B33**, 711.

56 Klepp, K.O. (1987) *J. Less-Common Met.*, **128**, 79.

57 Savelsberg, G. and Schäfer, H. (1981) *Mater. Res. Bull.*, **16**, 1291.

58 Park, Y., Degroot, D.C., Schindler, J., Kannewurf, C.R., and Kanatzidis, M.G. (1991) *Angew. Chem., Int. Ed. Engl.*, **30**, 1325.

59 (a) Park, Y. and Kanatzidis, M.G. (1991) *J. Chem. Mater.*, **3**, 781; (b) Zhang, X., Park, Y., Hogan, T., Schindler, J.L., Kannewurf, C.R., Seong, S., Albright, T., and Kanatzidis, M.G. (1995) *J. Am. Chem. Soc.*, **117**, 10300.

60 Emirdag, M., Schimek, G.L., and Kolis, J.W. (1999) *J. Chem. Soc., Dalton Trans.*, 1531.

61 DeGroot, M.W., Cockburn, M.W., Workentin, M.S., and Corrigan, J.F. (2001) *Inorg. Chem.*, **40**, 4678.

62 (a) Frank, F.C. and Kasper, J.S. (1958) *Acta Crystallogr.*, **11**, 184; (b) Frank, F.C. and Kasper, J.S. (1959) *Acta Crystallogr.*, **12**, 483.

63 (a) Kölmel, C. and Ahlrichs, R. (1990) *J. Phys. Chem.*, **94**, 5536; (b) Cotton, F.A., Feng, X., Matusz, M., and Poli, R. (1988) *J. Am. Chem. Soc.*, **110**, 7077; (c) Merz, K.M. and Hoffmann, R. (1988) *Inorg. Chem.*, **27**, 2120.

64 Müller, U. (1991) *Anorganische Strukturchemie*, 2nd edn, B.G. Teubner, Stuttgart, p. 101f.

65 (a) Jellinek, F. (1972) *MTP International Review of Science*, Inorganic Chemistry Series One (ed. D.W. Sharp), vol. 5, Butterworths, London, p. 339; (b) Folmer, C.W. and Jellinek, F. (1980) *J. Less-Common Met.*, **76**, 153.

66 Ahlrichs, R., Besinger, J., Eichhöfer, A., Fenske, D., and Gbureck, A. (2000) *Angew. Chem.*, **112**, 4089.

67 Wang, X.-J., Langetepe, T., Fenske, D., and Kang, B.-S. (2002) *Z. Anorg. Allg. Chem.*, **628**, 1158–1167.

68 (a) Chitsaz, S., Fenske, D., and Fuhr, O. (2006) *Angew. Chem.*, **118**, 8224–8228; (b) Chitsaz, S., Fenske, D., and Fuhr, O. (2006) *Angew. Chem., Int. Ed.*, **45**, 8055–8059.

69 (a) Fenske, D., Persau, C., Dehnen, S., and Anson, C.E. (2004) *Angew. Chem.*, **116**, 309–313; (b) Fenske, D., Persau, C., Dehnen, S., and Anson, C.E. (2004) *Angew. Chem., Int. Ed.*, **43**, 305–309.

70 (a) Wang, X.J., Langetepe, T., Persau, C., Kang, B.-S., Sheldrick, G.M., and Fenske, D. (2002) *Angew. Chem.*, **114**, 3972–3977; (b) Wang, X.J., Langetepe, T., Persau, C., Kang, B.-S., Sheldrick, G.M., and Fenske, D. (2002) *Angew. Chem., Int. Ed.*, **41**, 3818–3822.

71 (a) Fenske, D., Anson, C.E., Eichhöfer, A., Fuhr, O., Ingendoh, A., Persau, C., and Richert, C. (2005) *Angew. Chem.*, **117**, 5376–5381; (b) Fenske, D., Anson, C.E., Eichhöfer, A., Fuhr, O., Ingendoh, A., Persau, C., and Richert, C. (2005) *Angew. Chem., Int. Ed.*, **44**, 5242–5246.

72 Jin, X., Xie, X., Qian, H., Tang, K., Liu, C., Wang, X., and Gong, Q. (2002) *Chem. Commun.*, 600–601.

73 Jin, X., Tang, K., Liu, W., Zeng, H., Zhao, H., Ouyang, Y., and Tang, Y. (1996) *Polyhedron*, **15**, 1207–1211.

74 Tang, K., Xie, X., Zhao, L., Zhang, Y., and Jin, X. (2004) *Eur. J. Inorg. Chem.*, 78–85.

75 (a) Anson, C.E., Eichhöfer, A., Issac, I., Fenske, D., Fuhr, O., Persau, C., Sevillano, P., Stalke, D., and Zhang, J. (2008) *Angew. Chem.*, **120**, 1346–1351; (b) Anson, C.E., Eichhöfer, A., Issac, I., Fenske, D., Fuhr, O., Persau, C., Sevillano, P., Stalke, D., and Zhang, J. (2008) *Angew. Chem., Int. Ed.*, **47**, 1326–1331.

76 (a) Wiggers, G.A. (1971) *Am. Mineral.*, **56**, 1882–1888; (b) Pinsker, Z.G., Chiang Ling, C., Imamov, R.M., and Lapidus, E.L. (1965) *Sov. Phys. Cristallogr. (Engl. Transl.)*, **7**, 225–233; (c) Kracek, F.C. (1946) *Trans. Am. Geophys. Union.*, **27**, 364; (d) Djurle, S. (1958) *Acta Chem. Scand.*, **12**, 1427; (e) Rahlfs, P. (1936) *Z. Phys. Chem. Abt. B.*, **31**, 157–194; (f) Rickert, E. (1960) *Z. Phys. Chem.*, **24**, 418–421; (g) Böttcher, A., Haase, G., and Treupel, H. (1955) *Z. Angew. Phys.*, **7**, 478–487; (h) Junod, P. (1959) *Helv. Phys. Acta.*, **32**, 567–600; (i) FrPh, A.J. (1958) *Z. Kristallogr.*, **110**, 136–144.

77 (a) Ahlrichs, R., Fenske, D., Rothenberger, A., Wieber, S., and Schrodt, C. (2006) *Eur. J. Inorg. Chem.*, 1127–1129; (b) Ahlrichs, R., Fenske, D., McPartlin, M., Rothenberger, A., Schrodt, C., and Wieber, S. (2005) *Angew. Chem.*, **117**, 4002–4005; (c) Ahlrichs, R., Fenske, D., McPartlin, M., Rothenberger, A., Schrodt, C., and Wieber, S. (2005) *Angew. Chem., Int. Ed.*, **44**, 3932–3936; (d) Ahlrichs, R., Anson, C.E., Clérac, R., Fenske, D., Rothenberger, A., Sierka, M., and Wieber, S. (2004) *Eur. J. Inorg. Chem.*, 2933–2936.

78 (a) Maier, J. (2002) *Solid State Ionics*, **154–155**, 291–301; (b) Maier, J. (2002) *Solid State Ionics*, **148**, 367–374; (c) Münch, W. (2000) *Z. Anorg. Allg. Chem.*, **626**, 264–269.

79 Fenske, D., Zhu, N., and Langetepe, T. (1998) *Angew. Chem., Int. Ed. Engl.*, **37**, 2640.

80 Fenske, D. and Langetepe, T. (2002) *Angew. Chem., Int. Ed. Engl.*, **41**, 300.

81 Corrigan, J.F. and Fenske, D. (1996) *Chem. Commun.*, 943.

82 Corrigan, J.F., Fenske, D., and Power, W.P. (1997) *Angew. Chem., Int. Ed. Engl.*, **36**, 1176.

83 Corrigan, J.F. and Fenske, D. (1997) *Chem. Commun.*, 1837.

84 Langetepe, T. and Fenske, D. (2001) *Z. Anorg. Allg. Chem.*, **627**, 820.

85 (a) Schmidbaur, H. (1999) *Gold: Progress in Chemistry, Biochemistry, and Technology*, John Wiley & Sons, New York; (b) Yam, V.W.-W., Chan, C.-L., Li, C.-K., and Wong, K.M.-C. (2001) *Coord. Chem. Rev.*, **216**, 173 and references therein.

86 Fenske, D., Langetepe, T., Kappes, M.M., Hampe, O., and Weis, P. (2000) *Angew. Chem., Int. Ed. Engl.*, **39**, 1857.

3.3
Synthesis of Metal Nanoparticles

This sub-chapter is divided into two parts. In the first part, attention is focused on particles, except for a few examples that consist mainly of noble metals. Nanoparticles of noble metals, in particular, have a long tradition and the fundamental synthetic strategies associated with these materials have long been known. Thus, the aim here is not to describe the vast number of synthetic pathways for metal nanoparticles, but rather to summarize the recognized syntheses (although some novel procedures have been included in so far as they complement traditional procedures).

The second part of the sub-chapter deals exclusively with magnetic particles, which have attracted much attention during the past decade. The reason for this relates to the manifold future applications in nanodevices. This second part also describes the most important properties of these particles.

3.3.1
Noble Metal Nanoparticles
Günter Schmid

3.3.1.1 Introduction
Advancements of the understanding of small-particle science, and the potential for new materials sciences based on the chemistry and physics of nanoscale metal clusters, rest on the measurement and application of useful size-dependent properties of small metal nanoparticles. Ideally, this requires the preparation and isolation of monodispersed metal particles with a great degree of control over size, shape, and composition. Although metals in varying levels of dispersion have a long history in many technological applications, the methods used for their preparation are usually determined by the intended use for the metal; they are not necessarily developed with the aim of preparing well-defined metal particles. For example, highly dispersed metal catalysts are normally prepared by the high-temperature reduction of metal precursors on oxide supports. These severe preparative conditions tend to yield metal particles with ranges of sizes and morphologies broad enough to hide the potentially unique properties that a monodispersed nanoscale sample could exhibit.

With these considerations in mind, synthetic chemists have begun to address the needs of metal particle research by developing the synthetic chemistry of nanosized metals, with a view to using the strategies of molecular chemistry to prepare well-defined metal nanoparticles. The goal may be stated as "…the search for synthetic methods for metal nanoparticles of narrow size distribution and, if possible, with shape-control." Furthermore, bimetallic species will be considered, either with core–shell architecture or in alloyed form.

3.3.1.2 History and Background
Metal colloid science can be said to have begun with the experiments of Michael Faraday on gold sols in the mid-nineteenth century [1]. In this case, deep red

solutions of colloidal gold were prepared by the reduction of chloroaurate solutions, using phosphorus as a reducing agent. The later studies conducted by Zsigmondy, at the turn of the century, put the physical investigation of colloidal solutions on a firm basis; indeed, Zsigmondy's invention of the ultramicroscope enabled careful studies to be carried out on the effect of preparation conditions on particle size [2]. Many methods for the preparation of colloidal metals have been developed over the years since Faraday's experiments. A summary of the research into the preparation of metal sols during the first half of the century is available in the 1951 report of Turkevitch and coworkers [3], which describes detailed synthesis conditions and the early electron microscopic characterization of gold sols prepared using the methods of Faraday, and also by several other related methods involving the action of reducing agents on chloroaurate solutions. These preparations introduce many of the methods still used today for preparative metal colloid chemistry. Turkevitch's extended studies yielded a detailed knowledge of the nucleation, growth, and agglomeration of colloidal metals, and subsequently allowed the reproducible preparation of gold sols with narrow particle size distributions over a range of mean sizes [4]. In the following text, the term "colloid" is, in most cases, substituted by the term "nanoparticle" (as is usual in the literature). A transmission electron microscopy (TEM) image of typical, very narrowly size-distributed 18 nm gold nanoparticles (AuNPs), prepared using the method of Turkevitch, is shown in Figure 3.104.

In general, the particle sizes found for these AuNPs and related metal nanoparticles may vary between about 2 to 100 nm, depending on preparative conditions, with most examples falling in the large-diameter range.

3.3.1.3 Stabilization of Metal Nanoparticles

Before beginning a description of synthetic methods, a general and crucial aspect should be considered – that is, the means by which metal particles are stabilized in a dispersing medium, as small metal particles are unstable with respect to agglomeration to the bulk. At short interparticle distances, two particles would be attracted to each other by van der Waals forces and, in the absence of any repulsive forces to counteract this attraction, an unprotected sol would coagulate. This counteraction can be achieved by two methods, namely *electrostatic stabilization* and *steric stabilization* [5]. In classical gold sols, for example, prepared by the reduction of aqueous $[AuCl_4]^-$ by sodium citrate, the colloidal gold particles are surrounded by an electrical double layer formed from the adsorbed citrate, chloride ions and cations that are attracted to the particles. This results in a coulombic repulsion between the particles, the net result of which is shown schematically in Figure 3.105.

The weak minimum in potential energy at moderate interparticle distance defines a stable arrangement of particles which is easily disrupted by medium effects and, at normal temperatures, by the thermal motion of the particles. Thus, if the electric potential associated with the double layer is sufficiently high, electrostatic repulsion will prevent particle agglomeration. However, an electrostatically stabilized sol can be coagulated if the ionic strength of the dispersing medium is increased sufficiently. If the surface charge is reduced by the displacement of adsorbed anions by a more

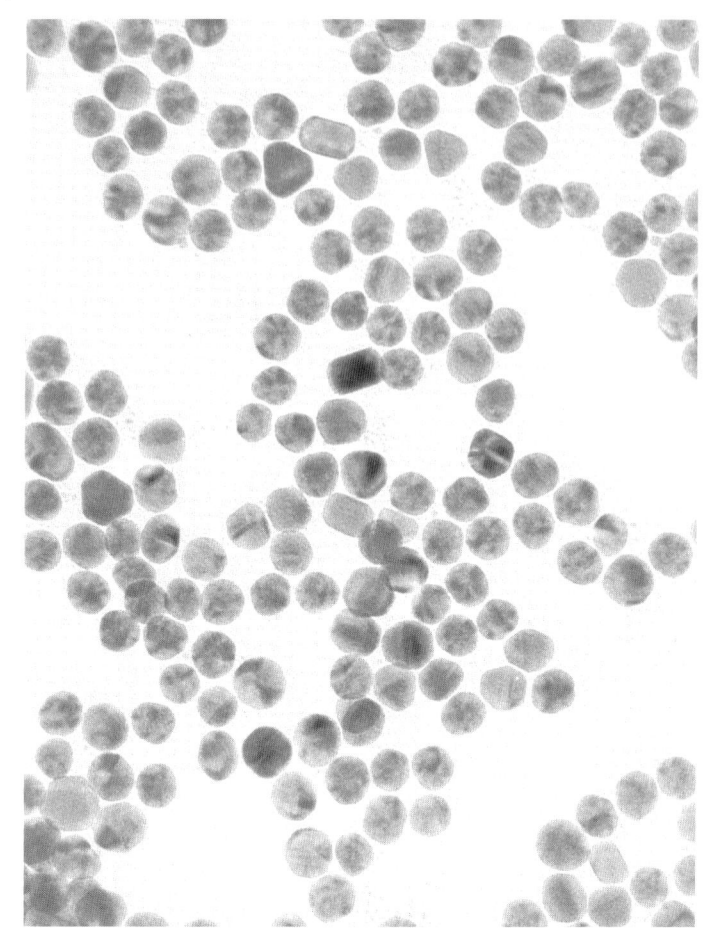

Figure 3.104 TEM image of monodispersed 18 nm gold particles, prepared by citrate reduction of $HAuCl_4$.

strongly binding neutral adsorbate, the particles can now collide and agglomerate under the influence of the van der Waals attractive forces. This phenomenon can easily be demonstrated by the addition of pyridine to a gold sol of the type mentioned above [6].

Even in organic media, in which electrostatic effects might not normally be considered to be important, the development of charge has been demonstrated on inorganic surfaces, including metals, in contact with organic phases such as solvents and polymers [7] For example, the acquisition of charge by gold particles in organic liquids has been demonstrated, and the sign and magnitude of the charge found to vary as a function of the donor properties of the liquid [8]. Thus, even for colloidal metals in suspension in relatively nonpolar liquids, the possibility cannot be excluded that electrostatic stabilization contributes to the stability of the sol. Nevertheless, electrostatic stabilization of metal nanoparticles is rather weak and never allows the

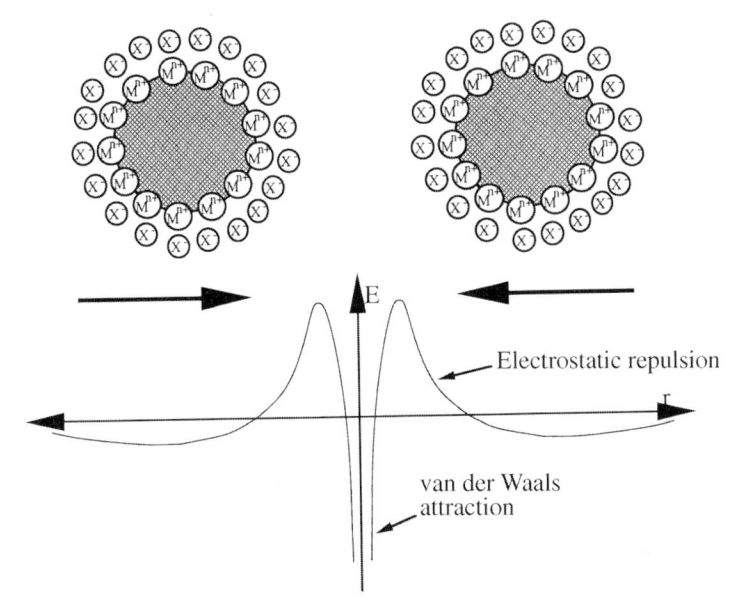

Figure 3.105 Electrostatic stabilization of metal nanoparticles. Attractive van der Waals forces are outweighed by repulsive electrostatic forces between adsorbed ions and associated counterions at moderate interparticle separation.

isolation of those nanoparticles as solids without extended aggregation and even formation of the bulk metal.

A second, much better, means by which nanoparticles can be prevented from aggregating is based on steric effects, and occurs by the addition of polymers and surfactants at the surface of the particles. Polymers are widely used, and it is clear that the protectant, in order to function effectively, must not only coordinate to the particle surface but also be adequately solvated by the dispersing fluid – such polymers are termed "amphiphilic." The choice of polymer is determined by consideration of the solubility of the metal precursor, the solvent of choice, and the ability of the polymer to stabilize the metal nanoparticles. Before the advent of synthetic polymer chemistry, natural polymers such as gelatin and agar were often used and, indeed, related stabilizers such as cellulose acetate, cellulose nitrate [9] and cyclodextrins [10] have been used more recently. Thiele [11] proposed the "Protective Value" as a measure of the ability of a polymer to stabilize colloidal metals. This was defined, similarly to the older "Gold Number" of Zsigmondy, as the weight of the polymer that would stabilize 1 g of a standard red gold sol containing 50 mg l^{-1} gold against the coagulating effect of 1% sodium chloride solution. Several other studies have been performed on the relative ability of polymers to act as steric stabilizers [12–14]. Despite the fact that these quite subjective studies focus on very specific (and quite different) sol systems it seems that, among the synthetic polymers considered, vinyl polymers with polar side groups such as poly(vinylpyrrolidone) (PVP) and poly(vinyl alcohol) (PVA) are especially useful in this respect.

Electrostatic and steric stabilization are, in a sense, combined in the use of long-chain alkylammonium cations and surfactants, either in single-phase sols or in the reverse micelle synthesis of colloidal metals.

Subsequently, ligand molecules – which are traditional in complex chemistry – were found to stabilize metal nanoparticles in such a way as to allow the generation of metal nanoparticles from less than 1 nm up to 100 nm and more, in a very specific manner. The predominantly covalent interactions between the ligand molecule and the nanoparticle usually stabilize the particles to such an extent that they may be isolated in solid state, without aggregation. Furthermore, the varying ligand strengths enable ligand exchange reactions with important consequences with regards to solubility and chemical properties. Amines, phosphines and thiols have dominated the chemistry of metal nanoparticles for the past two decades, and some examples are provided in the following sections. Metal nanoparticles, when protected by appropriate molecular ligands, can be considered as the natural continuation of classical complex chemistry and molecular metal clusters.

3.3.1.4 Synthetic Methods

The successful establishment of this important part of nanoscience and nanotechnology depends first of all on the control of particle size, morphology and the composition of metal nanoparticles in syntheses, which produce chemically significant quantities. Thus, for present purposes the preparation of size-selected particles in a molecular beam experiment, although elegant, is less relevant to materials chemistry. The problem of preparing isolable yields of uniform samples is also one which was central to the development of synthetic organometallic cluster chemistry. The obvious difference between molecular synthesis and molecular approaches to nanosized metals is that, with the former, it is the constant aim of the organometallic chemist to maintain metals in complexed molecular form, whereas with the latter the goal is reversed, and controlled methods must be sought for metal formation from molecular or ionic precursors. Whilst a reaction that results in the formation of elemental metal in the laboratory of the synthetic molecular chemist is considered a failure, it is also a potential source of metal nanoparticles. Such a process would normally lead to the formation of a metallic precipitate or mirror but, in the presence in solution of an appropriate stabilizing agent, aggregation of the metal atoms can be arrested at an early stage, such that the metal particles will be preserved in suspension in the liquid.

A survey of the literature has revealed that many of the various methods reported for the preparation of metal nanoparticles are applicable to a number of metals across the Periodic Table. For example, salt reduction using main group hydride-reducing agents has been used for the preparation of many metals in nanosized form (not only those cited in this subchapter). It is not the goal here to provide a directory of all reports of nanoparticles syntheses, but rather to provide examples of the principal types of preparative methods that can be used.

The synthetic methods used have included modern versions of established methods of metal colloid preparation, such as the mild chemical reduction of solutions of transition metal salts and complexes. Some of these reactions have been in use for many years, while some are the results of research stimulated by the current resurgence in metal colloid chemistry. Yet, the list of preparative methods is being extended daily, and as examples of these methods are described below, the reader will quickly become aware that almost any organometallic reaction or physical process which results in the deposition of a metal is in fact a resource for the metal colloid chemist. The acquisition of new methods requires only the opportunism of the synthetic chemist to turn a previously negative result into a synthetic possibility.

The synthetic methods described in the following sections all follow the so-called "bottom-up" strategy. ("Top-down" techniques, which are mechanically or anywise based on diminishing processes start from bulk metals, and will not be considered here.) Bottom-up strategies predominantly employ the chemical or electrochemical reduction of metal salts in solution. Thermolytic, photolytic, radiolytic or sonochemical procedures, as well as the decomposition of appropriate metastable precursor molecules, complete the wide spectrum of synthetic pathways to nanosized metal particles [15]. In any case, the formation of nanoparticles occurs when individual, neutral atoms, are subjected to collide with other atoms, resulting in a nucleation to smaller or larger particles. The particles growth can be influenced by concentration and solvent, and especially by the presence of a stabilizing agent. Short nucleation times support the formation of monodisperse particles, since freshly generated atoms will be trapped by existing nuclei in the diffusion-controlled growth; consequently, no new particles will be formed and a broad particle size distribution will be prevented. An important factor of particle formation is the specific surface energy of the corresponding metal. If such energy is significantly higher than the loss of entropy, then "Ostwald ripening" [16] will occur; that is, larger particles will grow to the debit of smaller particles, such that a broad size distribution usually results. Passivation of the particles' surfaces due to the presence of appropriate ligand molecules may also help in the size-selective generation of nanoparticles, since not only particle growth (from a distinct size on) but also agglomeration will be prevented. The so-called "digestive ripening" occurs if a polydisperse solution of AuNPs is heated in the presence of excess ligand molecules [17–21]. The reason for this occurring is that the larger particles will be reduced in size, whereas the smaller particles will grow. This is a reverse process of the Ostwald ripening, and is caused by the thermodynamically favored reduction of nonstabilized surfaces. The control of size is only one factor to be considered, however, with shape control becoming increasingly of interest for possible applications in various areas of nanotechnology. In this sense, structures such as spheres, cubes, rods, hollow cages, and wires are attracting ever-more interest; indeed, the creation of these structures frequently requires the use of special conditions, a situation which will be discussed by means of examples in the following sections.

3.3.1.4.1 **Salt Reduction** The reduction of transition metal salts in solution is the most widely practiced method of generating metal nanoparticles. Although, in aqueous systems, the reducing agent must be added or generated *in situ*, in nonaqueous systems the solvent and reducing agent may be one and the same.

Easily oxidized solvents such as alcohols can thus function both as the reducing agent and as the stabilizer, and have been used widely in nanoparticle preparations. Hirai and Toshima developed a so-called "alcohol reduction process" which represented a very general process to produce metal nanoparticles, and was often stabilized by organic polymers such as PVP, PVA and poly(methyl vinyl ether) (PMVE) [13, 22–28]. In general, those alcohols which serve as useful reducing agents contain α-hydrogen (*tert*-butanol, for example, is ineffective) and are oxidized to the corresponding carbonyl compound (methanol to formaldehyde, ethanol to acetaldehyde, *iso*-propanol to acetone) [see Eq. (3.4)].

$$RhCl_3 + 3/2\ R_1R_2CHOH \rightarrow Rh(0) + 3/2\ R_1R_2C=O + 3\ HCl \qquad (3.4)$$

Palladium organosols have been prepared by reduction of palladium acetate in refluxing methanol, and also at a higher temperature in 2-ethoxyethanol, in the presence of PVP, to yield particles of about 6.5 nm [29, 30]. Platinum and palladium nanoparticles have been synthesized by heating salts of the metals in long-chain aliphatic alcohols such as 1-decanol, which not only reduce the salts to the metal but also act as stabilizers for the particles [31]. Esumi has reported the use of several alcohols in a comparative synthetic study, in which C_1–C_5 aliphatic alcohols, α-phenylethanol, β-phenylethanol and benzyl alcohol were used as solvent and reducing agent for palladium acetate in the presence of PVP, giving particle sizes in the 1–4 nm range [32]. The generation of bimetallic nanoparticles has also been reviewed [22].

The so-called "polyol method" is based on the reduction power of alcohols with more than one OH function [33]. Nanoparticles in a solid state of Pb, Co, Ir, Os, Cu, Ag, Au, Pd, and Cd have been prepared and generated using $HO-CH_2-CH_2-OH$, which is oxidized via $CH_3-C(H)O$ to $CH_3-CO-CO-CH_3$, as shown in Eq. (3.5) [34].

$$H_2C-CH_2 \underset{\underset{H}{\overset{|}{O}}\ \underset{H}{\overset{|}{O}}}{|\ |} \xrightarrow{-2H_2O} H_3C-\underset{H}{\overset{\overset{O}{\parallel}}{C}} \xrightarrow{H_2PtCl_6} H_3C-\underset{\overset{\parallel}{O}}{C}-\underset{\overset{\parallel}{O}}{C}-CH_3 + H_2O + HCl + Pt_{Particle} \qquad (3.5)$$

Without special conditions, the size distribution of as-prepared particles is broad. However, by varying the temperature the use of only partially soluble precursor complexes supersaturation of metal atoms in solution can be avoided, such that the solid material will serve as a type of reservoir. Long-chain polyalcohols such as 1,2-hexanediol [35] can also act as protectors, similar to other materials such as PVP, to help control particle size distribution [36–38]. Previously, PVP was applied to the generation of numerous monodisperse metal nanoparticles of Pd [39, 40], Pt [40], and Au [41]. A series of alloy-like nanoparticles has also been prepared using the polyol method [35, 42–54]. Monodisperse particles of Pd and Ag could be generated if they were to be deposited on alumina in *status nascendi*, a technique that may be helpful for the generation of heterogeneous catalysts.

The reduction of transition metal salts by the addition of a reducing agent in a nonreducing solvent represents one of the oldest established procedures for the preparation of metal nanoparticles solutions. A wide range of reducing agents has been used, and they are frequently interchangeable from metal to metal.

Faraday reduced aqueous solutions of $[AuCl_4]^-$ with phosphorus vapor to produce gold hydrosols [1], and a variety of reducing agents have subsequently been used with tetrachloroaurate, both with and without protective polymers, to produce colloidal gold with particles ranging from one to several hundred nanometers in diameter. Turkevitch and coworkers reproduced many of these methods, and have established reliable preparative procedures for the preparation of gold sols with quite precisely defined particle sizes. The 20 nm gold sol prepared by the reduction of $[AuCl_4]^-$ by sodium citrate has become a standard for histological staining applications. The mechanism of metal salt reduction by citrate was also investigated by the same authors who showed that, in the reduction of $[AuCl_4]^-$, an induction period that was present when citrate was the reducing agent was absent when $[AuCl_4]^-$ was reduced by acetone dicarboxylate (an oxidation product of citrate), such that a rapid formation of colloidal gold resulted.

$$\begin{array}{c} CH_2\text{-COOH} \\ | \\ HOC\text{-COOH} \\ | \\ CH_2\text{-COOH} \end{array} \xrightarrow{[AuCl_4]} \begin{array}{c} CH_2\text{-COOH} \\ | \\ O=C \\ | \\ CH_2\text{-COOH} \end{array} \longrightarrow CO_2 + Au(0) \qquad (3.6)$$

A similar mechanism may be postulated for other reductions with citrate, and this method has been widely used for the preparation of other metal sols, such as platinum [55–57]; citrate can also be added as an ionic stabilizer in preparations which require an additional reducing agent. The use of formate, citrate and acetone dicarboxylate as reducing agents at various pH values was reported to give good control over particle size in the preparation of a series of platinum hydrosols [58]. Turkevitch's pioneering studies of the 1950s were to some extent continued by Henglein et al. during the late 1970s [59, 60].

Sodium borohydride, hydroxylamine hydrochloride, dimethylamine borane, sodium citrate, hydrazine monohydrate, sodium formate, trimethylamine borane, sodium trimethoxyborohydride and formaldehyde have all become routine reducing agents for the generation of metal nanoparticles.

Among the various reducing agents, BH_4^- meanwhile plays the dominant role. Brust and Shiffrin introduced this method to reduce gold salts in the presence of alkanethiols, the result being AuNPs of 1–5 nm. The Brust–Schiffrin method has since become well established for nanoparticles syntheses, and especially of AuNPs [61–63]. Stabilization with thiols guarantees a high stability of such species, due to the extraordinary strength of the Au–S bond. Not only do simple alkylthiols find applications, but so too (and to the same extent) do thiol moieties such as cysteine, glutathione (GSH) or more complex sulfur-containing biomolecules, which protect AuNPs of different sizes [64]. Variations in the thiol concentrations have provided a good size control of the particles between 2 and 5 nm [65]. Thiol-stabilized AuNPs were also prepared by following a seeding growth approach, starting with 3.5 nm colloids [66]. Subsequently, AuNPS in the size range of 10 to about 30 nm have been prepared by using mercaptosuccinic acid for stabilization and BH_4^- as the reducing compound. Variations in the $[AuCl_4]^-$: succinic acid ratio allow the specific synthesis of 10.2, 10.8, 12.8, 19.4, and 33.6 nm species [67].

The generation of reducing agents *in situ* can also be used for the preparation of gold hydrosols from $[AuCl_4]^-$. For instance, under alkaline conditions tetrakis (hydroxymethyl)phosphonium chloride (THPC) is hydrolyzed so as to eliminate formaldehyde, while under conditions of hydrolysis $[AuCl_4]^-$ oxidizes THPC with the formation of small (1–2 nm) gold clusters [68, 69].

Gaseous hydrogen is another long-standing and very effective reducing agent for metal salts. In this case, hydrosols of palladium, platinum, rhodium, and iridium, stabilized with PVA, were prepared by a hydrogen reduction of the metal hydroxides [70–74].

Moiseev *et al.* used hydrogen to synthesize Pd nanoclusters of very uniform size and composition, such that the particles could be described by a rather precise formula: $Pd_{561}L_{\approx 60}(OAc)_{\approx 180}$ (L = phenanthroline, bipyridine) [75–80]. Pd nanoclusters, which also were of a discrete size, shape and composition, were described by Schmid *et al.*, who also used H_2 to reduce Pd salts to produce $Pd_{561}phen_{36}O_{\approx 200}$, $Pd_{1415}phen_{60}O_{\approx 1100}$, $Pd_{2057}phen_{84}O_{\approx 1600}$ (phen = phenanthroline) [81–83].

These nanoclusters belong to the family of so-called "full-shell" clusters, which means that their metal nuclei consist of a number of atoms that follow hexagonal or cubic close-packing. A coordination number of 12 in such metallic structures leads to a first full-shell cluster of $1 + 12 = 13$ atoms, followed by another shell of 42 atoms, resulting in 55 atoms, and so on. Hence, the above-mentioned Pd clusters correspond to five shells (561 atoms), seven shells (1415 atoms), and eight shells (2057 atoms), although of course, by increasing their size, the exact number of atoms can in practice deviate from the ideal number. However, it must be stated that the full-shell clusters known to date are indeed more or less monodisperse, especially when compared to the huge number of larger nanoparticles which normally show size-distributions of at least ±10%. An elucidation of the organization of full-shell clusters is shown in Figure 3.106, while Figure 3.107 shows the TEM images of a Pt four-shell and a Pd seven-shell cluster, where 9 $(1 + 4 + 4)$ and 17 $(1 + 8 + 8)$ atomic planes of dense packed atoms can be counted.

Full-shell cluster chemistry was first devised in 1981, when ligand-protected Au_{13} [84, 85] and Au_{55} [86] clusters were initially synthesized. Whereas, the icosahedral Au_{13} species could be described as a typical molecular cluster, the cubic close-packed (ccp) Au_{55} nucleus in $Au_{55}(PPh_3)_{12}Cl_6$ and the numerous derivatives represented exactly the transition from molecule to bulk, and was therefore of great significance in many respects. Notably, it represented the prototype of a metallic quantum dot (QD) [87–131]. An excellent review on Au_{55} clusters is provided in Ref. [132]., while further details of these cluster types are provided in Chapters 4 and 5.

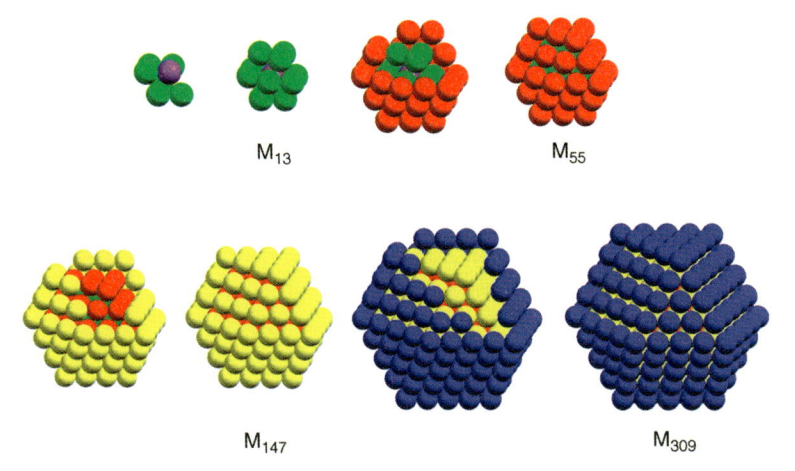

Figure 3.106 Organization of cubo-octahedral full-shell clusters with $10n^2 + 2$ atoms in the nth shell.

The synthesis of $Au_{55}(PPh_3)_{12}Cl_6$ occurs via a very special reduction method. First, diborane is allowed to react with Ph_3PAuCl in warm benzene or toluene [86, 133], while B_2H_6 serves not only as a reducing agent but also to trap any excess of PPh_3 so as to yield $Ph_3P–BH_3$. Numerous derivatives of the PPh_3-stabilized cluster are available by ligand-exchange reactions. The details of some ligand-exchange reactions are shown schematically in Figure 3.108, while in Figure 3.109 is shown a space-filling atom model of $Au_{55}(PPh_3)_{12}Cl_6$ together with a scanning electron microscopy (STM) image of an individual cluster in the same orientation [128].

For instance, $Ph_2PC_6H_4SO_3Na$ removes PPh_3 (Figure 3.108, pathway **a**) quantitatively in a phase-transfer process from CH_2Cl_2 to H_2O, rendering the compound water-soluble (Figure 3.108, pathway **b**) [89]. Thiols are best suited to substitute PPh_3, due to their stronger Au–S bonds compared to the weaker Au–P bonds [118, 134, 135].

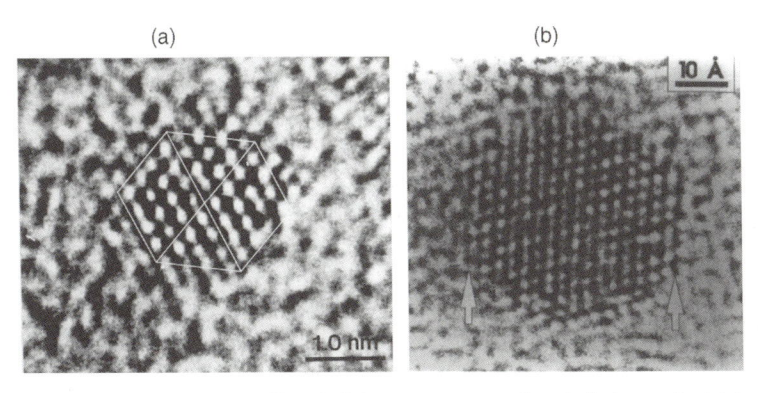

Figure 3.107 TEM images of full-shell clusters. (a) A Pt_{309} four-shell cluster; (b) A Pd_{561} five-shell cluster.

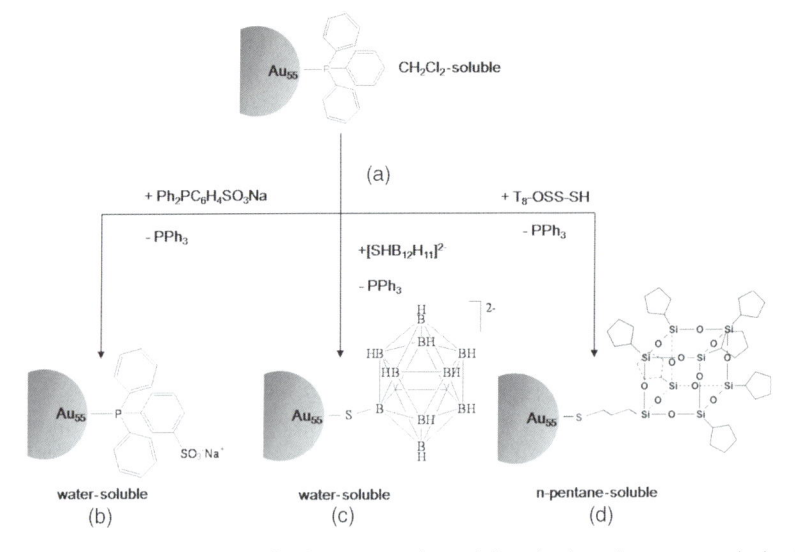

Figure 3.108 Substitution of PPh$_3$ in Au$_{55}$(PPh$_3$)$_{12}$Cl$_6$ by other ligands, generating hydrophilic (**b** + **c**) and hydrophobic (**d**) derivatives.

Then, the sodium salt of the closo-borate [SHB$_{12}$H$_{11}$]$^{2-}$ transfers the mother cluster to the water-soluble 24-fold negatively charged derivative (Figure 3.108, pathway **c**), whereas the thiol-substituted silsesquioxane T$_8$-OSS–SH yields the totally unpolar derivative (Figure 3.108, pathway **d**), which is soluble in *n*-pentane due to the existence in total of 72 cyclopentyle substituents. B$_2$H$_6$ has also been used to synthesize ligand-stabilized Pt$_{55}$, Rh$_{55}$ and Ru$_{55}$ clusters [136]. The only four-shell cluster to have become well known is Pt$_{309}$phen$_{36}$O$_{30\pm10}$ (phen = 4,7-C$_6$H$_4$SO$_3$Na-substituted 1,10-phenanthroline) [81]. This is prepared from Pt(II) acetate, which is

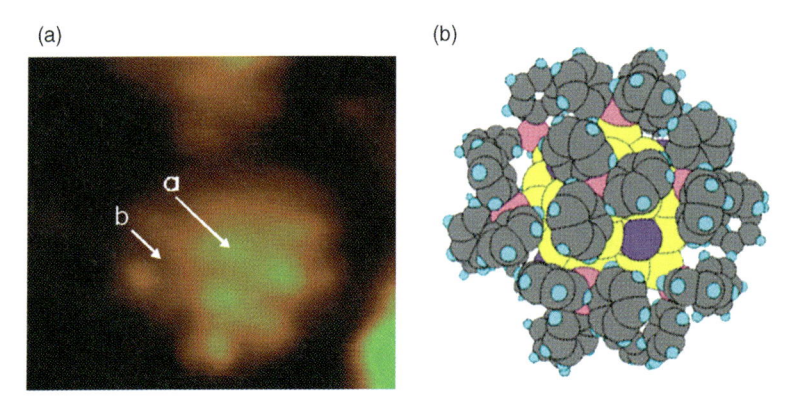

Figure 3.109 (a) STM image of Au$_{55}$(PPh$_3$)$_{12}$Cl$_6$; (b) The corresponding space-filling atom model. Arrow 'a' in the STM image indicates a phenyl ring, and arrow 'b' a neighbored non-coordinated position.

reduced by hydrogen at room temperature in acetic acid solution in the presence of equivalent amounts of ligand. Under these conditions, the oxygen-free cluster is formed first, and then carefully treated with oxygen to provide the air-stable final product. The nature of the oxygen is most likely (as in the above-mentioned Pd full-shell clusters) either superoxidic or peroxidic, and this can easily be removed from the surfaces by hydrogen. The structure of the Pt_{309} core is ccp, as in the case of the Au_{55}, the Pd_{561}, the Pd_{1415}, and the Pd_{2057} nanoclusters; this has been demonstrated using X-ray diffraction (XRD) and high-resolution transmission electron microscopy (HR-TEM) studies. Additional structural details are provided in Ref. [137].

Remarkable contributions to the synthetic field of transition metal nanoparticles were provided by the group of Bönnemann [28, 138–151], some details of which will be discussed in the following sections. The use of tetra-alkylammoniumhydrotriorganoborates $[R_4N]^+[BR'_3H]^-$, as reducing species for transition metal salts, was found to result in numerous valuable $[NR_4]^+$-stabilized metal nanoparticles. The general formation of metal nanoparticles following this route is shown in Eq. (3.7):

$$MX_n + n[R_4N] \rightarrow M(0) + n[R_4N]X + nBEt_3 + n/2\, H_2$$
$$(X = Cl, Br, R = alkyl, C_6-C_{20}; M = metals\ of\ groups\ VI\text{-}XI) \quad (3.7)$$

The ammonium salts accomplish two functions: (i) the hydride reduces the positively charged metal in MX_n; and (ii) the relatively bulky $[R_4N]^+$ cations stabilize the metal nanoparticles, as indicated in Figure 3.110.

Figure 3.110 Sketch of an ammonium salt-protected metal nanoparticle.

Nanoparticles of numerous transition metals could be generated by this so-called tetra-alkylammonium method (the particle size, in nm, is shown in parentheses): Cr (2–3); Fe (3.0); Ru (1.3); Co (2.8); Rh (2.1); Ir (1.5); Ni (2.8); Pd (2.5); Pt (2.8); Cu (8.3); Ag (2–13); and Au (10). The $[BEt_3H]^-$ anion as a potassium salt even reduces salts of the early transition metals that are only stabilized by THF; consequently, Ti, V, and Nb nanoparticles can be acquired [152–155]. A modified version of the creation of tetra-alkylammonium-stabilized metal nanoparticles involves the use of tetra-alkylammonium carboxylates, acting as both the reductant and stabilizer [156]:

$$M^+ + R'COO^- NR_4^+ \rightarrow M^0(R'COO^- NR_4)_x + CO_2 + R-R' \quad (3.8)$$
$$(M^+ = \text{metal ion}, R = \text{octyl}, R' = \text{alkyl, aryl, H})$$

A rather unexpected reaction occurs between metal salts MX_n (M = metals of groups VI–XI, X = halogen, acetylacatonate, $n = 2$–4, R = C_1–C_8 alkyl) and AlR_3, giving particles in the size range of 1–12 nm [157, 158]. It has been suggested that a nonprecisely determinable layer of organoaluminum species covers the nanoparticles.

Instead of chemically generated electrons as reducers, the electrolytic reduction of metal salts represents a successful alternative in connection with the fabrication of tetra-alkylammonium salt-stabilized nanoparticles [159–165]. The first step comprises an oxidative anodic dissolution of the corresponding metal, followed by the formation of zerovalent metal atoms at the cathode. Nucleation and particle growth then follow, some of which is stopped by the addition of a tetra-alkylammonium salt. This technique not only prevents the formation of byproducts but also allows a rather good size-selectivity: high current densities lead to small particles, while low current densities yield larger species. Ammonium salt-protected nanoparticles of Ti, Fe, Co, Ni, Pd, Pt, Ag, and Au may also be prepared in this way.

3.3.1.4.2 **Controlled Decomposition** Since many organometallic compounds of transition metals decompose thermally to their respective metals under relatively mild conditions, these compounds provide a rich source of nanoparticle precursors. Moreover, this method is widely applicable.

The syntheses of organosols of palladium and platinum have been reported by the thermolysis of precursors such as palladium acetate, palladium acetylacetonate and platinum acetylacetonate in high-boiling organic solvents such as methyl-*iso*butylketone, and bimetallic colloids of copper and palladium from the thermolysis of mixtures of their acetates in similar solvents [166–171]. These preparations were performed in the absence of stabilizing polymers and, as a result, relatively broad size distributions and large particles were observed. A preparation of bimetallic PVP-stabilized copper–palladium particles with diameters of 1–4 nm was reported by Toshima, in which palladium acetate and cupric sulfate were transformed first to colloidal hydroxides by treatment of an ethylene glycol solution of the salts with sodium hydroxide solution, in the presence of PVP [172]. The colloidal mixed hydroxide was then heated to 198 °C under nitrogen, to yield a stable bimetallic sol.

In the meantime, the method of thermal decomposition of organometallics has undergone further development, notably under guidance from the group of Chaudret. By following this approach, numerous tetra-alkylammonium salt-stabilized nanoparticles have been obtained [30, 173–201].

In view of the extensive literature on the syntheses and structures of bimetallic metal carbonyl clusters, it is even more surprising that few attempts have been reported to employ the facile decomposition of these compounds in the preparation of bimetallic nanoparticles. Although this would be an unnecessarily complicated route for nanoparticles of miscible metals if their salts could be co-reduced to the same end, there remains the possibility of preparing bimetallic particles of immiscible metals starting from well-defined molecular bimetallic clusters of immiscible metals. Such clusters are known to exist, despite the immiscibility of their constituent metals in the bulk; indeed, there is reason to believe that bulk immiscibility might not apply to very small particles, due to the importance of surface effects. Microwave heating is appropriate to generate metal nanoparticles of a narrow size distribution, and polymer-stabilized Pt particles of 2–4 nm have already been fabricated in this way [202–205]. The same is valid for sonochemical techniques, using organometallic complexes [205–207].

Although, photolysis, radiolysis and laser irradiation have also been applied for metal nanoparticles synthesis, they are of limited application value due to the need for low concentrations [208–210].

The final method used to generate size-controlled metal nanoparticles involves ligand displacement from organometallic complexes with zero-valent metals.

Since the 1970s, palladium and platinum complexes with dibenzylideneacetone Pd(dba)$_2$ and M$_2$(dba)$_3$ (M = Pd, Pt) have been known to react under mild conditions with either hydrogen or carbon monoxide, with the formation of a metal [211]. Indeed, there exists a long series of examples where CO and H$_2$ have been used to decompose organometallic precursor molecules [173–177, 179–181, 183, 184, 186–188, 212–216]. As an example, the decomposition Ru(COD)(COT) (COD = cyclo-octadiene; COT = cyclo-octatriene) in an atmosphere of hydrogen is worthy of mention [189, 190]. (Scheme 3.20). In this case, the precursor molecule is dissolved in a methanol–THF mixture and is contacted with H$_2$ (3 bar pressure) at room temperature for at least 45 min. Depending on the nature of the MeOH–THF mixture, the Ru particle size can be designed between 3 and 86 nm.

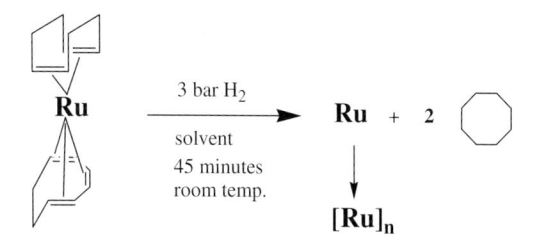

Scheme 3.20 Formation of Ru nanoparticles from Ru(COD)COT).

3.3.1.5 Shape Control

In the past, metal nanoparticle synthesis has been mainly focused on a limited size distribution, with the value of any new procedure consisting mainly of improvements in size distribution. Except for a few examples of discrete size (as in the full-shell clusters M_{55}, M_{309} or M_{561}), particles of ±10% in size are generally also referred to as "monodisperse."

Meanwhile, for particles in the upper size region another criterion has become quite important – that of shape control. With the increasing practical use of metal nanoparticles in nanotechnologies, it is not only the size and size distribution of the particles which has become more important, but also their shape. Consequently, shapes such as cubes, tetrahedrons, octahedrons, icosahedrons, cylinders, wires and many other forms have been recognized, mainly of noble metals. As with size control, the main point is to identify a fabrication technique that produces a defined geometry for the resultant species. The reason behind this increasing interest in metal nanoparticles of distinct shape is to be seen in the numerous shape-dependent properties, especially in the field of optics. It has long been known that the optical properties of metal nanoparticles are determined by the collective oscillations of the conduction electrons, caused in turn by the interaction with electromagnetic radiation, especially in the case of the metals Cu, Ag, and Au. This oscillation depends, among others factors, on the particles' size and shape. For example, in contrast to spherical nanoparticles, nonspherical particles (such as rods) exhibit two oscillations: transverse and longitudinal.

Particles of distinct shape are frequently referred to as "nanocrystals," the definition of which states that such species must be limited to 1–100 nm in at least one dimension [217]. The targeted synthesis of shape-controlled metal nanoparticles has developed greatly during the past decade [218, 219], and although in the ideal case they are single-crystalline, they are invariably polycrystalline. Both, thermodynamic and kinetic factors must be considered to achieve such a goal. In a thermodynamically controlled reaction the most stable product will be formed, whereas a kinetically controlled particle formation will be influenced by the speed of reduction or decomposition of the starting compound during crystal germ formation. Although the thermodynamically most stable shape would be a truncated octahedron, cubes, cubo-octahedrons, octahedrons and tetrahedrons can also be formed [220]. In practice, kinetic factors play the dominant role, where reaction conditions such as the reaction time, protecting molecules, concentrations and temperature influence both the nucleation and crystal growth [219].

The organization of metal atoms to a nanocrystal occurs, as with any other case of crystallization, via a decisive nucleus that is difficult to observe due to its small size. Various experimental data have been acquired for specialized cases, by using electrospray mass spectrometry [221, 222], complemented by *ab initio* calculations [223]. Based on the existence of a series of full-shell clusters M_{13}, M_{55}, M_{309}, M_{561}, M_{1415}, and M_{2057} (see Section 3.3.1.4.1), it is assumed that such nanoclusters could serve as nucleation centers in crystallization processes, culminating in crystal germs and finally in nanocrystals [218]. Whilst crystal germs may simultaneously possess monocrystalline and multiple-twinned structures, in order to obtain a single type of nanocrystal it is essential first to control the population of the crystal germ of interest.

In general, there exists a correlation between the different types of crystal germs and the resultant nanocrystals. Single-crystalline crystal germs of fcc metals generally result in octahedrons, cubo-octahedra or cubes, though which of these geometries is ultimately realized depends on the relative rate of growth along the (111)- and (100)-directions [224]. If crystal growth along a single axis can be induced, then cubo-octahedral and cubic crystal germs will develop to form eight-angled sticks or rectangulars [225], while single-twinned crystal germs will develop to regularly shaped bipyramids [226, 227] or, if growth occurs in only one direction, to nanocylinders [228]. Multiple-twinned crystal germs form icosahedrons, decahedrons or pentagonal nanosticks and nanowires, depending on whether the {100}-faces are stabilized, or not [219, 225, 229–234]. Triangular and hexagonal thin platelets are formed from crystal germs with stacking faults [235–244].

The formation of Ag triangular nanoprisms from spherical nanoparticles in the presence of trisodium citrate and bis(p-sulfonatophenyl)phenylphosphine (BSPP) (dipotassium salt) with fluorescent light [245] has been described, as has that of Au nanoprisms [246].

Shape control can be directed by the use of molecular capping agents, which adsorb to specific crystal planes such that growth is limited on those crystal planes without capping molecules, or which have only weakly coordinated molecules. Surfactants, polymers and biomolecules, small gas molecules and even different metal ions have each been shown capable of controlling nanocrystal growth [219].

Shape-inducing cationic surfactants and other additives have been recognized not only for direct gold-rod formation during the electrochemical reduction of $HAuCl_4$, but also because they act simultaneously as stabilizers for the resultant species. When using cetyltrimethylammonium bromide (CTAB) and tetradecylammonium bromide (TDAB), the ratio of the two surfactants can be applied to control the aspect ratio of the generated Au nanorods [247–250]. Cu nanorods have also been generated using bis(2-ethylhexyl)sulfosuccinate (AOT) [251]. If capping ligands provide very strong bonds to metals (as do thiol functions to noble metals), then spherical shapes are generally formed [252, 253]. Primary amines, amine-terminated dendrimers and ammonium-terminated surfactants have each been shown to support anisotropic growth that resulted in rod, bipod, tripod, or tetrapod AuNPs [254]. At present, however, the details of nanocrystal formation are far from understood.

Besides the use of directive coordinating surfactants or other species, the polyol method (see above) in connection with the application of PVP as the surface capping agent, has been shown to function very successfully with respect to shape control. If metal salt reduction by various polyols occurs in the presence of PVP, the polymer not only prevents aggregation of metal nanoparticles but also acts a "crystal-habit modifier," by directing the addition of metal atoms to specific faces [219]. Based on spectroscopic studies, it is known that PVP and noble (fcc) metals interact preferentially via the carbonyl group of the pyrrolidone ring mainly on top of the {111} and {100} planes, although the details of this are currently under consideration. The formation of a silver octahedron via different geometries, beginning with a cube

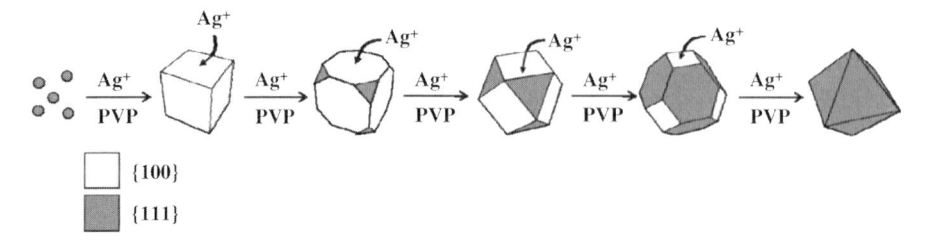

Figure 3.111 Shape evolution of polyhedral Ag nanocrystals.

and assuming that new Ag atoms are exclusively added onto {100} facets, is illustrated schematically in Figure 3.111.

Recently, PVP has been used to decisively control the shape of Ag [255–258], Au [257, 258] and Pt [259] nanoparticles. Each of these three metals can be generated in a variety of shapes such as rods, cubes, wires, octahedra, and icosahedra. A selection of as-prepared AgNPs and AuNPs is shown (as TEM images) in Figure 3.112. Clearly, the Ag and Au cubes, cubo-octahedra and octahedra can each be tuned by careful control of the reaction conditions [256, 259, 260].

As noted above, in addition to large surfactants and even polymers, selective shape-directing plane-blocking can also be induced by small molecules, and even by atomic species [261]. For example, nitrogen dioxide (NO_2), if present during the deposition of Pd on small Pt germs, will cause an increased growth of {111} faces compared to {100} faces. Likewise, the presence of traces of Ag^+ ions has been shown to lead to variations in the {111}:{100} ratio during Au and Pt nanocrystal growth in polyol reactions [257]. Ag nanocrystal growth be influenced in a similar manner [259]. Pd cubes, bars and pentagonal rods may be generated using KBr in micromolar amounts as capping agents during the reduction of Na_2PdCl_4 with ascorbic acid [262], while

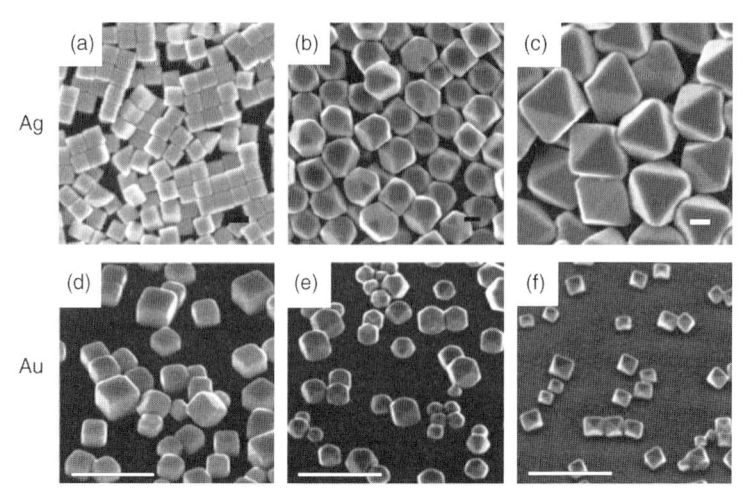

Figure 3.112 Silver cubes (a), cubo-octahedra (b), and octahedra (c). (Scale bar = 100 nm); (d–f) The same shapes, but with gold nanoparticles (scale bar = 1 μm).

citric acid may also serve as a capping agent (in addition to its reducing function) towards Na_2PdCl_4 to produce truncated octahedra, icosahedra, octahedral, and decahedra [262].

Finally, bioinspired techniques to control shape of metal nanocrystals are briefly described. Proteins, peptides or nucleic acids invariably possess sulfide, amide or carbonyl functions, while polypeptide sequences may initiate Au particle growth, resulting in thin platelets [263]. The shape control of AgNPs has also been observed with phage-display peptides [264].

One very special type of "shape control" results in hollow nanoparticles of different geometries. In principle, the process is rather simple, in that a particle of any structure may serve as template that is coated by a layer of the desired metal. Subsequent removal of the inner template, either by calcination or by selective dissolution, will result in the formation of hollow nanostructures. Whilst this general technique can be applied to many types of material, one special application that involves only metals is as follows. In this case, silver nanocrystals of different shapes are coated by gold layers, via a simple chemical process based on the reductive power of elementary silver towards Au^{3+}.

By using this procedure, which is shown schematically in Figure 3.113 [265], numerous shapes of AgNPs are available. Typically, the reduction of $HAuCl_4$ on the Ag surface causes small gold particles to be deposited on the silver surface (pathway A), enabling the diffusion of $HAuCl_4$ and AgCl through the imperfect shell (pathway B). The process is discontinued (pathway C) when the silver template is completely consumed such that, if the final step is performed at 100 °C, the imperfect gold coating will be transferred onto a crystalline material. The advantage of this technique is to be seen in the transformation of an original silver shape into a hollow gold architecture. This process can also be applied to other metals, examples being the redox systems $Ag–Pd^{2+}$ and $Ag–Pt^{2+}$, with corresponding results.

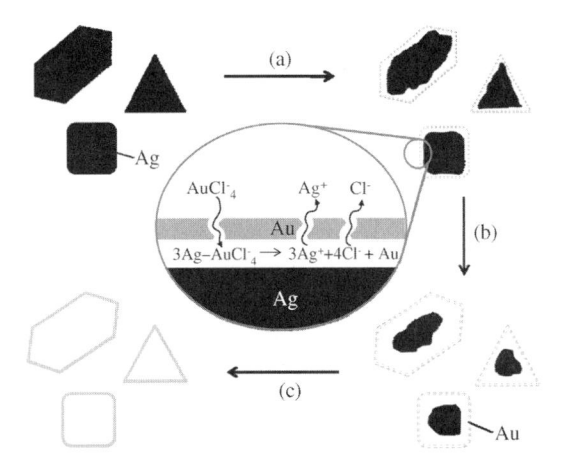

Figure 3.113 Formation of hollow gold nanostructures. For details, see the text.

Reference

1. Faraday, M. (1857) *Philos. Trans. R. Soc.*, **147**, 145.
2. Zsigmondy, R. (1909) *Colloids and the Ultramicroscope (English Edition)*, John Wiley, New York.
3. Turkevitch, J., Stevenson, P.C., and Hillier, J. (1951) *Disc. Faraday Soc.*, **11**, 55.
4. Turkevitch, J. (1985) *Gold Bull.*, **18**, 86.
5. Overbeek, J.T.G. (1981) *Colloidal Dispersions* (ed. J.W. Goodwin), Royal Society of Chemistry, London, p. 1.
6. Blatschford, C.G., Campbell, J.R., and Creighton, J.A. (1982) *Surface Sci.*, **120**, 435.
7. Labib, M.E. (1988) *Colloids Surf.*, **29**, 293.
8. Labib, M.E. and Williams, R. (1984) *J. Colloid Interface Sci.*, **97**, 356.
9. Duteil, A., Quéau, R., Chaudret, B., Mazel, R., Roucau, C., and Bradley, J.S. (1993) *Chem. Mater.*, **5**, 341.
10. Komiyama, M. and Hirai, H. (1983) *Bull. Chem. Soc. Jpn*, **56**, 2833.
11. Thiele, V.H. and Kowallik, J. (1965) *J. Colloid Sci.*, **20**, 679.
12. Hess, P.H. and Parker, P.H. (1966) *J. Appl. Polymer. Sci.*, **10**, 1915.
13. Hirai, H., Nakao, Y., and Toshima, N. (1979) *J. Macromol. Sci. Chem.*, **13**, 727.
14. Hirai, H. (1985) *Macromol. Chem. Suppl.*, **14**, 55.
15. Bönnemann, H. and Nagabhushana, K.S. (2008) *Metal Nanoclusters in Catalysis and Materials Science. The Issue of Size Control* (eds B. Corain, G. Schmid, and N. Toshima), Elsevier, Amsterdam, p. 21.
16. El Nakat, J.H., Dance, J.G., Fisher, K.J., and Willet, G.D. (1991) *Inorg. Chem.*, **30**, 2957.
17. Lin, X.M., Sorensen, C.M., and Klabunde, K.J. (2000) *J. Nanoparticle Res.*, **2**, 157.
18. Stoeva, S.I., Klabunde, K.J., Sorensen, C.M., and Dragieva, I. (2002) *J. Am. Chem. Soc.*, **124**, 2305.
19. Stoeva, S.I., Prasad, B.L.V., Uma, S., Stoimenov, P.K., Zaikovski, V., Sorensen, C.M., and Klabunde, K.J. (2003) *J. Phys. Chem. B*, **107**, 7441.
20. Prasad, B.L.V., Stoeva, S.I., Sorensen, C.M., and Klabunde, K.J. (2003) *Chem. Mater.*, **15**, 935.
21. Stoeva, S.I., Zaikovski, V., Prasad, B.L.V., Stoimenov, P.K., Sorensen, C.M., and Klabunde, K.J. (2005) *Langmuir*, **21**, 10280.
22. Toshima, N. and Yonezawa, T. (1998) *New J. Chem.*, **22**, 1179.
23. Hirai, H., Nakao, Y., Toshima, N., and Adachi, K. (1976) *Chem. Lett.*, 905.
24. Hirai, H., Nakao, Y., and Toshima, N. (1978) *Chem. Lett.*, 545.
25. Hirai, H., Nakao, Y., and Toshima, N. (1978) *J. Macromol. Sci. Chem. A*, **112**, 1117.
26. Toshima, N. and Hirakawa, K. (1999) *Polymer J.*, **31**, 1127.
27. Lu, P., Teramishi, T., Asakura, K., Miyake, M., and Toshima, N. (1999) *J. Phys. Chem. B*, **103**, 9673.
28. Bönnemann, H. and Richards, R.M. (2001) *Eur. J. Inorg. Chem.*, 2455.
29. Bradley, J.S., Millar, J.M., and Hill, E.W. (1991) *J. Am. Chem. Soc.*, **113**, 4016.
30. Bradley, J.S., Hill, E.W., Behal, S., Klein, C., Chaudret, B., and Duteil, A. (1992) *Chem. Mater.*, **4**, 1234.
31. Mandler, D. and Willner, I. (1987) *J. Phys. Chem.*, **91**, 3600.
32. Esumi, E., Itakura, T., and Torigoe, K. (1994) *Colloids Surf. A*, **82**, 111.
33. Figlarz, M. (1989) *Mater. Res. Soc. Bull.*, **14**, 29.
34. Bonet, F., Delmas, V., Grugeon, S., Urbina, R.H., Silvert, P.-Y., and Elhsissen, K.T. (1999) *Nanostruct. Mater.*, **11**, 1277.
35. Liu, C., Wu, X., Klemmer, T., Shukla, N., Yang, X., Weller, D., Roy, A.G., Tanase, M., and Laughlin, D. (2004) *J. Phys. Chem. B*, **108**, 6121.
36. Silvert, P.-Y., Vijayakrishnan, P., Vibert, P., Urbina, R.H., and Elhsissen, K.T. (1996) *Nanostruct. Mater.*, **7**, 611.
37. Silvert, P.-Y., Urbina, R.H., Duvauchelle, N., Vijayakrishnan, V., and Elhsissen, K.T. (1996) *J. Mater. Chem.*, **6**, 573.
38. Silvert, P.-Y. and Urbina, R.H. (1997) *J. Mater. Chem.*, **7**, 293.
39. Sanguesa, C.D., Urbina, R.H., and Figlarz, M. (1993) *Solid State Ionics*, **63–65**, 25.

40 Elhsissen, T.K., Bonet, F., Grugeon, S., Lambert, S., and Urbina, R.H. (1999) *J. Mater. Res.*, **14**, 3707.

41 Silvert, P.-Y. and Elhsissen, K.T. (1995) *Solid State Ionics*, **82**, 53.

42 Elhsissen, T.K., Silvert, P.-Y., and Urbina, R.H. (1999) *J. Alloys Compd.*, **292**, 96.

43 Viau, G., Vincent, F.F., and Fiévet, F. (1996) *Solid State Ionics*, **84**, 259.

44 Viau, G., Vincent, F.F., and Fiévet, F. (1996) *J. Mater. Chem.*, **6**, 1047.

45 Toneguzzo, P., Acher, O., Viau, G., Pierrard, A., Vincent, F.F., Fiévet, F., and Rosenman, I. (1999) *IEEE Trans. Magn.*, **235**, 3469.

46 Viau, G., Toneguzzo, P., Pierrard, A., Acher, O., Vincent, F.F., and Fiévet, F. (2001) *Scripta Mater.*, **44**, 2263.

47 Bonet, F., Grugeon, S., Dupont, L., Urbina, R.H., Huery, C., and Tarascon, J.M. (2003) *J. Solid State Chem.*, **172**, 111.

48 Brayner, R., Viau, G., da Cruz, G.M., Vincent, F.F., Fiévet, F., and Verduraz, F.B. (2000) *Catal. Today*, **57**, 187.

49 Viau, G., Brayner, R., Poul, L., Chakroune, N., Lacaze, E., Vincent, F.F., and Fiévet, F. (2003) *Chem. Mater.*, **15**, 486.

50 Chakroune, N., Viau, G., Ricolleau, C., Vincent, F.F., and Fiévet, F. (2003) *J. Mater. Chem.*, **13**, 312.

51 Sun, S., Murray, C.B., Weller, D., Folks, L., and Moser, A. (2000) *Science*, **287**, 1989.

52 Elumalai, P., Vasan, H.N., Verelst, M., Lcante, P., Carles, V., and Tailhades, P. (2002) *Mater. Res. Bull.*, **37**, 353.

53 Elumalai, P., Vasan, H.N., Munichandraiah, N., and Shivashankar, S.A. (2002) *J. Appl. Electrochem.*, **32**, 1005.

54 Elhsissen, T.K., Bonet, F., Grugeon, S., Lambert, S., and Urbina, R.H. (1999) *J. Alloys Compd.*, **292**, 96.

55 Furlong, D.N., Launikonis, A., Sasse, W.H.F., and Saunders, J.V. (1984) *J. Chem. Soc., Faraday Trans.*, **80**, 571.

56 Harriman, A., Millward, G.R., Neta, P., Richoux, M.C., and Thomas, J.M. (1988) *J. Phys. Chem.*, **92**, 1286.

57 Turkevitch, J., Miner, R.S.J., Okura, I., and Namba, S. (1981) *Proc. Swed. Symp. Catal.*, 111.

58 Turkevitch, J. and Kim, G. (1970) *Science*, **169**, 873.

59 Tausch-Treml, R., Henglein, A., and Lilie, J. (1978) *Ber. Bunsenges. Phys. Chem.*, **82**, 1335.

60 Henglein, A. (1985) *Modern Trends in Colloid Science in Chemistry and Biology*, Birkhauser Verlag, Stuttgart.

61 Brust, M., Walker, A., Bethell, D., Schiffrin, D.J., and Whyman, R. (1994) *Chem. Commun.*, 801.

62 Brust, M., Bethell, D., Schiffrin, D.J., and Kiely, C.J. (1995) *Adv. Mater.*, **7**, 795.

63 Brust, M., Fink, J., Bethell, D., Schiffrin, D.J., and Kiely, C.J. (1995) *Chem. Commun.*, 1655.

64 Levy, R., Thanh, N.T.K., Doty, R.C., Hussain, I., Nichols, R.J., Schiffrin, D.J., Brust, M., and Fernig, D.G. (2004) *J. Am. Chem. Soc.*, **126**, 10076.

65 Hostetler, M.J., Wingate, J.E., Zhong, C.J., Harris, J.E., Vachet, R.W., Clark, M.R., Londono, J.D., Green, S.J., Stokes, J.J., Wignale, G.D., Glish, G.L., Porter, M.D., Evans, N.D., and Muarry, R.W. (1998) *Langmuir*, **14**, 17.

66 Jana, N.R., Gearheart, L., and Murphy, C.J. (2001) *Langmuir*, **17**, 6782.

67 Chen, S. and Kimura, K. (1999) *Langmuir*, **15**, 1075.

68 Duff, D.G., Baiker, A., and Edwards, P.P. (1993) *Langmuir*, **9**, 2301.

69 Duff, D.G., Baiker, A., and Edwards, P.P. (1993) *Langmuir*, **9**, 2310.

70 Rampino, L.D. and Nord, F.F. (1941) *J. Am. Chem. Soc.*, **63**, 2745.

71 Rampino, L.D. and Nord, F.F. (1941) *J. Am. Chem. Soc.*, **63**, 3268.

72 Kavanagh, K.E. and Nord, F.F. (1943) *J. Am. Chem. Soc.*, **65**, 2121.

73 Hernandez, L. and Nord, F.F. (1948) *J. Colloid Sci.*, **3**, 363.

74 Dunsworth, W.P. and Nord, F.F. (1950) *J. Am. Chem. Soc.*, **72**, 4197.

75 Vargaftik, M.N., Zargorodnikov, V.P., Stolarov, I.P., Moiseev, I.I., Likholobov, V.A., Kochubey, D.I., Chuvilin, A.L., Zaikovsky, V.I., Zamaraev, K.I., and Timofeeva, G.I. (1985) *Chem. Commun.*, 937.

76 Vargaftik, M.N., Zargorodnikov, V.P., Stolarov, I.P., Moiseev, I.I., Kochubey, D.I., Likholobov, V.A., Chuvilin, A.L., and Zarnaraev, K.I. (1989) *J. Mol. Catal.*, **53**, 315.

77 Volkov, V.V., Tendeloo, G.v., Tsirkov, G.A., Cherkashain, N.V., Vargaftik, M.N., Moiseev, I.I., Novotortsev, V.M., Krit, A.V., and Chuvilin, A.L. (1996) *J. Cryst. Growth*, **163**, 377.

78 Moiseev, I.I., Vargaftik, M.N., Volkov, V.V., Tsirkov, G.A., Charkashina, N.V., Novotortsev, V.M., Ellett, O.G., Petrunenka, I.A., Chuvilin, A.L., and Krit, A.V. (1995) *Mendeleev Commun.*, 87.

79 Oleshko, V., Volkov, V.V., Jacob, W., Vargaftik, M.N., Moiseev, I.I., and Tendeloo, G.v. (1995) *Z. Physik D*, **34**, 283.

80 Moiseev, I.I., Vargaftik, M.N., Chernysheva, T.V., Stromnova, T.A., Gekham, A.E., Tsirkov, G.A., and Makhlina, A.M. (1996) *J. Mol. Catal. A: Chem.*, **108**, 77.

81 Schmid, G., Morun, B., and Malm, J.-O. (1989) *Angew. Chem., Int. Ed. Engl.*, **28**, 778.

82 Schmid, G. (1991) *Mater. Chem. Phys.*, **29**, 133.

83 Schmid, G., Harms, M., Malm, J.-O., Bovin, J.-O., Ruitenbeck, J.v., Zandbergen, H.W., and Fu, W.T. (1993) *J. Am. Chem. Soc.*, **115**, 2046.

84 Briant, C.E., Theobald, B.R.C., White, J.W., Beil, C.K., and Mingos, D.M.P. (1981) *Chem. Commun.*, 201.

85 Velden, J.W.A.v.d., Vollenbroek, F.A., Bour, J.J., Beurskens, P.I., Smits, J.M.M., and Bosmann, W.P. (1981) *Rec. J. R. Neth. Chem. Soc.*, **100**, 148.

86 Schmid, G., Boese, R., Pfeil, R., Bandermann, F., Meyer, S., Calis, G.H.M., and Velden, J.W.A.v.d. (1981) *Chem. Ber.*, **114**, 3634.

87 Staveren, M.P.J.v., Brom, H.B., Jongh, L.J.d., and Schmid, G. (1986) *Solid State Commun.*, **60**, 319.

88 Smit, H.H.A., Thiel, R.C., Jongh, L.J.d., Schmid, G., and Klein, N. (1988) *Solid State Commun.*, **65**, 915.

89 Schmid, G., Klein, N., Korste, L., Kreibig, U., and Schönauer, D. (1988) *Polyhedron*, 7, 605.

90 Schmid, G. (1988) *Polyhedron*, 7, 2321.

91 Benfield, R.E., Creighton, J.A., Eadon, D.G., and Schmid, G. (1989) *Z. Phys. D*, **12**, 533.

92 Fairbanks, M.C., Benfield, R.E., Newport, R.J., and Schmid, G. (1990) *Solid State Commun.*, **73**, 431.

93 Feld, H., Leute, A., Rading, D., Benninghoven, A., and Schmid, G. (1990) *Z. Phys. D*, **17**, 73.

94 Schmid, G. (1990) *Endeavour*, **14**, 172.

95 Feld, H., Leute, A., Rading, D., Benninghoven, A., and Schmid, G. (1990) *J. Am. Chem. Soc.*, **112**, 8166.

96 Becker, C., Fries, T., Wandelt, K., Kreibig, U., and Schmid, G. (1991) *J. Vac. Sci. Technol.*, **B9**, 810.

97 Kreibig, U., Fauth, K., Granqvist, C.-G., and Schmid, G. (1990) *Z. Phys. Chem. Neue Folge*, **169**, 11.

98 Fauth, K., Kreibig, U., and Schmid, G. (1991) *Z. Phys. D*, **20**, 297.

99 Quinten, M., Sander, I., Steiner, P., Kreibig, U., Fauth, K., and Schmid, G. (1991) *Z. Phys. D*, **20**, 377.

100 Leemput, L.E.C.v.d., Gerritsen, J.W., Rongen, P.H.H., Smokers, R.T.M., Wieringa, H.A., Kempen, H.v., and Schmid, G. (1991) *J. Vac. Sci. Technol.*, **B9**, 814.

101 Dusemund, B., Hoffmann, A., Salzmann, T., Kreibig, U., and Schmid, G. (1991) *Z. Phys. D*, **20**, 305.

102 Schmid, G. (1992) *Chem. Rev.*, **92**, 1709.

103 Schmid, G. (1994) *Physics and Chemistry of Metal Cluster Compounds* (ed. J.d. Jongh), Kluwer, p. 107.

104 Simon, U., Schmid, G., and Schön, G. (1993) *Angew. Chem., Int. Ed.*, **32**, 250.

105 Mulder, F.M., Zeeuw, E.A.v.d., Thiel, R.C., and Schmid, G. (1993) *Solid State Commun.*, **85**, 93.

106 Herrmann, M., Kreibig, U., and Schmid, G. (1993) *Z. Phys. D*, **26**, 1.

107 Baak, J., Brom, H.B., Jongh, L.J.d., and Schmid, G. (1993) *Z. Phys. D*, **26**, 30.

108 Brom, H.B., Baak, J., Jongh, L.J.d., Mulder, F.M., Thiel, R.C., and Schmid, G. (1993) *Z. Phys. D*, **26**, 27.

109 Cluskey, P.D., Newport, R.J., Benfield, R.E., Gurmann, S.J., and Schmid, G. (1993) *Z. Phys. D*, **26**, 8.

110 Houbertz, R., Feigenspan, T., Mielke, F., Memmert, U., Hartmann, U., Simon, U., Schön, G., and Schmid, G. (1994) *Europhys. Lett.*, **28**, 641.

111 Smith, B.A., Zhang, J.Z., Giebel, U., and Schmid, G. (1997) *Chem. Phys. Lett.*, **270**, 139.

112 Tominaga, T., Tenma, S., Watanabe, H., Giebel, U., and Schmid, G. (1996) *Chem. Lett.*, 1033.
113 Ruffieux, V., Schmid, G., Braunstein, P., and Rosé, J. (1997) *Chem. Eur. J.*, **3**, 900.
114 Bezryadin, A., Dekker, C., and Schmid, G. (1997) *Appl. Phys. Lett.*, **71**, 1273.
115 Chi, L.F., Hartig, M., Drechsler, T., Schwaak, T., Seidel, C., Fuchs, H., and Schmid, G. (1998) *Appl. Phys. A*, **66**, 187.
116 Schmid, G. and Chi, L.F. (1998) *Adv. Mater.*, **10**, 515.
117 Schmid, G. (1998) *J. Chem. Soc., Dalton Trans.*, 1077.
118 Schmid, G., Pugin, R., Sawitowski, T., Simon, U., and Marler, B. (1999) *Chem. Commun.*, 303.
119 Schmid, G., Meyer-Zaika, W., Pugin, R., Sawitowski, T., Majoral, J.-P., Caminade, A.-M., and Turrin, C.-O. (2000) *Chem. Eur. J.*, **6**, 1693.
120 Herrmann, M., Koltun, R., Kreibig, U., Schmid, G., and Güntherodt, G. (2001) *Adv. Funct. Mater.*, **11**, 202.
121 Torma, V., Schmid, G., and Simon, U. (2001) *Chem. Phys. Chem.*, **5**, 321.
122 Torma, V., Reuter, T., Vidoni, O., Schumann, M., Radehaus, C., and Schmid, G. (2001) *Chem. Phys. Chem.*, **8–9**, 546.
123 Liu, Y., Schumann, M., Raschke, T., Radehaus, C., and Schmid, G. (2001) *Nano Lett.*, **8**, 405.
124 Sawitowski, T., Miquel, Y., Heilmann, A., and Schmid, G. (2001) *Adv. Funct. Mater.*, **11**, 435.
125 Boyen, H.-G., Kästle, G., Weigl, F., Ziemann, P., Schmid, G., Garnier, M.G., and Oelhafen, P. (2001) *Phys. Rev. Lett.*, **87**, 276401.
126 Boyen, H.-G., Kästle, G., Weigl, F., Koslowski, B., Dietrich, C., Ziemann, P., Spatz, J.P., Riethmüller, S., Hartmann, C., Möller, M., Schmid, G., Garnier, M.G., and Oelhafen, P. (2002) *Science*, **297**, 1533.
127 Torma, V., Vidoni, O., Simon, U., and Schmid, G. (2003) *Eur. J. Inorg. Chem.*, 1121.
128 Zhang, H., Schmid, G., and Hartmann, U. (2003) *Nano Lett.*, **3**, 305.
129 Schön, G. and Simon, U. (1995) *Colloid Polymer Sci.*, **273**, 101.
130 Schön, G. and Simon, U. (1995) *Colloid Polymer Sci.*, **273**, 202.
131 Simon, U. (1998) *Adv. Mater.*, **10**, 1487.
132 Schmid, G. (2008) *Chem. Soc. Rev.*, **37**, 1909.
133 Schmid, G. (1990) *Inorg. Synth.*, **7**, 214.
134 Schmid, G., Pugin, R., Malm, J.-O., and Bovin, J.-O. (1998) *Eur. J. Inorg. Chem.*, 813.
135 Schmid, G., Pugin, R., Meyer-Zaika, W., and Simon, U. (1999) *Eur. J. Inorg. Chem.*, 2051.
136 Schmid, G. and Huster, W. (1986) *Z. Naturforsch.*, **41b**, 1028.
137 Schmid, G. (1994) *Clusters and Colloids – From Theory to Applications* (ed. G. Schmid), VCH, Weinheim.
138 Bönnemann, H., Brijoux, W., Brinkmann, R., Dinjus, E., Joussen, T., and Korall, B. (1991) *Angew. Chem., Int. Ed.*, **30**, 1312.
139 Bönnemann, H., Brijoux, W., Brinkmann, R., Dinjus, E., Joussen, T., Fretzen, R., and Korall, B. (1992) *J. Mol. Catal.*, **74**, 323.
140 Bönnemann, H., Brijoux, W., Brinkmann, R., Fretzen, R., Joussen, T., Köppler, R., Neiteler, P., and Richter, J. (1994) *J. Mol. Catal.*, **86**, 129.
141 Bönnemann, H., Braun, G., Brijoux, W., Brinkmann, R., Tilling, A.S.S., Seevogel, K., and Siepen, K. (1996) *J. Organomet. Chem.*, **520**, 143.
142 Bönnemann, H., and Brijoux, W. (1996) *Active Metals* (ed. A. Fürster), VCH, Weinheim, p. 339.
143 Bönnemann, H. and Brijoux, W. (1999) *Metal Clusters in Chemistry* (ed. P. Braunstein), Wiley-VCH, Weinheim, p. 913.
144 Bönnemann, H. and Brijoux, W. (1996) *Advanced Catalysts and Nanostructured Materials*, (ed W. Moser), Academic Press, San Diego, p. 165.
145 Bönnemann, H. and Nagabhushana, K.S. (2004) *Dekker Encyclopedia of Nanoscience and Nanotechnology* (eds J.A. Schwarz, C.I. Contescu, and K. Putyera), Marcel Dekker, New York, p. 739.
146 Bönnemann, H. and Nagabhushana, K.S. (2004) *Encyclopedia of Nanoscience and Nanotechnology* (ed. H.S. Nalwa),

American Scientific Publishers, Stevenson Ranch, p. 777.
147 Nagabhushana, K.S. and Bönnemann, H. (2004) *Nanotechnology in Catalysis* (eds B. Zhou, S. Hermans, and G.A. Somorjai), Kluwer Academic–Plenum Publishers, New York, p. 51.
148 Bönnemann, H. and Richards, R. (2003) *Catalysis and Electrocatalysis at Nanoparticle Surfaces* (eds A. Weikowski, E.R. Savinova, and C.G. Vayenas), Marcel Dekker, New York, p. 343.
149 Bönnemann, H., Nagabhushana, K.S., and Richards, R.M. (2007) *Nanoparticles and Catalysis* (ed. D. Astruc), Wiley-VCH, Weinheim, p. 49.
150 Bönnemann, H., Braun, G., Brijoux, W., Brinkmann, R., Schulze, A., Tilling, A.S., Seevogel, K., and Siepen, K. (1996) *J. Organometal. Chem.*, **520**, 143.
151 Bönnemann, H., Brijoux, W., Brinkmann, R., Dinjus, E., Joussen, T., and Korall, B. (1991) *Angew. Chem., Int. Ed. Engl.*, **30**, 1312.
152 Aiken, J.D. III and Finke, R.G. (1999) *J. Mol. Catal. A*, **145**, 1.
153 Bönnemann, H. and Brijoux, W. (1995) *Nanostruct. Mater.*, **5**, 135.
154 Franke, R., Rothe, J., Pollmann, J., Hormes, J., Bönnemann, H., and Brijoux, W. (1996) *J. Am. Chem. Soc.*, **118**, 12090.
155 Bönnemann, H. and Korall, B. (1992) *Angew. Chem., Int. Ed.*, **31**, 1490.
156 Reetz, M.T. and Maase, M. (1999) *Adv. Mater.*, **11**, 773.
157 Angermund, K., Bühl, M., Endruschat, U., Mauschick, F.T., Mörtel, R., Mynott, R., Tesche, B., Waldöfner, N., Bönnemann, H., Köhl, G., Modrow, H., Hormes, J., Dinjus, E., Gassner, F., Haubold, H.-G., Vad, T., and Kaupp, M. (2003) *J. Phys. Chem. B*, **107**, 7507.
158 Angermund, K., Bühl, M., Endruschat, U., Mauschick, F.T., Mörtel, R., Mynott, R., Tesche, B., Waldöfner, N., Bönnemann, H., Köhl, G., Modrow, H., Hormes, J., Dinjus, E., Gassner, F., Haubold, H.-G., and Vad, T. (2002) *Angew. Chem., Int. Ed.*, **41**, 4041.
159 Reetz, M.T. and Helbig, W. (1994) *J. Am. Chem. Soc.*, **116**, 7401.
160 Reetz, M.T., Helbig, W., and Quaiser, S.A. (1996) *Active Metals* (ed. A. Fürstner), VCH, Weinheim, p. 279.
161 Reetz, M.T., Quaiser, S.A., and Merk, C. (1996) *Chem. Ber.*, **129**, 741.
162 Reetz, M.T., Helbig, W., and Quaiser, S.A. (1995) *Chem. Mater.*, **7**, 2227.
163 Reetz, M.T. and Quaiser, S.A. (1995) *Angew. Chem., Int. Ed. Engl.*, **34**, 2240.
164 Kolb, U., Quaiser, S.A., Winter, M., and Reetz, M.T. (1996) *Chem. Mater.*, **8**, 1889.
165 Becker, J.A., Schäfer, R., Festag, W., Ruland, W., Wendorf, J.H., Pebler, J., Quaiser, S.A., Helbig, W., and Reetz, M.T. (1995) *J. Chem. Phys.*, **103**, 2520.
166 Tano, T., Esumi, K., and Meguro, K. (1989) *J. Colloid Interface Sci.*, **133**, 530.
167 Esumi, K., Tano, T., and Meguro, K. (1989) *Langmuir*, **5**, 268.
168 Esumi, K., Tano, T., Torigoe, K., and Meguro, K. (1990) *Chem. Mater.*, **2**, 564.
169 Esumi, K., Suzuki, M., Tano, T., Torigoe, K., and Meguro, K. (1991) *Colloid Surf.*, **55**, 9.
170 Esumi, K., Sato, N., Torigoe, K., and Meguro, K. (1992) *J. Colloid Interface Sci.*, **149**, 295.
171 Esumi, K., Sadakane, O., Torigoe, K., and Meguro, K. (1992) *Colloids Surf.*, **62**, 255.
172 Toshima, N. and Wang, Y. (1993) *Chem. Lett.*, 1611.
173 Amiens, C., Caro, D.D., Chaudret, B., and Bradley, J.S. (1993) *J. Am. Chem. Soc.*, **115**, 11638.
174 deCaro, D., Wally, H., Amiens, C., and Chaudret, B. (1994) *Chem. Commun.*, **24**, 1891.
175 Rodriguez, A., Amiens, C., Chaudret, B., Casanove, M.-J., Lecante, P., and Bradley, J.S. (1996) *Chem. Mater.*, **8**, 1978.
176 Bardaji, M., Vidoni, O., Rodriguez, A., Amiens, C., Chaudret, B., Casanove, M.-J., and Lecante, P. (1997) *New J. Chem.*, **21**, 1243.
177 Duteil, A., Quéau, R., Chaudret, B.M., Roucau, C., and Bradley, J.S. (1993) *Chem. Mater.*, **5**, 341.
178 deCaro, D., Agelou, V., Duteil, A., Chaudret, B., Mazel, R., Roucau, C., and Bradley, J.S. (1995) *New J. Chem.*, **19**, 1265.
179 Dassenoy, F., Philippot, K., Ely, T.O., Amiens, C., Lecante, P., Snoeck, E.,

Mosset, A., Casanove, M.-J., and Chaudret, B. (1998) *New J. Chem.*, **19**, 703.

180 Osuna, J., Caro, D.D., Amiens, C., Chaudret, B., Snoeck, E., Respaud, M., Broto, J.-M., and Fert, A. (1996) *J. Phys. Chem.*, **100**, 14571.

181 Ely, T.O., Amiens, C., Chaudret, B., Snoeck, E., Verelst, M., Respaud, M., and Broto, J.M. (1999) *Chem. Mater.*, **11**, 526.

182 Bradley, J.S., Hill, E.W., Chaudret, B., and Duteil, A. (1995) *Langmuir*, **11**, 693.

183 Dassenoy, F., Casanove, M.-J., Lecante, P., Verelst, M., Snoeck, E., Mosset, A., Ely, T.O., Amiens, C., and Chaudret, B. (2000) *J. Chem. Phys.*, **112**, 8137.

184 Verelst, M., Ely, T.O., Amiens, C., Snoeck, E., Lecante, P., Mosset, A., Respaud, M., Broto, J.-M., and Chaudret, B. (1999) *Chem. Mater.*, **11**, 2702.

185 Soulantica, K., Maisonnat, A., Fromen, M., Casanove, M., Lecante, P., and Chaudret, B. (2001) *Angew. Chem., Int. Ed.*, **40**, 448.

186 Nayrak, C., Viala, E., Fau, P., Senocq, F., Jumas, J., Maisonnet, A., and Chaudret, B. (2000) *Chem. Eur. J.*, **6**, 4082.

187 Gomez, S., Philippot, K., Colliere, V., Chaudret, B., Senocq, F., and Lecante, P. (2000) *Chem. Commun.*, **19**, 1945.

188 Pan, C., Dassenoy, F., Casanove, M., Philippot, K., Amiens, C., Lecante, P., Mosset, A., and Chaudret, B. (1999) *J. Phys. Chem. B*, **103**, 10098.

189 Pelzer, K., Vidoni, O., Philippot, K., and Chaudret, B. (2003) *Adv. Funct. Mater.*, **13**, 118.

190 Pelzer, K., Philippot, K., Chaudret, B., Meyer-Zaika, W., and Schmid, G. (2003) *Z. Anorg. Allg. Chem.*, **629**, 1217.

191 Chaudret, B. (2005) *C. R. Phys.*, **6**, 117.

192 Chaudret, B. (2005) *Surface and Interfacial Organometallic Chemistry and Catalysis: Topics in Organometallic Chemistry*, (ed C. Coperet and B. Chaudret), Springer, p. 233.

193 Ely, T.O., Pan, C., Amiens, C., Chaudret, B., Dassenoy, F., Lecante, P., Casanove, M.J., Mosset, A., Respaud, M., and Broto, J.M. (2000) *J. Phys. Chem. B*, **104**, 695.

194 Philippot, K. and Chaudret, B. (2003) *C. R. Chim.*, **6**, 1019.

195 Pan, C., Pelzer, K., Philippot, K., Chaudret, B., Dassenoy, F., Lecante, B., and Casanove, M.J. (2001) *J. Am. Chem. Soc.*, **123**, 7584.

196 Pery, T., Pelzer, K., Mathes, J., Buntkowski, G., Philippot, K., Limbach, H.-H., and Chaudret, B. (2005) *Chem. Phys. Chem.*, **6**, 605.

197 Mevellec, V., Roucoux, A., Ramirez, E., Philippot, K., and Chaudret, B. (2004) *Adv. Synth. Catal.*, **346**, 72.

198 Jansat, S., Go'mez, M., Philippot, K., Muller, G., Guiu, E., Claver, E., Castillo'n, S., and Chaudret, B. (2004) *J. Am. Chem. Soc.*, **126**, 1592.

199 Gomez, M., Philippot, K., Colliere, V., Lecante, P., Muller, G., and Chaudret, B. (2003) *New J. Chem.*, **27**, 114.

200 Pelzer, K., Philippot, K., and Chaudret, B. (2003) *Z. Phys. Chem.*, **217**, 1539.

201 Jansat, S., Picurelli, D., Pelzer, K., Philippot, K., Gomez, M., Muller, G., Lecante, P., and Chaudret, B. (2006) *New J. Chem.*, **30**, 115.

202 Boxall, D., Deluga, G., Kenik, E., King, W., and Lukehart, C. (2001) *Chem. Mater.*, **13**, 891.

203 Komarneni, S., Li, D., Newalkar, B., Katsuki, H., and Bhalla, A.S. (2002) *Langmuir*, **18**, 5959.

204 Gonsalves, K.E., Li, H., Perez, R., Santiago, P., and Yacaman, M.J. (2000) *Coord. Chem. Rev.*, **206–207**, 607.

205 Suslick, K.S. and Prince, J. (1998) *J. Mater. Chem.*, **8**, 445.

206 Dhas, A. and Gedanken, A. (1998) *J. Mater. Chem.*, **8**, 445.

207 Koltypin, Y., Fernandez, A., Rojas, C., Campora, J., Palma, P., Prozorov, R., and Gedanken, A. (1999) *Chem. Mater.*, **11**, 1331.

208 Henglein, A. (2000) *J. Phys. Chem. B*, **104**, 2201.

209 Henglein, A. and Meisel, D. (1998) *Langmuir*, **14**, 7392.

210 Belloni, J., Mostafavi, M., Remita, H., Marignier, J.L., and Delcourt, M.O. (1998) *New J. Chem.*, **22**, 1239.

211 Takahashi, Y., Ito, T., Sakai, S., and Ishii, Y. (1970) *Chem. Commun.*, 1065.

212 de Caro, D., Wally, H., Amiens, C., and Chaudret, B. (1994) *Chem. Commun.*, 1891.

213 Bradley, J.S., Hill, E.W., Behal, S., Klein, C., Chaudret, B., and Duteil, A. (1993) *Chem. Mater.*, **4**, 1234.

214 de Caro, D., Agelou, V., Duteil, A., Chaudret, B., Mazel, R., Roucau, C., and Bradley, J.S. (1995) *New J. Chem.*, **19**, 1265.

215 Bradley, J.S., Hill, E.W., Chaudret, B., and Duteil, A. (1995) *Langmuir*, **11**, 693.

216 Soulantica, K., Maisonnat, A., Fromen, M., Casanove, M., Lecante, P., and Chaudret, B. (2001) *Angew. Chem., Int. Ed.*, **40**, 448.

217 Fahlmann, B.D. (2007) *Materials Chemistry*, Springer, Mount Pleasant, p. 282.

218 Xia, Y., Xiong, Y., Lim, B., and Skrabalak, S.E. (2009) *Angew. Chem., Int. Ed.*, **48**, 60.

219 Tao, A.R., Habas, S., and Yang, P. (2008) *Small*, **4**, 310.

220 Frenken, J.W.M. and Stoltze, P. (1999) *Phys. Rev. Lett.*, **82**, 3500.

221 Kéki, S., Nagy, L., Deák, G., Zsuga, M., Somogyi, L., and Lévai, A. (2004) *J. Am. Soc. Mass Spectrom.*, **15**, 879.

222 McLean, J.A., Stumpo, K.A., and Russel, D.H. (2005) *J. Am. Chem. Soc.*, **127**, 5304.

223 Li, X., Kuznetsov, A.E., Zhang, H.F., Boldyrev, A.I., and Wang, L.S. (2001) *Science*, **291**, 859.

224 Wang, Z.L. (2000) *J. Phys. Chem. B*, **104**, 1153.

225 Xiong, Y. and Xia, Y. (2007) *Adv. Mater.*, **19**, 3385.

226 Wiley, B.J., Xiong, Y., Li, Z.-Y., Yin, Y., and Xia, Y. (2006) *Nano Lett.*, **6**, 765.

227 Xiong, Y., Cai, H., Yin, Y., and Xia, Y. (2007) *Chem. Phys. Lett.*, **440**, 273.

228 Wiley, B.J., Wang, Z., Wei, J., Yin, Y., Cobden, D.H., and Xia, Y. (2006) *Nano Lett.*, **6**, 2273.

229 Wiley, B.J., Sun, Y., Mayers, B., and Xia, Y. (2005) *Chem. Eur. J.*, **11**, 454.

230 Wiley, B.J., Sun, Y., Chen, J., Cang, H., Li, Z.-Y., Li, X., and Xia, Y. (2005) *MRS Bull.*, **30**, 356.

231 Wiley, B., Sun, Y., and Xia, Y. (2007) *Acc. Chem. Res.*, **40**, 1067.

232 Murphy, C.J. and Jana, N.R. (2002) *Adv. Mater.*, **14**, 80.

233 Murphy, C.J., Sau, T.K., Gole, A.M., Orendorff, C.J., Gao, J., Gao, L., Hunyadi, S.E., and Li, T. (2005) *J. Phys. Chem. B*, **109**, 13857.

234 Murphy, C.J., Gole, A.M., Hunyadi, S.E., and Orendorff, C.J. (2006) *Inorg. Chem.*, **45**, 7544.

235 Kirkland, A.I., Jefferson, D.A., Duff, D.G., Edwards, P.P., Gameson, I., Johnson, B.F.G., and Smith, D.J. (1993) *Proc. R. Soc. London Ser. A*, **440**, 589.

236 Germain, V., Li, J., Ingert, D., Wang, Z.L., and Pileni, M.P. (2003) *J. Phys. Chem. B*, **107**, 8717.

237 Ho, P.-F. and Chi, K.-M. (2004) *Nanotechnology*, **15**, 1059.

238 Xiong, Y., Siekkinen, A.R., Wang, J., Yin, Y., Kim, M.J., and Xia, Y. (2007) *J. Mater. Chem.*, **2006**, 2600.

239 Washio, I., Xiong, Y., Yin, Y., and Xia, Y. (2006) *Adv. Mater.*, **18**, 1745.

240 Xiong, Y., Washio, I., Chen, J., Cai, H., Li, Z.-Y., and Xia, Y. (2006) *Langmuir*, **22**, 8563.

241 Lim, B., Camargo, P.H.C., and Xia, Y. (2008) *Langmuir*, **24**, 10437.

242 Xiong, Y., McLellan, J.M., Chen, J., Yin, Y., Li, Z.-Y., and Xia, Y. (2005) *J. Am. Chem. Soc.*, **127**, 17118.

243 Sun, Y. and Xia, Y. (2003) *Adv. Mater.*, **15**, 695.

244 Sun, Y., Mayers, B., and Xia, Y. (2003) *Nano Lett.*, **3**, 675.

245 Jin, R., Cao, Y., Mirkin, C.A., Kelly, K.L., Schatz, G.C., and Zheng, J.G. (2001) *Science*, **294**, 1901.

246 Millstone, J.E., Hurst, S.J., Metraux, G.S., Cutler, J.I., and Mirkin, C.A. (2009) *Small*, **5**, 646.

247 Martin, C.R. (1994) *Science*, **266**, 1961.

248 Schönenberger, C., van der Zande, B.M.I., Fokkink, L.G.J., Henny, M., Schmid, C., Krüger, M., Bachtold, A., Huber, R., Birk, H., and Staufer, U. (1997) *J. Phys. Chem. B*, **101**, 5497.

249 Yu, Y.-Y., Chang, S.-S., Lee, C.-L., and Wang, C.R.C. (1997) *J. Phys. Chem. B*, **101**, 6661.

250 Liz-Marzán, L.M. (2004) *Materials Today*, **7**, 26.

251 Pileni, M.P. (2003) *Nat. Mater.*, **2**, 145.

252 Brust, M., Walker, M., Bethell, D., Schiffrin, B.J., and Whyman, R. (1994) *Chem. Commun.*, 801.

253 Sarathy, K.V., Kulkarni, G.U., and Rao, C.N.R. (1997) *Chem. Commun.*, 537.

254 Chen, S., Wang, Z.L., Ballato, J., Foulger, S.H., and Carroll, D.L. (2003) *J. Am. Chem. Soc.*, **125**, 16286.
255 Sun, Y. and Xia, Y. (2002) *Science*, **298**, 2176.
256 Tao, A.R., Sinsermsuksakul, P., and Yang, P. (2006) *Angew. Chem., Int. Ed.*, **45**, 4597.
257 Kim, F., Connor, S., Song, H., Kuykendall, T., and Yang, P. (2004) *Angew. Chem., Int. Ed.*, **43**, 3673.
258 Wiley, B.J., Sun, Y., and Xia, Y. (2007) *Acc. Chem. Res.*, **40**, 1067.
259 Song, H., Kim, F., Connor, S., Somorjai, G.A., and Yang, P. (2005) *J. Phys. Chem. B*, **109**, 188.
260 Jun, Y.-W., Choi, J.-S., and Cheon, J. (2006) *Angew. Chem., Int. Ed.*, **45**, 3414.
261 Habas, S.E., Lee, H., Radmilovic, V., Somorjai, G.A., and Yang, P. (2007) *Nat. Mater.*, **6**, 692.
262 Lim, B., Jiang, M., Tao, J., Camargo, P.H.C., Zhu, Y., and Xia, Y. (2009) *Adv. Funct. Mater.*, **19**, 189.
263 Brown, S., Sarikaya, M., and Johnson, E. (2000) *J. Mol. Biol.*, **299**, 725.
264 Naik, R.R., Stringer, S.J., Agarwal, G., Jones, S.E., and Stone, M.O. (2002) *Nat. Mater.*, **1**, 169.
265 Sun, Y., Mayers, B.T., and Xia, Y. (2002) *Nano Lett.*, **2**, 481.

3.3.2
Synthesis, Properties and Applications of Magnetic Nanoparticles
Galyna Krylova, Maryna I. Bodnarchuk, Ulrich I. Tromsdorf, Elena V. Shevchenko, Dmitri V. Talapin, and Horst Weller

3.3.2.1 Introduction
Today, magnetic particles of micrometer and nanometer size continue to attract a growing fundamental and technological interest. The use of magnets is important in the development of electric motors, in electrical power transformation, for data storage devices and electronics in general. Moreover, miniaturization – which today represents one of the most important streams of modern technology – calls for extremely small magnets of micrometer, or even nanometer, size. In this regard, it is difficult to see how such materials might be prepared, and also what properties might be expected from an extremely small piece of material which is ferromagnetic in its bulk state. In this chapter, attention will be focused on the nanoparticles of ferromagnetic metals, their preparation, and structural and magnetic characterization. In nature, there are only three ferromagnetic elements at room temperature – iron, cobalt, and nickel (although gadolinium, which is ferromagnetic below 16 °C, might also be included in this series). In addition, some compounds and alloys of manganese, chromium and europium can exhibit ferromagnetic or ferrimagnetic behavior [1]. The exciting discovery that rhodium clusters (Rh_n, where $n = 12-32$) are ferromagnetic [2] not only demonstrates the nontrivial scaling behavior of magnetism, but also shows that magnetic nanomaterials can reveal novel, unexpected phenomena.

Magnetic metal nanoparticles are considered as promising candidates for applications as catalysts [3], as single-electron devices [4–7], and as contrast agents for biomedical imaging [8], among other roles. Monodisperse magnetic nanoparticles can organize themselves into highly ordered superstructures, and may even form macroscopic "crystals" that consist of nanoparticles as building blocks. Thin granular films of ferromagnetic nanocrystals already serve as the basis of conventional magnetic storage media (hard drives), and it is expected that advanced magnetic media, based on ordered arrays of monodisperse nanocrystals, might soon achieve magnetic recording densities of between 100 Gb per in^2 and 1 Tb per in^2 (15 Gb per cm^2 and 150 Tb per cm^2) [9]. Spin-dependent tunneling electron transport has been recently demonstrated in an array of close-packed cobalt [4] and magnetite [10] nanoparticles. Granular materials consisting of nanometer-sized magnetic particles in a dielectric or nonmagnetic metal matrix have been shown to exhibit giant magnetoresistive properties [11–15].

Magnetic colloids, or ferrofluids, are based on magnetic nanoparticles coated with a layer of surfactant dispersed in a carrying solvent. The molecules of surfactant (hereafter referred to as "stabilizing agents" or "stabilizers") bind to the nanoparticle surface, preventing coagulation and providing solubility and the desired surface properties. Moreover, stabilizers protect the particle from oxidation and can strongly affect the magnetic properties [16–18]. Ferrofluids have been extensively studied and harnessed in a variety of applications [19–21]; for example, the magnetorheological properties of magnetic colloids have been exploited in high-performance bearings and seals [22–24]. Several reviews are available on the properties and technological applications of ferrofluids [19–24].

Within the nanoparticle, the electron spins can be aligned in a certain direction by applying an external magnetic field. Thus, the possibility of manipulating nanoparticles across a distance by applying an external magnetic field opens a broad field for their biomedical application. Specific surface groups of the stabilizers may be used to link different species to the nanoparticle surface; for example, drugs and/or antibodies may be attached to the surface of a colloidal magnetic nanoparticle, and steered into regions of the body where they are required, by the application of a magnetic field [1]. Similarly, functionalized magnetic nanoparticles can be attached specifically to a tumor and then heated by the application of an AC magnetic field; this results in tumor thermoablation (this process referred to as "magnetic fluid hyperthermia") [25]. In addition, magnetic nanoparticles coupled with oligonucleotides have been used as nanosensors for the detection and separation of complementary specific oligonucleotides [26].

The preparation of nanoparticles of desired sizes is the first, and very important, requisite for their further investigation and use in practice. A narrow size distribution, or monodispersity, is important because the magnetic properties become strongly size-dependent within the nanometer size range (Section 3.3.2.5). The magnetic behavior of nanoparticles depends not only on their chemical composition and size, but also on their crystalline modification and the presence of structural defects such as stacking faults or twin planes [9]. Murray *et al.* suggested using the term "nanocrystals" for crystalline particles with low concentrations of defects, whereas the more general term "nanoparticle" would involve particles containing gross internal grain boundaries, fractures, or internal disorders [27]. In the case of granular films composed of magnetic nanoparticles, the interparticle spacing strongly affects the magnetic and charge transport properties [28].

As the different preparation methods lead to magnetic nanoparticles with differences in crystalline structure, surface chemistry, shape, and so on, the fabrication technique will have a major influence on the magnetic properties of the materials obtained [29]. Indeed, many examples have been reported where nanocrystalline materials prepared via different routes but with similar grain sizes have exhibited differing magnetic properties [30]. In the past, many physical and chemical methods have been employed to produce magnetic nanostructures, including molecular beam epitaxy (MBE) [31], chemical vapor deposition (CVD) [32], pulsed laser deposition [33, 34], sputtering [35], and electrodeposition [36, 37], especially in the preparation of thin magnetic films.

- *Inert gas condensation technique*: Separated nanometer-sized nanoparticles of Fe, Co and Ni can be prepared using this technique, in which metal is evaporated at a very high temperature (~1500 °C) into a high-purity gas (e.g., helium) [38, 39]. Upon collision with the inert gas, the metal atoms lose their kinetic energy and condense onto a cold finger in the form of ultrafine powder. A modification of this method is based on the evaporation of metals by arc discharge into a circulating gas mixture of H_2 and Ar; this approach is used for the large-scale preparation of 20–30 nm particles of Fe, Ni, and Fe–Ni alloys. Usually, the nanoparticles synthesized by this approach are amorphous and some amount of oxide presents at their surface [38, 39].
- *Arc discharge technique*: This is similar to that used for fullerene synthesis, was applied for the preparation of carbon-encapsulated Ni, Co, or Fe nanoparticles.
- *Laser pyrolysis*: This is based on the heating with a laser beam of an organometallic reactant vapor (e.g., a metal carbonyl), mixed with an inert gas. The reactant vapors rapidly decompose to the atoms, which form clusters upon collision with inert gas molecules [29]. Nanoparticles of iron and iron carbides have been prepared in this way.

Further detailed information on each of these methods is available in Ref. [29].

Solution-phase chemical syntheses are commonly used for the preparation of colloids of magnetic metal nanoparticles. Several general approaches can be used to control the size and shape of nanoparticles in solution-phase synthesis.

3.3.2.1.1 **Reverse Micelles Technique** The water-in-oil microemulsion ("reverse micelles") technique is based on using droplets of water sustained in an organic phase by a surfactant. Inside these nanometer-sized reactors, a metal salt dissolved in the nanodroplets can be reduced to form metal nanoparticles. The amount of metal precursor available for particle growth is limited by the micelle volume, which can be varied by tailoring the water-to surfactant ratio. This technique has been used to prepare nanoparticles of magnetic metals [40, 41], metal alloys [42], core–shell structures [43] and oxides [44, 45]. Subsequently, Pileni et al. synthesized nanoparticles of Co, Fe or Fe–Cu alloy in a water-in-*iso*-octane microemulsion by the reduction of metal salts, using hydrazine or sodium borohydride in the presence of sodium *bis* (2-ethylhexyl)sulfosuccinate [Na(AOT)] as a surfactant [40], where the size of the metal nanoparticles increased with the increasing water-to-surfactant ratio ($[H_2O]$:[AOT]) [46]. Although synthesis using reverse micelles is a simple method, and does not require any special costly equipment, it does have certain drawbacks. First, the size distribution of nanoparticles is usually rather broad (~20%; [47]), which means that monodisperse colloids can be obtained only after applying costly post-preparative size-selective fractionation procedures [48]. Another important problem is that the reactions in reverse micelles occur at relatively low temperatures (<100 °C). The nanoparticles of magnetic metals do not crystallize well at room temperature [49], and the as-synthesized material is often poorly crystalline [50]. Consequently, annealing at 200–300 °C is generally required in order to prepare virtually defect-free magnetic nanocrystals [51].

3.3.2.1.2 **Sonochemical Synthesis** Sonochemical synthesis is based on the decomposition of a metal precursor (usually a metal carbonyl complexes), dissolved in a high-boiling solvent under irradiation with high-intensity ultrasound. An ultrasonic treatment initiates acoustic cavitation – that is, the formation, growth and collapse of bubbles within a liquid. The collapse of these bubbles causes a local heating of the media, up to thousands of Kelvin. Consequently, volatile organometallic compounds rapidly decompose inside the collapsing bubbles, yielding the individual metal atoms which, in the presence of a stabilizing agent, agglomerate to form nanosized particles [52]. This technique was developed by Suslick *et al.* for the preparation of Ni [53], nanostructured silica-supported Fe and Fe–Co alloys [54] and Fe [52] nanoparticles. The sonochemical synthesis usually yields amorphous, rather polydisperse nanoparticles, and a further high-temperature heating is required for their crystallization [52–54].

3.3.2.1.3 **Colloidal Syntheses** Among various solution-phase approaches, colloidal syntheses in high-boiling solvents appear to be the most promising routes for the preparation of high-quality magnetic nanocrystals. This methodology is based on the reduction of metal salts or the thermolysis of zero-valent organometallic precursors in the medium of a high-boiling solvent, in the presence of specialized stabilizing agents that adsorb reversibly to the nanoparticle surface, thus mediating the growth rate. The correct choice not only of the surfactants and precursors, but also of the temperature regime, allows the successful preparation of monodisperse colloids of highly crystalline magnetic nanoparticles; details of these materials are provided in the following section.

3.3.2.2 Colloidal Synthesis of Magnetic Metal Nanoparticles

The majority of high-quality colloidal magnetic nanoparticles have been synthesized via the thermolysis or reduction of organometallic precursors, in the presence of ligand molecules (so-called "stabilizing agents") which bind reversibly to the nanoparticle surface and prevent its further growth and coagulation with other particles. The correct choice of stabilizing agents is vital, as attractive magnetic interactions between nanocrystals tend to reduce the overall stability of a magnetic colloid. A high reaction temperature (150–300 °C) allows the annealing of crystalline defects and, consequently, is important when preparing highly crystalline nanoparticles. The main peculiarities of the organometallic synthetic approach, using Co and $CoPt_3$ nanocrystals as typical examples, are discussed in the following sections. In the past, cobalt nanocrystals have most likely received the most attention among nanocrystalline elemental magnetic materials, and CoPt$_3$ may be considered as a model system for magnetic alloy nanocrystals.

3.3.2.2.1 **General Remarks on the Synthesis of Co and CoPt$_3$** Nanocrystals The organometallic synthesis of magnetic metal nanocrystals usually requires a rigorous air-free atmosphere, and is carried out under dry inert gas using a Schlenk line technique or a glovebox. The nanoparticle size and width of size distribution are usually determined by the kinetics of particle nucleation and growth [55]. A narrow particle size distribution can be obtained if a temporally discrete nucleation event is followed by controlled growth of the preformed nuclei [56]. Temporally discrete nucleation in the organometallic synthesis can be attained by the hot injection technique, shown schematically in Figure 3.114. For this, a solution of organometallic precursors or a reducing agent (solution A) is rapidly injected into a hot solution containing the stabilizing agents (solution B); this results in an explosive nucleation, followed by a fall in temperature and a slow growth of preformed nuclei. The optimized injection temperature, duration of heating and correct choice of precursors and stabilizing agents are prerequisites for the successful synthesis of magnetic nanoparticles with controllable size and shape, and a narrow size distribution.

If the synthesis does not yield nanoparticles of a desired monodispersity, the size distribution of as-prepared nanoparticles can be narrowed by applying a size-selective precipitation procedure that exploits the differences in solubility of smaller and larger particles [48, 57]. In a typical size-selective precipitation of a nanoparticle colloid, a sample of the as-prepared nanoparticles with a broad size distribution is dispersed in an appropriate solvent. A nonsolvent is then added dropwise; this gradually destabilizes the colloidal dispersion, causing the solution to become slightly turbid. The largest nanoparticles in the sample exhibit the greatest attractive van der Waals and magnetic forces, and tend to aggregate before the smaller particles. Aggregates of the

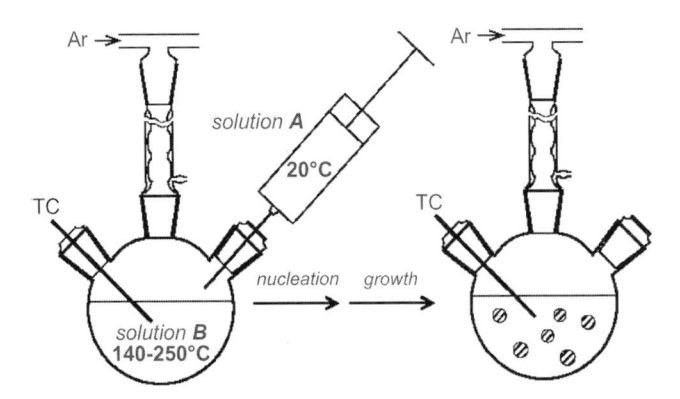

Figure 3.114 Schematic representation of the hot injection technique usually employed in the organometallic synthesis of magnetic nanocrystals. A thermocouple (TC) controls the temperature in the reaction vessel. The composition of the solutions A and B is detailed in the text for each particular recipe.

largest nanoparticles can be isolated by centrifugation or filtration, and redissolved in an appropriate solvent. Subsequently, the next portion of the nonsolvent is added to the supernatant to isolate the second size-selected fraction, and so on. For magnetic nanoparticles, a size-selective precipitation can also be achieved by applying an external magnetic field, as the larger particles will precipitate in the magnetic field before their smaller counterparts [27]. These procedures, if repeated several times, will allow a series of monodisperse fractions to be obtained with a near-Gaussian size distribution and a standard deviation below 5% ($\pm$1 atomic layer).

3.3.2.2.2 Synthesis of Cobalt Nanoparticles with Different Crystalline Modification

Cobalt nanoparticles can be synthesized in three different crystalline modifications: hexagonal close-packed (hcp), face-centered cubic (fcc)' and epsilon (ε). The ε-Co structure, which is not observed in the bulk, consists of a 20-atom unit cell with cubic symmetry similar to the structure of β-Mn [58]. The hcp structure is the most stable phase for bulk cobalt at room temperature. Despite this fact, experimental data have repeatedly shown that ε-Co is the most often found crystalline structure in nanoparticles prepared by wet chemistry [59].

The ε-Co nanoparticles can be synthesized via the reduction of a cobalt salt by a superhydride (LiBEt$_3$H), in the presence of stabilizing agents [27]. Injection of the solution of superhydride in dioctylether (solution A) into the hot (200 °C) solution of $CoCl_2$ in dioctylether containing also oleic acid and trioctylphosphine (TOP) as stabilizing agents (solution B) (Figure 3.114), yields relatively monodisperse ε-Co nanoparticles (Figure 3.115a) [60]. Examinations using high-resolution transmission electron microscopy (HR-TEM) (Figure 3.116b) and simulations of X-ray powder data indicate that the ε-Co nanoparticles are near-perfect single crystals [60]. The width of the X-ray diffraction (XRD) reflexes decreases coarsely as the nanoparticle size increases, as shown in Figure 3.115a for the (221), (310), and (311) reflections. The average nanocrystal size can be tuned from $\sim$2 nm up to 6 nm by tailoring the concentrations of the stabilizers; typically, increase in the concentrations of oleic acid and TOP will yield a smaller mean particle size. In order to prepare ε-Co nanocrystals larger than 6 nm, it is necessary to use less-bulky stabilizers which can effectively cap particles with a lower surface curvature. Thus, by substituting tributylphosphine (TBP) for TOP, larger (7–11 nm) ε-Co nanocrystals will be formed [27, 60, 61]. The fine-tuning of particle size can be achieved by a size-fractionation of crude solutions, yielding a series of near-monodisperse ε-Co nanocrystals (Figure 3.116a).

An alternative method of preparing cobalt nanoparticles is based on the thermal decomposition of $Co_2(CO)_8$ in the presence of a suitable surfactant [27]. A typical recipe involves the injection of a solution of $Co_2(CO)_8$ in diphenylether (solution A) into the hot (200 °C) mixture of oleic acid and TOP or TBP, dissolved also in diphenylether (solution B). The mixture is then heated at 200 °C for 15–20 min. Decomposition of the cobalt carbonyl, under the conditions described above, results in Co nanoparticles with a so-called multiply-twinned face-centered cubic (mt-fcc) lattice (Figures 3.115b and (3.116)d). The multiply-twinned particles are composed of domains with a distorted fcc lattice, this structure being similar to that observed in multiply-twinned, icosahedral gold particles [27]. Post-preparative size-selection

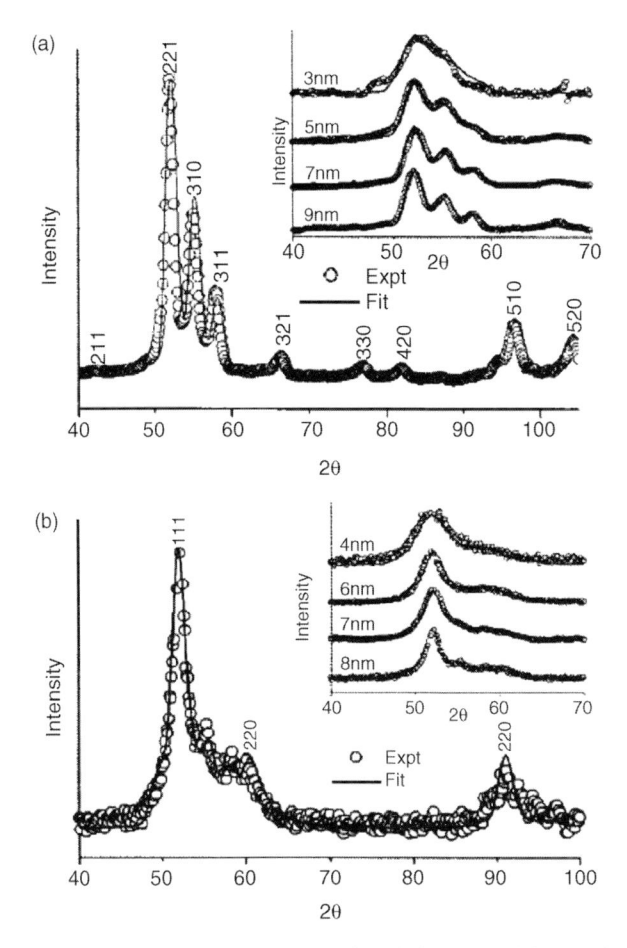

Figure 3.115 Powder XRD patterns for cobalt nanoparticles (circles) and their corresponding simulations (solids lines) using atomic-level models of the particles. (a) ε-Co; (b) multiply-twinned fcc particles. The insets show X-ray patterns and simulations for several particle sizes. Reproduced with permission from Ref. [60]; © 2001, American Chemical Society.

procedures allow the isolation of fairly monodisperse fractions of mt-fcc Co nanocrystals (Figure 3.116c). Depending on the reaction medium and stabilizers, the thermal decomposition of cobalt carbonyl can also yield nanoparticles of other crystalline phases [59, 62–64].

The annealing of ε-Co nanocrystals at 300 °C under vacuum causes them to be converted into hcp-Co phase (Figure 3.117), which has a higher magnetocrystalline anisotropy compared to the ε-Co and mt-fcc Co modifications [61]. Alternatively, the hcp-Co nanoparticles can be synthesized in solution by the reduction of cobalt acetate by 1,2-dodecanediol at 250 °C in a diphenylether medium in the presence of oleic acid and TOP as stabilizing agents [27, 61].

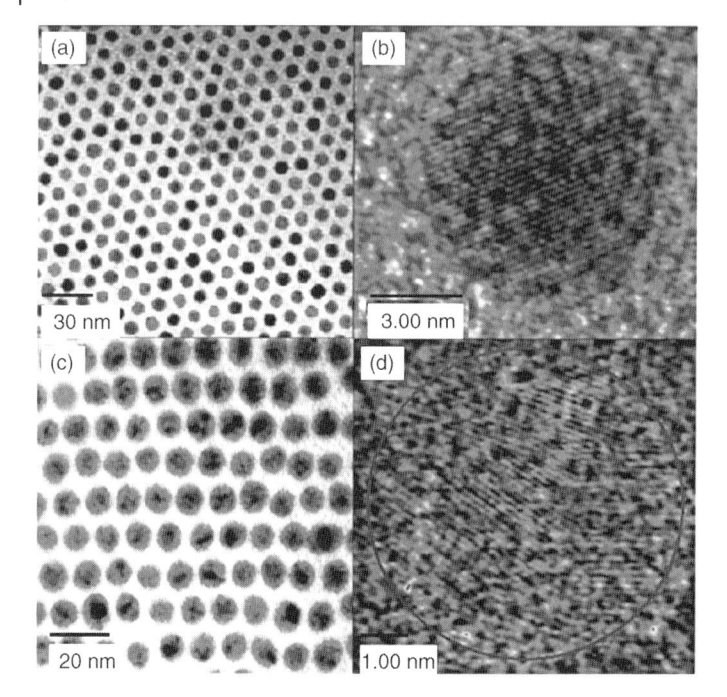

Figure 3.116 TEM images of cobalt nanoparticles. (a) Low-magnification image of ε-Co. The even contrast across the individual particles implies a uniform crystalline structure for this phase; (b) HR-TEM image of ε-Co, showing perfect crystallographic coherence and discrete faceting; (c) Low-magnification image of multiply-twinned fcc structure. The contrast changes within particles indicate that they are composed of crystal domains; (d) HR-TEM image showing the domains within a single multiply-twinned fcc nanoparticle. Reproduced with permission from Ref. [60]; © 2001, American Chemical Society.

As shown above, the optimized combination of precursors ($CoCl_2$, $Co(CH_3COO)_2$ or $Co_2(CO)_8$) and reducing agents (superhydride or polyalcohol) allows the selective preparation of monodisperse cobalt nanoparticles with a desired crystalline phase. Such behavior shows that the solution-phase chemical synthesis of magnetic nanocrystals is not thermodynamically controlled, and thus can allow the preparation of crystal phases that are metastable, such as the ε-Co structure [9]. The control of size and the crystalline phase of nanoparticles is important, as these parameters greatly affect the magnetic properties (see Section 3.3.2.5).

One serious drawback of cobalt and other elemental magnetic nanoparticles (Fe, Ni, Nd, etc.) is their relatively low stability against oxidation under ambient conditions. Thus, the exposure of cobalt nanoparticles to air results in the formation of a shell of approximately one to two monolayers of cobalt oxide. This shell of antiferromagnetic cobalt oxide causes considerable disturbance of the magnetic properties of the cobalt core. The intrinsic magnetic coercivity of elemental (Co, Fe) magnetic nanoparticles is too low for some applications, an example being advanced magnetic recording media, and nanoparticles of more stable materials with a higher

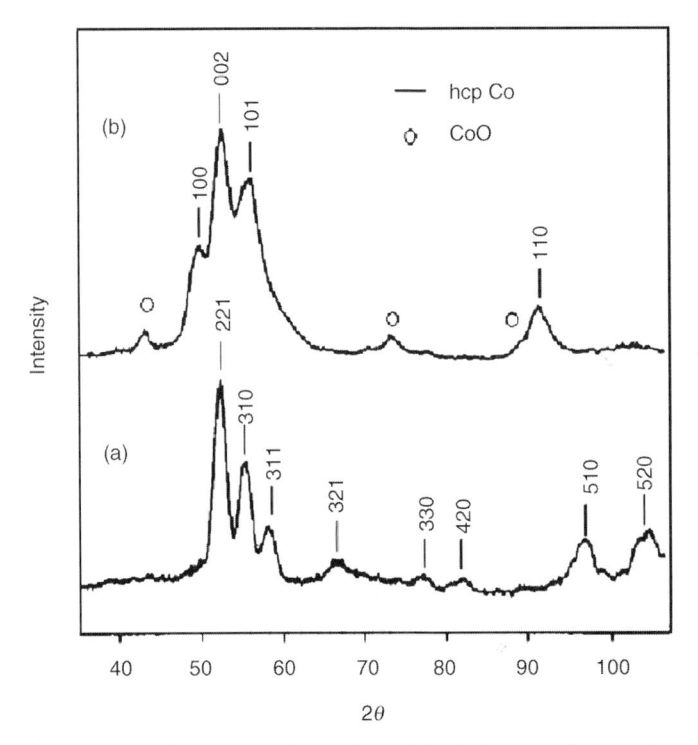

Figure 3.117 XRD patterns of the ε-Co particles before and after annealing. (a) As-synthesized 9 nm particles; (b) After annealing at 300 °C under vacuum for 3 h. Reproduced with permission from Ref. [9]; © 1999, American Institute of Physics.

coercivity are strongly desired for this role. As a convenient solution, intermetallic materials such as Co_xPt_{1-x} and Fe_xPt_{1-x} are considerably more stable than elemental cobalt or iron, and have $L1_0$ crystalline modification with very high magnetocrystalline anisotropy, leading to a large coercivity of the nanoparticles.

3.3.2.2.3 Synthesis of CoPt₃ Magnetic Alloy Nanocrystals The synthetic approach developed for the preparation of elemental nanoparticles can be further extended to intermetallic compounds. Thus, high-quality $CoPt_3$ nanocrystals can be synthesized via the simultaneous reduction of platinum acetylacetonate and the thermal decomposition of cobalt carbonyl in the presence of 1-adamantanecarboxylic acid (ACA) and hexadecylamine (HDA) as stabilizing agents [65].

In a typical preparation, the mixture of platinum acetylacetonate, 1,2-hexadecandiol, ACA and HDA (solution B in Figure 3.114) were heated to a certain temperature in the range from 140 to 220 °C; solution A, prepared by dissolving cobalt carbonyl in 1,2-dichlorobenzene, was swiftly injected into solution B, with vigorous stirring. The temperature of injection of the stock solution will be further referred to as the "reaction temperature." After injection, the color changed from pale yellow to black, indicating the formation of $CoPt_3$ nanocrystals. Further heating was normally continued for 1 h at the injection temperature, followed by annealing at the refluxing temperature (~275–285 °C) for 1 h. The annealing stage was necessary to improve the crystallinity of the as-formed $CoPt_3$ nanoparticles.

Information regarding nanoparticle crystallinity can be obtained by a comparison of the average particle sizes estimated by powder XRD and TEM techniques. The width of the XRD reflections provides information about the X-ray coherence length, which is close to the average size of the single crystalline domain inside the nanocrystal, whereas the TEM images show the total size of a nanoparticle. (Nanocrystal sizes estimated by these methods will be further referred to as "XRD-size" and "TEM-size," correspondingly.) For $CoPt_3$ nanoparticles prepared at 140–200 °C, the XRD-size was considerably smaller than the TEM-size (Figure 3.118a), due to the internal structural defects (stacking faults, twinned planes, etc.) present within as-formed particles. Annealing of $CoPt_3$ nanoparticles at the boiling point of the crude solution (~275–285 °C) for ~30 min resulted in an increase of the XRD-size, whereas no considerable change in TEM-size was observed. The annealing process required a relatively high temperature (200–300 °C), which was necessary to trigger atom diffusion inside the nanocrystals. After annealing, the XRD-size became almost equal to the TEM-size (Figure 3.118b), indicating that the most of nanocrystals were single crystallites with a near-perfect lattice. Subsequent HR-TEM investigations confirmed the excellent crystallinity of annealed $CoPt_3$ nanocrystals (Figure 3.119), with almost all particles possessing the lattice fringes without stacking faults and other defects. From a practical viewpoint, the perfect crystallinity of magnetic nanoparticles is important because the structural defects cause a considerable reduction in coercivity, the parameter which determines the applicability of a material for magnetic recording and storage [9].

Tuning of the mean size of $CoPt_3$ nanoparticles can be achieved by exerting control over the nucleation and growth rates, as illustrated in Figure 3.120. Fast nucleation provides a high concentration of nuclei finally yielding smaller nanocrystals, whereas

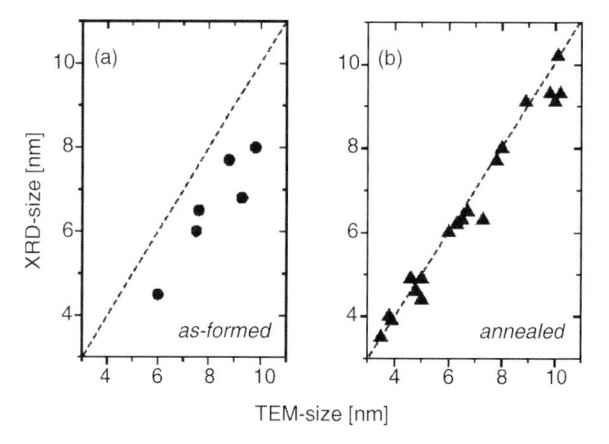

Figure 3.118 *In situ* annealing of the $CoPt_3$ nanoparticles. Comparison of the XRD- and TEM-sizes before and after annealing. Reproduced with permission from Ref. [55]; © 2003, American Chemical Society.

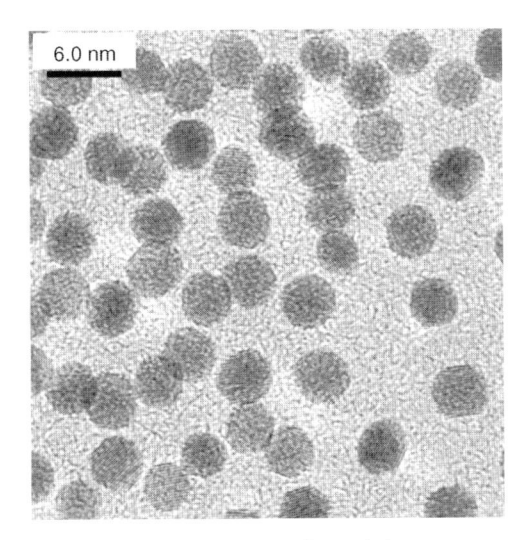

Figure 3.119 HR-TEM image of annealed CoPt$_3$ nanocrystals. Note that almost all nanoparticles possess the lattice fringes without stacking faults and other defects.

slow nucleation provides a low concentration of seeds that consume the same amount of precursors and thus result in larger particles [55].

The balance between nucleation and growth rates can be tuned by adjusting the reaction temperature, since the activation energy for the homogeneous nucleation is typically much higher than that for particle growth [66]. This means that the nucleation rate is much more sensitive to changes in temperature compared to the growth rate. At a higher reaction temperature more nuclei are formed and, according to the scheme in Figure 3.120, the final particle size will be smaller. Indeed, an increase in temperature from 145 to 220 °C allows a reduction in the average size of the CoPt$_3$ nanocrystals, from ∼10 nm to 3 nm – that resulting in the preparation of nanocrystals over a wide range of sizes (Figure 3.121a). This concept of size turning implies that Ostwald ripening – that is, the growth of large particles

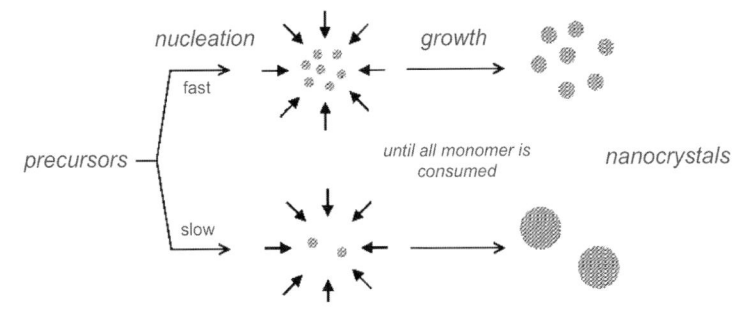

Figure 3.120 Schematic representation of the synthesis of CoPt$_3$ nanocrystals. Reproduced with permission from Ref. [55]; © 2003, American Chemical Society.

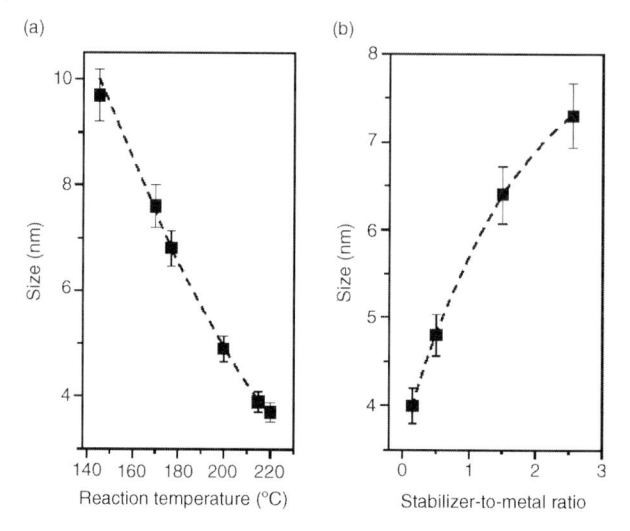

Figure 3.121 (a) Dependence of $CoPt_3$ nanocrystals size on the reaction temperature. (b) Influence of initial amount of the stabilizing agent (1-adamantancarboxylic acid) on nanocrystal size. Reproduced with permission from Ref. [55]; © 2003, American Chemical Society.

under the concomitant dissolution of small particles – is very slow. As reported elsewhere [49, 55], this is indeed the case in most organometallic preparations of metal particles.

The nanoparticle size can also be tuned by tailoring the concentrations of the stabilizing agents, although such an effect can be different for various systems. Thus, an increase in stabilizer concentration resulted in smaller nanocrystals in the case of Co [27, 67] and Ni [27], whereas an opposite behavior was observed for Fe [68] and $CoPt_3$ [55] nanocrystals (Figure 3.121b).

For the kinetically driven synthesis of nanoparticles, a decrease in the mean particle size with increasing stabilizer concentration can occur if a strong passivation of the nanocrystal surface results in a slowing of the growth rate, whereas the nucleation rate remains much less influenced (see Figure 3.120). However, the opposite situation can also be realized, as was observed recently for $CoPt_3$ [55] and Fe [68] nanocrystals. Stabilizing agents such as ACA and primary amines can also form stable complexes with individual metal atoms of a molecular precursor. Thus, whilst the platinum precursor Pt acetylacetonate was decomposed in a HDA/diphenyl ether/1,2-hexadecandiol mixture at $\sim$130 °C, the addition of ACA raised the decomposition temperature to $\sim$220 °C. A similar behavior was also observed for the cobalt precursor: here, the addition of ACA drastically enhanced the stability of the cobalt carbonyl solution against thermal decomposition. The formation of these complexes precedes the nucleation step [69], decreases the monomer reactivity and, therefore, an increase of the concentration of ligand is expected to suppress the nucleation rate. In accordance with the scheme shown in Figure 3.120, a slowing of the nucleation rate will result in a larger final nanocrystal size, as was also observed experimentally (Figure 3.121b).

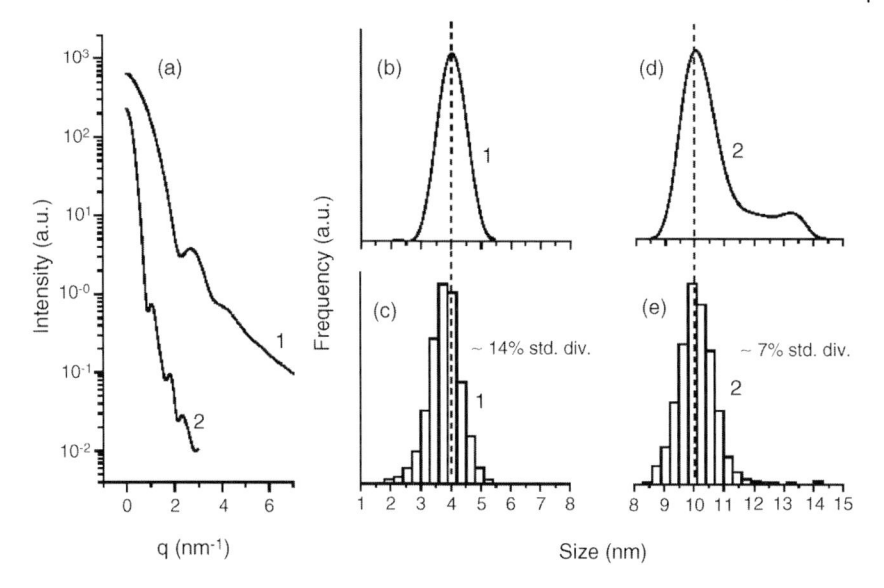

Figure 3.122 (a) The angular dependence of scattered X-ray intensity (SAXS) from two samples of as-prepared CoPt$_3$ nanocrystals with mean size (1) ∼4 nm and (2) ∼10 nm; (b and d) Particle volume fraction versus size curves calculated from the SAXS data shown in part (a); (c and e) Particle size distribution histograms estimated for the same samples from TEM images. Reproduced with permission from Ref. [55]; © 2003, American Chemical Society.

Information regarding the *in situ* size distribution of CoPt$_3$ nanoparticles can be obtained from a statistical evaluation of all nanocrystals in a representative TEM image, or from small-angle X-ray scattering (SAXS) data by means of the indirect Fourier transformation technique [70, 71]. The latter method provides the information averaged over a huge number of nanoparticles and, moreover, allows the *in situ* investigation of particle aggregation. Figure 3.122a shows SAXS patterns for two samples of as-prepared CoPt$_3$ nanocrystals with mean sizes of ∼4 and 10 nm, where the nanocrystal size was tuned by varying the reaction temperature (215 and 145 °C). Figure 3.122b and c show the particle size distribution functions of as-prepared ∼4 nm CoPt$_3$ nanocrystals as obtained from the SAXS and TEM data, correspondingly. The SAXS data show that small CoPt$_3$ nanocrystals form stable suspensions of isolated nanoparticles (single-particle population) [72, 73]. Both, the SAXS and TEM measurements confirmed a relatively narrow and near-symmetric particle size distribution for the as-prepared ∼4 nm CoPt$_3$ nanocrystals (Figure 3.122b and c), whereas only a minor deviation was observed between the mean sizes estimated by these methods. HR-TEM always provided smaller estimates for particle size compared to SAXS; this was because the size distribution from SAXS is a volume (or mass) distribution, whereas that from microscopy is a number distribution. However, the difference between volume and number distribution is negligible when the distribution is narrow. Nonetheless, in broad distributions, or in the presence of a few aggregates, the larger particles count visibly more in volume distributions (SAXS) than in number distributions (TEM).

In the case of ~10 nm $CoPt_3$ nanocrystals, the SAXS size distribution curve was superimposed by a peak which corresponded to the single particle population, with a tail in the large size region (Figure 3.122d). Subsequent TEM investigations of the same sample revealed a symmetric size distribution curve without any features around 12–13 nm (Figure 3.122e). This allows a partial aggregation of ~10 nm $CoPt_3$ at room temperature to be assumed [72]. The long-term stability of colloidal solutions, in addition to an absence of agglomerates in the TEM images, provides evidence that the aggregation is both dynamic and reversible.

The absolute width of the particle size distribution is almost constant for both, the ~4 nm and the ~10 nm samples (Figure 3.122). This means that the relative size distribution of as-prepared ~4 nm $CoPt_3$ nanocrystals is much broader compared to that of the ~10 nm nanocrystals (~14% and ~7% of standard deviation, respectively). The general tendency here is a narrowing of the size distribution with an increase in the particle size, as observed for all samples irrespective of the method of size control. The progressive narrowing of the particle size distribution with an increasing mean size is due to the more rapid growth of smaller particles than larger particles [66]. This growth regime was recently observed for ensembles of CdSe nanocrystals, and was referred to as the "focusing" of size distribution [74].

A combination of the methods of size control described above allows series of near-monodisperse $CoPt_3$ nanocrystals to be obtained which range from ~3 nm up to 17 nm in size. Subsequently, it was found that different experimental conditions (e.g., reaction temperature, ratio between the cobalt and platinum precursors, etc.) would yield the nanocrystals with identical chemical compositions. The systematic evolution of XRD patterns and the HR-TEM images of $CoPt_3$ nanocrystals, on increasing their size, are shown in Figures 3.123 and 3.124, respectively.

Although, $CoPt_3$ nanocrystals with sizes less than ~7 nm are usually spherical (see Figures 3.119 and 3.124), any further increase in the nanocrystal size results in a rather abrupt transition from spherical to cubic, truncated cubic or, in some cases, plate-like shapes (see Figure 3.125). Annealing at ~275 °C results in a smoothing of the edges of cubic nanocrystals formed at 145 °C (cf. Figures 3.124d and 3.125). Here, each side of the cubic crystal lattices corresponds to the {100} direction of the fcc nanocrystal lattice. Cubic particles can form if the growth rate in the {111} direction is higher than that in the {100} direction [75].

3.3.2.2.4 Shape-Controlled Synthesis of Magnetic Nanoparticles Generally, the shape anisotropy of magnetic nanocrystals can result in advanced magnetic properties, and is therefore of great practical interest [30]. The shape of a colloidal nanoparticle can be controlled by the selective adsorption of organic surfactants onto particular crystallographic facets, in order to inhibit particle growth in a particular crystallographic direction. This approach to shape control is based on the presence of structurally and chemically dissimilar crystal lattice facets, for example, the (011) and (001) facets of the hcp lattice. Shape-controlled colloidal syntheses were recently

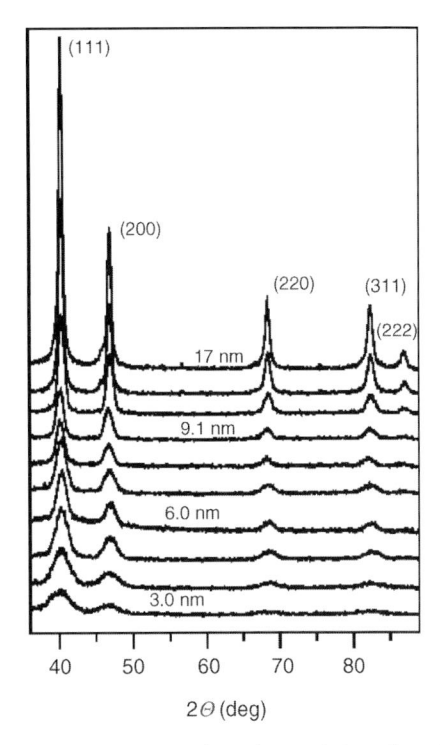

Figure 3.123 Size-dependent evolution of powder XRD patterns for CoPt$_3$ nanocrystals. The average nanocrystal sizes were calculated using the Debye–Sherrer equation. Reproduced with permission from Ref. [55]; © 2003, American Chemical Society.

developed for semiconductor (e.g., CdSe [76, 77], PbS [78], ZnO [79]) and noble metal (Pt [48], Au [49]) nanoparticles.

The possibility of carrying out shape control has also been reported for Co nanocrystals [59, 63, 64]. Thus, the decomposition of cobalt carbonyl in the presence of primary amines (e.g., octadecylamine), together with TOPO or oleic acid, yielded nanodisks of hcp-Co [64] (Figure 3.126). Alkyl amines were shown to inhibit the growth of the unique (001) face of hcp-Co upon any increase in particle size in the kinetic regime. Both, the length and diameter of the nanodisks can be controlled by tailoring the reaction time, as well as by varying the precursor-to-surfactant ratios. The long-term heating of hcp-Co nanodisks (to the thermodynamic limit) resulted in a transformation of the thermodynamically stable spherical nanoparticles of the ε-Co phase. The nanodisks were ferromagnetic and capable of self-assembling spontaneously into long ribbons from stacked nanodisks lying perpendicular to the substrate. A side view of the nanodisks in TEM images (Figure 3.126) strongly resembled nanorods, while TEM tilting experiments were necessary to verify the disk shape [64].

The synthesis of hcp-Co nanorods and nanowires was also reported [80]. Thermal decomposition of the organometallic cobalt precursor, [Co(η^3-C$_8$H$_{13}$)(η^4-C$_8$H$_{12}$)], in

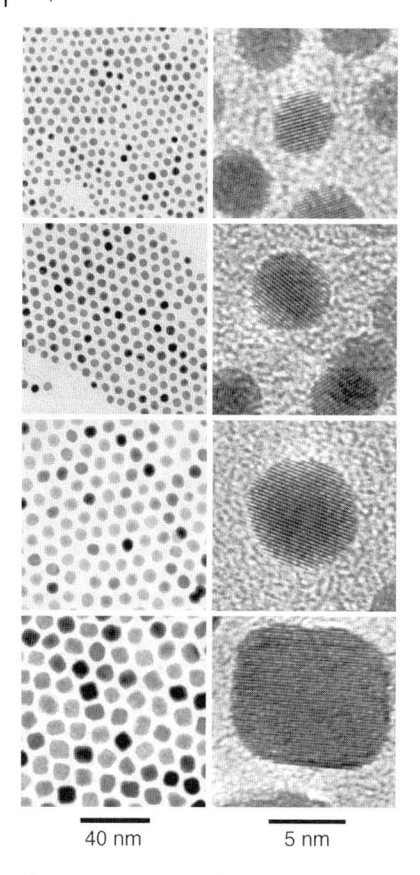

Figure 3.124 TEM and HR-TEM images showing the effect of the reaction temperature on the mean size and size distribution of 3.7 nm, 4.9 nm, 6.3 nm and 9.3 nm CoPt$_3$ nanocrystals prepared at 220, 200, 170 and 145 °C. The molar ratio of Pt: Co: ACA was 1:3:6 for all samples. Reproduced with permission from Ref. [55]; © 2003, American Chemical Society.

the presence of oleic acid and oleylamine under a pressure of 3 bar H$_2$ yielded ~3 nm spherical hcp-Co nanoparticles which were further transformed into uniform nanorods (Figure 3.127) after 48 h of heating at 150 °C. Both, the length and diameters of the hcp-Co nanorods could be varied by tailoring the nature and concentration of the stabilizing agents. Thus, increasing the concentration of oleic acid allowed the preparation of very long (micron range) nanowires of 4 nm diameter. Conversely, the replacement of oleylamine with octylamine resulted in smaller and wider Co nanoparticles (17 × 10 nm). The mechanism of transformation of the initially formed ~3 nm spherical nanoparticles into rods and wires remains unclear although, in principle, the nanorods might be obtained either through a coalescence of spherical particles followed by fusion [79, 81], or upon using the initial particles as nuclei for an anisotropic growth process [76].

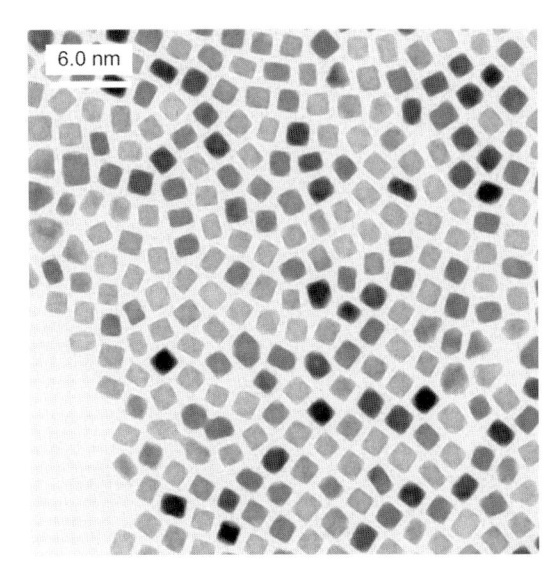

Figure 3.125 TEM overview image of faceted CoPt$_3$ nanocrystals prepared at 145 °C.

3.3.2.2.5 Other Metal Magnetic Nanoparticles Synthesized by Methods of Colloidal Chemistry In the previous sections, the colloidal synthesis of Co and CoPt$_3$ nanocrystals and the great potential of an organometallic approach in the preparation of high-quality magnetic nanoparticles have been discussed. Yet, these synthetic

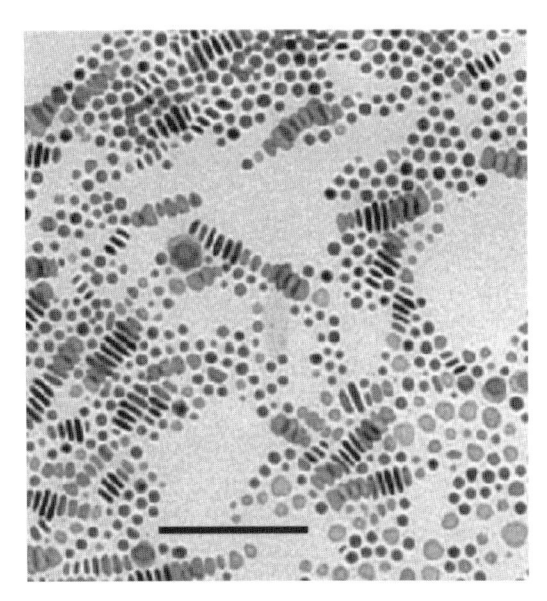

Figure 3.126 TEM partially self-assembled Co nanodisks. Scale bar = 100 nm. Reproduced with permission from Ref. [64]; © 2002, American Chemical Society.

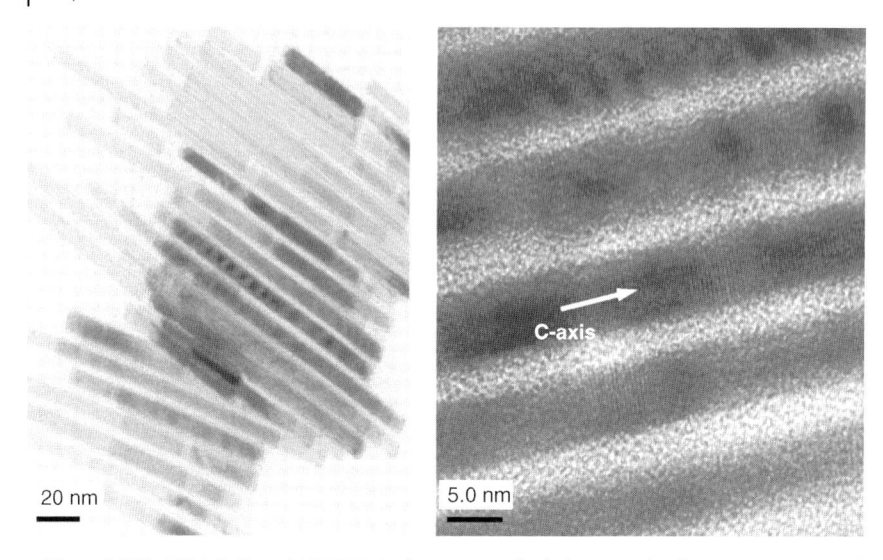

Figure 3.127 TEM (left) and HR-TEM (right) images of cobalt nanorods. Illustration courtesy of Dr. B. Chaudret.

schemes may also be adopted for the preparation of nanoparticles with different magnetic materials, as detailed in the following sections.

Murray et al. described the synthesis of monodisperse iron nanoparticles by the thermal decomposition of $Fe(CO)_5$, under an inert atmosphere in the presence of oleic acid and tri-butyl phosphine [27]. These ~6 nm Fe nanoparticles had a structure that resembled the bulk bcc lattice, albeit with some disorder, and were very sensitive to oxidation (a very brief contact of the nanoparticle surface with air led to the formation of ~2 nm-thick oxide layer [27]). The thermal decomposition of iron carbonyl in tri-n-octylphosphine oxide at 320–340 °C yielded ~2 nm spherical iron nanoparticles [64] that Hyeon et al. used as building blocks for iron nanorods [82]. Thus, refluxing the 2 nm nanoparticles in pyridine, in the presence of didodecyldimethylammonium bromide (DDAB), led to the formation of uniform rod-shaped iron particles with diameter of 2 nm and a length that varied from 11 to 27 nm, depending on the DDAB concentration employed. The electron diffraction pattern of the nanorods obtained exhibited a body center cubic (bcc) structure of α-Fe. Transformation of the nanospheres to nanorods was explained by the oriented attachment of several 2 nm spherical particles, followed by their fusion (the shape-dependent magnetic properties of iron nanoparticles are discussed in section 3.3.2.5). Although the practical applications of iron nanoparticles are strongly hindered due to their chemical instability under ambient conditions, monodisperse iron particles may be gently oxidized by trimethylamine N-oxide into stable monodisperse magnetic γ-Fe_2O_3 nanocrystals, as described below [68]. Chaudret et al. showed that the reaction of the metal–organic precursor $Fe[N(SiMe_3)_2]_2$ with H_2, in the presence of a long-chain acid and a long-chain amine in various proportions, produced relatively stable monodisperse, zerovalent iron nanocubes [83]. Both, magnetic and Mössbauer measurements revealed that these Fe particles displayed magnetic properties that matched those of bulk iron.

Spherical nickel nanoparticles have been synthesized by the reduction of nickel salts by polyalcohols [27], with a mixture of oleic acid, tri-butylphosphine (TBP) and tri-butylamine (TBA) as stabilizing agents, favoring the formation of $\sim$12–13 nm nanoparticles. An increase of the alkyl chain length in the stabilizing agents resulted in a decreased nanoparticle size. Thus, the replacement of TBP and TBA with TOP and tri-octylamine (TOA) resulted in 8–10 nm Ni nanoparticles. The decomposition of Ni(cyclo-octa-1,5-diene)$_2$, in the presence of HDA as a stabilizing and shape-controlling agent, yielded Ni nanorods [84]. The simultaneous reduction of nickel and cobalt salt was used to prepare cobalt–nickel alloy nanoparticles with fcc structure [27]. As nickel was more readily incorporated into the growing nanoparticles, the use of equal amounts of nickel and cobalt precursors would lead to the formation of a Ni-rich alloy ($\sim$Ni$_{60}$Co$_{40}$).

Currently, the iron–platinum alloys are among the most well-studied magnetic nanoparticle materials [85]. The classical synthesis by Sun *et al.* is based on the simultaneous high-temperature reduction of platinum acetylacetonate and the decomposition of iron pentacarbonyl in the presence of oleic acid and oleylamine as stabilizing agents [86]. The size of the FePt nanocrystals can be controlled by first growing $\sim$2.5–3 nm monodisperse seed particles, followed by the addition of more reagents so as to enlarge the existing seeds to the desired size of up to $\sim$10 nm [86]. The chemical composition of Fe$_x$Pt$_{1-x}$ binary alloy nanocrystals can be slightly tuned between $x = 0.48$ and $x = 0.7$ by controlling the molar ratio of iron carbonyl to the platinum salt.

The as-prepared FePt nanoparticles exhibited a chemically disordered fcc structure, with relatively poor magnetic properties. Thus, fcc FePt nanoparticles were superparamagnetic at room temperature, and could be handled in the form of a colloidal solution for casting films, and producing self-assembled arrays, for example (Figure 3.128). Further, the FePt nanoparticles could be annealed at 550–600 °C

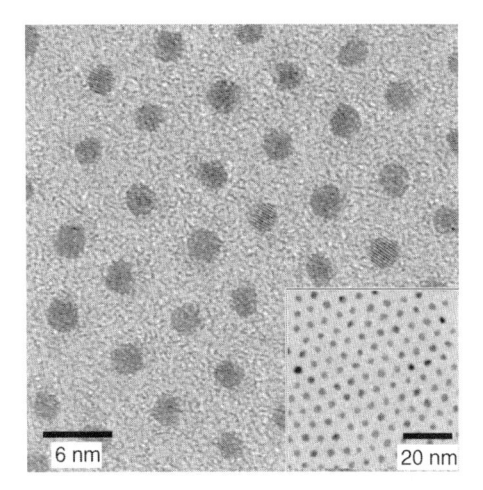

Figure 3.128 Hexagonal close-packed 2-D array of FePt nanocrystals on a TEM grid.

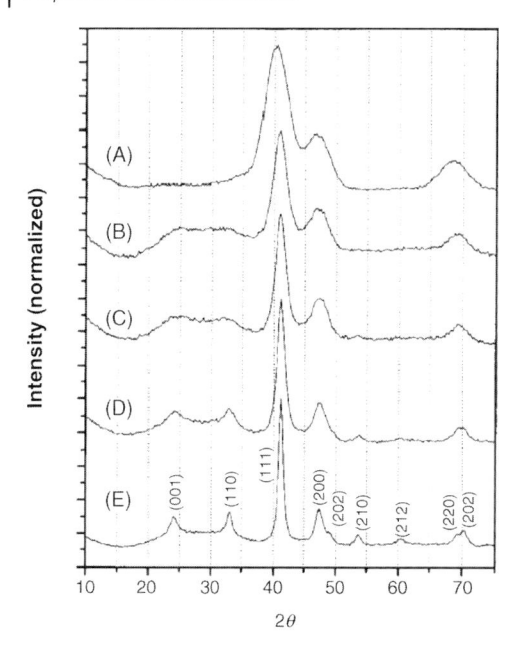

Figure 3.129 XRD patterns (A) of as-synthesized 4 nm $Fe_{52}Pt_{48}$ particle assemblies and a series of similar assemblies annealed under atmospheric N_2 gas for 30 min at temperatures of (B) 450 °C, (C) 500 °C, (D) 550 °C, and (E) 600 °C. The indexing is based on tabulated fct FePt reflections (25). Reproduced with permission from Ref. [86]; © 2000, American Association for the Advancement of Science.

under an inert atmosphere; this induced the iron and platinum atoms to rearrange and for the particles to be converted to the chemically ordered face-centered tetragonal (fct) phase (Figure 3.129). The particles were also transformed into nanoscale ferromagnets with a room temperature coercivity sufficiently high for the arrays of fct FePt nanocrystals to be used in a data storage device [86]. Depending on Fe_xPt_{1-x} stoichiometry and nanocrystal size, the transition temperature was in the range of 500–700 °C [87], whereas for the bulk material the reported value was 1300 °C [88]. In order to further reduce the phase-transition temperature, it was proposed to introduce silver atoms into the lattice of FePt nanocrystals during their synthesis [89]. Such Ag incorporation would promote the transition from fcc to tetragonal phase, reducing the transition temperature by ∼100–150 °C when compared to the pure FePt nanocrystals. It was also suggested that Ag atoms might leave the FePt lattice at a temperature below 400 °C; the vacancies thus formed would increase the mobility of the Fe, while the Pt atoms would accelerate the phase transformation.

The synthetic approach developed for the synthesis of platinum–iron binary alloy nanoparticles was subsequently adopted for the preparation of several other binary alloy nanoparticles, such as FePd [90] or MnPt [91]. However, these alloys required further structural characterization, as well as further developments of the methods for controlling the particle size, shape, and composition.

The preparation of "core–shell"-type magnetic nanoparticles was reported [92, 93] as a two-step synthesis in which nanoparticles of one metal served as the seeds for growth of the shell from another metal. Thus, the reduction of platinum salts in the presence of Co nanoparticles would allow the preparation of air-stable $Co_{core}Pt_{shell}$ nanoparticles [92]. The thermal decomposition of cobalt carbonyl in the presence of silver salt led to the formation of $Ag_{core}Co_{shell}$ nanoparticles [93]. The syntheses of these multicomponent magnetic nanostructures are discussed in Section 3.3.2.4.

3.3.2.3 Iron Oxide-Based Magnetic Nanocrystals

Iron oxides, of which 16 types exist as either oxides, oxide–hydroxides or hydroxides, are highly abundant in nature where, in most compounds, the iron exists in its trivalent state [94]. The majority of iron oxide phases exist naturally, except for β-Fe_2O_3 and high-pressure FeOOH phases, which have been synthesized in the laboratory. Hematite (α-Fe_2O_3), magnetite (Fe_3O_4) and maghemite (γ-Fe_2O_3) represent some of the most abundant iron-based minerals. Wüstite (FeO) and goethite (α-FeOOH) are antiferromagnetic, whereas magnetite, maghemite and metal ferrites with the general composition MFe_2O_4 (M = Co, Mn, Zn, Ni, Bi, etc.) are ferrimagnetic. Elemental iron is a ferromagnet. The crystal lattice of iron oxides consists of close-packed arrays (fcc or hcp) of anions (O^{2-}) in which the interstices are partially filled with Fe^{2+} or Fe^{3+}, in either octahedral (FeO_6), or tetrahedral (FeO_4) coordinations. The differences between the various oxides arise from the various spatial arrangements of the FeO_6 and FeO_4 units.

3.3.2.3.1 Maghemite and Magnetite Nanocrystals
One of the first syntheses of magnetite nanocrystals employed aqueous solutions of $FeCl_2$ and $FeCl_3$, and was reported by Massart [95]. The chemical reaction leading to the Fe_3O_4 phase occurred at pH 8–14 under airless conditions; the process can be expressed as:

$$Fe^{2+} + 2Fe^{3+} + 8OH^- \rightarrow Fe_3O_4 + 4H_2O \tag{3.9}$$

The size, shape, and composition of iron oxide nanocrystals depends somewhat on the type of salt, the Fe^{2+} : Fe^{3+} ratio, the temperature, the pH value, and the ionic strength of the reaction mixture. Rigorous adjustments of the aqueous preparations yielded magnetite nanocrystals with sizes tunable in the range of 4.2–16 nm [96]. The composition of nanocrystals can be varied by the Fe^{2+} : Fe^{3+} ratio [97]; for example, if the ratio is very small then goethite (α-FeOOH) formed. This synthesis method is both inexpensive and straightforward, but permits only a very limited control over the nanocrystal size and morphology. Magnetite can be readily oxidized by oxygen present in the air and, as a consequence, is slowly converted to maghemite (γ-Fe_2O_3).

The thermal decomposition of organometallic compounds in high-boiling solvents in the presence of stabilizers has been widely investigated, and is currently used for the synthesis of Fe_3O_4 and γ-Fe_2O_3 nanocrystals. Iron carbonyls [68, 98], acetylacetonate [99] and iron fatty acid salts [100, 101] are the typical precursors, while oleic acid and other fatty acids, as well as oleylamine and other aliphatic amines, are the most frequently used surfactants. The important parameters for controlling the size and morphology include the precursor-to-stabilizer and stabilizer-to-solvent ratios, the reaction temperature the duration, and the heating rate.

Iron pentacarbonyl has an historic role with respect to the synthesis of various iron-based nanocrystals. The decomposition of metal carbonyls has long been used to prepare ferrofluids based on metallic ultrafine powders (now known as nanocrystals). In 1966, for example, monodisperse, 20 nm Co nanocrystals were prepared by decomposing dicobalt octacarbonyl in boiling toluene, containing polymers as stabilizers [102]. In 1979, a similar approach was reported for iron nanoparticles [103]; in this case, iron pentacarbonyl was decomposed at 150 °C in nonpolar organic solvents (e.g., decalin, xylene, o-dichlorobenzene) in the presence of various vinyl polymers, and under an inert atmosphere. Remarkably, the particle size was tunable in the range between 1.5 and 20 nm, with 20% standard deviation. The nanoparticles of iron oxide were easily formed upon exposure of the as-synthesized iron particles to air (Figure 3.130a).

Interest in iron carbonyl as a precursor for nanomaterials has been revived during the past decade such that, in 2001, Hyeon *et al.* reported the synthesis of monodisperse 4–16 nm γ-Fe_2O_3 nanocrystals from iron pentacarbonyl in octyl ether in the presence of either oleic or lauric acids as surface ligands [68]. In this approach, highly reactive $Fe(CO)_5$ first formed iron–oleate complexes, the subsequent decomposition of which at 300 °C led to the formation of transient metallic iron species or strongly reduced iron oxocomplexes, followed by their oxidation to the γ-Fe_2O_3 phase by introducing $(CH_3)_3NO$ into the reaction mixture.

A simple, inexpensive and reproducible method for producing monodisperse Fe_3O_4 nanocrystals by the pyrolysis of iron-oleates was independently proposed in 2004 by the groups of Peng [100], Hyeon [101], and Colvin [104]. In this case, the

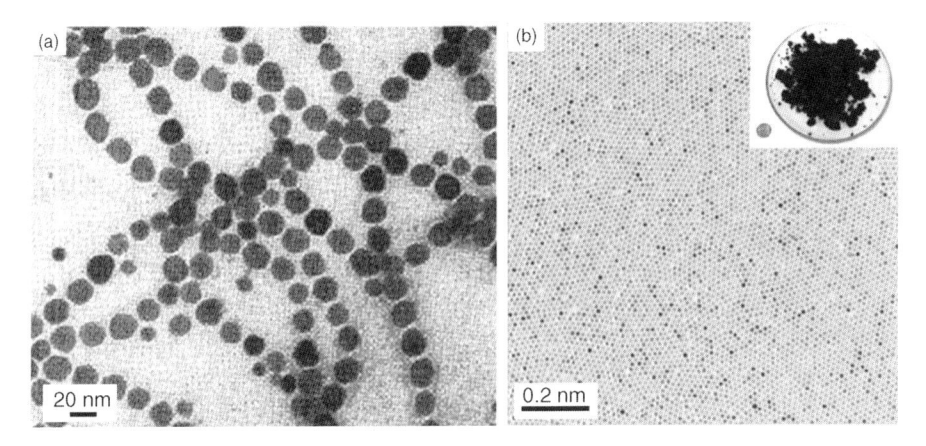

Figure 3.130 (a) Iron oxide nanocrystals synthesized from iron pentacarbonyl by Griffiths *et al.* in 1979. Reproduced with permission from Ref. [103]; © 1979, American Institute of Physics; (b) TEM image of monodisperse magnetite nanocrystals synthesized by decomposing iron oleate. The inset shows a Petri dish containing multi-gram amounts of this nanomaterial. Reproduced with permission from Ref. [101]; © 2004, Nature Publishing Group.

iron–oleate complexes were prepared by the reaction of ferric and/or ferrous salts with sodium hydroxide and oleic acid [100], or with sodium oleate [101]. The iron oleate was then decomposed in noncoordinating solvents (e.g., octadecene, n-eicosane, tetracosane) at 300–360 °C. By varying the oleic acid: oleate ratio and the decomposition temperature, the size of nanoparticles could be precisely tuned from 8 to 50 nm. While this synthesis provided a remarkably large yield of material, on the multigram scale (Figure 3.130b), a precise size control also allowed the detailed study of the nanocrystals' size-dependent magnetic properties (see also Section 3.3.2.7).

In a colloidal synthesis, the choice of an appropriate surfactant is crucially important for controlling the nucleation, growth, and shape of the nanocrystal product. A simple approach to controlling the crystal shape of iron oxide (Fe_3O_4) nanocrystals, independently of their size, was recently proposed [105–107] where, by using sodium oleate as the stabilizer and adjusting the reaction temperature profile, monodisperse spherical, cubic, and bipyramidal nanocrystals (Figure 3.131a,b) could be prepared. The different behaviors of oleic acid and its sodium salt were attributed to their electrolytic properties at the temperature of synthesis. For example, sodium oleate undergoes a strong dissociation, providing a significantly higher concentration of "free" oleate ions that in turn enhances their preferential adsorption onto {100} facets. Sun et al. used iron(III) acetylacetonate, $Fe(acac)_3$, to synthesize 4–20 nm magnetite nanocrystals [99] via the thermal decomposition of $Fe(acac)_3$ in the presence of 1,2-hexadecanediol, oleic acid and

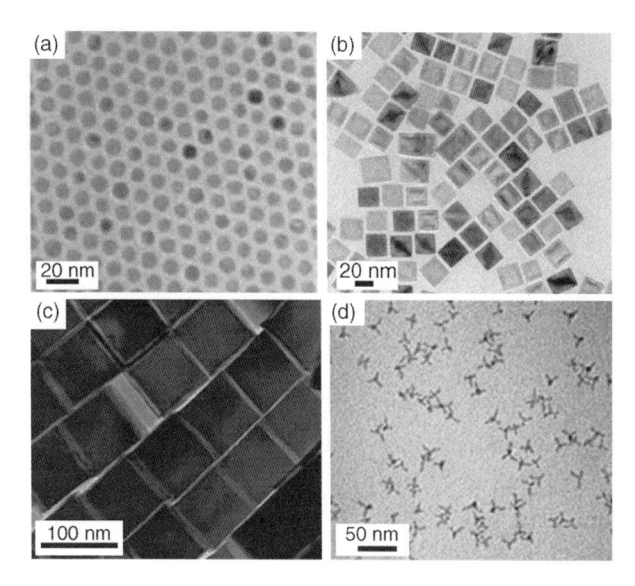

Figure 3.131 TEM images of various shaped iron oxide nanocrystals. (a) 12 nm Fe_3O_4 spheres. Reproduced with permission from Ref. [101]; © 2004, Nature Publishing Group; (b) 22 nm Fe_3O_4 cubes. Reproduced with permission from Ref. [106]; © 2007, American Chemical Society; (c) 80 nm Fe_3O_4 cubes. Reproduced with permission from Ref. [108]; © 2009, American Chemical Society; (d) γ-Fe_2O_3 tetrapods. Reproduced with permission from Ref. [109]; © 2006, American Chemical Society.

oleylamine in diphenyl ether. Recently, Hyeon and coworkers reported the synthesis of uniform 30–160 nm nanocubes of magnetite via the decomposition (Fe(acac)$_3$) in a mixture of oleic acid and benzyl ether (Figure 3.131c) [108].

Cheon et al. reported the synthesis of γ-Fe$_2$O$_3$ nanocrystals of various shapes, including hexagonal plates, diamonds, triangle plates and spheres, via the aerobic decomposition of iron pentacarbonyl in a hot solution of dichlorobenzene (180 °C) that contained dodecylamine as the capping ligand [98]. The observed size and shape evolution was attributed to a modulation of the growth rate along different crystallographic directions by selective adhesion of the alkylamine ligand. Tetrapod-shaped iron oxide nanocrystals were synthesized by Cozzoli et al. via the decomposition of Fe(CO)$_5$ in octadecene, in the presence of a complex mixture of surfactants (oleic acid, oleylamine, hexadecane-1,2-diol) at 240 °C, followed by oxidation under air at 80 °C (Figure 3.131d) [109]. Unlike the tetrapods of group II–VI materials [110–112], the formation of tetrapods in the Fe$_3$O$_4$ system cannot be rationalized based on the polymorphism of the crystal structure.

3.3.2.3.2 Nanocrystals of Other Iron Oxides (Hematite, Wüstite, Goethite)

Under ambient conditions, the most thermodynamically stable iron oxide is α-Fe$_2$O$_3$ (hematite), and which is used widely as a catalyst, a gas sensor, a pigment, and a magnetic material. At this point, some reports are highlighted that have been selected from many synthetic recipes for this material. Near-monodisperse cubic α-Fe$_2$O$_3$ nanocrystals stabilized with oleic acid can be obtained via a hydrothermal method [113, 114] in which an iron(III) salt such as FeCl$_3$, sodium hydroxide (or sodium oleate) and oleic acid in an ethanol/water mixture are loaded into an autoclave and heated for 10–12 h at 180 °C. In this case, the Fe(oleate)$_3$ complex would be expected to form as an intermediate, and to decompose slowly to the iron oxide. Although the as-synthesized nanocrystals were rather polydisperse, quite uniform 15 nm nanocubes were obtained after size-selection. Larger α-Fe$_2$O$_3$ nanocubes (70–800 nm) were produced by using iron chloride and an amino acid (α-amino-δ-guanidovaleric acid) as the precursor and capping agent, respectively [115]. Rhombohedrally shaped hematite nanocrystals were synthesized following the slow (ca. 10 days) hydrolysis of an acidic ferric chloride solution at 95 °C [116]. Hematite nanocrystals of 20–60 nm were also obtained using ethylene oxide and iron chloride, via a sol–gel process [117].

Relatively few investigations have been conducted into the synthesis of Fe$_x$O (wüstite, $0.84 < x < 0.95$) nanocrystals, most likely due to the metastability of the Fe$_x$O phase at room temperature, its high tendency to disproportionate into Fe and Fe$_3$O$_4$, and its easy oxidation under ambient conditions. To the best of the present authors' knowledge, no reports have been made on the aqueous-based preparation of Fe$_x$O nanocrystals. Perhaps the first organometallic synthesis of wüstite nanocrystals was reported by Redl et al. [118], in which the decomposition of iron acetate was effected at 250–300 °C in trioctylamine or octyl ether in the presence of oleic acid, followed by gentle oxidation with pyridine N-oxide. The findings of Redl and coworkers showed clearly that the synthesis of iron oxide nanocrystals in organic solvents is a complex process with a rich chemistry that leads to the diverse compositions of iron oxides. In 2007, using a very similar process, Sun et al.

synthesized single-phase Fe_xO nanocrystals with well-defined spherical or truncated octahedral morphologies [119]. Recently, Fe_xO nanocrystals stabilized by hexadecylamine were synthesized by the aerobic aging of $Fe[N(SiMe_3)_2]_2$ in tetrahydrofuran for one week at room temperature, yielding 5 nm crystalline Fe_xO nanocrystals with a rather broad size distribution [120].

Goethite, α-FeOOH, is another thermodynamically stable iron oxide phase that has an important role as a pigment. Nanorods of goethite (α-FeOOH) nanocrystals can be prepared by the dropwise addition of $NaHCO_3$ solution to aqueous $Fe(NO_3)_3$, followed by aging at pH 12 for 24 h at 90 °C [121, 122]. Recently, Hyeon *et al.* synthesized monodisperse and uniform goethite nanotubes via the reaction of iron oleate with hydrazine immobilized in reverse micelles (Figure 3.132) [123]. With its high affinity to various contaminants and heavy-metal ions, goethite represents an important component of environmental remediation processes. Recently, small nanotubes have provided useful applications in biosensing and molecular separation techniques.

3.3.2.3.3 Nanocrystals of Metal Ferrites
Similar to Fe_3O_4, metal ferrite nanocrystals can be obtained by a variety of methods, including aqueous coprecipitation, microemulsion technique, and the thermal decomposition of organometallic com-

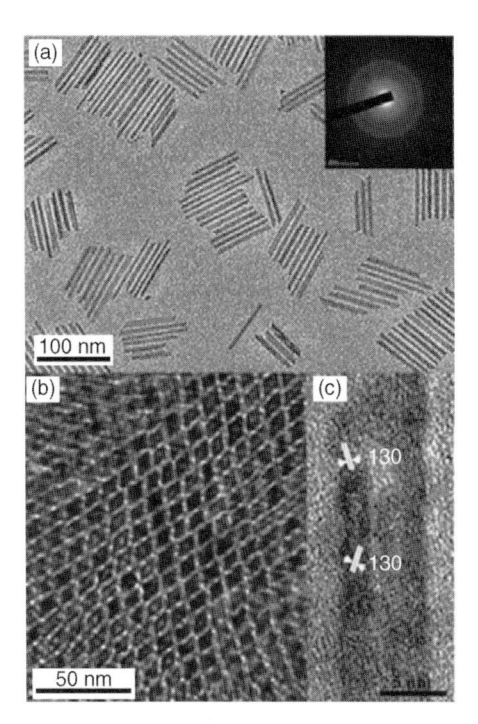

Figure 3.132 (a) TEM image of goethite (α-FeOOH) nanotubes; (b) Superlattice of ordered goethite nanotubes; (c) HR-TEM image of goethite nanotubes. Reproduced with permission from Ref. [123]; © 2007, American Chemical Society.

pounds. However, the main problem associated with all of these techniques is controlling the size, shape and exact stoichiometry of the nanocrystals.

In a simple coprecipitation approach, $MnFe_2O_4$ nanocrystals were synthesized by mixing either ferric or ferrous salts with $MnCl_2$ and sodium hydroxide, followed by aging at 100 °C [124]. Ferric salts yielded 5–25 nm $MnFe_2O_4$ nanocrystals, while ferrous salts led to 180 nm $Mn_xF_{3-x}O_4$ nanocrystals ($0.2 < x < 0.7$). Moreover, the particle size was found to depend heavily on the metal ion: NaOH ratio. Using this technique, 2–6 nm $CoFe_2O_4$ nanocrystals were prepared from $FeCl_3$ and $CoCl_2$ by adding ammonium hydroxide under an argon atmosphere at 80 °C [125]. After washing, the $CoFe_2O_4$ nanocrystals could be functionalized with oleic acid, rendering them soluble in various nonpolar solvents.

The high-temperature decomposition of organometallic compounds allows the creation of high-quality metal ferrite nanocrystals. As an example, monodisperse $CoFe_2O_4$ nanocrystals were synthesized via a high-temperature (305 °C) reaction of iron(III) and cobalt acetylacetonates with 1,2-hexadecanediol in the presence of oleic acid and oleylamine [126]. By adjusting the reaction parameters, and by using a seeded growth initialized by smaller nanocrystals, it proved possible to tune the nanocrystal size within the range of 3 to 20 nm. The organometallic synthesis of monodisperse $CoFe_2O_4$ nanocrystals using single-source precursors was also reported by Hyeon et al. [127]. In this case, $(\eta_5\text{-}C_2H_5)\text{-}CoFe_2(CO)_9$ as a single-source precursor was mixed with oleic acid in dioctyl ether and heated at 300 °C for about 1 h. The FeCo alloy nanocrystals, which formed at the initial stage, were then oxidized with trimethylamine N-oxide to yield uniform 6 nm $CoFe_2O_4$. Notably, larger nanocrystals could be obtained by using mixtures of lauric and oleic acids as stabilizers. $MnFe_2O_4$ nanocrystals were also synthesized using the same approach [128]. More recently, Cheon et al. described a generalized synthesis of MFe_2O_4 (M = Co, Mn, Ni) nanocrystals via a high-temperature reaction between metal dichloride and iron-2,4-pentadionate in the presence oleylamine and oleic acid as stabilizers [129].

3.3.2.4 Multicomponent Magnetic Nanocrystals

During the past decade, significant progress has been made in the synthesis of multicomponent magnetic nanostructures. The assembly of several materials into one tiny piece of solid represents an attractive approach for designing complex systems with diverse physical and chemical properties. In multicomponent systems, it might also be expected that unusual properties would emerge, originating from collective interactions between the constituents at the nanoscale. Investigations into the design and creation of magnetic multicomponent nanocrystals are driven by the potentially interesting properties of these materials, both from a fundamental point of view and from their great promise in applied sciences. Indeed, these materials have already demonstrated enhanced optical, magnetic and catalytic properties compared to their individual single-component analogs [130, 131].

The morphologies of magnetic multicomponent nanocrystals can be subdivided into several groups, namely *core–shell nanoparticles, dumbbell nanoparticles,* and the more recently discovered *hollow nanostructures.* The general approach towards the preparation of nanoscale heterostructures is based on either a two-step or a multistep synthesis. The first step is to create monodisperse nanocrystals of one type, which then serve as the nucleation centers for the deposition of a second component. The latter deposition may result in the production of a continuous uniform shell, leading either to the formation of core–shell structures, or to the nucleation and growth of a second component at a certain facet of the "seed" material, producing in turn a dumbbell-like structure. The most common challenge in the synthesis of multicomponent materials is to achieve compositional and structural uniformity. However, the successful synthesis of multicomponent structures requires the suppression of all possible side-processes, such as the homogeneous nucleation of the secondary material.

3.3.2.4.1 **Magnetic Core–Shell Nanoparticles** Recent advances in colloidal synthesis have led to the preparation of a variety of magnetic core–shell nanosized structures, where the magnetic component can be encapsulated within a semiconducting or plasmonic shell, or it can form an outer shell around a magnetic or semiconducting nanocrystal. Co–CdSe nanocrystals [132] are examples of nanomaterials that combine both semiconductor and magnetic properties. In these structures, both the magnetic and semiconducting properties of the constituents were shown to be conserved. The core–shell geometry also offers a convenient approach for combining different magnetic phases and to achieve magnetic exchange coupling of the components (Figure 3.133). The combination of magnetically hard and soft materials in the form of a core–shell provides an elegant approach towards the design of very strong, permanent magnets [133], where a direct contact between the core and the shell ensures a strong magnetic exchange coupling [134]. The figure of merit by which permanent magnetic materials are judged is termed the "energy product"; this is a measure of the maximum magnetostatic energy that would be stored in free space

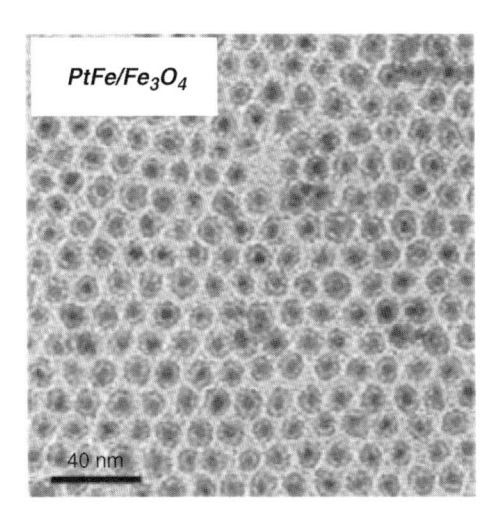

Figure 3.133 TEM image of PtFe–Fe_3O_4 core–shell nanocrystals. Reproduced with permission from Ref. [134]; © 2004, American Chemical Society.

between the pole pieces of a magnet made from the material in question [135]. To obtain a large energy product requires a large magnetization and a large coercivity, but this can be achieved by "exchange coupling" between a hard (high-coercivity) material and a soft (low-coercivity) material with a large magnetization. In a two-phase mixture of such materials, the exchange forces between the phases mean that the resultant magnetization and coercivity of the material will be an average of the properties of the two constituent phases [135]. The exchange-coupled $FePt$–Fe_3O_4 nanocrystals can be used as building blocks for such exchange-coupled magnets. Upon reductive annealing under 5% H_2 + 95% Ar at 650 °C, the assembly of $Fe_{58}Pt_{42}$–Fe_3O_4 nanocrystals was transformed into a nanocomposite with the energy product of ~18 MGOe, providing a 38% increase over the theoretical maximum value of 13 MGOe for the non-exchange-coupled isotropic Pt-transition metal bulk materials [136]. A similar behavior was also reported for $FePt$–Fe_3Pt nanocomposites obtained by the reductive sintering of FePt and Fe_3O_4 nanoparticle mixtures [133].

The shape-controlled synthesis of exchange-coupled Fe_xO–MFe_2O_4 core–shell nanocrystals has been recently reported by Bodnarchuk et al. [137]. The Fe_xO and $MeFe_2O_4$ components are highly compatible not only compositionally, but also structurally [138]. The NaCl-type Fe_xO and the spinel-type $MeFe_2O_4$ are both based on a cubic, close-packed lattice of oxygen atoms with a small lattice mismatch (~3%) between the fcc oxygen sublattices. The decomposition of the mixed $Fe(3+)Co(2+)$ oleate occurred in high-boiling organic solvents at 300–340 °C in the presence of oleic acid or sodium oleate acting as shape-regulating stabilizing agents. Nanocrystals sized between 11 and 25 nm, with a narrow size distribution, have been prepared by tuning the decomposition temperature, as well as by adjusting the stabilizer: precursor molar ratio. Using magnetically hard $CoFe_2O_4$ as the nanocrystal shell material, the strong exchange coupling between the antiferromagnetic core and the ferrimagnetic shell was demonstrated. The possibility of shape control, as well as the use of other MFe_2O_4 (M = Zn, Mn) components as shell materials, has been reported; this has highlighted the possibility of tuning by the choice of M in MFe_2O_4, where magnetically hard $CoFe_2O_4$ led to the most pronounced exchange coupling.

3.3.2.4.2 **Dumbbell-Like Nanoparticles** The dumbbell-like (or heterodimer) structure is another example of a typical nanoscale heterostructure. This structure offers two functional surfaces suitable, for example, for the attachment of different types of molecule, and rendering such species especially attractive as multifunctional probes for diagnostic and therapeutic applications. A chemical approach allows the synthesis of dumbbells with various compositions, where magnetic nanoparticles can be combined with other magnetic, plasmonic, or semiconducting components. The magnetic constituent can be used as a "seed" for the growth of second components, or it can be grown on the top of preformed nanocrystals. Thus, Au–Fe_xO_y, Pt–Fe_3O_4 and Pd–Fe_3O_4 nanodumbbells were synthesized by the thermal decomposition of $Fe(CO)_5$ in the presence of Au, Pt, and Pd seeds, respectively [139–141]. Some examples of Au–Fe_3O_4 nanocrystals with different sizes of Au and Fe_3O_4 are shown in Figure 3.134 [139]. Since both Fe_3O_4 and Au phases have cubic crystalline lattices with lattice parameters of 4.08 Å and 8.35 Å, respectively, the Au–Fe_3O_4 structures

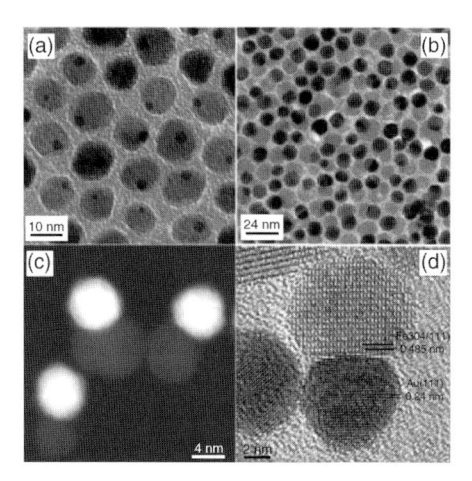

Figure 3.134 TEM images of (a) 3–14 nm and (b) 8–14 nm Au–Fe$_3$O$_4$ nanodumbbells; (c) High-angle annular dark-field scanning electron micrograph of the 8–9 nm Au–Fe$_3$O$_4$ particles; (D) HR-TEM image of a 8–12 nm Au–Fe$_3$O$_4$ nanodumbbell. Reproduced with permission from Ref. [139]; © 2005, American Chemical Society.

were assumed to derive from the epitaxial growth of Fe$_3$O$_4$ on the Au nanocrystals. The dimers of CoPt$_3$–Au (Figure 3.135) and Ag–CoFe$_2$O$_3$ can be prepared by the nucleation of Au on preformed CoPt$_3$ and Ag nanocrystals, respectively [142, 143]. The dumbbells combining plasmonic and magnetic components showed

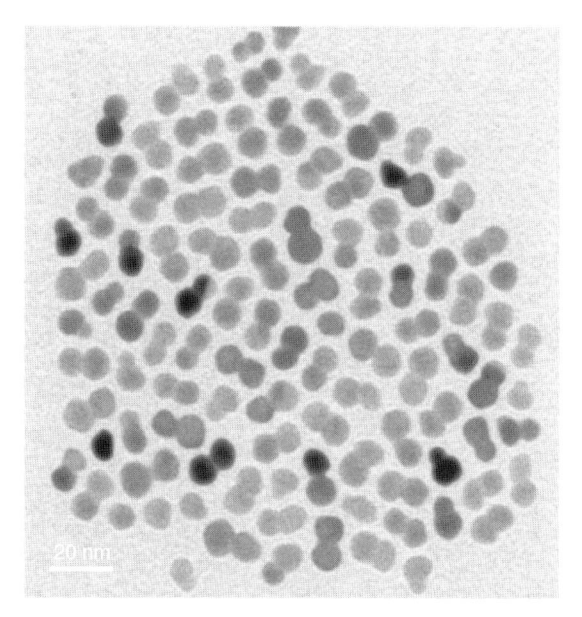

Figure 3.135 TEM image of symmetric Au–CoPt$_3$ nanocrystals synthesized by a modified procedure as described in Ref. [142].

interesting magneto-optic properties, with measurements of Faraday rotation for Ag–CoFe$_2$O$_3$ and CoFe$_2$O$_3$ revealing the materials to have not only a similar magnitude of the rotation but also shape of the hysteresis loops at short wavelength. At longer wavelengths, however, the rotation became strongly enhanced for Ag–CoFe$_2$O$_3$; indeed, at 633 nm this difference approached almost an order of magnitude [143].

CdS–FePt represents an example of a magnetic-semiconductor dumbbell structure, where the proposed mechanism of formation assumes, first, the deposition of an intermediate sulfur layer and subsequent growth of a thin CdS shell. The latter is dewetted from the FePt core during the later reaction stages, forming dumbbell-like nanocrystals (Figure 3.136) [144]. A set of parameters, such as the reaction temperature, size of the seed material, capping ligands and concentrations of reagents, allows control over the heterodimer structures [142]. It is clear from Figures 3.134–3.136 that dumbbells may be both symmetric and highly asymmetric.

3.3.2.4.3 Hollow Magnetic Nanocrystals

Hollow nanoscale structures were first obtained by Y. Yin during the sulfurization of cobalt nanocrystals at elevated temperatures [145]. This process was found to lead to the formation of hollow cobalt sulfide nanocrystals such that, depending on the size of the cobalt nanocrystals and the cobalt: sulfur molar ratio, different stoichiometries of hollow cobalt sulfide could be obtained. Hollow nanostructures are usually formed through the nanoscale Kirkendall effect, which is based on the difference in diffusion rates of two species, and results in an accumulation and condensation of vacancies [146]. This phenomenon was first observed by Kirkendall at the interface of copper and zinc in brass in 1947 [147]. As a typical example of the nano-Kirkendall effect, the controllable oxidation of iron nanoparticles by air can lead to the formation of hollow iron oxide nanostructures, as shown in Figure 3.137. During the course of metal nanoparticle oxidation, the outward diffusion of metal occurs much faster in

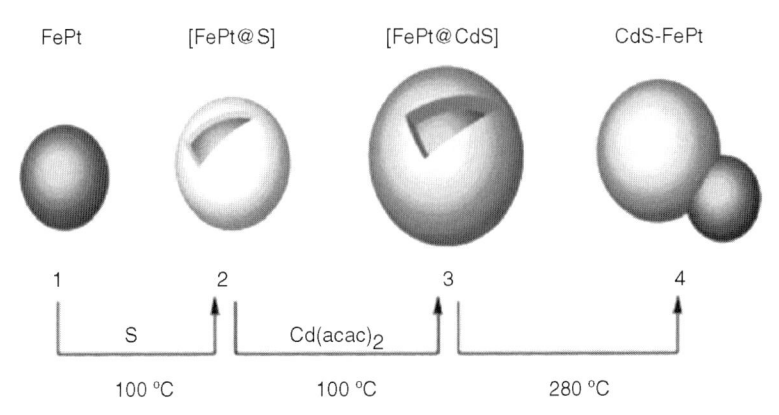

Figure 3.136 Proposed mechanism for the formation of CdS–PtFe nanodumbbells. Reproduced with permission from Ref. [144]; © 2004, American Chemical Society.

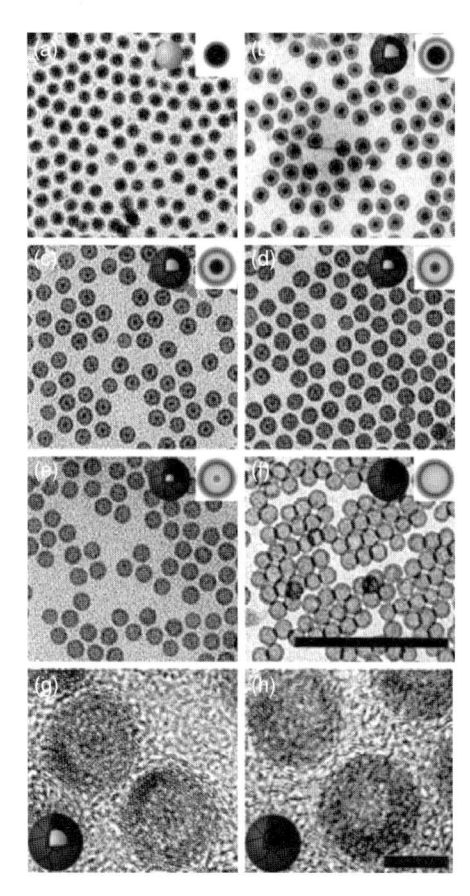

Figure 3.137 Hollow iron–iron oxide nanocrystals formed by the nano-Kirkendall effect. TEM images of iron–iron oxide nanoparticles exposed to dry 20% oxygen. (a) <1 min at room temperature; (b) 1 h at 80 °C; (c) 12 h at 80 °C; (d) 5 min at 150 °C; (e) 1 h at 150 °C; (f) 1 h at 350 °C on a substrate; (g,h) High-resolution of partial and fully oxidized iron nanoparticles. The low- and high-resolution scale bars correspond to 100 and 6 nm, respectively. Two-dimensional projections of the cross-sections for electron scattering in iron–iron(III) oxide core–void–shell nanospheres are shown as insets. The simulated evolution of the particle size, core diameter, and shell thickness corresponds to an oxide growth at the oxide–solution interface. Reproduced with permission from Ref. [149, 150]; © 2007, American Chemical Society.

the oxide layer than does the inward diffusion of oxygen, and this leads to the formation of a nanoscale void in the center of a nanoparticle [148, 149]. Both, the reaction temperature and oxidation time allow a precise tuning of the thickness of an initial oxide shell. Subsequent TEM images of the partially oxidized particles showed three differentiated contrast regions (Figure 3.137), where the darker inner region corresponded to the iron core and the outermost shell (which corresponded to a lower-density material) was the iron oxide. As the chemical transformation proceeded, the oxide shell grew thicker due to the continual appearance and

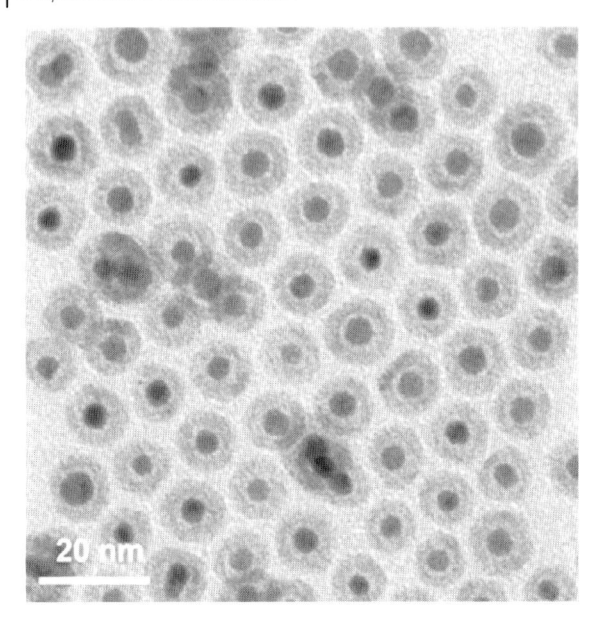

Figure 3.138 TEM image of gold core–iron oxide hollow-shell nanocrystals.

subsequent oxidation of iron atoms on the outermost surface of the oxide. The disappearance of the spherical iron core in the center of the hollow particle was clearly observed. When the iron atoms diffused outwards, the vacancies left behind ultimately coalesced into a single central void (Figure 3.137f). Since the reaction was carried out at a relatively high temperature (80–150 °C), hollow nanocrystals could be synthesized within several minutes, whereas the growth of a 4 nm-thick oxide layer in iron films could take about 600 years at room temperature [149].

The formation of hollow nanostructures via the Kirkendall effect can be used for the design of materials with complex morphologies, such as hollow Au–Fe_3O_4 (Figure 3.138) and Pt–CoO core–shells [145, 151]. These core–hollow shell nanoparticles have been prepared by the deposition and subsequent oxidation of the transition metal shell. By using this approach, catalytically active transition metal nanoparticles can be encapsulated inside oxide shells so as to protect them against coalescence at high temperatures. Both, HR-TEM and electron energy loss spectroscopy (EELS), in combination with scanning transmission electron microscopy (STEM), have revealed a polycrystalline character and nanoscale pores in the outer shell of the Au–Fe_3O_4 samples. Selected small molecules are able to enter the internal cavity via grain boundaries and also via small average diameter pinholes that permit reactant access and product removal, providing in turn a means for molecular shape selectivity. Both, size and shape selectivity is to be expected for molecules with an average cross-section that is less than the pinhole diameter. Compared to catalysts on open surfaces, or in the channels of porous structures, catalysts confined in this manner may reduce secondary reactions and deliver only the desired reaction products, in the desired amounts. In one investigation, these confined catalysts efficiently promoted the reaction of ethylene (C_2H_4) and hydrogen to form ethane (C_2H_6). It was assumed that these small molecules would travel along grain boundaries through the shell [145]. The combination of magnetic and high-contrast X-ray or plasmonic materials may represent a promising route towards the design of dual agents that allow both MRI–X-ray and MRI–optical detection.

To summarize, a combination of organometallic and colloidal chemistry provides a powerful method for the preparation of monodisperse magnetic nanoparticles with controllable size, shape, composition, and crystalline modification. At present, the synthesis of magnetic nanoparticles provides a palette of high-quality materials not only for investigations into nanomagnetism but also for a wide variety of applications. Despite much progress having been made in the synthesis of magnetic nanocrystals, there remains a clear lack of any theoretical understanding of the processes that occur during nanoparticle growth in colloidal solutions, and in maintaining a narrow particle size distribution. Clearly, future synthetic progress will depend on an ability to understand and control those parameters that govern the properties of colloidally grown nanocrystals.

3.3.2.5 Size- and Shape-Dependent Magnetic Properties of Magnetic Metal Nanoparticles

The phenomenon of magnetism occurs as a result of moving charges. However, elementary particles – such as electrons – also have an intrinsic magnetic moment (spin) that determines their quantum state. The magnetic properties of materials arise mainly from the orbital motion and the spin motion of their electrons, while other contributions, such as nuclear magnetic effects, are normally much smaller compared to the electronic effect. As the electronic structure of a nanometer-sized metal particle is heavily size-dependent, size-dependency should be also noted among the magnetic properties. Many physical phenomena – such as the magnetic domain size and exchange-coupling effects – which determine the experimentally observable magnetic properties of materials, have natural length scales that lay in the nanometer and micrometer size range. The questions remain, therefore, as to what happens if the size of a ferromagnetic material is shrunk below 10 nm? The aim of this section is to demonstrate the effects of particle size, shape and structure on the magnetic properties of nanomaterials. More detailed descriptions of the physical aspects of nanoscale magnetism are available in several excellent books [1, 29, 152] and journal reviews [27, 30].

The first step is to introduce very briefly the commonly measured magnetic parameters. All materials interact with a magnetic field, and this interaction can be either *attractive* towards a magnetic pole (ferromagnetism and paramagnetism) or *repulsive* (diamagnetism). The application of a magnetic field (H) results in the magnetization (M) of a sample which can be measured by, for example, using a superconductive quantum interference device (SQUID); this is one of the most popular and sensitive techniques used to investigate magnetic properties. When a ferromagnetic material is magnetized by an increasing applied field, and the field is then decreased, the magnetization does not follow the initial magnetization curve obtained during the increase. This irreversibility is called *hysteresis* [1]; a typical

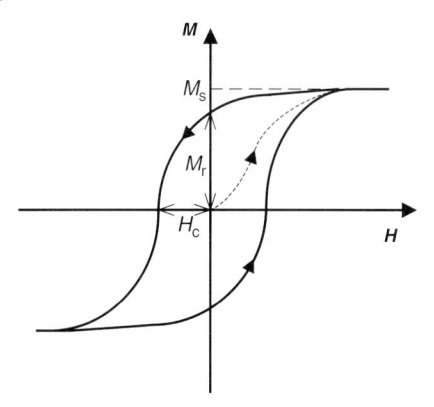

Figure 3.139 A typical hysteresis loop for a ferromagnetic material and the important magnetic parameters which can be obtained from the hysteresis loop. The dotted line shows the first scan.

hysteresis loop arises from measuring the magnetization of the material as a function of magnetic field applied in positive and negative directions: the response of the materials will contain two distinct paths on *magnetization* and *demagnetization* (Figure 3.139). With large fields, the magnetization approaches the maximum value termed the *saturation magnetization* (M_s). Magnetic materials in a ferromagnetic state have a residual magnetization at zero external field, called *remanent magnetization* (M_r), while *coercivity* (H_c) characterizes the reverse-field strength needed to reduce the magnetization to zero. Thus, hysteresis measurements allow information to be obtained regarding the coercivity, remanent magnetization, and saturation magnetization of a given material (Figure 3.139).

Today, it is still common in scientific and engineering literature to use the old cgs (centimeter, gram, second) system for magnetic units and, as a consequence, there are two widely used systems of units, the cgs and SI (Système Internationale) systems. Thus, the magnetic field strength H can be measured either in oersteds (Oe; for cgs) or in amperes per meter (A m^{-1}; for SI). Moreover, the applied field can be multiplied with the induction constant μ_0, and given in units of Gauss (cgs) or Tesla (SI). The magnetization M is measured in emu cm^{-3} (cgs) or A m^{-1} (SI) or, being multiplied with μ_0, can also be given in Gauss or Tesla. It should also be noted that, in the cgs system, magnetization can be written per gram of substance. The magnetic units used in the cgs and SI systems, and some unit conversions, are listed in Table 3.2.

In order to minimize their energy, macroscopic ferromagnetic materials are broken into domains of parallel magnetic moments. Within a domain, the magnetic moments orient in one direction, while the alignment of spins in neighboring domains is usually antiparallel. The oppositely aligned magnetic domains are separated from each other by a domain wall (the Bloch wall). As the particle size decreases below some critical value, the formation of domain walls becomes energetically unfavorable and the ferromagnetic particle can support only a single domain. This critical size depends on the material, and is usually on the order of tens of nanometers varying from ∼14 nm for Fe up to ∼170 nm for γ-Fe$_2$O$_3$. Magnetic particles of nanometer size are usually single domain in nature [27].

Table 3.2 Magnetic units and unit conversions.

Magnetic units		cgs	SI	Conversion
Magnetic field strength	H	Oe (oersted)	A m^{-1}	1 Oe = $\frac{10^3}{4\pi}$ A m^{-1}
Magnetization	M	emu cm^{-3}	A m^{-1}	1 emu cm^{-3} = 10^3 A m^{-1}
Magnetic flux density (induction)	B	G (gauss)	T (tesla)	1 T = 10^4 G
		B = H + 4πM	B = μ_0(H + M)	
Induction constant	μ_0	G–Oe	T-(A m^{-1})	μ_0 = 1 G·Oe
				4π × 10^{-7} T·(A m^{-1})

The dimensionless parameter – *magnetic susceptibility* (χ) – defined as $\chi = \partial M/\partial H$, is used to estimate the efficiency of the applied magnetic field for magnetizing a material. In many cases, the susceptibility of a material will depend on the direction in which it is measured; such a situation is termed *magnetic anisotropy*. The physical origin of magnetic anisotropy may be, for example, the symmetry of the crystalline lattice or the shape of a particular piece of magnetic material (this will be discussed below in more detail). At this point, it will only be emphasized that anisotropy is an extremely important parameter that determines the behavior of a ferromagnetic material. When magnetic anisotropy exists, the total magnetization of a system will prefer to lie along a special direction, termed the *easy axis of magnetization*. For a small, single-domain particle the energy associated with this alignment is called the *anisotropy energy*, and this can be written in the simplest uniaxial approximation as $E_a = KV \sin^2 \theta$, where K – is the anisotropy constant, V the particle volume, and θ is the angle between the moment and the easy axis (Figure 3.140) [27].

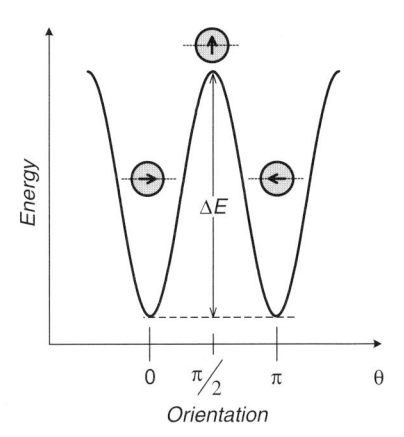

Figure 3.140 The energy associated with rotation of a magnetization direction for a uniaxial magnetic particle. The minima correspond to alignment of the magnetization along the easy axis schematically shown by dotted lines. ΔE represents the energy barrier to the rotation of magnetization.

When an external magnetic field interacts with the single-domain particle, an additional potential energy of $E_f = -\vec{M}_s \cdot \vec{H}$ is supplied. Depending on the direction of the magnetic field with respect to the particle easy axis, the response of the particle moment to an applied field is different. Thus, if the field is applied perpendicular to the easy axis, the dependence of the particle potential energy on the magnetization direction evolves with the strength of applied field, as shown in Figure 3.141a. The equilibrium direction of the magnetic moment, corresponding to the minimum of the total alignment energy $E_{tot} = E_a + E_f$, turns toward the field to the angle $\theta = \arcsin(M_s H/(2\,KV))$. At $H = 2\,KV/M_s$, the moment aligns perpendicular to the easy axis (i.e., along the applied field) and magnetization approaches the saturation. In the M versus H scans, the magnetization component parallel to the applied field changes linearly from $-M_s$ to $+M_s$, as shown in Figure 3.141c. Note that no magnetization hysteresis is observed if the magnetic field is applied perpendicular to the easy axis of an uniaxial, single-domain particle.

If a magnetic field is applied parallel to the easy axis, the particle moment can align either parallel or antiparallel to the field direction; the latter case is considered in Figure 3.141b. Although the alignment of particle magnetization along the field is energetically favorable, in a relatively weak field the energy barrier blocks the magnetization reversal. The height of the barrier decreases with increasing field strength (Figure 3.141b) such that, at $H = 2\,KV/M_s$ the barrier vanishes and the particle magnetic moment jumps from the antiparallel to the parallel alignment. This behavior leads to a square hysteresis loop with coercivity $H_c = 2\,KV/M_s$, as shown in Figure 3.141d. The effect of magnetic field applied at any other angle to the particle easy axis can be evaluated in a similar manner.

As can be seen, the response of an individual single-domain particle crucially depends on the direction of the applied magnetic field. The two situations considered above represent the extreme cases of the possible hysteresis curves: totally closed (no hysteresis), and totally open (square). Other orientations yield hysteresis curves between these limits. In reality, there is usually an ensemble of many particles with randomly oriented easy axes (Figure 3.142a), and the measured hysteresis loop is the result of averaging over all possible orientations. In an extremely strong magnetic field, all particles align their moments along the field approaching the largest (saturation) magnetization. If the particles are fixed – for example, if they are dispersed in a solid nonmagnetic matrix – then the easy axes are randomly aligned and magnetization occurs through rotation of the particle moments, also known as *Neel rotation* (Figure 3.142b). The time scale of the Neel rotation process is approximately 10^{-9} s. However, the particles dispersed in a liquid can considerably decrease the potential energy aligning their easy axes along the external field (Figure 3.142c). This alignment occurs via the Brownian motion process when the particles strike with solvent molecules, and is much slower ($\sim 10^{-7}$ s for solvents such as hexane) than the rotation of magnetic moment inside the particle.

More explicitly, the Neel (τ_N) and Brownian (τ_B) magnetic relaxation times of a magnetic particle are described as:

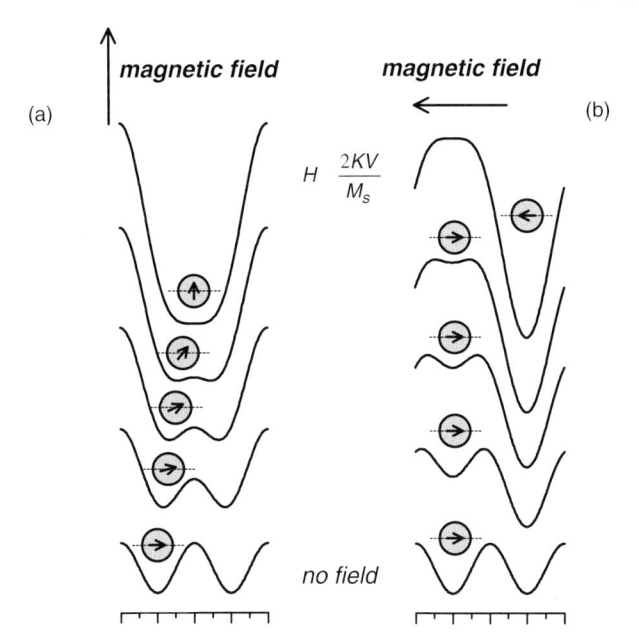

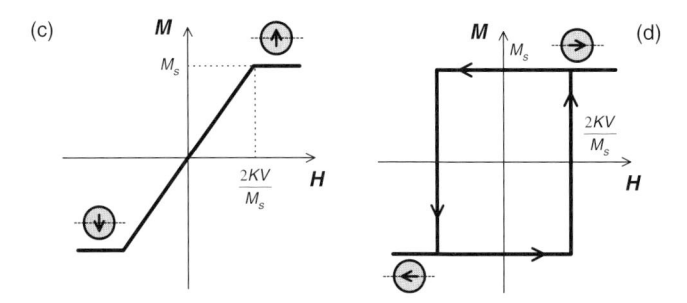

Figure 3.141 (a,b) Dependence of potential energy on the direction of a particle magnetic moment in presence of external magnetic fields applied (a) perpendicular and (b) parallel to the easy magnetization axis. The lower curves correspond to zero magnetic field, and the upper curves to the saturation magnetization when the magnetic moment align along the field; (c,d) Hysteresis loops for applied fields perpendicular (c) and parallel (d) to the easy axis of an uniaxial single-domain particle. The direction of easy magnetization axis is shown by the dotted lines.

$$\tau_N = \tau_0 \exp\frac{KV}{k_B T} \qquad (3.10)$$

and

$$\tau_B = \frac{3\eta V_H}{k_B T}, \qquad (3.11)$$

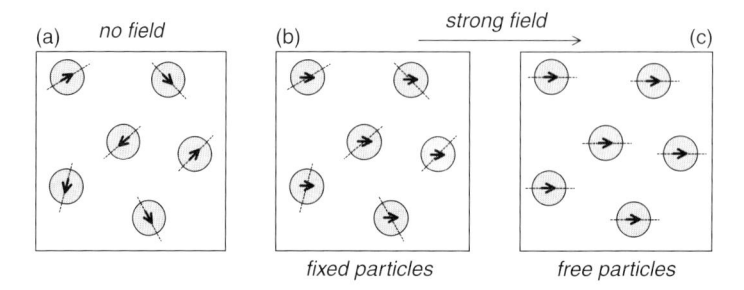

Figure 3.142 (a) An ensemble of magnetic particles with randomly distributed easy magnetization axis; (b) Magnetization of particle ensemble in a strong field via rotation of the particle magnetic moments; (c) Magnetic particles dispersed in a liquid can align their easy axes along the applied field. The directions of particle easy magnetization axes are shown by the dotted lines.

where τ_0 is the attempt frequency which is on the order of $\sim 10^{-9}$–10^{-10} s, K is the magnetic anisotropy constant, V is the volume of the magnetic particle, k_B is Boltzmann's constant, T is the temperature, η is the viscosity, and V_H is the hydrodynamic particle volume. These parameters will be discussed in more detail in Section 3.2.2.7, with regards to the biomedical applications of magnetic nanomaterials.

When considering the role of thermal fluctuations in the behavior of very small magnetic particles, in a zero magnetic field the energy barrier ΔE must be overcome in order to rotate the magnetization of a single domain particle (see Figure 3.140). The height of this barrier, $\Delta E = KV$, is proportional to the particle volume and, as the particle size decreases, ΔE may become comparable to the thermal energy ($k_B T$). In this case, the energy barrier can no longer pin the magnetization direction on the time scale of observation, and rotation of the magnetization direction will occur due to thermal fluctuations. Such a particle is then said to be *superparamagnetic*. The coercivity of a superparamagnetic particle is zero ($H_c = 0$), because thermal fluctuations prevent the existence of a stable magnetization. The cooling of a superparamagnetic particle reduces the energy of thermal fluctuations such that, at a certain temperature, the free movement of magnetization becomes blocked by anisotropy. The temperature of the transition from superparamagnetic to ferromagnetic state, termed the *blocking temperature* (T_B), is related to the particle volume and the anisotropy constant. The latter can be calculated as $K = 25 k_B T_B / V$ [1].

The coercivity of a magnetic particle depends heavily on its size. Thus, if the particle is large enough to support a multidomain structure, magnetization reversal occurs through a domain walls motion [30]; this is relatively easy and, hence, the coercivity is low. On the other hand, in a single-domain particle the change of magnetization direction can occur only by coherent rotation of spins; this results in a considerably higher coercivity of single-domain particles in comparison to multidomain particles (Figure 3.143) [1]. Upon further decrease of the particle size, the coercivity falls off due to the progressively increasing role of the thermal fluctuations leading to superparamagnetism, with $H_c = 0$ (Figure 3.143).

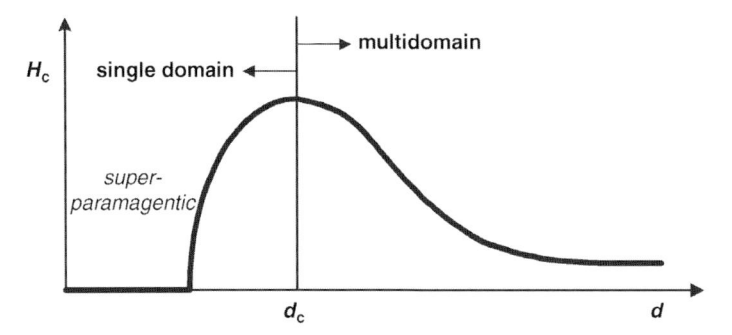

Figure 3.143 Particle coercivity versus size. The largest coercivity is observed at the particle size d_c, corresponding to the transition from multidomain to single-domain structure.

In general, a magnetic nanoparticle is a single-domain magnet. Thus, in order to investigate the magnetic properties inherent to independent nanoparticles, all measurements must be performed on an ensemble of uniform, noninteracting nanoparticles. Moreover, all sources of sample inhomogeneity (particle size and shape distribution, presence of particles with different crystalline modifications, etc.) must be minimized. Interparticle interactions can be reduced by dissolving the particles in an appropriate solvent at a low concentration [27]. However, if a distribution of particle sizes is present, then the initial susceptibility will be more sensitive to the larger particles present, whereas the approach to saturation will be sensitive to the smallest particles of the sample [28]. In addition to the hysteresis measurements, which allow estimation of the coercivity and specific magnetization of a sample, zero-field-cooled (ZFC) and field-cooled (FC) measurements provide information about the blocking temperature (T_B).

In ZFC scans, a sample is cooled under zero applied magnetic field to a temperature well below the suspected T_B. The system is then warmed up and the magnetization measured as a function of temperature, by applying a relatively low external magnetic field. As the thermal energy increases, the nanoparticles become aligned with the applied field and the magnetization is increased. At T_B, the magnetization is maximal, while further growth of temperatures above T_B will result in a decrease of the magnetization due to the effect of thermal energy causing fluctuations of the magnetic moments of nanoparticles.

In the FC scans, a sample is cooled in a small magnetic field, thus freezing-in a net alignment of the nanoparticle moments. The field is then removed, and the magnetization measured as the sample is slowly warmed up. Thermal energy unpins and randomizes the nanoparticle moments, lowering the sample's net magnetization. Below T_B, the free movement of magnetic moments is "blocked" by the anisotropy, and so the particles are in a ferromagnetic state; above T_B, however, the particles are characterized by superparamagnetic behavior [47]. At T_B, the ZFC and FC curves will converge [27]. The T_B in a superparamagnetic system decreases with increasing applied measuring fields, and is proportional to $H^{2/3}$ at large magnetic fields and to H^2 at lower fields, respectively [30].

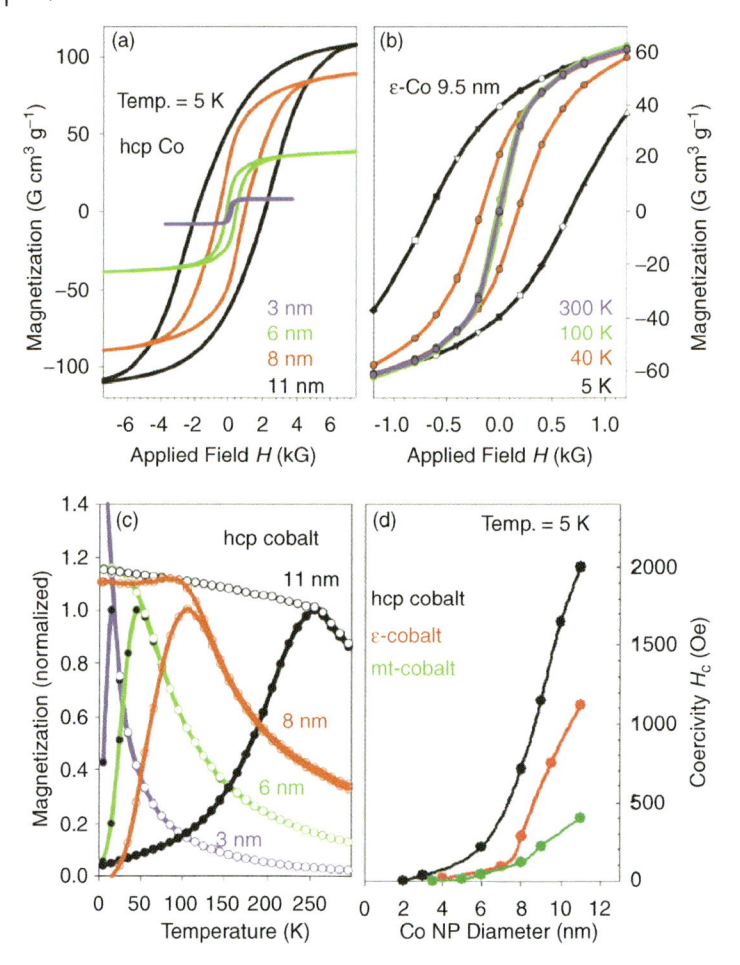

Figure 3.144 (a) Magnetization versus applied-field hysteresis loops at 5 K for 3, 6, 8, and 11 nm hcp cobalt nanocrystals; (b) Comparison of the hysteresis loops for 9.5 nm ε-Co nanocrystals measured at 5, 40, 100, and 300 K; (c) A series of zero-field-cooled (ZFC) and field-cooled (FC) magnetization scans for hcp-Co nanocrystals of different size (3, 6, 8, and 11 nm); (d) Dependence of coercivity of the nanoparticle size for a collection of hcp-Co, ε-Co, and mt-fcc Co samples. Reproduced with permission from Ref. [27]; © 2001, Materials Research Society.

Figure 3.144 shows a set of measurements performed to characterize the magnetic properties of cobalt nanoparticles [27]. As discussed in Section 3.3.2.2.2, a size series of monodisperse cobalt nanoparticles can be synthesized with three different crystal structures: hexagonal (hcp), ε-, and mt-fcc. All investigated sizes are far below the critical single domain size which is, for example, 70 nm for the hcp-Co phase, and the particles are small enough to be superparamagnetic at room temperature. The hysteresis measurements performed at 5 K on hcp-Co nanoparticles of different size revealed a strong decrease in coercivity with decreasing particle size (Figure 3.144a and d). As already discussed, the number of spins coupled by exchange interactions decreases with the size of the single-domain particle, leading to an easier spontaneous reorientation of magnetization at a given temperature, and thus to a decrease in coercivity [27]. The drop in saturation magnetization with decreasing nanoparticle size (Figure 3.144a) was attributed to an increase in the nanoparticles' surface-to-volume ratio, and the presence of about one to two monolayers of cobalt oxide [27]. Measurements of the hysteresis loop at different temperatures demonstrate the transition from ferromagnetic to superparamagnetic behavior in 9.5 nm ε-Co nanocrystals (Figure 3.144b). Increasing the thermal energy results in fluctuations of the nanoparticle magnetization direction, reducing coercivity and leading to superparamagnetic behavior at temperatures above 125 K. More precise information regarding T_B can be obtained from the ZFC and FC scans shown in Figure 3.144c for a size series of hcp-Co nanocrystals. The blocking temperature, as estimated from the peak magnetization on the ZCF scans, increases with the particle size from 20 K for 3 nm up to 240 K for 11 nm hcp-Co nanoparticles. The increase in T_B, magnetization and coercivity with the increase in nanoparticle size was observed for all crystalline modifications. A similar behavior was also observed for $CoPt_3$ nanocrystals.

The magnetic anisotropy of nanocrystalline materials originates from the influence of crystal symmetry, shape and surface (or grain boundaries) effects. The most common anisotropies in the case of magnetic nanoparticles are the magnetocrystalline and shape anisotropies. Magnetocrystalline anisotropy arises from a coupling of the electron spins to the electronic orbit which, in turn, is coupled to the crystal lattice. The magnetization is energetically favorable to align along a specific crystallographic direction called the "easy axis" of the material. In the case of hexagonal cobalt, the easy axis of magnetization is the c axis. Magnetocrystalline anisotropy is intrinsic to a given material and is independent of particle shape [30], although it does depend on temperature. In a cubic lattice, symmetry creates multiple easy axes; typically, at room temperature the easy magnetization axes of cubic Ni and Fe are <111> and <100> axes, respectively [153].

The highest magnetization and coercivity were achieved in the case of hcp cobalt nanoparticles (Figure 3.144d), because this structure has a larger magnetocrystalline anisotropy as compared to the mt-fcc and ε-Co phases [27]. For each system, the coercivity is lower than that predicted for idealized nanocrystals, as internal structural defects (stacking faults, twinned planes) reduce H_c [9]. Future ultra-high-density recording media will require uniform particles (size distribution less than 10%) with an average diameter of <6 nm and a room temperature H_c of 3000–5000 Oe [27]. As the coercivity is proportional to anisotropy constant, monodisperse nanoparticles of materials with a high intrinsic magnetocrystalline anisotropy (e.g., platinum–iron, cobalt–platinum binary alloys) will certainly be required [30].

Cobalt nanoparticles have a great tendency to undergo oxidation in air, such that their surfaces are usually covered with a layer of antiferromagnetic cobalt oxide, which influences the magnetic properties [30]. This influence of the oxide layer is especially noticeable for small nanoparticles with an extremely high surface-to-volume ratio. Thus, the coercivity of the smallest nanoparticles in the size series shown in Figure 3.144d is dominated by coupling to the surface CoO layer [27, 30].

The phenomenon, termed *exchange anisotropy* [154], occurs when a ferromagnet is in close proximity to an antiferromagnet or ferrimagnet [30]. Exchange anisotropy arises as a consequence of the interfacial, exchange coupling. Although first discovered in oxide-coated cobalt particles (with sizes in the range from 10 to 100 nm), this effect has since been observed in a variety of other systems.

Some molecular species, when adsorbed onto the surfaces of nanoparticles, can also significantly influence the magnetic behavior through quenching of the surface atom contribution. Complete quenching of the magnetic moment of the surface nickel atoms was reported for small nickel particles and $Ni_{38}Pt_6$ clusters coated by carbonyl ligands [17].

Nonspherical nanoparticles can also possess shape anisotropy, which accounts for the preferential orientation of magnetization along the long axis of an elongated particle. The demagnetization field is less in the long direction, because the induced poles at the surface are farther apart [1]. Shape anisotropy is comparable to magnetocrystalline anisotropy, and can produce large coercive forces. Thus, the T_B of Fe nanorods (2 × 11 nm) is almost one order of magnitude higher than that of 2 nm spherical nanoparticles (Figure 3.145) [82]. The advanced magnetic properties of nonspherical nanoparticles attract much attention to the shape-controlled colloidal synthesis, and the syntheses of Co nanodisks [64], Co nanorods [80], Fe nanorods [82], $CoPt_3$ nanocubes [55] and $CoPt_3$ nanowires [65] have been recently reported. Progress in the synthesis of monodisperse nonspherical magnetic particles is an important prerequisite for the investigation of shape-dependent magnetic properties.

If the concentration of nanoparticles in a sample is high, the magnetic moments of individual nanoparticles can interact, such that these dipolar interactions will greatly affect the magnetic behavior of the sample [155]. Interparticle interactions in both two-dimensional (2-D) and three-dimensional (3-D) assemblies of magnetic nanoparticles result in the T_B being shifted to higher values. Thus, an increase of the T_B by

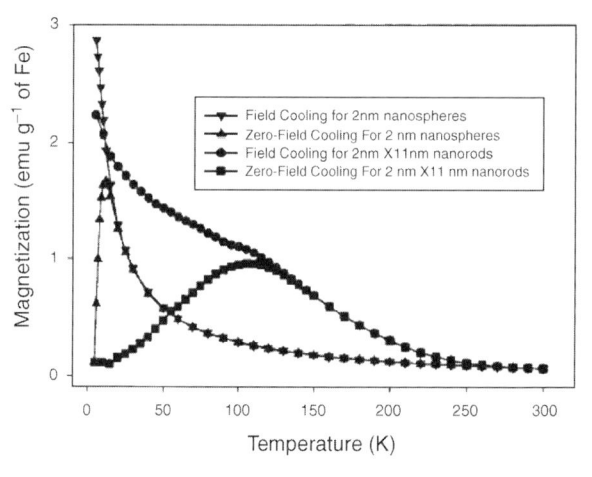

Figure 3.145 Zero-field-cooled (ZFC) and field-cooled (FC) magnetization scans for the 2 nm spherical iron nanoparticles and the 2 nm × 11 nm iron nanorods at the applied magnetic field of 100 Oe. Reproduced with permission from Ref. [82]; © 2000, American Chemical Society.

5 K was observed for 5.8 nm cobalt particles arranged in a 2-D network, compared to the isolated, noninteracting particles [47]. An increase in T_B and a broadening of the ZFC–FC scans were found in close-packed assemblies of monodisperse nickel [27] and CoPt$_3$ nanoparticles. Dilution of the close-packed film of 9 nm cobalt particles ($T_B = 165$ K) with 1-octadecylamine reduced the magnetostatic coupling of particles and resulted in a sharper transition from superparamagnetic to ferromagnetic state at 105 K [9].

3.3.2.6 Magnetic Nanocrystals for Data Storage Applications

The use of colloidal nanocrystals for magnetic data storage applications was first proposed by Sun and Murray, in 2000 [86]. As the electron spins inside a magnetic nanocrystal are all aligned parallel to each other, their orientations can be manipulated by applying an external magnetic field. A ferromagnetic nanocrystal can also store one bit of information, encoded in its magnetic moment. In the most optimistic scenario, where magnetic nanocrystals are assembled in a long-range ordered array and each is used to store information, these advanced magnetic media could provide enormous magnetic recording densities of up to 1 Tb in^{-2}, although the technical realization of this great idea requires many difficult problems to be addressed. As the detailed discussion of magnetic data storage is beyond the scope of this book, the achievements and challenges encountered on the way to using magnetic nanocrystals for data storage will be only briefly described. Further detailed information on this topic is available in various reviews and book chapters [156–159].

In order to be used as a magnetic storage medium, the magnetization direction in a material must be very stable and not be reversed due to thermal fluctuations. In other words, the nanocrystals should have a high magnetic coercivity. As the coercivity of a superparamagnetic particle is zero, such nanocrystals could not be used for data storage; hence, in order to overcome the superparamagnetic limit, materials with a very large magnetic anisotropy must be used.

Figure 3.146 shows a standard write–read test acquired for an annealed film of 4 nm FePt nanocrystals [86]. The as-synthesized fcc FePt nanoparticles are superparamagnetic at room temperature, and can be handled in the form of a colloidal solution for casting films and preparing self-assembled arrays. To render them ferromagnetic, FePt nanoparticles must be annealed at 500–700 °C under an inert atmosphere; this annealing induces the iron and platinum atoms to rearrange, and converts the particles to the chemically ordered face-centered tetragonal (fct) phase. It also causes them to be transformed into nanoscale ferromagnets with a room-temperature coercivity sufficiently high that the arrays of fct FePt nanocrystals can be used in a data storage device [156, 160, 161]. The annealed FePt nanocrystals array (Figure 3.146a) can also support stable magnetization reversal transitions (bits) at room temperature [86].

The read-back sensor voltage signals (Figure 3.146c) from written data tracks (Figure 3.146b) correspond to linear densities of 500, 1040, 2140, and 5000 flux changes per millimeter (fc mm^{-1}). The results of these write–read experiments showed that a 4 nm FePt ferromagnetic nanocrystal assembly could support magnetization reversal transitions at moderate linear densities that could be read back

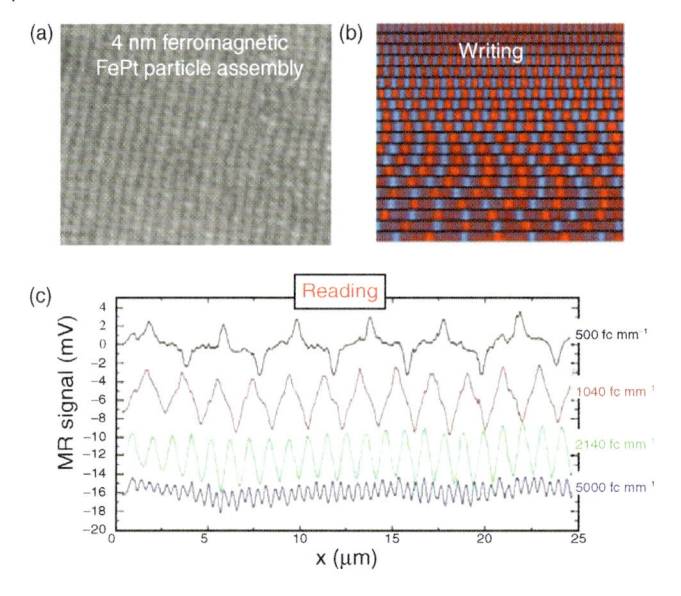

Figure 3.146 Magneto-resistive (MR) readback signals from written bit transitions in a 120 nm-thick assembly of 4 nm-diameter $Fe_{48}Pt_{52}$ nanocrystals. The individual line scans reveal magnetization reversal transitions at linear densities of 500, 1040, 2140, and 5000 flux changes per millimeter (fc mm^{-1}) [86]. Courtesy of C. B. Murray.

nondestructively. To achieve much higher recording densities, however, it is not only the size and magnetic properties of the individual nanocrystals that must be controlled; rather, the orientation of the individual nanocrystals in the array must be pinned so as to provide a unidirectional alignment of the easy axes of magnetization (i.e., the preferable magnetization direction, typically linked to a certain crystallographic axis) for individual nanocrystals. This is a major challenge, especially considering that such alignment must be achieved throughout the entire hard disk area by using high-throughput fabrication techniques. The deposition of magnetic nanocrystals in the presence of a magnetic field, or the design of nanocrystals with an anisotropic shape, have been proposed to facilitate the orientation of individual particles [162]. Future perspectives for colloidal nanocrystals for magnetic data storage will depend heavily on the development of novel syntheses for materials with a very high magnetic anisotropy (e.g., Co_5Sm), in addition to the fast and reliable assembly of chemically synthesized nanocrystals into uniform and, ideally, long-range ordered 2-D arrays.

3.3.2.7 Biomedical Applications of Magnetic Nanoparticles

The combination of unique properties of magnetic nanoparticles has led to a variety of applications in biomedicine. First, the size of a nanoparticle is smaller, or at least comparable, to those of cells (10–100 µm), viruses (20–450 nm), proteins (5–50 nm), or DNA (2 nm wide and ~10–100 nm long). Second, the existence of magnetic dipoles allows the manipulation of magnetic nanoparticles by an external magnetic field that have an intrinsic ability to penetrate into human tissues. Third, the surface of nanoparticles can be modified with constituents carrying certain functionalities. Thus, magnetic nanoparticles can be used as contrast agents in magnetic resonance imaging (MRI), or as a platform for the precise delivery of anticancer drugs or biological species. In this section, the major application areas of magnetic nanoparticles in biomedicine, such as MRI, magnetic hyperthermia, magnetotransport, and drug and gene delivery, are discussed.

3.3.2.7.1 **Design of Magnetic Particles for Biomedical Applications** In order to be useful for any biological application, magnetic nanoparticles must first be compatible with biological species. Next, the particle surface coating should allow further functionalization and provide stability against agglomeration and oxidation [163]. Both, magnetite (Fe_3O_4) and maghemite (γ-Fe_2O_3) are considered to be among the most suitable materials for a variety of biomedical applications, such as drug delivery, magnetic hyperthermia, and MRI contrast agents [164]. Since both Fe(III) and Fe(II) ions are present in the human body in significant concentrations, their presence in iron oxide nanomaterials is unlikely to cause any serious toxicological issues. Yet, in parallel with the development of iron-oxide based systems, various research groups have continued to seek other materials. For example, whilst nanocrystals of Fe, Co, Ni FeAu, FePt, $SmCo_5$, $CoFe_2O_4$, $NiFe_2O_4$, and $MnFe_2O_4$ all have a higher saturation magnetization compared to iron oxides, their biomedical uses have been limited by potential hazards associated with leaching of the metal ions. As Co^{2+}, Ni^{2+}, and Mn^{2+} are cytotoxic, the use of corresponding materials for *in vivo* applications would require a nontoxic and protective coating, such as a noble metal shell [165, 166].

Typically, as-synthesized magnetic nanoparticles are hydrophobic due to the layer of hydrocarbon organic molecules that is attached to their surface, preventing their aggregation and protecting their surfaces against oxidation. For bioapplications, nanoparticles must be transferred from a nonpolar environment into an aqueous medium, an effect which can be achieved by a ligand exchange based on replacement of the native hydrophobic surfactants with hydrophilic counterparts. New ligands consist of a strong binding group that attaches to the nanoparticle surface, and a hydrophilic region with a low surface affinity (Figure 3.147). Frequently used anchor groups for binding to the surfaces of metal oxide nanoparticles include dopamine [167–169], carboxylic acids [170, 171], phosphine oxides [172, 173], and phosphates [174]. The major challenge during the phase transfer from organic solvent to the aqueous medium is to achieve a complete exchange of the original surface ligands, since an incomplete surface exchange may lead to an uncontrolled aggregation of the nanoparticles. The encapsulation of hydrophobic nanoparticles into an amphiphilic polymer shell represents another strategy to create water-soluble nanoparticles (see Figure 3.147). In this case, the nonpolar part of the amphiphilic polymer interacts with hydrophobic surfactants, while the hydrophilic part aligns to water molecules, providing solubility in an aqueous environment. The nanoparticles may also be trapped inside biocompatible micelles (Figure 3.147).

For many applications, the surface ligands must possess reactive groups available for chemical reactions with biomolecules (e.g., proteins, antibodies). The current

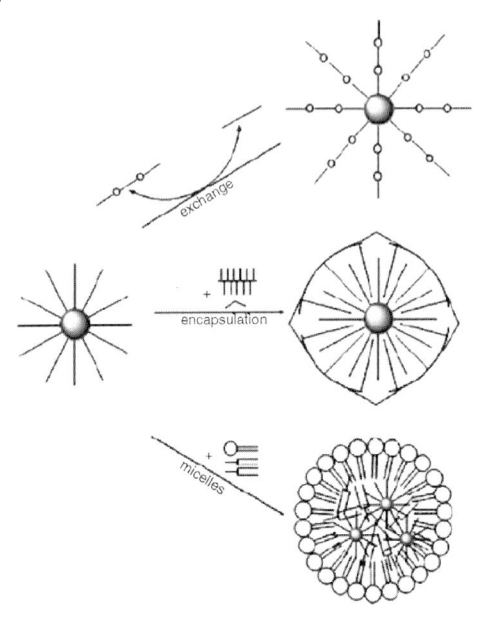

Figure 3.147 Typical strategies used for surface modification and hydrophilization of hydrophobic nanocrystals. Reproduced with permission from Ref. [150]; © 2007, American Chemical Society.

most popular bioconjugation techniques are summarized in Table 3.3. The coating of magnetic nanoparticles with biocompatible natural polymers, such as carbohydrates or proteins, is common practice [175–181], and dextran-coated magnetic nanoparticles are available commercially today. In this case, the hydroxyl groups on the carbohydrate skeletons simplify their further functionalization [182]. The main limiting factors for most natural polymer coatings are their low mechanical strength, porosity, nonselective adsorption, and water solubility, although the latter problem can be overcome by crosslinking with neighboring molecules. Although synthetic polymers such as poly(ethyleneglycol) (PEG), poly(vinyl alcohol) (PVA) and poly-L-lactic acid (PLA) have demonstrated a better mechanical strength, their porosity at the molecular level would not prevent corrosion of the magnetic core and the leaching of potentially harmful metal ions. PEG can provide high stability under physiological conditions, together with a resistance against protein-induced aggregation. PEG is also capable of suppressing unspecific phagocytosis [183], which is crucial for targeted imaging. *Silica coatings* offer a high mechanical strength and a high stability in aqueous media, due to electrostatic stabilization at pH ≥ 2. *Silanol surface groups* ($-Si-OH$) can be used for various surface modifications, using linker groups such as NH_2, NHR, COOH, CHO, and SH, that can be further attached to the desired biological or therapeutic molecules [184, 185]. *Noble metal coatings* may also offer a good alternative; the main advantages of a gold coating are its high biocompatibility, ease of surface functionalization by thiol linker groups [186], and high stability [187].

In order to enhance the magnetic moments beyond that of the individual nanoparticles, it is possible to use agglomerates of superparamagnetic particles.

Table 3.3 Properties of natural and synthetic polymers for coating magnetic nanoparticles [189].

Polymer	Surface hydrophobicity	Application areas	Reference(s)
Natural polymers			
Carbohydrates:			
Dextran	Hydrophilic	Drug delivery	[190]
		Radioimmunoassay	[191]
		MRI	[192]
		Hyperthermia	[193]
Starch	Hydrophilic	Tumor targeting, MRI, X-ray imaging	[194]
Proteins:			
Albumin	Hydrophilic	MRI	[195]
RGD	Hydrophilic	Fluorescent imaging and MRI	[196]
Lipids	Hydrophobic	Immunoassay	[197]
Synthetic polymers			
Poly(ethyleneglycol) (PEG)	Hydrophilic	MRI	[198, 199]
		Drug delivery	[200]
Polyvinyl alcohol (PVA)	Hydrophilic	Drug delivery	[201, 202]
Silica	Hydrophilic	Fluorescent imaging MRI and drug delivery	[203]
Gold	Hydrophobic (Surface modification can render hydrophilicity)	Hyperthermia	[204]
		MRI	[205, 206]

Very often, magnetic nanoparticles used in *in vitro* and *in vivo* (animal model) experiments are represented by large ~100 nm aggregates of superparamagnetic particles (e.g., ~5 nm magnetite), covered with starch polymers. In practical terms, these may be more strongly attracted by an applied external magnetic field, and also better withstand the flow dynamics within the circulatory system (veins, arteries) [188]. The aggregation of superparamagnetic particles also greatly influences the r_2 relaxivity of the surrounding water protons, which is an important parameter for the design of MRI contrast agents. Unfortunately, however, the use of substantially large aggregates may cause thromboembolism.

3.3.2.7.2 **Drug Delivery** The concept of using of magnetic microparticles and nanoparticles for drug delivery was proposed during the late 1970s [207–209]. Unfortunately, many currently used chemotherapeutic agents are not only highly toxic but also nonspecific, which means that they will attack healthy cells on an equal basis. Moreover, as chemotherapy is very often a long-term procedure, it is associated with multiple adverse side effects, including nephrotoxicity, hepatitis, or gastrointestinal ulceration. In the ideal situation, the chemotherapeutic agent would be delivered precisely to the target area, a situation made possible by the use of a composite of magnetic nanoparticles and drugs. By focusing a high magnetic field and field gradient over the target site, the therapeutic agent can be directed to the tumor area, where it is captured and concentrated for effective therapy.

In the presence of a magnetic field gradient, a translational force will be exerted on the particle–drug complex, effectively trapping it in the field at the target site and pulling it towards the magnet [210, 211]. This magnetic force (F_{mag}) can be described as:

$$F_{mag} = (\chi_2 - \chi_1) V \frac{1}{\mu_0} H(\nabla H) \qquad (3.12)$$

where H is the magnetic field strength, ∇H is the field gradient (which can be reduced to ∂H–∂x, ∂H–∂y, ∂H–∂z), χ_2 is the magnetic susceptibility of the magnetic particle, χ_1 is the magnetic susceptibility of the medium ($\chi_1 \ll \chi_2$), μ_0 is the vacuum permeability, and V is the particle volume.

As it follows on from Eq. (3.12), the magnetic properties and particle volume, magnetic field strength, and magnetic field gradient determine the effectiveness of magnetic capture of the nanoparticles. Simple theoretical studies and preliminary experimental results have indicated that an efficient capturing of magnetic particles at the target site can be achieved at fields of ~200–700 mT, with gradients along the z-axis of ~8 Tm^{-1} for femoral arteries, and greater than 100 Tm^{-1} for carotid arteries [212, 213]. In most cases, the magnetic field gradient is generated by a strong permanent magnet, such as Nd–Fe–B, which is fixed outside the body over the target site.

Magnetic drug delivery agents are required to have the following building units: (i) a magnetic core; (ii) surface molecules; and (iii) therapeutic molecules (drugs, DNA, etc.), as shown in Figure 3.148. Once attached, the particle–therapeutic agent complex is injected into the bloodstream, generally using a catheter to position the injection site close to the target. This approach is very efficient for targets close to the body surface, as the magnetic field strength falls off rapidly with distance, and deeper tissues are more difficult to target. To overcome this problem, it was suggested that the magnetic complexes should be implanted close to the target site, within the body [214, 215]. When the drug–carrier complex has been concentrated at the target, the drug can be released either via enzymatic activity or via changes in physiological conditions, such as pH, osmosis, or temperature, and taken up by the tumor cells [194]. The drug molecules should be linked to the nanoparticle in a way that prevents their instant release during delivery, but which provides control over the release mechanism inside the targeted cell or tissue. Very often, the attachment of the drug molecules is based on electrostatic interactions; in this case, drug release can be effected by changing the ionic surroundings, by altering the pH of the media or by raising the temperature. In some other cases, catalytic or redox reactions have been involved in the release process.

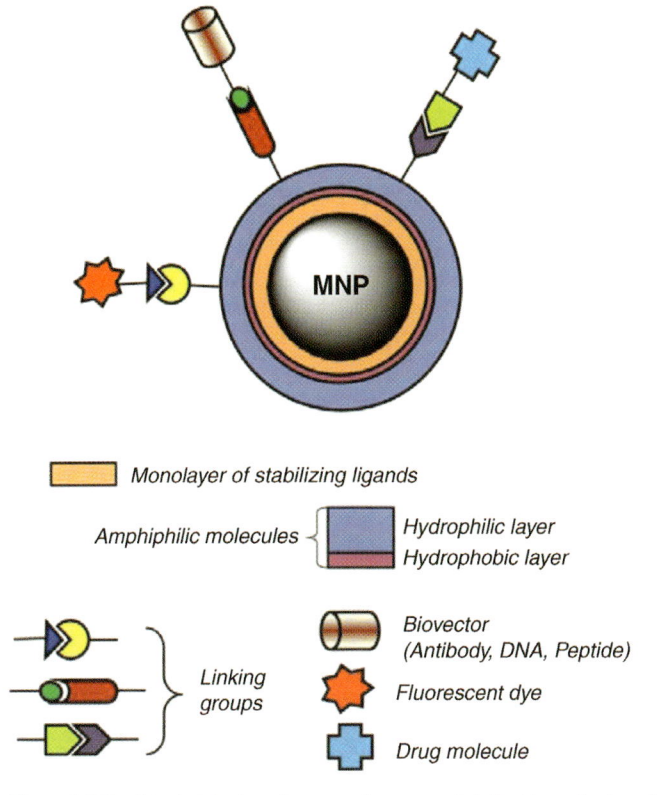

Figure 3.148 A typical design of a magnetic nanoparticle for biomedical applications. Reproduced with permission from Ref. [189]; © 2008, Dove Medical Press.

The design of the magnet–drug composites may differ, however. For example, therapeutic agents can be attached to the surface of magnetic core or encapsulated with magnetic microparticles or nanoparticles inside the porous shell. Cytotoxic drugs used for chemotherapy, and therapeutic DNA used for the correction of a genetic defect, are often attached to the polymer or metal coating of a magnetic core.

Kohler *et al.* [198, 216] proposed a simple strategy for magnetic drug carriers, based on magnetite nanoparticles modified with the cytotoxic anticancer agent, methotrexate (MTX). Although MTX is an analog of folic acid (vitamin B$_9$), unlike folic acid it binds irreversibly to the enzyme called dihydrofolate reductase (DHFR), "switching off" the enzyme in the folic acid cycle. The conjugation of MTX with the nanoparticle surface was achieved through an amide bond by covalent binding between the self-assembled PEG monolayer and the glutamic acid residue of the MTX molecule (Figure 3.149). Such complexes are stable under intravenous conditions. The nanoparticles are taken up via folic acid receptor-mediated endocytosis (Figure 3.150) and transported to the early endosomes; the latter fuse with low-pH lysosomes that contain proteases and which, during normal cellular metabolism, are responsible for

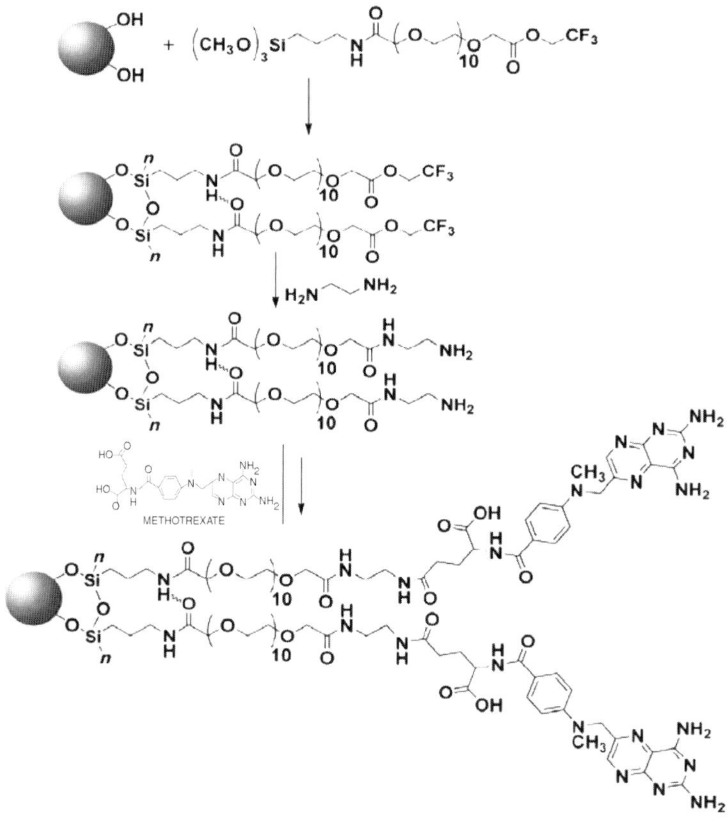

Figure 3.149 A strategy for the immobilization of methotrexate (anticancer agent) at the surface of magnetic nanoparticles. Reproduced with permission from Ref. [198]; © 2006, Wiley-VCH Verlag GmbH & Co. KGaA, Weinheim.

the acidic breakdown of proteins and other exogenous materials brought into the cell. The proteases cleave the peptide bond between the MTX and the nanoparticle, such that the MTX is released from the particle surface inside the target cell. Once free from the nanoparticle surface, the MTX can enter the cytosol where it is free to inhibit DHFR, stopping the folic acid cycle and reducing cell viability. In addition, a prolonged particle retention may allow the MRI imaging of tumor cells exposed to the nanoparticle–PEG–drug conjugate over an extended therapeutic course [198].

Another example of drug delivery using magnetotransport was in the *in vivo* treatment of the VX-2 squamous cell carcinoma which had been implanted into rabbits. Following the administration of mitoxantrone (MX)–magnetite conjugates and the application of a ~1.7 T magnetic field for 60 min, the drug was shown to have been concentrated in the area of the tumor [217, 218]. Importantly, within the first 10 min – when concentration of the drug delivery agents occurs – there had been no release of MX according to UV spectroscopy data for *in vitro* experiment at the physiological conditions. A subsequent histological examination revealed a uniform

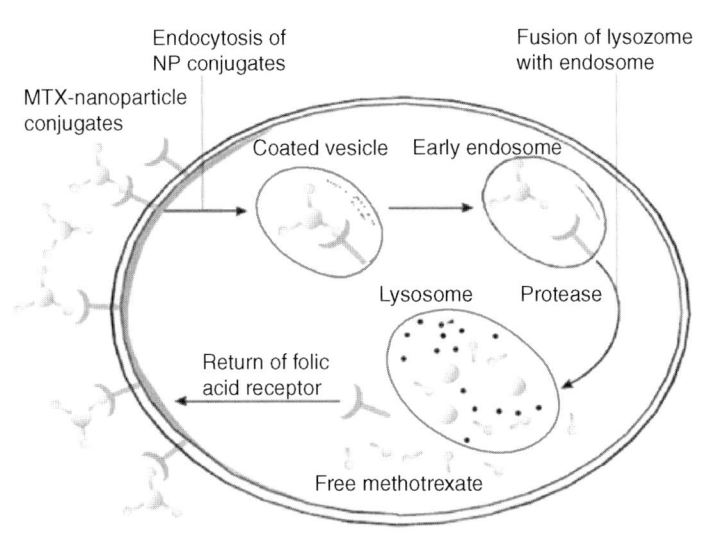

Figure 3.150 Schematic representation of the intracellular uptake of methotrexate-modified nanoparticles into the breast cancer cells. Reproduced with permission from Ref. [216]; © 2005, American Chemical Society.

distribution of the conjugates throughout the entire tumor in 1.5 h after the drug injection. After a three-month observation period, however, no viable tumor tissue was histologically evident in the animals, and no metastases were identified in the regional lymph nodes, nor in any other organs. Some residual particles were found in the animals' spleens, but none was evident in the liver, lungs, or brain, nor at the implantation site and surrounding musculature and skin. Neither were any other macroscopic or histological pathological changes found in any of the investigated organs.

Only a limited number of clinical trials have been conducted with drug delivery using magnetic particles [219–222]. Thus, the treatment of 32 patients with particle–doxorubicin hydrochloride complexes showed the tumor to be effectively targeted in 30 cases [221]. In other studies, 64–91% of the tumor volume of hepatocellular carcinoma was affected by the drug doxorubicin [222]. The use of magnetic carriers may also allow a significant reduction in therapeutic doses of the drug. For example, a magnetic composite with a fivefold lower concentration of MX had the same therapeutic effect as when MX was used alone [217].

Notably, the nanoscale dimensions of particles allow them not only to pass through the narrowest blood vessels, but also to penetrate the cell membranes when necessary [223]. More specific targeting can be achieved by the conjugation of drug-carrying magnetic nanocomposite with active biomolecules showing specificity towards certain molecules on the cell membrane. Each type of cell has its own set of highly selective receptors embedded into the phospholipid bilayer of the cell membrane. Antibodies can bind specifically to such receptors, and hence can be used for the biovectorization of nanoparticles. Avidin and its derivatives (neutravidin, streptavidin) are common examples of proteins that allow the bioconjugation of nanoparticles that can be recognized by biotin. Likewise, oligosaccharides have a high affinity towards lectins, which are highly specific sugar-binding proteins. Among other examples of biomolecules used for the biofunctionalization of nanoparticles, nucleic acids and folic acids are included nucleic acids and folic acid [164]. Folic acid itself has often been used as a biovector [203], because the α-folate receptor is known to be upregulated in various types of human cancer [224, 225].

Many parameters must be taken into account when conducting clinical studies with these materials, including the magnetic properties and size of the carrier particles, the field strength and geometry, the drug/gene binding capacity, the depth to target, the rate of blood flow, vascular supply, and body weight [226]. Serious difficulties have been encountered when scaling-up from small animal trials with near-surface targets to large animals and humans; hence, the development of a detailed theory is required to optimize the design of drug-delivery agents [210].

Figure 3.151 shows that magnetic field intensity falls drastically with distance from the actual source, which makes magnetic capture difficult when tumors are located in deep tissues or organs. As permanent magnetic fields up to 9.4 T have been shown to have no significant adverse effects on the human body [227], further progress in the delivery of magnetic fields to deeper tissues will undoubtedly promote the spread of this technique in medicine. Progress is also expected in the design of biocompatible magnetic particles with high magnetic moments, which will lead to the use of smaller composites and a lower risk of adverse effects, such as thromboembolism.

3.3.2.7.3 **Gene Delivery** The role of nucleic acids as encoders of cellular processes has led to the idea that any process within living cells can be purposefully influenced by the introduction of nucleic acids into living cells from outside. Introducing a foreign DNA into host cells (gene delivery) is one of the steps necessary for gene

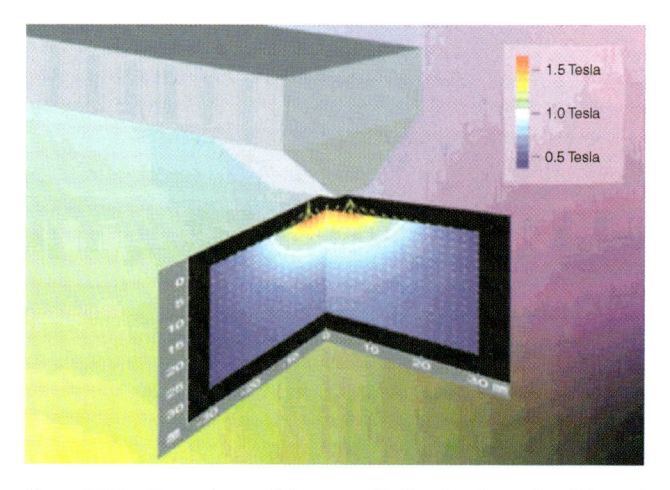

Figure 3.151 Dependence of the magnetic flux density on the distance to pole shoe with the electromagnet. Reproduced with permission from Ref. [218]; © 2000, American Association for Cancer Research.

therapy and genetic modification. The objectives of nucleic acid delivery in research and therapy are: (i) the overexpression of a particular gene; (ii) the silencing or knockdown of a selected gene; and (iii) the actual correction of genetic defects by DNA repair, mediated by transfected nucleic acid molecules. DNA can be delivered into the host cells from outside by both viral and nonviral methods. The viral approach to nucleic acid delivery into host cells, as devised by mother Nature herself, is based on the ability of a virus to inject its DNA inside host cells [228]. Among nonviral physical methods of gene delivery, one promising approach is based on the same principle as magnetic drug delivery [229]. Thus, a targeted delivery of DNA was achieved using magnetic particles coated with adenoassociated virus (AAV) encoding green fluorescent protein (GFP) adhered to the surface of magnetic particles by a heparin sulfate linker [229]. Unfortunately, however, the response of the immune system and removal of vectors from the target tissue by the bloodstream limits viral gene delivery. Scherer *et al.* [230] reported an alternative method, *magnetofection*, which utilizes gene-magnetic particle complexes. In this case, it was found that electrostatic interactions of the negatively charged phosphate backbone of DNA and of a positively charged nanoparticle coating, poly(ethyleneimine) (PEI), led to the formation of stable conjugate that allowed a rapid *in vitro* DNA delivery. Since such DNA–particle complexes are held as an entity by electrostatic interactions, the release of DNA can be efficiently achieved in low-pH lysosomes, whereas in the case of viral delivery the DNA release requires the outer viral shell to be ruptured [231]. A similar approach was used successfully to deliver antisense oligonucleotides into human umbilical vein endothelial cells (HUVECs) [232], and small interfering RNA (siRNA) to downregulate gene expression in HeLa cells [233]. Notably, the overall transfection efficiency could be improved by using oscillating magnetic fields [234]. Carbon nanotubes (CNTs) have also been used as a carrier for both magnetic nanoparticles (inside the CNT) and DNA (on a surface). The application of an external magnetic field will orient, in very strong fashion, the magnetic moments of nanoparticles inside the CNTs along the applied field, such that the CNTs "spear" the cell membrane and promote delivery of the target DNA [235].

3.3.2.7.4 **Magnetic Separation** Magnetic separation is used to extract a magnetically susceptible material from a mixture by applying external magnetic fields. The development of these approaches began during the late 1970s [236, 237], at the same time as magnetic drug delivery techniques. Originally, the use of magnetic separation was limited to a few iron-containing cells (e.g., erythrocytes) [238], but more recent progress in the conjugation of magnetic particles with biomolecules has made possible the extraction of proteins, antibodies and nucleic acids [239–243], as well as the capture of specific bacteria [244] and viruses [245]. In the case of magnetic separations, both individual superparamagnetic nanoparticles and larger magnetic composites (e.g., microspheres represented by sub-micron-sized magnetic particles incorporated in a polymeric binder) can be used [246].

Magnetic separation is a two-step process that includes: (i) labeling of the desired biological species with magnetic nanoparticles; and (ii) their separation by passing the fluid mixture through regions with a magnetic field gradient. The high mobility of magnetic nanoparticles allows a shorter reaction time and a greater volume of reagent to be used, when compared to a standard immunoassay where the antibody is bound to a plate. The modification of magnetic particles with fluorescent dye or enzyme [247, 248] allows the simultaneous performance of magnetic separation, sample concentration and cell detection using sensitive optical readout techniques. A higher throughput of solution in flow design separators [249–251] improves the efficiency and rate of separation of magnetically labeled biospecies as compared to the application of permanent magnetic fields to a static solution [252].

Magnetic separation has been used as a highly sensitive technique for the separation of lung cancer cells [253], urological cancer cells [254], lymphoblastic leukemia cells [255] and neuroblastoma cells in the bone marrow [256], and also for malarial parasites in blood samples [257, 258], by enhancing the selectivity of their detection. The selectivity of these methods is provided by modifying the nanoparticles' surfaces with specific antibodies. Recently, magnetic separation was used to "catch" metastatic rare tumor cells in the bloodstream of cancer patients [259], thus improving (at least, potentially) their long-term survival.

Although magnetic separation is used extensively in the laboratory for diagnostic purposes, its use *in vivo* faces the same problems as encountered by magnetic drug delivery, namely the shallow penetration of the magnetic field. Nonetheless, magnetic separation is currently considered to show much promise in the treatment of soft forms of cancer (e.g., leukemia) that are the most difficult to cure [260].

3.3.2.7.5 Magnetic Hyperthermia This method is based on the generation of heat by a magnetic material when subjected to an alternating current (AC)-generated magnetic field. The main concept of hyperthermia, which is to effect the local heating of malignant cells, but not of healthy cells [261–263], is achieved by the local delivery of an AC magnetic field into the tumor tissues. Although, when heated above 50 °C, the tumor cells are killed by necrosis, even a short-term (minutes) treatment can be associated with critical adverse side effects, such as "shock syndrome" due to a sudden release of large amounts of necrotic tumor material and a major inflammatory response [264]. At the same time, raising the temperature to 42–45 °C will significantly improve the efficacy of many anti-cancer drugs. Magnetic materials were first used for hyperthermia in 1957 by Gilchrist *et al.* [265], when it was shown that various tissue samples could be heated by using 20–100 nm-sized particles of γ-Fe_2O_3 via the application of a 1.2 MHz magnetic field. Since then, a number of reports have described a variety of schemes with different types of magnetic material, and different field strengths and frequencies [164, 266–271]. Magnetic hyperthermia has been tested on human patients in Germany [272], whilst research using animal models is currently ongoing in many research groups worldwide [273].

The delivery of an adequate amount of magnetic particles necessary to generate enough heat in the target under clinically acceptable conditions, represents the main challenge for this technique. As the heating of healthy tissues is undesirable, it is important to take into account and to minimize heat transfer from the tumor to its surrounding tissues [274, 275]. The optimal heat deposition rate has been estimated at approximately 100 mW cm^{-3}, while the frequency and strength of the external AC magnetic field is limited by deleterious physiological responses to high-frequency magnetic fields [276, 277], such as an undesirable stimulation of the peripheral and skeletal muscles, cardiac stimulation and arrhythmia, as well as the nonspecific inductive heating of tissues. The ranges of frequency f and amplitude H considered to be safe for humans are $f = 0.05$–1.2 MHz and $H = 0$–15 k A m^{-1} (up to ~ 18.85 T). Experimentally, it has been shown that AC fields with $H \times f < 4.85 \times 10^8$ A m^{-1} s^{-1} should be considered safe. The heating efficiency depends heavily on the magnetic properties of the particles, but both single- or multi-domain ferromagnetic particles and superparamagnetic particles can be used for this purpose. The coupling of antibodies to magnetic nanoparticles facilitates targeting of the cancer cells.

Magnetic particles generate heat under AC magnetic fields due to the magnetic losses that arise from the various processes of magnetization reversal: (i) hysteresis; (ii) Neel or Brown relaxation; and (iii) frictional losses in viscous suspensions [264]. Hysteresis losses depend heavily on the field amplitude H, and can be estimated as $A \times f$, where A is the integrated area of the hysteresis loop and f is the frequency of the magnetic field. Hysteresis losses of particles with sizes above the single domain (i.e., multidomain particles) depend, in a complicated manner, on the type and configuration of domain wall pinning, as determined by the particle structure. In the case of ferromagnetic particles, heat is dissipated only when the external magnetic field strength is larger than the coercive force for a given frequency. It should be noted that the coercivity of ferromagnetic particles increases with frequency and, as a result, the external field required to generate heat through the hysteresis losses will become larger [264] and may even exceed the safety limit. Thus, magnetic particles with high-saturation magnetization and a relatively low coercivity are preferred for hyperthermic applications.

With decreasing particle size, the energy barrier for magnetization reversal is also decreased, which means that the magnetic moments of superparamagnetic particles can easily switch direction under the influence of thermal energy, such that zero coercivity will be observed at room temperature. In the case of superparamagnetic particles, the Neel and Brown relaxation processes provide the major contribution to the heat dissipation: the magnetic moment of a single-domain particle can relax through Brownian rotation of the particle itself and through Neel rotation of the magnetic moment within the particle. The characteristic times for Neel and Brownian relaxation were described in Section 3.3.2.5. Typically, the resonant frequency of Brownian relaxation is below 100 KHz, whereas that based on Neel relaxation is in the GHz range [278]. In real nanoparticle systems, the relative contributions of Neel and Brownian relaxations depend on the magnetic properties and nanoparticle size.

The energy absorbed by the magnetic particles under an applied external AC magnetic field is described by the specific absorption rate [SAR; the term specific loss power (SLP) may also be used in some cases]. The SAR can be determined as:

$$\text{SAR} = c(\Delta T/\Delta t), \tag{3.13}$$

where c is the sample specific heat capacity and $(\Delta T/\Delta t)$ is the initial slope of temperature increase versus the heating time [279]. SAR is measured in units of

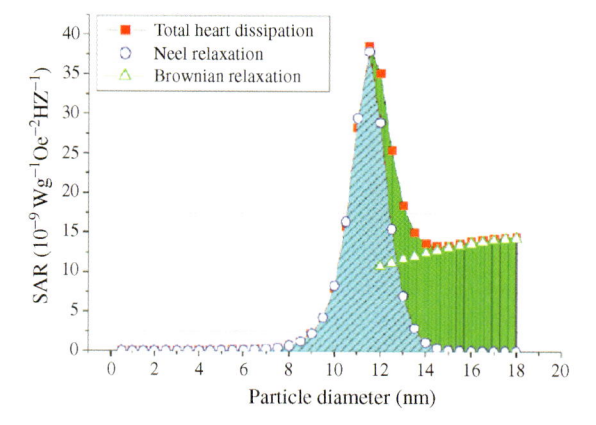

Figure 3.152 Theoretical estimation of the heating rates as a function of particle diameter. The calculations used surfactant layer thickness of 3.2 nm, the anisotropy constant 30 kJ m^{-3}, temperature 300 K, viscosity 0.00089 kg m^{-1} s^{-1}, applied AC magnetic field 40 Oe, and frequency 600 KHz. Reproduced with permission from Ref. [286]; © 2009, Elsevier.

W g^{-1} [280–282]. the data in Figure 3.152 show that the relative contribution of Neel and Brownian relaxations to SAR depends on the size of the particle, and that the heat dissipated through Neel relaxation is not influenced by the viscosity of the medium, as compared to Brownian relaxation. Taking into account the fact that the viscosity of the *in vivo* medium is rather high, and that particles can be immobilized on the cell membranes due to interactions with hydrophilic protective coatings, the Brownian rotation is often suppressed. Thus, it is essential to have particles that have high magnetic moments and are capable of relaxing mostly through the Neel rotation.

When the applied magnetic field is smaller than the saturation magnetization, within the validity of linear response theory (i.e., magnetization depends linearly on the applied magnetic field), the SAR can be determined as the loss power density P(f,H) normalized by the mass density of the particles (Wm^{-3}), expressed as:

$$P(f, H) = \mu_0 \pi \chi''(f) H^2 \tag{3.14}$$

where μ_0 is the permeability of free space, χ'' is the imaginary part of magnetic susceptibility indicating dissipative processes in the sample, and H and f are amplitude and frequency of the applied magnetic field, respectively.

Frictional losses in viscous suspensions generate heat due to viscous friction between rotating particles and the surrounding medium. This loss type is not restricted to superparamagnetic particles [264]. The increase of SAR by about an order of magnitude has been demonstrated by the example of ∼100 nm magnetite particles modified with gelatin and suspended in water. Such a significant enhancement of the SAR was the result of gelatin melting above 30 °C, and has been attributed to the effect of viscous losses [283]. The highest SAR value reported for chemically synthesized iron oxide nanoparticles materials was 600 W g^{-1} at 400 KHz and 11 kA m^{-1} [283], although iron oxide nanoparticles synthesized by bacteria can provide a higher SAR value of ~960 W g^{-1} at 410 KHz and 10 kA m^{-1} [283]. The particle size distribution substantially reduces the SAR, and should be avoided.

The maximum SAR calculated for pure Fe- and Co-based systems can be significantly higher compared to iron oxide nanoparticles [204, 284, 285]. However, the main concern for use for *in vivo* applications is their potential toxicity associated with the release of cytotoxic Co^{2+} ions.

3.3.2.7.6 Magnetic Resonance Imaging

The technique of magnetic resonance imaging (MRI) allows the visualization of soft tissues and metabolic processes in a human body [287]. Basically, the MRI signal originates from the nuclear magnetic resonance (NMR) of protons and their relaxation in an external magnetic field [288]. The interaction of the protons with an external magnetic field, B_0, leads to a precession of proton spins about the field direction (defined as z-direction). The precession angular frequency ω_0 is given by the Larmor equation:

$$\omega_0 = \gamma \cdot B_0, \tag{3.15}$$

where γ is the proton gyromagnetic ratio ($\gamma = 42.6$ MHz T^{-1}). Strictly speaking, B defines magnetic induction, and is related to the field strength through the induction constant μ_0 (as discussed in Section 3.3.2.5). In the following, the units and terminology commonly accepted in the MRI-associated literature will be used.

The protons can align either parallel or antiparallel to the magnetic field B_0. Although the energy difference between states with different spin alignments is quite small, there is an excess of proton spins aligned parallel to the field, and these cause a net magnetization along B_0 [287]:

$$M_0 = \frac{\varrho_0 \gamma^2 \hbar^2}{4 k_B T} B_0, \tag{3.16}$$

where M_0 is the net magnetization (parallel to B_0) and ϱ_0 is the proton spin density. Both, NMR and MRI are based on the ability to manipulate and detect this net magnetization caused by the bulk precession of proton spins in a sample or a tissue, respectively.

When a radiofrequency (RF) pulse of a certain duration is applied to the spin system, the protons can absorb energy and, as a consequence, will undergo a transition from the parallel to the non-parallel alignment. Since the energy associated with the Larmor frequency, ω_0, is very small, the transition runs into saturation – that is, both states are equally populated and, as a result, the net magnetization along B_0 vanishes (Figure 3.153). The applied RF-field is circularly polarized in the transverse plane; this means that its vector rotates with the Larmor frequency about the z-axis. Immediately after the RF pulse, the spin system starts to relax. In order to understand the NMR experiment, it is important to distinguish between two types of relaxation of the proton spin. The further discussions will deal with one component that describes changes of the magnetization along the longitudinal z-direction (M_z), and the other part which describes the corresponding transversal xy component (M_{xy}). The first relaxation mechanism, which is referred to as "longitudinal" or "spin lattice"

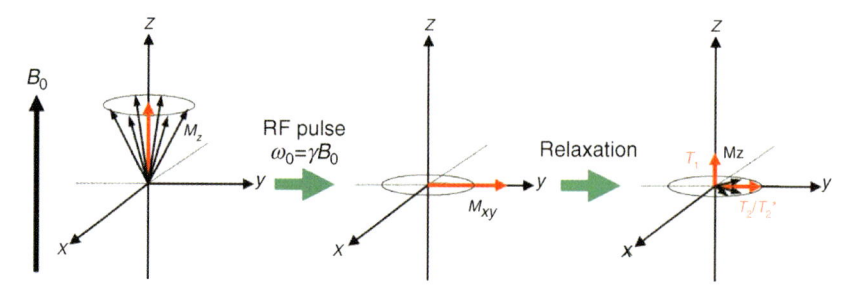

Figure 3.153 In a typical NMR experiment, when an ensemble of protons is placed in an external magnetic field B_0, a net magnetization along B_0 is caused. After applying a 90° RF pulse, the net magnetization is flipped to the xy plane. The longitudinal relaxation process (T_1) describes the re-growth of the magnetization along B_0 to its initial value, whereas the transverse relaxation process (T_2 or T_2^*) is characterized by the phase loss of protons in the xy plane (see text for details).

relaxation, describes the return of the longitudinal magnetization from $M_z = 0$ to its initial equilibrium state ($M_z = M_0$). This process is associated with the relaxation time T_1, and arises from a continuous growth of the spin excess corresponding to the parallel alignment. Therefore, the z-component of the magnetization is given by:

$$M_z(t) = M_0(1 - e^{-t/T_1}) \qquad (3.17)$$

where T_1 is the spin lattice relaxation time.

In order to discuss the second relaxation mechanism, first it is necessary to review several aspects of the NMR experiment. So far, only the z-component of the net magnetization has been mentioned and, before applying the RF pulse, the z-component is the only component as the spins are rotating randomly about the field direction – that is, their vector sum is zero and therefore the transverse magnetization is also zero. In a typical NMR experiment, however, the RF pulse is a so-called 90° pulse, because it tips the net magnetization away from the longitudinal z-direction to the perpendicular transverse xy plane. The 90° pulse causes all the spins to rotate about the z-axis with the Larmor frequency ω_0, and all spins are in-phase with their vector components having the same x and y components (Figure 3.153). In order to detect the as-generated transverse magnetization, M_{xy}, a detector coil is placed along the y direction. Due to the precession of the net magnetization about the z-axis, a detectable periodic signal is induced in the coil, such that the second relaxation mechanism comes into play.

The initial phase coherence of the spins in the xy plane gradually fades with time, a process which is reflected in the time-dependent disappearance of the transverse magnetization M_{xy}; that is, a time-dependent signal decay in the detection coil. The spin dephasing is caused by two main processes. First, fluctuating inhomogeneities in the microscopic environment of the individual protons induce phase shifts [289]. The Larmor frequency of the spins remains unaltered at the same time. This phase loss is characterized by the transverse relaxation time T_2. Second, static local inhomogeneities in the external field B_0, as well as in the tissue of the body, cause additional so-called "susceptibility effects" that cause clusters of spins to experience different Larmor frequencies. These additional inhomogeneities result in an increased decay of the transverse magnetization, and are often summarized in a separate time constant T_2'. As a consequence, the observable signal decay in the detection coil is a superposition of both processes, and is referred to as free induction decay (FID). It is given by:

$$M_{xy}(t) = M_0 e^{-t/T_2^*}, \tag{3.18}$$

where T_2^* is the effective transverse relaxation time. T_2^* can thus be expressed as:

$$\frac{1}{T_2^*} = \frac{1}{T_2} + \frac{1}{T_2'}. \tag{3.19}$$

The simple NMR experiment described so far is not suitable for MRI purposes. Although MRI relies on the NMR effect, it has some important differences that allow an image of the NMR signals to be obtained over large objects, such as a human body. In other words, the signals provide spatial information, and for this purpose the use of magnetic gradient fields is necessary, although unfortunately these additional magnetic gradients greatly shorten the transverse relaxation process. To compensate for this additional shortening, two main pulse sequences have been developed for MRI. The gradient echo sequences allow the measurement of T_2^* – that is, the MR signal mainly depends on T_2^* – while the spin echo sequences compensate the additional static inhomogeneities which are summarized in T_2'. As a result, the MRI signal is mainly dependent on T_2 in this case. In other words, the use of spin-echo sequences leads to a complete recovery of the T_2' process, and the spins are refocused with respect to their Larmor frequency.

3.3.2.7.7 Tomographic Imaging
The key step to obtaining a 3-D image of the NMR signals is to ensure that each region of spins experiences a unique magnetic field. In other words, it is necessary to correlate a series of signal measurements with the spatial locations. To accomplish this, a spatially changing magnetic field across the sample superimposes the static field B_0; that is, a magnetic field gradient in direction z is applied:

$$\omega(z) = \gamma B(z) = \gamma \cdot (B_0 + G_z z). \tag{3.20}$$

Here, the field gradient G_z is produced by an additional coil that changes the magnetic field linearly. For three spatial directions there are three gradient fields required. The first spatial encoding step is the slice selection in the z-direction, as shown in Figure 3.154. During the first 90° excitation pulse, which flips the net magnetization down to the transverse plane, a linear slice selection gradient G_z is applied. Therefore, the Larmor frequency also changes linearly with respect to the proton position in the z-direction. The 90° pulse contains a certain bandwidth of Larmor frequencies $\Delta\omega$ that correlate with a particular slice thickness Δz. All other

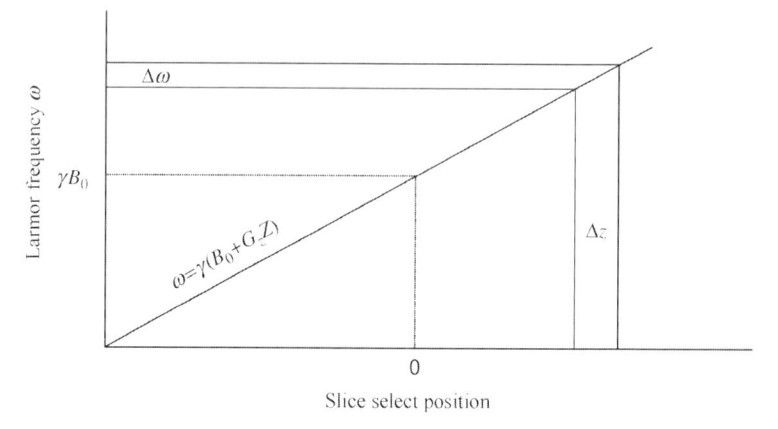

Figure 3.154 To obtain spatial information, a linear gradient is applied along the slice select axis. The spectral bandwidth of the RF pulse is adjusted in a way that the protons within the slice of thickness Δz are uniformly excited. Reproduced with permission from Ref. [287]; © 1999, John Wiley & Sons, New York.

Larmor frequencies are not addressed by the 90° pulse, such that only protons in the slice Δz are excited.

For encoding of the y-direction, a phase-encoding gradient is applied for a certain duration Δt. The protons in the y-direction thus experience an additional field of G_y. After the time interval Δt, the gradient is turned off and the protons exhibit their initial Larmor frequencies again. Hence, the field causes a linear phase shift $\Delta \varphi$ with respect to the y-position. The process of phase encoding is thus comparable to the pure T_2 process, caused by fluctuating inhomogeneities in the microscopic environment of the individual protons. The x-direction is encoded using a frequency-encoding gradient, which is applied during the data acquisition, such that each region of protons along the x-axis possesses its own Larmor frequency. The obtained signal is then a composition of all Larmor frequencies along the slice which, subsequently, can be related to their image position through the use of Fourier transform. This process is somehow comparable to the additional T_2' dephasing caused by static inhomogeneities.

3.3.2.7.8 The Role of Magnetic Nanoparticle-Based Contrast Agents in MRI

A basic requirement for diagnostic MRI is the ability to distinguish between diseased and normal tissues. Consequently, the two tissues must exhibit different MR signals which depend on a variety of parameters and are closely related to the term "contrast." More precisely, contrast is defined as the MR signal difference between two tissues A and B:

$$C_{AB} = S_A - S_B \tag{3.21}$$

In an image, a region of high signal intensity appears bright, whereas a region of low intensity appears dark. In general, MRI has the ability to manipulate the tissue signal via various contrast mechanisms. The most frequently used mechanisms are based on differences in the spin density ϱ_0, which directly determines the magnitude of M_0 according to Eq. (3.15) and the relaxation times T_1 and T_2 (T_2^*). Although the aim here is to weight the image according to one of these three parameters, in order to generate sufficient contrast the signal should be a function of all parameters – a fact that must be borne in mind when selecting the experimental set-up. Under particular circumstances, if two tissues have very similar properties, they may not be distinguishable with any of the parameters, and in these cases the use of an MR contrast agent might be helpful. The role of a contrast agent is to selectively alter the signal of the tissue that takes up the contrast agent. Here, the aim is to deliver the contrast agent specifically into the tissue of interest (A), where it will alter the NMR signal. In the surrounding tissue (B), the NMR signal will remain unaltered when the same imaging sequences are applied, such that the contrast between A and B might be increased dramatically.

MRI contrast agents consist not only of superparamagnetic nanoparticles, but also of particular types of paramagnetic complexes. In general, MRI contrast agents are divided into two major types:

- *Positive* contrast agents mainly consist of paramagnetic complexes, in particular of Gd^{3+}. They are applied to shorten the longitudinal relaxation time T_1, and cause an increased signal intensity in the region of their accumulation, causing the region to appear bright.
- *Negative* contrast agents on the other hand are used to shorten the transverse relaxation time T_2, the result of which is a signal loss leading to a dark image.

In order to describe, quantitatively, the effectiveness of a contrast agent, the relaxivity coefficients r_i are introduced [290]. The relaxivities describe the shortening of a certain relaxation time per millimole of contrast agent. They are, therefore, concentration-independent values, as the inverse relaxation time ($R_i = 1/T_i$) – that is, the relaxation rate – increases linearly with the contrast agent concentration:

$$\frac{1}{T_i} = \frac{1}{T_{i,0}} + r_i[CA] \equiv R_i = R_{i,0} + r_i[CA] \qquad (3.22)$$

In the further discussion, the term relaxivity will be used solely to describe the properties of the contrast agents, and attention will be focused on the r_2 and r_2^* process. For more than 20 years, T_2 contrast agents have been based on superparamagnetic iron oxide particles (SPIOs) [291], for which a few products are currently in clinical use [292]. These conventional contrast agents can be classified according to their hydrodynamic sizes, which in turn greatly determines their biological properties. SPIOs with hydrodynamic diameters of 60–150 nm are commonly applied to image cells of the reticuloendothelial system (RES) that are located primarily in the liver and spleen; hence, they are mainly used for the diagnosis of liver disease. These conventional contrast agents are synthesized by coprecipitation reactions in aqueous medium; the particles produced consist of multiple iron oxide cores within a dextran coating, which provides water solubility. However, due to their extensive uptake by the cells of the RES, these SPIOs will have a short blood half-life.

Smaller iron oxide particles (20–40 nm hydrodynamic diameter), which usually are referred to as ultrasmall superparamagnetic iron oxides (USPIOs) are selected to image lymph nodes and bone marrow. USPIO-based agents of approximately 20 nm mean diameter have been applied to magnetic resonance angiography (MRA), as they demonstrate much longer half-lives in the blood.

The above-described conventional iron oxide-based contrast agents are rather polydisperse and polycrystalline particles, and have limitations in terms of targeted imaging. For future MRI applications, such as targeted molecular and cellular imaging, a careful engineering of the nanoparticle magnetic properties, and also of the surface ligands, will be required. Consequently, the crystalline core will need to be carefully tuned so as to provide optimal properties for the contrast generation.

In general, the overall relaxivity coefficient r_i is the sum of two contributions which determine their relaxation behavior [290, 293]. The inner-sphere relaxivity, r_i^{IS}, arises from the water molecules that are directly coordinated to the contrast agent, and is dominant in T_1 contrast agents. On the other hand, the outer-sphere relaxivity, r_i^{OS}, describes the relaxation behavior of water molecules diffusing in the surrounding of the contrast agent, and is the most important contribution for T_2 shortening agents. The outer-sphere term for a superparamagnetic particle depends on a number of parameters such as the magnetic moment, the degree of aggregation, and the total hydrodynamic radius. Two main aspects must be considered when designing efficient T_2 contrast agents, namely the composition and dimensions. Although iron oxide is still the most frequently used material, based mainly on its low toxicity, investigations are ongoing to seek potentially better materials. For example, Cheon and coworkers have systematically investigated the magnetic and relaxometric properties of various monodisperse spinel ferrite (MFe_2O_4, M = Ni, Co, Fe, Mn) nanoparticles [173]. In this case, the $MnFe_2O_4$ nanoparticles showed the highest magnetic moment of $5\mu_B$, and thus the highest r_2 relaxivity (see Figure 3.155). The magnetic moments of the other ferrites are lower, and correlate strongly with the transverse relaxivities. The clustering of individual nanoparticles will greatly increase the r_2 relaxation coefficient [294].

The size and shape of nanoparticles must be adjusted for the desired application since, on the nanometer scale, magnetic properties such as saturation magnetization and magnetic anisotropy strongly size- and shape-dependent. In particular, M_S is greatly decreased with decreasing size of a superparamagnetic nanoparticle. As M_S directly determines the r_2 relaxivity, size control for the crystalline core is the key requirement.

Magnetic nanoparticles offer a powerful platform for the design of multipurpose nanocomposites suitable for both diagnostic and treatment purposes. This can be achieved by combining the building blocks necessary, for example, to provide drug delivery and hyperthermia [295], MRI and drug delivery [21], MRI and gene delivery [296], MRI and photothermal therapy [205], and cell separation and imaging [297]. In clinical research with human patients, much more attention is now paid to prior mathematical simulations and their approach to "real" conditions [298]. Indeed, the most challenging tasks for future investigations in this field are the rendering of safety and biocompatibility to nanoparticles of a wider range of magnetic

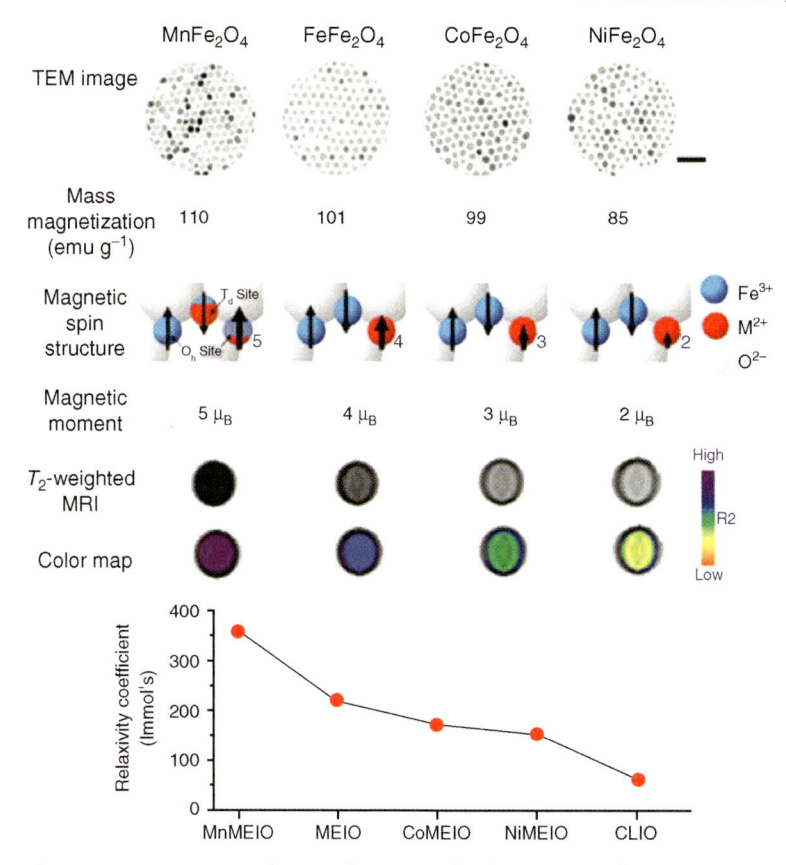

Figure 3.155 Transverse relaxivity of various uniformly sized spinel ferrite nanoparticles with respect to composition. r_2 correlates strongly with the saturation magnetization M_S [129]. Reproduced with permission from Ref. [129]; © 2007, Nature Publishing Group.

materials, and in particular to those with a high magnetic moment. Finally, long-term toxicity tests and possible biodegradation pathways studies should be carried out beyond proof-of-concept studies. The synthesis of tailored block copolymers should allow the formation of micelles with a narrow size distribution and with a controllable amount of iron oxide nanoparticles embedded in the hydrophobic core [299–301]. Not surprisingly, these micelles represent promising candidates as future T_2 MRI contrast agents.

One important aspect in the development of future MRI contrast agents is the aim to label certain regions, for example a certain type of cell, in specific fashion. If diseased regions could be labeled selectively, then the obtained contrast would be optimal and diagnostic reliability improved. With this in mind, research currently aims on the modification of nanoparticles with specific biomolecules. Consequently, whilst MRI can be expected to become increasingly important in the selective diagnosis of early stages of diseases, it will continue to be dependent on the development of innovative and sophisticated contrast agents.

Reference

1 Sorensen, C.M. (2001) *Nanoscale Materials in Chemistry* (ed. K.J. Klabunde), John Wiley & Sons, Inc., p. 162.
2 Cox, A.J., Louderback, J.G., and Bloomfield, L.A. (1993) *Phys. Rev. Lett.*, **71**, 923.
3 Paulus, U.A., Wokaun, A., Scherer, G.G., Schmidt, T.J., Stamenkovic, V., Radmilovic, V., Markovic, N.M., and Ross, P.N. (2002) *J. Phys. Chem. B*, **106**, 4181.
4 Black, C.T., Murray, C.B., Sandstrom, R.L., and Sun, S.H. (2000) *Science*, **290**, 1131.
5 Doty, R.C., Yu, H.B., Shih, C.K., and Korgel, B.A. (2001) *J. Phys. Chem. B*, **105**, 8291.
6 Schmid, G. and Liu, Y.P. (2001) *Nano Lett.*, **1**, 405.
7 Parthasarathy, R., Lin, X.M., and Jaeger, H.M. (2001) *Phys. Rev. Lett.*, **87**, 186807.
8 Cheon, J. and Lee, J.H. (2008) *Acc. Chem. Res.*, **41**, 1630.
9 Sun, S.H. and Murray, C.B. (1999) *J. Appl. Phys.*, **85**, 4325.
10 Zeng, H., Black, C.T., Sandstrom, R.L., Rice, P.M., Murray, C.B., and Sun, S.H. (2006) *Phys. Rev. B*, **73**, 020402.
11 Chien, C.L. (1991) *J. Appl. Phys.*, **69**, 5267.
12 Berkowitz, A.E., Mitchell, J.R., Carey, M.J., Young, A.P., Zhang, S., Spada, F.E., Parker, F.T., Hutten, A., and Thomas, G. (1992) *Phys. Rev. Lett.*, **68**, 3745.
13 Xiao, J.Q., Jiang, J.S., and Chien, C.L. (1992) *Phys. Rev. Lett.*, **68**, 3749.
14 Lipkin, H.J. (1987) *Phys. Rev. Lett.*, **58**, 425.
15 Takahashi, S. and Maekawa, S. (1998) *Phys. Rev. Lett.*, **80**, 1758.
16 Russier, V., Petit, C., Legrand, J., and Pileni, M.P. (2000) *Phys. Rev. B*, **62**, 3910.
17 Vanleeuwen, D.A., Vanruitenbeek, J.M., Dejongh, L.J., Ceriotti, A., Pacchioni, G., Haberlen, O.D., and Rosch, N. (1994) *Phys. Rev. Lett.*, **73**, 1432.
18 Bodker, F., Morup, S., and Linderoth, S. (1994) *Phys. Rev. Lett.*, **72**, 282.
19 Zahn, M. and Shenton, K.E. (1980) *IEEE Trans. Magn.*, **16**, 387.
20 O'Grady, K. and Chantrell, R. (1992) *Magnetic Properties of Fine Particles*, Elsevier, Amsterdam.
21 Aharony, A. (1996) *Introduction to the Theory of Ferromagnetism*, Oxford University Press, New York.
22 Blums, E., Cebers, A., and Maiorv, M.M. (1997) *Magnetic Fluids*, Walter de Gruyter, New York.
23 Rosenwieg, R.E. (1998) *Ferrohydrodynamics*, Dover Publishing, New York.
24 Berkovskii, B.M. and Bashtovoi, V.G. (1996) *Magnetic Fluids and Applications Handbook*, Begell House, New York.
25 Jordan, A., Scholz, R., Wust, P., Fahling, H., and Felix, R. (1999) *J. Magn. Magn. Mater.*, **201**, 413.
26 Josephson, L., Perez, J.M., and Weissleder, R. (2001) *Angew. Chem., Int. Ed.*, **40**, 3204.
27 Murray, C.B., Sun, S.H., Doyle, H., and Betley, T. (2001) *MRS Bull.*, **26**, 985.
28 Collier, C.P., Saykally, R.J., Shiang, J.J., Henrichs, S.E., and Heath, J.R. (1997) *Science*, **277**, 1978.
29 O'Connor, C.J., Tang, J., and Zhang, J. (2001) *Magnetism: Molecules to Materials III* (eds J.S. Miller, and M. Drillon), Wiley-VCH Verlag GmbH, Weinheim, p. 1.
30 Leslie-Pelecky, D.L. and Rieke, R.D. (1996) *Chem. Mater.*, **8**, 1770.
31 Dumm, M., Zolfl, M., Moosbuhler, R., Brockmann, M., Schmidt, T., and Bayreuther, G. (2000) *J. Appl. Phys.*, **87**, 5457.
32 Choi, S.W.K., and Puddephatt, R.J. (1997) *Chem. Mater.*, **9**, 1191.
33 Jeon, I.J., Kim, D., Song, J.S., Her, J.H., Lee, D.R., and Lee, K.B. (2000) *Appl. Phys. A – Mater. Sci. Process*, **70**, 235.
34 Jeon, I.J., Kang, D.W., Kim, D.E., Kim, D.H., Choe, S.B., and Shin, S.C. (2002) *Adv. Mater.*, **14**, 1116.
35 Falicov, L.M., Pierce, D.T., Bader, S.D., Gronsky, R., Hathaway, K.B., Hopster, H.J., Lambeth, D.N., Parkin, S.S.P., Prinz, G., Salamon, M., Schuller, I.K., and Victora, R.H. (1990) *J. Mater. Res.*, **5**, 1299.

36 Cheung, C., Nolan, P., and Erb, U. (1994) *Mater. Lett.*, **20**, 135.
37 Delplancke, J.L., Bouesnard, O., Reisse, J., and Winand, R. (1997) *Mater. Res. Soc. Symp. Proc.*, **451**, 383.
38 Loffler, J.F., Meier, J.P., Doudin, B., Ansermet, J.P., and Wagner, W. (1998) *Phys. Rev. B*, **57**, 2915.
39 Prados, C., Multigner, M., Hernando, A., Sanchez, J.C., Fernandez, A., Conde, C.F., and Conde, A. (1999) *J. Appl. Phys.*, **85**, 6118.
40 Tanori, J., Duxin, N., Petit, C., Lisiecki, I., Veillet, P., and Pileni, M.P. (1995) *Colloid Polymer Sci.*, **273**, 886.
41 Duxin, N., Stephan, O., Petit, C., Bonville, P., Colliex, C., and Pileni, M.P. (1997) *Chem. Mater.*, **9**, 2096.
42 Carpenter, E.E., Seip, C.T., and O'Connor, C.J. (1999) *J. Appl. Phys.*, **85**, 5184.
43 Carpenter, E.E., Sangregorio, C., and O'Connor, C.J. (1999) *IEEE Trans. Magn.*, **35**, 3496.
44 Li, S.C., John, V.T., Rachakonda, S.H., Irvin, G.C., McPherson, G.L., and O'Connor, C.J. (1999) *J. Appl. Phys.*, **85**, 5178.
45 Moumen, N. and Pileni, M.P. (1996) *Chem. Mater.*, **8**, 1128.
46 Tanori, J. and Pileni, M.P. (1997) *Langmuir*, **13**, 639.
47 Petit, C., Taleb, A., and Pileni, M.P. (1998) *Adv. Mater.*, **10**, 259.
48 Chemseddine, A. and Weller, H. (1993) *Ber. Bunsenges. Phys. Chem. Chem. Phys.*, **97**, 636.
49 Murray, C.B., Kagan, C.R., and Bawendi, M.G. (2000) *Annu. Rev. Mater. Sci.*, **30**, 545.
50 Petit, C., Taleb, A., and Pileni, M.P. (1999) *J. Phys. Chem. B*, **103**, 1805.
51 Yang, M.H. and Flynn, C.P. (1989) *Phys. Rev. Lett.*, **62**, 2476.
52 Suslick, K.S., Fang, M.M., and Hyeon, T. (1996) *J. Am. Chem. Soc.*, **118**, 11960.
53 Ramesh, S., Koltypin, Y., Prozorov, R., and Gedanken, A. (1997) *Chem. Mater.*, **9**, 546.
54 Suslick, K.S., Hyeon, T.W., and Fang, M.M. (1996) *Chem. Mater.*, **8**, 2172.
55 Shevchenko, E.V., Talapin, D.V., Schnablegger, H., Kornowski, A., Festin, O., Svedlindh, P., Haase, M., and Weller, H. (2003) *J. Am. Chem. Soc.*, **125**, 9090.
56 Lamer, V.K. and Dinegar, R.H. (1950) *J. Am. Chem. Soc.*, **72**, 4847.
57 Murray, C.B., Norris, D.J., and Bawendi, M.G. (1993) *J. Am. Chem. Soc.*, **115**, 8706.
58 Dinega, D.P. and Bawendi, M.G. (1999) *Angew. Chem., Int. Ed.*, **38**, 1788.
59 Puntes, V.F., Krishnan, K.M., and Alivisatos, A.P. (2001) *Science*, **291**, 2115.
60 Diehl, M.R., Yu, J.Y., Heath, J.R., Held, G.A., Doyle, H., Sun, S.H., and Murray, C.B. (2001) *J. Phys. Chem. B*, **105**, 7913.
61 Sun, S., Murray, C.B., and Doyle, H. (1999) *Mater. Res. Soc. Symp. Proc.*, **577**, 385.
62 Kitakami, O., Sato, H., Shimada, Y., Sato, F., and Tanaka, M. (1997) *Phys. Rev. B*, **56**, 13849.
63 Puntes, V.F., Krishnan, K., and Alivisatos, A.P. (2002) International Symposium on Nanomaterials – Synthesis, Characterization and Catalysis, Honolulu, HI, p. 145.
64 Puntes, V.F., Zanchet, D., Erdonmez, C.K., and Alivisatos, A.P. (2002) *J. Am. Chem. Soc.*, **124**, 12874.
65 Shevchenko, E.V., Talapin, D.V., Rogach, A.L., Kornowski, A., Haase, M., and Weller, H. (2002) *J. Am. Chem. Soc.*, **124**, 11480.
66 Sugimoto, T. (2001) *Monodisperse Particles*, Elsevier.
67 Park, J.I., Kang, N.J., Jun, Y.W., Oh, S.J., Ri, H.C., and Cheon, J. (2002) *Chem. Phys. Chem.*, **3**, 543.
68 Hyeon, T., Lee, S.S., Park, J., Chung, Y., and Na, H.B. (2001) *J. Am. Chem. Soc.*, **123**, 12798.
69 Yu, W.W. and Peng, X.G. (2002) *Angew. Chem., Int. Ed.*, **41**, 2368.
70 Glatter, O. (1977) *J. Appl. Crystallogr.*, **10**, 415.
71 Glatter, O. (1980) *J. Appl. Crystallogr.*, **13**, 7.
72 Korgel, B.A., Fullam, S., Connolly, S., and Fitzmaurice, D. (1998) *J. Phys. Chem. B*, **102**, 8379.
73 Mattoussi, H., Radzilowski, L.H., Dabbousi, B.O., Thomas, E.L., Bawendi, M.G., and Rubner, M.F. (1998) *J. Appl. Phys.*, **83**, 7965.
74 Peng, X., Wickham, J., and Alivisatos, A.P. (1998) *J. Am. Chem. Soc.*, **120**, 5343.
75 Wang, Z.L. (1998) *Adv. Mater.*, 10, 13.

76. Peng, X.G., Manna, L., Yang, W.D., Wickham, J., Scher, E., Kadavanich, A., and Alivisatos, A.P. (2000) *Nature*, **404**, 59.
77. Manna, L., Scher, E.C., and Alivisatos, A.P. (2000) *J. Am. Chem. Soc.*, **122**, 12700.
78. Lee, S.-M., Jun, Y.-W., Cho, S.-N., and Cheon, J. (2002) *J. Am. Chem. Soc.*, **124**, 11244.
79. Pacholski, C., Kornowski, A., and Weller, H. (2002) *Angew. Chem., Int. Ed.*, **41**, 1188.
80. Dumestre, F., Chaudret, B., Amiens, C., Fromen, M.C., Casanove, M.J., Renaud, P., and Zurcher, P. (2002) *Angew. Chem., Int. Ed.*, **41**, 4286.
81. Tang, Z., Kotov, N.A., and Giersig, M. (2002) *Science*, **297**, 237.
82. Park, S.J., Kim, S., Lee, S., Khim, Z.G., Char, K., and Hyeon, T. (2000) *J. Am. Chem. Soc.*, **122**, 8581.
83. Dumestre, F., Chaudret, B., Amiens, C., Renaud, A., and Fejes, P. (2004) *Science*, **303**, 821.
84. Cordente, N., Respaud, M., Senocq, F., Casanove, M.J., Amiens, C., and Chaudret, B. (2001) *Nano Lett.*, **1**, 565.
85. Service, R.F. (2000) *Science*, **287**, 1902.
86. Sun, S., Murray, C.B., Weller, D., Folks, L., and Moser, A. (2000) *Science*, **287**, 1989.
87. Dai, Z.R., Sun, S.H., and Wang, Z.L. (2001) *Nano Lett.*, **1**, 443.
88. Hansen, M. (1958) *Constitution of Binary Alloys*, McGraw-Hill, New York, Toronto, London.
89. Kang, S., Harrell, J.W., and Nikles, D.E. (2002) *Nano Lett.*, **2**, 1033.
90. Chen, M. and Nikles, D.E. (2002) *J. Appl. Phys.*, **91**, 8477.
91. Ono, K., Okuda, R., Ishii, Y., Kamimura, S., and Oshima, M. (2003) *J. Phys. Chem. B*, **107**, 1941.
92. Park, J.I. and Cheon, J. (2001) *J. Am. Chem. Soc.*, **123**, 5743.
93. Sobal, N.S., Hilgendorff, M., Mohwald, H., Giersig, M., Spasova, M., Radetic, T., and Farle, M. (2002) *Nano Lett.*, **2**, 621.
94. Cornell, R.M. and Schwertmann, U. (1996) *The Iron Oxides: Structure, Properties, Reactions, Occurrence and Uses*, VCH, Weinheim, Germany.
95. Massart, R. (1981) *IEEE Trans. Magn.*, **17**, 1247.
96. Massart, R. and Cabuil, V. (1987) *J. Chim. Phys. Phys.-Chim. Biol.*, **84**, 967.
97. Jolivet, J.P. (1994) De la solution à l'Oxyde, 2nd edn, InterEdition/CNRS Editions.
98. Cheon, J., Kang, N.-J., Lee, S.-M., Lee, J.-H., Yoon, J.-H., and Oh, S.J. (2004) *J. Am. Chem. Soc.*, **126**, 1950.
99. Sun, S. and Zeng, H. (2002) *J. Am. Chem. Soc.*, **124**, 8204.
100. Jana, N.R., Chen, Y., and Peng, X. (2004) *Chem. Mater.*, **16**, 3931.
101. Park, J., An, K.J., Hwang, Y.S., Park, J.G., Noh, H.J., Kim, J.Y., Park, J.H., Hwang, N.M., and Hyeon, T. (2004) *Nature Mater.*, **3**, 891.
102. Thomas, J.R. (1966) *J. Appl. Phys.*, **37**, 2914.
103. Griffiths, C.H., Ohoro, M.P., and Smith, T.W. (1979) *J. Appl. Phys.*, **50**, 7108.
104. Yu, W.W., Falkner, J.C., Yavuz, C.T., and Colvin, V.L. (2004) *Chem. Commun.*, 2306.
105. Shavel, A. and Liz-Marzan, L.M. (2009) *Phys. Chem. Chem. Phys.*, **11**, 3762.
106. Kovalenko, M.V., Bodnarchuk, M.I., Lechner, R.T., Hesser, G., Schaffler, F., and Heiss, W. (2007) *J. Am. Chem. Soc.*, **129**, 6352.
107. Ahniyaz, A., Sakamoto, Y., and Bergstrom, L. (2007) *Proc. Natl Acad. Sci. USA*, **104**, 17570.
108. Kim, D., Lee, N., Park, M., Kim, B.H., An, K., and Hyeon, T. (2009) *J. Am. Chem. Soc.*, **131**, 454.
109. Cozzoli, P.D., Snoeck, E., Garcia, M.A., Giannini, C., Guagliardi, A., Cervellino, A., Gozzo, F., Hernando, A., Achterhold, K., Ciobanu, N., Parak, F.G., Cingolani, R., and Manna, L. (2006) *Nano Lett.*, **6**, 1966.
110. Manna, L., Milliron, D.J., Meisel, A., Scher, E.C., and Alivisatos, A.P. (2003) *Nat. Mater.*, **2**, 382.
111. Talapin, D.V., Nelson, J.H., Shevchenko, E.V., Aloni, S., Sadtler, B., and Alivisatos, A.P. (2007) *Nano Lett.*, **7**, 2951.
112. Carbone, L., Nobile, C., De Giorgi, M., Sala, F.D., Morello, G., Pompa, P., Hytch, M., Snoeck, E., Fiore, A., Franchini, I.R., Nadasan, M., Silvestre, A.F., Chiodo, L., Kudera, S., Cingolani, R., Krahne, R., and Manna, L. (2007) *Nano Lett.*, **7**, 2942.

113 Liang, X., Wang, X., Zhuang, J., Chen, Y.T., Wang, D.S., and Li, Y.D. (2006) *Adv. Funct. Mater.*, **16**, 1805.
114 Wang, S.B., Min, Y.L., and Yu, S.H. (2007) *J. Phys. Chem. C*, **111**, 3551.
115 Cao, H.Q., Wang, G.Z., Warner, J.H., and Watt, A.A.R. (2008) *Appl. Phys. Lett.*, 92, 013110.
116 Rodriguez, R.D., Demaille, D., Lacaze, E., Jupille, J., Chaneac, C., and Jolivet, J.P. (2007) *J. Phys. Chem. C*, **111**, 16866.
117 Dong, W.T. and Zhu, C.S. (2002) *J. Mater. Chem.*, **12**, 1676.
118 Redl, F.X., Black, C.T., Papaefthymiou, G.C., Sandstrom, R.L., Yin, M., Zeng, H., Murray, C.B., and O'Brien, S.P. (2004) *J. Am. Chem. Soc.*, **126**, 14583.
119 Hou, Y.L., Xu, Z.C., and Sun, S.H. (2007) *Angew. Chem., Int. Ed.*, **46**, 6329.
120 Glaria, A., Kahn, M.L., Lecante, P., Barbara, B., and Chaudret, B. (2008) *Chem. Phys. Chem.*, **9**, 776.
121 Cwiertny, D.M., Handler, R.M., Schaefer, M.V., Grassian, V.H., and Scherer, M.M. (2008) *Geochim. Cosmochim. Acta*, **72**, 1365.
122 Burleson, D.J. and Penn, R.L. (2006) *Langmuir*, **22**, 402.
123 Yu, T., Park, J., Moon, J., An, K., Piao, Y., and Hyeon, T. (2007) *J. Am. Chem. Soc.*, **129**, 14558.
124 Tang, Z.X., Sorensen, C.M., Klabunde, K.J., and Hadjipanayis, G.C. (1991) *J. Colloid Interface Sci.*, **146**, 38.
125 Fried, T., Shemer, G., and Markovich, G. (2001) *Adv. Mater.*, **13**, 1158.
126 Sun, S., Zeng, H., Robinson, D.B., Raoux, S., Rice, P.M., Wang, S.X., and Li, G. (2004) *J. Am. Chem. Soc.*, **126**, 273.
127 Hyeon, T., Chung, Y., Park, J., Lee, S.S., Kim, Y.-W., and Park, B.H. (2002) *J. Phys. Chem. B*, **106**, 6831.
128 Kang, E., Park, J., Hwang, Y., Kang, M., Park, J.G., and Hyeon, T. (2004) *J. Phys. Chem. B*, **108**, 13932.
129 Lee, J.H., Huh, Y.M., Jun, Y., Seo, J., Jang, J., Song, H.T., Kim, S., Cho, E.J., Yoon, H.G., Suh, J.S., and Cheon, J. (2007) *Nat. Med.*, **13**, 95.
130 Gao, J., Zhang, B., Gao, Y., Pan, Y., Zhang, X., and Xu, B. (2007) *J. Am. Chem. Soc.*, **129**, 11928.
131 Kim, H., Achermann, M., Balet, L.P., Hollingsworth, J.A., and Klimov, V.I. (2005) *J. Am. Chem. Soc.*, **127**, 544.
132 Klimov, V.I., Ivanov, S.A., Nanda, J., Achermann, M., Bezel, I., McGuire, J.A., and Piryatinski, A. (2007) *Nature*, **447**, 441.
133 Zeng, H., Li, J., Liu, J.P., Wang, Z.L., and Sun, S. (2002) *Nature*, **420**, 395.
134 Zeng, H., Li, J., Wang, Z.L., Liu, J.P., and Sun, S. (2004) *Nano Lett.*, **4**, 187.
135 Sellmyer, D.J. (2002) *Nature*, **420**, 374.
136 Klemmer, T., Hoydick, D., Okumura, H., Zhang, B., and Soffa, W.A. (1995) *Scr. Metall.*, **33**, 1793.
137 Bodnarchuk, M.I., Kovalenko, M.V., Groiss, H., Resel, R., Reissner, M., Hesser, G., Lechner, R.T., Steiner, W., Schäffler, F., and Heiss, W. (2009) *Small*, 5, 2247.
138 Ketteler, G., Weiss, W., Ranke, W., and Schlogl, R. (2001) *Phys. Chem. Chem. Phys.*, **3**, 1114.
139 Yu, H., Chen, M., Rice, P.M., Wang, S.X., White, R.L., and Sun, S. (2005) *Nano Lett.*, **5**, 379.
140 Wang, C., Daimon, H., and Sun, S. (2009) *Nano Lett.*, **9**, 1493.
141 Wang, C., Xu, C., Zeng, H., and Sun, S. (2009) *Adv. Mater.*, 21, 3045.
142 Pellegrino, T.E.A. (2006) *J. Am. Chem. Soc.*, **128**, 6690.
143 Li, Y., Zhang, Q., Nurmikko, Q.A., and Sun, S. (2005) *Nano Lett.*, **5**, 1689.
144 Gu, H., Zheng, R., Zhang, X., and Xu, B. (2004) *J. Am. Chem. Soc.*, **126**, 5664.
145 Yin, Y., Rioux, R.M., Erdonmez, C.K., Hughes, S., Somorjai, G.A., and Alivisatos, A.P. (2004) *Science*, **304**, 711.
146 Smigelskas, A.D. and Kirkendall, E.O. (1947) *Trans. AIME*, **171**, 130.
147 Smigelskas, A.D. and Kirkendall, E.O. (1947) *Trans. AIME*, **171**, 130.
148 Peng, S. and Sun, S. (2007) *Angew Chem., Int. Ed.*, **46**, 4155.
149 Cabot, A., Puntes, V.F., Shevchenko, E., Yin, Y., Balcells, L., Marcus, M.A., Hughes, S.M., and Alivisatos, A.P. (2007) *J. Am. Chem. Soc.*, **129**, 10358.
150 Tromsdorf, U.I., Bigall, N.C., Kaul, M.G., Bruns, O.T., Nikolic, M.S., Mollwitz, B., Sperling, R.A., Reimer, R., Hohenberg, H., Parak, W.J., Forster, S., Beisiegel, U.,

Adam, G., and Weller, H. (2007) *Nano Lett.,* **7**, 2422.

151 Shevchenko, E.V., Bodnarchuk, M.I., Kovalenko, M.V., Talapin, D.V., Smith, R.K., Aloni, S., Heiss, W., and Alivisatos, A.P. (2008) *Adv. Mater.,* **20**, 4323.

152 deHeer, W.A. (2000) *Characterisation of Nanophase Materials* (ed. Z.L. Wang,), Wiley-VCH Verlag GmbH, Weinheim, p. 289.

153 Kittel, C. (1996) *Introduction to Solid State Physics,* John Wiley & Sons, New York, Singapore.

154 Meiklejohn, W.H. and Bean, C.P. (1957) *Phys. Rev.,* **105**, 904.

155 Held, G.A., Grinstein, G., Doyle, H., Sun, S.H., and Murray, C.B. (2001) *Phys. Rev. B,* 64, 012408.

156 Sun, S. (2006) *Adv. Mater.,* **18**, 393.

157 Tsoukalas, D., Dimitrakis, P., Kolliopoulou, S., and Normand, P. (2005) *Mater. Sci. Eng. B – Solid State Mater. Adv. Tech.,* **124**, 93.

158 Chappert, C., Fert, A., and Van Dau, F.N. (2007) *Nat. Mater,* **6**, 813.

159 Terris, B.D. and Thomson, T. (2005) *J. Phys. D – Appl. Phys.,* **38**, R199.

160 Teng, X.W. and Yang, H. (2003) *J. Am. Chem. Soc.,* **125**, 14559.

161 Nguyen, H.L., Howard, L.E.M., Giblin, S.R., Tanner, B.K., Terry, I., Hughes, A.K., Ross, I.M., Serres, A., Burckstummer, H., and Evans, J.S.O. (2005) *J. Mater. Chem.,* **15**, 5136.

162 Chen, M., Kim, J., Liu, J.P., Fan, H., and Sun, S. (2006) *J. Am. Chem. Soc.,* **128**, 7132.

163 Zhang, D.J., Klabunde, K.J., Sorensen, C.M., and Hadjipanayis, G.C. (1996) *High Temp. Mater. Sci.,* **36**, 135.

164 Laurent, S., Forge, D., Port, M., Roch, A., Robic, C., Elst, L.V., and Muller, R.N. (2008) *Chem. Rev.,* **108**, 2064.

165 Lee, W.R., Kim, M.G., Choi, J.R., Park, J.I., Ko, S.J., Oh, S.J., and Cheon, J. (2005) *J. Am. Chem. Soc.,* **127**, 16090.

166 Bao, Y.P. and Krishnan, K.M. (2005) *J. Magn. Magn. Mater.,* **293**, 15.

167 Xu, C.J., Xu, K.M., Gu, H.W., Zheng, R.K., Liu, H., Zhang, X.X., Guo, Z.H., and Xu, B. (2004) *J. Am. Chem. Soc.,* **126**, 9938.

168 Xie, J., Xu, C.J., Xu, Z.H., Hou, Y.L., Young, K.L.L., Wang, S.X., Pourmand, N., and Sun, S.H. (2007) *Chem. Mater.,* **19**, 1202.

169 Wu, H., Zhu, H., Zhuang, J., Yang, S., Liu, C., and Cao, Y.C. (2008) *Angew. Chem., Int. Edit.,* **47**, 3730.

170 Jun, Y.W., Huh, Y.M., Choi, J.S., Lee, J.H., Song, H.T., Kim, S., Yoon, S., Kim, K.S., Shin, J.S., Suh, J.S., and Cheon, J. (2005) *J. Am. Chem. Soc.,* **127**, 5732.

171 Zhang, T.R., Ge, J.P., Hu, Y.P., and Yin, Y.D. (2007) *Nano Lett.,* **7**, 3203.

172 Kim, S.W., Kim, S., Tracy, J.B., Jasanoff, A., and Bawendi, M.G. (2005) *J. Am. Chem. Soc.,* **127**, 4556.

173 Bin Na, H., Lee, I.S., Seo, H., Il Park, Y., Lee, J.H., Kim, S.W., and Hyeon, T. (2007) *Chem. Commun.,* 5167.

174 White, M.A., Johnson, J.A., Koberstein, J.T., and Turro, N.J. (2007) *J. Am. Chem. Soc.,* **129**, 4504.

175 Schroder, U., Segren, S., Gemmefors, C., Hedlund, G., Jansson, B., Sjogren, H.O., and Borrebaeck, C.A.K. (1986) *J. Immunol. Methods,* **93**, 45.

176 Berry, C.C., Wells, S., Charles, S., and Curtis, A.S.G. (2003) *Biomaterials,* **24**, 4551.

177 Nitin, N., LaConte, L.E.W., Zurkiya, O., Hu, X., and Bao, G. (2004) *J. Biol. Inorg. Chem.,* **9**, 706.

178 Ito, A., Ino, K., Kobayashi, T., and Honda, H. (2005) *Biomaterials,* **26**, 6185.

179 De la Fuente, J.M. and Penades, S. (2006) *BBA.-Gen. Subjects,* **1760**, 636.

180 Liang, X., Wang, X., Zhuang, J., Chen, Y.T., Wang, D.S., and Li, Y.D. (2006) *Adv. Funct. Mater.,* **16**, 1805.

181 McDonald, M.A. and Watkin, K.L. (2006) *Acad. Radiol.,* **13**, 421.

182 Heinze, T., Liebert, T., Heublein, B., and Hornig, S. (2006) *Polysaccharides Ii,* vol. **205**, Springer-Verlag, Berlin, p. 199.

183 Xie, J., Xu, C., Kohler, N., Hou, Y., and Sun, S. (2007) *Adv. Mater.,* **19**, 3163.

184 Yiu, H.H.P. and Wright, P.A. (2005) *J. Mater. Chem.,* **15**, 3690.

185 Yiu, H.H.P., Wright, P.A., and Botting, N.P. (2001) *J. Mol. Catal. B: Enzym.,* **15**, 81.

186 Hu, M., Chen, J.Y., Li, Z.Y., Au, L., Hartland, G.V., Li, X.D., Marquez, M., and Xia, Y.N. (2006) *Chem. Soc. Rev.,* **35**, 1084.

187 Bertilsson, L. and Liedberg, B. (1993) *Langmuir,* **9**, 141.

188 Lubbe, A.S., Bergemann, C., Brock, J., and McClure, D.G. (1998) *2nd International Conference on Scientific and Clinical Applications of Magnetic Carriers (SCAMC2)*, Elsevier Science BV, Cleveland, Ohio, p. 149.

189 McBain, S.C., Yiu, H.H.P., and Dobson, J. (2008) *Int. J. Nanomedicine*, **3**, 169.

190 Yuan, X.B., Li, H., Zhu, X.X., and Woo, H.G. (2006) *J. Chem. Technol. Biotechnol.*, **81**, 746.

191 Li, X.H., Zhang, D.H., and Chen, J.S. (2006) *J. Am. Chem. Soc.*, **128**, 8382.

192 Morales, M.P., Bomati-Miguel, O., de Alejo, R.P., Ruiz-Cabello, J., Veintemillas-Verdaguer, S., and O'Grady, K. (2003) *J. Magn. Magn. Mater.*, **266**, 102.

193 Jordan, A., Scholz, R., Wust, P., Schirra, H., Schiestel, T., Schmidt, H., and Felix, R. (1999) *J. Magn. Magn. Mater.*, **194**, 185.

194 Alexiou, C., Arnold, W., Hulin, P., Klein, R.J., Renz, H., Parak, F.G., Bergemann, C., and Lubbe, A.S. (2001) *J. Magn. Magn. Mater.*, **225**, 187.

195 Roser, M., Fischer, D., and Kissel, T. (1998) *Eur. J. Pharm. Biopharm.*, **46**, 255.

196 Montet, X., Montet-Abou, K., Reynolds, F., Weissleder, R., and Josephson, L. (2006) *Neoplasia*, **8**, 214.

197 Matsunaga, T. and Takeyama, H. (1998) *Supramol. Sci.*, **5**, 391.

198 Kohler, N., Sun, C., Fichtenholtz, A., Gunn, J., Fang, C., and Zhang, M.Q. (2006) *Small*, **2**, 785.

199 Veiseh, O., Sun, C., Gunn, J., Kohler, N., Gabikian, P., Lee, D., Bhattarai, N., Ellenbogen, R., Sze, R., Hallahan, A., Olson, J., and Zhang, M.Q. (2005) *Nano Lett.*, **5**, 1003.

200 Gupta, A.K. and Wells, S. (2004) *IEEE Trans. Nanobiosci.*, **3**, 66.

201 Schulze, K., Koch, A., Schopf, B., Petri, A., Steitz, B., Chastellain, M., Hofmann, M., Hofmann, H., and von Rechenberg, B. (2005) *J. Magn. Magn. Mater.*, **293**, 419.

202 Schulze, K., Koch, A., Petri-Fink, A., Steitz, B., Kamau, S., Hottiger, M., Hilbe, M., Vaughan, L., Hofmann, M., Hofmann, H., and von Rechenberg, B. (2006) *J. Nanosci. Nanotechnol.*, **6**, 2829.

203 Liong, M., Lu, J., Kovochich, M., Xia, T., Ruehm, S.G., Nel, A.E., Tamanoi, F., and Zink, J.I. (2008) *ACS Nano*, **2**, 889.

204 Kline, T.L., Xu, Y.H., Jing, Y., and Wang, J.P. (2009) *J. Magn. Magn. Mater.*, **321**, 1525.

205 Ji, X.J., Shao, R.P., Elliott, A.M., Stafford, R.J., Esparza-Coss, E., Bankson, J.A., Liang, G., Luo, Z.P., Park, K., Markert, J.T., and Li, C. (2007) *J. Phys. Chem. C*, **111**, 6245.

206 Choi, J.-s., Choi, H.J., Jung, D.C., Lee, J.-H., and Cheon, J. (2008) *Chem. Commun.*, 2197.

207 Widder, K.J., Senyei, A.E., and Scarpelli, D.G. (1978) *Proc. Soc. Exp. Biol. Med.*, **158**, 141.

208 Senyei, A., Widder, K., and Czerlinski, G. (1978) *J. Appl. Phys.*, **49**, 3578.

209 Mosbach, K. and Schroder, U. (1979) *FEBS Lett.*, **102**, 112.

210 Grief, A.D. and Richardson, G. (2005) *J. Magn. Magn. Mater.*, **293**, 455.

211 Pankhurst, Q.A., Connolly, J., Jones, S.K., and Dobson, J. (2003) *J. Phys. D Appl. Phys.*, **36**, R167.

212 Voltairas, P.A., Fotiadis, D.I., and Michalis, L.K. (2002) *J. Biomech.*, **35**, 813.

213 Ruuge, E.K. and Rusetski, A.N. (1992) *6th International Conference on Magnetic Fluids (ICMF 6)*, Elsevier Science Bv, Paris, France, p. 335.

214 Kubo, T., Sugita, T., Shimose, S., Nitta, Y., Ikuta, Y., and Murakami, T. (2000) *Int. J. Oncol.*, **17**, 309.

215 Yellen, B.B., Forbes, Z.G., Halverson, D.S., Fridman, G., Barbee, K.A., Chorny, M., Levy, R., and Friedman, G. (2004) *J. Magn. Magn. Mater.*, **293**, 647.

216 Kohler, N., Sun, C., Wang, J., and Zhang, M.Q. (2005) *Langmuir*, **21**, 8858.

217 Alexiou, C., Schmid, R.J., Jurgons, R., Kremer, M., Wanner, G., Bergemann, C., Huenges, E., Nawroth, T., Arnold, W., and Parak, F.G. (2006) *Eur. Biophys. J. Biophys.*, **35**, 446.

218 Alexiou, C., Arnold, W., Klein, R.J., Parak, F.G., Hulin, P., Bergemann, C., Erhardt, W., Wagenpfeil, S., and Lubbe, A.S. (2000) *Cancer Res*, **60**, 6641.

219 Lubbe, A.S., Bergemann, C., Riess, H., Schriever, F., Reichardt, P., Possinger, K., Matthias, M., Dorken, B., Herrmann, F., Gurtler, R., Hohenberger, P., Haas, N., Sohr, R., Sander, B., Lemke, A.J.,

Ohlendorf, D., Huhnt, W., and Huhn, D. (1996) *Cancer Res.*, **56**, 4686.

220 Lubbe, A.S., Bergemann, C., Huhnt, W., Fricke, T., Riess, H., Brock, J.W., and Huhn, D. (1996) *Cancer Res.*, **56**, 4694.

221 Koda, J., Venook, A., Walser, E., and Goodwin, S. (2002) *14th EORTC–NCI–AACR Symposium on Molecular Targets and Cancer Therapeutics*, Pergamon-Elsevier Science Ltd, Frankfurt, Germany, p. 43.

222 Wilson, M.W., Kerlan, R.K., Fidelman, N.A., Venook, A.P., LaBerge, J.M., Koda, J., and Gordon, R.L. (2004) *Radiology*, **230**, 287.

223 Willard, M.A., Kurihara, L.K., Carpenter, E.E., Calvin, S., and Harris, V.G. (2004) *Int. Mater. Rev.*, **49**, 125.

224 Sudimack, J. and Lee, R.J. (2000) *Adv. Drug Delivery Rev.*, **41**, 147.

225 Kam, N.W.S., O'Connell, M., Wisdom, J.A., and Dai, H.J. (2005) *Proc. Natl Acad. Sci. USA*, **102**, 11600.

226 Neuberger, T., Schopf, B., Hofmann, H., Hofmann, M., and von Rechenberg, B. (2005) *J. Magn. Magn. Mater.*, **293**, 483.

227 Browne, M. and Semelka, R.C. (1999) *MRI: Basic Principles and Applications*, Wiley, New York.

228 Luo, D. and Saltzman, W.M. (2000) *Nat. Biotechnol.*, **18**, 893.

229 Mah, C., Fraites, T.J., Zolotukhin, I., Song, S.H., Flotte, T.R., Dobson, J., Batich, C., and Byrne, B.J. (2002) *Mol. Ther.*, **6**, 106.

230 Scherer, F., Anton, M., Schillinger, U., Henkel, J., Bergemann, C., Kruger, A., Gansbacher, B., and Plank, C. (2002) *Gene Ther.*, **9**, 102.

231 Akinc, A., Thomas, M., Klibanov, A.M., and Langer, R. (2005) *J. Gene. Med.*, **7**, 657.

232 Krotz, F., de Wit, C., Sohn, H.Y., Zahler, S., Gloe, T., Pohl, U., and Plank, C. (2003) *Mol. Ther.*, **7**, 700.

233 Schillinger, U., Brill, T., Rudolph, C., Huth, S., Gersting, S., Krotz, F., Hirschberger, J., Bergemann, C., and Plank, C. (2004) *5th International Conference on Scientific and Clinical Applications of Magnetic Carriers*, Elsevier Science Bv, Lyon, France, p. 501.

234 McBain, S.C., Yiu, H.H.P., El Haj, A., and Dobson, J. (2007) *J. Mater. Chem.*, **17**, 2561.

235 Cai, D., Mataraza, J.M., Qin, Z.H., Huang, Z.P., Huang, J.Y., Chiles, T.C., Carnahan, D., Kempa, K., and Ren, Z.F. (2005) *Nat. Methods*, **2**, 449.

236 Kronick, P.L., Campbell, G.L., and Joseph, K. (1978) *Science*, **200**, 1074.

237 Widder, K.J., Senyei, A.E., Ovadia, H., and Paterson, P.Y. (1979) *Clin. Immunol. Immunopathol.*, **14**, 395.

238 Kemshead, J.T. and Ugelstad, J. (1985) *Mol. Cell. Biochem.*, **67**, 11.

239 Safarik, I., Safarikova, M., and Forsythe, S.J. (1995) *J. Appl. Bacteriol.*, **78**, 575.

240 Olsvik, O., Popovic, T., Skjerve, E., Cudjoe, K.S., Hornes, E., Ugelstad, J., and Uhlen, M. (1994) *Clin. Microbiol. Rev.*, **7**, 43.

241 Sauzedde, F. (1997) Université Claude Bernard Lyon 1 (Lyon).

242 Ding, X.B., Sun, Z.H., Zhang, W.C., Peng, Y.X., Wan, G.X., and Jiang, Y.Y. (2000) *J. Appl. Polym. Sci.*, **77**, 2915.

243 Oster, J., Parker, J., and Brassard, L.A. (2000) in *3rd International Conference on Scientific and Clinical Applications of Magnetic Carriers*, Elsevier Science Bv, Rostock, Germany, p. 145.

244 Morisada, S., Miyata, N., and Iwahori, K. (2002) *J. Microbiol. Methods*, **51**, 141.

245 Veyret, R., Elaissari, A., Marianneau, P., Sall, A.A., and Delair, T. (2005) *Anal. Biochem.*, **346**, 59.

246 Ugelstad, J., Prestvik, W.S., Stenstad, P., Kilaas, L., and Kvalheim, G. (1998) *Magnetism in Medicine* (ed. H. Nowak), Wiley-VCH, Berlin, p. 471.

247 Kala, M., Bajaj, K., and Sinha, S. (1997) *Anal. Biochem.*, **254**, 263.

248 Tibbe, A.G.J., de Grooth, B.G., Greve, J., Liberti, P.A., Dolan, G.J., and Terstappen, L. (1999) *Nat. Biotechnol.*, **17**, 1210.

249 Rheinlander, T., Kotitz, R., Weitschies, W., and Semmler, W. (2000) *J. Magn. Magn. Mater.*, **219**, 219.

250 Moore, L.R., Rodriguez, A.R., Williams, P.S., McCloskey, K., Bolwell, B.J., Nakamura, M., Chalmers, J.J., and Zborowski, M. (2001) *J. Magn. Magn. Mater.*, **225**, 277.

251 Rheinlander, T., Kotitz, R., Weitschies, W., and Semmler, W. (2000) *J. Magn. Magn. Mater.*, **219**, 219.

252 Owen, C.S. (1983) *Cell Separation: Methods and Selected Applications*, vol. 2 (eds T.G. Pretlowand T.G. Pretlow), Academic Press, New York, p. 327.

253 Kularatne, B.Y., Lorigan, P., Browne, S., Suvarna, S.K., Smith, M.O., and Lawry, J. (2002) *Cytometry B: Clin. Cytom.*, **50**, 160.

254 Zigeuner, R.E., Riesenberg, R., Pohla, H., Hofstetter, A., and Oberneder, R. (2003) *J. Urol.*, **169**, 701.

255 Poynton, C.H., Dicke, K.A., Culbert, S., Frankel, L.S., Jagannath, S., and Reading, C.L. (1983) *Lancet*, **1**, 524.

256 Kemshead, J.T., Treleaven, J.G., Gibson, F.M., Ugelstad, J., Rembaum, A., and Philip, T. (1985) *Prog. Exp. Tumor Res.*, **29**, 249.

257 Paul, F., Melville, D., Roath, S., and Warhurst, D.C. (1981) *IEEE Trans. Magn.*, **17**, 2822.

258 Seesod, N., Nopparat, P., Hedrum, A., Holder, A., Thaithong, S., Uhlen, M., and Lundeberg, J. (1997) *Am. J. Trop. Med. Hyg.*, **56**, 322.

259 Liberti, P.A., Rao, C.G., and Terstappen, L. (2000) *3rd International Conference on Scientific and Clinical Applications of Magnetic Carriers*, Elsevier Science Bv, Rostock, Germany, p. 301.

260 Scarberry, K.E., Dickerson, E.B., McDonald, J.F., and Zhang, Z.J. (2008) *J. Am. Chem. Soc.*, **130**, 10258.

261 van der Zee, J. (2002) *Ann. Oncol.*, **13**, 1173.

262 Moroz, P., Jones, S.K., and Gray, B.N. (2001) *J. Surg. Oncol.*, **77**, 259.

263 Wust, P., Hildebrandt, B., Sreenivasa, G., Rau, B., Gellermann, J., Riess, H., Felix, R., and Schlag, P.M. (2002) *Lancet Oncol.*, **3**, 487.

264 Hergt, R., Dutz, S., Muller, R., and Zeisberger, M. (2006) *J. Phys. - Condens. Mater.*, **18**, S2919.

265 Gilchrist, R.K., Medal, R., Shorey, W.D., Hanselman, R.C., Parrott, J.C., and Taylor, C.B. (1957) *Ann. Surg.*, **146**, 596.

266 Gordon, R.T., Hines, J.R., and Gordon, D. (1979) *Med. Hypotheses*, **5**, 83.

267 Rand, R.W., Snow, H.D., Lagasse, L., and Brown, W.J. (1983) *Radiat Res.*, **94**, 567.

268 Chan, D.C.F., Kirpotin, D.B., and Bunn, P.A. (1993) *J. Magn. Magn. Mater.*, **122**, 374.

269 Suzuki, M., Shinkai, M., Kamihira, M., and Kobayashi, T. (1995) *Biotechnol. Appl. Biochem.*, **21**, 335.

270 Jordan, A., Wust, P., Scholz, R., Faehling, H., Krause, J., and Felix, R. (1997) *Sci. Clin. Appl. Magn. Carriers*, 569.

271 Hilger, I., Fruhauf, K., Andra, W., Hiergeist, R., Hergt, R., and Kaiser, W.A. (2002) *Acad. Radiol.*, **9**, 198.

272 Jordan, A., Scholz, R., Maier-Hauff, K., Johannsen, M., Wust, P., Nadobny, J., Schirra, H., Schmidt, H., Deger, S., Loening, S., Lanksch, W., and Felix, R. (2000) *J. Magn. Magn. Mater.*, **225**, 118.

273 Moroz, P., Jones, S.K., and Gray, B.N. (2002) *Int. J. Hyperthermia*, **18**, 267.

274 Rabin, Y. (2002) *Int. J. Hyperthermia*, **18**, 194.

275 Salloum, M., Ma, R.H., and Zhu, L. (2008) *Int. J. Hyperthermia*, **24**, 589.

276 Reilly, J.P. (1992) *Ann. N. Y. Acad. Sci.*, **649**, 96.

277 Oleson, J.R., Heusinkveld, R.S., and Manning, M.R. (1983) *Int. J. Radiat. Oncol.*, **9**, 549.

278 Jordan, A., Scholz, R., Wust, P., Fahling, H., and Felix, R. (1999) *J. Magn. Magn. Mater.*, **201**, 413.

279 Ramanujan, R.V., Ang, K.L., and Venkatraman, S. (2007) *Mater. Sci. Eng. C: Biomim. Supramol. Syst.*, **27**, 347.

280 Jones, S.K., Gray, B.N., Burton, M.A., Codde, J.P., and Street, R. (1992) *Phys. Med. Biol.*, **37**, 293.

281 Maehara, T., Konishi, K., Kamimori, T., Aono, H., Naohara, T., Kikkawa, H., Watanabe, Y., and Kawachi, K. (2002) *Jpn. J. Appl. Phys. 1*, **41**, 1620.

282 Hergt, R., Andra, W., d'Ambly, C.G., Hilger, I., Kaiser, W.A., Richter, U., and Schmidt, H.G. (1998) *IEEE Trans. Magn.*, **34**, 3745.

283 Hergt, R., Hiergeist, R., Zeisberger, M., Schuler, D., Heyen, U., Hilger, I., and Kaiser, W.A. (2005) *J. Magn. Magn. Mater.*, **293**, 80.

284 Kim, D.-H., Nikles, D.E., Johnson, D.T., and Brazel, C.S. (2008) *J. Magn. Magn. Mater.*, **320**, 2390.

285 Choi, H.S., Liu, W., Misra, P., Tanaka, E., Zimmer, J.P., Ipe, B.I., Bawendi, M.G., and Frangioni, J.V. (2007) *Nat. Biotechnol.*, **25**, 1165.

286 Suto, M., Hirota, Y., Mamiya, H., Fujita, A., Kasuya, R., Tohji, K., and Jeyadevan, B. (2009) *J. Magn. Magn. Mater.*, **321**, 1493.

287 Haacke, E.M., Brown, R.W., Thompson, M.R., and Venkatesan, R. (1999) *Magnetic Resonance Imaging – Physical Principles and Sequence Design*, John Wiley & Sons, New York.

288 Reiser, M. and Semmler, W. (2002) *Magnetresonanztomographie*, Springer-Verlag, Berlin.

289 Dilcher, L., Venator, M., and Dilcher, S. (2002) *Handbuch der Kernspintomographie*, 2nd edn, Edwin Ferger Verlag, Bergisch Gladbach.

290 Toth, E., Helm, L., and Merbach, A.E. (2002) *Contrast Agents I*, **221**, 61.

291 Bulte, J.W.M. and Kraitchman, D.L. (2004) *NMR Biomed.*, **17**, 484.

292 Wang, Y.X.J., Hussain, S.M., and Krestin, G.P. (2001) *Eur. Radiol.*, **11**, 2319.

293 Caravan, P., Ellison, J.J., McMurry, T.J., and Lauffer, R.B. (1999) *Chem. Rev.*, **99**, 2293.

294 Roch, A., Gossuin, Y., Muller, R.N., and Gillis, P. (2005) *J. Magn. Magn. Mater.*, **293**, 532.

295 Kim, D.-H., Rozhkova, E.A., Rajh, T., Bader, S.D., and Novosad, V. (2009) *IEEE. Trans. Magn.*, 45, 4821.

296 Huh, Y.M., Lee, E.S., Lee, J.H., Jun, Y.W., Kim, P.H., Yun, C.O., Kim, J.H., Suh, J.S., and Cheon, J. (2007) *Adv. Mater.*, **19**, 3109.

297 Quarta, A., Di Corato, R., Manna, L., Argentiere, S., Cingolani, R., Barbarella, G., and Pellegrino, T. (2008) *J. Am. Chem. Soc.*, **130**, 10545.

298 Salloum, M., Ma, R., and Zhu, L. (2009) *Int. J. Hyperthermia*, **25**, 309.

299 Kim, B.S., Qiu, J.M., Wang, J.P., and Taton, T.A. (2005) *Nano Lett.*, **5**, 1987.

300 Nasongkla, N., Bey, E., Ren, J.M., Ai, H., Khemtong, C., Guthi, J.S., Chin, S.F., Sherry, A.D., Boothman, D.A., and Gao, J.M. (2006) *Nano Lett.*, **6**, 2427.

301 Ai, H., Flask, C., Weinberg, B., Shuai, X., Pagel, M.D., Farrell, D., Duerk, J., and Gao, J.M. (2005) *Adv. Mater.*, **17**, 1949.

4
Organization of Nanoparticles

4.1
Semiconductor Nanoparticles

Nikolai Gaponik and Alexander Eychmüller

Once formed in solution or as powders, semiconductor nanocrystals may become building blocks for larger supramolecular units. These units may form "automatically," that is, by crystallization or self-assembly, or they may be formed "intentionally" using one of several techniques. The latter methods of formation are discussed here. Structures of semiconductor nanoparticles up to several hundreds of microns in size can be formed as crystals or superlattices (see Section 4.1.1), depending on the fine adjustment of the experimental conditions. Several experimental methods, such as the Langmuir–Blodgett and layer-by-layer (LbL) techniques – which have in common the build-up of layered structures consisting of disordered particles – are described in Section 4.1.2, while the coupling (either ionic or covalent) of semiconductor particles is detailed in Section 4.1.3.

4.1.1
Molecular Crystals and Superlattices

One of the most interesting developments of the past decade has been the merging of the colloidal preparation of nanostructures with synthetic approaches that stem from advanced inorganic chemical routes (see, e.g., Section 3.2.3), both of which lead to very similar nanoparticles. Both routes yield species that crystallize into superstructures with sizes that may reach several hundreds of microns. When single-crystal X-ray diffraction (XRD) is performed on these macroscopic crystals, the data obtained yield the exact positions of every single atom in the superstructures.

Based on the studies of Strickler [1] and Dance and coworkers [2–5] in 1993, Herron *et al.* reported on the crystallization of $Cd_{32}S_{14}(SPh)_{36}$·4DMF (Ph = C_6H_5; DMF = dimethyl formamide) as pale yellow cubes [6]. The structure unraveled by single-crystal XRD studies consisted of an 82-atom CdS core that was a piece of the

Nanoparticles: From Theory to Application. Edited by Günter Schmid
Copyright © 2010 WILEY-VCH Verlag GmbH & Co. KGaA, Weinheim
ISBN: 978-3-527-32589-4

bulk cubic lattice appearing as a tetrahedron, the points of which were capped by DMF solvent molecules. The cluster remained soluble, and as such retained the properties of a discrete molecule. From the point of view of a nanocrystal chemist interested in studying the properties of particles as a function of size, the "Cd_{32}-cluster" produced is exceptional because, unlike all other colloidally prepared samples, it has no size distribution; rather, its structure is a larger homolog of the $[Cd_{17}S_4(SPh)_{28}]^{2-}$ cluster prepared by Lee et al. [5]. As in the case of the "Cd_{32}-cluster", this "Cd_{17}-cluster" appears as a noncovalently bound cluster in the crystal. An identical cluster core was prepared by Vossmeyer et al., but via a distinctly different route and with another superstructure in the evolving crystals [7]. The preparative route was essentially described earlier as being colloid–chemical in nature, in which $Cd(ClO_4)_2$ and 2-mercaptoethanol were dissolved in water, followed by H_2S injection in alkaline solution [8]. Exhaustive dialysis against water yielded crystalline material which first appeared as needles and evolved as blocky crystals after standing in the mother liquor for some weeks. In contrast to the above-mentioned clusters, in this case intermolecular bridges were formed that built up a superlattice framework. Smaller thiophenolate-capped clusters containing zinc – for example, $[Zn_4(\mu_2\text{-}SPh)_6(SPh)_4]^{2-}$, $[Zn_{10}(\mu_3\text{-}S)_4(\mu_2\text{-}SPh)_{12}]$, and $[Zn_{10}(\mu_3\text{-}S)_4(\mu_2\text{-}SPh)_{12}(SPh)_4]^{4-}$ – have been prepared as supertetrahedral fragments and possible molecular models for cubic ZnS, and their electronic structure has been evaluated theoretically [9]. Very similar structures may be obtained via organometallic synthetic routes; clusters derived in this manner include $[Cd_{32}Se_{14}(SePh)_{36}(PPh_3)_4]$ [10] and its mercury analog, which display exactly the same internal structure as the above-mentioned sulfur-containing clusters. Other crystalline material consists of $Cd_{17}Se_4(SePh)_{24}(PPh_2Pr)_4]^{2+}[Cd_8Se(SePh)_{12}Cl_4]^{2-}$ [11] ($Pr = n$-propyl) and the smaller cluster $[Cd_{10}Se_4(SPh)_{12}(PPr_3)_4]$ [12]. The latter studies also included results relating to the cluster compound $[Cd_{16}(SePh)_{32}(PPh_3)_2]$, which lacks a central atom, in contrast to the "Cd_{17}" cluster, thus forming a structure similar to that of Koch pyramids [13].

Together with the Banin group, the authors of Refs [10–12] carried out optical spectroscopy investigations on some of the cluster molecules obtained [14–16]. These materials were treated as the molecular limit of the bulk semiconductor CdSe, and issues such as oscillator strength, steady-state and time-resolved photoluminescence and photoluminescence excitation were addressed. In addition, emission-mediating vibrational modes were detected, and photobleaching effects observed.

In an investigation of particle–particle interaction in semiconductor nanocrystal assemblies, Döllefeld et al. [17] examined (among other structures) the crystalline superstructure of $[Cd_{17}S_4(SCH_2CH_2OH)_{26}]$ (as described in Ref. [7]). In the UV-visible absorption spectrum of this compound in solution, the first electronic transition was shifted to higher energies by about 150 meV as compared to the reflection spectrum of the crystalline material. In addition, the transition was broadened from a full-width at half maximum of approximately 390 meV to about 520 meV. Most likely, a complete description of the interaction of semiconductor nanocrystals in crystalline superstructures would include both electronic and dipole–dipole interactions. The electronic coupling may be introduced by covalent bridging of the nanocrystals in the crystalline superstructure, though detailed theoretical studies would provide a more quantitative picture and, indeed, the first attempts in this direction were reported only recently [18].

Yet another development of remarkable nanostructured materials yields superlattices of nanosized objects. As there is no clear distinction between "molecular crystals" and "superlattices" formed from nanoparticles, at this point reference will be made to structures composed of very similar (but most likely not exactly identical) nanoparticles, namely colloidal particles in the size range 2 to 10 nm. Two excellent reviews by leading experts in the field were produced in 1998 and 2000 [19, 20], the titles of which contained the terms "nanocrystal superlattices" and "close-packed nanocrystal assemblies." These are in line with the above-outlined delimitation, although Collier *et al.* have also reported on molecular crystals (as above). The two reviews comprised approximately 100 pages with some 300 references, and summarized the state of the art at that time in exemplary fashion. The topics included preparative aspects of the formation of monodisperse nanoparticles of various compositions including metals, the superlattice formation itself with some theoretical background, covalent linking of nanocrystals (see below), and an appropriate description of the physical properties and characterization of the nanocrystal superlattices.

Naturally, both reviews referred to the pioneering studies of Murray *et al.* from 1995, in which the self-organization of CdSe nanocrystals into three-dimensional (3-D) colloidal crystals was demonstrated [21] (see also Refs [22, 23]). A prerequisite for superlattice formation is a high degree of monodispersity of the particles, as a result of synthetic advances. Characterization of the evolving structures may be carried out by using absorption and emission spectroscopy, high-resolution scanning electron microscopy (HR-SEM) and transmission electron microscopy (TEM), electron diffraction and, very impressively, XRD, so as to provide clear-cut evidence of both the long-range order and tunability of the interdot spacing by varying the size and nature of the capping agents.

One very fine example of superlattice formation was reported by Talapin *et al.* [24, 25]. The concept of the nucleation procedure is depicted in Figure 4.1, where a nonsolvent (methanol) for the CdSe particles dispersed in toluene diffuses either directly (left tube in Figure 4.1a) or is slowed down through a buffer layer (propan-2-ol, right tube) into the colloidal solution. Over weeks, this leads to irregularly shaped crystals in the case of fast diffusion (Figure 4.1b), or to perfectly faceted hexagonal colloidal crystals with sizes of about 100 μm (Figure 4.1c) consisting of about 10^{12} individual 3.5 nm CdSe particles.

Figure 4.2 shows the TEM images of these materials, in which a close packing of up to seven monolayers of CdSe particles is displayed (Figure 4.2a), demonstrating the earliest stages of 3-D superlattice formation starting from a two-dimensional (2-D) lattice. Figure 4.2b and c, with their corresponding fast Fourier transforms (insets) display differently oriented segments of the evolving superlattices, leading to an assignment of the 3-D alignment of the particles to fcc-like.

Similar methods were later successfully applied for the fabrication of PbS, PbSe, PbTe, and γ-Fe_2O_3 superlattices [26–28]. Moreover, binary superlattices built from variations of PbS, PbSe, $CoPt_3$, Fe_2O_3, Au, Ag, and Pd nanocrystals exhibiting one of

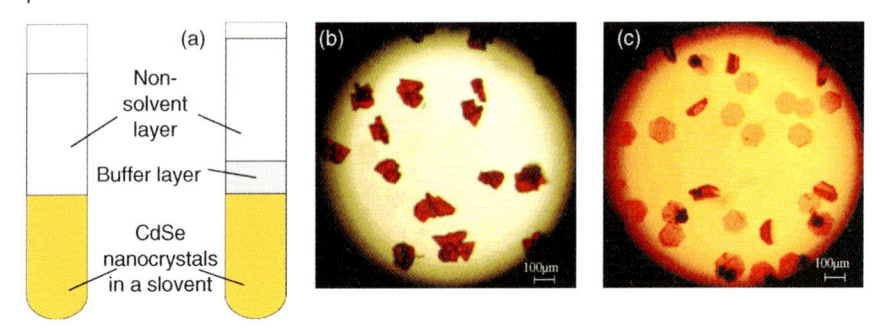

Figure 4.1 (a) Schematic illustration of superlattice formation from colloidal solutions of 3.5 nm CdSe nanocrystals in toluene; (b,c) Optical micrographs of the evolving nanocrystal superlattices after fast (b) and slow (c) nucleation. Reproduced with permission from Ref. [24].

the AB, AB_2, AB_3, AB_4, AB_5, AB_6, or AB_{13} stoichiometries with cubic, hexagonal, tetragonal and orthorhombic symmetries have been realized and identified [29, 30]. The self-organization of CdSe and CdSe/CdS nanorods into nematic, smectic, and crystalline solids by a slow destabilization of the respective nanoparticle solution (the diffusion of a nonsolvent into the colloidal solution of nanorods) was demonstrated [31]. The colloidal crystals of nanorods obtained showed a characteristic birefringence which was assigned to a specific spherulite-like texture of each nanorod assembly. The fact that the nanocrystal building blocks present in the superlattices were in the closest possible proximity may have led to an efficient energy transfer as well as electronic interactions, and to the formation of collective electronic states related to the bulk supercrystalline structure, and to crystal defects as well as to surface states. Nonetheless, the physico-chemical characterization of these promising novel materials remains in its infancy. Among others, an efficient energy transfer in the CdSe closely packed solids has been reported, as has been a weak fluorescence from the surface of CdSe supercrystals [32], energy transfer and fluorescence quenching in binary CdSe:Au superlattices [33], fluorescence from organized

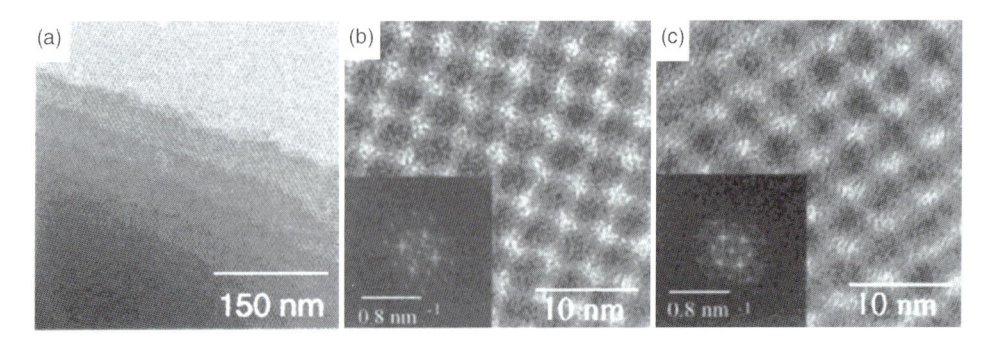

Figure 4.2 Transmission electron microscopy (TEM) images of superlattices formed from CdSe nanocrystals. Different orientations of the CdSe nanocrystal superlattice with corresponding fast Fourier transforms as insets leading to the assignment of fcc-like packing of the particles. Reproduced with permission from Ref. [24].

assemblies of CdSe/CdS heterostructure nanorods [31], as well as enhanced thermopower in PbSe superlattices [34].

In addition to the above-described self-organization, there exist other approaches to the formation of superlattices, including the use of bacterial S-layers as templates for the nucleation of ordered 2-D arrays of CdS nanocrystals [35], crystallization on other periodically patterned surfaces [36], or the "lost-wax approach" (as reported recently), in which polymer templates are first formed from silica colloidal crystals, with subsequent filling of the evolving voids of the porous polymer with a variety of colloidal particles [37]. Filling of the voids of "artificial opals" (periodic structures formed from, for example, SiO_2, latex, or polymethylmethacrylate spheres) with semiconductor nanocrystals such as CdSe [38], followed by destruction of the template, leads to so-called "inverse opals," which are of eminent interest in the evolving field of photonics. Another procedure yields CdS nanoparticles inside 3-D hexagonal mesoporous silica films [39]. Dimethylaminoethanethiol-stabilized CdTe nanocrystals may self-assemble due to charge–dipole interactions, to form free-floating 2-D sheets [40, 41]. One area still to be approached with group II–VI semiconductor materials is the "nanotectonic" approach, in which 3-D objects are formed via a template-directed assembly, using nanocrystals as the building units [42, 43].

Based on studies of the formation of amorphous 3-D arrays of CdSe nanocrystals [44] and of compact round disks of similar particles [45], a step forward was proposed by Ge and Brus, who reported planar monolayer aggregates that consisted of 80–100 identical 4 nm CdSe nanocrystals which diffused on planar graphite surfaces [46]. Subsequent investigations into the collective motion of a very small nanocrystal superlattice have opened up yet another new field of research.

4.1.2
Layers of Semiconductor Nanocrystals

Layers of semiconductor nanoparticles may be prepared, in a variety of ways, either from preformed colloids via techniques such as spin-coating or LbL assembly, or from ionic, atomic or molecular precursors via, for example, electrodeposition, chemical vapor deposition (CVD), or metal-organic chemical vapor deposition (MOCVD). Selected examples of layered structures, derived from a wealth of reports, are described here as examples, together with some attempts at their characterization.

At a relatively early stage in the development of the field, the Langmuir–Blodgett (LB) technique was recognized as a useful tool for preparing thin layers of nanocrystalline materials, as well as for controlling parameters such as film composition and thickness. Based on the results of earlier studies performed in 1992 [47–51], Grieser et al. used the LB technique to generate not only multilayer films of CdS, CdSe and CdTe, but also mixed crystal compounds such as CdS_xCdSe_{1-x}, CdS_xCdTe_{1-x} and $CdSe_xCdTe_{1-x}$ [52]. Most of the films are formed from multilayers of cadmium arachidate that are exposed to H_2X (X = S, Se, Te), and which results in a formation of the respective CdX particles within the matrix of the LB films. The best characterization in terms of particle size is based on the relationship between absorption energy and size, whereas direct methods such as atomic force microscopy (AFM) and TEM do not yield unambiguous results. One fine example of the early application of a local probe technique on colloido-chemically derived nanostructures was reported by Facci and coworkers on similarly derived LB films of CdS particles [53]. The film formation was monitored using XRD, absorption spectroscopy, nanogravimetry, and scanning tunneling microscopy (STM).

The organometallically derived TOP/TOPO-capped CdSe particles (see Section 3.2.1.3) are also suitable for application as monolayer-forming elements on the water surface of an LB trough, as demonstrated by Bawendi and coworkers [54]. Whilst the results of absorption and luminescence studies have indicated that the monolayers would retain the optical properties of the isolated crystallites, TEM investigations have revealed the formation of 2-D hexagonal close-packed regions. Fendler *et al.* also described studies with LB-derived films of CdS nanoparticles [55] and other sulfides and selenides [56]. Characterization of the films included the creation of surface pressure versus surface area isotherms, in addition to Brewster angle microscopy, while Ref. [55] focused on the emission properties (spectral and time-resolved) of the films, and Ref. [56] on methodic issues of the *in situ* generation of size-quantized semiconductor particulate films. The emission properties of LB-derived nanocrystal monolayers were also investigated by Cordero *et al.* [57] who, by comparing wet and dry films of CdSe particles with and without illumination, suggested that water molecules, when adsorbed onto the surface of the quantum dots (QDs), would serve to: (i) passivate the surface traps, accompanied by an increased emission; and (ii) oxidize the surface species, so as to introduce new surface defects that would result in a reduction of the luminescence quantum yield. Xu *et al.* have demonstrated the possibility to derive an average limiting nanoparticle area from the π-A isotherms of the CdSe and CdSe/ZnS TOPO-capped QDs, which could be used to establish an average size of the QDs if the thickness of the TOPO shell were to be counted [58]. Klimov and coworkers studied the spectrally resolved dynamics of Förster resonant energy transfer (FRET) in single monolayers and bilayers of semiconductor nanocrystal QDs assembled using LB techniques [59]. For a single monolayer, a distribution of transfer times was observed from approximately 50 ps to about 10 ns. This could be quantitatively modeled, assuming that the energy transfer was dominated by interactions of a donor nanocrystal with acceptor nanocrystals from the first three shells surrounding the donor. These authors also detected an effective enhancement of the absorption cross-section (by up to a factor of 4) for larger nanocrystals on the red side of the size distribution; this resulted from a strong interdot electrostatic coupling in the LB film (the "light-harvesting antenna effect").

As demonstrated by Heath *et al.*, application of the LB technique in conjunction with semiconductor nanoparticles may lead to the generation of tunnel diodes [60]. These devices consist of a monolayer of 3.8 nm CdSe nanocrystals and an insulating bilayer of eicosanoic acid, sandwiched between an Au and an Al electrode. Advanced spectroscopic techniques such as attenuated low-energy photoelectron spectroscopy were also applied to LB-derived multilayered nanostructured assemblies of differently sized CdS particles [61]. Recent examples of applications of the LB technique have included conjugated polymer–CdSe QD nanocomposites suitable for photovoltaics [62], the micropatterning of InP and PbSe nanocrystals on silicon [63], and biosensing based on avidin-modified and LB-assembled ZnS–CdSe QD layers [64].

Both, chemical solution deposition (CD) and electrodeposition (ED) – two alternative techniques of film formation – make use of neutral and charged species in solution. These types of film making are inseparably linked to the name of Gary Hodes. Although a long list of reports could be cited here, reference is made instead to a review [65] and to a few examples demonstrating the basic concepts [66–68]. One method of film formation which has attracted considerable interest in recent years is the LbL assembly. This method, which was first proposed by G. Decher [69], is based on the alternating adsorption of oppositely charged species, and was originally developed for positively and negatively charged polyelectrolytes. It also provides a useful tool for the assembly of nanoparticles and nanoparticle–polymer composites on planar substrates, and on other surfaces of different geometries and curvatures [70–72]. As seen from Figure 4.3, the formation of monolayers of deposited material is based on the electrostatic interaction between the nanocrystals and the surface. Alternation of the sign of the charges of the species to be deposited allows the multilayers to grow quite thickly, while the introduction of new components in one of the layers provides the opportunity of virtually unlimited, but controllable, variations of LbL-multistructure compositions.

Based on its universality, simplicity and low cost, the LbL method has become very widely used, and the list of reports on the subject is very long; hence, only a few recent examples will be described at this point. One of the most successful applications of

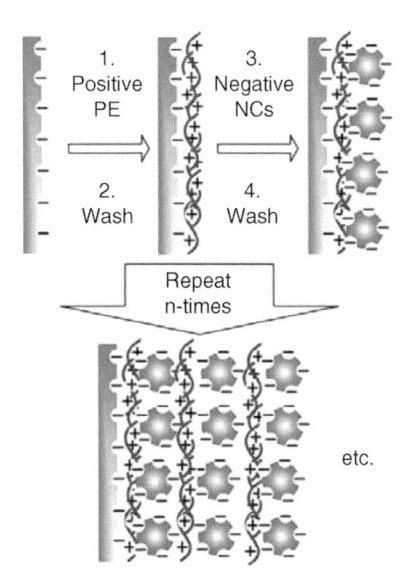

Figure 4.3 Schematic representation of the LbL assembly involving polyelectrolyte (PE) molecules and oppositely charged nanoparticles. The procedures 1 to 4 can be repeated to assemble more polyelectrolyte/nanoparticle bilayers.

the LbL technique is in the fabrication of thin film light-emitting diodes (LEDs). The LbL-assembled nanocrystal-based LEDs, which originate from the studies of M. Gao et al., were first built from negatively charged CdSe particles embedded in poly(p-phenylene vinylene) [73]. A broad, almost white and relatively weak electroluminescence was observed from devices of this type. As the next step, a color control of the LbL electroluminescent devices was achieved by using thioglycolic acid-capped CdTe nanoparticles of different sizes, alternately assembled with poly(diallyldimethylammonium chloride) (PDDA) [74]. Further developments of CdTe-based LEDs have resulted in devices with an external quantum efficiency of up to 0.51%, and with a reasonably low turn-on voltage of 2.5 V [75, 76]. The typical electroluminescence characterization, together with a photographic image, of the best-performing red-emitting LED built from 30 PDDA/CdTe bilayers is shown in Figure 4.4. The perspectives of this type of device have been discussed recently by Rogach et al. [77].

Another fascinating area of LbL applications is that of spherical assemblies, where nanoparticles are assembled on the surfaces of larger polymer or inorganic beads [78]. Examples of these assemblies have shown promise for encoding combinatorial libraries and bio-recognition [79], in drug-delivery systems [80–82], photonic crystals [83], spherical resonators [84, 85], subwavelength emitters [86], and photonic molecules [87]. Ultralow-threshold, continuous-wave lasing has been achieved in the near-infrared (NIR) spectral region (tunable between 1240–1780 nm) at room

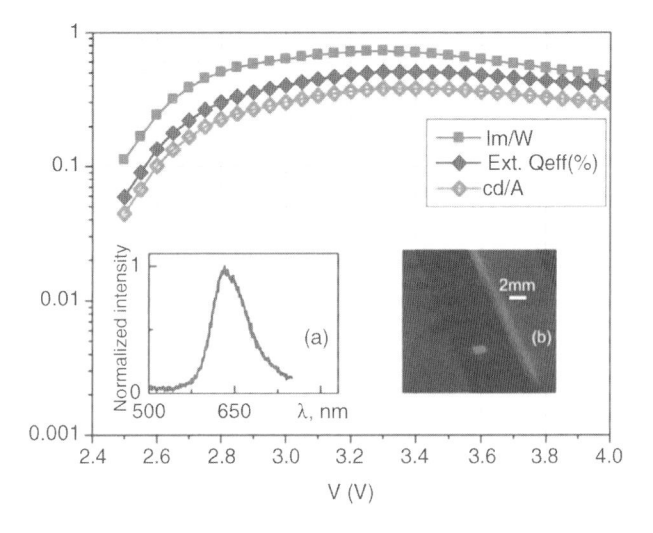

Figure 4.4 Efficiency values for a CdTe/PDDA light-emitting diode (LED). The device showed an electroluminescence turn on at 2.5 V, and maximum light output was obtained at 3.3 V and 350 mA cm^{-2}, with a peak radiated power of 141 nW, corresponding to an external quantum efficiency of 0.51%. The luminous efficiencies reached 0.4 cd A^{-1} and 0.81 Lm W^{-1}. The insets show (a) the emission spectrum of a 30-bilayer device and (b) a view of the emissive area while the device is operated under standard laboratory conditions, without sealing or packaging. Reprinted with permission from Ref. [75]; © 2007, American Institute of Physics.

temperature in a fused-silica microsphere that has been coated with HgTe QDs [85]. Small polystyrene beads covered with CdTe nanocrystals, when mounted at the end of an optical tip, were used to enhance the signal and improve the resolution of scanning near-field optical microscopy (SNOM), due to Förster resonance energy transfer (FRET) effects [88].

One of the major advantages of the LbL technique is that it can be applied to very different (i.e., both multifunctional and multicomponent) assemblies, so as to create a variety of fascinating architectures that allow efficient FRET. For example, LbL films built from CdTe nanocrystals of two different sizes – one (the donor) with an emission maximum matching the first absorption maximum of another (the acceptor) – shows a FRET rate of up to 251 ps^{-1} [89]. In the above-mentioned case, PDDA was used as positively charged polyelectrolyte to assemble negatively (thioglycolic acid-capped) charged nanocrystals. In this case, the PDDA layer increased the distance between the donor and acceptor such that was not optimal for efficient FRET. Subsequently, a further improvement was achieved by using an alternating deposition of negatively and positively (cysteamine-capped) CdTe nanocrystals, without any polyelectrolyte in between [90]. As a result, the FRET rate in the thus-formed nanocrystal bilayers exceeded 71 ps^{-1}! This efficient energy transfer, combined with a possibility of building graded semiconductor nanocrystal films [91], has resulted in the LbL fabrication of exciton recycling structures that possess great potential as building blocks for energy scavengers [92]. The principle of such an exciton recycling is shown in Figure 4.5.

Multicomponent composites built by LbL assembly may consist not only of nanocrystals but also of nanowires, biomolecules, dyes, and functional polymers. In addition, metal nanoparticles may be employed with potential applications in the fields of drug delivery and biodetection, energy harvesting, optical signal processing, and emission enhancement or quenching [70, 93].

Generally, the LbL deposition has been shown possible for all types of semiconductor nanoparticles synthesized in water, including ZnSe [94], CdS [95],

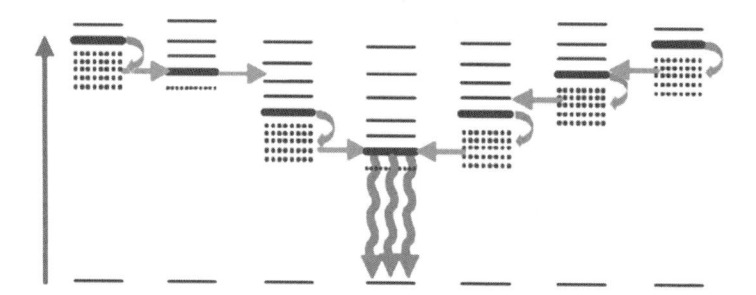

Figure 4.5 Exciton recycling in a LbL-assembled graded CdTe nanostructure. Excitons in the six layers of smaller-sized nanocrystals (NCs) may be trapped in surface states. However, the trapped excitons can be transferred very efficiently from layers with smaller NCs to layers with larger NCs. Most importantly, these recycled excitons finally reach the center layer of red-emitting NCs with only low excess energy, which reduces the probability of being trapped. Adapted with permission from Ref. [92].

CdSe [73, 96], and HgTe [97, 98], because these types of particles are charged. In order to apply LbL deposition to nanoparticles synthesized in organic solutions and stabilized by nonpolar molecules, a surface capping exchange must be utilized whereby, among others, nonpolar surfactants can be efficiently exchanged with mercaptopropionic or mercaptoundecanoic acids, with the resultant nanoparticles dissolving readily in aqueous solution for further use in LbL assemblies.

A report by Gao *et al.* [99] was devoted to the lateral patterning of CdTe nanocrystal films by a modification of the conventional LbL technique, namely by using an electric field. The idea here is that the electric field will direct the spatially selective deposition of the LbL films, such that binary patterned arrays of films containing two differently sized CdTe nanocrystals and a polyelectrolyte can be produced on conductive indium–tin oxide electrodes in a multistep process. The impressive results of these operations are displayed in Figure 4.6. In this case, Figure 4.6a and b show the differently sized negatively charged CdTe nanocrystals, which are deposited on either the upper (immersion of the biased electrode into the solution of the red emitting particles) or the lower (immersion into a solution of the green emitting particles) biased electrode. Both types of particle are also found in the unbiased marginal regions; this finally leads to an orange emission in the case of successive deposition of both sorts (cf. Figure 4.6c and the corresponding emission spectrum on the left-hand side of the figure).

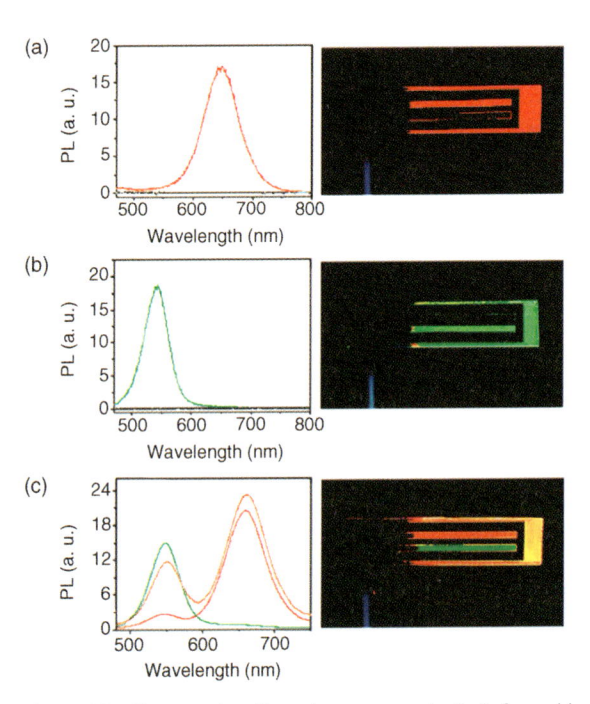

Figure 4.6 Photographs of lateral structures selectively formed by green- and red-luminescing CdTe nanocrystals via an electric field-directed LbL assembly, and the corresponding emission spectra. Reproduced with permission from Ref. [99].

Several other techniques have also been utilized for film formation, including *drop casting* (e.g., CdSe on a Pt electrode in order to study electrochromic effects in nanocrystal films [100]), *spin casting* (e.g., composite films of CdSe and electron-transporting block copolymers [101]), *photochemical deposition* (e.g., CdSe films from aqueous solution [102]), and *electrospray organometallic CVD* for the formation of QD composites [103].

Thiol-capped nanocrystals (e.g., CdTe or ZnSe) may be self-assembled with at least 50% coverage efficiency on freshly cleaned gold surfaces [104].

Regardless of the film-formation procedure, the question of the interaction of semiconductor nanocrystal assemblies still arises, and has been the subject of much investigation by research groups during recent years. In 1994, Vossmeyer *et al.* reported on the absorption properties of very small CdS nanoparticles [8], where a red shift of the first electronic transition with respect to the transition seen in solution was observed when compact layers were formed from the material by spin coating. A comparatively small red shift in the emission spectra of the crystalline superstructure of CdSe nanoparticles (see Section 4.1.1) with respect to the emission peaks in solution, can be explained by the small-sized deviations of particles within the sample, resulting in an energy transfer from the smaller to the larger particles [21, 105, 106]. Similar observations were made for group III–V semiconductor nanocrystals of InP [107]. Artemyev *et al.* prepared unstructured solids also from CdSe nanoparticles with polymers acting as spacers, in order to control the distance between the particles [108–110]. The pure solid that was built up by drop-casting solutions of semiconductor nanoparticles showed very broad transitions, and bore a resemblance to a bulk spectrum rather than to isolated absorption bands. Micic *et al.* obtained similar results in disordered films of InP nanoparticles [111]. Following studies on the photoconductivity in CdSe QD solids [112], Leatherdale *et al.* reported on CdSe particles in different dielectrics [113]. In agreement with the results in Ref. [8], a red shift of the first electronic transition was found that exhibited no significant broadening. Kim *et al.* presented the results of studies monitoring pressure dependency in solutions and in compact layers of semiconductor nanoparticles of CdSe [114]. In this case, different stabilizing agents were used and, in contrast to the findings of Leatherdale *et al.*, no red shift was found for the compact layers of TOP/TOPO-stabilized CdSe particles. Only for those particles stabilized with pyridine was observed a shift to a lower transition energy. Again, no significant broadening of the electronic transitions was noted. As noted in Section 4.1.1, reflection spectroscopy on the cluster crystal compound $Cd_{17}S_4(SCH_2CH_2OH)_{26}$ revealed a substantial broadening and a low-energy shift of the first absorption band which, as a first step, may be explained by both dipole–dipole interaction and electronic coupling of the nanocrystals in the crystalline superstructure. When comparing these results with those from disordered solids obtained by spin-coating concentrated solutions of the same material, it was noted that extreme care must be taken in order to avoid the measurement of artifacts [115]. The misinterpretation of observed changes in transition energies in these structures can be avoided by selecting an integrating sphere as the experimental set-up. The resultant "real" red shift of the transition energy can be explained in a simple picture, by the dipole–dipole interactions of the semiconductor nanocrystals in the films. More elaborate, but theoretical, studies have been undertaken to examine topics such as exciton dissociation and interdot transport [116], to take into account the dielectric medium surrounding the nanoparticles [117], and to explain any observed solvatochromism in CdSe colloidal QDs. This also takes into account the screening of the ligand shell [113], and uses pair potentials to predict the phase behavior and thermodynamics of small particles [118]. Because of its potential applications (based also on experimental studies of, for example, fast kinetics [119] and electronic transport in films of colloidal particles [120–124]), this area of research will surely remain active in the years to come.

4.1.3
Coupling of Semiconductor Nanocrystals

Perhaps one of the most demanding challenges in current preparational research is the directed coupling of semiconductor nanoparticles. If the QDs were called "artificial atoms," this endeavor would yield "artificial molecules," and the challenge would be to create such QD molecules in a desired manner to form, for example, dimers ("two-atomic QD molecules") of various compositions, AB_4 molecules, and so on. As the field is about to emerge from its infant status, the very first attempts in this direction are reported in the following sections.

The coupling of semiconductor nanocrystals to, for example, small particles of metals [125] or insulators [126, 127], is beyond the scope of this chapter. In addition, the coupling of semiconductor nanoparticles to DNA [128–133] and to other biological systems [134–140], as well as to organic [141] or metal-organic fluorophores [142] and functional polymers [143], have been excluded at this point, so that the literature can be more clearly arranged.

A report by Kolny, Kornowski and Weller described the use of an electrostatic interaction between oppositely charged CdS nanocrystals [144]. It has been shown, by using absorption spectroscopy, XRD and high-resolution electron microscopy, that two types of aggregate may be formed: (i) complex-like structures, which are obtained when a large excess of one type of particle is used; and (ii) three-dimensionally ordered solids, when comparable amounts of the constituents are mixed. In addition, the strength of interaction between the particles may be controlled by adjusting the ionic strength of the solution. Another approach relies on the covalent bonding of different [145, 146] or similar [147] semiconductor nanocrystals. The latter studies describe the synthesis, isolation and characterization of CdSe-homodimers. When monodisperse CdSe particles capped with TOPO were treated with various surface-active species, thiols proved to be a good choice to passivate the surfaces. For coupling of the nanocrystals, bis(acylhydrazine) was used, which can bridge two semiconductor nanocrystals via its two terminating NH_2-groups. Size-selective precipitation finally leaves mostly homodimers in solution. The covalent coupling of different CdTe nanocrystals has been reported [145]. In this case, the CdTe particles are stabilized with thioacids or aminothiols, so as to form an amide bond using carbodiimide as a mediating agent. Thus, the stabilizing ligands can be used directly for the coupling, without a bridging molecule, in contrast to the above-mentioned case and the findings of Trindade et al. [146], in which the coupling of PbS particles to CdS or CdSe nanocrystals, via the bridging ligand 2,2′-bipyrimidine, was outlined. A careful analysis of the IR transmission spectra of a sample containing only a mixture of the two types of nanocrystal, and of a sample of the coupled particles, revealed the existence of N–H vibrations within the amide group, and a diminishment of the O–H vibration in the carbonyl group due to the coupling reaction [145]. These authors also stated that a possible alternative interpretation should be considered. In their own words:

> "Since the nanocrystalline powders are washed with buffer solution and dried in a vacuum in order to perform the IR measurements, the existence of coupled stabilizers without connection to at least one nanocrystalline surface can be excluded. However, it is conceivable that some of these 'amide bridges' fall off the nanocrystal on one side. As the S–H bond vibrations are detected in neither Raman nor IR spectroscopy, this hypothesis cannot yet be verified."

The study of interactions in this type of covalent assembly, as compared to those built electrostatically (due to opposite surface charges of thioacid- and thioamine-stabilized nanoparticles), showed FRET rates between nanocrystals in both types of assembly in the range of 10^{10}–10^{11} s^{-1}, this being relatively faster in the case of covalent-coupled assemblies due to the shorter interparticle distances and direct covalent interactions [148]. Divalent, positively charged Ca^{2+} ions can serve as electrostatic chelating linkers capable of inducing the clustering of CdTe nanocrystals (NCs) through coordination to the negatively charged carboxylic groups of thioglycolic or mercaptopropionic acid ligands [149]. Moreover, the addition of sodium carbonate, which binds Ca^{2+}, reverses the clustering process. An efficient FRET was found to be responsible for changes in the emission spectra and the luminescence decay times of both binary- and ternary-sized CdTe NC clusters. Together with theoretical investigations [18, 150, 151], this subfield of nanoscience is expected to continue and to grow in stature as more sophisticated preparational attempts are launched and new demands from the applications area are imposed.

References

1 Strickler, P. (1969) *J. Chem. Soc., Chem. Commun.*, 655.
2 Dance, I.G., Scudder, M.L., and Secomb, R. (1983) *Inorg. Chem.*, **22**, 1794.
3 Dance, I.G., Choy, A., and Scudder, M.L. (1984) *J. Am. Chem. Soc.*, **106**, 6285.
4 Dance, I.G. (1986) *Polyhedron*, **5**, 1037.
5 Lee, G.S.H., Craig, D.C., Ma, I., Scudder, M.L., Bailey, T.D., and Dance, I.G. (1988) *J. Am. Chem. Soc.*, **110**, 4863.
6 Herron, N., Calabrese, J.C., Farneth, W.E., and Wang, Y. (1993) *Science*, **259**, 1426.
7 Vossmeyer, T., Reck, G., Katsikas, L., Haupt, E.T.K., Schulz, B., and Weller, H. (1995) *Science*, **267**, 1476.
8 Vossmeyer, T., Katsikas, L., Giersig, M., Popovic, I.G., Diesner, K., Chemseddine, A., Eychmüller, A., and Weller, H. (1994) *J. Phys. Chem.*, **98**, 7665.

9 Bertoncello, R., Bettinelli, M., Casarin, M., Maccato, C., Pandolfo, L., and Vittadini, A. (1997) *Inorg. Chem.*, **36**, 4707.

10 Behrens, S., Bettenhausen, M., Deveson, A.C., Eichhoefer, A., Fenske, D., Lohde, A., and Woggon, U. (1996) *Angew. Chem., Int. Ed. Engl.*, **35**, 2215.

11 Behrens, S. and Fenske, D. (1997) *Ber. Bunsenges. Phys. Chem.*, **101**, 1588.

12 Behrens, S., Bettenhausen, M., Eichhofer, A., and Fenske, D. (1997) *Angew. Chem., Int. Ed. Engl.*, **36**, 2797.

13 Ptatschek, V., Schmidt, T., Lerch, M., Mueller, G., Spanhel, L., Emmerling, A., Fricke, J., Foitzik, A.H., and Langer, E. (1998) *Ber. Bunsenges. Phys. Chem.*, **102**, 85.

14 Soloviev, V.N., Eichhofer, A., Fenske, D., and Banin, U. (2000) *J. Am. Chem. Soc.*, **122**, 2673.

15 Soloviev, V.N., Eichhofer, A., Fenske, D., and Banin, U. (2001) *Phys. Status Solidi B*, **224**, 285.

16 Soloviev, N.V., Eichhofer, A., Fenske, D., and Banin, U. (2001) *J. Am. Chem. Soc.*, **123**, 2354.

17 Döllefeld, H., Weller, H., and Eychmüller, A. (2001) *Nano Lett.*, **1**, 267.

18 Nosaka, Y. and Tanaka, H. (2002) *J. Phys. Chem. B*, **106**, 3389.

19 Collier, C.P., Vossmeyer, T., and Heath, J.R. (1998) *Annu. Rev. Phys. Chem.*, **49**, 371.

20 Murray, C.B., Kagan, C.R., and Bawendi, M.G. (2000) *Annu. Rev. Mater. Sci.*, **30**, 545.

21 Murray, C.B., Kagan, C.R., and Bawendi, M.G. (1995) *Science*, **270**, 1335.

22 Heath, J.R. (1995) *Science*, **270**, 1315.

23 Weller, H. (1993) *Angew. Chem., Int. Ed. Engl.*, **32**, 41.

24 Talapin, D.V., Shevchenko, E.V., Kornowski, A., Gaponik, N., Haase, M., Rogach, A.L., and Weller, H. (2001) *Adv. Mater.*, **13**, 1868.

25 Rogach, A.L., Talapin, D.V., Shevchenko, E.V., Kornowski, A., Haase, M., and Weller, H. (2002) *Adv. Funct. Mater.*, **12**, 653.

26 Nagel, M., Hickey, S.G., Froemsdorf, A., Kornowski, A., and Weller, H. (2007) *Z. Phys. Chem.*, **221**, 427.

27 Urban, J.J., Talapin, D.V., Shevchenko, E.V., and Murray, C.B. (2006) *J. Am. Chem. Soc.*, **128**, 3248.

28 Talapin, D.V., Shevchenko, E.V., Murray, C.B., Titov, A.V., and Kral, P. (2007) *Nano Lett.*, **7**, 1213.

29 Shevchenko, E.V., Talapin, D.V., Murray, C.B., and O'Brien, S. (2006) *J. Am. Chem. Soc.*, **128**, 3620.

30 Shevchenko, E.V., Talapin, D.V., O'Brien, S., and Murray, C.B. (2005) *J. Am. Chem. Soc.*, **127**, 8741.

31 Talapin, D.V., Shevchenko, E.V., Murray, C.B., Kornowski, A., Förster, S., and Weller, H. (2004) *J. Am. Chem. Soc.*, **126**, 12984.

32 Talapin, D.V., Shevchenko, E.V., Gaponik, N., Radtchenko, I.L., Kornowski, A., Haase, M., Rogach, A.L., and Weller, H. (2005) *Adv. Mater.*, **17**, 1325.

33 Shevchenko, E.V., Ringler, M., Schwemer, A., Talapin, D.V., Klar, T.A., Rogach, A.L., Feldmann, J., and Alivisatos, A.P. (2008) *J. Am. Chem. Soc.*, **130**, 3274.

34 Wang, R.Y., Feser, J.P., Lee, J.-S., Talapin, D.V., Segalman, R., and Majumdar, A. (2008) *Nano Lett.*, **8**, 2283.

35 Shenton, W., Pum, D., Sleytr, U.B., and Mann, S. (1997) *Nature*, **389**, 585.

36 Lin, K.-h., Crocker, J.C., Prasad, V., Schofield, A., Weitz, D.A., Lubensky, T.C., and Yodh, A.G. (2000) *Phys. Rev. Lett.*, **85**, 1770.

37 Jiang, P., Bertone, J.F., and Colvin, V.L. (2001) *Science*, **291**, 453.

38 Vlasov, Y.A., Yao, N., and Norris, D.J. (1999) *Adv. Mater.*, **11**, 165.

39 Besson, S., Gacoin, T., Ricolleau, C., Jacquiod, C., and Boilot, J.-P. (2002) *Nano Lett.*, **2**, 409.

40 Tang, Z., Zhang, Z., Wang, Y., Glotzer, S.C., and Kotov, N.A. (2006) *Science*, **314**, 274.

41 Zhang, Z., Tang, Z., Kotov, N.A., and Glotzer, S.C. (2007) *Nano Lett.*, **7**, 1670.

42 Davis, S.A., Breulmann, M., Rhodes, K.H., Zhang, B., and Mann, S. (2001) *Chem. Mater.*, **13**, 3218.

43 Meldrum, F.C. and Colfen, H. (2008) *Chem. Rev.*, **108**, 4332.

44 Kim, B.S., Avila, L., Brus, L.E., and Herman, I.P. (2000) *Appl. Phys. Lett.*, **76**, 3715.

45 Ge, G. and Brus, L. (2000) *J. Phys. Chem. B*, **104**, 9573.

46 Ge, G. and Brus, L.E. (2001) *Nano Lett.*, **1**, 219.
47 Smotkin, E.S., Lee, C., Bard, A.J., Campion, A., Fox, M.A., Mallouk, T.E., Webber, S.E., and White, J.M. (1988) *Chem. Phys. Lett.*, **152**, 265.
48 Zylberajch, C., Ruaudel-Teixier, A., and Barraud, A. (1989) *Thin Solid Films*, **179**, 9.
49 Chen, H., Chai, X., Wei, Q., Jiang, Y., and Li, T. (1989) *Thin Solid Films*, **178**, 535.
50 Zhao, X.K., Yuan, Y., and Fendler, J.H. (1990) *J. Chem. Soc., Chem. Commun.*, 1248.
51 Scoberg, D.J., Grieser, F., and Furlong, D.N. (1991) *J. Chem. Soc., Chem. Commun.*, 515.
52 Grieser, F., Furlong, D.N., Scoberg, D., Ichinose, I., Kimizuka, N., and Kunitake, T. (1992) *J. Chem. Soc., Faraday Trans.*, **88**, 2207.
53 Facci, P., Erokhin, V., Tronin, A., and Nicolini, C. (1994) *J. Phys. Chem.*, **98**, 13323.
54 Dabbousi, B.O., Murray, C.B., Rubner, M.F., and Bawendi, M.G. (1994) *Chem. Mater.*, **6**, 216.
55 Tian, Y., Wu, C., and Fendler, J.H. (1994) *J. Phys. Chem.*, **98**, 4913.
56 Fendler, J.H. (1993) *Isr. J. Chem.*, **33**, 41.
57 Cordero, S.R., Carson, P.J., Estabrook, R.A., Strouse, G.F., and Buratto, S.K. (2000) *J. Phys. Chem. B*, **104**, 12137.
58 Xu, J., Ji, X., Gattas-Asfura, K.M., Wang, C., and Leblanc, R.M. (2006) *Colloids Surf. A*, **284+285**, 35.
59 Achermann, M., Petruska, M.A., Crooker, S.A., and Klimov, V.I. (2003) *J. Phys. Chem. B*, **107**, 13782.
60 Kim, S.H., Markovich, G., Rezvani, S., Choi, S.H., Wang, K.L., and Heath, J.R. (1999) *Appl. Phys. Lett.*, **74**, 317.
61 Samokhvalov, A., Berfeld, M., Lahav, M., Naaman, R., and Rabani, E. (2000) *J. Phys. Chem. B*, **104**, 8631.
62 Goodman, M.D., Xu, J., Wang, J., and Lin, Z. (2009) *Chem. Mater.*, **21**, 934.
63 Lambert, K., Moreels, I., Van Thourhout, D., and Hens, Z. (2008) *Langmuir*, **24**, 5961.
64 Gole, A., Jana, N.R., Selvan, S.T., and Ying, J.Y. (2008) *Langmuir*, **24**, 8181.
65 Hodes, G. (1994) *Sol. Energy Mater. Sol. Cells*, **32**, 323.
66 Hodes, G. (1993) *Isr. J. Chem.*, **33**, 95.
67 Baranski, A.S. and Fawcett, W.R. (1980) *J. Electrochem. Soc.*, **127**, 766.
68 Hodes, G. and Albu-Yaron, A. (1988) *Proc. Electrochem. Soc.*, **88-14**, 298.
69 Decher, G. (1997) *Science*, **277**, 1232.
70 Hammond, P.T. (2004) *Adv. Mater.*, **16**, 1271.
71 Shavel, A., Gaponik, N., and Eychmüller, A. (2005) *Eur. J. Inorg. Chem.*, 3613.
72 Kotov, N.A., Dekany, I., and Fendler, J.H. (1995) *J. Phys. Chem. US*, **99**, 13065.
73 Gao, M., Richter, B., Kirstein, S., and Möhwald, H. (1998) *J. Phys. Chem. B*, **102**, 4096.
74 Gao, M., Lesser, C., Kirstein, S., Möhwald, H., Rogach, A.L., and Weller, H. (2000) *J. Appl. Phys.*, **87**, 2297.
75 Bertoni, C., Gallardo, D., Dunn, S., Gaponik, N., and Eychmüller, A. (2007) *Appl. Phys. Lett.*, **90**, 034107.
76 Gallardo, D.E., Bertoni, C., Dunn, S., Gaponik, N., and Eychmüller, A. (2007) *Adv. Mater.*, **19**, 3364.
77 Rogach, A.L., Gaponik, N., Lupton, J.M., Bertoni, C., Gallardo, D.E., Dunn, S., Li Pira, N., Paderi, M., Repetto, P., Romanov, S.G., O'Dwyer, C., Sotomayor Torres, C.M., and Eychmüller, A. (2008) *Angew. Chem., Int. Ed.*, **47**, 6538.
78 Susha, A.S., Caruso, F., Rogach, A.L., Sukhorukov, G.B., Kornowski, A., Möhwald, H., Giersig, M., Eychmüller, A., and Weller, H. (2000) *Colloids Surf. A*, **163**, 39.
79 Gaponik, N., Radtchenko, I.L., Sukhorukov, G.B., Weller, H., and Rogach, A.L. (2002) *Adv. Mater.*, **14**, 879.
80 Gaponik, N., Radtchenko, I.L., Gerstenberger, M.R., Fedutik, Y.A., Sukhorukov, G.B., and Rogach, A.L. (2003) *Nano Lett.*, **3**, 369.
81 Gaponik, N., Radtchenko, I.L., Sukhorukov, G.B., and Rogach, A.L. (2004) *Langmuir*, **20**, 1449.
82 Sukhorukov, G.B., Rogach, A.L., Zebli, B., Liedl, T., Skirtach, A.G., Koehler, K., Antipov, A.A., Gaponik, N., Susha, A.S.,

Winterhalter, M., and Parak, W.J. (2005) *Small*, **1**, 194.

83 Rogach, A., Susha, A., Caruso, F., Sukhorukov, G., Kornowski, A., Kershaw, S., Möhwald, H., Eychmüller, A., and Weller, H. (2000) *Adv. Mater.*, **12**, 333.

84 Rakovich, Y.P., Donegan, J.F., Gaponik, N., and Rogach, A.L. (2003) *Appl. Phys. Lett.*, **83**, 2539.

85 Shopova, S.I., Farca, G., Rosenberger, A.T., Wickramanayake, W.M.S., and Kotov, N.A. (2004) *Appl. Phys. Lett.*, **85**, 6101.

86 Olk, P., Buchler, B.C., Sandoghdar, V., Gaponik, N., Eychmüller, A., and Rogach, A.L. (2004) *Appl. Phys. Lett.*, **84**, 4732.

87 Rakovich, Y.P., Donegan, J.F., Gerlach, M., Bradley, A.L., Connolly, T.M., Boland, J.J., Gaponik, N., and Rogach, A. (2004) *Phys. Rev. A*, **70**, 051801.

88 Müller, F., Götzinger, S., Gaponik, N., Weller, H., Mlynek, J., and Benson, O. (2004) *J. Phys. Chem. B.*, **108**, 14527.

89 Franzl, T., Koktysh, D.S., Klar, T.A., Rogach, A.L., Feldmann, J., and Gaponik, N. (2004) *Appl. Phys. Lett.*, **84**, 2904.

90 Franzl, T., Shavel, A., Rogach, A.L., Gaponik, N., Klar, T.A., Eychmüller, A., and Feldmann, J. (2005) *Small*, **1**, 392.

91 Mamedov, A.A., Belov, A., Giersig, M., Mamedova, N.N., and Kotov, N.A. (2001) *J. Am. Chem. Soc.*, **123**, 7738.

92 Franzl, T., Klar, T.A., Schietinger, S., Rogach, A.L., and Feldmann, J. (2004) *Nano Lett.*, **4**, 1599.

93 Srivastava, S. and Kotov, N.A. (2008) *Acc. Chem. Res.*, **41**, 1831.

94 Shavel, A., Gaponik, N., and Eychmüller, A. (2004) *J. Phys. Chem. B*, **108**, 5905.

95 Kim, D., Okahara, S., Shimura, K., and Nakayama, M. (2009) *J. Phys. Chem. C*, **113**, 7015.

96 Das, B.C., Batabyal, S.K., and Pal, A.J. (2007) *Adv. Mater.*, **19**, 4172.

97 Rogach, A.L., Koktysh, D.S., Harrison, M., and Kotov, N.A. (2000) *Chem. Mater.*, **12**, 1526.

98 Rogach, A.L., Kotov, N.A., Koktysh, D.S., Susha, A.S., and Caruso, F. (2002) *Colloids Surf. A*, **202**, 135.

99 Gao, M., Sun, J., Dulkeith, E., Gaponik, N., Lemmer, U., and Feldmann, J. (2002) *Langmuir*, **18**, 4098.

100 Wang, C., Shim, M., and Guyot-Sionnest, P. (2002) *Appl. Phys. Lett.*, **80**, 4.

101 Mattoussi, H., Radzilowski, L.H., Dabbousi, B.O., Fogg, D.E., Schrock, R.R., Thomas, E.L., Rubner, M.F., and Bawendi, M.G. (1999) *J. Appl. Phys.*, **86**, 4390.

102 Ichimura, M., Takeuchi, K., Nakamura, A., and Arai, E. (2001) *Thin Solid Films*, **384**, 157.

103 Danek, M., Jensen, K.F., Murray, C.B., and Bawendi, M.G. (1996) *Chem. Mater.*, **8**, 173.

104 Poznyak, S.K., Osipovich, N.P., Shavel, A., Talapin, D.V., Gao, M., Eychmüller, A., and Gaponik, N. (2005) *J. Phys. Chem. B*, **109**, 1094.

105 Kagan, C.R., Murray, C.B., and Bawendi, M.G. (1996) *Phys. Rev. B*, **54**, 8633.

106 Kagan, C.R., Murray, C.B., Nirmal, M., and Bawendi, M.G. (1996) *Phys. Rev. Lett.*, **76**, 1517.

107 Micic, O.I., Jones, K.M., Cahill, A., and Nozik, A.J. (1998) *J. Phys. Chem. B*, **102**, 9791.

108 Artemyev, M.V., Bibik, A.I., Gurinovich, L.I., Gaponenko, S.V., and Woggon, U. (1999) *Phys. Rev. B*, **60**, 1504.

109 Artemyev, M.V., Woggon, U., Jaschinski, H., Gurinovich, L.I., and Gaponenko, S.V. (2000) *J. Phys. Chem. B*, **104**, 11617.

110 Artemyev, M.V., Bibik, A.I., Gurinovich, L.I., Gaponenko, S.V., Jaschinski, H., and Woggon, U. (2001) *Phys. Status Solidi B*, **224**, 393.

111 Micic, O.I., Ahrenkiel, S.P., and Nozik, A.J. (2001) *Appl. Phys. Lett.*, **78**, 4022.

112 Leatherdale, C.A., Kagan, C.R., Morgan, N.Y., Empedocles, S.A., Kastner, M.A., and Bawendi, M.G. (2000) *Phys. Rev. B*, **62**, 2669.

113 Leatherdale, C.A. and Bawendi, M.G. (2001) *Phys. Rev. B*, **63**, 165315/1.

114 Kim, B.S., Islam, M.A., Brus, L.E., and Herman, I.P. (2001) *J. Appl. Phys.*, **89**, 8127.

115 Döllefeld, H., Weller, H., and Eychmüller, A. (2002) *J. Phys. Chem. B*, **106**, 5604.

116 Franceschetti, A. and Zunger, A. (2001) *Phys. Rev. B*, **63**, 153304/1.

117 Rabani, E. (2001) *J. Chem. Phys.*, **115**, 1493.
118 Ramakrishnan, S. and Zukoski, C.F. (2000) *J. Chem. Phys.*, **113**, 1237.
119 Maly, P., Kudrna, J., Trojanek, F., Mikes, D., Nemec, P., Maciel, A.C., and Ryan, J.F. (2000) *Appl. Phys. Lett.*, **77**, 2352.
120 Vanmaekelbergh, D. and De Jongh, P.E. (1999) *J. Phys. Chem. B*, **103**, 747.
121 Meulenkamp, E.A. (1999) *J. Phys. Chem. B*, **103**, 7831.
122 Solbrand, A., Henningsson, A., Södergren, S., Lindström, H., Hagfeldt, A., and Lindquist, S.-E. (1999) *J. Phys. Chem. B*, **103**, 1078.
123 Noack, V., Weller, H., and Eychmüller, A. (2002) *J. Phys. Chem. B*, **106**, 8514.
124 Morgan, N.Y., Leatherdale, C.A., Drndic, M., Jarosz, M.V., Kastner, M.A., and Bawendi, M. (2002) *Phys. Rev. B*, **66**, 075339/1.
125 Brust, M., Bethell, D., Schiffrin, D.J., and Kiely, C.J. (1995) *Adv. Mater.*, **7**, 795.
126 Gopidas, K.R., Bohorquez, M., and Kamat, P.V. (1990) *J. Phys. Chem.*, **94**, 6435.
127 Lawless, D., Kapoor, S., and Meisel, D. (1995) *J. Phys. Chem.*, **99**, 10329.
128 Torimoto, T., Yamashita, M., Kuwabata, S., Sakata, T., Mori, H., and Yoneyama, H. (1999) *J. Phys. Chem. B*, **103**, 8799.
129 Shipway, A.N., Katz, E., and Willner, I. (2000) *ChemPhysChem*, **1**, 18.
130 Mirkin, C.A. (2000) *Inorg. Chem.*, **39**, 2258.
131 Niemeyer, C.M. (2001) *Angew. Chem., Int. Ed.*, **40**, 4128.
132 Parak, W.J., Gerion, D., Zanchet, D., Woerz, A.S., Pellegrino, T., Micheel, C., Williams, S.C., Seitz, M., Bruehl, R.E., Bryant, Z., Bustamante, C., Bertozzi, C.R. and Alivisatos, A.P. (2002) *Chem. Mater.*, **14**, 2113.
133 Medintz, I.L., Berti, L., Pons, T., Grimes, A.F., English, D.S., Alessandrini, A., Facci, P., and Mattoussi, H. (2007) *Nano Lett.*, **7**, 1741.
134 Bruchez, M. Jr, Moronne, M., Gin, P., Weiss, S., and Alivisatos, A.P. (1998) *Science*, **281**, 2013.

135 Medintz, I.L., Uyeda, H.T., Goldman, E.R., and Mattoussi, H. (2005) *Nat. Mater.*, **4**, 435.
136 Sapsford, K.E., Pons, T., Medintz, I.L., Higashiya, S., Brunel, F.M., Dawson, P.E., and Mattoussi, H. (2007) *J. Phys. Chem. C*, **111**, 11528.
137 Mamedova, N.N., Kotov, N.A., Rogach, A.L., and Studer, J. (2001) *Nano Lett.*, **1**, 281.
138 Rosenthal, S.J. (2001) *Nat. Biotechnol.*, **19**, 621.
139 Smith, A.M., Duan, H., Mohs, A.M., and Nie, S. (2008) *Adv. Drug Delivery Rev.*, **60**, 1226.
140 Michalet, X., Pinaud, F.F., Bentolila, L.A., Tsay, J.M., Doose, S., Li, J.J., Sundaresan, G., Wu, A.M., Gambhir, S.S., and Weiss, S. (2005) *Science*, **307**, 538.
141 Schmelz, O., Mews, A., Basche, T., Herrmann, A., and Muellen, K. (2001) *Langmuir*, **17**, 2861.
142 Ni, T., Nagesha, D.K., Robles, J., Materer, N.F., Muessig, S., and Kotov, N.A. (2002) *J. Am. Chem. Soc.*, **124**, 3980.
143 Agrawal, M., Rubio-Retama, J., Zafeiropoulos, N.E., Gaponik, N., Gupta, S., Cimrova, V., Lesnyak, V., Lopez-Cabarcos, E., Tzavalas, S., Rojas-Reyna, R., Eychmüller, A., and Stamm, M. (2008) *Langmuir*, **24**, 9820.
144 Kolny, J., Kornowski, A., and Weller, H. (2002) *Nano Lett.*, **2**, 361.
145 Hoppe, K., Geidel, E., Weller, H., and Eychmüller, A. (2002) *Phys. Chem. Chem. Phys.*, **4**, 1704.
146 Trindade, T., O'Brien, P., and Zhang, X.-M. (1999) *J. Mater. Res.*, **14**, 4140.
147 Peng, X., Wilson, T.E., Alivisatos, A.P., and Schultz, P.G. (1997) *Angew. Chem., Int. Ed. Engl.*, **36**, 145.
148 Osovsky, R., Shavel, A., Gaponik, N., Amirav, L., Eychmüller, A., Weller, H., and Lifshitz, E. (2005) *J. Phys. Chem. B*, **109**, 20244.
149 Mayilo, S., Hilhorst, J., Susha, A.S., Hoehl, C., Franzl, T., Klar, T.A., Rogach, A.L., and Feldmann, J. (2008) *J. Phys. Chem. C*, **112**, 14589.
150 Landman, U., Barnett, R.N., Scherbakov, A.G., and Avouris, P. (2000) *Phys. Rev. Lett.*, **85**, 1958.
151 Rabani, E. and Egorov, S.A. (2002) *Nano Lett.*, **2**, 69.

4.2
Metal Nanoparticles

Günter Schmid, Dmitri V. Talapin, and Elena V. Shevchenko

The organization of metal nanoparticles is, like that of semiconductor nanoparticles, of vital relevance. The ultimate goal of all efforts in producing and investigating semiconducting or metal nanoparticles is their use in optoelectronics, nanoelectronics, storage systems, and so on, based on the size-quantized properties of these miniaturized materials. The fact that the electronic properties of metal particles are dramatically changed if their size falls below a well-defined limit, makes them very promising candidates for components such as single-electron switches and transistors. The organization of these particles into three-dimensional (3-D), two-dimensional (2-D) or even one-dimensional (1-D) systems seems indispensable if specific functions are be called upon. Consequently, this chapter describes the present state of the art with respect to 3-D, 2-D, and 1-D arrangements of metal nanoparticles.

4.2.1
Three-Dimensional Organization of Metal Nanoparticles

In principle, the 3-D organization of building blocks of matter occurs simply by nature - referred to as "crystallization." So, why should one subchapter be devoted to the 3-D organization of metal nanoparticles? Typically, ions and molecules forming 3-D crystals are identical in every respect, and this enables them to be arranged periodically in three dimensions. This is usually not the case for building blocks at the nanometer scale. Nanoparticles, which traditionally are also called "colloids," usually differ not only in size but also in structure. The surprising discovery with respect to the crystal formation of nanoparticles during the past decade was that, in spite of not having fully identical building blocks, 3-D organizations and (as will be shown below) 2-D and even 1-D arrangements become available if specific conditions are fulfilled. Of course, there is a continuous transition from 1-D to 2-D, and from 2-D to 3-D. In the following sections, the assemblies of some monolayers will be considered as quasi-2-D systems or thin films, rather than as 3-D architectures. In fact, 3-D crystals are considered to exhibit dimensions in the a, b, and c directions of the same order of magnitude.

Usually, nanoparticles to be 3-D assembled to superlattices are covered by a shell of stabilizing ligand molecules; these protect the particles from coalescence and also make them soluble in appropriate solvents.

A variety of 3-D superlattices formed from metal nanoparticles has been reported [1–8]. For instance, gold nanoparticles, when stabilized by alkylthiolates RS^- ($R = n\text{-}C_{12}H_{25}$) and rather narrowly distributed in size, could be obtained by the repeated fractional crystallization of a mixture of 1.5–3.5 nm particles from unpolar solvents. The success of fractionation was controlled by mass spectroscopy [6]. Based on these data, definite numbers of atoms could be assigned to the particles of distinct fractions (e.g., 140, 225, 314, and 459), all of which consisted of fcc-structured octahedra. Solutions of these remarkably well-defined particles can be used to generate 2-D assemblies (see Section 4.2.2), and also for growing 3-D crystals of considerable extension. The Au_{459} fraction forms well-shaped crystals of up to approximately 40 μm in length [6]. Clearly, no coalescence occurs during crystal formation as the crystals are highly soluble in nonpolar solvents. Assembly of the nanoparticles causes a change in color, from the well-known purple color of individual colloids to a metallic-like luster.

Superlattices of 5.0 nm silver nanoparticles have also become available from alkylthiolate-protected colloids [7–9]. Subsequent examination using high-resolution transmission electron microscopy (HR-TEM) permitted quite precise information regarding the architecture of the crystals to be obtained, and confirmed that these materials consisted of highly ordered truncated octahedrons (see Figure 4.7).

Crystals of up to ~5 μm edge length could be grown on the TEM copper grid from solution, while 3-D superlattices of thiol-stabilized 5–6 nm gold particles have also been described [10]. Micron-sized 3-D crystals of the well-known gold cluster compound, $Au_{55}(PPh_3)_{12}Cl_6$, were formed from dichloromethane (DCM) solution, simply by evaporating the solvent [11], although this was not surprising as these very small nanoparticles are indeed identical in size and shape. However, the growth of larger crystals fails due to a rapid decomposition of the clusters in solution, over the course of only hours. Although the rapid removal of solvent results in small crystals, they are more or less perfectly organized, as indicated by TEM imaging and small-angle X-ray diffraction (SAXRD) measurements. In this case, the 2θ angle of 4.3° corresponds with a d-value of 2.05 nm, in agreement with an fcc arrangement of the clusters of an effective size of 2.3–2.4 nm, including the ligand shell. Based on TEM studies, the hexagonally close-packed arrangement of the clusters could also be deduced, in agreement with the SAXRD results.

Independent of the dimensionality, the organization of metal nanoparticles seemed to be linked with the existence of a protecting ligand shell that enveloped the particles' surfaces. The most important reason for this condition was seen as a necessity to grow the arrangements from solution.

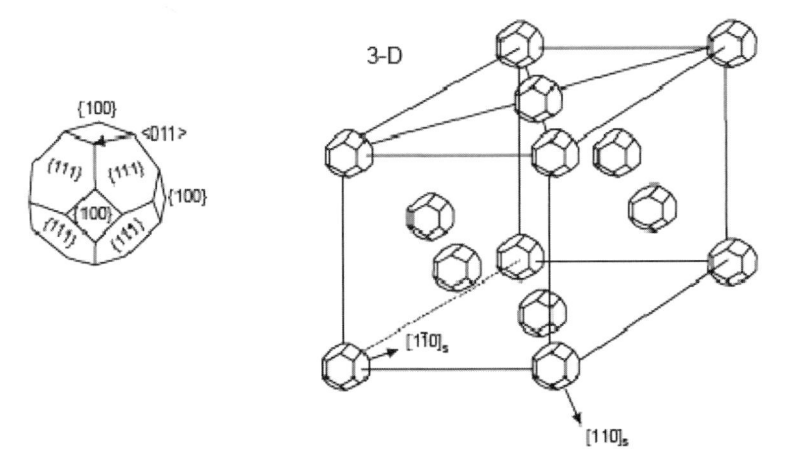

Figure 4.7 Sketch of a superlattice of truncated octahedral 5.0 nm Ag nanoparticles.

The soft removal of ligands from protected particles in solution might represent one way of realizing the generation of defined, "naked" nanoparticles. On the other hand, it is well known that the quantitative elimination of a ligand shell cannot be induced by heat treatment. As early as in 1986, it first emerged that naked metal clusters could organize themselves in an ordered manner, when the mild electrochemical decomposition of $Au_{55}(PPh_3)_{12}Cl_6$ in solution resulted in the formation of naked Au_{13} clusters (the inner cores of Au_{55}), together with Ph_3PAuCl and bulk gold. The Au_{13} clusters were found to build up clusters of clusters of the type $(Au_{13})_n$ [12, 13]. Subsequently, secondary ion mass spectrometry (SIMS) investigations of $Au_{55}(PPh_3)_{12}Cl_6$ resulted in similar species, as identified by masses which corresponded to $(Au_{13})_n$, with n-values of up to 55 [14, 15].

The generation of bare Au_{55} clusters under mild conditions later became possible by a surprising reaction: a thiol-terminated fourth-generation dendrimer with 96 SH groups (G4-SH) on its surface reacts with $Au_{55}(PPh_3)_{12}Cl_6$ in solution with quantitative elimination of PPh_3 and Cl ligands [16].

The reason for removing the ligand shell is to be seen in a formal substitution of the phosphines by the stronger thiol groups of the dendrimers, which are used in large excess. Ligand-free clusters on the dendrimers' surface clearly are not irreversibly fixed, but can move on the equivalent surface positions so as to finally achieve contact with a second, third, fourth... other bare cluster, with the formation of strong metal–metal bonds. The process can be described as a crystal growth in a matrix of organic material, with growth continuing until the solubility is exceeded. The possible formation of isolated microcrystals consisting of only Au_{55} building blocks is illustrated, in very simplified manner, in Figure 4.8. One of these perfectly shaped crystals is shown in Figure 4.9.

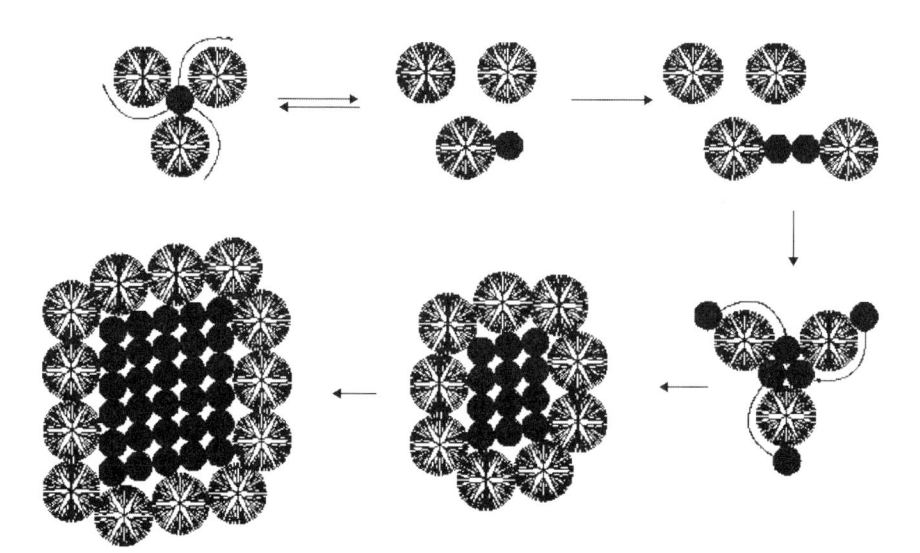

Figure 4.8 Schematic representation of the formation of Au_{55} microcrystals inside a matrix of G4-SH dendrimers.

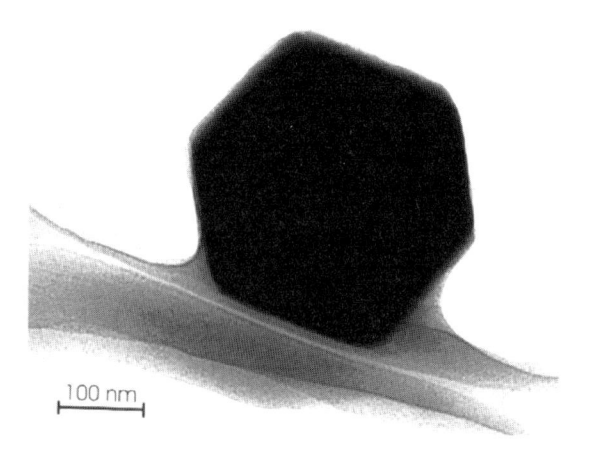

Figure 4.9 TEM image of an Au$_{55}$ microcrystal with a skin of G4-SH dendrimer molecules.

Surprisingly, not only are the PPh$_3$ ligands removed so as become detectable in solution, but also the Cl atoms have disappeared, as demonstrated by the lack of a characteristic Au–Cl vibration at 274 cm^{-1} in the infra-red (IR) spectrum.

The most important question is, how do the bare Au$_{55}$ clusters organize themselves to produce these perfect-looking microcrystals? A combination of detailed TEM, SAXRD and wide-angle X-ray diffraction (WAXRD) showed that the cubo-octahedrally shaped Au$_{55}$ clusters clearly touched each other via their edges, in a very regular manner, so as to build up superlattices of an architecture (as shown schematically in Figure 4.10).

These results show, impressively, that the organization of metal nanoparticles need not be limited to ligand-protected cases, but that ligand-free particles can also be organized in 3-D fashion to build up new modifications of metals – in this case, gold.

In contrast to the microcrystals formed of Au$_{55}$(PPh$_3$)$_{12}$Cl$_6$, which have been intensively investigated with respect to their electronic characteristics, the 3-D assemblies of bare Au$_{55}$ clusters have not yet been subjected to physical investigations.

It has been shown that 3-D superlattices of larger, ligand-protected Ag–Au alloy nanoparticles of narrow size-distribution are formed on solid substrates during solvent evaporation [17]. Truncated cubo-octahedrons, as well as multiply twinned icosahedrons, show the same tendency to grow three-dimensionally under appropriate experimental conditions. The truncated cubo-octahedrons alloy particles are available either with different compositions of the same size or of same composition, but with different sizes. Single crystalline forms of truncated cubo-octahedrons exist in three different packing patterns, whereas the icosahedral species form a hexagonally close-packed structure. The TEM images in Figure 4.11 show such icosahedral Ag–Au alloy nanoparticles as multi-layer (Figure 4.11a) and two-layer (Figure 4.11c) versions, together with a FFT pattern (Figure 4.11b) and a model of the superlattice (Figure 4.11d).

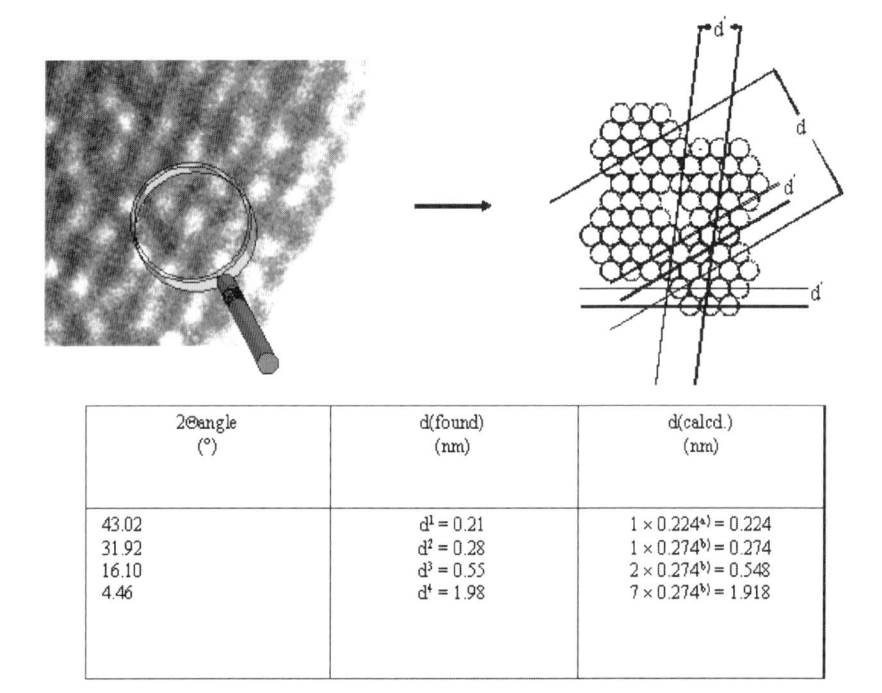

2θangle (°)	d(found) (nm)	d(calcd.) (nm)
43.02	$d^1 = 0.21$	$1 \times 0.224^{a)} = 0.224$
31.92	$d^2 = 0.28$	$1 \times 0.274^{b)} = 0.274$
16.10	$d^3 = 0.55$	$2 \times 0.274^{b)} = 0.548$
4.46	$d^4 = 1.98$	$7 \times 0.274^{b)} = 1.918$

(a) hkl : 111 = 0.2355 − 5% = 0.224
(b) hkl : 110 = 0.2884 − 5% = 0.274

Figure 4.10 Magnified TEM image of an Au_{55} microcrystal and crystallographic data from SAXRD and WAXRD studies. The hkl values 111 and 110 are reduced by 5% due to the size-determined contraction of Au–Au distances in small particles.

The self-assembly of monodisperse magnetic nanoparticles into ordered 3-D superstructures is usually achieved by the slow destabilization of a colloidal dispersion. Indeed, if the evaporation rate of a solvent is relatively high, close-packed glassy films of nanoparticles with a short-range order are formed [18]. On the other hand, superlattices with a low concentration of lattice defects and a long-range order of nanocrystal arrangement can be formed if the particles are mobile, and have sufficient time to find their lowest energy sites in the superstructure. For example, the slow evaporation of a high-boiling carrier solvent at increased temperatures (50–60 °C) was used to form ordered superstructures of monodisperse cobalt [18–21], nickel [19], and magnetite [22] nanoparticles. Recently, a technique has been developed for growing macroscopic crystals (so-called colloidal crystals) that utilizes monodisperse semiconductor or metal nanoparticles as building blocks [23–25]. The method is based on the gentle destabilization of a colloidal solution, induced by the slow diffusion of a nonsolvent into a concentrated solution of monodisperse nanoparticles inside a vertically positioned glass tube (see Figure 4.12). The colloidal crystals nucleate and grow preferentially on the walls of a glass tube, with the growing process normally taking several weeks. The spatial distribution of local oversatura-

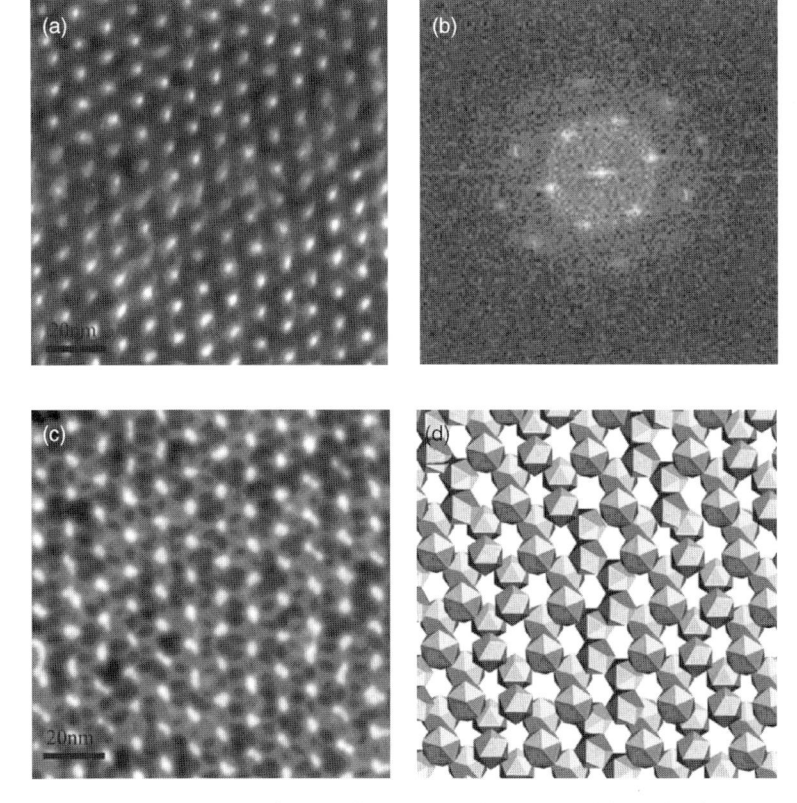

Figure 4.11 TEM images of Ag–Au alloy nanoparticles. (a) A multilayer; (b) The FFT pattern; (c) A two-layered sample; (d) The calculated model.

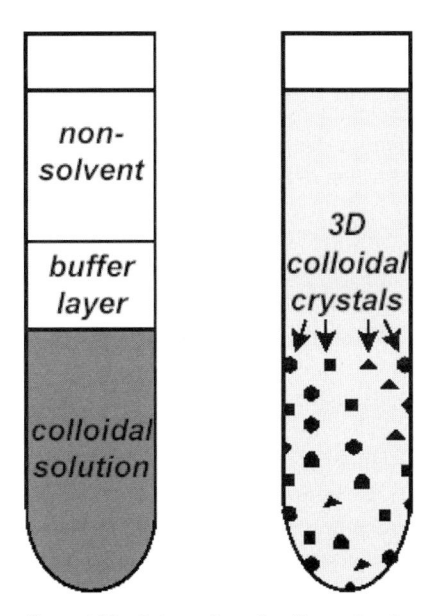

Figure 4.12 Schematic outline illustrating the concept of the three-layer oversaturation technique used for the crystallization of monodisperse FePt and CoPt3 nanoparticles.

tions caused by the diffusion of a nonsolvent determines the quality of the colloidal crystals. In order to make the oversaturation front less sharp, as in the case of a direct solvent–nonsolvent contact, a buffer layer with low (but not negligible) solubility for nanocrystals was inserted between the solvent and nonsolvent layers. This allowed the preparation of perfectly faceted colloidal crystals of monodisperse CdSe [23], FePt [24] and CoPt3 [26] nanocrystals, with sizes ranging from tens up to hundreds of microns.

The optical images in Figure 4.13a and b show the FePt colloidal crystals adhering to the glass. The crystals were grown on a slide, preferably in the form of faceted triangular or hexagonal platelets, and reached 10–30 µm in size, and crystals of a pyramidal shape were also observed. Subsequent studies using scanning electron microscopy (SEM) allowed a closer examination of the morphology of these colloidal crystals. In the case of FePt, faceted triangular platelets, incomplete hexagonal platelets and tetrahedronal colloidal crystals with imperfect sides showing terraces and cleaved ledges (Figure 4.13c–e) were clearly distinguishable. Figure 4.13f shows a TEM image taken from a small fragment of the FePt colloidal crystal obtained by mechanical grinding and treatment in an ultrasonic bath. Hexagonal arrangements of nanocrystals (i.e., the building blocks of the colloidal crystals) can be clearly seen at the edges of the crystalline pieces. Figure 4.14 shows the SEM images of well-facetted hexagonal, triangular and pyramidal colloidal crystals built from $CoPt_3$ nanocrystals.

The arrangement of nanocrystals at the surface of colloidal crystals can be investigated using HR-SEM (Figure 4.15). Thus, the ledges and terraces of ordered $CoPt_3$ nanoparticles are clearly seen at the surface of colloidal crystal. These 6.3 nm $CoPt_3$ nanoparticles are superparamagnetic, with the blocking temperature of 19 K. However, in 3-D ordered structures, dipolar interactions between adjacent particles can orient the magnetic moments of individual nanocrystals, such that ferromagnetic domains can appear within the supercrystal, resulting in a considerable increase in the blocking temperature [21, 27].

The self-assembly of magnetic nanocrystals can also be affected by applying an external magnetic field. Indeed, it has been found that relatively large external magnetic fields can disturb the crystallization of magnetic nanoparticles into macroscopic colloidal crystals, causing the formation of glass-like solids rather than 3-D ordered super-lattices.

Moreover, the direction of the applied magnetic field will influence the morphology of self-assembled structures of magnetic nanoparticles [19, 28, 29]. For example, when the field was applied in plane of the substrate, the deposition of Co nanoparticles led to spindle-shaped superlattices, with their long axes aligned along the external magnetic field [19, 30]. Within these spindle-shaped aggregates, the individual magnetic nanoparticles showed a near-perfect long-range order [19]. Yet, when $CoPt_3$ nanocrystals were slowly precipitated in an external magnetic field of 0.5 T, applied parallel to the substrate, they organized themselves into parallel, equidistant stripes (see Figure 4.16a). In these structures the periodic stripes of close-packed $CoPt_3$ nanocrystals were separated from each other by regions where no nanocrystals were deposited. Moreover, the period of these self-assembled gratings was shown to decrease with in line with increases in the strength of the external magnetic field.

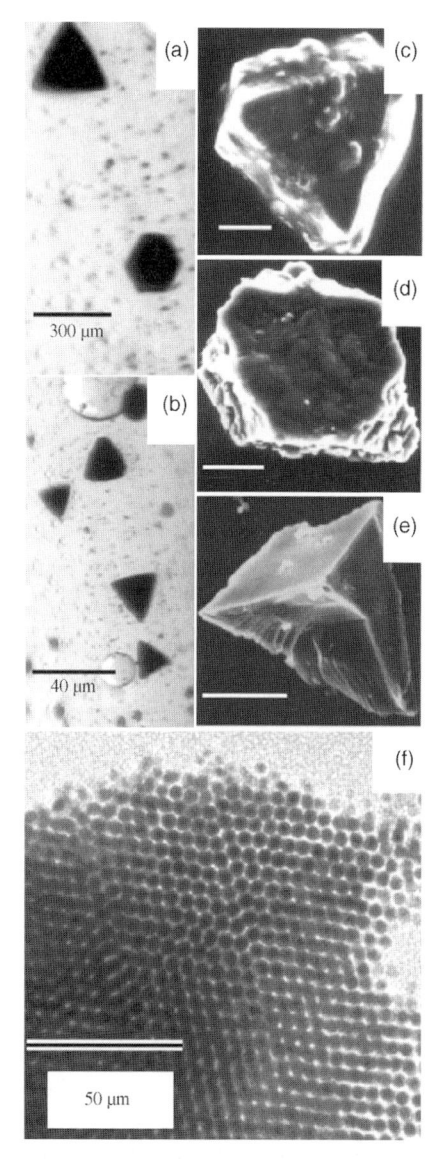

Figure 4.13 Optical micrographs (a,b) and SEM images (c–e) of colloidal crystals of FePt nanocrystals, illustrating their different shapes; (f) TEM image of the fragment of an FePt colloidal crystal. Reproduced with permission from Ref. [24]; © 2001, Wiley-VCH.

A magnetic field, when applied perpendicularly to the substrate, induces the formation of 2-D hexagonal superlattices of individual Co nanoparticles [30] Pileni et al. [28] observed the formation of a hexagonal network of about 1 μm dot-shaped aggregates made from 8 nm cobalt nanoparticles. CoPt$_3$ nanocrystals (4 nm), when deposited under a magnetic field of 0.9 T applied perpendicularly to the substrate, can

Figure 4.14 SEM images of colloidal crystals of CoPt$_3$ nanoparticles, illustrating their different shapes. Reproduced with permission from Ref. [24]; © 2002, American Chemical Society.

self-assemble into needles of diameter 0.1–2 µm and length up to 100 µm (Figure 4.16b). HR-SEM investigations revealed no long-range ordering of the CoPt$_3$ nanoparticles inside these needles.

The self-organization phenomena in colloids of magnetic nanoparticles (ferrofluids) induced by external magnetic fields has attracted considerable attention on the basis of importance from both fundamental and applied perspectives [31]. However, the lack of systematic investigations, combined with some contradictory findings, have led to these materials remaining a relative novelty. Wirtz *et al.* [32] have suggested that the formation of macroscopic 1-D periodic patterns composed of

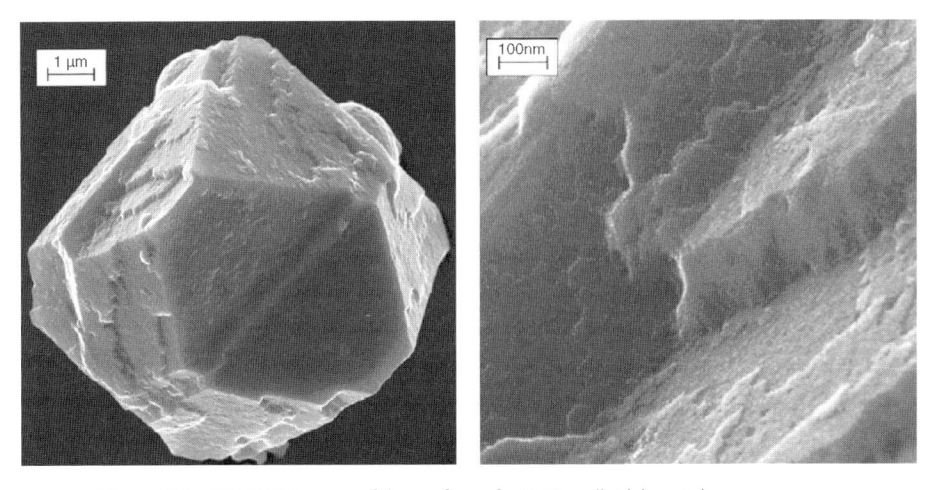

Figure 4.15 HR-SEM images of the surface of a CoPt$_3$ colloidal crystal.

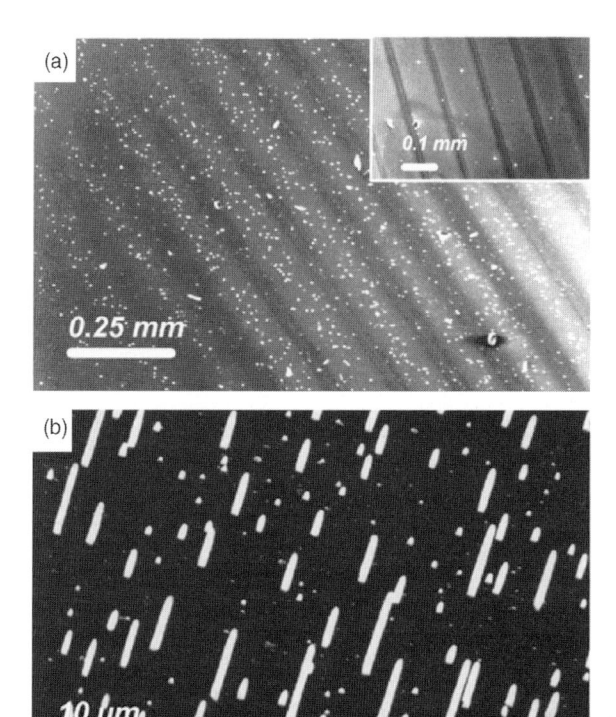

Figure 4.16 (a) Stripes of 9 nm CoPt$_3$ nanocrystals deposited in a magnetic field of 0.5 T applied parallel to a glass substrate; (b) Needles composed of 4.0 nm CoPt$_3$ nanocrystals deposited in the magnetic field of 0.9 T, applied perpendicular to a glass substrate. The image was acquired at 40° tilt of the substrate.

high- and low-concentration regions in a ferrofluid would be possible only in an oscillating magnetic field. However, Wang et al. have succeeded in forming periodic branched structures (similar to those shown in Figure 4.16a) under a constant magnetic field [33]. Recently, Wen et al. modeled the self-assembly of planar magnetic colloidal crystals under an external magnetic field using glass microspheres coated with nickel floating at the interface of two immiscible liquids of different densities [34]. A hexagonal lattice, in which the lattice constant increased steeply in line with the increased magnetic field strength, was observed when the magnetic field was applied perpendicular to the plane of magnetic particles. Tilting the magnetic field at an angle 60° away from the normal projection transformed the hexagonally ordered monolayer into equidistant straight chains of particles, aligned along the field [34]. The study of self-organization phenomena in ferrofluids requires extremely monodisperse samples of magnetic nanoparticles, mainly because the magnetic dipole moment of a single-domain nanoparticle is proportional to the third power of its radius; hence, particles of different sizes will have different susceptibilities to the magnetic field. Nonetheless, the considerable progress achieved during the past few years in the preparation of monodisperse magnetic nanocrystals will surely help towards any further understanding of self-organization and the phase behavior of ferrofluids.

4.2.2
Two- and One-Dimensional Structures of Metal Nanoparticles

Nature tends to organize materials in 3-D fashion, which is why 2-D and 1-D assemblies usually do not grow, except under special experimental conditions. As noted above, the formation of 2-D assemblies is also possible without having identical building blocks. Moreover, tolerance with respect to deviations from size and shape has been found to be even greater than in the 3-D case. Consequently, the formation of 2-D structures of metal nanoparticles has developed into a major field in the nanoscience, with promising aspects regarding future applications.

The generation of 2-D and 1-D structures of metal nanoparticles can be achieved using very different techniques of: (i) self-assembly; (ii) guided self-assembly; and (iii) aimed structures. Examples of these three principal routes are presented and discussed in the following subsections.

4.2.2.1 Self-Assembly

Although the self-assembly of nanoparticles on appropriate substrates is, without doubt, the simplest way to produce monolayers, the method suffers from several disadvantages, mainly that little or no influence can be exerted on the structure and extension of the monolayers. The formation of self-assembled monolayers (SAMs) depends on experimental conditions, making reproducibility difficult. Nevertheless, many quite impressive 2-D structures of metal nanoparticles have become known during the course of the past decade.

One of the first hints that gold nanoparticles (AuNPs) could be deposited in an ordered manner was provided by Schiffrin et al. [35], when 8 nm gold particles, which had been prepared in a two-phase liquid–liquid system and stabilized by dithiols [36], provided crosslinked assemblies for which the quality of ordering was rather limited compared to recent results. Nonetheless, these preliminary findings confirmed that 2-D self-assembled arrays could be produced.

The same research group, and others, have subsequently contributed considerably to the improvement of 2-D architectures by varying the experimental conditions [37]. Now, instead of dithiols, tetra-alkylammonium bromides [38], $[R_4N]^+Br^-$, are used, whereby the separation of the clusters depends on the chain length of R in $[R_4N]^+$. Bilayered structures were also prepared using these particles, and formed unusual patterns owing to the occupancy of two gold saddle sites, leading to linear and circular arrangements. The reason for this unexpected structure is to be seen in the balance between local electrostatic repulsion and dispersion forces between the ammonium-stabilized particles.

Schiffrin et al., also for the first time, made another important discovery when they fabricated bimodal ensembles of nanosized gold particles. Alkanethiol-stabilized AuNPs of different (but well-defined) sizes can organize in 2-D fashion, namely with size segregation, with each region consisting of hexagonally close-packed structures, randomly positioned particles and, most importantly, also ordered bimodal arrays.

"Alloy-like" bimodal 2-D structures, consisting of two different metals, are likewise available, as shown for both silver and gold [39, 40]. There appears to be a relationship between the geometric packing of these sphere-like particles and that of intermetallic alloys on an atomic scale. This follows the well-known Laves rules [41], which describes the ratio of the atomic radii r_A and r_B of two metallic species r_A/r_B (where $r_A > r_B$), and of the local number ratio of the two metals n_A/n_B. Furthermore, there are polarizability components to be considered. In the case of only one metal of two different particle sizes (<10 nm) the 2-D lattice structures mainly depend on the ratios n_A/n_B and r_A/r_B, since polarization effects can be neglected. AB_2 structures are formed if $n_A/n_B = 0.5$, and if $0.482 < r_A/r_B < 0.624$. If, however, $n_A/n_B = 1$ and $0.27 < r_A/r_B < 0.425$, then AB lattices with NaCl structure result. Both alternatives are sketched in Figure 4.17.

An Au/Ag mixture, the particles of which are approximately the same size, gives only randomly oriented colloids. However, an Au:Ag mixture (10:1) with 8 nm Au and 5 nm Ag particles results in an AB_2 superlattice, in agreement with the Laves alloy rules. It should be stated that these comparisons of true alloys with mixtures of nanoparticles of different sizes are not really permissible, as the conditions do not correspond. For instance, the fact that the nanoparticles are ligand-stabilized means that the metal cores do not actually touch each other; however, as a simple model to understand the different structures of the 2-D superlattices, this is quite satisfactory.

When two monodisperse colloids of $CoPt_3$ nanocrystals (4.5 nm and 2.6 nm diameter) were mixed together, followed by a slow evaporation of the solvent, an AB_5-type superlattice analogous to the structure of intermetallic compound $CaCu_5$ [42] was obtained (Figure 4.18) [26]. A similar structure was observed for binary mixtures of latex spheres of two different sizes [43, 44]. In the first plane of this lattice (Figure 4.18), each 4.5 nm $CoPt_3$ nanocrystal is surrounded with a hexagon formed by 2.6 nm nanocrystals. The second plane consists only of hexagons of small particles, while the third plane repeats the first.

Well-ordered and largely extended (up to 1 μm) 2-D superlattices by the self-organization of silver and gold nanoparticles, protected by alkylthiolates, were identified by Whetten et al. in connection with the formation of 3-D assemblies (see Section 4.2.1) [6, 7]. The Ag particles were produced using a special aerosol technique in which elementary silver was evaporated at 1200–1500 K into an atmosphere of helium, followed by a cooling step to produce 4–5 nm Ag particles that were finally co-condensed with an excess of alkylthiols (C_{12} chains) [7].

The AuNPs were the same as those used for the fabrication of 3-D crystals [6]. In this case, fcc-structured monolayers and multilayers were formed, from solution, on carbon-coated copper grids (detailed structural investigations are provided in the original reference).

Notably, several other groups also contributed to this development during the late 1990s, mostly also using the noble metals silver and gold [10, 45–47].

A spontaneous self-assembly was observed when colloidal solutions of $CoPt_3$ nanocrystals were spread onto a substrate with subsequent slow evaporation of the

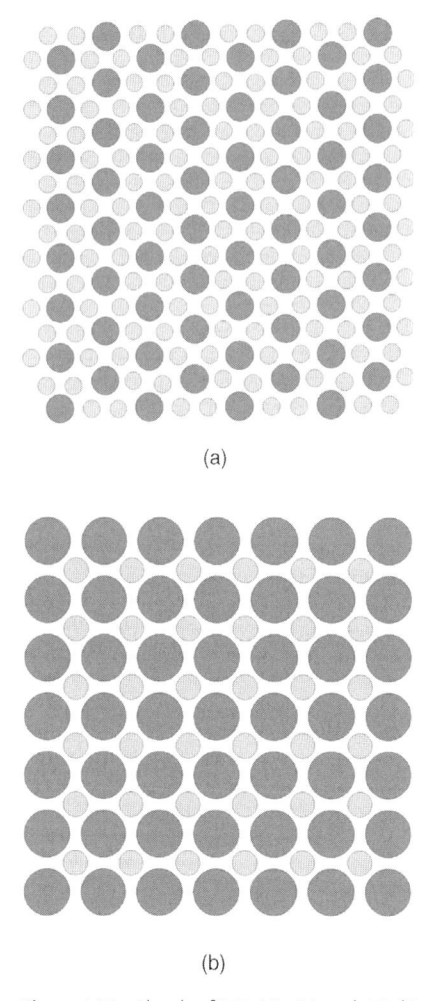

Figure 4.17 Sketch of 2-D AB_2 (a) and AB (b) superlattices.

carrier solvent [25, 26]. Depending on the particle size and conditions of solvent evaporation, several types of self-organized superstructure were obtained. Figure 4.19a shows a TEM overview image of a close-packed monolayer of 4.8 nm $CoPt_3$ nanocrystals. If the surface coverage with nanocrystals was thicker than one monolayer, then nanocrystals of the second layer occupied positions between the nanocrystals in the first layer (Figure 4.19b and c). Owing to the relatively large interparticle spacing (ca. 2.5 nm) maintained by the bulky 1-adamantanecarboxylic acid (ACA) capping ligands, two layers of $CoPt_3$ nanocrystals form the abovementioned structure, were observable for the different nanocrystal sizes. The third layer of nanocrystals occupied the positions typical for the cubic close-packed (ccp) structure (Figure 4.19d and e). Figure 4.19e shows a magnified image of a three-layered assembly of $CoPt_3$ nanocrystals about 4.0 nm in size. The difference between

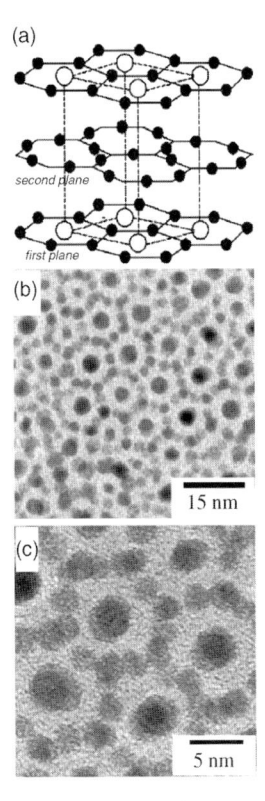

Figure 4.18 (a) Schematic view of an intermetallic compound $CaCu_5$; (b,c) TEM images illustrating an AB_5-type superstructure formed by $CoPt_3$ nanocrystals with bimodal size distribution. Reproduced with permission from Ref. [26]; © 2002, American Chemical Society.

the phase contrast of the two underlying layers and that of the darker third layer, allows each nanocrystal to be attributed to the layer in which it is placed, and leads to the conclusion that $CoPt_3$ nanocrystals are packed into a ccp-like superlattice in which the nanocrystals are separated from each other by thick (2.5 nm) organic shells. Figure 4.19f shows a TEM image of multilayer 3-D superlattice, where the 4.5 nm $CoPt_3$ nanocrystals are arranged in a nearly defect-free 3-D structure exhibiting long-range order. The graphical illustration (Figure 4.19g) clarifies the three-layer arrangement of the $CoPt_3$ nanocrystals.

The shape of individual nanocrystals can greatly affect the morphology of self-assembled structures [48–50]. In the case of spherical particles, the lattices of individual nanocrystals are randomly oriented within a superstructure, as a rule. However, faceted – and, in particular, cubic – nanocrystals can form "lattice-matched superstructures" where each nanocrystal has the same orientation as its neighbors (Figure 4.20). Such superstructures are of interest for creating materials with high magnetic anisotropy constants, due to the possibility of aligning the easy magnetization axes of individual nanocrystals. The high magnetic anisotropy of $CoPt_3$

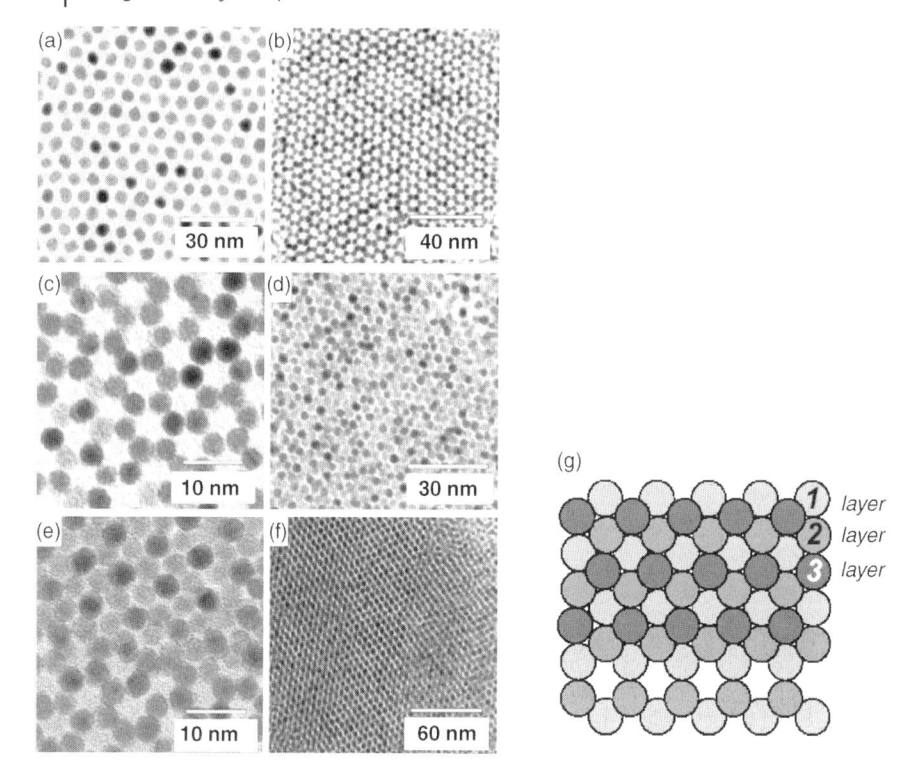

Figure 4.19 TEM images. (a) A monolayer of 4.8 nm $CoPt_3$ nanoparticles; (b,c) Two layers of 3.6 and 4.0 nm $CoPt_3$ nanoparticles; (d) Three layers of 3.6 nm $CoPt_3$ nanoparticles; (e) Three layers of 4.0 nm CoPt3 nanoparticles; (f) More than five layers of 4.5 nm $CoPt_3$ nanoparticles; (g) Graphical illustration of a three-layer arrangement of $CoPt_3$ nanocrystals. Reproduced with permission from Ref. [26]; © 2002, American Chemical Society.

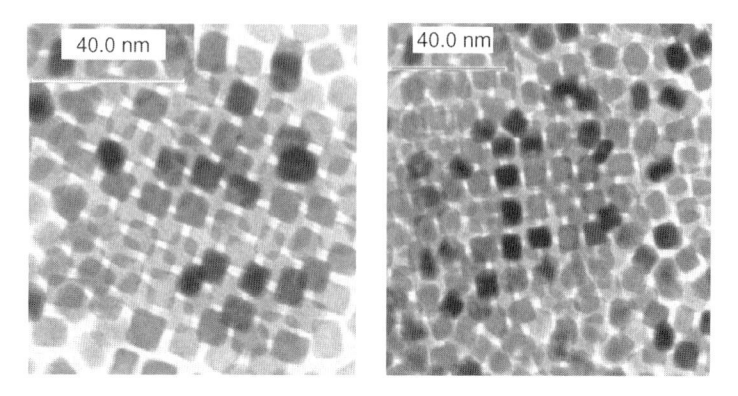

Figure 4.20 TEM images of self-assembled structures of cubic $CoPt_3$ nanocrystals.

nanocrystals [51] makes them superior candidates for realizing this new type of artificial solid.

Another exciting example of the effect of particle shape on self-assembly is the formation of the cobalt nanorod superstructures, as observed by Chaudret et al. [52]. In this case, exceptionally monodisperse cobalt nanorods were synthesized by the thermal decomposition of a Co(η_3-C_8H_{13})(η_4-C_8H_{12}) organometallic precursor in anisole at 150 °C, in the presence of hexadecylamine and stearic acid under hydrogen at a pressure of 3 bar. Upon aging the colloidal solution, the Co nanorods became stuck to each other so as to form a superstructure which consisted of perfectly aligned nanorods (Figure 4.21). It should be noted that, as the easy magnetization of individual nanorods were aligned in one direction, a high magnetic anisotropy and very high coercivity would be expected in the material obtained.

The wetting properties of the substrate can influence the superlattice morphology [19]. For example, if the colloidal solution of nanoparticles wets the substrate, then a 2-D superlattice will grow preferentially, forming a monolayer. As the surface coverage increases, however, the nanocrystals will adsorb to the ledges and kinks of the growing structure, so as to form terraces that extend laterally across the substrate.

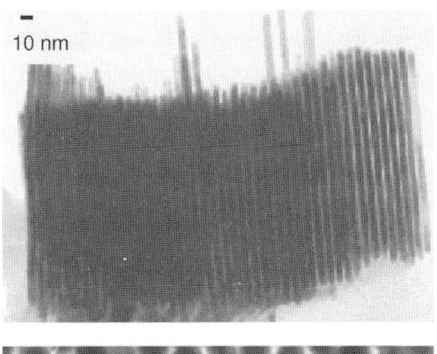

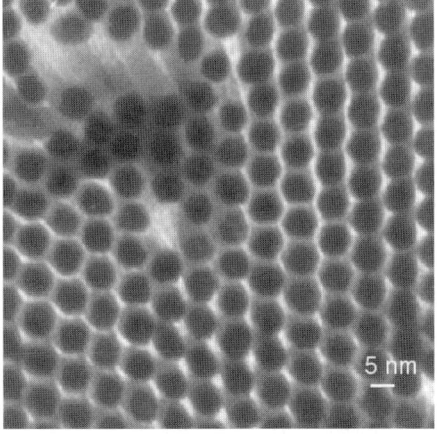

Figure 4.21 Three-dimensional organization of hcp cobalt nanorods. (a) Side view; (b) Top view. Illustration courtesy of Dr B. Chaudret.

In contrast, if the colloidal solution does not wet the substrate, a 3-D superlattice will grows preferentially, with facets that reflect the packing symmetry inherent in the nanocrystals. Annealing of the superlattice can alter the packing symmetry of self-assembled magnetic nanocrystals. When annealing 2-D arrays of ε-Co nanoparticles under vacuum or in an inert atmosphere at about 300 °C, the packing symmetry of 3-D ordered regions was converted from originally hexagonal to cubic [20]. The internal transition of the nanocrystal structure from cubic ε-Co to hcp-Co was responsible for this change. Further annealing of Co nanoparticles led to a dramatic change of the crystalline core, with fusing and sintering of the nanoparticles and destruction of the superlattice. The temperature of nanocrystal coalescence may depend on the substrate. Thus, no obvious aggregation of PtFe nanocrystals in the monolayer was observed even at 700 °C on a SiO_2 substrate, whereas on an amorphous carbon surface the particles coalesced at about 530 °C [48, 53].

The monolayers described so far have all been grown by self-assembly processes, where ligand-stabilized metal nanoparticles are deposited on "innocent" surfaces from solution, followed by a more or less rapid evaporation of the solvent. Depending on the concentration and other experimental conditions, those procedures may occasionally also result in multilayers, or even in small 3-D microcrystals. Self-assembly processes can, however, also be supported by chemical, electrochemical, magnetic, or mechanical effects; some examples of these are provided in the following subsections.

The organization of metal nanoparticles on surfaces is based not only on the presence of almost identical building blocks and attractive forces between them, but also on the interaction between particles and surface. Strong bonds between the nanoparticles and the surface do not normally lead to ordered structures of any appreciable extension. The reason for this is that chemically bound particles cannot move sufficiently to support the organization process. For instance, only very small areas of ordered $Au_{55}(PPh_2C_6H_4SO_3H)_{12}Cl_6$ clusters could be obtained on poly(ethyleneimine) (PEI)-modified surfaces, as the relatively strong acid–base interactions between clusters and surface disturb the organization process [54]. On the other hand, the lack of any interaction between the surface and the particle may also be disadvantageous, as there is no reason to deposit from solution as long as solvent is present. These conditions often end result in the formation of circles of particles, owing to the drying mechanism of the solvent. Ultimately, the moderate attraction between nanoparticles and surface were found to be best suited to generate ordered areas of remarkable extent; the following examples illustrate these findings.

Polymer films on a water surface with an extended π electron system, interacting weakly with the electronic π system of the phenyl rings of the PPh_3 ligands, give perfectly organized 2-D structures of $Au_{55}(PPh_3)_{12}Cl_6$. Examples are poly(p-phenylene-ethynylene) (PPE) derivatives, although poly(vinylpyrolidone) (PVP) and several other polymers have produced comparable results [55, 56]. In addition, these polymer films can act as a type of "carpet," supporting the transfer to solid substrates.

A TEM image of a cutout of cubically organized $Au_{55}(PPh_3)_{12}Cl_6$ on a PVP surface is shown in Figure 4.22 [56]. PVP also supports the formation of quasi-1-D Au_{55} cluster arrangements. When distributed in a thin DCM film on a water surface in a LB

4.2 Metal Nanoparticles | 345

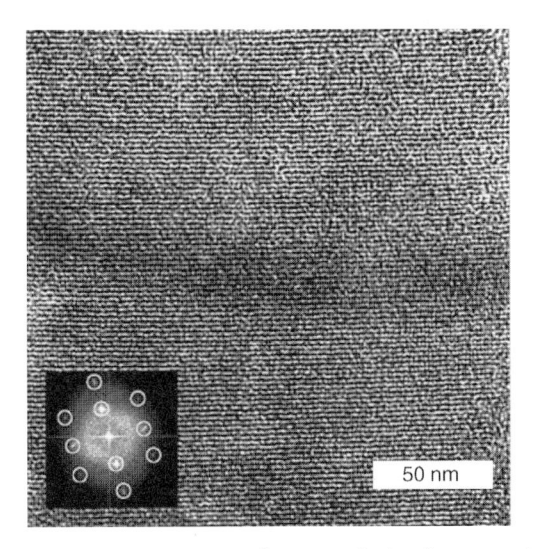

Figure 4.22 TEM image of a cutout of cubically organized $Au_{55}(PPh_3)_{12}Cl_6$ on PVP. The insert shows the diffraction pattern calculated from the recorded image.

trough, cluster-covered individual polymer molecules can be imaged using AFM. When the surface pressure is increased, the PVP–nanoparticle wires continuously form dens filaments, which show numerous knots, leading to a 2-D network [57]. It is assumed that these knots consist of AuNPs since PVP, when imaged under the same conditions, does not exhibit corresponding behavior. Figure 4.23 shows the

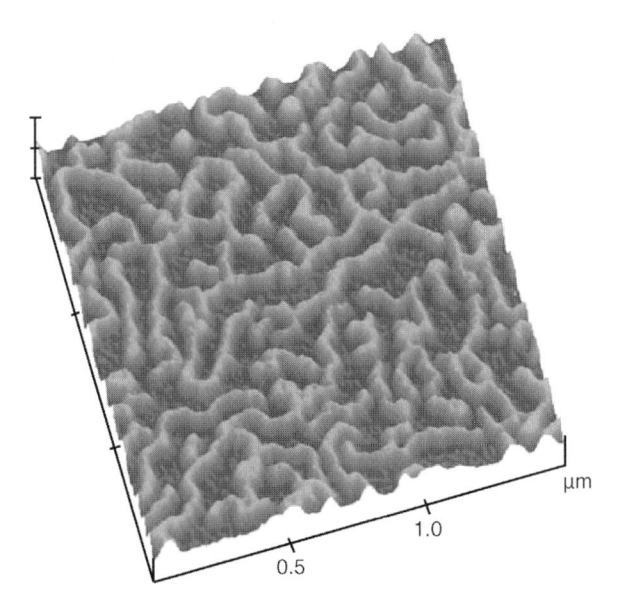

Figure 4.23 AFM image of $Au_{55}(PPh_3)_{12}Cl_6$ clusters decorating chains of PVP, forming a 2-D network.

formation of these filaments, which can be transferred to solid substrates without degradation because of the linkage via nanoparticles.

The relevance of the weak particle–surface interaction could also be demonstrated when alkyl- instead of aryl-substituted ligands were used, when there was no ordering process at all. However, if the same clusters with an "alkyl surface" were used in combination with saturated polymers such as poly(methylmethacrylate) (PMMA), then a reasonable ordering could be observed again.

Independently of the polymer, two different arrangements of $Au_{55}(PPh_3)_{12}Cl_6$ were observed, namely hexagonal and square. Although the reason for this is not quite clear, it can be postulated that the very first contact between a few clusters will be decisive for the formation of the one or the other structure, as the orientation of the PPh_3 ligands between different clusters could determine further growth. This situation is shown schematically in Figure 4.24a, while a simulation as a full-space model, elucidating the different densities, is shown in Figure 4.24b (simulations performed by M. Bühl and

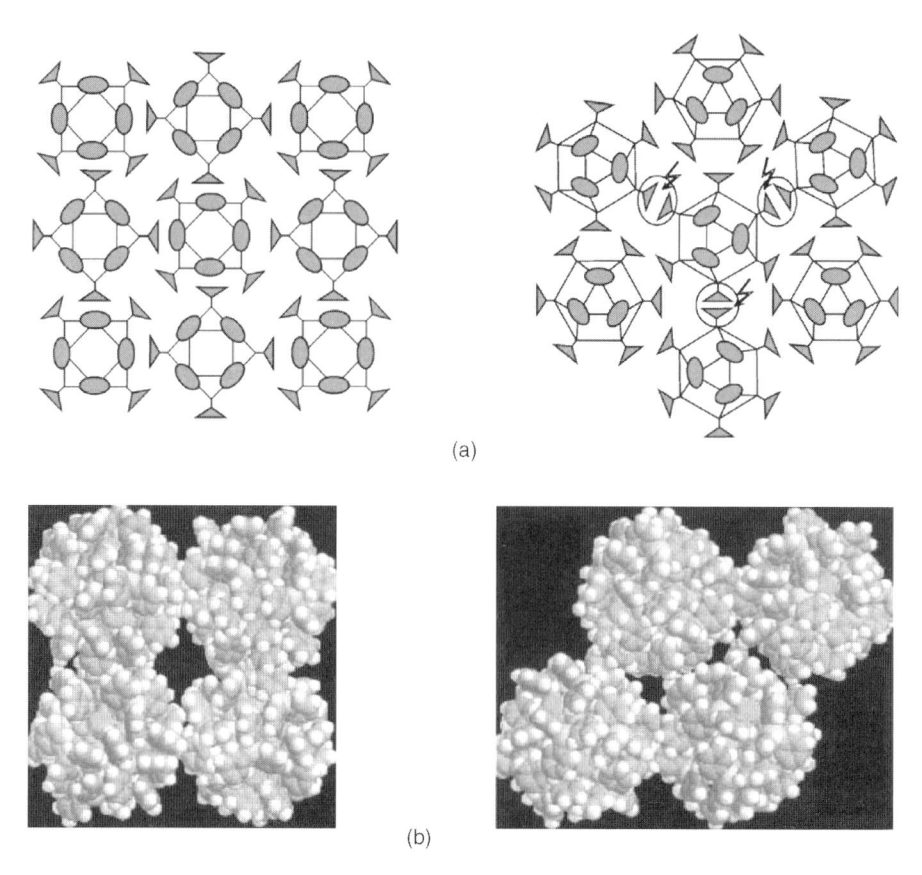

Figure 4.24 (a) Sketches of variously oriented $Au_{55}(PPh_3)_{12}Cl_6$ clusters, indicating the PPh_3 shapes and their positioning towards each other, resulting in square or hexagonal closed-packed structures; (b) Full-space models of both orientations.

Figure 4.25 TEM image of a magnified cutout of perfectly square-packed $Au_{55}(Ph_2PC_6H_4SO_3H)_{12}Cl_6$ clusters.

F. Terstegen, Max Planck Institut für Kohlenforschung, Mülheim, Germany). Indeed, statistical data acquired from many different TEM images clearly show different cluster–cluster distances, depending on the structure of the monolayers (as indicated in Figure 4.24a). The TEM image in Figure 4.25 shows a cutout of a perfectly square-packed monolayer of $Au_{55}(Ph_2PC_6H_4SO_3H)_{12}Cl_6$ clusters [55].

DNA is a natural "polymer" molecule that attracts appropriate metal nanoparticles by coulombic interactions. Recently, a rather interesting way to ordered arrangements of AuNPs in combination with DNA has been recognized [58], based on the coulombic interactions between the negatively charged phosphate backbone of synthetic and of natural DNA and positively charged 3.5 nm AuNPs, prepared from $HAuCl_4$ and $NaBH_4$ and capped with lysine molecules. Subsequent TEM investigations of these hybrid systems confirmed the formation of equidistant rows of nanoparticles between parallel DNA sequences, as shown schematically in Figure 4.26.

A unique type of 1-D organized metal nanoparticle, in connection with DNA as the templating material, was observed when λ-DNA was reacted with $Au_{55}(Ph_2PC_6H_4SO_3H)_{12}Cl_6$ [59]. It appeared that these clusters not only interacted with DNA (as did the above-mentioned 2–3 nm Pt particles) by assembling along the double-stranded DNA, but also occupied the major grooves of B-DNA, the preferred DNA version in aqueous solution. Subsequent molecular dynamics (MD) simulations indicated that bare Au_{55} clusters of 1.4 nm diameter fitted perfectly into the grooves. The loss of parts (or even all) of the original phosphine ligands can be understood as a substitution reaction of phosphines by phosphates of the DNA backbone, acting as a type of polydentate system; this can be seen in Figure 4.27a. However, the surprising effect is the degradation of these 1.4 nm particles in the B-DNA major grooves to 0.6–0.7 nm particles in ultrahigh-vacuum, most likely due to shrinkage of

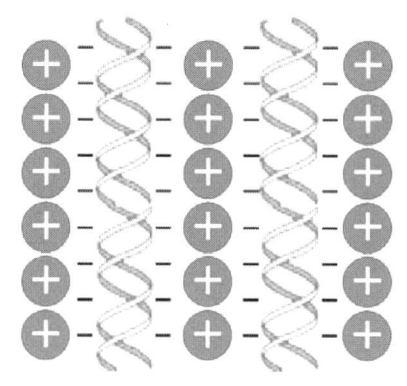

Figure 4.26 Sketch of an arrangement of DNA/Au nanoparticles. Positively charged 3.5 nm gold nanoparticles, capped with lysine, add to the negatively charged phosphate backbones of DNA to build up equidistant rows.

the major grooves of B-DNA, which changes to A-DNA under these conditions. Again, the grooves of this type of DNA fit perfectly to the observed size of particles, which agree reasonably with Au_{13} (see Figure 4.27b). Moreover, these Au_{13}–DNA hybrid systems become organized in such a way that equidistant wires of nanoparticles result, most likely via gold–gold interactions,. The ordering parameter is, without doubt, the A-DNA. The simulation and experimental result are shown in Figure 4.27c.

Whilst the formation of nanowires clearly occurs during drying on the grid used for TEM, Au_{55} clusters have never been observed to degrade under comparable conditions. This is clearly a unique example of a change in molecular conformation leading to a controlled degradation of nanoparticles. Under the conditions of electron beam irradiation, however, the cluster–DNA hybrid system is rather unstable such that, during the course of only seconds, the lines break down to form larger gold particles, the formation of which can be observed microscopically. It is unclear whether a beam-conditioned decomposition of DNA, of the wires, or of both, causes the degradation. Figure 4.28b shows the situation following decomposition of the same area shown in Figure 4.28a.

4.2.2.2 Guided Self-Assembly

A "guided" self-assembly of AuNPs on patterned GaAs surfaces has been elaborated by Andres et al., for the development of electronic arrays to create semiconductor device layers [60]. In this case, a structured Be-doped GaAs layer is first covered by a monolayer of xylyl dithiol, $HS-CH_2-C_6H_4-CH_2-SH$. Then, if a solution of dodecanethiol-protected 5 nm AuNPs in hexane is spread onto a convex water surface, a uniform monolayer of nanoparticles will be formed, in accordance with the above-described procedure. If the modified GaAs surface is allowed briefly to touch the nanoparticle monolayer, it will be transferred to the substrate. In this case, it is assumed that the dodecanethiol ligand molecules are partially substituted by the xylyl dithiol covering the surface. The quality of the hexagonal close-packed monolayer of

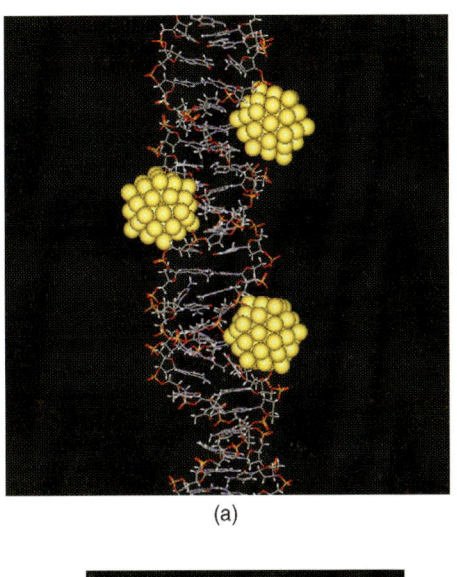

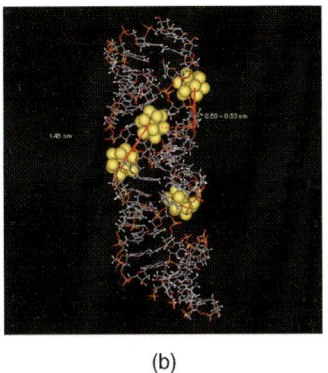

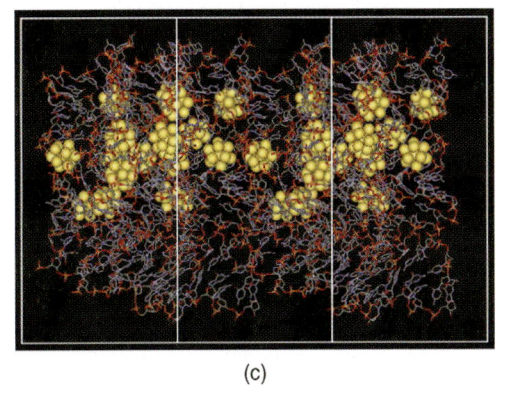

Figure 4.27 (a) Molecular modeling resulting from the combination of B-DNA with bare Au_{55} clusters trapped in the major grooves; (b) Transition from B- to A-DNA, linked with shrinking of the major groves from 1.4 to 0.7 nm, causes degradation of the Au_{55} clusters to 0.6–0.7 nm particles; (c) These hybrid systems then combine to produce equidistant wires.

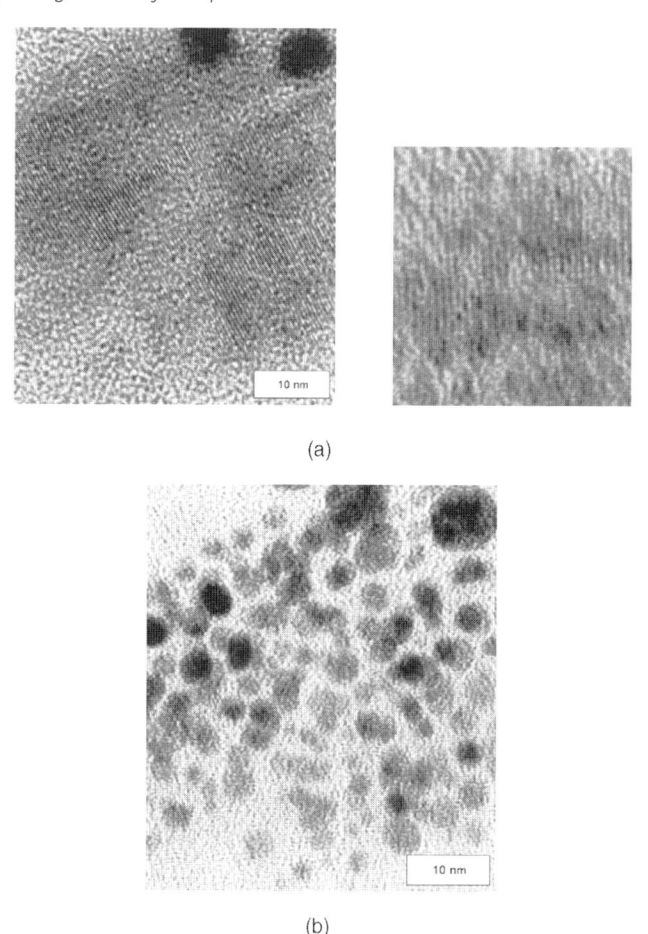

Figure 4.28 (a) TEM image of gold nanowires, consisting of 0.6–0.7 nm particles (see magnified cutout) arranged equidistantly (0.5 nm); (b) The same area after continued electron-beam irradiation, showing the formation of large gold nanoparticles from the wires.

AuNPs can be controlled by transferring samples to the TEM grids. It is assumed that the quality of the monolayer does not change if the thiol-modified GaAs surface is used instead of the grid. This procedure might be of particular interest, since when self-organization occurs on a water surface, the monolayer will become strongly fixed to the surface by covalent bonds. Subsequent rinsing of the surface with hexane will remove any untethered particles, but not those that are chemically fixed.

Recently, LB techniques have also been used to fabricate ordered 2-D structures of metal nanoparticles, with several approaches having been identified using particles in the size range of 3 to 5 nm [61–67]. For example, Heath et al. reported [62] that success in the organization of alkanethiolate-protected nanoparticles with LB techniques depends characteristically on the ratio of ligand length and the surface pressure.

Ligand-stabilized Au_{55} clusters of only about 2 nm in size were first organized in LB films in 2000 [68], but only relatively small ordered areas were observed. Much better results were obtained when the surface pressure was increased; under these conditions, Au_{55} clusters coated with PPh_3 ligands (some of which had been substituted by hydrophilic derivatives to render the particles amphiphilic) formed much larger areas of ordered particles [69]. Repeated pressing of the films at up to 35 mN m^{-2}, with alternating relaxation, led to the production of quite sufficiently ordered films of $0.25 \times 0.35\,\mu m$.

From the moment the photonic bandgap concept was introduced by Jablonovitch and John [70, 71], materials possessing a 3-D periodicity of the dielectric constant have attracted increasing attention. The periodically varying index of refraction in photonic crystals causes a redistribution of the density of photonic states, due to Bragg diffraction, which is associated with stop bands for light propagation. Chemical self-assembly (bottom-up) methods have been used to prepare 3-D photonic crystals (so-called "artificial opals") from colloidal silica or micron-sized polymer spheres. By using artificial opals as templates, so-called "inverse opals" that possess a pronounced photonic bandgap can be prepared by the impregnation of voids with different materials, followed by removal of the spheres via annealing or dissolution, so as to produce a complete photonic bandgap [72]. Magnetic nanocrystals can be used for the fabrication of metal–dielectric inverse opals that are addressable through an external magnetic field. The SEM image of an inverse FePt colloidal crystal, obtained by complete impregnation of a 3-D template of monodisperse latex microspheres with FePt nanocrystals, followed by dissolution of the polymer spheres in toluene, is shown in Figure 4.29.

Self-assembled structures are quite suitable for studying the electronic behavior of ordered monomolecular films, and have contributed considerably to the provision of a better understanding of the quantum size nature of metal nanoparticles. On the other hand, it is quite clear that metal particles in future nanoelectronic devices should be in structures in which the QDs should not be self-arranged, but rather are

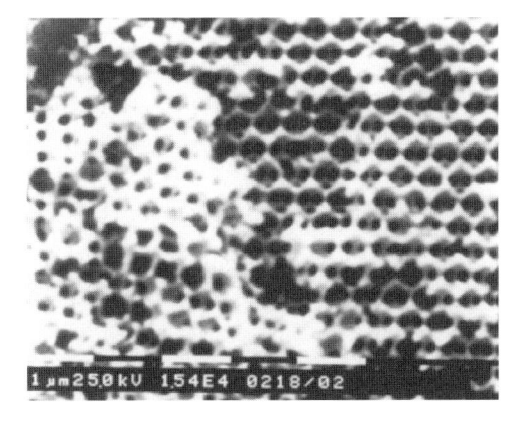

Figure 4.29 SEM image of an inverse colloidal crystal made from PtFe nanocrystals. Illustration courtesy of Dr N. Gaponik.

placed at definite positions. This is a major problem that will only be solved by techniques other than self-assembling.

Whilst the use of an appropriate template represents one possible route to success, the templates have still to be developed. Recently, remarkable progress has been made in the organization of metal nanoparticles to form given structures that has involved the use of *micelles*. For example, Möller *et al.* have used micellar structures in block copolymers as nano-sized reaction vessels to generate and to organize metal nanoparticles. Although this technique is not based on the self-assembly of metal nanoparticles, it is linked with the organization of micelles in the polymers. Whilst the technique differs from the self-assembly processes described above, it cannot be considered to solve the problem of creating artificial architectures. However, compared to the classical self-assembly procedures, the application of preformed micelles will open up novel routes for handling nanoparticles.

A-B diblock copolymers can be used to form micelles that allow the preparation of thin, coherent films of well-developed structure [73–79]. For example, both poly(styrene)-block (*b*)-poly(ethylene oxide) and poly(styrene)-*b*-poly(2-vinylpyridine) have been identified as useful combinations for generating well-ordered compartments in which metal nanoparticles can be fabricated in a variety of ways. The formation of diblock copolymers from polymer A and polymer B, ultimately producing nanospheres that can be transferred to films of highly ordered micelles, is illustrated in Figure 4.30a. The formation of metal nanoparticles inside the micelles is shown schematically in Figure 4.30b.

As a typical example, the formation of AuNPs is described here [80]. Both of the above-mentioned block copolymers can be treated with $HAuCl_4$ or $LiAuCl_4$; in this case, the tetrachloroaurates penetrate into the micelles in nonpolar solution to form thermodynamically stable dispersions inside the holes, with the solubilization process being supported by ultrasound irradiation (if necessary). A subsequent reduction of the gold salts results in the formation of nanoparticles, with suitable reducers including hydrazine and $NaBH_4$. Dispersion of the generated nanoparticles can be controlled by the domain structure of the block copolymer [74, 75]. Because of the reduction mechanisms used, it is not possible to prepare a single gold monocrystal per micelle via this method; rather, multiple nanoparticle formation in each domain is observed. A TEM image of an ordered array of micelles, each loaded with several gold particles, is shown in Figure 4.31 [75]. In this way, one single nanoparticle per micelle can be obtained, even with systematic variation of the particle size, if the process is performed at the glass-transition temperature (90 °C) of the poly(ethylene oxide). Particles of 6, 4, and 2.5 nm diameter have been prepared by varying the loading ratios and the micelle size (see Figure 4.32a–c).

Self-assembly processes leading to film formation usually occur by evaporation of the solvent from solutions containing loaded micelles. Micellar film formation can also be brought about by dipping platelets of different materials, such as GaAs, InP, glass or carbon-coated copper grids, into a corresponding solution, followed by their withdrawal [81]. Consequently, by using four kinetically controlled steps – reduction, mineralization, coagulation, and film formation – the structure of nanoparticle-loaded block copolymer composites is possible, the further use of which can be expected.

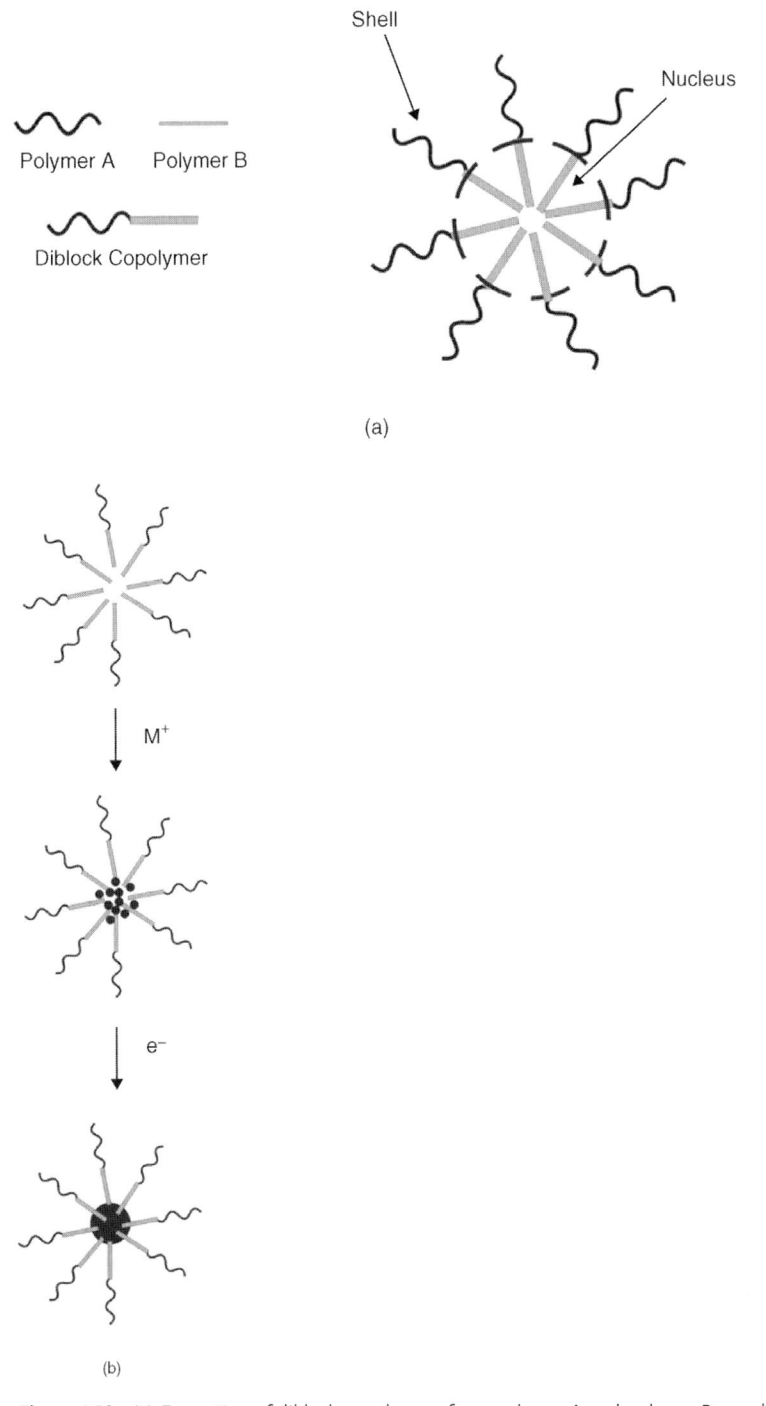

Figure 4.30 (a) Formation of diblock copolymers from polymer A and polymer B, resulting in micelles; (b) Formation of metal nanoparticles inside the micelles.

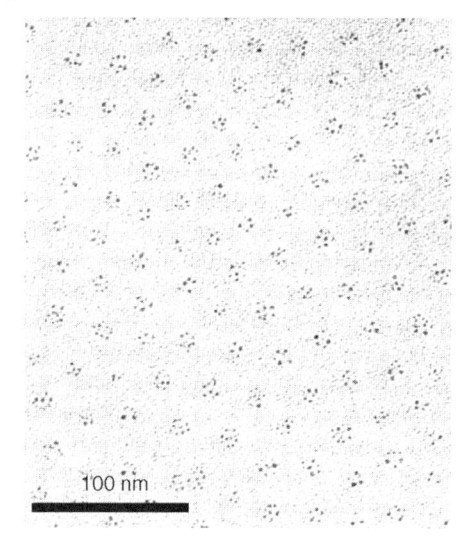

Figure 4.31 Gold particles inside of an ordered array of micelles.

Notably, gold is by no means the only material that can be incorporated into block copolymer micelles. Indeed, other metals such as palladium, silver, and cobalt have been successfully applied, and even nanoparticles of CdS and PbS have been generated inside micelles [82].

One interesting and important continuation of this development is the possibility to remove the diblock copolymer material without losing any ordering of the nanoparticles, thus generating nanostructures on different surfaces and with varying inter-particle distances. For this, Möller et al. have developed an elegant method to generate these materials on a routine basis, via the use of an oxygen plasma [83–85]. If ordered micellar films on glass or mica, containing either AuNPs or even the nonreduced tetrachloroaurates, are treated by an oxygen plasma at 200 W for 20 min, the bare nanoparticles will be deposited on the surfaces, without any loss of the original order. The formation of nanoparticles from $AuCl_4^-$ during the plasma process can be traced back to the existence of intermediate oxidation products of the polymer (e.g., CO), and also free electrons. Recent X-ray photoelectron spectroscopy

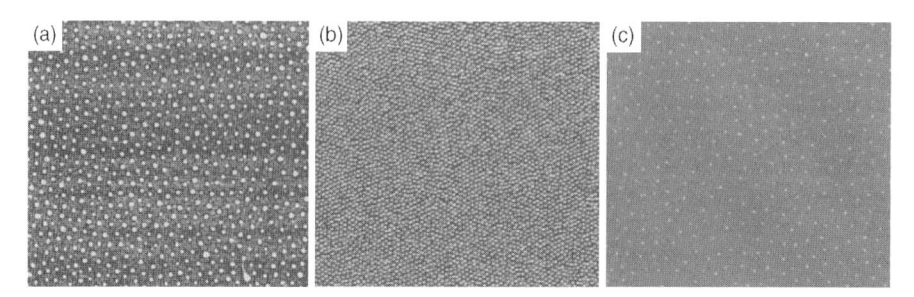

Figure 4.32 TEM images of gold nanoparticles of various sizes generated in micelles of different sizes and by different loading ratios. (a) 6 nm, (b) 4 nm, (c) 2.5 nm.

(XPS) studies have proved the absence of any residual polymer. The as-prepared particles are strongly fixed onto the surfaces, and cannot be removed by rinsing with solvents, nor by rubbing with a soft tissue. Another remarkable advantage associated with this method is the ability to deposit metal nanoparticles onto surfaces. In this case, the particle size, inter-particle distances – and, thereby, the particle density – can be varied. In addition, extension of the covered surfaces is, in principle, not limited. The three examples shown in Figure 4.33 are cutouts of 3×3 cm areas, demonstrating the versatility of the method.

The important point here is to use diblock copolymers of different molecular weights, and of different gold contents. For example, a poly[styrene(800)-b-2-vinyl-pyridine(HAuCl$_4$)$_{0.5}$(860)] (numbers in brackets = monomer units) gives gold colloids of 12 nm in height (by AFM) and a periodicity of 80 nm (Figure 4.33a). Typically, a styrene: vinylpyridine ratio of 325 : 75, with the same equivalent of HAuCl$_4$, results in 2 nm particles with 30 nm distances (Figure 4.33b), whereas changing the ratio to 1700 : 450 with only 0.1 equivalent of gold acid gives 1 nm particles of 140 nm periodicity (Figure 4.33c) [83].

Other diblock copolymers and metal compounds such as H_2PtCl_6, $Pd(Ac)_2$, $TiCl_4$ or $FeCl_3$ have been used successfully in this process. In this case, the less-noble metals Ti and Fe are deposited as oxides following plasma treatment.

Gold nanoparticles containing films of diblock copolymers can be used for the nanolithographic structuring of surfaces, and so may contribute to the search for techniques that allow the routine fabrication of nanostructured surfaces. Semiconductor substrates such as GaAs, InP, or layered systems can be covered by micellar films containing AuNPs. Subsequently, if such composites are etched by using an Ar^+ beam, then nanostructured surfaces will be generated, owing to the fact that the nanoparticulate gold is much more rapidly etched than either bulk gold or pure poly(styrene) and poly(vinylpyridine). Hence, all positions containing gold can be transferred into holes in the underlying semiconductor (as illustrated in Figure 4.34).

In contrast, if a hydrogen plasma is used instead of argon, then all organic material will first be removed, so that only the metal nanoparticles remain. On subsequent etching with an ion-beam chlorine plasma, GaAs cylinders of up to 80 nm in height were prepared [86]. Gold nanoparticles, generated from a gold-loaded diblock copolymer film by an oxygen plasma, also serves as a mask during continued etching by oxygen [87], such that gold-capped diamond pillars were generated with a diameter that roughly corresponded to the size of the gold particles.

The use of so-called "S-layers" is a combination of self-organization and spatial patterning [88]. S-layers consist of 2-D protein crystals that are formed naturally as the outermost cell surface layer (S-layer) of prokaryotic organisms. The subunits can recrystallize into nanoporous monolayers in suspension, at liquid–surface interfaces, on lipid films, or on solid substrates. The S-layers of *Bacillus sphaericus* CCM 2177 have been used to generate ordered arrays of 4–5 nm gold particles, with a 13.1 nm repeat distance, from $AuCl^-$ ions [89]. The spontaneous self-assembly of 5 nm AuNPs was shown to occur at the S-layer of *Deinococcus radiodurans*, to produce micrometer-sized ordered domains [90]. Arrays of 1.9 nm platinum particles were achieved from Pt salts in the S-layer of *Sporosarcina ureae* [90]; these were of square symmetry and had a lattice constant of 13.2 nm.

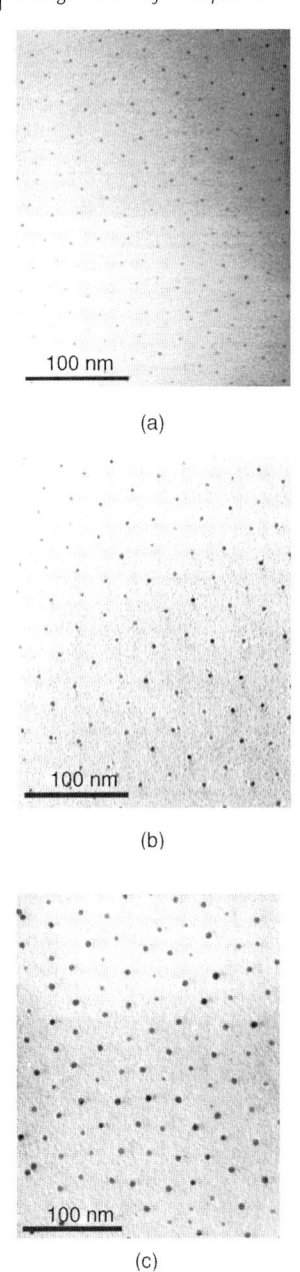

Figure 4.33 TEM images of cutouts of 3 × 3 cm areas of gold nanoparticles produced in diblock copolymers of different molecular weight, and of different gold contents.

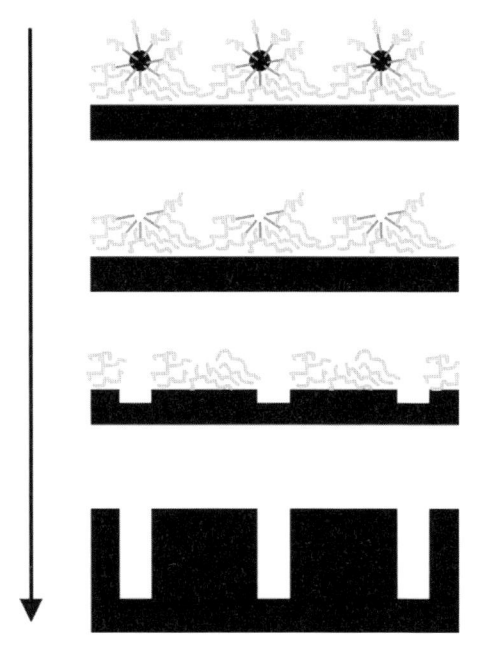

Figure 4.34 Scheme of etching semiconductor substrates, starting with gold nanoparticles containing micellar films, and followed by Ar$^+$ beam etching to remove the gold nanoparticles. This forms holes in the substrate with the same diameter as that of the gold nanoparticles.

4.2.2.3 Aimed Structures

Until now, the discussions have centered only on the arrangements of metal nanoparticles that were based on self-assembly mechanisms even if, as in the case of micellar films, some influence on particle size and periodicity was possible. The arrangements of nanoparticles completely free from natural self-assembly processes, thus allowing the formation of arbitrary arrays, has seemed possible since Mirkin *et al.* developed the process of dip-pen nanolithography (DPN) [91, 92]. This technique is based on the transport of molecules from the tip of an atomic force microscope, through a water meniscus, onto a flat surface that attracts the molecules by chemical bonding. The water meniscus is formed naturally in the laboratory atmosphere such that, when the tip is moved over the surface, the corresponding molecules are transported to the surface so as to form programmed patterns. Gold surfaces are very much suited to decoration with thiol molecules from the tip. The general method used to deposit regularly oriented nanoparticles onto a surface is shown in Figure 4.35 [93].

In a second step, an array of dots of molecules that will attract the nanoparticles are deposited onto the substrate. Although more recently, 16-thiohexadecanoic acid has been used to bind positively charged particles such as protonated amine- or amidine-modified polystyrene spheres as building blocks from solution, the unpatterned regions of the gold substrate previously were passivated using an alkanethiol.

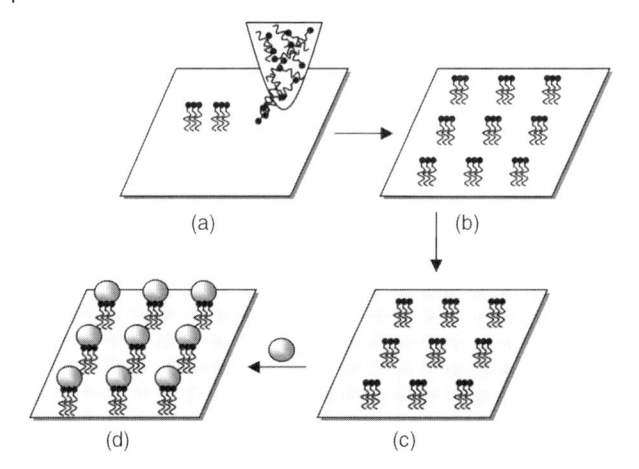

Figure 4.35 Schematic representation of dip-pen nanolithography. An AFM tip is loaded by thiol molecules (in this case, 16-thiohexadecanoic acid). Because of the presence of a "natural" water meniscus, the thiols are transported onto a gold surface via the moving tip (a) to form any kind of pattern (b). Passivation of the unpatterned regions by alkylthiols (c) enables deposition of nanospheres with basically modified surfaces by interactions with the carboxylic groups (d).

Orthogonal gold assemblies have been fabricated by varying this method in such a way that, after patterning the gold surface with 16-thiohexadecanoic acid and passivation of the unpatterned regions with 1-octadecanethiol, alkylamine-modified oligonucleotide ($a = $ TCTCAACTCGTAA$_{10}$) becomes chemically bound to the carboxylic groups. Several steps of rinsing with buffer solutions, washing with deionized water, and drying then follow. The superiority of the DPN technique has been impressively demonstrated by adding a second orthogonal structure to the first. In fact, under normal writing conditions (0.5 nN, 55% humidity, 23 °C), it is possible to replace 1-octadecanethiol by 16-thiohexadecanoic acid molecules emitting from the microscope tip. A second, quite different, alkylamine-modified oligonucleotide ($b = $ A$_{10}$CGCATTCAGGAT) is then combined with the acid functions (as in the first step), such that two nanopatterns consisting of two different oligonucleotide sequences can be created. The a,b-modified substrate is then treated with $a'b' = $ TACGAGTTGA-GAATCCTGAATGCG, where a' and b' are complementary to a and b, respectively. The deposition of gold nanoparticles of two different sizes occurs by adding a- and b-substituted 13 nm and 30 nm particles. The complementary forms of a and a' and b and b', respectively, result in a regular pattern of both types of nanoparticle [94] although clearly, the fabrication of 2-D arrangements of only one type of particle is also possible. The formation of nanostructures via complementary oligonucleotides is illustrated schematically in Figure 4.36a, while Figure 4.36b shows an AFM topographic image of the different Au particles after an orthogonal assembly process.

It can be foreseen that not only two different sizes of the same type of metal can be used to generate those or other structures, but also that different metals can be used, leading to many possibilities with respect to future applications.

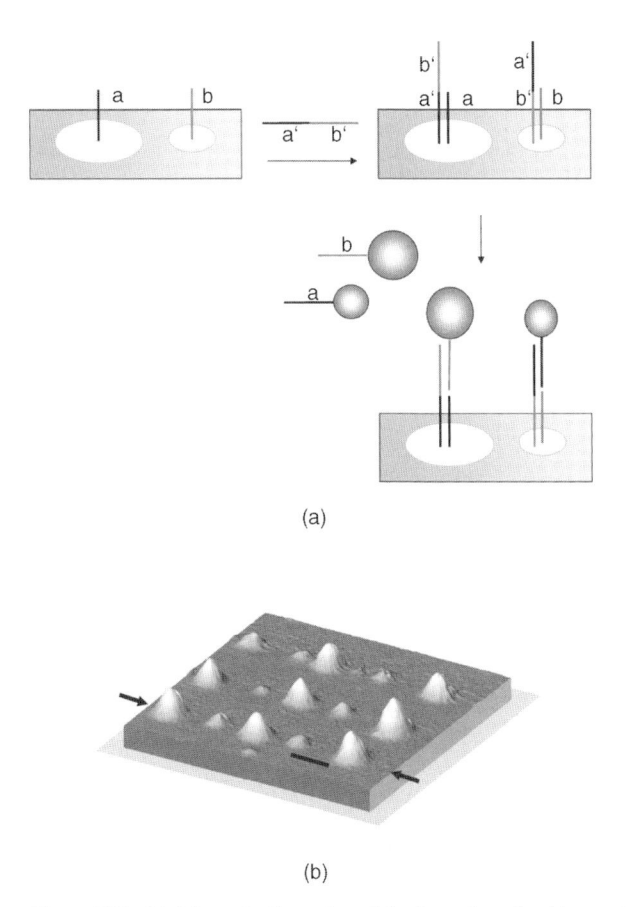

Figure 4.36 (a) Schematic illustration of the formation of gold nanostructures via complementary oligonucleotides; (b) AFM image of as-prepared patterns consisting of gold nanoparticles of different sizes.

A template-free fabrication of one-dimensionally organized Au_{55} clusters using a controlled degradation of ordered monolayers has been identified. As described in Section 4.2.2.1, islands of square and hexagonal structures of $Au_{55}(PPh_3)_{12}Cl_6$ clusters are formed on the surface of water, which was previously covered with a thin film of clusters in DCM, when the solvent is evaporated. If these islands are then transferred to solid substrates under well-defined conditions, they will be degraded into strictly parallel stripes, each consisting of three to four rows of clusters [95]. The optimum transfer conditions include a 20° withdrawal angle and a speed of 10 cm min^{-1}. The formation of these patterns is explained by the oscillation of the water meniscus at the water–substrate boundary; the complex mechanism is illustrated in Figure 4.37, where the monolayer is fractioned parallel to the transfer direction. A TEM image of an as-prepared pattern of stripes of Au_{55} clusters is shown in Figure 4.38.

This form of nanoparticles organization is related to a method which is based on the wetting instabilities of monomolecular layers when transferred onto solid

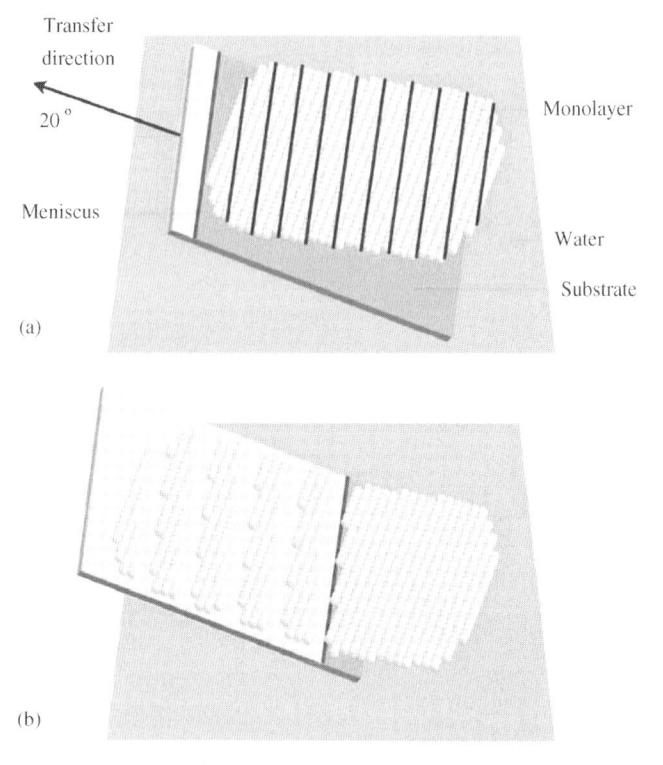

Figure 4.37 Sketch of the mechanism generating stripes of $Au_{55}(PPh_3)_{12}Cl_6$ chains by the division of ordered islands into sections caused by the oscillating meniscus at the water–substrate boundary.

substrates. When low-pressure monolayers of L-α-dipalmitoyl-phosphatidylcholine (DPPC) are placed on water, they will degrade so as to form a pattern with parallel hydrophilic channels that are 200–300 nm wide, when transferred to mica at a speed of 1 mm s^{-1} [96]. The channels of such structures can be used to deposit nanoparticles in one direction. Subsequently, thiol-stabilized Au_{55} clusters, dissolved in 1-phenyloctane, were used to fill the channels, such that aggregates were formed during removal of the solvent. The DPPC areas were almost completely free of clusters, owing to their nonwettability.

Because of the magnetic dipolar interactions, a 1-D self-assembly is rather typical for magnetic nanoparticles. For a magnetic single-domain nanoparticle of radius r, the magnetic dipole moment arising from the alignment of electron spins, is $\mu = 4\pi r^3 M_s/3$, where M_s is the saturation magnetization of the respective bulk material. The dipole–dipole interaction between two magnetic particles at contact is proportional to $\mu^2/\sigma^3 \propto r^6/\sigma^3$, where σ is the effective hard sphere diameter, consisting of the magnetic core diameter and the thickness of the surfactant layer [97]. These dipolar interactions, which favor the head-to-tail arrangement of dipoles, can compete with nondirectional van der Waals forces and steric repulsion, inducing an anisotropic agglomeration of nanoparticles. In 1970, de Gennes and Pincus [98] predicted, on a theoretical basis, that colloidal particles with a magnetic dipole

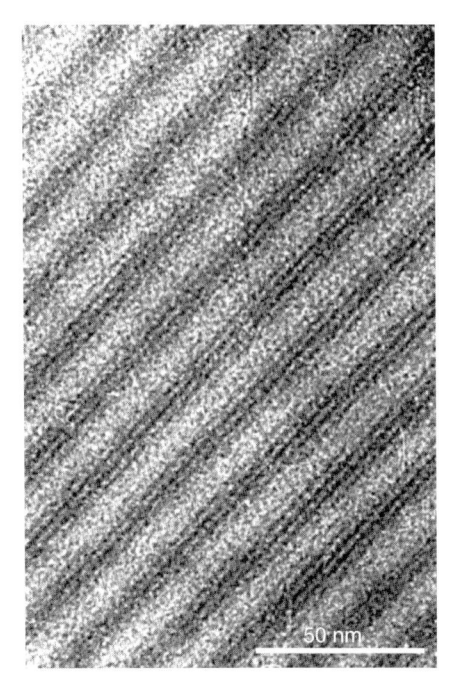

Figure 4.38 TEM image of equidistant stripes of $Au_{55}(PPh_3)_{12}Cl_6$ clusters consisting of three to four individual 1-D chains.

moment should be able to minimize the magnetostatic energy via self-assembly into flexible chains (see Figure 4.39a). Simulations have shown that such chains may be present in colloids of superparamagnetic particles, even in a zero magnetic field [99, 100]. Only in 2003, did Butter et al. [101] report the direct in situ observation of such linear chains of particles present in a dispersion of iron nanoparticles at zero field. These authors used a cryogenic TEM technique based on a rapid vitrification of the colloidal solution, followed by low-dose TEM imaging of metal nanoparticles inside the vitrified film. A clear transition from a single-particle population observed for 4 nm iron nanoparticles to a system of worm-like chains for particles larger than about 12 nm was observed, and this was explained by a rapid increase in dipolar interaction with particle size.

Unpaired dipoles present at the ends of a linear chain of nanoparticles can pair, yielding the closed ring (as shown in Figure 4.39b). These structures were observed by Tripp et al. [102] for 27 nm cobalt nanoparticles (weakly ferromagnetic at room temperature), which self-assemble into bracelet-like rings. Typically, the rings consisted of between five and 12 particles and were 50–100 nm in diameter. The key point for selective formation of the nanoparticle rings rather than large aggregates was an increased viscosity of the solvent during an earlier stage of drying the colloidal solution on a TEM grid.

An external magnetic field orients the magnetic dipoles of individual nanoparticles in one direction, facilitating the formation of linear head-to-tail superstructures. The chain of dipole–dipole-interacting nanoparticles will align itself in the direction of an

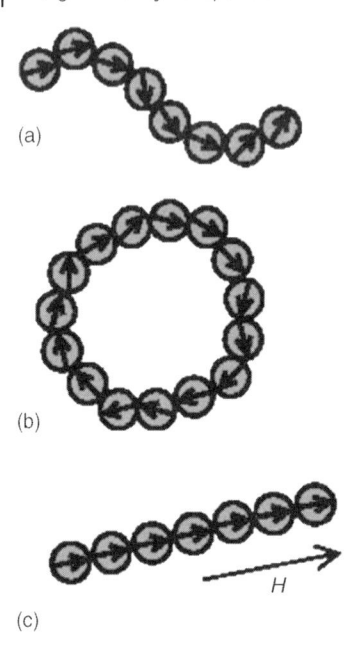

Figure 4.39 Dipolar chains formed by magnetic particles. (a) A semi-flexible chain of magnetic particles showing the head-to-tail, paired dipoles in the middle section and the unpaired dipoles at the ends; (b) A self-assembled closed ring of magnetic particles allows all magnetic dipoles to be paired; (c) Straight chains of magnetic particles can form and align in the magnetic field.

externally applied magnetic field (as shown in Figure 4.39c). Another striking effect of the magnetic field is the dependence of the lateral width of 1-D nanoparticle aggregates on the strength of the applied field, as was demonstrated by Giersig et al. using the example of cobalt nanoparticles (Figure 4.40) [30]. A progressive increase in the width of nanoparticle chains with increasing magnetic field can be explained by the attractive chain–chain interactions between parallel dipolar chains of particles aligned by a magnetic field [101, 103]. In the case of a substrate positioned vertically, the self-assembly of magnetic nanoparticles into 1-D chains of variable width has been explained as a cooperative effect of the external magnetic field and the force of gravitation [104].

The shape of the magnetic nanoparticles can also affect the preferential dimensionality of self-assembled superstructures. Thus, hcp-Co nanodisks, as synthesized by Puntes et al., showed a clear predisposition to stack face-to-face and to remain perpendicular to the substrate, forming long ribbons (Figure 4.41) [105]. In the TEM images, these resembled ribbons of nanorods. However, by tilting the TEM grid the standing disks were seen to vary in thickness when tilted in the short direction, but to remain of a constant length when tilted in the long direction (Figure 4.41), thus verifying the disk shape.

Currently, the best tool for fabricating aimed nanostructures is, without doubt, the atomic force microscope tip; this was used by Mirkin et al. to develop DPN for

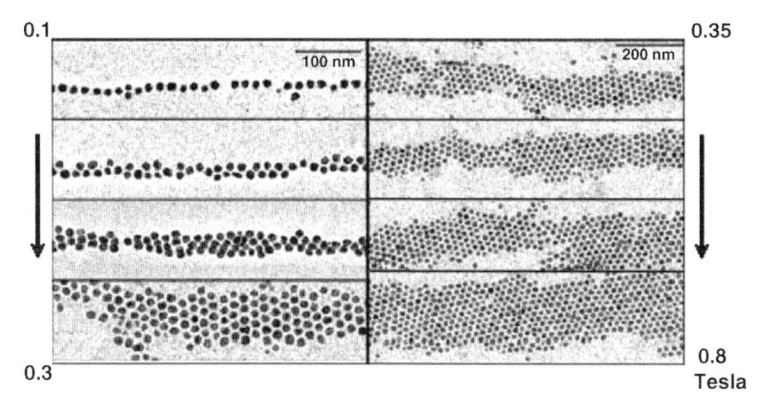

Figure 4.40 TEM images of 1-D ordered cobalt nanoparticles self-assembled in the magnetic fields of different strengths. Reproduced with permission from Ref. [30]; © 2002, Elsevier Science.

depositing thiol molecules onto gold surfaces [91, 92]. A similar technique that uses a position-specific deposition of metal nanoparticles has also been developed, using the tip of a *scanning tunneling microscope* to which a voltage has been applied [106]. By using a potential which is less negative than would be required to generally reduce metal ions in the relevant solution, only the microscope tip will become loaded with metal ions. If the tip is moved close to a surface, a small metal bridge between the tip and surface will be generated; subsequently, on withdrawing the tip, metal nanoparticles will be deposited at the corresponding positions. This rapid and easily reproducible process is suited to numerous types of metal ions.

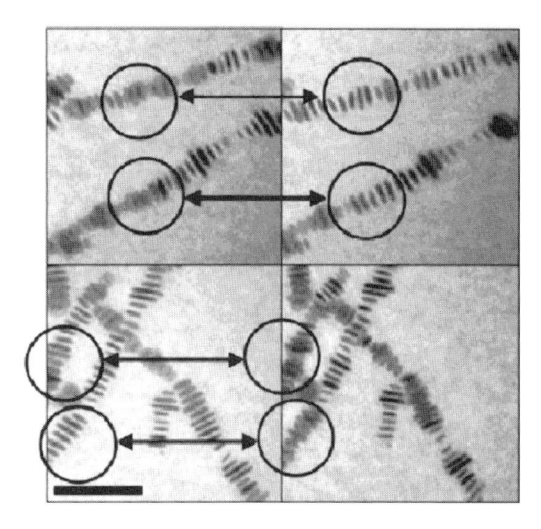

Figure 4.41 TEM image of stacked face-to-face hcp-Co nanodisks. The pictures on the left correspond to no tilt and those on the right the pictures on the right 25° tilt. The bar is 200 nm. Tilting direction corresponds with the double arrows. Reproduced from Ref. [105], Copyright 2002, with permission from American Chemical Society.

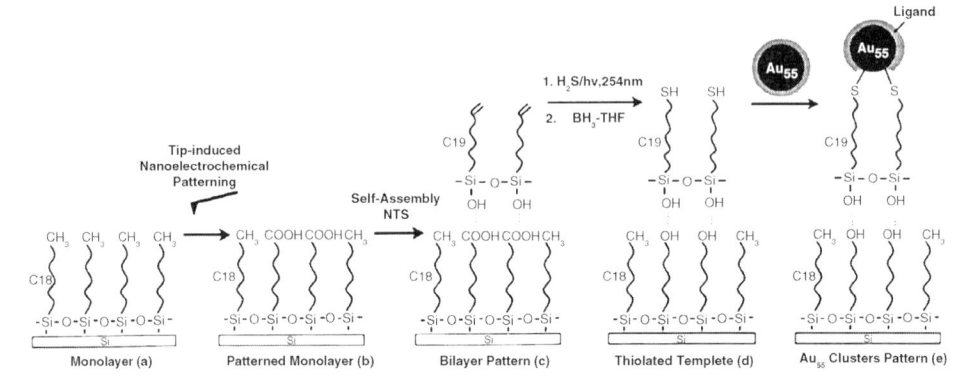

Figure 4.42 Generation of Au_{55} patterns on a prestructured surface. An alkyl-terminated, self-assembled monolayer (a) is oxidized along traces by a metalized AFM tip, inducing electric pulses (b). Alkene-terminated silanes are then added specifically to the COOH functions (c). This is followed by thiolization of the C=C-double bonds (d) and the addition of water-soluble $Au_{55}(Ph_2PC_6H_4SO_3Na)_{12}Cl_6$ clusters (e).

Conductive AFM tips can be used to emit electrical pulses to affect the top functions of ordered monolayers [107–109]. Here, the tip operates as a nanoelectrical pen, inscribing chemical information into appropriate surfaces. These traces, when chemically modified, can then be used to deposit nanoparticles for further applications.

As can be seen from Figure 4.42, a C_{18} alkyl monolayer on silicon has been oxidized by a tip-induced nanoelectrical reaction, generating carboxylic functions [110]. A second self-assembled monolayer, produced by nonadecenyltrichlorosilane, on top of the COOH groups, functionalizes the carboxylic tracks by C=C double bonds that can be attacked photolytically by H_2S to give the corresponding thiol-functionalized pattern. The purpose of using BH_3 is to reduce any S–S bonds that might be formed during the process. When the water-soluble gold cluster $Au_{55}(Ph_2PC_6H_4SO_3Na)_{12}Cl_6$ was used for deposition on the thiolated traces, the structures were remarkably thermally stable, as demonstrated by annealing up to 120 °C.

By using this technique, J. Sagiv *et al.* generated a series of planned patterns of 1.4 nm gold particles, the electronic properties of which are known to render them promising candidates as future single-electron switches. Some of these structures are shown in Figures 4.43–4.45. Subsequent AFM analyses of the patterns indicated the existence of single cluster layers; a pattern of parallel wires and the corresponding distance–height profiles is shown in Figure 4.43. The wires in Figure 4.43a consist of the bilayer system only (step d in Figure 4.42), whereas Figure 4.43b images the same pattern after the addition of the gold clusters (step e in Figure 4.42). As can be seen from the distance–height profiles, there is an increase in height from about 2 nm in Figure 4.43a to about 5 nm in Figure 4.43b, corresponding to a particle height of about 3 nm. Considering the inherent limitations of the AFM height determination, this value agrees quite well with the cluster size of about 2.5 nm.

Although details cannot be distinguished in Figure 4.43, gold clusters placed in two densely packed parallel rows can be observed in the magnified cutouts in Figure 4.44,

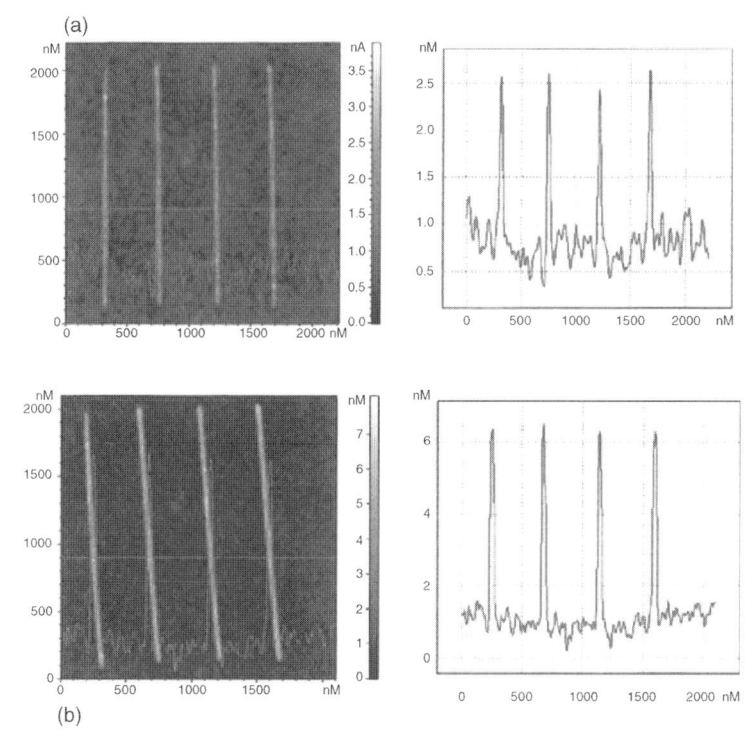

Figure 4.43 (a) Lines of SH-functionalized traces corresponding to step (d) in Figure 4.42 and height profile; (b) Au_{55}-decorated lines corresponding to step (e) in Figure 4.42. The height profile agrees well with the addition of about 2.5 nm to the 2 nm height of the thiol-functionalized traces.

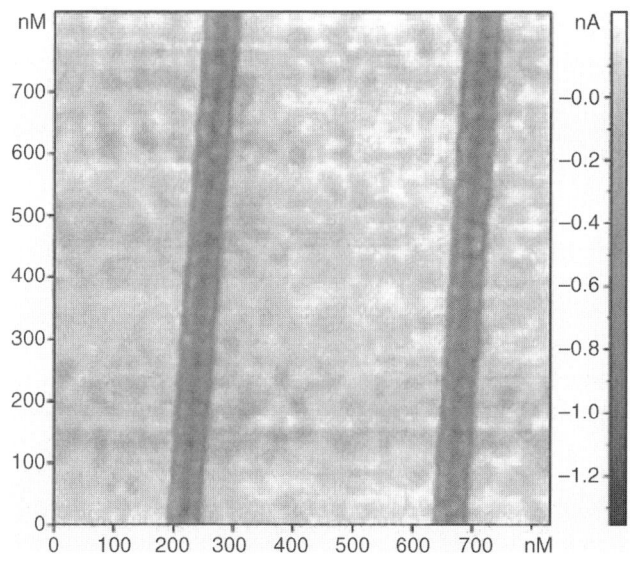

Figure 4.44 Magnified pieces of Au_{55} cluster-decorated lines of Figure 4.43, indicating two rows of clusters per line.

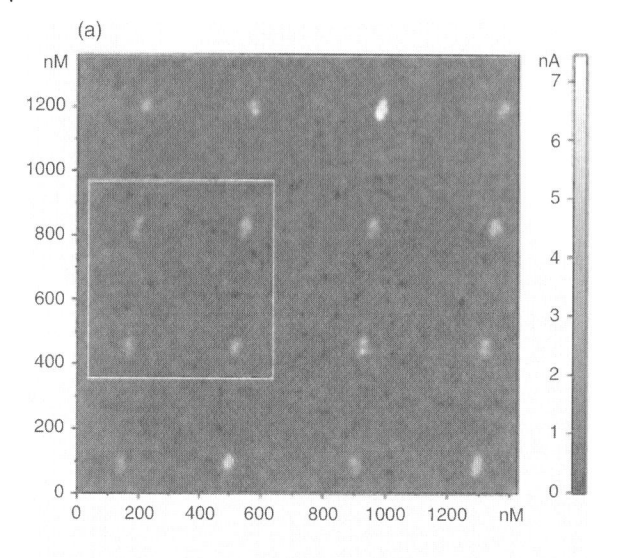

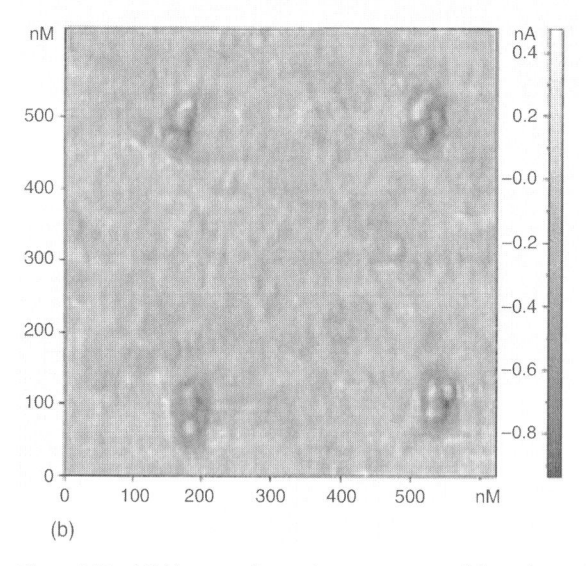

Figure 4.45 AFM images of a regular arrangement of dots of Au$_{55}$ clusters, each dot consisting of two to three clusters.

apparently arranged along the periphery of the tracks. By comparison, Figure 4.45 shows a pattern of dots, each consisting of assemblies of two to three partially resolved clusters.

The capabilities of this method, provided that the experimental conditions are optimized, can be seen from Figure 4.46, where ideal 1-D wires of individual particles have become available. Indeed, it should even be possible to place single clusters into predefined intrawire gaps.

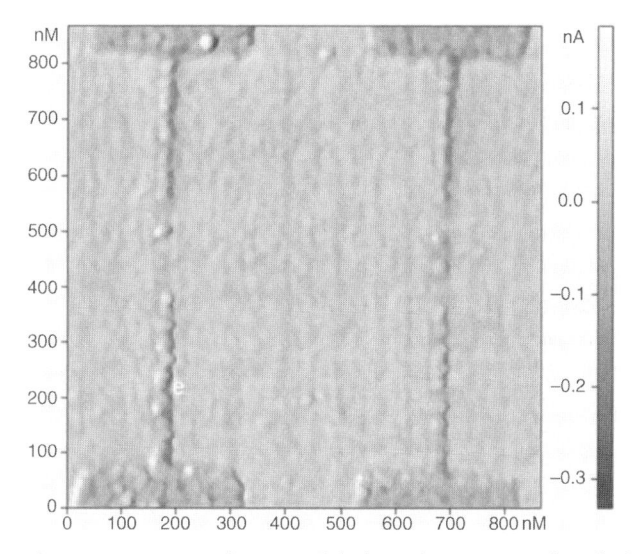

Figure 4.46 Optimized pattern of ideal 1-D cluster wires and single clusters in predefined intrawire gaps.

To date, this route represents the most promising recipe for the creation of planned architectures of QD although, of course there is still a long way to go before practical applications become realizable. On the other hand, there exists at present no other system which could be built up from particles as small as the 1.4 nm gold particles mentioned above, and which is electronically exposed in such a manner that single electrons can be used to switch, even at room temperature. This vision justifies any effort to further improve the techniques to build up working constructions.

The route to generate aimed structures in general, and those of metal nanoparticles in particular, via the electro-oxidation of SAMs shows great promise. A series of examples other than those discussed here is available in Ref. [111]. Further developments of this strategy will be necessary to render it competitive by comparison with current lithographic methods. Although the generation of extended nanostructured surfaces by means of automated techniques has already been achieved [112], the use of multi-tip systems should lead to remarkable improvements in the efficiency of the technique.

At this point, it can be concluded that the organization of metal nanoparticles has developed in a very promising manner. From very simple self-organization via guided self-assembly to the controlled patterning of surfaces, a wide variety of routes is currently available that should, in principle, make applications in nanoelectronics possible. It should be stated, however, that not all of the above-mentioned organized particles can be used as QDs, as they are still too large to exhibit the very special electronic properties required for single-electron transitions at room temperature. These quantum-sized characteristics are discussed in detail in Section 5.2.

References

1 Motte, L., Billoudet, F., and Pileni, M.P. (1995) *J. Phys. Chem.*, **99**, 16425.
2 Motte, L., Billoudet, F., Lacaze, E., Douin, J., and Pileni, M.P. (1997) *J. Phys. Chem. B*, **101**, 138.
3 Collier, C.P., Vossmeyer, T., and Heath, J.R. (1998) *Annu. Rev. Phys. Chem.*, **49**, 371.
4 Wang, Z.L., Harfenist, S.A., Vezmar, I., Whetten, R.L., Bentley, J., Evans, N.D., and Alexander, K.B. (1998) *Adv. Mater.*, **10**, 808.
5 Courty, A., Fermon, C., and Pileni, M.P. (2001) *Adv. Mater.*, **13**, 254.
6 Whetten, R.L., Khoury, J.T., Alvarez, M.M., Murthy, S., Vezmar, I., Wang, Z.L., Stephens, P.W., Cleveland, C.L., Luedtke, W.D., and Landmann, U. (1996) *Adv. Mater.*, **8**, 428.
7 Harfenist, St.A., Wang, Z.L., Alvarez, M.M., Vezmar, I., and Whetten, R.L. (1996) *J. Phys. Chem.*, **100**, 13904.
8 Harfenist, St.A., Wang, Z.L., Whetten, R.L., Vezmar, I., and Alvarez, M.M. (1997) *Adv. Mater.*, **9**, 817.
9 Wang, Z.L. (1998) *Adv. Mater.*, **10**, 13.
10 Martin, J.E., Wilcoxon, J.P., Odinek, J., and Provencio, P. (2000) *J. Phys. Chem.*, **104**, 9475.
11 Schmid, G., Pugin, R., Sawitowski, T., Simon, U., and Marler, B. (1999) *Chem. Commun.*, 1303.
12 Schmid, G. and Klein, N. (1986) *Angew. Chem., Int. Ed. Engl.*, **25**, 922.
13 Schmid, G. (1988) *Polyhedron*, **7**, 2321.
14 Feld, H., Leute, A., Rading, D., Benninghoven, A., and Schmid, G. (1990) *Z. Phys. D*, **17**, 73.
15 Feld, H., Leute, A., Rading, Dr., Benninghoven, A., and Schmid, G. (1990) *J. Am. Chem. Soc.*, **112**, 8166.
16 Schmid, G., Meyer-Zaika, W., Pugin, R., Sawitowski, T., Majoral, J.-P., Caminade, A.-M., and Turrin, C.-O. (2000) *Chem. Eur. J.*, **6**, 1693.
17 Zhang, Q., Xie, J., Liang, J., and Lee, J.Y. (2009) *Adv. Funct. Mater.*, **19**, 1387.
18 Murray, C.B., Kagan, C.R., and Bawendi, M.G. (2000) *Annu. Rev. Mater. Sci.*, **30**, 545.
19 Murray, C.B., Sun, S., Doyle, H., and Betley, T. (2001) *MRS Bull.*, **26**, 985.
20 Sun, S., Murray, C.B., and Doyle, H. (1999) *Mater. Res. Soc. Symp. Proc.*, **577**, 385.
21 Sun, S. and Murray, C.B. (1999) *J. Appl. Phys.*, **85**, 4325.
22 Sun, S. and Zeng, H. (2002) *J. Am. Chem. Soc.*, **124**, 8204.
23 Talapin, D.V., Shevchenko, E.V., Kornowski, A., Gaponik, N.P., Haase, M., Rogach, A.L., and Weller, H. (2001) *Adv. Mater.*, **13**, 1868.
24 Shevchenko, E.V., Talapin, D.V., Kornowski, A., Wiekhorst, F., Kötzler, J., Haase, M., Rogach, A.L., and Weller, H. (2001) *Adv. Mater.*, **14**, 287.
25 Rogach, A.L., Talapin, D.V., Shevchenko, E.V., Kornowski, A., Haase, M., and Weller, H. (2002) *Adv. Funct. Mater.*, **12**, 653.
26 Shevchenko, E.V., Talapin, D.V., Rogach, A.L., Kornowski, A., Haase, M., and Weller, H. (2002) *J. Am. Chem. Soc.*, **124**, 11480.
27 Petit, C., Taleb, A., and Pileni, M.P. (1998) *Adv. Mater.*, **10**, 259.
28 Pileni, M.P. (2001) *J. Phys. Chem. B*, **105**, 3358.
29 Ngo, A.T. and Pileni, M.P. (2000) *Adv. Mater.*, **12**, 276.
30 Giersig, M. and Hilgendorff, M. (2002) *Colloids Surf. A*, **202**, 207.
31 Safran, S.A. (2003) *Nat. Mater.*, **2**, 71.
32 Wirtz, D. and Fermigier, M. (1994) *Phys. Rev. Lett.*, **72**, 2294.
33 Wang, H., Zhu, Y., Boyd, C., Luo, W., Cebers, A., and Rosensweig, R.E. (1994) *Phys. Rev. Lett.*, **72**, 1929.
34 Wen, W., Zhang, L., and Sheng, P. (2000) *Phys. Rev. Lett.*, **85**, 5464.
35 Brust, M., Bethell, D., Shiffrin, D.J., and Kiely, C.J. (1995) *Adv. Mater.*, **7**, 795.
36 Brust, M., Walker, M., Bethell, D., Schiffrin, D.J., and Whyman, R. (1994) *J. Chem. Soc., Chem. Commun.*, 801.
37 Fink, J., Kiely, C.J., Bethell, D., and Schiffrin, D.J. (1998) *Chem. Mater.*, **10**, 922.
38 Reetz, M.T., Winter, M., and Tesche, B. (1997) *Chem. Commun.*, 147.

39 Kiely, C.J., Fink, J., Brust, M., Bethell, D., and Schiffrin, D.J. (1998) *Nature*, **396**, 444.

40 Kiely, C.J., Zheng, J.G., Fink, J., Brust, M., Bethell, D., and Schiffrin, D.J. (2000) *Adv. Mater.*, **12**, 640.

41 Laves, F. (1956) *The Theory of Alloy Phases*, American Society for Metals, Cleveland, OH, pp. 124–129.

42 Pearson, W.B. (1972) *Crystal Chemistry and Physics of Metals and Alloys*, Wiley-Interscience, London.

43 Hachisu, S. and Youshimura, S. (1980) *Nature*, **283**, 188.

44 Youshimura, S. and Hachisu, S. (1983) *Prog. Colloid Polym. Sci.*, **68**, 59.

45 Andres, R.P., Bielefeld, J.D., Henderson, J.I., Janes, D.B., Kolagunta, V.R., Kubiak, C.P., Mahoney, W.J., and Osifchin, R.G. (1996) *Science*, **273**, 1690.

46 Korgel, B.A. and Fitzmaurice, D. (1998) *Adv. Mater.*, **10**, 661.

47 Wilcoxon, J.P., Martin, J.E., and Provencio, P. (2001) *J. Chem. Phys.*, **115**, 998.

48 Dai, Z.R., Sun, S., and Wang, Z.L. (2001) *Nano Lett.*, **1**, 443.

49 Wang, Z.L. (1998) *Adv. Mater.*, **10**, 13.

50 Wang, Z.L., Dai, Z., and Sun, S. (2000) *Adv. Mater.*, **12**, 1944.

51 Wiekhorst, F., Shevchenko, E.V., Weller, H., and Kötzler, J. (2003) *Phys. Rev. B*, **67**, 224416.

52 Dumestre, F., Chaudret, B., Amiens, C., Fromen, M.-C., Casanove, M.-J., Renaud, P., and Zurcher, P. (2002) *Angew. Chem., Int. Ed.*, **41**, 4286–4289.

53 Sun, S., Murray, C.B., Weller, D., Folks, L., and Moser, A. (2000) *Science*, **287**, 1989.

54 Schmid, G. and Peschel, St. (1998) *New J. Chem.*, **7**, 669.

55 (a) Schmid, G., Bäumle, M., and Beyer, N. (2000) *Angew. Chem.*, **112**, 187; (b) Schmid, G., Bäumle, M., and Beyer, N. (2000) *Angew. Chem., Int. Ed. Engl.*, **39**, 181.

56 Schmid, G. and Beyer, N. (2000) *Eur. J. Inorg. Chem.*, 835.

57 Reuter, T., Vidoni, O., Torma, V., Schmid, G., Nan, L., Oleiche, M., Chi, L., and Fuchs, H. (2002) *Nano Lett.*, **2**, 709.

58 Kumar, A., Pattarkine, M., Bhadbhade, M., Mandale, A.B., Ganesh, K.N., Datar, S.S., Dharmadhikari, C.V., and Sastry, M. (2002) *Adv. Mater.*, **13**, 341.

59 (a) Liu, Y., Meyer-Zaika, W., Schmid, G., Leis, M., and Kuhn, H. (2003) *Angew. Chem.*, **115**, 2959; (b) Liu, Y., Meyer-Zaika, W., Schmid, G., Leis, M., and Kuhn, H. (2003) *Angew. Chem., Int. Ed.*, **42**, 2853.

60 Liu, J., Lee, T., Janes, D.B., Walsh, B.L., Melloch, M.R., Woodall, J.M., Reifenberger, R., and Andres, R.P. (2000) *Appl. Phys. Lett.*, **77**, 373.

61 (a) Ohara, P.C., Heath, J.R., and Gelbart, W.M. (1997) *Angew. Chem.*, **109**, 1120; (b) Ohara, P.C., Heath, J.R., and Gelbart, W.M. (1997) *Angew. Chem., Int. Ed. Engl.*, **36**, 1078.

62 Heath, J.R., Knobler, C.M., and Leff, D.V. (1997) *J. Phys. Chem. B*, **101**, 189.

63 Gelart, W.M., Sear, R.P., and Heath, J.R. (1999) *Faraday Discuss.*, **112**, 299.

64 Sear, R.P., Chung, S.W., and Markovich, G. (1999) *Phys. Rev. E*, **59**, R 6255.

65 Whetten, R.L., Shafigullin, M.N., and Khoury, J.T. (1999) *Acc. Chem. Res.*, **32**, 397.

66 Gutierrez-Wing, C., Santiago, P., Ascencio, J.A., Comacho, A., and Jose-Yacaman, M. (2000) *Appl. Phys. A*, **71**, 237.

67 Ohara, P.C., Leff, D.V., Heath, J.R., and Gelbart, W.M. (1995) *Phys. Rev. Lett.*, **75**, 3466.

68 Chi, L.F., Rakers, S., Hartig, M., Gleiche, M., Fuchs, H., and Schmid, G. (2000) *Colloids Surf.*, **171**, 241–248.

69 Brown, J.J., Porter, J.A., Daghlian, C.P., and Gibson, U.J. (2001) *Langmuir*, **17**, 7966.

70 Jablonovitch, E. (1987) *Phys. Rev. Lett.*, **58**, 2059.

71 John, S. (1987) *Phys. Rev. Lett.*, **58**, 2486.

72 Blanco, A., Chomski, E., Grabtchak, S., Ibisate, M., John, S., Leonard, S.W., Lopez, C., Meseguer, F., Miguez, H., Mondia, J.P., Ozin, G.A., Toader, O., and van Driel, H.M. (2000) *Nature*, **405**, 437.

73 Spatz, J.P., Roescher, A., and Möller, M. (1996) *Polymer Reprints, Am. Chem. Soc.*, **36**, 409.

74 Spatz, J.P., Sheiko, S., and Möller, M. (1996) *Macromolecules*, **29**, 3220.

75 Spatz, J.P., Roescher, A., and Möller, M. (1996) *Adv. Mater*, **8**, 337.

76 (a) Spatz, J.P., Mößner, S., and Möller, M. (1996) *Angew. Chem.*, **108**, 1673; (b) Spatz, J.P., Mößner, S., and Möller, M.

77 Spatz, J.P., Sheiko, S., and Möller, M. (1996) *Adv. Mater.*, **8**, 513.
78 Spatz, J.P., Möller, M., and Ziemann, P. (1999) *Phys. Bl.*, **55**, 1.
79 Spatz, J.P., Eibeck, P., Mößner, S., Möller, M., Herzog, T., and Ziemann, P. (1998) *Adv. Mater.*, **10**, 849.
80 Möller, M., Spatz, J.P., Roescher, A., Mößner, S., Selvan, S.T., and Klok, H.-A. (1997) *Macromol. Symp.*, **117**, 207.
81 Möller, M. and Spatz, J.P. (1997) *Curr. Opin. Colloid Interface Sci.*, **2**, 177.
82 Spatz, J.P., Herzog, T., Mößner, S., Ziemann, P., and Möller, M. (1999) *Adv. Mater.*, **11**, 149.
83 Spatz, J.P., Herzog, T., Mößner, S., Ziemann, P., and Möller, M. (1997) Micro- and Nanopatterning Polymers, in *ACS Symposium Series*, vol. **706** (eds H. Ito, E. Reichmanis, O. Nalamasu, and T. Ueno), American Chemical Society, p. 12.
84 Spatz, J.P., Mößner, S., Möller, M., Herzog, T., Plettl, A., and Ziemann, P. (1998) *J. Lumin.*, **7677**, 168.
85 Spatz, J.P., Mößner, S., Hartmann, C., Möller, M., Herzog, T., Krieger, M., Boyen, H.-G., Ziemann, P., and Kabius, B. (2000) *Langmuir*, **16**, 407.
86 Haupt, M., Miller, S., Bitzer, K., Thonke, K., Sauer, R., Spatz, J.P., Mößner, S., Hartmann, C., and Möller, M. (2001) *Phys. Status Solidi*, **224**, 867.
87 Koslowski, B., Strobel, S., Herzog, T., Heinz, B., Boyen, H.G., Notz, R., Ziemann, P., Spatz, J.P., and Möller, M. (2000) *J. Appl. Phys.*, **87**, 7533.
88 Dieluweit, S., Pum, D., and Sleytr, U.B. (1998) *Supramol. Sci.*, **5**, 15.
89 Hall, S.R., Shenton, W., Engelhardt, H., and Mann, S. (2001) *ChemPhysChem*, **2**, 194.
90 Mertig, M., Kirsch, R., Pompe, W., and Engelhardt, H. (1999) *Eur. Phys. J. D*, **9**, 45.
91 Piner, R.D., Zhu, J., Xu, F., Hong, S., and Mirkin, C.A. (1999) *Science*, **283**, 661.
92 Hong, S., Zhu, J., and Mirkin, C.A. (1999) *Science*, **286**, 523.
93 (a) Demers, L.M., and Mirkin, C.A. (2001) *Angew. Chem*, **113**, 3159; (b) Demers, L.M., and Mirkin, C.A. (2001) *Angew. Chem., Int. Ed.*, **40**, 3069.
94 (a) Demers, L.M., Perk, S.-J., Taton, T.A., Li, Z., and Mirkin, C.A. (2001) *Angew. Chem.*, **113**, 3161; (b) Demers, L.M., Perk, S.-J., Taton, T.A., Li, Z., and Mirkin, C.A. (2001) *Angew. Chem., Int. Ed.*, **40**, 3071.
95 Vidoni, O., Reuter, T., Torma, V., Meyer-Zaika, W., and Schmid, G. (2001) *J. Mater. Chem.*, **11**, 3188.
96 Gleiche, M., Chi, L.F., and Fuchs, H. (2000) *Nature*, **403**, 173.
97 Scholten, P.C. (1992) in *Magnetic Properties of Fine Particles* (eds J.L. Dortmann and D. Fiorani), Elsevier, Amsterdam, The Netherlands.
98 de Gennes, P.G. and Pincus, P.A. (1970) *Phys. Kondens. Mater.*, **11**, 189.
99 Chantrell, R.W., Bradbury, A., Popplewell, J., and Charles, S.W. (1982) *J. Appl. Phys.*, **53**, 2742.
100 Tavares, J.M., Weis, J.J., and Telo da Gama, M.M. (2002) *Phys. Rev. E*, **65**, 061201.
101 Butter, K., Pomans, P.H.H., Frederik, P.M., Vroege, G.J., and Philipse, A.P. (2003) *Nat. Mater.*, **2**, 8891.
102 Tripp, S.L., Pusztay, S.V., Ribbe, A.E., and Wei, A. (2002) *J. Am. Chem. Soc.*, **124**, 7914.
103 Halsey, T.C. (1992) *Science*, **258**, 761.
104 Hilgendorff, M., Tesche, B., and Giersig, M. (2001) *Aust. J. Chem.*, **54**, 497.
105 Puntes, V.F., Zanchet, D., Erdonmez, C.K., and Alivisato, A.P. (2002) *J. Am. Chem. Soc.*, **124**, 12874.
106 Kolb, D.M. and Simeone, F.C. (2005) *Electrochim. Acta*, **50**, 2989.
107 Maoz, R., Cohen, S.R., and Sagiv, J. (1999) *Adv. Mater.*, **11**, 55.
108 Maoz, R., Frydman, E., Cohen, S.R., and Sagiv, J. (2000) *Adv. Mater.*, **12**, 424.
109 Maoz, R., Frydman, E., Cohen, S.R., and Sagiv, J. (2000) *Adv. Mater.*, **12**, 725.
110 Liu, S., Maoz, R., Schmid, G., and Sagiv, J. (2002) *Nano Lett.*, **2**, 1055.
111 (a) Wouters, D., Hoeppener, St., and Schubert, U.S. (2009) *Angew. Chem.*, **121**, 1762; (b) Wouters, D., Hoeppener, St., and Schubert, U.S. (2009) *Angew. Chem., Int. Ed.*, **48**, 1732.
112 Wouters, D. and Schubert, U.S. (2007) *Nanotechnology*, **18**, 485306.

5
Properties

5.1
Semiconductor Nanoparticles

5.1.1
Optical and Electronic Properties of Semiconductor Nanocrystals
Uri Banin and Oded Millo

5.1.1.1 Introduction
Semiconductor nanocrystals are novel materials lying between the molecular and solid-state regime, with the unique feature of properties controlled by size and shape [1–8]. Nanocrystals contain hundreds to thousands of atoms, are typically 20 to 200 Å in diameter, and maintain a crystalline core with periodicity of the bulk semiconductor. However, as the wavefunctions of electrons and holes are confined by the physical nanometric dimensions of the nanocrystals, the electronic level structure and the resultant optical and electrical properties are greatly modified. Upon reducing the semiconductor size to the nanocrystal regime, a characteristic blue shift of the band gap appears, and discrete level structure develops as a result of the "quantum size effect" in these quantum structures [9]. In addition, because of their small size, the charging energy associated with the addition or removal of a single electron is very high, leading to pronounced single-electron tunneling effects [10–12]. These phenomena provide rich grounds for basic scientific research that has attracted – and continues to attract – considerable attention. Due to the unique optical and electrical properties, nanocrystals may play a key role in the emerging new field of nanotechnology in applications ranging from lasers [13, 14] and other optoelectronic devices [15–17], to biological fluorescence marking [18–20]. At this point, discuss the optical and electronic properties of colloidal grown quantum dots (QDs) and quantum rods (QRs) are discussed. Of particular interest here is the intuitive and appealing "artificial atom" picture for the electronic structure of semiconductor nanocrystals. In this respect, the combination of tunneling and optical spectroscopy methods has proven to be a powerful approach for the study of size- and shape-dependent level structure and single-electron charging in these systems.

Nanoparticles: From Theory to Application. Edited by Günter Schmid
Copyright © 2010 WILEY-VCH Verlag GmbH & Co. KGaA, Weinheim
ISBN: 978-3-527-32589-4

Colloidal grown nanocrystals have several favorable features, such as continuous size control and chemical accessibility due to their overcoating with organic ligands. This chemical compatibility enables the use of powerful chemical or biochemical means to assemble nanocrystals in controlled manner [21–27]. Further powerful control, in addition to size, on optical and electronic properties of nanocrystals has been achieved using colloidal synthesis. Heterostructured nanocrystals were developed, consisting of semiconductor shells grown on a core [28–36]. Enhanced fluorescence and increased stability can be achieved in these particles, compared to cores overcoated by organic ligands. Shape control was also achieved by correct modification of the synthesis, allowing the preparation of rod-shaped particles – QRs [6, 37, 38]. Such QRs manifest the transition from zero-dimensional (0-D) QDs to one-dimensional (1-D) quantum wires, in the sense that a strong quantum confinement holds for the radial direction, while a weak confinement holds for the axial direction [39, 40]. Quantum rods exhibit electronic and optical properties that differ from those of QDs, such as polarized emission [41] that leads to polarized lasing [14]. Additionally, the rod geometry was found to be advantageous in polymer–nanocrystal-based solar cells, due to improvements in charge-transport properties [42]. Further functionality is afforded by heterostructured seeded nanorods [43–46]. Such structures combine a 0-D QD seed embedded within a quasi-1-D QR shell. As in the spherical core–shells mentioned above, the band-alignment in these heterostructures determines the electronic and optical properties. Aside from the type I band-alignment, where the core band gap is embedded within that of the shell, a type II staggered band-offset is also possible, promoting an internal electron-hole charge separation which is important for solar energy harvesting. Thus, the determination of band-offsets in these systems is of critical importance for understanding their properties [47].

In this chapter, attention is focused on the electronic structure and optical and electrical properties of colloidal grown semiconductor nanocrystals, and also on the application of tunneling and optical spectroscopy to this problem. The application of optical and tunneling spectroscopy to the study of semiconductor nanocrystals, molding substances into the artificial atom analogy for their level structure, is detailed in Section 5.1.1.2, while the theoretical framework to describe the electronic level structure is described in Section 5.1.1.3. Modifications related to core–shell nanocrystals are discussed in Section 5.1.1.4, with particular attention paid to the atomic-like envelope wavefunctions that are visualized using scanning tunneling microscopy (STM). The electronic properties of QRs are discussed in Section 5.1.1.5, while Section 5.1.1.6 details the heterostructured seeded nanorods and introduces an approach to extract the band-offsets. The results on optical gain and lasing from semiconductor nanocrystals are outlined in Section 5.1.1.7.

5.1.1.2 Semiconductor Nanocrystals as Artificial Atoms

Since the early studies on the quantum confinement effect in colloidal semiconductor QDs, electronic levels have been assigned according to the spherical symmetry of the electron and hole envelope functions [9, 48]. The simplistic "artificial atom" model of a particle in a spherical box predicts discrete states with atomic-like state symmetries, for example, s and p. In order to probe the electronic structure of II–VI and III–V semiconductor nanocrystals, a variety of "size-selective" optical techniques has been used, mapping the size dependence of dipole-allowed transitions [49–53]. Theoretical models based on an effective mass approach with a varying degree of complexity [50, 54], as well as pseudopotentials [55, 56], were used to assign the levels. One model system that has been studied extensively in this context is that of CdSe nanocrystals, for which the spectroscopic mapping of electronic states using photoluminescence excitation (PLE) was performed by Bawendi and coworkers [49, 50]. A size-selective PLE method was also used to examine the size-dependent level structure of InAs QDs, an additional system that has attracted much interest [52, 53]. InAs is a narrow-gap semiconductor ($E_g = 0.418$ eV) with a large Bohr radius a_0 of 340 Å, as compared to CdSe, which has a significantly larger band gap, $E_g = 1.84$ eV, and a smaller Bohr radius of 55 Å. Therefore, both systems may serve as a prototypical framework for the study of quantum confinement effects.

Figure 5.1a shows the typical spectral features measured at 10 K for a sample of InAs nanocrystals with a mean radius of 2.5 nm. The absorption onset exhibits a ~0.8 eV blue shift from the bulk band gap. A pronounced first peak and several features at higher energies are observed in the absorption spectra. Band-edge luminescence is observed with no significant red-shifted (deep trapped) emission, while in the PLE spectrum a set of up to eight transitions is resolved. The full size-dependence was measured by changing the detection window, and by using different samples and a representative set of such PLE spectra, as shown in Figure 5.1b [52].

The map of excited transitions for InAs nanocrystals was extracted from such PLE data, and the levels were assigned using an eight-band effective mass model [52]. Briefly, in this approach each electron (e) and hole (h) state is characterized by its parity and its total angular momentum, $\mathbf{F} = \mathbf{J} + \mathbf{L}$, where $\mathbf{J}$ is the Bloch band-edge angular momentum (1/2 for the CB, 3/2 for the heavy and light hole bands, and 1/2 for the split-off band), and $\mathbf{L}$ is the angular momentum associated with the envelope function. The standard notation for the electron and hole states is nQ_F, where n is the main quantum number, and $Q =$ S,P,D,..., denotes the lowest L in the envelope wave function. The calculated level separations were found to closely reproduce the observed strong transitions.

Tunneling transport through semiconductor nanocrystals can yield complementary information on their electronic properties, which cannot be probed via optical measurements. Whilst in the optical spectra, allowed valence band (VB) to conduction band (CB) transitions are detected, in tunneling spectroscopy the CB and VB states can be separately probed. In addition, the tunneling spectra may show the effects of single-electron charging of the QD. Such interplay between single-electron charging and resonant tunneling through the QD states can provide unique information on the degeneracy, and therefore the symmetry of the levels.

The interplay between single-electron tunneling (SET) effects and quantum size effects in isolated nanoparticles can be experimentally observed most clearly when the charging energy of the dot by a single electron, E_c, is comparable to the electronic

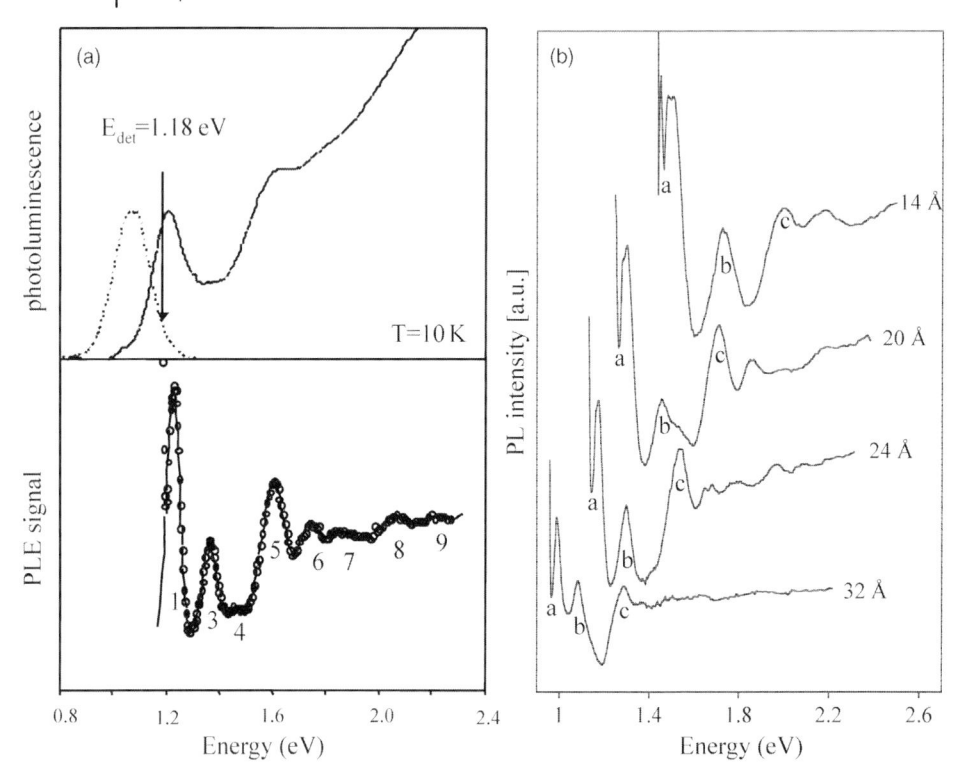

Figure 5.1 (a) Optical spectroscopy of InAs nanocrystals, with mean radius of 2.5 nm. The top frame shows the absorption (solid line), and the photoluminescence (dotted line) for the sample. The lower frame shows a size-selected photoluminescence excitation (PLE) spectrum, where eight transitions are resolved, measured with a narrow detection window positioned as indicated by the arrow in the top panel. The weak transition E_2 is not detected in this QD size; (b) Size-dependent PLE spectra for four representative InAs QD radii. The band gap transition (a), and two strong excited transitions (b, c) are indicated.

level separation ΔE_L, and both energy scales are larger than $k_B T$ [10, 57, 58]. These conditions are met by semiconductor nanocrystals in the strong quantum confinement regime, even at room temperature, whereas for metallic nanoparticles, E_c is typically much larger than ΔE_L. SET effects are relevant to the development of nanoscale electronic devices, such as single-electron transistors [59, 60].

However, for small colloidal nanocrystals, the task of wiring up the QD between electrodes for transport studies is exceptionally challenging. To this end, various mesoscopic tunnel junction configurations have been employed, such as the double barrier tunnel junction (DBTJ) geometry, where a QD is coupled via two tunnel junctions to two macroscopic electrodes [10, 11, 61, 62]. Klein *et al.* achieved this by attaching CdSe QDs to two lithographically prepared electrodes, and observing SET effects [63]. In this device, a gate voltage can be applied to modify the transport properties; alternatively, electrical transport through single QDs can be achieved by using scanning probe methods. Alperson *et al.* observed SET effects at room temperature in electrochemically deposited CdSe nanocrystals by using conductive atomic force microscopy (AFM) [64].

One of the most useful approaches for tunneling-transport spectroscopy on single nanocrystals is based on the use of STM. For such measurements, the nanocrystals can be linked to a gold film via hexane dithiol (DT) molecules [12, 65], which enables the realization of a DBTJ, as shown schematically in Figure 5.2a. An STM topographic image of an isolated InAs QD, 32 Å in radius, is presented in Figure 5.2a (left inset). Also shown is a tunneling current–voltage (I–V) curve that was acquired after positioning the STM tip above the QD. A region of suppressed tunneling current is observed around zero bias, followed by a series of steps at both negative and positive bias. Figure 5.2b shows the plot of dI/dV versus V, tunneling conductance spectrum,

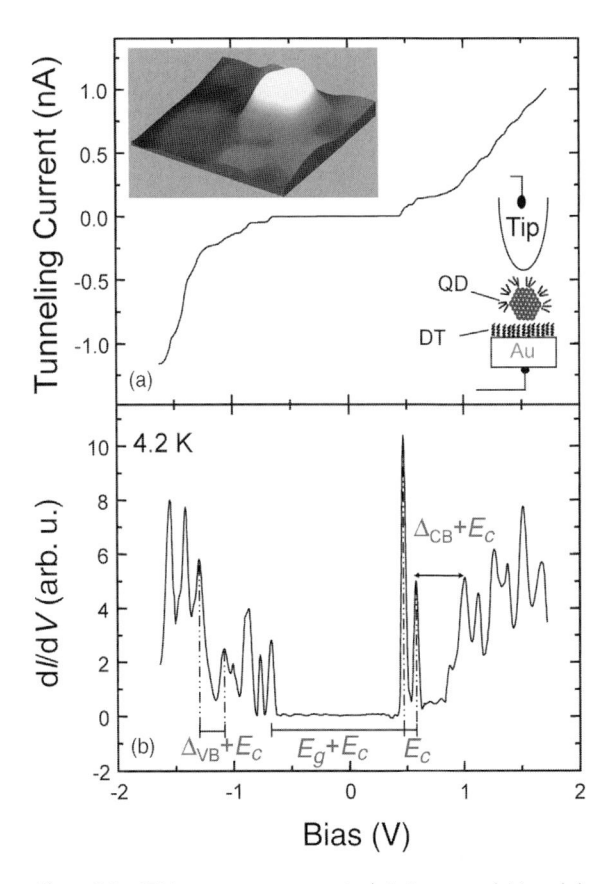

Figure 5.2 STM measurements on single InAs nanocrystal 3.2 nm in radius, acquired at 4.2 K. The QDs are linked to the gold substrate by dithiol (DT) molecules, as shown schematically in the right inset. The left inset presents a 10 × 10 nm STM topographic image of the QD. The tunneling I–V characteristic is presented in panel (a) and the tunneling conductance spectrum is shown in panel (b). The arrows depict the main energy separations: E_c is the single electron charging energy, E_g is the nanocrystal band-gap, and Δ_{VB} and Δ_{CB} are the spacing between levels in the valence and conduction bands, respectively.

which is proportional to the tunneling density of state (DOS). A series of discrete SET peaks is clearly observed, where the separations are determined by both E_c (addition spectrum) and the discrete level spacings (excitation spectrum) of the QD. The I–V characteristics were acquired with the tip retracted from the QD to a distance, where the bias predominantly drops on the tip–QD junction. Under these conditions, which represent an asymmetric DBTJ, the CB (VB) states appear at positive (negative) sample bias, and the excitation peak separations are almost equal to the real QD level spacings [12, 66].

On the positive bias side of Figure 5.2b, two closely spaced peaks are observed immediately after current onset, followed by a larger spacing and a group of six almost equidistant peaks. The doublet is assigned to tunneling through the lowest CB QD state, where the spacing corresponds to the single-electron charging energy, E_c. The observed doublet is consistent with the degeneracy of the envelope function of the first CB level, $1S_e$ (here, we revert to a simpler notation for the CB states), which has s character. In this case, a direct relationship between the degeneracy of a QD level and the number of addition peaks is expected. This is further substantiated by the observation that the second group consists of six peaks, corresponding to the degeneracy of the $1P_e$ state, spaced by E_c. This sequential level filling resembles the Aufbau principle of building up the lowest energy electron configuration of an atom, directly demonstrating the atomic-like nature of the QD.

The separation between the two groups of peaks is a sum of the level spacing $\Delta_{CB} = 1P_e − 1S_e$ and the charging energy, E_c. On the negative bias side, tunneling through filled dot levels takes place, reflecting the tunneling DOS of the QD valence band. Again, two groups of peaks are observed, although the multiplicity in this case, in contrast to the CB, cannot be clearly assigned to specific angular momentum degeneracy. In a manner similar to that described above for the CB states, a value of Δ_{VB} is extracted for the level separation between the two observed VB states. In the region of null current around zero bias, the tip and substrate Fermi energies are located within the QD band gap where the tunneling DOS is zero, and the energy gap E_g can be extracted from the observed spacing between the highest VB and the lowest CB peaks. The gap region appears to be free of spectral features (see also Figure 5.3). On removing the coating ligands by pyridine treatment, sub-gap states could be observed [67], which were attributed to surface states [68].

The tunneling conductance spectra for single InAs nanocrystals spanning a size range of 10–35 Å in radius are presented in Figure 5.3. Two groups of peaks are observed in the positive bias side (CB). The first is always a doublet, consistent with the expected s symmetry of the $1S_e$ level, while the second has higher multiplicity of up to six, consistent with $1P_e$. The separation between the two groups, as well as the spacing of peaks within each multiplet, increase with decreasing QD radius. This reflects quantum size effects, on both the nanocrystal energy levels and its charging energy, respectively. On the negative bias side, two groups of peaks – both of which exhibit similar quantum confinement effects – are also generally observed. Here, variations in the group multiplicities between QDs of different size, as well as variations of the peak energy spacings within each group, are observed. This behavior is partly due to the fact that E_c is very close to Δ_{VB}, the level spacing in the valence

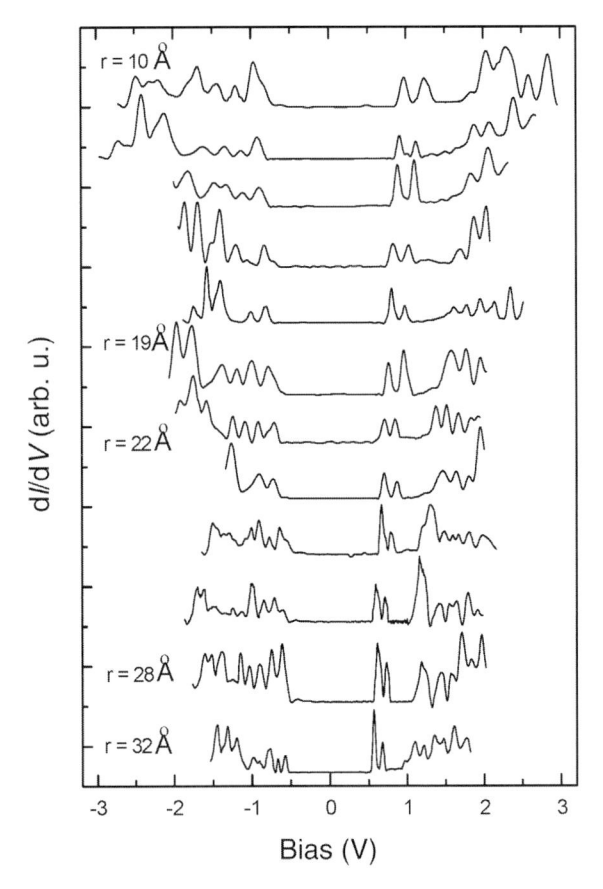

Figure 5.3 Size evolution of the tunneling spectra at 4.2 K of single InAs QDs, displaced vertically for clarity. The position of the centers of the zero-current gap showed nonsystematic variations with respect to the zero bias, of the order of 0.2 eV, probably due to variations of local offset potentials. For clarity of presentation, the spectra are offset along the V-direction to center them at zero bias. Representative nanocrystal radii are denoted.

band, as shown in Figure 5.4b. In this case, sequential SET may be either an addition to the same VB state, or excitation with no extra charging to the next state. An atomic analogy for this situation can be found in the changing order of electron occupation when moving from the transition to the noble metals within a row of the Periodic Table.

The comparison between tunneling and PLE data can be used to decipher the complex QD level structure. This correlation is also important for examining possible effects of charging and tip-induced electric field in the tunneling measurements on the nanocrystal level structure. First, the size dependence of the band gap E_g, as extracted from the tunneling data, can be compared with the nanocrystal sizing curve (Figure 5.4a). The sizing curve (open diamonds) was obtained by correlating the average nanocrystal size, measured using transmission electron microscopy (TEM), with the excitonic band gap of the same sample [53]. To compare these data with the

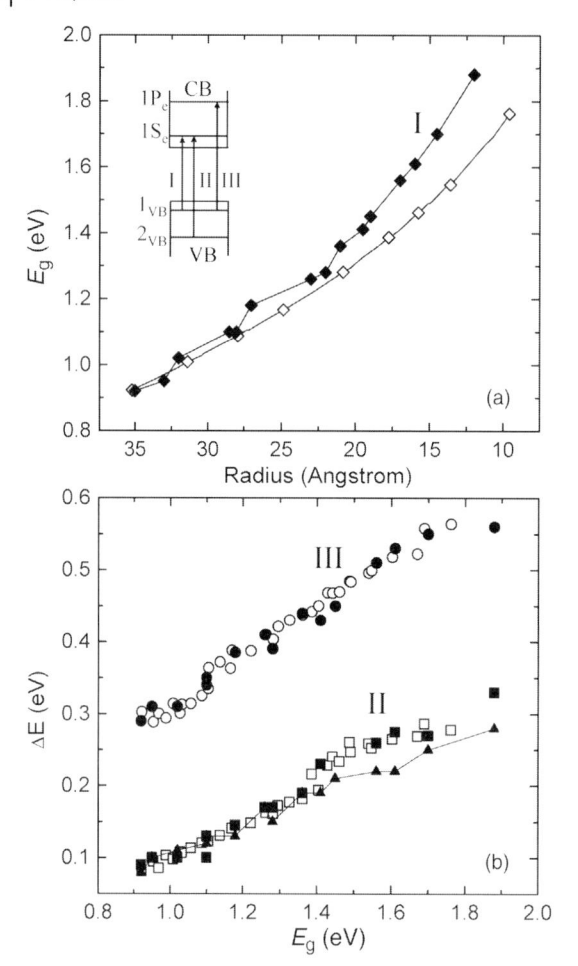

Figure 5.4 Correlation of optical and tunneling spectroscopy data for InAs QDs. The inset shows a schematic of the CB and VB level structure and optical transitions I, II, and III (a,b and c in Figure 5.3). (a) Comparison of the size dependence of the low-temperature optical band gap (transition I) after Coulomb correction (open diamonds), with the band-gap measured by the STM (filled diamonds); (b) Excited transitions plotted versus the band gap for tunneling and optical spectroscopy. The two lower data sets (II) depict the correlation between $\Delta_{VB} = 1_{VB} - 2_{VB}$, detected by tunneling spectroscopy (full squares), with the difference between transition II and the band gap transition I (open squares). The two upper data sets (III) depict the correlation between $\Delta_{CB} = 1P_e - 1S_e$, determined from the tunneling spectra (full circles), with the difference between optical transition III and I (open circles). Also shown is the size dependence of E_c from the tunneling data (full triangles).

tunneling results, a correction term of $1.8e^2/\varepsilon r$ was added to compensate for the electron–hole excitonic coulombic interaction that is absent from the tunneling data [9]. The agreement was good for the larger nanocrystal radii, with an increasing deviation for smaller nanocrystals that may be due to the difference in the estimation of QD size between the experiments, which should be more pronounced in the small size regime [12].

In Figure 5.4b, the size dependence of the higher strongly allowed optical transitions is compared with the level spacings measured by tunneling spectroscopy. The two lower data sets (II) compare the difference between the first strong excited optical transition and the band gap from PLE (b–a in Figure 5.1), with the separation $\Delta_{VB} = 2_{VB} - 1_{VB}$ in the tunneling data (open and full squares, respectively). The excellent correlation observed enables assignment of this first excited transition in the PLE to a $2_{VB} - 1S_e$ excitation, as shown schematically in the inset of Figure 5.4a. Strong optical transitions are allowed only between electron and hole states with the same envelope function symmetry. Thus, it is inferred that the envelope function for state 2_{VB} should have s character, and this state can be directly identified as the $2S_{3/2}$ valence band level.

Another important comparison is depicted by the higher pair of curves, set III. The second strong excited optical transition relative to the band gap (c–a in Figure 5.1), is plotted along with the spacing $\Delta_{CB} = 1P_e - 1S_e$ from the tunneling spectra. Again, an excellent correlation is observed, which allows this peak in the PLE to be assigned to the $1_{VB} - 1P_e$ transition (Figure 5.4a, inset). The top-most VB level, 1_{VB}, should thus have some p character for this transition to be allowed. From this, and considering that the band gap optical transition $1_{VB} - 1S_e$ is also allowed, it can be concluded that 1_{VB} has mixed s and p character, in agreement with theory [52, 69]. The good agreement between the tunneling and optical measurements indicates that charging and the tip-induced Stark effect do not significantly perturb the QD level structure.

With regards to the tunneling data, it is important to note that a detailed understanding of the role played by the DBTJ geometry, and the ability to control it, are essential for the correct interpretation of tunneling characteristics. In particular, the ratio between the junction capacitances determines the voltage division, and therefore the observed level spacing. By varying the tip–QD distance, it is possible to modify the voltage division (as discussed in Refs [70, 71]). Additionally, the degree of single-electron charging effects depends on the ratio of the tunneling rates between the two junctions. When the rate of one of the junctions is significantly higher, charging effects are suppressed, as was achieved by depositing InAs nanocrystals on highly oriented pyrolytic graphite (HOPG) without linker molecules, leading to a vanishing of the charging multiplets [66] (as seen in Figure 5.5). A similar effect was also observed for CdSe and CdS nanocrystals by Bakkers *et al.* [72].

5.1.1.3 Theoretical Descriptions of the Electronic Structure

The theoretical treatments for both optical and tunneling experiments on QDs require, first, a calculation of the level structure. Various approaches have been developed to treat this problem, including effective mass-based models, with various degrees of band-mixing effects [50, 52, 54, 73], and a more atomistic approach based on pseudo-potentials [69, 74]. Both approaches have been successfully applied to various nanocrystal systems [55, 56, 74–77]. In order to model the PLE data, it is necessary to calculate the oscillator strength of possible transitions, and to take into account the

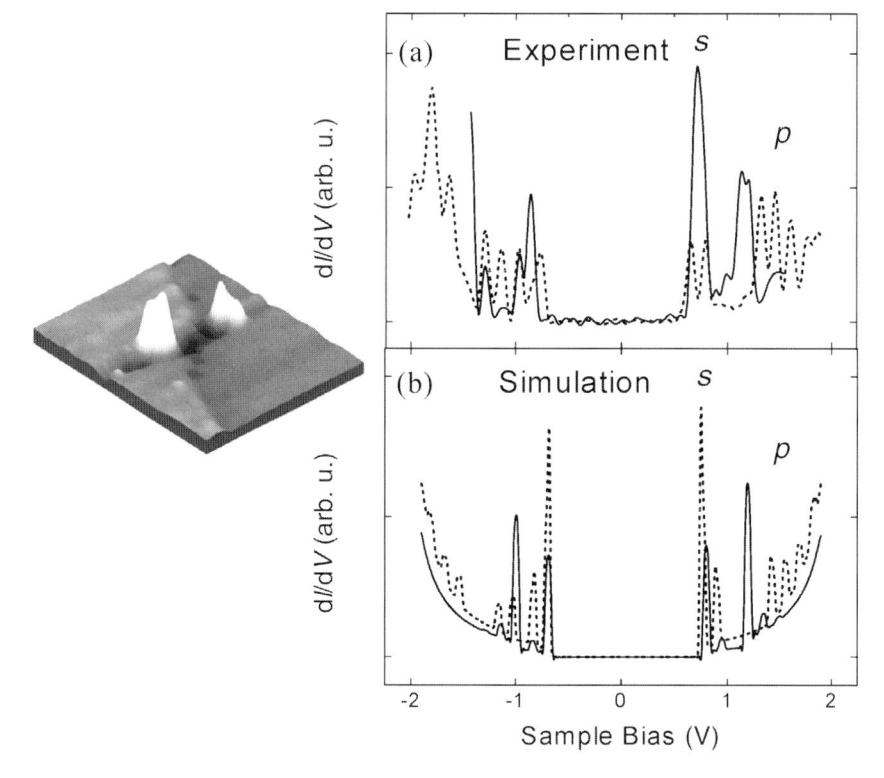

Figure 5.5 (a) Tunneling spectra measured on InAs QDs ~2 nm in radius. The solid curve was measured in the QD/HOPG geometry, and the dashed curve in the QD/DT/Au geometry; (b) Calculated spectra showing the effect of tunneling rate ratio. The dashed and solid curves were calculated with tunneling rate ratios, $\Gamma_2/\Gamma_1 \sim 1$ and 10, respectively. The image shows two QDs positioned near a step on HOPG.

electron–hole coulombic interaction which modifies the observed (excitonic) band gap. In the tunneling case, as discussed above, the device geometry should be carefully modeled and, in addition, the effects of charging and the tip-bias on the level structure must be considered. The charging may affect the intrinsic level structure and also determine the single-electron addition energy, while the tip-bias may perturb the levels via the Stark effect.

Zunger and coworkers treated the effects of electron charging for a QD embedded in a homogeneous dielectric medium characterized by ε_{out}. The addition energies and quasi-particle gap were calculated as a function of ε_{out} [78, 79]. While this isotropic model does not represent the experimental geometry of the tunneling measurements, the authors were able to find a good agreement between the energetic positions of the peaks for several QD sizes, using one value of ε_{out}. These authors also noted that the charging energy contribution associated with the band-gap transition may be different from that within the charging multiplets in the excited states. This difference was, however, on the order of the peak width in the spectra.

In another approach, Niquet et al. modeled the junction parameters (capacitances and tunneling rates), and used a tight-binding model for the level structure [80, 81]. The tunneling spectra were calculated using a rate equation method. The authors were able to reproduce the experimental tunneling spectra, attributing part of the tunneling peaks at negative bias to tunneling through the CB.

5.1.1.4 Atomic-Like States in Core–Shell Nanocrystals: Spectroscopy and Imaging

Additional complexity and functionality is afforded for semiconductor nanocrystals through the growth of shells. Of particular interest are the core–shell nanocrystals that exhibit an enhanced emission quantum yield and improved stability. In such composite core–shell structures, the shell type and thickness provide further control for tailoring the optical, electronic, electrical, and chemical properties of semiconductor nanocrystals. The growth of InAs–shell nanocrystals with high quantum yield (QY) in the technologically important near-infrared (NIR) spectral range are discussed in Section 3.1.3.4 and in Ref. [34]. These nanocrystals were successfully implemented as chromophores in polymer–nanocrystal light emitting diodes (LEDs) active in the telecommunications range [17]. They also constitute an important system for further investigation of the evolution of level structure in QDs. The combined tunneling and optical spectroscopy approach has been further applied to investigate core–shell nanocrystals.

Figure 5.6 shows tunneling-conductance spectra measured on two InAs–ZnSe core–shell nanocrystals with two- and six-monolayer (ML) shells, along with a typical curve for an InAs QD of radius similar to the nominal core radius ~1.7 nm, deposited on Au with DT linkers. The general appearance of the spectra of the core and of the core–shell nanocrystals was similar, and the band-gap near-identical, as observed with optical absorption measurements [34]. In contrast, the s-p level separation is substantially reduced. Both effects were consistent with a model in which the s state was confined to the InAs core region, while the p level extended to the ZnSe shell.

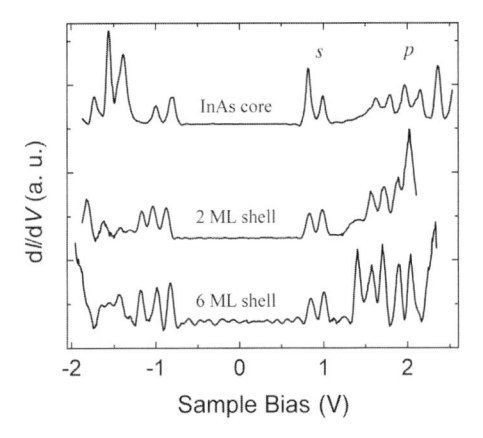

Figure 5.6 Tunneling conductance spectra of an InAs core QD and two core–shell nanocrystals with two and six monolayer (ML) shells with nominal core radii ~1.7 nm. The spectra were offset along the V direction to center the observed zero current gaps at zero bias.

In this case, the *p* state was red-shifted upon increasing shell thickness, whereas the *s* level did not shift, but yielded a closure of the CB *s–p* gap.

The complementary optical spectroscopy also provides evidence for the reduction of the *s–p* spacing upon shell growth, as manifested by the PLE spectra presented in Figure 5.7. The three spectra, for cores (solid line), and core–shells of 4 ML and 6 ML shell thickness (dotted and dashed lines, respectively), were measured using the same detection window (970 nm), which corresponded to the excitonic band-gap energy for InAs cores of 1.7 nm radius. The peak labeled III which, as discussed earlier, corresponds in the cores to the transition from the VB edge state to the CB $1P_e$ state, is red-shifted monotonically upon shell growth. The dependence of the difference between peak III and the band-gap transition I on shell thickness is depicted in the inset of Figure 5.7 (solid circles), along with the $1S_e - 1P_e$ level spacing extracted from the tunneling spectra (solid squares). While the qualitative trend of red shift is similar for both data sets, there is in fact a quantitative difference, with the optical shift being considerably smaller. This is in contrast to the good correlation between the optical and tunneling spectra observed for InAs cores, and provides an opportunity to examine the intricate differences between these two

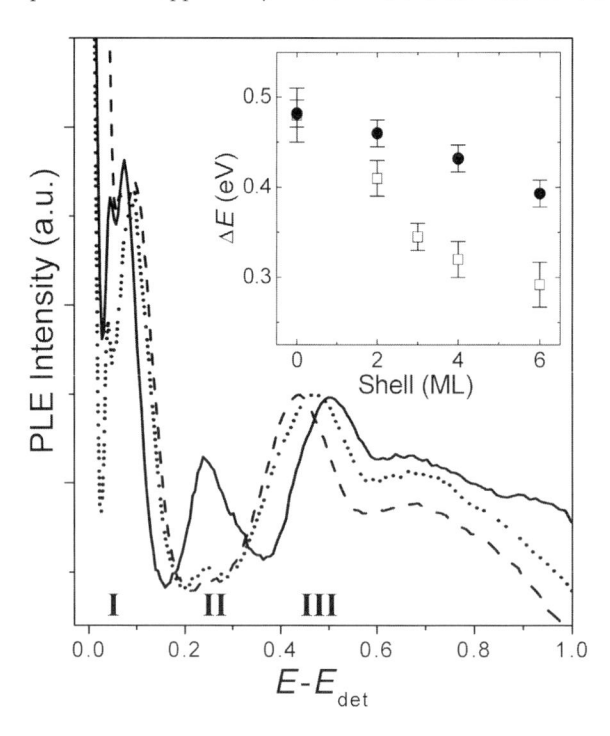

Figure 5.7 PLE spectra, normalized to peak III, for InAs cores (solid line) and InAs–ZnSe core–shell nanocrystals of four-ML (dotted line) and six-ML (dashed line) shells, with the zero of the energy scale taken at the detection window (970 nm). The inset depicts the dependence on shell thickness of the *s-p* level closure, as determined by tunneling (open squares) and PLE (solid circles). The PLE data points represent the difference between transition III and transition I (that hardly shifts upon shell growth).

complementary methods. While the tunneling data directly depict the spacing between the two CB states, the PLE data in the inset of Figure 5.7 represent the energy difference between two VB to CB optical transitions. Therefore, evolution of the complex VB edge states upon shell growth will inevitably affect the PLE spectra. In particular, a blue-shift of the *p*-like component of the VB edge state upon shell growth, will reduce the net observed PLE shift compared to the tunneling data, consistent with the experimental observations.

The elegant artificial atom analogy for QDs, borne out from optical and tunneling spectroscopy, can be tested directly by observing the shapes of the QD electronic wavefunctions, as demonstrated by various techniques for molecular beam epitaxy (MBE)-grown QDs. For colloidal, free-standing nanocrystal-QDs, the unique sensitivity of STM to the electronic density of states on the nanometer scale, seems to provide an ideal probe of the wavefunctions. A demonstration of this capability was provided recently for InAs–ZnSe core–shells, discussed above, where the different extent of the CB *s* and *p* states, as implied by the spectroscopic results, could be directly verified by using STM to image the QD atomic-like wavefunctions.

Bias-dependent current imaging measurements were performed (as shown in Figure 5.8) for a core–shell nanocrystal with a 6 ML shell. The dI/dV versus V spectrum is shown in Figure 5.8a, and the bias values for tunneling to the *s* and *p* states are indicated. A topographic image was measured at a bias value above the *s* and *p* states, $V_B = 2.1$ V (Figure 5.8b), simultaneously with three current images. At each point along the topography scan, the STM feedback circuit was disconnected momentarily, and the current measured at three different V_B values: at 0.9 V, corresponding to the CB *s* state (Figure 5.8c); at 1.4 V, within the *p* multiplet (Figure 5.8d); and at 1.9 V, above the *p* multiplet (Figure 5.8e). With this measurement procedure, the topographic and current images were all measured with the same constant local tip–QD separation. Thus, the main factor determining each current image was the local (bias-dependent) DOS, reflecting the shape of the QD electronic wavefunctions.

Upon comparing the current images, pronounced differences could be observed in the extent and shape of the *s* and *p* wavefunctions. The image corresponding to the *s*-like wavefunction (Figure 5.8c) is localized to the central region of the core–shell nanocrystal, while the images corresponding to the *p*-like wavefunctions extend out to the shell (Figure 5.8d and e), consistent with the model discussed above. This can also be seen in the cross-sections presented in Figure $5.8f_1$, taken along a common line through the center of each current image, and most clearly in Figure $5.8f_2$, which shows the current normalized to its maximum value along the same cuts. Figure 5.8e, which was recorded at a voltage above the *p* multiplet, manifests a near-spherical geometry similar to that of Figure 5.8c for the *s* state, but has a larger spatial extent. Figure 5.8d, which was recorded with V_B near the middle of the *p* multiplet, is also extended but has a truncated top with a small dent in its central region.

An illustrative model aids the interpretation of the current images, assuming a spherical QD shape, with a radial core–shell potential, as shown in the inset of Figure 5.8a [82]. The energy calculated for the *s* state is lower than the barrier height at the core–shell interface, and has about the same values for core and core–shell QDs. In

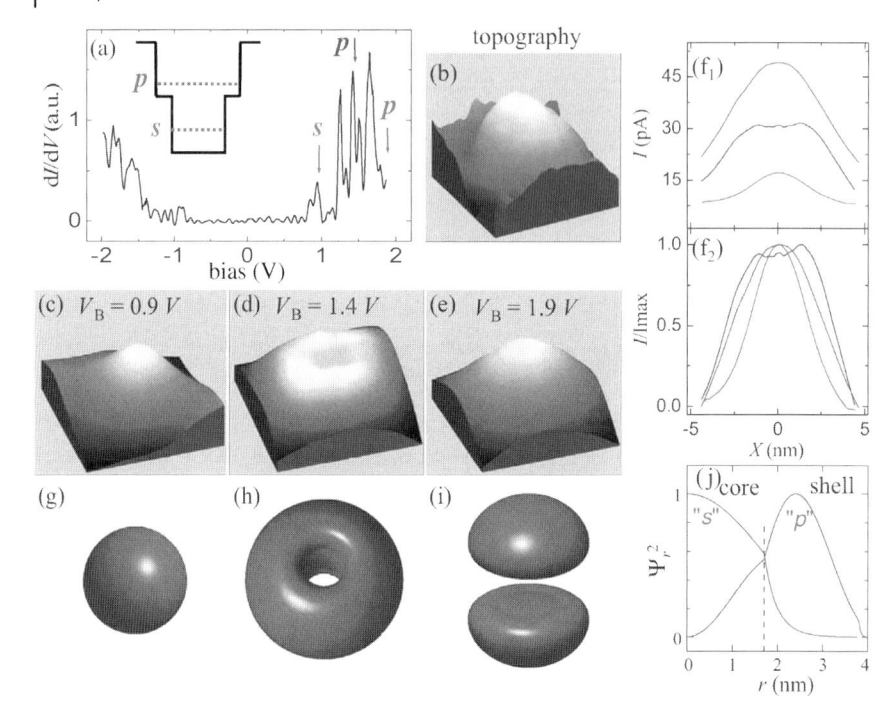

Figure 5.8 Wavefunction imaging and calculation for an InAs–ZnSe core–shell QD having a six ML shell. (a) A tunneling spectrum acquired for the nanocrystal; (b) 8 × 8 nm² topographic image taken at $V_B = 2.1\,\text{V}$ and $I_s = 0.1\,\text{nA}$; (c–e) Current images obtained simultaneously with the topographic scan at three different bias values denoted by arrows in panel (a); (f$_1$) Cross-sections taken along the diagonal of the current images at 0.9 V (lower curve), 1.4 V (middle curve) and 1.9 V (upper curve); (f$_2$) The same cross-sections normalized to their maximum current values; (g–j) Envelope wavefunctions calculated within a "particle in a sphere" model. The radial potential and the energies of the s and p states are illustrated in the inset of panel (a); (g–i) Isoprobability surfaces, showing s^2 (g), $p_x^2 + p_y^2$ (h), and p_z^2 (i); (j) The square of the radial parts of the s and p wavefunctions normalized to their maximum values.

contrast, the energy of the p state is above the core–shell barrier and red shifts with shell growth, in qualitative agreement with the spectroscopic findings discussed in the previous section. Isoprobability surfaces for the different wavefunctions are presented in Figure 5.8g–i, with Figure 5.8g showing the s state, Figure 5.8h the in-plane component of the p wavefunctions, $p_x^2 + p_y^2$, that has a torus-like shape, and Figure 5.8i depicting the perpendicular component, p_z^2. The square of the radial parts of the s and p wavefunctions are presented in Figure 5.8j. The calculated probability density for the s state is spherical in shape and mostly localized in the core, whereas the p components extend much further to the shell, consistent with the bias-dependent current images. Moreover, the different shapes observed in the current images can be assigned to different combinations of the probability density of the p components.

A filled torus shape, similar to the current Figure 5.8d, recorded at the middle of the p multiplet, can be obtained by a combination with larger weight of the $(p_x^2 + p_y^2)$ component, parallel to the gold substrate, and a smaller contribution of the p_z component. The non-equal weights reflect preferential tunneling through the in-plane components, but this may result from a tip-induced Stark effect leading to a small degeneracy lifting, as shown theoretically by Tews and Pfannkuche [83]. A spherical shape for the isoprobability surfaces results from summing all the p components with equal weights, consistent with the current image measured at a bias above the p manifold (Figure 5.8e). This example of wavefunction imaging, combined with the tunneling and optical spectra, provides visualization of the atomic-like character of nanocrystal QDs.

5.1.1.5 Level Structure of CdSe Quantum Rods

An additional system prepared by colloidal chemistry which continues to attract considerable interest is that of QRs that exhibit electronic and optical properties that differ from those of QDs. For example, due to their cylindrical symmetry, QRs have a linearly polarized emission, as demonstrated by fluorescence measurements on single rods [41], leading to polarized lasing [14]. The powerful combination of optical and tunneling spectroscopy discussed here was also applied to investigations of the length- and diameter-dependence of the electronic level structure of CdSe QRs [40].

The central conclusion drawn from the spectroscopic studies, as well as from calculations employing a multiband effective mass model, is that the QR level spectrum depends mainly on the rod diameter, and not on its length for CdSe rods with lengths above ~10 nm [40]. This is demonstrated in Figure 5.9, by both PLE and tunneling data. The PLE spectra (Figure 5.9a) measured on QR samples which differ considerably in length but, by having similar diameters, appear very similar; in contrast, the spectra measured on QRs of different radii appeared to differ. This becomes evident when comparing the upper two PLE curves measured on thin rods (radii ~1.7 nm) with the lower curves measured on thicker rods (radii ~3.1 nm). The tunneling spectra (Figure 5.9b), acquired in the regime where charging effects were suppressed, revealed a similar behavior. It can be seen here also that the band-gap depends mainly on – and decreases with – the QR radius, in agreement with the optical measurements (see Ref. [39]).

5.1.1.6 Level Structure and Band-Offsets in Heterostructured Seeded Quantum Rods

By tailoring the potential landscape in core–shell nanocrystals through material choice, the wavefunctions of the electron and hole ground states can each be either confined to the core or extended out to the shell, thus affecting the degree of overlap and, consequently, their optoelectronic properties. For example, core–shells with type I band-offsets exhibit a bright and stable fluorescence that is widely applied in LEDs, as both the electron and hole can be confined to the core. Multiexciton properties, which are important in optical gain applications, are also controlled by the band offsets where type II alignment, leading to charge separation, may prove advantageous. It is therefore clear that a knowledge of band-offsets in colloidal nanostruc-

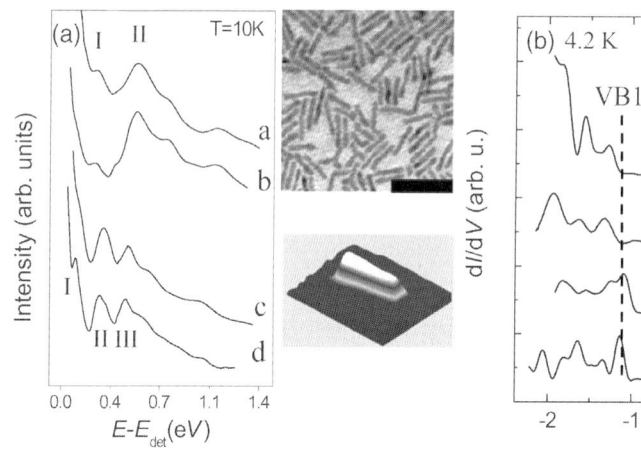

Figure 5.9 (a) Optical and tunneling spectroscopy of CdSe QRs. PLE spectra for CdSe QRs: spectrum (a) 31 × 1.9 (length times radius) nm; spectrum (b) 11 × 1.6 nm; spectrum (c) 60 × 3.3 nm; spectrum (d) 11 × 2.9 nm, with the zero energy representing the position of the detection window. Relevant optical transitions are marked. The structure above 0.7 eV is overlapping peaks of the excitation lamp that could not be completely normalized out; (b) Tunneling spectra for QRs of different sizes. For clarity, the spectra were shifted horizontally to align the CB1 peaks. Between panels (a) and (b) are shown a TEM image of a QR sample 31 × 3.9 nm (top, scale bar = 50 nm) and an STM topographic image of a single QR 25 × 2 nm (bottom).

tures is necessary for predictive heterostructure design. Yet, with a clear lack of direct experimental measurements, a common approach taken in this context has been to assume band-offset values of the bulk materials or of epitaxial superlattices although, obviously, in nanoscale colloidal heterostructures the band offsets can differ considerably. While optical measurements can provide energies of allowed transitions between the bands, scanning tunneling spectroscopy (STS) can provide information on the conduction and valence states separately. This, along with the capability to map the local DOS with nanometer resolution in composite nanocrystals was used to extract level offsets between the different components of colloidal heterostructures. Along with theoretical modeling, these data allow determination of the band offsets [47].

This approach was applied to heterostructured seeded nanorods, both for CdSe–CdS (type I) and for ZnSe–CdS QD/QR core–shell nanocrystals. The former system was studied extensively, using a variety of optical spectroscopy methods; the data acquired suggested a charge separation, where the hole is located in the CdSe core [positioned close to one end of the nanorod (NR)] and the electron extends over the CdS shell [84]. This picture is consistent with a small value of the conduction band offset, typically $\Delta_C < 0.2$ eV, extracted from the bulk regime. However, a direct measurement of the band offsets and the consequent charge distribution in such nanocrystals is of major interest.

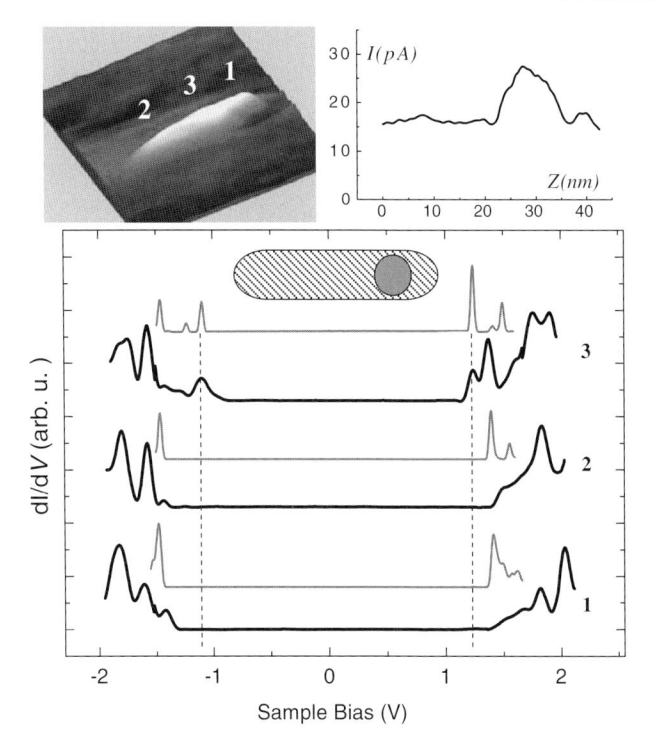

Figure 5.10 Three tunneling spectra (black curves) acquired on a CdSe/CdS nanocrystal at locations marked on the topographic image shown to the top-left. The corresponding theoretical curves are drawn above the measured spectra (thin red curves). A cross-section of a current image taken at a voltage of 1.2 V is shown to the upper-right. The inset illustrates the CdSe–CdS QD/NR core–shell structure.

The dI/dV spectra acquired at 4.2 K on a single CdSe/CdS core/shell rod are presented in Figure 5.10. The spectra were measured at different locations along the rod, as marked in the topographic image shown at the top-left, and the corresponding theoretical curves are shown above the experimental spectra (thin red curves). Curve 1 was measured near one apex of the NR, and curve 2 on the other side of the QR, but closer to the middle. Both curves exhibit a single-particle band-gap (between the electron and hole ground states) of about 2.9 eV. The band gap measured along the CdS QR, far enough from the CdSe core, was almost constant, and varied between 2.8 and 2.9 eV. This measured gap is in very good agreement with the computed value (2.85 eV), and larger than the bulk CdS gap due to quantization effects in the CdS rod. Curve 3, however, which was acquired about 15 nm from the right edge of the NC, exhibits a significantly smaller band gap ($\sim$2.3 eV), a value consistent with the measured optical band-gap of the system. At this position, the first tunneling peak on the negative bias side, corresponding to the valence band ground state, is red-shifted with respect to curves 1 and 2, consistent with the expected large Δ_V between the CdSe core and CdS shell. It was thus concluded that spectrum 3 was measured above the CdSe core. This position is also consistent with the location of the core deduced previously from TEM data [85], and the STM topographic image indeed portrays there a slight broadening of the NR diameter, as frequently observed in the TEM images.

Surprisingly, a red shift with respect to the CdS is seen also for the ground-state of the electron, indicating that the electron ground state (and not only the hole ground state) is localized in the core. Further experimental evidence for the above electron localization scenario is provided by current imaging tunneling spectroscopy measurements, which yield information on the shape and extent of the electronic wavefunctions, as described above. This is shown in the upper-right panel of Figure 5.10, where a cross-section of a current image acquired at a bias of 1.2 V is presented. This voltage corresponds to the energetic position of the conduction-band (electron) ground state, as depicted by the upper curve presented in the lower panel of Figure 5.10. The peak in this current cross-section curve, located about 15 nm from one end of the NR, clearly manifests the localized electron ground state wavefunction.

In order to determine the conduction-band offset that is sufficient for localizing the electron ground state in the core, calculation of the level structure (and the corresponding tunneling spectra) along the nanorods was performed using an effective mass-based model. The theoretical spectra (presented by the red curves in the lower panel of Figure 5.10), which reproduce very well the measured band-gaps and the level offsets, have been obtained using $\Delta_C = 0.30$ eV ($\Delta_V = 0.44$ eV). This conduction band-offset, which is somewhat larger than the maximal reported "bulk" value, is found to be sufficient to allow localization of the electronic ground state in the CdSe seed. This combined STS-theory approach was also successfully applied for the type II ZnSe–CdS QD–NR hybrid nanocrystals, clearly revealing their type II band-gap. The band-offsets extracted by this approach can be applied to core–shell structures of similar composition and of different shapes.

5.1.1.7 Optical Gain and Lasing in Semiconductor Nanocrystals

The broad spectral coverage for luminescence from the ultraviolet (UV) to the NIR offered via the tunability of composition, size and shape in semiconductor nanocrystals, presents an obvious advantage for the use of such materials as tunable optical gain media, and in laser applications. Furthermore, low lasing thresholds are predicted for QDs and quantum wires, compared with two-dimensionally confined quantum wells, form the basis of the present semiconductor diode laser devices which are ubiquitous in information and telecommunication technologies [86, 87].

Amplified spontaneous emission was recently observed by Bawendi, Klimov and coworkers for spherical colloidal CdSe QDs in close-packed films where pumping with an amplified femtosecond laser source was used to compete with fast non-radiative Auger decay processes [13, 88]. In further studies, films with CdSe nanocrystals were deposited on a distributed feed-back grating structure to yield optically pumped lasing that was tunable through the visible range by changing the nanocrystal size [89, 90].

Lasing was also observed for semiconductor nanocrystals in solution at room temperature, by using a cylindrical microcavity with nanosecond excitation. For this, a simple microcavity set-up was used which provided an uncomplicated means for screening colloidal nanocrystals as potential laser chromophores directly in solution [14].

Figure 5.11 presents the results of lasing for CdSe–ZnS QRs, along with their structural and optical characteristics. The QRs were grown using the well-developed methods of colloidal nanocrystal. The rod–shell configuration was chosen as the growth of ZnS on organically coated CdSe QRs enhances the fluorescence quantum yield, from ~2% to 14%. A TEM image of the QRs, deposited on a grid from hexanes solution, is presented in Figure 5.11a. A good size and shape monodispersivity is observed, with the average nanorod length being 25 nm and the average diameter 4 nm. The monodispersivity is deemed essential for achieving lasing, by reducing the

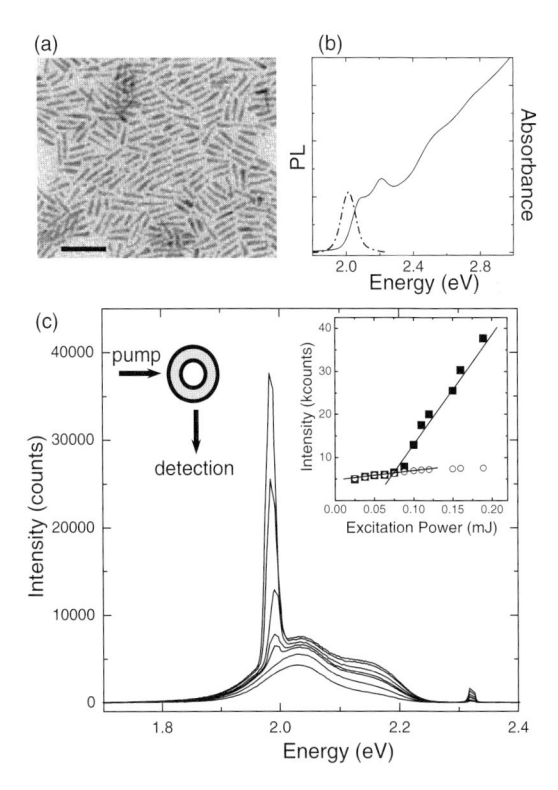

Figure 5.11 (a) A TEM image of the quantum rod sample (scale bar = 50 nm); (b) The absorption (solid line) and emission (dot-dashed line) for the quantum rods (QRs) in solution' (c) Emission spectra for the QRs loaded in the microcavity at different pump powers. The peak at 2.33 eV is the residual scattering of the pump laser. The pump intensities from low to high were: 0.016, 0.038, 0.075, 0.088, 0.10, 0.15, 0.19 mJ. The narrow peak at 1.98 eV emerging above the fluorescence is the laser signal that appears at a threshold pump level of ~0.08 mJ. The right inset of panel (c) shows the intensity of the lasing peak (filled squares), and the fluorescence peak (empty circles), versus the pump power. Lasing shows a clear threshold behavior as accentuated by the solid lines. The left inset in panel (c) is a schematic top view of the cylindrical microcavity experimental set-up, with the gray shaded area representing the QR solution occupying the volume between the fiber and inner capillary walls.

inhomogeneous spectral broadening of the gain profile. The optical characteristics of the nanorod sample are summarized in Figure 5.11b.

In Figure 5.11c, emission spectra collected from the cylindrical microcavity, loaded with a QR sample in hexanes, is shown for several laser excitation intensities where each spectrum corresponds to a single laser shot. The peak at 2.33 eV is due to scattering of the pump laser. At low excitation intensities (e.g., 0.016 mJ), the spectrum resembles the emission spectrum measured for the QRs in solution (Figure 5.11b), but at higher intensities changes in the spectral shape are observed. Most significantly, above ~0.08 mJ a narrow lasing peak clearly emerges at 1.98 eV, to the red of the fluorescence peak. The dependence of the intensity of the lasing peak on pump power is shown in the right-hand inset of Figure 5.11c (filled squares), along with the dependence of the fluorescence peak intensity (open circles). The lasing peak intensity exhibits an abrupt change of slope at the onset of laser action whilst, at the same time, the peak fluorescence intensity is in effect saturated. This threshold behavior of the laser peak intensity provides an additional important signature of lasing [91]. Higher-resolution spectral measurements revealed that lasing occurs in a sequence of discrete modes, corresponding to the "whispering gallery" modes of the cylindrical microcavity. Polarized measurements of the lasing for rods versus CdSe dots in solution showed that, whilst for QDs the lasing is not polarized, for QRs a linearly polarized laser signal is detected that is directly related to their symmetry [14]. These studies on gain and lasing in semiconductor nanocrystals provide examples that may lead to further technological developments and the implementation of colloidal nanocrystals in laser and optical amplifier applications.

References

1 Alivisatos, A.P. (1996) *Science*, **271**, 933.
2 Brus, L.E. (1991) *Appl. Phys. A*, **53**, 465.
3 Weller, H. (1993) *Angew. Chem., Int. Ed. Engl.*, **32**, 41.
4 Nirmal, M. and Brus, L.E. (1999) *Acc. Chem. Res.*, **32**, 407.
5 Collier, C.P., Vossmeyer, T., and Heath, J.R. (1998) *Annu. Rev. Phys. Chem.*, **49**, 371.
6 Peng, X.G., Manna, L., Yang, W.D., Wickham, J., Scher, E., Kadavanich, A.V., and Alivisatos, A.P. (2000) *Nature*, **404**, 59.
7 Murray, C.B., Norris, D.J., and Bawendi, M.G. (1993) *J. Am. Chem. Soc.*, **115**, 8706.
8 Guzelian, A.A., Banin, U., Kadavanich, A.V., Peng, X., and Alivisatos, A.P. (1996) *Appl. Phys. Lett.*, **69**, 1432.
9 Brus, L.E. (1984) *J. Chem. Phys.*, **80**, 4403.
10 Grabert, H. and Devoret, M.H. (eds) (1992) *Single Charge Tunneling*, Plenum, New York.
11 Averin, D.V. and Likharev, K.K. (1991) *Mesoscopic Phenomena in Solids* (eds B.L. Altshuler, P.A. Lee, and R.A. Webb), Elsevier, Amsterdam, p. 173.
12 Banin, U., Cao, Y.W., Katz, D., and Millo, O. (1999) *Nature*, **400**, 542.
13 Klimov, V.I., Mikhaelovsky, A.A., Xu, S., Malko, A., Hollingsworth, J.A. *et al.* (2000) *Science*, **290**, 314.
14 Kazes, M., Lewis, D.Y., Ebenstein, Y., Mokari, T., and Banin, U. (2002) *Adv. Mater.*, **14**, 317.
15 Colvin, V.L., Schlamp, M.C., and Alivisatos, A.P. (1994) *Nature*, **370**, 354.
16 Dabboussi, B.O., Bawendi, M.G., Onitsuka, O., and Rubner, M.F. (1995) *Appl. Phys. Lett.*, **66**, 1316.
17 Tessler, N., Medvedev, V., Kazes, M., Kan, S.H., and Banin, U. (2002) *Science*, **295**, 1506.

18 Bruchez, M.P., Moronne, M., Gin, P., Weiss, S., and Alivisatos, A.P. (1998) *Science*, **281**, 2013.
19 Chan, W.C.W. and Nie, S. (1998) *Science*, **281**, 2016.
20 Mitchell, G.P., Mirkin, C.A., and Letsinger, R.L. (1999) *J. Am. Chem. Soc.*, **121**, 8122.
21 Bowen-Katari, J.E., Colvin, V.L., and Alivisatos, A.P. (1994) *J. Phys. Chem.*, **98**, 109.
22 Murray, C.B., Kagan, C.R., and Bawendi, M.G. (1995) *Science*, **270**, 1335.
23 Collier, C.P., Vossmeyer, T., and Heath, J.R. (1998) *Annu. Rev. Phys. Chem.*, **49**, 371.
24 Whetten, R.L., Khoury, J.T., Alvarez, M.M., Murthy, S., Vezmar, I. et al. (1996) *Adv. Mater.*, **8**, 428.
25 Black, C.T., Murray, C.B., Sandstrom, R.L., and Sun, S. (2000) *Science*, **260**, 1131.
26 Pileni, M.P. (2001) *J. Phys. Chem. B*, **105**, 3358.
27 Alivisatos, A.P., Johnson, K.P., Peng, X., Wilson, T.E., Loweth, C.J. et al. (1996) *Nature*, **382**, 609.
28 Mews, A., Eychmüller, A., Giersig, M., Schoos, D., and Weller, H. (1994) *J. Phys. Chem.*, **98**, 934.
29 Cao, Y.W. and Banin, U. (1999) *Angew. Chem., Int. Ed. Engl.*, **38**, 3692.
30 Hines, M.A. and Guyot-Sionnest, P.J. (1996) *J. Phys. Chem.*, **100**, 468.
31 Peng, X., Schlamp, M.C., Kadavanich, A.V., and Alivisatos, A.P. (1997) *J. Am. Chem. Soc.*, **119**, 7019.
32 Dabbousi, B.O., Rodriguez-Viejo, J., Mikulec, F.V., Heine, J.R., Mattoussi, H. et al. (1997) *J. Phys. Chem. B*, **101**, 9463.
33 Tian, Y., Newton, T., Kotov, N.A., Guldi, D.M., and Fendler, J.H. (1996) *J. Phys. Chem.*, **100**, 8927.
34 Cao, Y.W. and Banin, U. (2000) *J. Am. Chem. Soc.*, **122**, 9692.
35 Kershaw, S.V., Burt, M., Harrison, M., Rogach, A., Weller, H., and Eychmuller, A. (1999) *Appl. Phys. Lett.*, **75**, 1694.
36 Harrison, M.T., Kershaw, S.V., Rogach, A.L., Kornowski, A., Eychmuller, A., and Weller, H. (2000) *Adv. Mater.*, **12**, 123.
37 Manna, L., Scher, E.C., and Alivisatos, A.P. (2000) *J. Am. Chem. Soc.*, **122**, 12700.
38 Peng, Z.A. and Peng, X. (2001) *J. Am. Chem. Soc.*, **123**, 1389.
39 Li, L.S., Hu, T.J., Yang, W.D., and Alivisatos, A.P. (2001) *Nano Lett.*, **1**, 349.
40 Katz, D., Wizansky, T., Millo, O., Rothenberg, E., Mokari, T., and Banin, U. (2002) *Phys. Rev. Lett.*, **89**, 86801.
41 Hu, J., Li, L.S., Yang, W., Manna, L., Wang, W., and Alivisatos, A.P. (2001) *Science*, **292**, 2060.
42 Huynh, W.U., Dittmer, J.J., and Alivisatos, A.P. (2002) *Science*, **295**, 2425.
43 Talapin, D.V., Koeppe, R., Götzinger, S., Kornowski, A., Lupton, J.M., Rogach, A.L., Benson, O., Feldmann, J., and Weller, H. (2003) *Nano Lett.*, **3**, 1677.
44 Carbone, L. et al. (2007) *Nano Lett.*, **7**, 2942.
45 Talapin, D.V., Nelson, J.H., Shevchenko, E.V., Aloni, S., Sadtler, B., and Alivisatos, A.P. (2007) *Nano Lett.*, **7**, 2951.
46 Dorfs, D., Popov, I., Salant, A., and Banin, U. (2008) *Small*, **5**, 1319.
47 Steiner, D., Dorfs, D., Banin, U., Della Sala, F., Manna, L., and Millo, O. (2008) *Nano Lett.*, **8**, 2954.
48 Vahala, K.J. and Sercel, P.C. (1990) *Phys. Rev. Lett.*, **65**, 239.
49 Norris, D.J., Sacra, A., Murray, C.B., and Bawendi, M.G. (1994) *Phys. Rev. Lett.*, **72**, 2612.
50 Norris, D.J. and Bawendi, M.G. (1996) *Phys. Rev. B*, **53**, 16338.
51 Bertram, D., Micic, O.I., and Nozik, A.J. (1998) *Phys. Rev. B*, **57**, R4265.
52 Banin, U., Lee, J.C., Guzelian, A.A., Kadavanich, A.V., Alivisatos, A.P. et al. (1998) *J. Chem. Phys.*, **109**, 2306.
53 Banin, U., Lee, J.C., Guzelian, A.A., Kadavanich, A.V., and Alivisatos, A.P. (1997) *Superlattices Microstruct.*, **22**, 559–568.
54 Ekimov, A.I., Hache, F., Schanne-Klein, M.C., Ricard, D., Flytzanis, C. et al. (1993) *J. Opt. Soc. Am. B*, **10**, 100.
55 Fu, H., Wang, L.W., and Zunger, A. (1997) *Appl. Phys. Lett.*, **71**, 3433.
56 Williamson, A.J. and Zunger, A. (2000) *Phys. Rev. B*, **61**, 1978.
57 Porath, D., Levi, Y., Tarabiah, M., and Millo, O. (1997) *Phys. Rev. B*, **56**, 9829.
58 Porath, D. and Millo, O. (1997) *J. Appl. Phys.*, **85**, 2241.
59 Kastner, M.A. (1993) *Phys. Today*, **46** (1), 24.
60 Kouwenhoven, L. (1997) *Science*, **257**, 1896.

61 Amman, M., Mullen, K., and Ben-Jacob, E. (1989) *J. Appl. Phys.*, **65**, 339.
62 Hanna, A.E. and Tinkham, M. (1991) *Phys. Rev. B*, **44**, 5919.
63 (a) Klein, D.L., Roth, R., Lim, A.K.L., Alivisatos, A.P., and McEuen, P.L. (1996) *Nature*, **389**, 699; (b) Klein, D.L. *et al.* (1996) *Appl. Phys. Lett.*, **68**, 2574.
64 Alperson, B., Cohen, S., Rubinstein, I., and Hodes, G. (1995) *Phys. Rev. B*, **52**, R17017.
65 Colvin, V.L., Goldstein, A.N., and Alivisatos, A.P. (1992) *J. Am. Chem. Soc.*, **114**, 5221.
66 Katz, D., Millo, O., Kan, S.H., and Banin, U. (2001) *Appl. Phys. Lett.*, **79**, 117.
67 Millo, O., Katz, D., Cao, Y.W., and Banin, U. (2000) *J. Low Temp. Phys.*, **118**, 365.
68 Alperson, B., Hodes, G., Rubinstein, I., Porath, D., and Millo, O. (1999) *Appl. Phys. Lett.*, **75**, 1751.
69 (a) Fu, H. and Zunger, A. (1998) *Phys. Rev. B*, **57**, R15064;(b) Fu, H. *et al.* (1998) *Phys. Rev. B*, **57**, 9971.
70 Bakkers, E.P.A.M. and Vanmaekelbergh, D. (2000) *Phys. Rev. B*, **62**, R7743.
71 Katz, D., Kan, S.H., Banin, U., and Millo, O. (2001) *Physica A*, **302**, 328.
72 Bakkers, E.P.A.M., Hens, Z., Zunger, A., Franceschetti, A., Kouwenhoven, L.P. *et al.* (2001) *Nano Lett.*, **1**, 551.
73 Efros, A.L. and Rosen, M. (2000) *Annu. Rev. Phys. Chem.*, **30**, 475.
74 Franceschetti, A., Fu, H., Wang, L.W., and Zunger, A. (1999) *Phys. Rev. B*, **60**, 1819.
75 Rabani, E., Hetenyi, B., Berne, B.J., and Brus, L.E. (1999) *J. Chem. Phys.*, **110**, 5355.
76 Zunger, A. (1998) *MRS Bull.*, **23**, 35.
77 Hu, J.T. *et al.* (2002) *J. Phys. Chem.*, **106**, 2447.
78 Franceschetti, A. and Zunger, A. (2000) *Phys. Rev. B*, **62**, 2614.
79 Franceschetti, A. and Zunger, A. (2000) *Appl. Phys. Lett.*, **76**, 1731.
80 Niquet, Y.M., Delerue, C., Lannoo, M., and Allan, G. (2001) *Phys. Rev. B*, **64**, 113305.
81 Niquet, Y.M., Delerue, C., Allan, G., and Lannoo, M. (2002) *Phys. Rev. B*, **65**, 165334.
82 Schooss, D., Mews, A., Eychmuller, A., and Weller, H. (1994) *Phys. Rev. B*, **49**, 17072.
83 Tews, M. and Pfannkuche, D. (2002) *Phys. Rev. B*, **65**, 73307.
84 Müller, J. *et al.* (2005) *Nano Lett.*, **5**, 2044.
85 Carbone, L. *et al.* (2007) *Nano Lett.*, **7**, 2942.
86 Asada, M., Miyamoto, Y., and Suematsu, Y. (1986) *IEEE J. Quantum Electron.*, **QE-22**, 1915.
87 Grundmann, E. (2000) *Physica E*, **5**, 167.
88 Klimov, V.I., Schwarz, Ch.J., McBranch, D.W., Leatherdale, C.A., and Bawendi, M.G. (1999) *Phys. Rev. B*, **60**, R2177.
89 Eisler, H.J., Sundar, V.C., Bawendi, M.G., Walsh, M., Smith, H.I., and Klimov, V.I. (2002) *Appl. Phys. Lett.*, **80**, 4614.
90 Malko, A.V., Mikhailovsky, A.A., Petruska, M.A., Hollingsworth, J.A., Htoon, H., Bawendi, M.G., and Klimov, V.I. (2002) *Appl. Phys. Lett.*, **81**, 1303.
91 Yariv, A. (1986) *Quantum Electronics*, 3rd edn, John Wiley & Sons, New York.

5.1.2
Optical and Thermal Properties of Ib–VI Nanoparticles

Stefanie Dehnen, Andreas Eichhöfer, John F. Corrigan, Olaf Fuhr, and Dieter Fenske

5.1.2.1 Optical Spectra of Selenium-Bridged and Tellurium-Bridged Copper Clusters

The discrete molecular cluster species presented are ideal subjects for the study of the molecular quantum size effect [1], which is one of the most important questions prompting research in this area of cluster chemistry. It is necessary, therefore to explore the electronic properties of the cluster species, beginning with the interplay of cluster size and HOMO–LUMO gap. A number of theoretical investigations have provided the preliminary answers to this question [2, 3]. The first dipole- and spin-allowed singlet or triplet excitations of selected "naked" copper selenide clusters [$Cu_{2n}Se_n$] (up to $n=15$) were calculated on the CIS level (CIS = configuration interaction considering single excitations) [4, 5] for structures previously optimized for the same symmetry employing the CCSD(T) [6a] method [CCSD(T) = coupled cluster approximation up to single and double excitation with additional consideration of third excitations by means of perturbation theory] ($n=1, 2$), or the MP2 [6b] method [MP2 = second-order Møller–Plesset perturbation theory] ($n>2$). Triplet excitations were investigated in order to take into account the fact that relativistic effects (spin–orbit coupling) [7] are noticeably present in Cu_2Se compounds. The results are provided graphically in Figure 5.12.

The development of the energy values for singlet or triplet excitations with increasing cluster size shows the same trend, with the triplet excitation energies being 0.2–0.7 eV lower than the singlet excitation energies. The excitation energies increase with slight oscillation to a value around 6 eV for the hexamer ($n=3$), and decrease afterwards down to a value of approximately 3.8 eV for the [$Cu_{30}Se_{15}$] species. The initial increase in excitation energy values can be explained first, by a stabilization of selenium p-orbitals by an admixture of s-, p-, or d-contributions of copper atoms and, second, by an increasing completion of the selenium shell around the copper substructure. The latter reduces the space for electrons in excited states with s-character. Thus, for [$Cu_{12}Se_6$], significant p-type contributions are observed for the calculated transitions. The two exceptions ($n=2; 4$) of the described trend feature less-compact molecular structures that lead to lower excitation energies. Having reached a maximum coverage of selenium atoms on the copper framework, the spatial extension of the cluster molecules gains more relevance. This confirms the assumption of the splitting of the Se-p and Cu-s states to form a valence and a conduction band. For clusters of ever-increasing size, and finally bulk Cu_2Se, the first excitation energy should approach a value of ca. 1 eV [8, 9]. According to quantum

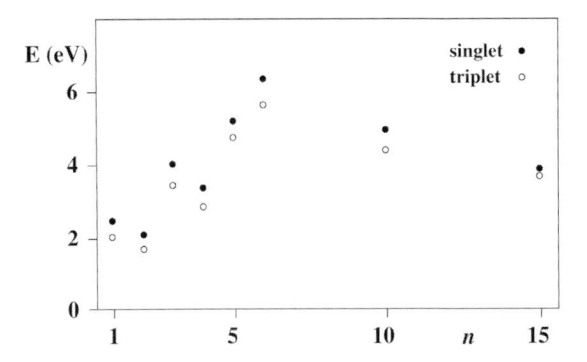

Figure 5.12 First dipole- and spin-allowed singlet or triplet excitations of selected "naked" copper selenide clusters [$Cu_{2n}Se_n$] (up to $n=15$) calculated on the CIS level (CIS = configuration interaction with single excitations) for structures previously optimized for the same symmetry employing the CCSD(T) method [CCSD(T) coupled cluster up to single and double excitation with additional consideration of third excitations by means of perturbation theory] ($n=1, 2$) or MP2 method ($n>2$).

chemical investigations, "naked" copper selenide clusters of medium size should be colorless insulators that gradually adopt semiconductor properties with increasing cluster size. Calculations on small PH_3- or PMe_3-ligated copper selenide clusters have shown an increase in the first excitation energies. Estimating an error of about 1.5 eV, the respective transitions could indeed absorb violet light – in agreement with the red color of small experimentally observed copper selenide clusters.

The crystals that are obtained from the cluster formation reactions are intensely colored. In fact, the intensity of the color increases when going from sulfur- to selenium- to tellurium-bridged compounds (see below), as might be expected for an increase in the covalent or (semi-) metallic binding properties. Small copper sulfide and selenide clusters form light red, orange, or purple crystals, but with increasing cluster size the color varies from dark red to reddish-black to (finally) black with a metallic sheen. The optical spectra of some copper selenide cluster compounds have been studied by means of solid-state UV-visible spectroscopy.

Figure 5.13 shows the absorption spectra of PEt_2Ph- or PEt_3-ligated compounds $[Cu_{26}Se_{13}(PEt_2Ph)_{14}]$ **(1)** [10], $[Cu_{44}Se_{22}(PEt_2Ph)_{18}]$ **(3)** [11], $[Cu_{70}Se_{35}(PEt_2Ph)_{23}]$ **(4)** [12], and $[Cu_{140}Se_{70}(PEt_3)_{34}]$ **(6)** [13]. The influence of the phosphine ligands should be approximately the same for all four clusters, and thus the increments in the measured shift of the absorption band can be ascribed solely to the increasing size and structural changes of the Cu_2Se framework.

A remarkable shift can be observed in the onset of absorption to higher wavelengths, by approximately 0.8 eV, with increasing cluster size (this is related to the increase in the number of copper and selenium cluster core atoms) in this narrow size regime, from approximately 600 nm (2.1 eV) for **1** (39 cluster core atoms) to

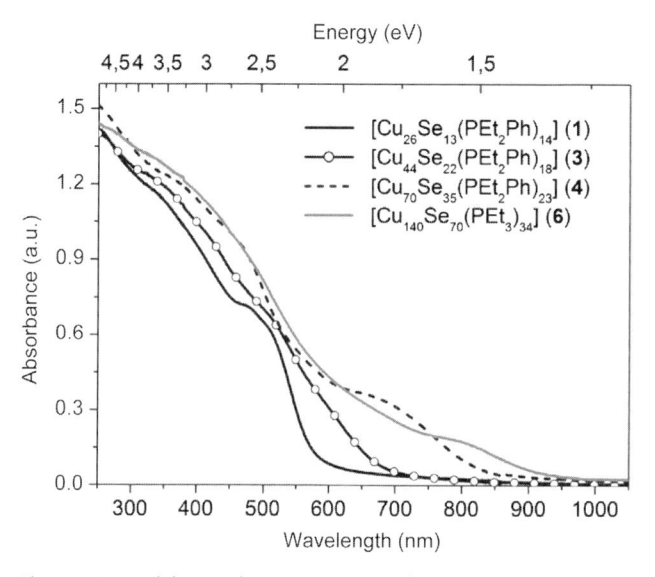

Figure 5.13 Solid state absorption spectra of PEt_2Ph- or PEt_3-ligated compounds **1**, **3**, **4**, and **6**. For the measurements, single crystalline samples, previously pulverized in mineral oil, were placed between two quartz plates.

950 nm (1.3 eV) for **6** (210 cluster core atoms). This agrees in principle with a decrease of the HOMO–LUMO gap in the same direction. Although, the experimental value for the band-gap in Cu_2Se (1.1 eV, i.e., 1127 nm, from optical measurements [9]; 0.37 eV, i.e., 3350 nm, from electrochemical measurements [14]) is not reached, the tendency can thus in fact be interpreted in view of the quantum size effect as an approximation of the frontier orbitals with increasing number of atoms.

With regards to the photoluminescence properties of copper selenide cluster molecules, it has been found that the CuSe clusters luminesce with a relatively high efficiency in the red to near-infrared (NIR) spectral range at cryogenic temperatures [15]. The photoluminescence excitation spectra of the copper selenide cluster molecules $[Cu_{26}Se_{13}(PEt_2Ph)_{14}]$ (**1**), $[Cu_{44}Se_{22}(PEt_2Ph)_{18}]$ (**3**) and $[Cu_{70}Se_{35}(PEt_2Ph)_{23}]$ (**4**) recorded at 16 K resemble more or less the room-temperature absorption spectra, but with a shift of the absorption onset to a higher wavelength by an increase in cluster size; however, they display more pronounced features (Figure 5.14). Interestingly, their PL spectra and kinetics demonstrate also size-dependent effects. For example, **1**, **3**, and **4** all show a relatively weak NIR photoluminescence at room temperature, the intensity of which increases strongly as the temperature is decreased to 16 K. The photolumi-

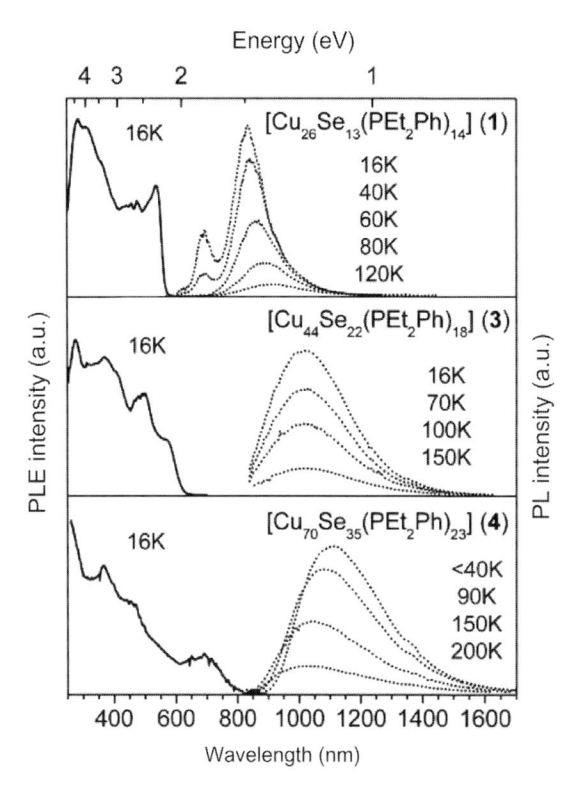

Figure 5.14 Photoluminescence excitation (PLE) spectra (solid lines) and photoluminescence (PL) spectra (dotted lines) of $[Cu_{26}Se_{13}(PEt_2Ph)_{14}]$ (**1**), $[Cu_{44}Se_{22}(PEt_2Ph)_{18}]$ (**3**), and $[Cu_{70}Se_{35}(PEt_2Ph)_{23}]$ (**4**) measured at different temperatures.

nescence maximum shifts thereby for **1** and **4**, but remains almost constant for **3**. For the 16 K spectra, a shift in the photoluminescence maximum is detected by the increase of cluster size ongoing from **1** at 835 nm (1.5 eV) to **3** at 1010 nm (1.2 eV) to **4** at 1120 nm (1.1 eV). An unusual feature here is a second photoluminescence peak that begins to develop at ~690 nm at temperatures below ~50 K. The photoluminescence excitation (PLE) spectrum indicates that this band is related to the low-lying excited electronic states (Figure 5.14). The origin of the dual photoluminescence is not clear at present, but it might be due to the existence of a minor isomer of **1**. The photoluminescence kinetics of the cluster molecules are quite complicated, and similar to those of cadmium selenide cluster molecules [16]. The photoluminescence decays relatively rapidly, within hundreds of nanoseconds, while the decay rates are increased as the temperature is increased. In general, the decays occur more rapidly at the "red" edge of the emission spectrum. The kinetics could be well fitted with the Williams–Watts function.

Likewise, for copper telluride clusters, the colors of the microcrystalline powders already reveal differences in the electronic properties of the compounds, with larger clusters being significantly darker in color than their smaller counterparts. In contrast to copper selenide cluster molecules, the existence of mixed-valence compounds has a pronounced influence on the spectra, and complicates the interpretation of size effects on the basis of a shift of the absorption onset. The UV-visible spectra (Figure 5.15) of a series of copper telluride cluster molecules reveal that most of the mixed-valence compounds, such as $[Cu_{16}Te_9(PPh_3)_8]$ (**7**), $[Cu_{23}Te_{13}(PPh_3)_{10}]$ (**8**), and $[Cu_{44}Te_{23}(PPh^nPr_2)_{15}]$ (**9**), belong – according to the classification of Robin and Day [17] – to

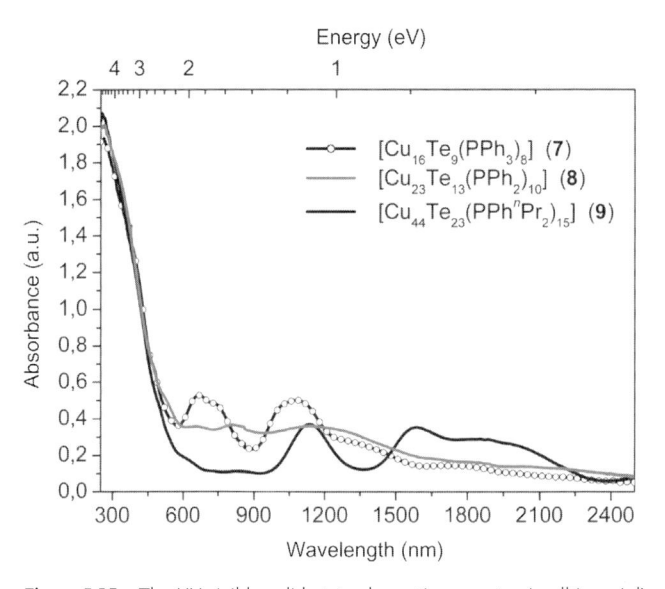

Figure 5.15 The UV-visible solid-state absorption spectra (mull in nujol) of the mixed-valence copper telluride clusters $[Cu_{16}Te_9(PPh_3)_8]$ (**7**), $[Cu_{23}Te_{13}(PPh_3)_{10}]$ (**8**), and $[Cu_{44}Te_{23}(PPhnPr_2)_{15}]$ (**9**).

class IIIa (metal clusters) [18]. In this class, metal centers with different valencies cannot be distinguished, and a total delocalization of the charge is achieved. Interestingly, as the energy of the mixed-valence transition is strongly correlated with the degree of delocalization, an even better charge delocalization – that is, a smaller transition energy – is observed for the larger cluster **9**.

5.1.2.2 Thermal Behavior of Selenium-Bridged Copper Clusters

As with the thermal properties of copper selenide cluster molecules in the solid state, investigations with compound **4** have revealed that the phosphine ligand shell can be cleaved at moderate temperatures (145 °C) in vacuum, in a one-step process that results in the formation of nanocrystalline copper selenide [12]. Powder diffraction measurements have shown that such cleavage of the ligand molecules is accompanied by an aggregation of the copper selenide cluster cores, which is in agreement with the observations made with X-ray photoelectron spectroscopy (XPS) measurements on a series of copper selenide clusters [19]. This corresponds with the results of quantum chemical investigations, which claimed that the clusters were metastable species relative to the binary chalcogenide Cu_2E, comprising very low Cu–P binding energies [20b, 20, 21].

Investigations into the thermal properties of a series of cluster molecules, namely $[Cu_{26}Se_{13}(PEt_2R)_{14}]$ (R = Ph (**1**) [10], Et (**2**)), $[Cu_{44}Se_{22}(PEt_2Ph)_{18}]$ (**3**) [11], $[Cu_{70}Se_{35}(PEt_2R)_{23}]$ (R = Ph (**4**) [12], Et(**5**) [22]), and $[Cu_{140}Se_{70}(PEt_3)_{34}]$ (**6**) [13], were focused on the cleavage behavior of the phosphine ligands, along with their cleavage temperatures under different experimental conditions and characterization of the residues [23]. Therefore, the thermogravimetric analyses of **1–6** were conducted in a vacuum and under an inert gas flow of helium. The total mass changes during the thermogravimetric analysis (TGA) were, for all compounds, in good agreement with the calculated weight loss for all of the phosphine ligands. The simultaneously recorded mass spectra of the volatile products confirmed the liberation of either PEt_3 or PEt_2Ph. These results suggested, in all cases, the formation of Cu_2Se, and this was confirmed by powder diffractometry and an elemental analysis of the black residue from the thermolysis. For both series of compounds, the TGA-curves of **1–6** in a helium atmosphere displayed complex multistage processes, whereas in a vacuum the differential thermogravimetry (DTG) graphs mostly showed a single peak that belonged to a one-step process, except those of **1** and **2**. However, for curves measured in helium gas flow, no intermediates could be stabilized by isothermal measurements at given temperatures. Apart from these clear observations, the different onsets and temperature ranges of the TGA of **1–6** in either a helium gas flow or in a vacuum suggested the presence of complex processes and influencing parameters which are not yet understood in full detail. With respect to the different types of phosphine ligand, the mean cleavage temperatures in vacuum were found to be significantly lower for **2** and **5** ligated by PEt_3 than were observed for the similarly sized clusters **1**, **3** and **4**, which are coordinated by PEt_2Ph. One explanation for this increased temperature might be a stronger Cu–P bonding for PEt_2Ph rather than for PEt_3, although it is more likely that such behavior originates from the different boiling points of PEt_3 (126–128 °C) and PEt_2Ph (222 °C).

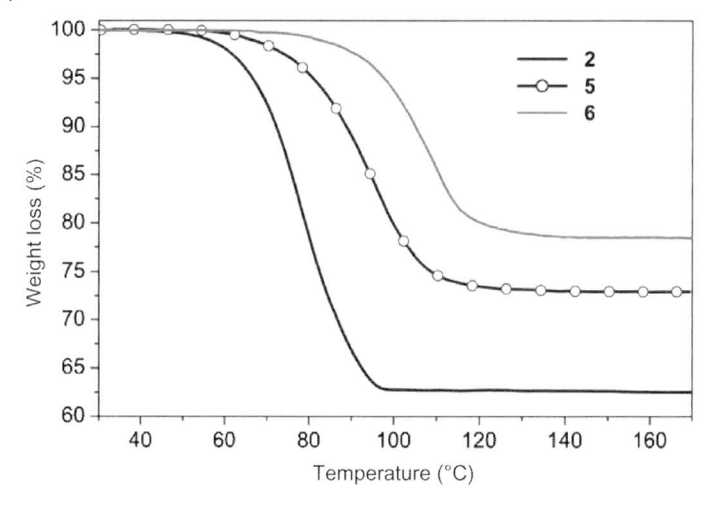

Figure 5.16 Thermogravimetric analysis of $[Cu_{26}Se_{13}(PEt_3)_{14}]$ (**2**), $[Cu_{70}Se_{35}(PEt_3)_{23}]$ (**5**) and $[Cu_{140}Se_{70}(PEt_3)_{34}]$ (**6**) under vacuum conditions.

Interestingly, for **2**, **5** and **6** – which are coordinated by the lower-boiling PEt_3 – under a vacuum atmosphere there was a distinct dependence of the mean cleavage temperature on cluster size, in that both parameters increased in line with each other (Figure 5.16). A similar behavior was also observed for the PEt_3-ligated cluster molecules measured under a helium gas flow. Assuming that these effects were not governed by the size and quality of the crystals of the precursor complexes, this fact might be rather explained by the stability of the different-sized cluster molecules than by any effects arising from different Cu–P binding energies. Calculations have revealed that the stability of these types of cluster should increase with increasing cluster size, which indicates that the tendency to form bulk material decreases in the same direction [20]. This would fit with the measured increase of the mean cleavage temperature of the PEt_3 ligands in **2**, **5** and **6**, if it is assumed that cleavage of the ligands is not determined by the Cu–P bond energy but rather by the stability of the cluster itself. In following this interpretation, the TGA curves for **2**, **5** and **6** depict the thermal stability of the different cluster molecules, whereas for **1**, **3** and **4**, with the higher-boiling PEt_2Ph, the shapes of the curves are determined more by the evaporation of the already liberated phosphine from the crucibles.

Powder X-ray diffraction (XRD) patterns of the black TGA residues have revealed the formation of Cu_2Se for all compounds **1–6**. Concomitant with the disappearance of the strong low 2θ reflections which illustrate a long-range ordering of the crystalline precursor cluster molecules, both weak and broad peaks appear at higher diffractions angles during the thermal treatment. The d-values of the powder patterns of the black residues following the cleavage process for all clusters were in good agreement with those found for monoclinic α-Cu_2Se (Figure 5.17) [24]. The final structural determination of the low-temperature α-Cu_2Se phase remains a point of discussion, however [25, 26].

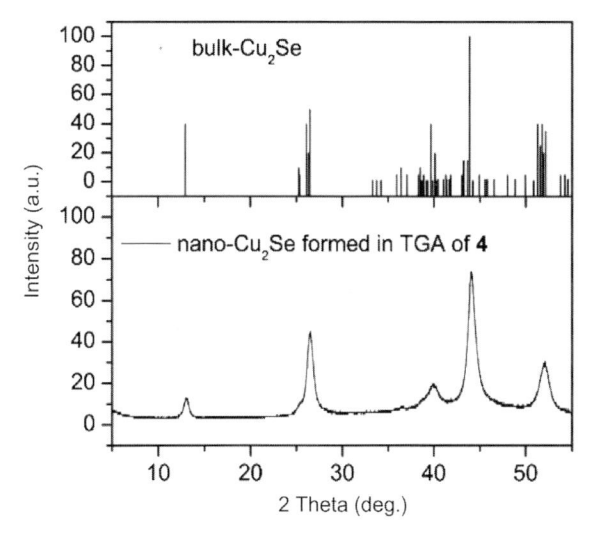

Figure 5.17 X-ray powder diffraction pattern of the black residue of the thermogravimetric analysis of [$Cu_{70}Se_{35}(PEt_2Ph)_{23}$] (**4**) in vacuum (lower graph) and of monoclinic α-Cu_2Se [115] (upper graph).

The relative broadness of the diffraction peaks indicates a nanostructuring of the material. Due to the Scherrer equation, the calculated mean particle size D (for the reflection at $2\theta = 13°$) yielded values between 12 and 16 nm for Cu_2Se powders resulting from a thermal treatment of **1–6** in vacuum up to 150 °C. The cluster cores of **1–6** have edge lengths of 0.8–1.5 nm, which reveals a distinct growth of the precursor cluster molecules with cleavage of the phosphine ligands. The differences in size are difficult to explain, and show no obvious relation to the size of the precursor cluster, although they may result from a difference in size and quality of the precursor crystals. Sintering of the samples at 150 °C for 5 h did not lead to any significant increase in particle size; however, the size of the crystalline domains doubled when the samples were heated to a final temperature of 300 °C.

Subsequent transmission electron microscopy (TEM) measurements of the nanostructured Cu_2Se powders as a suspension in tetrahydrofuran (THF) did not provide any valuable structural information, due to the relative thickness of the material that resulted from agglomeration on the grids. Thus, the thermally treated brittle crystals were embedded in resin and subsequently cut using a microtome. A small (but representative) sample section of a crystal of **4** heated to 150 °C in vacuum with two intergrown crystalline domains of copper selenide with domain sizes of approximately 15 nm, is shown in Figure 5.18. The d-values of the simultaneously recorded electron diffraction patterns were in good agreement with those calculated from the powder XRD patterns.

In conclusion, each of the copper selenide nanoclusters investigated lose their phosphine ligand shells at temperatures ranging from 60 to 200 °C, with the exact temperature depending on the experimental conditions, the type of phosphine ligand, and the size of the cluster itself. Nanostructured Cu_2Se with crystallite sizes

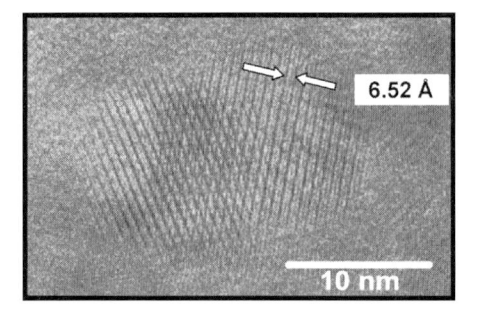

Figure 5.18 Section of a HR-TEM image of a microtome slice of a crystal of thermally treated [$Cu_{70}Se_{35}(PEt_2Ph)_{23}$] (**4**).

of approximately 15 nm are formed during the ligand-dissociation process. The results of TGA analyses suggest that cleavage of the phosphine ligands is most likely determined by the tendency of metastable clusters to form a bulk material.

References

1. (a) Nimtz, G., Marquard, P., and Gleiter, H. (1988) *J. Cryst. Growth*, **86**, 66; (b) Koutecky, J. and Fantucci, P. (1986) *Chem. Rev.*, **86**, 539; (c) Morse, M.D. (1986) *Chem. Rev.*, **86**, 1049; (d) Kappes, M.M. (1988) *Chem. Rev.*, **88**, 369; (e) Weller, H. (1993) *Angew. Chem.*, **105**, 43.
2. (a) Schäfer, A., Huber, C., Gauss, J., and Ahlrichs, R. (1993) *Theor. Chim. Acta*, **87**, 29; (b) Schäfer, A. and Ahlrichs, R. (1994) *J. Am. Chem. Soc.*, **116**, 10686.
3. Schäfer, A. (1994) PhD Thesis, University of Karlsruhe.
4. Foresman, J.B., Head-Gordon, M., Pople, J.A., and Frisch, M.J. (1992) *J. Phys. Chem.*, **96**, 135.
5. McWheeny, R. (1992) *Methods of Molecular Quantum Mechanics*, 2nd edn, Academic Press, London.
6. (a) Raghavachari, K., Trucks, G.W., Pople, J.A., and Head-Gordon, M. (1989) *Chem. Phys. Lett.*, **157**, 479; (b) Møller, C. and Plesset, S. (1934) *Phys. Rev.*, **46**, 618–622.
7. Schwabl, F. (1990) *Quantenmechanik*, Springer, Heidelberg, p. 189f.
8. Voskanyan, A.A., Inglizyan, P.N., Lalykin, S.P., Plyutto, I.A., and Shevchenko, Y.M. (1978) *Fiz. Tekh. Poluprovodn.*, **12**, 2096.
9. Garba, E.J.D. and Jacobs, R.L. (1986) *Physica B*, **138**, 5398.
10. Deveson, A., Dehnen, S., and Fenske, D. (1997) *J. Chem. Soc., Dalton Trans.*, 4491–4497.
11. (a) Dehnen, S. and Fenske, D. (1994) *Angew. Chem.*, **106**, 2369–2372; (b) Dehnen, S. and Fenske, D. (1994) *Angew. Chem., Int. Ed. Engl.*, **33**, 2287–2289.
12. Eichhöfer, A., Beckmann, E., Fenske, D., Herein, D., Krautscheid, H., and Schlögl, R. (2001) *Isr. J. Chem.*, **41**, 31–37.
13. Zhu, N. and Fenske, D. (1999) *J. Chem. Soc., Dalton Trans.*, 1067–1075.
14. Mostafa, S.N. and Schiman, S.A. (1983) *Ber. Bunsenges. Phys. Chem.*, **87**, 113.
15. Lebedkin, S., Eichhöfer, A., Fenske, D., and Kappes, M., unpublished results.
16. (a) Soloviev, V.N., Eichhöfer, A., Fenske, D., and Banin, U. (2001) *J. Am. Chem. Soc*, **123**, 2354; (b) Soloviev, V.N., Eichhöfer, A., Fenske, D., and Banin, U. (2001) *Phys. Status Solidi B*, **224**, 285.
17. Robin, M.B. and Day, P. (1967) *Adv. Inorg. Radiochem.*, **10**, 248.
18. Eichhöfer, A., Corrigan, J.F., Fenske, D., and Tröster, E. (2000) *Z. Anorg. Allg. Chem.*, **626**, 338.
19. van der Putten, D., Olevano, D., Zanoni, R., Krautscheid, H., and Fenske, D. (1995) *J. Electron. Spectrosc. Relat. Phenom.*, **76**, 207–211.

20 (a) Dehnen, S., Schäfer, A., Fenske, D., and Ahlrichs, R. (1994) *Angew. Chem.*, **106**, 786–790; (b) Dehnen, S., Schäfer, A., Fenske, D., and Ahlrichs, R. (1994) *Angew. Chem., Int. Ed. Engl.*, **33**, 746–748.
21 Dehnen, S., Schäfer, A., Ahlrichs, R., and Fenske, D. (1996) *Chem. Eur. J.*, **2**, 429–435.
22 (a) Fenske, D. and Krautscheid, H. (1990) *Angew. Chem.*, **102**, 1513–151;(b) Fenske, D. and Krautscheid, H. (1990) *Angew. Chem., Int. Ed. Engl.*, **29**, 1452–1454.
23 Cave, D., Corrigan, J.F., Eichhöfer, A., Fenske, D., Kowalchuk, C.M., Rösner, H., and Scheer, P. (2007) *J. Cluster Science*, **18**, 157–172.
24 Stevels, A.L.N. (1969) *Philips Res. Rep. Suppl.*, **9**, 39–44.
25 Kashida, S. and Akai, J. (1988) *J. Phys. C: Solid State Phys.*, **21**, 5329–5336.
26 Yamamoto, K. and Kashida, S. (1991) *J. Solid State Chem.*, **93**, 202–211.

5.2
Electrical Properties of Metal Nanoparticles

Kerstin Blech, Melanie Homberger, and Ulrich Simon

5.2.1
Introduction

Today, modern microelectronics tends continuously towards a higher degree of integration. Since the fabrication of the first transistor by Shockley, Brattain, and Bardeen during the late 1940s [1], transistors in microelectronic components have become increasingly smaller. Indeed, this is the basis of all the present-day key technologies that play roles in all areas of day-to-day living.

As miniaturization continues, shorter distances between transistors and related switching elements on a microchip will lead to an increased speed of performance. Likewise, the availability of nanolithographic fabrication techniques has permitted a scaling down to 50 nm or below [2, 3], which in turn has already had a major impact on the performance of traditional semiconductor circuits, and also opened up new possibilities based on quantum effects. But, the same is also true for so-called "metallic" electronics such that today, by exploiting charging effects or so-called "Coulomb effects" in metallic circuits that comprise tunnel junctions with submicron sizes, individual charge carriers can be handled. Today, this field has become known as "single electronics" (SE) [4].

Although charging effects have previously been observed in granular thin metal films [5, 6], SE was itself born during the late 1980s, when ultra-small metal-insulator–metal sandwich structures (tunnel junctions) – and simple systems that included them – were first studied intensely from both theoretical and experimental standpoints [7, 8]. In this situation, the discrete nature of the electric charge becomes essential, and the tunneling of electrons in a system of such junctions can be effected by the coulombic interaction of electrons, which may in turn be varied by the external application of a voltage, or by injected charges. As a consequence, a profound understanding of this phenomenon – termed single-electron tunneling (SET) – was developed. Interest in this field was not purely academic, however, as it would be very attractive to exploit the principles of SE to develop logic and memory cells that, in principle, could lead to the construction of a computer capable of operating on single electrons, realizing a "single-electron logic" (SEL). Today, as the physical limits of production technologies for device fabrication are rapidly being approached, it is clear that a continuous increase in integration density will require new concepts to produce components that are tens of nanometers or less in size, or ideally, at the molecular level.

Different approaches to bridge the size gap between conventionally fabricated circuit elements and truly atomic scale components have been discussed. These include the use of electron transport through single molecules, the concept of molecular rectification, and the design of atomic relays or molecular shuttles. Today, the approach that receives the most attention among the different concepts of futuristic circuit design is circuitry based on single-electron transistors.

The aim of this subchapter on the one hand is to acquaint the reader with charge transport phenomena in nanoparticles, the physical principles of SET, and basic SE structures where the charging effects occur as intended. On the other hand, it will be shown, by means of selected examples, that in further development of the ideas of SE, ligand-stabilized nanoparticles of noble metals have been recognized to be suitable building blocks for single-electron devices [9]. This topic has already been addressed from different views in the first edition [10] of this book, as well as in parts in more recent book chapters and reviews [11–16], respectively. The following sections will therefore summarize and update the present state of knowledge by means of selected examples. It should be emphasized that other academic and technical fields, such as electrochemical properties of metal nanoparticles, biochemical or chemical sensing properties, are not discussed in this subchapter, as these topics are far beyond its scope.

5.2.2
Physical Background and Quantum Size Effect

The availability of small solids such as transition metal clusters [17–19] offers a field of research that raises basic questions: How small must the number of atoms be for the properties of the original metal to be lost? How does an ordered accumulation of atoms behave when it is no longer under the influence of its ambient bulk matter? And, perhaps most important in the context of this subchapter: What types of applications of these new materials in microelectronic devices can be expected in the future?

To answer these questions, the following scenario should be considered. If a metal particle, initially having bulk properties, is reduced in size down to a few hundreds or dozens of atoms, the density of states in the valence and conduction bands decreases and the electronic properties change dramatically – that is, the conductivity, collective magnetism and optical plasmon resonance vanish such that the quasi-continuous density of states is replaced by quantized levels with a size-dependent spacing.

This "size-quantization" effect may be regarded as the onset of the metal–nonmetal transition such that, in another terminology, these quantum-size nanoparticles may also labeled as "artificial atoms" [20–22].

This is the great opportunity for the chemistry of nanoparticles, as the electronic properties are now tunable via the particle size. With respect to the electrical properties of nanoparticles and of all types of arrangement built from nanoparticles, the most important property is the amount of energy required to add one extra electron to an initially uncharged particle [23–26]. This energy – the so-called "charging energy" – scales roughly with $1/r$ (r = radius of the nanoparticle). Since it can be estimated from the charging energy of a macroscopic metallic sphere scaled down to the nanometer range, a physical effect such as the Coulomb blocking of tunneling events, which results from the huge charging energy of metal nanoparticles, is sometimes regarded as a "classical size effect." Based on the concept of charging energy, the physical principle of SET, in terms of classical considerations, is described in the following section.

5.2.2.1 Single-Electron Tunneling

The transfer of electric charges – or, in other words, an electric current – in a metallic conductor is associated with the motion of a huge amount of free electrons over the entire conductor. Because of electron–electron and electron–phonon interactions, the effective charge which passes a cross-section of the conductor per second is a result of the displacement of individual charges. As a result, in spite of the discrete nature of the charge carriers, the current flow in a metal is averaged over a huge amount of charge carriers, and is hence quasi-continuous. In contrast, when dealing with an isolated nanoscale piece of metal, the number of electrons in this is always integral. The number of electrons can be changed, for example, by means of direct connection, followed by disconnection of that nanoparticle to another charged object or a lead with a different electric potential. The resulting change in the number of electrons, which on the macroscopic scale is large and uncontrollable, clearly is also integral. Thus, the question arises as to how electrons may be transferred in a circuit which is built up from nanoscale metallic objects one by one, in a strictly controlled manner.

Physically, such transfer of electrons can be realized by means of quantum mechanical tunneling. The probability of such a tunneling event depends on the potential applied to the circuit, as well as on the distribution of excess electrons over the constituting sites. Therefore, an SET circuit should present a number of reservoirs for free electrons, which should be small and well-conducting islands separated by poorly permeable tunneling barriers (Figure 5.19). As long as the size of these so-called "mesoscopic islands" is larger than the atomic scale, they certainly comprise a huge amount of free charge carriers. However, the handling of individual

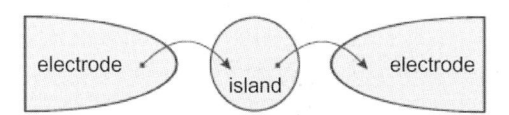

Figure 5.19 The tunneling of a single electron between two metal electrodes through an intermediate island can be blocked if the electrostatic energy of a single excess electron on the central island is large compared to the energy of thermal fluctuation.

charges is still possible if the characteristic electrical capacitance (C) of the island is small enough. In this case, the presence of one excess electron changes the electric potential ΔV of the island considerably, that is, $\Delta V = e/C$. The control of single tunneling events can then be realized in the following scenario. The enhanced potential could prevent further charging of the island, essentially reducing the probability of corresponding tunneling, as it would raise the energy of the whole system. If then, the potential of the neighboring islands and/or gates were to be changed, the probability of tunneling would be restored and the tunneling of strictly one electron would then occur. Simply, it might be concluded that SE deals with a small and defined amount of excess electrons on islands changing their distribution over the islands in time in a desirable way.

In order to realize this practically, the following two principal conditions must be fulfilled [27, 28]: First, the insulating barriers separating the conducting islands from each other, as well as from the electrodes, should be fairly opaque. The barrier properties determine the transport of electrons, which obey the rules of quantum mechanics and therefore are described by wavefunctions. The coordinate dependence of a wavefunction depends on the energy profile of the barrier. If the energy barrier separating the islands is high and thick enough (e.g., 1 eV and 2 nm, respectively), it provides an essential decay of the electron wavefunction outside the island and, as a result, only a weak overlapping of the wavefunctions of the neighboring islands will occur within the inter-island space. If, in addition, the number of electronic states contributing to tunneling is small enough – as is the case for nanoparticles – the total exchange of electrons between the islands becomes negligibly small. This situation is often referred to as the case of small quantum fluctuations of charge. In spite of a relatively complex rigorous quantum mechanical consideration, quantitatively this condition can illustratively be formulated using the tunneling resistance R_T as a characteristic of the tunneling junction. It should be much higher than the characteristic resistance expressed via fundamental constants, the so-called resistance quantum, $R_q = h/e^2 = 25.8\,\mathrm{k\Omega}$. The electrons in the island can then be considered to be localized, and their number already behaves classically, although they undergo thermodynamic fluctuations just like any statistical variable.

Second, in order to have these thermal fluctuations small enough, and consequently to make the exchange of electrons controllable, the energy associated with charging by one extra electron should be much greater than the characteristic thermal energy $k_B T$. This is the above-mentioned charging energy E_C, and it depends on the charge Q, on the size, and on the charge on the capacitances of any junctions, gates, conductors, and so on in the vicinity of the island. As long as the size of an island exceeds the electrical screening length (a few tenths of a nanometer), the geometric dependence of the charging energy is simply incorporated into the parameter C (the total island capacitance), and the charging energy takes the simple form $E_C = Q^2/2C$. Then, the condition for the Coulomb energy to take precedence over the thermal energy is quantitatively expressed by $E_C = e^2/2C \gg k_B T$. Stated simply, the smaller the island, the smaller the capacitance and the larger the E_C, as well as the temperature, at which single-electron charging can be observed experimentally.

When the above-mentioned two conditions are satisfied, the system is characterized by the set of configurations with various distributions of a small integral number of excess charges over the islands. These configurations, or states, of the system can be switched by means of tunneling of individual electrons. The characteristic time scale for the switching between the states due to externally induced single-electron transitions depends on temperature, and also on the energy distribution over the whole system. This is the so-called "recharging time" τ_R, which ranges from about 10^{-11} s for typical values of $R_T = 100\,k\Omega$ and $C = 10^{-16}$ F, up to several hours with extremely large R_T, in accordance with the relationship $\tau_R = R_T C$.

If such an array is connected to a voltage source, the charge Q on the junction will increase with increasing voltage unless a tunneling event takes place. As long as $|Q| < e/2$, the junction is in the Coulomb-blockaded state. At $|Q| \geq e/2$, tunneling is allowed, and will occur after a random amount of time because of the stochastic nature of the process. This is reflected in the current–voltage characteristic of the junction, as the Coulomb blockade appears at voltages below $0.5\,e/C$, and with a corresponding shift of the tunnel current characteristic.

If the array is biased by a constant current I, then at $Q \geq e/2$ tunneling will appear, making Q jump to $-e/2$, such that a new charging cycle starts again. This leads to an oscillation of the voltage across the junction in a saw tooth-like manner, with the fundamental frequency ν_{SET}:

$$\nu_{SET} = I/e$$

In this state, the tunneling of single electrons is correlated (Figure 5.20).

This gives rise to the manipulation of charges in a circuit on the single-electron level, and therefore to the employment of this for the creation of, for example, sensitive amplifiers and electrometers, switches, current standards, transistors, ultrafast oscillators, or, generally, digital electronic circuits in which the presence

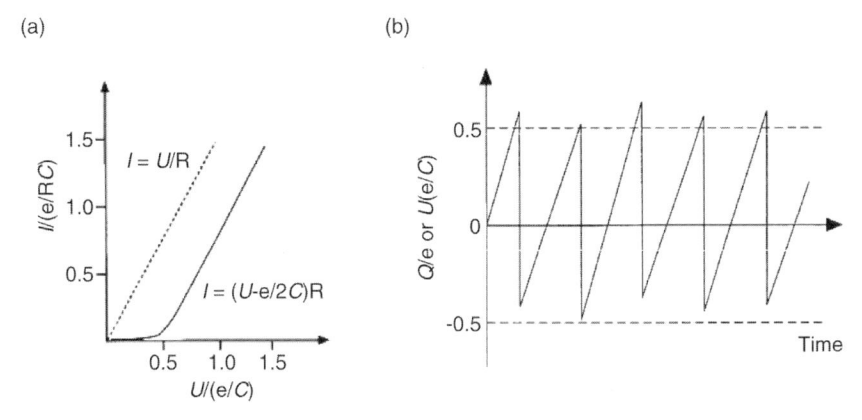

Figure 5.20 (a) The dependence of the time averaged current I on U (the $I(V)$ characteristic) for a single-electron junction. Coulomb blockade appears at voltages below e/C. The dottet line indicates Ohmic behavior; (b) Time development of the junction charge or voltage imposed by a constant current source.

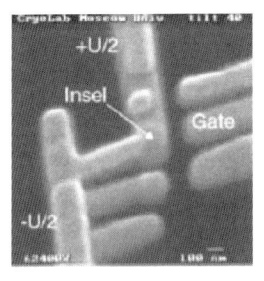

Figure 5.21 Scanning electron microscopy image of a single-electron transistor with source and drain ($+U/2$ and $-U/2$, respectively) feeding the central island (Insel), which is capacitively coupled to a gate electrode. The size of the central island is about 60×60 nm. Reproduced with kind permission from BMBF report INFO PHYS TECH, No. 22, 1998.

or absence of a single electron at a certain time and place provides the digital information.

5.2.2.2 The Single-Electron Transistor

The simplest circuit which reveals the peculiarities of SET comprises only one Coulomb island and two leads (electrodes) attached to it. This geometry presents the double-tunnel junction system; an example of where such geometry is realized is depicted in Figure 5.21. If this system is to function as a simple on/off switch, then a gate electrode must be capacitively coupled to the island (Figure 5.22). Such a system has external control of the state of charge, often called the single-electron transistor. The application of a voltage to the outer electrodes of this circuit may cause either the sequential transfer of electrons into and out of the central island, or no charge transport – that is, the transistor will remain in the nonconductive state. The result

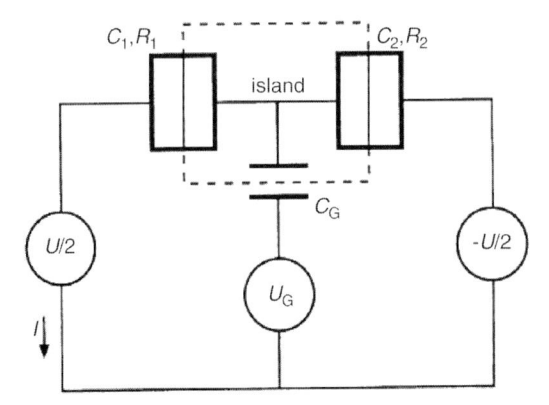

Figure 5.22 Circuit equivalent of a simple single-electron transistor, consisting of a two-junction arrangement and a gate electrode coupled capacitively to the central island. The transport voltage U induces a net flow of charge through the device, the value of current I being controlled be the gate voltage U_G.

depends on the voltage applied to the electrodes, U, as well as on the voltage applied to the gate, U_g.

Because of the small total capacitance of the island, $C = C_1 + C_2 + C_0 + C_{0'}$ (where C_1 and C_2 are the capacitances of the junctions, which for the sake of simplicity will be assumed equal, $C_1 = C_2 = C_J$, while C_0, the capacitance with respect to the gate and $C_{0'}$, the capacitance with respect to other remote conducting objects, will be assumed to be much less than C_J), the island has a large Coulomb energy, and $E_C \gg k_B T$. The total energy change of the system ΔE while one electron is tunneling in one of the junctions, consists of the charging energy of the island itself, as well as of the work done by the voltage source. For $U_g = 0$, this is simply given by:

$$\Delta E = Q_f^2/2C - Q_i^2/2C - qU,$$

where Q_i and Q_f are the initial and final charges of the island, respectively, and q is the charge (not necessarily integral) passed from the voltage source. From the elemental reasoning it is clear that if, say, $Q_i = 0$ and $Q_f = e$, then $q = \pm 2\,e/2$ (depending on the direction of electron tunneling) and, therefore, $\Delta E = e^2/2C \pm eU/2$. This means that for $-e/C$, the change of energy is positive when $-e/C \leq U \leq e/C$. Hence, electron tunneling would only increase the energy of the system, and this transition does not occur if the system cannot "borrow" some energy from its environment, when the temperature is assumed to be low enough. Therefore, there is a Coulomb-blocked state of the single-electron transistor when the voltage U is within the interval given previously and U_g is zero. Outside this range, the device conducts current by means of the sequential tunneling of electrons.

When the gate voltage U_g is finite, the calculation of the energy change due to one electron tunneling gives another value, because of additional polarization of the island electrostatically induced by the gate. The result of this consideration is illustrated in Figure 5.23, where the U versus U_g diagram of the Coulomb blockade is shown.

The diagram reflects an interesting feature: it is periodic with respect to the voltage U_g with a period of e/C_0. This results from the fact that every new "portion" of the gate

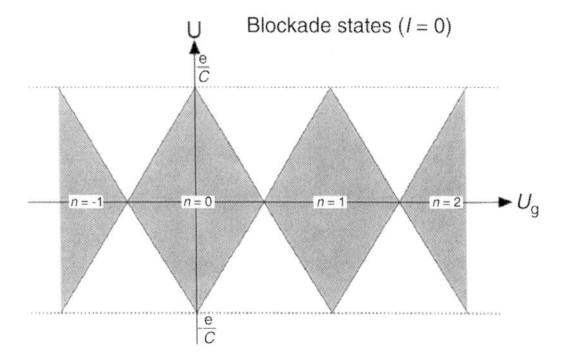

Figure 5.23 Idealized periodic rhomb pattern showing the Coulomb blockade region on the plane of voltages U and U_G. A transistor conducts only outside the rhombic-shaped regions. The value of the number n is the number of extra electrons trapped in the island in the blocked state.

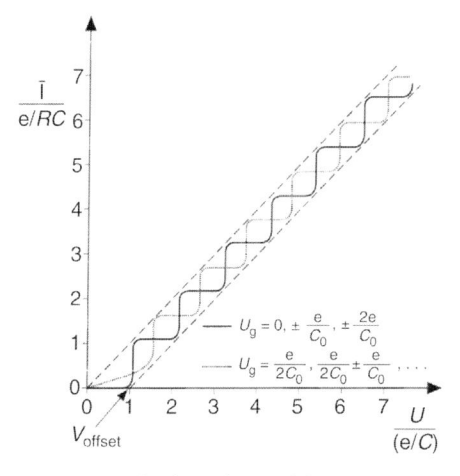

Figure 5.24 The dependence of the time-averaged current I versus U for an asymmetric single-electron transistor.

In the figure: $U_g = 0, \pm \frac{e}{C_0}, \pm \frac{2e}{C_0}$ and $U_g = \frac{e}{2C_0}, \frac{e}{2C_0} \pm \frac{e}{C_0}, \ldots$

voltage of e/C_0 is compensated by one extra electron on the island, and it returns to the previous conditions for electron tunneling. Therefore at constant bias U, the current through the device is alternately turning on and off with a sweep of voltage U_g.

The behavior of the transistor outside the Coulomb blockade region also shows the single-electron peculiarities, especially for the case of a highly asymmetrical junction, where $R_1 \gg R_2$, and $C_1 \gg C_2$. The $I(V)$-characteristics, with a step-like structure fading with decrease of U, can be seen in Figure 5.24. This so-called Coulomb "staircase" results from the fact that an increase in U increases the number of channels for tunneling in a step-like manner, allowing an increasing larger number of electrons to be present on the island. Another manifestation of charging effects in the SE transistor is the offset of the linear asymptotes by e/C.

The dependence of the transistor current on the gate voltage opens up the opportunity to fabricate a sensitive device which measures either directly an electric charge on the island or the charge induced on the island by the charges collected to the gate, that is, a highly sensitive electrometer. The sensitivity of such an electrometer, which has already been achieved in practice, is of the order of 10^{-4} to 10^{-5} parts of electronic charge, exceeding by far the charge sensitivity of conventional devices. It will be seen later that even the association of charges due to adsorption of molecules on the coulomb island causes the charge transport through such a single-electron transistor.

5.2.3
Thin Film Structures

It is almost two decades ago that the fabrication techniques to produce "traditional" single-electron circuits (e.g., transistors) were developed [29, 30], and these represented the first steps towards a new device concept. Respective techniques are based on the evaporation of metal (usually Al) films through a fine-pattern mask, which is often made from a polymer resist, by using an electron-beam lithographic process. The mask provides a number of thin splits which determine the sizes and shapes of the resulting wires and islands, while the bridges that interrupt them determine the position of the desired tunnel junctions and gates. Because of the composite polymer resist layer, which is chemically under-etched, the resultant mask is rigidly fixed above the substrate, while its bridges become suspended above. These arrangements of splits and bridges enable metals to be evaporated onto the substrate from various angles, which allows the resulting evaporated layers to be placed in different positions.

In order to form a set of islands connected by tunnel junctions, the evaporation is usually carried out from two or three different angles in two or three steps, respectively, with an intermediate oxidation process, when aluminum is used. By adjusting the angles, the linear sizes of the resulting overlapping area of the first and the second metallic layers may be even smaller than the width of the strips – that is, somewhat less than 100 nm. At the same time, a gate electrode coupled to the islands in only capacitive fashion should not overlap them, and this must be taken into account in the mask pattern design. After double evaporation, the mask and supporting layer are lifted off. Figure 5.21 shows such a single-electron transistor fabricated as an Al–Al_2O_3–Al stack.

The resulting barrier of Al_2O_3, with a desirable thickness of a few nanometers, is mechanically and chemically stable. The height of the corresponding energy barrier is about 1 eV, and this readily provides a tunnel resistance in the range of 100 kΩ for junctions of the above-mentioned size. The resulting capacitance of such tunnel junctions is about 10^{-16} F, which means that such circuits will operate reliably only at temperatures below 1 K which, although attainable, are reached only with immense technical effort.

For an essential decrease in tunnel junction size and corresponding increase in operating temperature, more sophisticated technologies are required for the fabrication of SE structures. Meanwhile, advances in lithographic methods have opened the way to much smaller structures. Nevertheless, with the decreasing size the technical effort is increased drastically. Thus, an alternative route would be to use ligand-stabilized metal nanoparticles as building blocks.

5.2.4
Single-Electron Tunneling in Metal Nanoparticles

The theoretical background of SET has achieved experimental manifestation in lithographically fabricated capacitors, as found in conventional computer circuits. These typically have capacitances in the pF range but, because of their extremely low charging energies, they would need to be cooled down to the sub-mK range for single-electron operation. Furthermore, such a capacitor is typically driven at an operating voltage of 10–100 mV, which would lead to the storage of a few tens of thousands of electrons per charging.

However, SET can also be observed on ligand-stabilized metal nanoparticles in a size range of a few nanometers [26]. In this case, the surrounding ligand shell, which typically consists of organic molecules, plays the role of an insulating layer in contact not only with neighboring clusters but also with conducting objects. The use of such particles as building blocks for single-electron circuits of different complexities requires the development of techniques for a controlled organization. To date, approaches to achieve this requirement have referred to self-assembly processes upon controlling the intermolecular interaction. The basic principles of these self-assembly processes represent one of the main interests of supramolecular chemistry, and appear as one of the key features in the development of SET devices utilizing nanoparticles.

Although, in general, the small size of nanoparticle makes the number of electrons countable, this requires a slight modification of the two principal conditions for "classical" SET:

- In the standard (macroscopic) electrostatic approach – where a conductor is treated as a continuous entity with an infinitely thin screening depth – will fail as the latter reaches almost the size of the object, and the number of weakly bound or metallic electrons in the object is small. As a result, the electron–electron interaction is not completely screened out, so that the concept of electrical capacitances cannot be strictly applied. In this case, calculation of the charging energy should be carried out by counting the energy of interacting charges. However, the Coulomb energy can still be roughly described by the same elemental formula for the charging energy of a capacitor. The symbol C will then denote a capacitance that generally depends on the total number of interacting electrons occupying the nanoparticle.
- Because of the small size of a nanoparticle, the energy spectrum is changed essentially because of the above-mentioned quantum size effect. In contrast to a macroscopic piece of metal, which has a quasi-continuous electron energy spectrum (as assumed in the discussion of charging effects in larger structures), a metal cluster presents for electrons a small potential energy well which, in accordance with quantum mechanics, has a set of distinctly discrete energy levels, the density and distribution of which depend on the size and shape of the cluster and, consequently, the tunnel junctions. When one of its electrodes is a nanoparticle, a tunnel junction can hardly be characterized by such a simple parameter as a constant tunneling resistance R_T. Such an approximation is valid when densities of levels in both electrodes are high, and independent of energy. In clusters of 1–2 nm, the level spacing appears, and is on the scale of the charging energy itself.

These peculiarities of metal nanoparticles do not, however, eliminate charging effects in them. Both, theoretical and experimental investigations have shown that the fundamental results of SET still hold true qualitatively. For example, the current–voltage characteristic of a double-barrier junction with a quantum dot (QD) possesses at low temperature, as well as the Coulomb staircase, a fine structure due to energy quantization inside the QD. Nevertheless, the interpretation of these results is complicated by the fact that the characteristic time $R_T C$ becomes as short as the characteristic time of the energy relaxation inside the QD, and consequently the electrons no longer have the pure Fermi distribution. Thus, the tunneling characteristics of single nanoparticles provide information about their electron energy spectrum, and hence the density of levels and their degeneracy.

The experimental set-ups that allow the probing of these tunneling characteristics of metal nanoparticles involve nanometer-separated electrodes, fabricated by lithographic techniques, scanning tunneling microscopy (STM), in which the microscope tip acts as one electrode, and electrochemical methods. Recent reviews have provided a detailed description of how electrochemical experiments can be used to provide data concerning the tunneling characteristics of metal nanoparticles that are analogous to that obtained by nanogap or STM configurations [31, 32]. Due to the limited space the electrochemical approach will not be discussed here and the following sections focus on STM and nanogap configurations.

5.2.4.1 STM Configurations

In STM measurements, a sharp metallic tip is brought close to a nonisolating sample until a tunneling current is detected. STM measurements are performed either by keeping the distance between the tip and sample constant while measuring the resulting tunneling current (constant-height mode), or by measuring changes in the tip-to-sample distance at a constant current (constant-current mode). The tunneling current depends exponentially on the tip-to-sample distance, and this forms the basis for imaging the topography of samples. For scanning tunneling spectroscopy (STS) investigations, the tip-to-sample distance is usually fixed such that, depending on the tip-to-sample bias, the tunneling current and differential conductance can be measured. The spectroscopic data obtained provide information concerning the local electronic characteristics of the sample, such as the density of states (DOS) [33].

In the STM configuration, metal nanoparticles are immobilized on a conducting substrate with a thin insulating spacer layer (e.g., represented either by a thin oxide layer, or organic ligand molecules) between the nanoparticle and the conducting substrate, with the tip positioned immediately above the metal nanoparticle. Thus, a double-barrier tunneling junction (DBTJ) (see Figure 5.25) is formed. In the case that the metal nanoparticle is electronically coupled only weakly to the leads (conducting substrate and tip) – that is, when the resistance of the barriers (the thin insulating layer between the conducting substrate, R_2; see Figure 5.25) is much higher than the quantum of resistance $h/2e^2$ (the resistance which must be overcome to add/remove a single electron to the cluster), then the physics of the electron transport can be described by the "orthodox" theory of SET. The high resistance of the barriers is essential to ensure that the nanoparticle has a well-defined number of added/removed electrons. Due to the geometry of the STM configuration, the tunneling rate between the particle and the substrate is fixed, while the resistance (R_1) – and therefore the tunneling rate between the tip and the nanoparticle – can be adjusted by moving the tip further or closer to the nanoparticle. In the limiting case that the tunneling rate between the tip and the nanoparticle is much higher than that between the nanoparticle and the substrate, the electrons tunnel through the nanoparticle one-by-one such that the number of electrons accumulating in the nanoparticle will

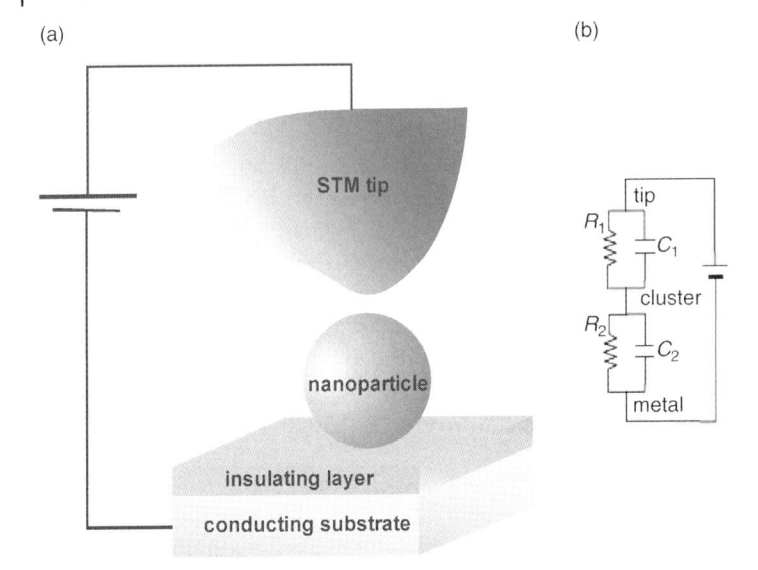

Figure 5.25 (a) General scheme of the STM configuration for the investigation of SET on individual metal nanoparticles; (b) Corresponding electronic circuit (C_1 = tip/nanoparticle capacitance, C_2 = nanoparticle/substrate capacitance). Reprinted with permission from Ref. [37]; © 2007, Elsevier.

be zero; this results in a linear current increase with the bias voltage in the $I(V)$ spectrum. In the case that the tunneling rate between the tip and nanoparticle is much higher than that between the nanoparticle and substrate, the electrons will accumulate in the nanoparticle such that the electron occupancy will depend on the bias voltage between the tip and the substrate. In this case, Coulomb charging effects become visible in the $I(V)$ characteristics. The tunneling resistance R_2 is determined by the resistive properties of the insulating layer and the contact area. The contact area may vary between different clusters of the same size and, as clusters have different facets (squares and triangles), the cluster capacitance C_2 will depend on the exact way in which the cluster lies on the substrate. Furthermore, in the case of STS measurements of ligand-stabilized clusters, the ligands may have different orientations that vary from cluster to cluster with respect to the underlying substrate. This will cause a different tunneling barrier between the cluster and the ground, and thus a different capacitance. Finally, residual water molecules from the solvent may be bound physically or chemically onto either the cluster surface or the ligand shell, such that the tunnel junction between the cluster and substrate may differ for different clusters.

One advantage of STM/STS measurements is, that the individual cluster can be imaged and analyzed spectroscopically in one single experiment, which ensures that reliable data are acquired from individual clusters.

The first investigations on SET events in metal nanoparticles, using the above-described STM configuration, date back to the 1990s. One of the earliest examples was an STM investigation of approximately 4 nm gold nanoparticles, obtained by metal evaporation and deposition onto a 1 nm-thick layer of ZrO_2 as insulating layer on a flat gold substrate [34]. The experimentally determined capacitance at room temperature of the nanoparticle–substrate junction in this case was about 10^{-18} F, a value which was in good agreement with the theoretically estimated value based on the model of a parallel-plate capacitor with ZrO_2 as the defined dielectric. Subsequently, Dorogi et al. [35] improved this STM set-up by using self-assembled monolayers (SAMs) of xylene-α,α′-dithiol as the insulating layer. This allowed the covalent attachment of approximately 2 nm gold clusters to the surface, via the formation of a sulfur–gold bond, such that the cluster position could be maintained in a fixed position during the measurements enabling reliable studies of the electrical transport. The tunneling current as a function of applied voltage, provided evidence of SET at room temperature, and by fitting the $I(V)$ curves to a Coulomb blockade model, the electrical resistance of the xylene-α,α′-dithiol molecules was estimated at about 9 MΩ, which was in good agreement with theoretical expectations of 12.5 MΩ for these molecules [36]. In a very recent example, van Haesendonck and coworkers [37] reported on the systematic and reproducible observations of Coulomb charging effects for individual naked gold clusters of approximately 2 nm, observed using low-temperature ($T = 78$ K) STS measurements. The gold clusters were produced by a laser vaporization source and, in analogy to the approach of Dorogi et al., deposited on a Au(111) film covered with a SAM of 1,4-benzenedimethanethiol (BDMT) as the insulating layer. Subsequently, the current–voltage spectra were obtained for several hundreds of individual gold clusters. Based on the formula of a parallel-plate capacitor, $C_2 = \varepsilon_0 \varepsilon_r A/d$, where ε_0 is the dielectric constant of the vacuum, ε_r the relative dielectric constant the BDMT layer, A the contact area, and d the interlayer thickness of 0.8 nm (determined by ellipsometry), the cluster surface capacitance was estimated to be in the order of 10^{-19} F, which fitted well with theoretical expectations. Almost all clusters showed the presence of Coulomb blockade effects, with a step width ranging between 200 and 600 mV (again, in the theoretical expected range). Furthermore, for a small fraction of the investigated gold clusters, apart from the equidistant Coulomb charging peaks, additional nonequidistant peaks in the order of 100 mV in the $dI/dV(V)$ data (Figure 5.26) were observed. These peaks, which could not be reproduced by the semiclassical orthodox theory, were considered to indicate the presence of discrete energy levels.

Very early results obtained on single ligand-stabilized nanoparticles were reported by van Kempen et al. in 1995 [38, 39], with details of STM measurements at 4.2 K on a $Pt_{309}phen_{36}O_{20}$ cluster synthesized by Schmid and coworkers (Figure 5.27). In this case, the $I(V)$ characteristics exhibited clear charging effects, which indicated that the ligands were acting as sufficiently insulating tunnel barriers between the cluster and the substrate. The experimentally observed charging energy was in the range of 50–500 meV, while a value of 140 meV was expected, assuming a continuous density of state in the particles and a dielectric constant of $\varepsilon_r = 10$ for the ligand molecules when the formula for the charging of a metallic sphere with $E_C = e^2/4\pi\varepsilon_0\varepsilon_r R$ (where R is the radius of the particle) was applied.

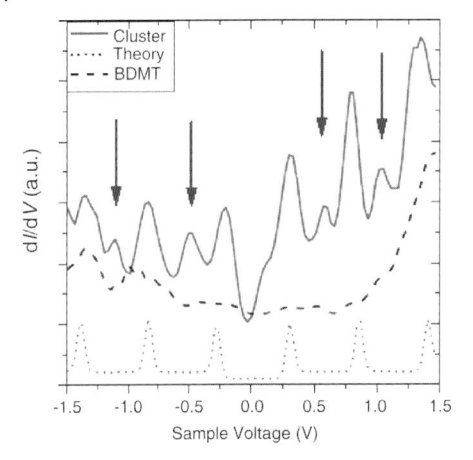

Figure 5.26 Conductance curves $dI/dV(V)$ taken on an individual cluster deposited on a self-assembled monolayer of 1,4-benzenedimethanethiol (BDMT) on Au(111). Both, equidistant Coulomb charging peaks and nonequidistant peaks (indicated by the arrows) are observed. The latter peaks are attributed to discrete electronic levels caused by the electron confinement inside the cluster ($T = 78$ K, $V_{set} = -1.5$ V, $I_{set} = 0.2$ nA). The lower curve shows the fitting result to the charging peaks of the experimental data by the orthodox theory based mode at 78 K. Parameters are: $R_1 = 6.5$ GΩ, $R_2 = 10$ MΩ, $C_1 = 2.9 \cdot 10^{-19}$ F, $C_2 = 2.5 \cdot 10^{-19}$ F. Reprinted with permission from Ref. [37]; © 2007, Elsevier.

In analogy to the above-mentioned results of van Haesendonck and coworkers, additional structures on the charging characteristics have been observed in some cases, which might be expected from a discrete electron level spectrum in the cluster caused by the quantum size effect. The effect of discrete levels on charging characteristics has been treated theoretically by Averin and Korotkov [40], who extended the existing "orthodox" theory of correlated SET in a double normal-metal tunnel junction to the case of a nanoscaled central electrode. According to this, the $I(V)$ characteristics should exhibit small-scale singularities reflecting the structure of the energy spectrum of the central electrode, that is, the nanoparticle. Furthermore, the energy relaxation rate becomes evident because of the small recharging time, τ, resulting from the small junction capacitance according to $\tau = R_T C$, where R_T is the

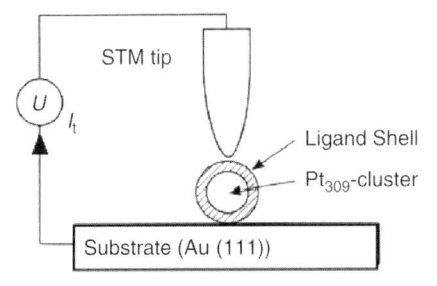

Figure 5.27 A single ligand-stabilized Pt_{309}-cluster between a STM tip and a Au(111) facet. The junction between the cluster and the substrate is built up by the ligand shell.

resistance of the respective tunnel barrier and C is the capacitance of the junction. While the theoretical prediction for the level splitting Δ according to $\Delta = 4E_F/3N$ or to a refined approach by Halperin [41] leads to an assumed splitting of about 8 meV (E_F is the Fermi energy of the metal, N is the number of free electrons), the fit of the spectra lies between 20 and 50 meV. Although different reasons were discussed in the cited report, the measurements reflected the discreteness of the level spectrum in the clusters as large as the 2.2 nm for Pt_{309}.

Experiments in the room temperature range have been performed on smaller ligand-stabilized metal clusters as for example Au55(PPh3)12Cl6 (1.4 nm). Already, cluster pellets (i.e., 3-D compacts of the clusters), have provided hints of single-electron transfer in the $I(V)$ curves [42] taken at room temperature. However, in these investigations neither the vertical nor the lateral arrangement of the clusters was well defined, as would be the case in a 1-D, 2-D or 3-D superlattice. As a consequence, the charging energy had a wider spread, as might be expected from the estimation of the charging energy of a single cluster according to the above-mentioned formula.

Chi et al. studied tunneling spectroscopy on Au_{55} monolayers prepared on various technically relevant substrates [43]. These samples were obtained using a two-step self-assembly (SA) process and a combined Langmuir–Blodgett/SA process. The spectroscopy provided clear evidence of the Coulomb blockade that originated from the double barrier at the ligand-stabilized cluster as the central electrode, up to the room temperature range.

From a fit of the experimental data at 90 K (Figure 5.28), the capacitance of the cluster was calculated as 3.9×10^{-19} F. This value, which is very sensitive towards residual charges and nearby background charges, is close to that of the microscopic

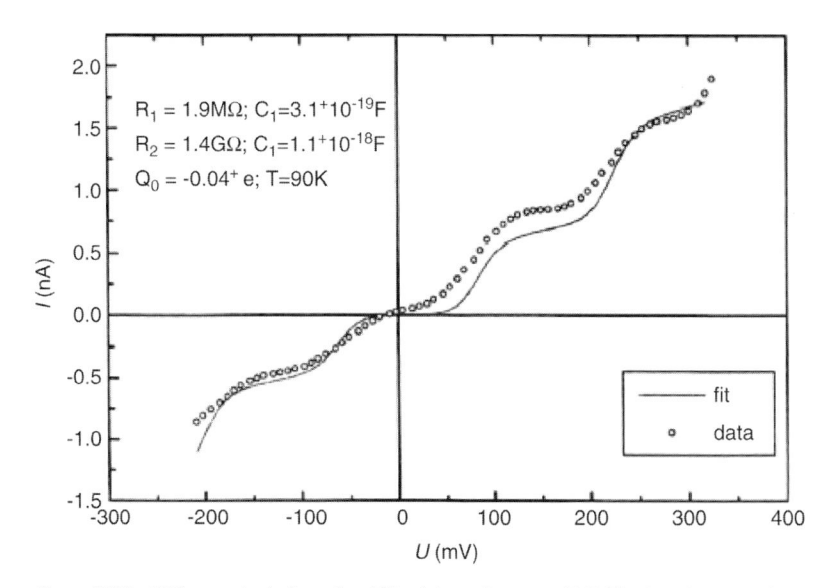

Figure 5.28 SET on a single ligand-stabilized Au_{55}-cluster at 90 K. The junction capacitance was calculated to be 3.9×10^{-19} F by fitting. Reprinted with kind permission from Ref. [43]; © 1998, Springer Science + Business Media.

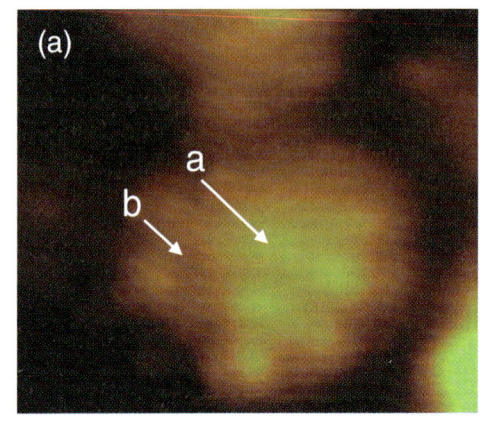

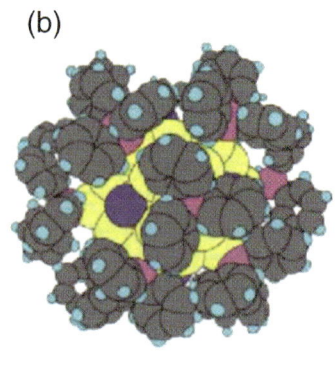

Figure 5.29 (a) STM image of a single Au$_{55}$[P(C$_6$H$_5$)$_3$]$_{12}$Cl$_6$ cluster on a Au(111) surface at 7 K, obtained at a bias of 2 V and a current of 100 pA using a Pt–Ir tip. Image size: 3.3 × 2.9 nm; (b) Space-filling model of the cluster compound: cubo-octahedral core with 55 Au atoms (yellow), 12 P(C$_6$H$_5$)$_3$ molecules (with P pink, C gray and H blue) bound to the 12 edges of the cubo-octahedron and six Cl atoms (violet) located in the center of the six square faces of the Au core. A comparison of (a) and (b) shows that the STM resolves the C$_6$H$_5$ rings. Spectroscopic data were acquired at the two distinct locations marked in panel (a). Reprinted with permission from Ref. [44]; © 2003, American Chemical Society.

capacitance determined previously via temperature-dependent impedance measurements [9].

Low-temperature tunneling spectroscopy on individual Au$_{55}$ clusters under ultra-high-vacuum conditions has also been reported [44]. In accordance with the above-mentioned results, clear evidence of the Coulomb blockade was given. In this regime, the conductivity appeared to be largely suppressed, but is not zero – a fact attributed to a certain probability of co-tunneling within the Coulomb gap at the finite temperature of 7 K. At this temperature, thermal motion would be sufficiently reduced, and the molecular structure of the ligand shell partly visible. The STM image fits with the space-filling model of the cluster fairly well; both are shown in Figure 5.29. In this figure, the two locations at which the tunneling spectra have been recorded are indicated. One location is just above a C$_6$H$_5$ ring of the PPh$_3$ ligand, and the other is next to the ring. The conductivity peaks, which precisely coincide for both spectra, are shown in Figure 5.30. This shows that the discrete energy levels of the cluster become visible in terms of conductivity oscillations with an average level spacing of 135 meV. If this value is compared to that expected for a simple free-electron model (as described above), the "electronical apparent" cluster diameter is about 1.0 nm – significantly smaller than the geometrically determined diameter of 1.4 nm for Au$_{55}$, and slightly larger than the expected value of 0.84 nm for Au$_{13}$. The authors claimed that these differences were most likely caused by the six Cl atoms located at the six square faces of the cubo-octahedral Au$_{55}$ surface. As Cl has a high electronegativity, it removes one electron from the Au$_{55}$ core; nevertheless, within the discrete energy levels the cluster exhibits a metallic behavior.

In a recent report, Xu *et al.* [45] described the STS investigation of hexanethiolate-protected gold nanoparticles sized between 3.2–6.3 nm and 11.8 nm, deposited on a

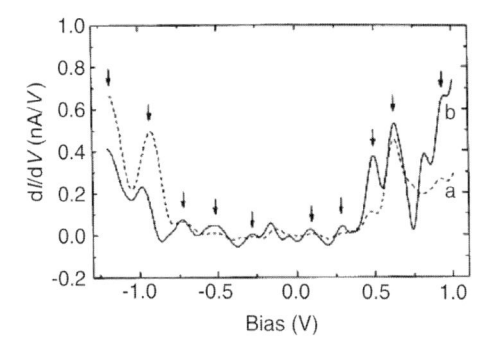

Figure 5.30 Tunneling spectra acquired at the two distinct locations marked in Figure 5.29a. The dashed curve was taken right above the C₆H₅ ring, and the solid curve next to the ring. The bias refers to the substrate potential. The arrows indicate conductivity peaks which coincide precisely for both spectra. Reprinted with permission from Ref. [44]; © 2003, American Chemical Society.

SAM of hexanethiol on a gold substrate. Also reported was the dependence of the tunneling resistance upon exposure of the nanoparticle/SAM/gold structure to organic vapors. The organic vapor molecules were shown to penetrate into the nanoparticle–SAM interface, and thus to modify the tunnel junction. The results indicated that, for particles with a core diameter of approximately 6 nm, the $I(V)$ profiles exhibited a rather sensitive variation upon exposure to various organic vapors, and this was reflected by a drastic enlargement of the Coulomb gap with increasing vapor concentration and decreasing vapor relative polarity. These results highlighted the strong link to the development of nanoparticle-based chemiresistors for chemical vapor detection, leading to the concept of chemical gating (as described in the following subsection).

5.2.4.2 Chemical Switching and Gating of Current Through Nanoparticles

One significant advantage of incorporating nanoparticles into single-electron devices is the opportunity to control the size of the particles chemically, as well as the thickness, composition, and state of charge of the ligand shell. The current flow through such a device will thus be very sensitive to any charges and impurities that reside on the nanoparticles or in the ligand shell. This sensitivity can be used to switch the "transparency" of the ligand shell, or to use the ligands as "chemical gates" to manipulate the SET current.

Schiffrin and coworkers have described how to control the transparency of the insulation barriers between a substrate and a nanoparticle in a STS experiment [46] (Figure 5.31). In this case, a bipyridyl moiety (viologen group, V^{2+}) was used as a redox group incorporated in the ligand shell of the particle. Electrons were incorporated into this group under electrochemical control, while the transparency of the insulating barrier was measured by STS. It was found that reduction of V^{2+} to the radical $V^{\bullet+}$ led to a significant decrease in the barrier height; any further reduction to V^0 resulted in a very large increase in the barrier height. This result reflected the supporting effect of a half-filled molecular orbital in $V^{\bullet+}$. This situation

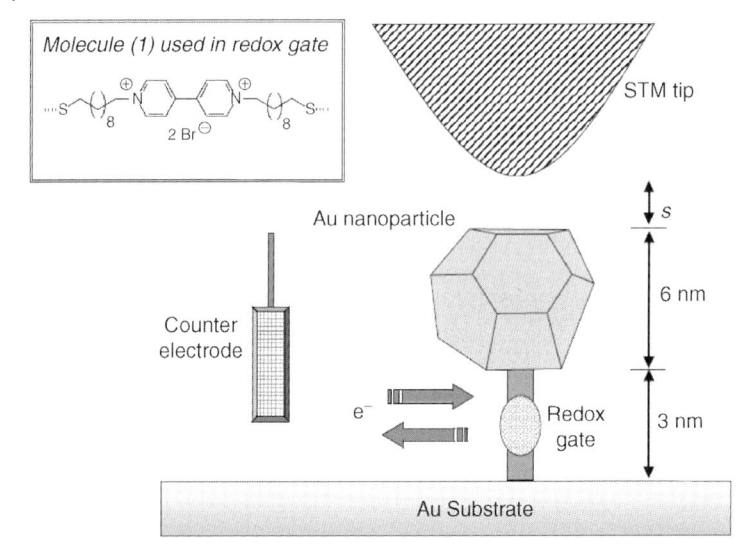

Figure 5.31 Scheme describing the redox switch, which is based on a viologen redox center incorporated within the nanoparticles ligand shell. Reprinted with permission from Ref. [46]; © 2000, Macmillan Publishers Ltd.

might lead to an extension of the electronic wave function from the nanoparticle to the substrate via the orbitals of the radical $V^{\bullet+}$. As soon as V^0 is formed, electron pairing in the LUMO (lowest unoccupied molecular orbital) suppresses the direct electronic interaction.

These results show impressively that switching of the SET current through a ligand-stabilized nanoparticle can be induced by electron injection into a specific redox group within the barriers of the tunnel junction. While the configuration, which was studied in these studies, requires reduction by at least 30 electrons to change the transparency of the barrier in the nanoparticle layer, a major challenge will be to integrate these switching elements into a self-assembled SET circuit.

Feldheim et al. have been considering the possibility of employing particle-capping ligands as "chemical gates" to control the SET current flow in an STS configuration [47]. These authors reported the $I(V)$ characteristics of ligand-stabilized Au nanoparticles in aqueous solutions, whereas octanethiol-stabilized 5 nm particles (C8-Au) and galvinol-stabilized 3 nm particles (Gal-Au) were used as the central island. Galvinol represents a pH-active probe with $pK_a \sim 12$, whereas about 15 galvinol ligands were introduced into the ligand shell. Upon increasing the pH, galvinol, however, was converted to the galvinoxide anion, causing the ligand shell to be charged negatively. Feldheim et al. have presented a collection of $I(V)$ characteristics for galvin and octanethiol nanoparticles in H_2O with several clear current steps and voltage plateaus, reflecting SET in the individual particles. While only small differences in the positions and magnitudes of the staircase voltage plateaus appeared for octanethiol-stabilized Au nanoparticles upon changing the pH, galvin stabilized Au nanoparticles was seen to react sensitively when the pH was changed from 5 to 8, 10,

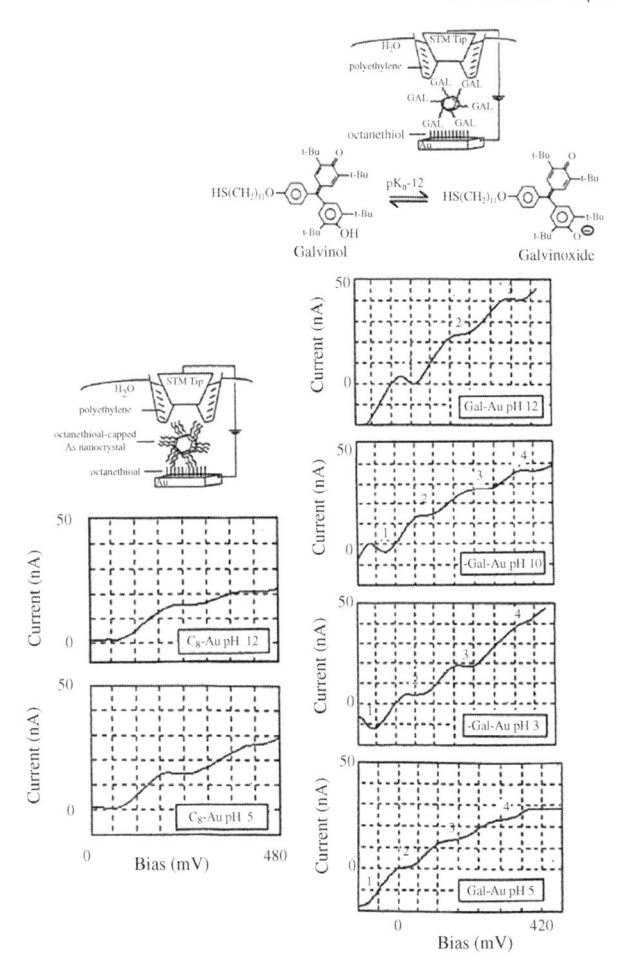

Figure 5.32 Current–voltage curves for C_8-Au (left) and Gal-Au (right) in H_2O as a function of pH (adjusted with phosphate buffer, see Ref. [6]). The numbers 1–4 in the Gal-Au data identify voltage plateaus. Cartoons of the experimental arrangements for measuring I–V curves of individual nanoparticles in solution are shown at the top of each data column. The insulated STM tip, ligand-capped Au nanoparticle and octanethiol-coated planar Au substrate are shown. Length and shapes are not to scale. Reprinted with permission from Ref. [47]; © 1998, American Chemical Society.

and finally to 12 (Figure 5.32). The authors pointed out that, first, a subtle shift (ca. 30 mV) in the entire staircase to positive bias potentials was noted from pH 5 to pH 8. The shift was even more pronounced in $I(V)$ curves obtained at pH 10 and 12 (from about 60 to 120 mV). Second, $\Delta V = e/C$ decreased in magnitude with increasing pH. Third, slight peaks were evident in the $I(V)$ curves at pH ≥ 8. These chemically induced changes, which led to an increase in the negative charge on the cluster, have two consequences: (i) the total self-capacitance of the nanoparticles decreases from about 2.2×10^{-18} F at pH 5 to circa 0.31×10^{-18} F at pH 12; and (ii) the induced negative charge, by forming galvanoxide, acts like a negatively charged gate in solid-state SET devices, causing a shift of the Coulomb blockade to positive bias potentials. It was concluded that, since potential shifts and capacitance changes were observed only for galvin-stabilized Au and not for octane thiol-stabilized Au, the concept of pH-gated SET would be supported.

In a recent study conducted by Albrecht et al. [48], monolayers of hexylmercaptan- and 4-mercaptopyridine-protected Au_{145} nanoparticles immobilized on single-crystal Pt(111) electrodes were investigated using differential pulse voltammetry (DPV; Figure 5.33b, red curve) and electrochemical STM (Figure 5.33a, black curve). In accordance with others, the authors observed clearly sequential charging events with an average peak spacing of 0.148 V, which led to a capacitance of 1.08 aF per nanoparticle. The charging steps were also observed in the tunneling spectroscopy investigations, though their position and intensity varied from scan to scan, as the STM tip slowly drifted over the substrate surface. However, the abundance distribution of peak positions (Figure 5.33b, inset) shows clearly that the peaks are not randomly distributed, but rather accumulate at certain potentials.

High particle charges (increasing overpotential η, with $\eta = E_s - E°$, whereby E_s = substrate potential, $E°$ = equilibrium potential of the "redox"-species) were observed, in addition to an increasing tunneling rate constant; this was explained by dielectric saturation effects in the immediate surroundings of the charged nanoparticles. Based on these findings, it was suggested that the individual nanoparticles could be operated as multistate switches in condensed media at room temperature.

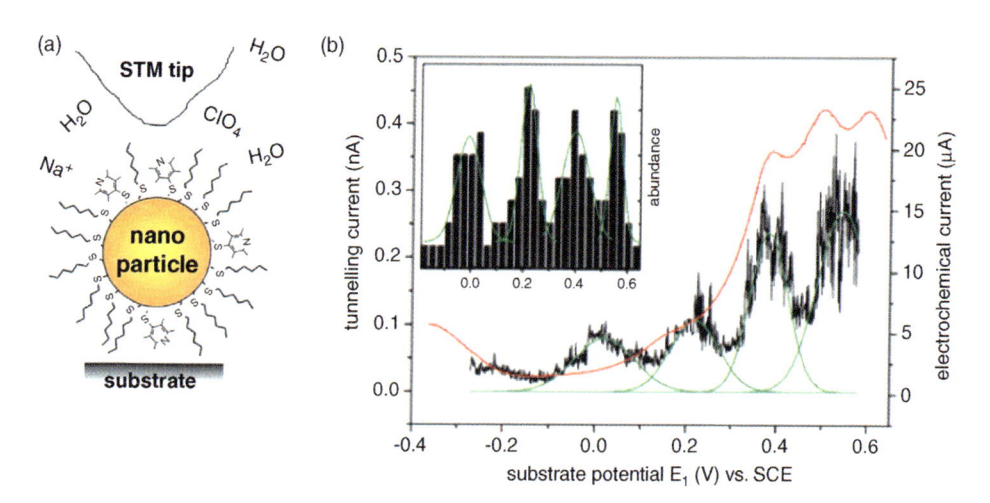

Figure 5.33 (a) Monolayer-protected nanoparticle in an electrochemical in situ STM configuration; (b) Sequential charging events observed in DPV (red) and $I_t(\eta)$ tunneling spectroscopy (black) at constant $V_{bias} = 0.05$ V ($I^0_{set} = 0.05$ nA). Inset: Abundance distribution of $I_t(\eta)$ peak positions (binning: 0.03 V), electrolyte: 0.1 M $KClO_4$ (tunneling spectroscopy) or 0.1 M $NaClO_4$ (DPV). Reprinted with permission from Ref. [48]; © 2007, American Chemical Society.

Taken together, these results showed that SET was not only dependent on the nature of the ligands stabilizing the nanoparticles, but also on the composition of the solution surrounding the function array. These points would be useful when studying chemical signal transduction, where SET currents should be sensitive to single redox or analysis events.

5.2.4.3 Individual Particles and 1-D Assemblies in Nanogap Configurations

Nanometer-spaced electrodes – so-called "nanogaps" – can be used to study the $I(V)$ characteristics of nanoparticles or nanoparticle assemblies. Nanogap configurations with gap separations of <10–100 nm can be fabricated using electron-beam lithography (EBL) and metal evaporation. In this case, the electrodes typically consist of an adhesion layer of Cr or Ti a few nanometers thick, and an overlayer of metal (typically Au) a few tens of nanometers thick. Furthermore, gaps of <10 nm can be created by, for example, mechanical breaking or electromigration [33]. In the electromigration process, a pair of thick electrodes is first fabricated by lithography, utilizing a shadow mask, after which the gap is bridged by a thin layer of a metal. On applying a gradually increasing current, a controlled break of the bridge is achieved, which results in a gap with a width of a few nanometers. Nevertheless, it should be noted that a number of challenges associated with the use of nanogap configurations for probing the tunneling characteristics of metal nanoparticles remain. First, due to the fabrication process involving the nanogap formation and the immobilization of the metal nanoparticles into the gap, device-to-device variations can be large and require the statistical analysis of a large number of samples. Furthermore, any contact between the electrode and the nanoparticle should be both robust and reproducible. Moreover, the atomic-scale details of the nanoparticle-electrode-contact are not well known.

An early example was provided by Alivisatos and coworkers in 1996, who realized an electrode structure scaled down to the level of a single Au nanoparticle [49]. To fabricate this structure, the techniques of optical lithography and angle evaporation were combined, such that a narrow gap of a few nanometers between two Au leads on a Si substrate could be defined. The Au leads were functionalized with hexane-1,6-dithiol, which binds linearly to the Au surface. These free ends which face the solution were used to immobilize 5.8 nm Au nanoparticles from solution between the leads. The resulting device revealed slight current steps in the $I(V)$ characteristic at 77 K (see Figure 5.34). By means of curve fitting to classical Coulomb blockade models, the junction capacitances were found to be 2.1×10^{-18} F, and the resistances 32 MΩ and 2 GΩ, respectively. In a further development of this device, a gate electrode to externally control the current flow through the central island, a CdSe nanoparticle, was applied to realize a SET transistor [50].

Schmid and Dekker reported a technique, which allows the controlled deposition of a single nanoparticle between two metal nanoelectrodes – that is, the technique of electrostatic trapping (ET) [51]. This method is based on the attraction of a polarized metal nanoparticle to the strongest point of the electric field, which is applied to two Pt electrodes (Figure 5.35). From solution, the particles can be immobilized in the gap between the Pt electrodes, which can be reduced down to 4 nm and were fabricated by

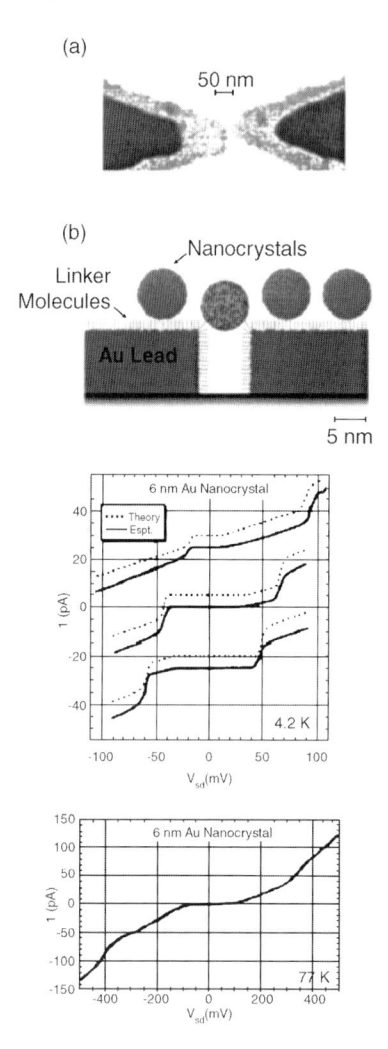

Figure 5.34 (a) Field-emission scanning electron microscopy image of a lead structure before the nanoparticles are introduced. The light gray region is formed by the angle evaporation, and is ~10 nm thick. The darker region is from a normal angle evaporation and is ~70 nm thick; (b) Schematic cross-section of nanoparticles bound via a bifunctional linker molecule to the leads. Transport between the leads occurs through the mottled nanoparticle bridging the gap; (c) $I(V_{sd})$ characteristics of a 5.8 nm-diameter Au nanoparticle measured at 4.2 K. The solid lines show three $I(V)$ curves measured over the course of several days. Each is offset for clarity. These different curves result from changes in the local charge distribution about the dot. The dashed lines are fits to the data using the orthodox Coulomb blockade model as discussed; (d) $I(V_{sd})$ characteristic of a 5.8 nm Au nanoparticle measured at 77 K; several Coulomb steps of periods ΔV_{sd} ~200 mV can be seen. Reprinted with permission from Ref. [49]; © 1996, American Institute of Physics.

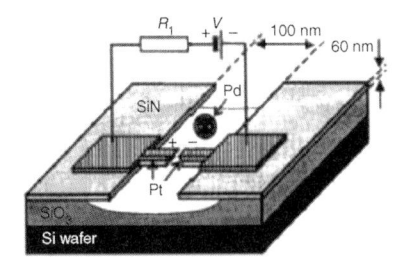

Figure 5.35 Schematic representation of the set-up for single particle measurements. Pt denotes two freestanding Pt electrodes (dashed region). A ligand-stabilized Pd cluster is polarized by the applied voltage and attracted to the gap between the Pt electrodes (electrostatic trapping; ET). Reprinted with permission from Ref. [51]; © 1997, American Institute of Physics.

thermal growth on Si. A 60 nm SiN film was then deposited on top, and EBL with poly(methylmethacrylate) (PMMA) and reactive ion etching used to open a 100 nm slit in the SiN film, with a local constriction with 20 nm spacing. Under-etching with HF enabled the formation of free-standing SiN "fingers," which were sputtered with Pt to reduce the gap down to 4 nm. Within this gap, Pd nanoparticles were trapped (Figure 5.36) to enable studies of the electrical transport properties of this double-barrier system. A typical $I(V)$ curve is shown in Figure 5.37. At 4.2 K, the most pronounced feature is the Coulomb gap at a voltage of about 55 mV, which disappears at 295 K. Furthermore, the $I(V)$ curve is not linear above the gap voltage, but increases exponentially; this was explained by a suppression of the effective tunnel barrier by the applied voltage.

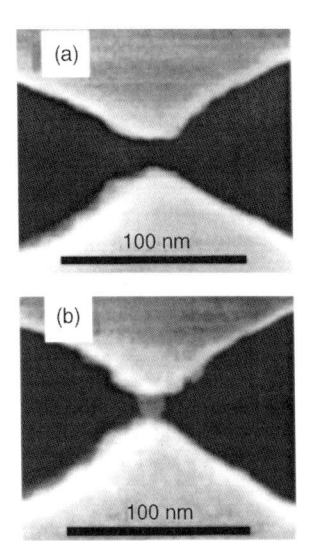

Figure 5.36 (a) Pt electrodes (white) separated by a ~14 nm gap; (b) After electrostatic trapping, the same electrodes are bridged by a single ~17 nm Pd particle. Reprinted with permission from Ref. [51]; © 1997, American Institute of Physics.

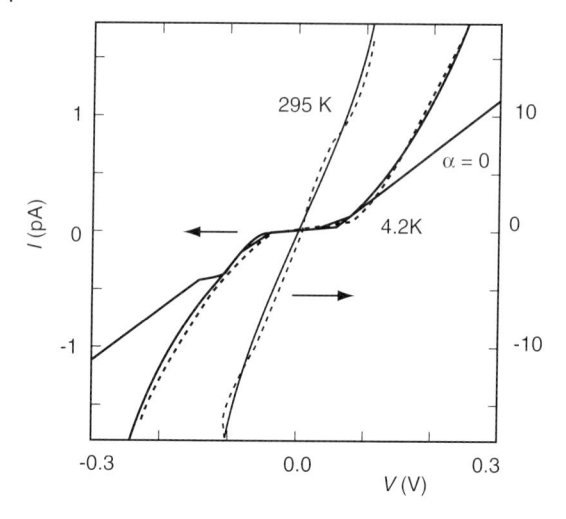

Figure 5.37 Current–voltage curves measured at 4.2 K (open squares) and at 295 K (solid squares). The solid curves denote fits of the Korotkov–Nazarov model, as described in Ref. [51]. Fitting parameters for these curves are $V_c = 55$ mV, $R_0 = 1.1 \times 10^{11}\,\Omega$, $q_0 = 0.15e$ (offset charge), and $a = E\,c/h = 0.5$. The dashed curve ($a = 0$) represents the conventional model which assumes a voltage-independent tunnel barrier. Reprinted with permission from Ref. [51]; © 1997, American Institute of Physics.

A biomolecular-based approach for the assembly of a nanogap/nanoparticle structure was reported by Chung et al. [52], who used direct-write dip-pen nanolithography (DPN) to add specific local chemical functionality in the form of specific DNA sequences to lithographically defined electrodes with a 20–100 nm gap distance. In this way, 20 to 30 nm oligonucleotide-functionalized gold nanoparticles were immobilized between the electrodes, using linker-oligonucleotide strands to induce circuit assembly (Figure 5.38a).

The $I(V)$-characteristics were measured for different temperatures, and a Coulomb gap could be observed at 4.2 K (Figure 5.38b), depending on nanoparticle size. The smaller particles had smaller capacitances and higher charging energies, and thus exhibited wider Coulomb gaps (≈ 73 mV for the 20 nm particles, ≈ 25 mV for the 30 nm particles), providing further evidence for the immobilization of a single electrically active cluster into the electrode junctions.

The structures mentioned up to this point have represented examples of quasi-zero-dimensional (0-D) systems, in which self-assembly or electrostatic trapping are used to place the individual (or at least a few) particles into a nanogap for electrical transport measurements. However, the processes are not necessarily suited to electrically address a well-defined number in a desired arrangement in a reliable manner, which in turn means that neither method can lead to strict and defect-free 1-D arrangements. In addition, inherent disorder cannot be avoided. Thus, until now, the electrical transport properties through a "perfect" 1-D array have been studied on a theoretical basis.

(a)

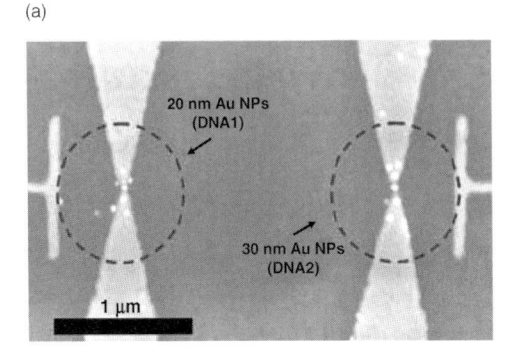

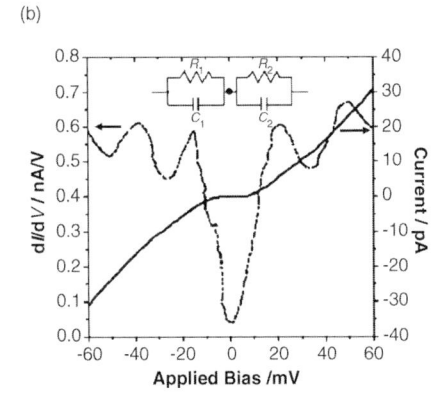

20 nm Au NPs
(DNA1)

30 nm Au NPs
(DNA2)

1 µm

(b)

Figure 5.38 (a) SEM image of single 20 and 30 nm-diameter Au nanoparticles assembled from solution and bridging the two adjacent nanoelectrode junctions; (b) Current–voltage (I(V)) characteristics and corresponding conductance (numerical dI/dV) plot for a DPN-generated nanogap device assembled with oligonucleotide-modified 30 nm diameter Au nanoparticles at $T = 4.2$ K. The inset shows a model circuit of the system with a double-barrier junction used for fitting the experimental data of solution-modified, Au nanoparticle devices. Reproduced with permission from Ref. [52]; © Wiley-VCH Verlag GmbH & Co. KGaA.

To analyze the effect of disorder, an exact analytical solution in terms of Green's function (GF) for the potential distribution in a finite 1-D array was shown to be appropriate (Figure 5.39) [53, 54]. The GF approach allows the formulation of the so-called partial "solitary" problem of small mesoscopic tunnel junctions, similar to the problem of the behavior of an electron in 1-D tight binding and in a set of random delta-function models.

The discussion of 1-D metal cluster arrangements of ligand-stabilized nanoparticles, as reported in Ref. [53, 54], assumes that: (i) either the capacitance C is the same for all junctions, whereas the self-capacitance C_0 can fluctuate from site to site because of a finite size distribution; or (ii) that the C_0 is the same, while C can fluctuate from site to site because of packing defects. It has further been assumed that the metal nanoparticles have a continuous density of states (DOS); that is, quantum

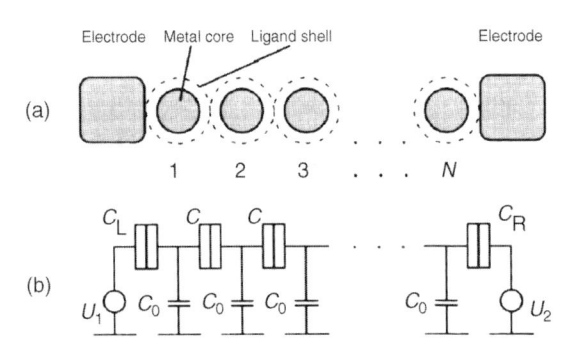

Figure 5.39 (a) 1-D array of N ligand-stabilized metal clusters; (b) The corresponding circuit equivalent. Reprinted with permission from Ref. [53]; © 1997, Elsevier.

size effects and their influence on the capacitance have been excluded from consideration. These simplifications illustrate the practical implications of disorder with respect to future applications in microelectronics, although in general they are not necessary assumptions for the method discussed.

The main results are that a decrease in particle size at one position of the array increases the potential at this point, and this may lead (at least) to localization – that is, the single excess electron in the array may be trapped. As a packing defect, which affects the inter-particle capacitance at one point, acts like an inhomogeneity, the soliton will interact with its mirror-image soliton (or anti-soliton), and will therefore be attracted. This method is of practical use because the total reflection amplitude obtained by these calculations is directly related to the Landauer resistance [55–57], which increases exponentially with the number of defects. For the sake of illustration, it is possible to consider a 1-D chain of 2 nm neutral metal particles in which a number N of the nanoparticles are replaced by smaller particles (1 nm). The 1 nm particles represent defects, which lead to a decreased transmission probability for electrons in the periodic chain, and in turn to an increase in the overall resistance. This situation is reflected in Figure 5.40, where the increase in resistance ΔG^{-1} is given for this selected example, according to Ref. [58].

A much more refined picture of the transport properties of 1-D arrays is given by Schoeller and coworkers [59], who calculated the quantum transport properties through a linear array of metallic nanoparticles, which were immobilized on DNA (an overview of the methods used to fabricate these assemblies is provided in Ref. [14, 60]). In detail, these authors determined the current and shot noise through such a system in variation of parameters such as the applied gate and bias voltage, the temperature, and the strength of dissipative effects. One special focus of the calculations was the geometry of the array, as it was intended to design the electron transport properties by controlling the shape of the device. In the model system used (see Figure 5.41), the gate, leads and nanoparticles were all treated as ideal conductors, whereas other parts of the system (substrate, DNA, ligand shell) were modeled as dielectrics.

As a basis, Schoeller et al. took a semi-classical Master equation approach to calculate the transport properties. Thereby, they assumed an incoherent tunneling, which was treated as a perturbation, while the Coulomb interaction between charged nanoparticles was taken into account nonpertubatively within a capacitance model. However, in contrast to the standard orthodox theory, they explicitly considered the discrete nature of the electronic spectrum of the nanoparticles. In the calculated $I(V)$

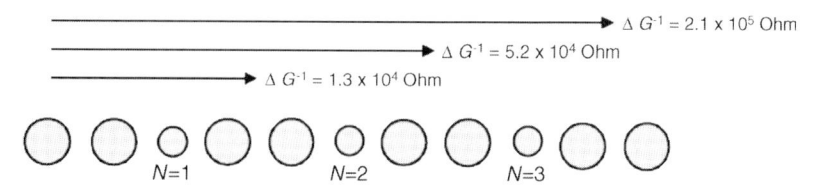

Figure 5.40 A 1-D chain of 2 nm metal particles with $N = 3$ "impurities" (i.e., 1 nm metal particles). The arrows indicate that the resistance increases exponentially with N [58].

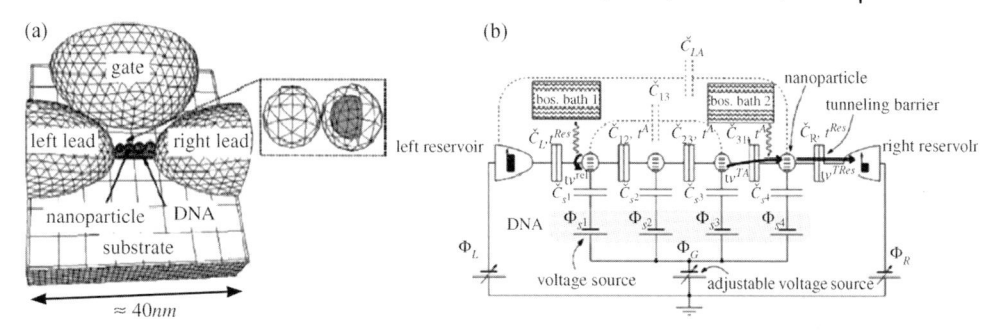

Figure 5.41 (a) Wireframe of the geometric set-up; (b) The model system. Reproduced with kind permission from Ref. [59]; © 2005, American Physical Society.

characteristics of a two-nanoparticle array, Schoeller et al. found a fine structure besides the Coulomb staircase which had originated from the level spacing. At increasing temperatures these fine structures were obliterated but the typical staircase remained.

The $I(V)$ characteristics of an array of nanoparticles with uniformly growing diameters illustrate a striking asymmetry in dependency of the gate voltage in the low bias regime (Figure 5.42). For higher bias voltages, the characteristics become symmetric and independent of the applied gate voltage. The mechanism behind this asymmetry is a combination of the asymmetric size of the nanoparticles, which leads to asymmetric capacitance matrices, and the Coulomb interaction. Additionally, the offset voltage increases with array length, whereas the asymptotic conductance can increase or decrease with array length, which can be tuned by different sizes of nanoparticles.

The negative differential conductance (NDC) effect occurs when nanoparticles are assumed to have discrete, equally spaced energy levels (Figure 5.43). This effect appears when a strong tunneling to the reservoirs is large, and induces a high-charge

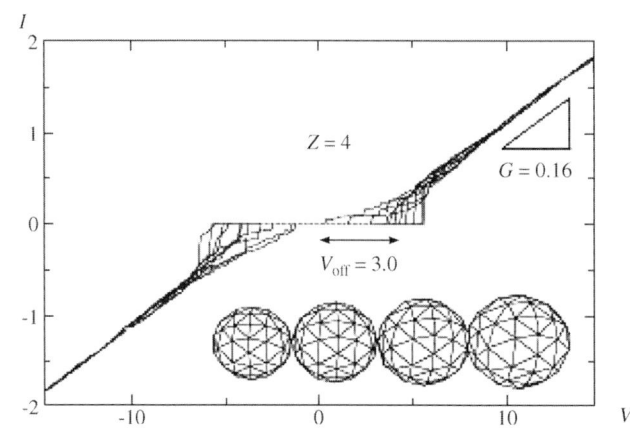

Figure 5.42 $I(V)$ characteristics for different gate voltages of a four-nanoparticle array with growing diameter. Reproduced with kind permission from Ref. [59]; © 2005, American Physical Society.

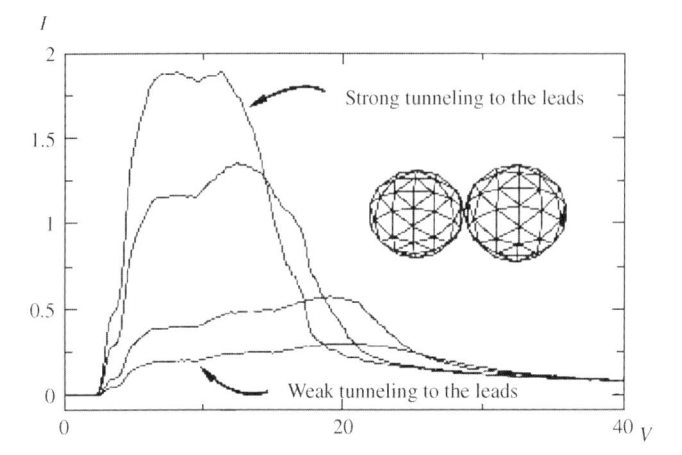

Figure 5.43 Increasing NDC effect for higher tunneling rates between array and reservoir in a two-nanoparticle array. Reproduced with kind permission from Ref. [59]; © 2005, American Physical Society.

gradient that leads to smaller transition rates between the particles, with increasing voltages and a large offset between the electronic spectra. By varying the distance between the array and the leads, the tunneling can be tuned and is heavily dependent on the geometry.

The first experimental attempts to achieve a 1-D array of nanoparticles were conducted by Sato et al. [61], who built up single electron transistors made from nanoparticles bridging a 30 nm gap between source and drain. The observed electrical conduction through these devices showed clear Coulomb staircases and periodic conductance oscillations in dependence of the gate voltage. The devices were prepared as follows. First, a metal electrode system consisting of source, drain and gate electrodes, with a gap of 30 nm between source and drain, was formed on thermally grown SiO_2 surfaces by using EBL. The surfaces were modified with alkanesiloxane molecules so as to form an adhesion layer. After functionalization with aminosilane, the samples were immersed into a gold nanoparticle solution to form a submonolayer coating of nanoparticles (this was caused by binding of the amino group to the gold surface). The submonolayer was treated with 1,6-hexane-dithiol to replace the citrate adsorbates on the gold nanoparticle surfaces, which led to sulfur-terminated gold nanoparticles. Following a second immersion step within a gold nanoparticle solution, the formation of nanoparticle chains was observed (Figure 5.44). In the main, two to four nanoparticles formed a chain in which the particles of the second deposition were bound covalently by dithiol molecules in the gaps of the first nanoparticle layer. This bifunctional organic molecule acted as a defined spacer between the particles and the particle–electrode connection, and provided the tunnel barriers.

Although, the number of bridging nanoparticles and the location of the chains varied from device to device, in general the electron conduction was dominated by electron-charging effects, as indicated by Coulomb gap formation. The $I(V)$ char-

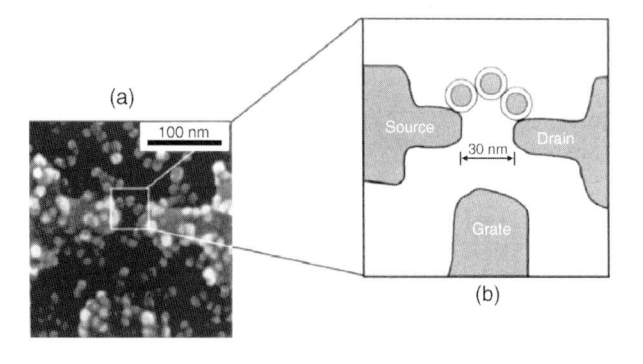

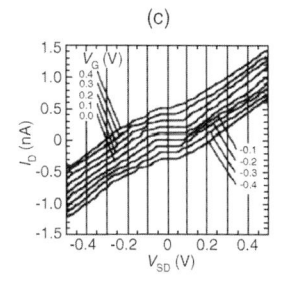

Figure 5.44 SET transistor, based on self-assembly of gold nanoparticles on electrodes fabricated by electron beam epitaxy. Reprinted with permission from Ref. [61]; © 1997, American Institute of Physics.

acteristics of a three-dot chain showed a clear Coulomb gap of ~150 mV, with a Coulomb staircase at 4.2 K that was smeared out at room temperature. The characteristics of a three-dot device measured at 4.2 K, in dependence of the gate voltage (−0.4 to 0.4 V), is shown in Figure 5.44c. With an increasing magnitude of the gate voltage, a squeezing of the Coulomb gap was observed. The plot of current through the device was clearly dependent on the gate voltage, and showed the typical current

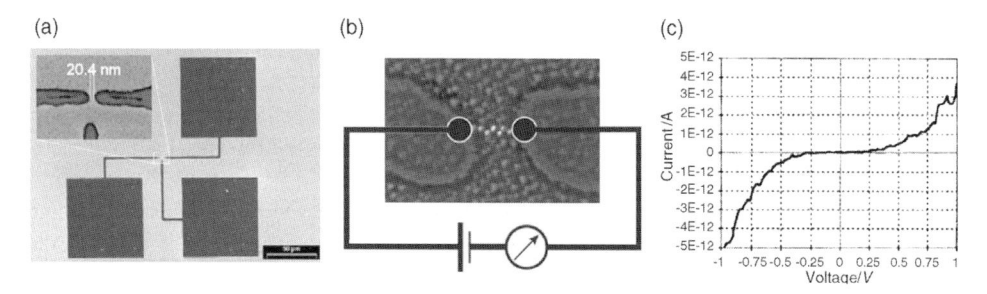

Figure 5.45 (a) SEM image of a three-electrode configuration with source, drain and gate electrodes. The inset shows (magnified) the gap in the tungsten electrodes; (b) High-resolution SEM image of Au_{55} clusters forming a quasi-1-D chain; (c) $I(V)$ characteristic of the device. Reprinted with permission from Ref. [63]; © 2001, American Chemical Society.

oscillations that proved that the desired function of the single-electron transistor had been achieved.

With Monte Carlo simulations based on a three-dot (four junction) single electron transistor, the measured $I(V)$ characteristics could be well reproduced. The calculated Coulomb gap was in good agreement with the measured value of ~ 150 mV, and all capacitances were determined as 1.8–2.0 aF.

According to Samanta et al. [62], who used a GF-based method to calculate the transmission function of electrons across the dithiol ligand, the resistance R per molecule was determined by the Landauer formula, $R = (h/2e^2)/T(E_g)$, where $T(E_g)$ is the transmission function, $T \sim \exp[2(mE_g)^{1/2}/h]$. For the dithiol molecule, the barrier height was estimated to be $E_g \sim 2.8$ eV, and the resulting resistance calculated as $R \sim 30$ GΩ.

Another strategy to form a 1-D array involves the electrostatic trapping of clusters by applying an electric field between electrodes. This method was used to generate cluster chains consisting of $Au_{55}(PPh_3)_{12}Cl_6$ clusters in a three-tungsten electrode configuration on an SiO_2 surface with a 30 nm gap (Fig. 5.45) [63]. These clusters had a 1.4 nm gold core, and were stabilized by a shell of 0.35 nm phosphine molecules. Deposition of the clusters was achieved by dipping the electrodes bearing the wafer sample in a cluster solution, and applying a voltage in the range of 0.5 to 2.0 V for 40–120 s. At room temperature, the current–voltage measurements of these QD wires showed equivalent Coulomb blockades in the region between -0.5 to 0.5 V. From this electronic behavior, it followed that the Au_{55} cluster would act as a single-electron transistor at room temperature (as already assumed for this type of nanoparticle).

A more selectively way to assemble nanoparticles into nanogaps was demonstrated by Lee et al. [64], who used 10 nm gold colloidal nanoparticles functionalized with thiol-modified single-stranded DNA (ssDNA) to fabricate single-electron transistors. These Au nanoelectrodes, which had a 30 nm gap and were prepared by EBL and lift-off techniques, were immersed into a solution of complementary thiol-modified ssDNA. In a second step, the DNA-modified nanoparticles were added to immobilize them on each side of the electrodes, through DNA hybridization. After washing and drying, another Au particle solution which had been functionalized with the same DNA strands as the electrode structure, was used to bridge the gap between the electrodes (Figure 5.46). In order to control the strength of the tunneling barrier, DNA was employed with different numbers of bases (11, 22, and 30).

The $I(V)$ characteristics of these devices showed a resistance of 700 MΩ at room temperature, and clear Coulomb staircases with a gap of 160 mV at 4.2 K (Figure 5.47a).

From the $I(V)$ characteristics of the device with 22-mer DNA strands, a larger Coulomb gap of 280 mV could be derived which had resulted from the longer DNA strands providing stronger tunneling barriers. In device III (Figure 5.47c), a larger room-temperature resistance of about 26 GΩ was calculated, and no characteristic Coulomb oscillation could be observed. This effect may be explained by the fact that the longer DNA strands act as resistors connected in series, rather than as tunneling barriers.

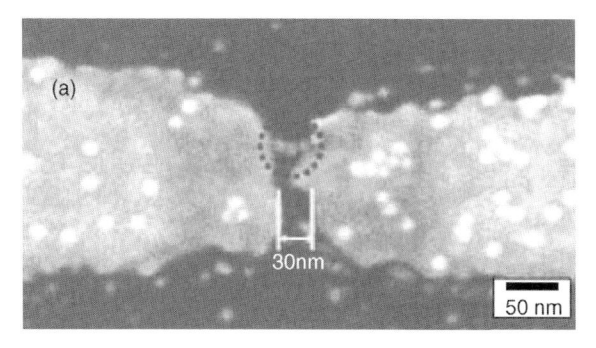

Figure 5.46 Field-emission SEM image of a fabricated device. Reprinted with permission from Ref. [64]; © 2005, American Institute of Physics.

In contrast to previously described techniques, Weiss et al. described the lithographic contacting of previously self-assembled nanoparticles [65]. While all tunnel junctions were similar to the previous examples, the main advantage of this method is the high ratio of gate capacitance to total capacitance. A device consisting of two 50 nm nanoparticles coated with octanethiol molecules is shown in Figure 5.48.

A differential conductance measurement of the double-island device at liquid He temperatures is shown in Figure 5.49a. In this case, due to Coulomb blockade the conductance is suppressed in the region of $V_{DS} = 0$. The Coulomb charging energy was calculated from the size of the diamonds, and had a value ~20 meV, which was in good agreement with that obtained from the simulation of the conductivity expected for such a described device. The basic model for this simulation and the conductance plot are shown in Figure 5.49b and c.

The high ratio of gate capacitance (Si substrate acts as gate electrode) and total capacitance was shown to be remarkable for this single-electron transistor, and was an order of magnitude larger than for previously described nanoparticle devices [50]. The authors suggested that this effect had resulted from a reduced capacitance between the outermost island and the electrode based on the electrode-incorporated nanoparticles. This effect would allow accessing to multiple charge states with moderate gate voltages, which is important for nanoelectronic devices. Additionally, in differential conductance measurements the investigated devices showed a high

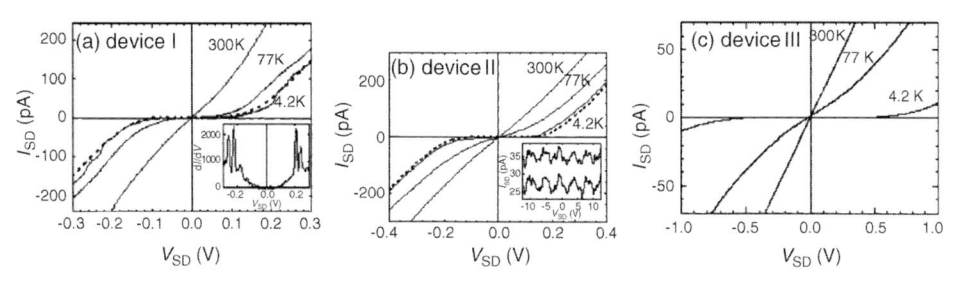

Figure 5.47 I(V) characteristics of devices I–III. (a) 11-mer; (b) 22-mer; (c) 30-mer DNA strands. Reprinted with permission from Ref. [64]; © 2005, American Institute of Physics.

Figure 5.48 Two-island device fabricated by contacting a self-assembled chain of nanoparticles. The two particles framing the two island particles form part of the electrodes, such that all three tunnel junctions are formed by self-assembly of the particles. Reprinted with permission from Ref. [65]; © 2006, American Institute of Physics.

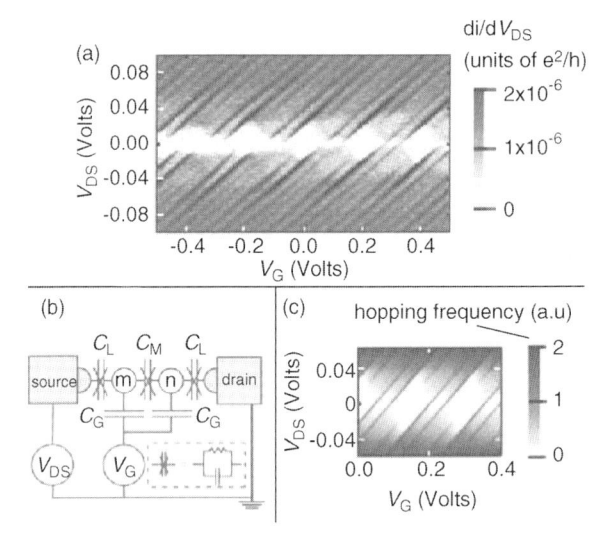

Figure 5.49 Experimental and predicted differential conductance plots of the double-island device of Figure 5.48. (a) Differential conductance measured at 4.2 K; four peaks are found per gate period. Above the threshold for the Coulomb blockade, the current can be described as linear with small oscillations superposed, which give the peaks in dI/dV_{DS}. The linear component corresponds to a resistance of $\sim$20 GΩ; (b) Electrical modeling of the device. The silicon substrate acts as a common gate electrode for both islands; (c) Monte Carlo simulation of a stability plot for the double-island device at 4.2 K with capacitance values obtained from finite-element modeling: $C_G = 0.84$ aF (island-gate capacitance), $C_M = 3.7$ aF (inter-island capacitance), $C_L = 4.9$ aF (lead-island capacitance); the left, middle, and right tunnel junction resistances were, respectively, set to 0.1, 10, and 10 GΩ to reproduce the experimental data. Reprinted with permission from Ref. [65]; © 2006, American Institute of Physics.

Figure 5.50 Scanning electron microscopy image of a pair of nanoelectrodes with gold nanoparticles immobilized into the gap using a mixture of large and small particles derivatized with RNA phosphate buffer containing NaCl in the presence of Mg^{2+}. Reprinted with permission from Ref. [66]; © 2006, American Chemical Society.

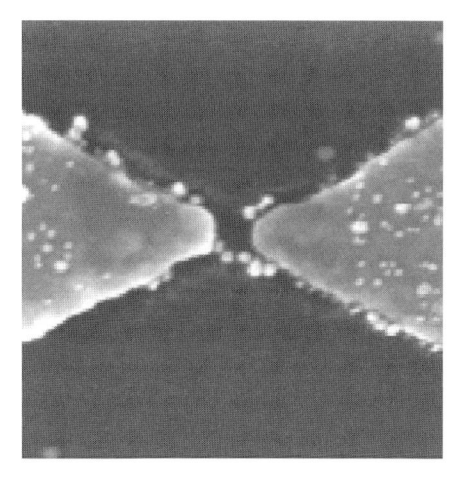

stability over 24 h, a property which would be important for future applications in nanoelectronics.

Besides the specific assembly of DNA-functionalized nanoparticles via the hybridization of DNA, an alternative selective working assembly is that of Mg^{2+}-mediated RNA–RNA loop–receptor interaction [66]. This method was demonstrated successfully by Bates et al., who prepared nanowires made from 15 or 30 nm gold nanoparticles functionalized with DNA, which hybridized with the extension sequences on RNA (Figure 5.50). A wire deposited between the lithographically fabricated nanoelectrodes exhibited an activated conduction, by electron hopping, over the temperature range of 150 to 300 K.

The conductivity (σ) decreased with decreasing temperature above 160 K in the low-bias regime, and was typical for an activated transport. The data were modeled as $\sigma = \sigma_0 \exp(-E_a/k_B T)$, where σ_0 is the maximum conductivity, E_a is the activation energy, and k is the Boltzmann constant. From these calculations, it was found that $E_a = 0.29$ eV and $\sigma_0 = 10$ mS (Figure 5.51). Below 160 K, the rate of change of conductivity with temperature was reduced significantly, indicating a deviating conductivity processes. The authors concluded that the transport was due to the gold nanoparticles and thermally activated electron hopping in-between.

Another interesting method for generating single-electron transistors is to pre-structure the substrate surface, followed by deposition of the nanoparticles, and finally to place two or more metal electrodes across the nanoparticle chain. This method was described by Coskun et al. [67], who first spin-coated a cleaned silicon substrate with PMMA, and then structured the surface with EBL to define the desired patterns, treated the substrate with a aminopropyltriethoxysilane (APTES) solution,

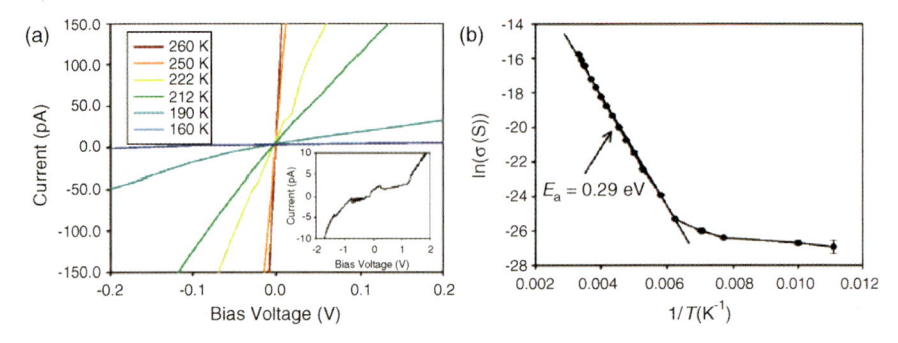

Figure 5.51 (a) $I(V)$ characteristics at low bias voltages of a self-assembled nanowire in the temperature range 160–260 K. The inset shows the characteristic at 130 K in the bias voltage range −2 to +2 V; (b) Arrhenius plot of the conductivity at lower bias voltages. E_a denotes the corresponding activation energy. Reprinted with permission from Ref. [66]; © 2006, American Chemical Society.

and deposited 13 nm and 50 nm gold nanoparticles before the self-assembled nanowires were contacted with metal electrodes (Figure 5.52).

The measured conductance (Figure 5.53) of a chain of 50 nm Au nanoparticles showed conductance oscillations that depended on the gate voltage. The characteristics indicated a good reproducibility by sweeping the gate voltage and the lack of the offset charges, an effect which was interpreted as an absence of contaminants. From the sizes of the Coulomb diamonds (Figure 5.53b), a nanoparticle-gate capacitance of 0.3 aF and a nanoparticle capacitance of ∼10–30 aF could be calculated; these values were in good agreement to previously reported values for similar structures [65].

In all previously envisaged electrical characterizations of single-electron transistors, lithographically prepared electrode configurations were used throughout. An *a priori* more flexible measuring method, based on a nanomanipulator system in a scanning electron microscope, was introduced to address low-dimensional nanostructures *in situ* [68]. The investigated samples were prepared first by processing PMMA-coated Si wafers using extreme-ultraviolet interference lithography (EUV-IL),

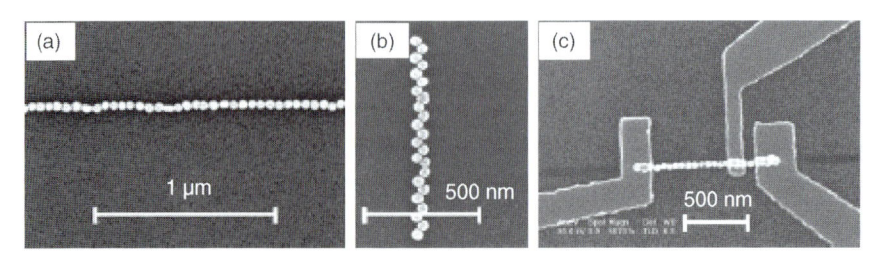

Figure 5.52 (a,b) Gold nanoparticle deposition along a patterned line with different widths; (c) Contacting the deposited nanoparticle chains by electron-beam lithography. Reprinted with permission from Ref. [67]; © 2008, American Institute of Physics.

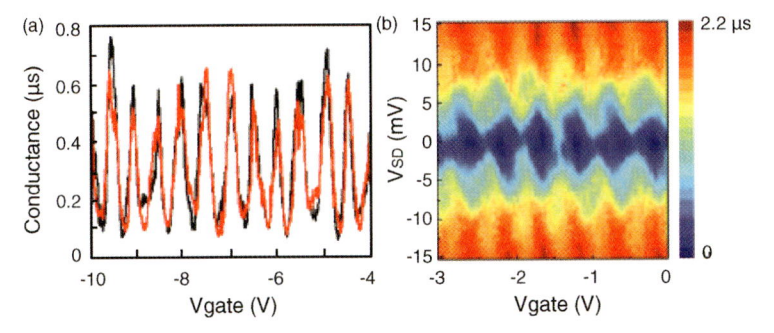

Figure 5.53 (a) Differential conductance in dependence of gate voltage measured at $V_{gate} = 4.2$ V; (b) "Coulomb diamonds" in the conductance measurement as a function of the gate voltage and the source-drain bias. Reprinted with permission from Ref. [67]; © 2008, American Institute of Physics.

and the resulting grooves filled with 44 nm Au gold nanoparticles by dip-coating the substrates into a concentrated nanoparticle suspension.

For the electrical measurements, two metallized tips were made to approach the surface of the sample, under simultaneous observation by scanning electron microscopy (SEM), and therefore represented a "flexible nanogap." After contacting the required chain of interest, the $I(V)$ characteristics were monitored. The first set of measurements for the as-prepared samples are shown in Figure 5.54, while ozone-cleaned samples were characterized electrically to estimate the contribution of the ligand shell to the overall resistance (see Figure 5.55).

From these $I(V)$ characteristics, a resistance per particle before and after the cleaning procedure could be calculated. Removal of the insulating ligand shell with ozone cleaning resulted in a higher conductivity and a lower resistance per particle, respectively. These results were in good agreement with previously reported lithographically fabricated structures, thus verifying this measurement set-up for the routine addressing of structures in the sub-10 nm range.

5.2.5
Collective Charge Transport in Nanoparticle Assemblies

The single-particle properties discussed above highlighted the significance of the high charging energy of metal nanoparticles as a prerequisite for SET events at elevated temperatures. Together with the limited number of free electrons, this may lead to these materials being regarded as "artificial atoms." Yet, this raises fundamental questions regarding the design of "artificial molecules" or "artificial solids" built up from these nanoscale subunits [69–72]. Remacle and Levine have reviewed these ideas, in association with the use of chemically fabricated QDs as building blocks for a new state of matter [73].

Students in their first chemistry course learn that the simplest molecule, H_2, is formed by the overlap of the electron wavefunctions centered on the individual hydrogen atoms. Correspondingly, the wavefunctions of artificial atoms can also

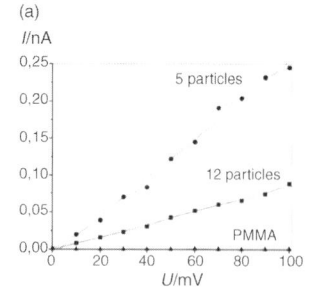

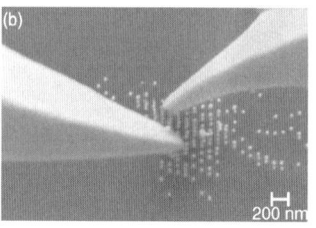

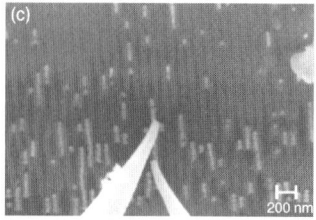

Figure 5.54 (a) $I(V)$ characteristics of as-prepared gold nanoparticle chains with different numbers of particles and a reference measurement on the PMMA surface; (b,c) Corresponding SEM images of the measurements performed with the nanomanipulating system. Reproduced with kind permission from Ref. [68]; © 2008, American Scientific Publishers.

overlap, and electrons can exchange coherently and reversibly between them – this is the basis of a covalent bond. By extending this idea further, the ordered assembly of identical nanoparticles in one, two, or three dimensions represents the formation of an artificial solid or superlattice. The fabrication of 2-D and 3-D ordered superlattices has been demonstrated on many occasions [74–82]. Furthermore, it is possible to mix nanoparticles of different chemical compositions, of bimodal size, or of different shapes so as to obtain tailored nanoalloys [83]. Levine termed these new products "designer materials," with tunable properties [73]. Such artificial solids were shown to exhibit delocalized electron states, depending on the strength of the electronic coupling between the adjacent nanoparticles. However, the latter effect would depend on the size of the nanoparticles, the nature and the covering density of the organic ligands, the particle spacing, and the symmetry of packing.

As it is known from solid-state physics, any type of disorder will affect the electronic structure of a solid. Accordingly, size distribution, packing defects, and

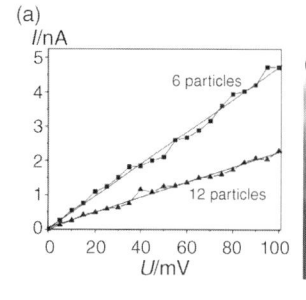

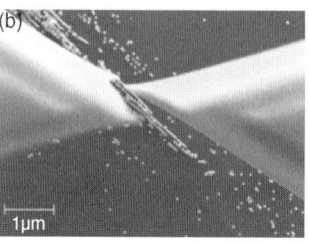

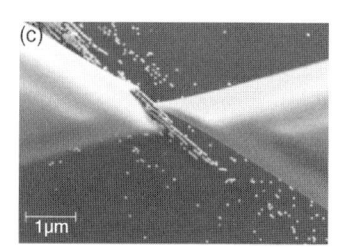

Figure 5.55 (a) $I(V)$ characteristics of O_3-cleaned samples; (b,c) Corresponding SEM images of the electrical transport measurements. Reproduced with kind permission from Ref. [68]; © 2008, American Scientific Publishers.

chemical impurities each lead to modifications of the electronic structure of the QD superlattice. Even in the case of identically sized nanoparticles over microdomains of the assembly, slight deviations in ligand density or orientation mean that the nanoparticles are not strictly the same at each lattice site, as would be the case for atoms. Therefore, solids which consist of chemically tailored nanoparticles are inherently disordered, and the loss of translational symmetry affects the extended states in the solid, which is immediately reflected in their optical and electrical properties. Remacle et al. illustrated the crucial point of symmetry breaking by comparing the experimental and theoretical second-order harmonic generation (SHG) response of a 2-D, hexagonal array of Ag particles ($R = 1.5$ nm) with variable inter-particle spacing D, where $D/2R$ is in the range 1.05–1.5 nm. At large inter-dot spacing ($D/2R = 1.5$), the wavefunction will be localized on the individual nanoparticles. If the 2-D layer is compressed, the wavefunction will be delocalized over small domains. A quantum phase transition to a fully delocalized state is achieved for closely packed nanoparticles at $D/2R = 1.05$ (see Figure 5.56).

A corresponding response has been obtained from complex impedance measurements on a monolayer of Ag nanoparticles with $2R = 3.5$ nm. Heath and coworkers, when analyzing the complex dielectric modulus (which is the inverse of the complex dielectric permittivity; for details, see Ref. [84]), reported a reversible metal–insulator transition. In this case, a transition was observed from RC relaxational behavior to an inductive metallic-like response, when the monolayer was compressed to decrease the inter-particle spacing to <0.6 nm (Figure 5.57). This suggests that a sequential transition from hopping transport between localized states via tunneling to metallic transport appears, accompanied by a rapid decay in the relaxation time,

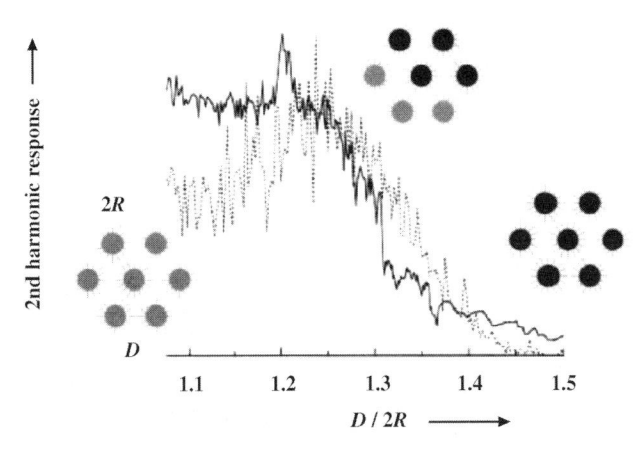

Figure 5.56 Transition from a localized to a delocalized state in a 2-D hexagonal array of ligand-stabilized 3 nm Ag particles upon mechanical compression. The transition is probed by the second-harmonic response, which is sensitive to the extent of delocalization of the electronic wavefunction. Reproduced with permission from Ref. [73]; © Wiley-VCH Verlag GmbH & Co. KGaA.

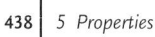

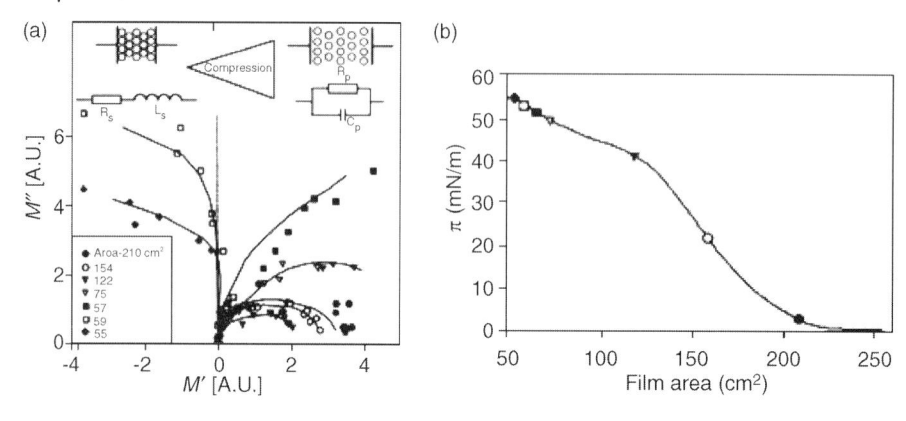

Figure 5.57 (a) Complex impedance response (plot of the imaginary part of the complex modulus M″ versus the real part M′ in the complex plane) of a monolayer (3.5 nm diameter) of propanethiol-capped Ag nanoparticles. The particle film response is characterized initially by an RC circuit equivalent, in conformity with a picture of capacitively determined hopping localized conductivity. As the particles are compressed to a separation of less than 0.6 nm, the film becomes inductive, indicating the presence of metal-like transport in the film; (b) The Langmuir isotherm of the monolayer compression with labels corresponding to the various curves in panel (a). Although the isotherm indicates a collapse near the 130 cm² film area, the collapse is heterogeneous, occurring only at the mobile barriers. The central region of the film, where transport measurements were carried out, remains an uncollapsed monolayer. Redrawn and modified from Ref. [84]; © 1998, American Physical Society.

when the layer is compressed. The latter effect may result from an increasing tunneling rate – that is, the resistance decreases between the particles as well as in extended domains with delocalized states, and this is essentially affected by the degree of disorder.

In order to discuss the signatures of localization and delocalization, and its significance to the application of nanoparticles in microelectronic devices, the following sections provide examples of the electrical properties of 2-D and 3-D nanoparticle arrays.

5.2.5.1 Two-Dimensional Arrangements

Janes *et al.* reported the electronic conduction across a network of 4 nm gold nanoparticles interconnected by diisonitrile ligands (1,4-di (4-isocyanophenylethynyl)-2ethylbenzene) to produce a conjugated, rigid molecule with an approximate inter-particle spacing of 2.2 nm [85]. The 2-D assembly results from the deposition of nanoparticles from a colloidal solution, followed by reactive addition of the linking molecules. In order to allow electrical characterization, the layers were deposited on a SiO_2-supported GaAs wafer with gold contacts, and with separations of 500 and 450 nm, respectively.

The $I(V)$-characteristics of the arrays were linear over a broad voltage range. As a dot-to-dot capacitance of 2×10^{-19} F may be assumed, the total dot capacitance will be

1.2×10^{-18} F, if each cluster is assumed to have six nearest-neighbors. The corresponding charging energy will thus be approximately 11 meV, which is only about half of the characteristic thermal energy at room temperature. Therefore, a Coulomb gap at room temperature is not apparent.

In order to improve the packing symmetry and reduce the defect concentration, many attempts were made to prepare monolayers or multilayers of near-monodisperse nanoparticles. In a straightforward approach, monolayers could be prepared almost defect-free on a water surface, and transferred via microcontact printing onto solid substrates [86] (Figure 5.58).

In an alternative approach, membranes prepared from a monolayer of dodecanethiol-stabilized gold nanoparticles with a diameter of 6 nm were obtained via a simple evaporation process on a silicon nitride "window" area [87] (Figure 5.59). Another example showed that nanoparticles could self-organize in 2-D or 3-D superlattices so as to form "supra" crystals [88] (Figure 5.60).

Another approach towards the development of molecular electronic circuits is the reversible formation of stable 2-D networks of molecular junctions [89]. This device was prepared by self-assembling 10 nm alkanethiol-capped gold nanoparticles, and transferring these particles onto a silicon wafer by microcontact printing, followed by an *in situ* ligand-exchange reaction. For the electrical transport measurements, contact pads were deposited on the top of the printed structures in which a TEM grid acted as shadow mask (Figure 5.61a).

The electrical characterization of the devices was carried out in air at room temperature, and showed a linear characteristic. In order to compare the resistances at different stages of the exchange experiment, the authors calculated the sheet resistances, defined as $R_s = Rwl^{-1}$, where R is the measured resistance and w and l are the width and length of the array, respectively. In the first step, the $I(V)$ characteristics of the as-prepared array were investigated (Figure 5.61b, curve 1) in which the particles were linked with octanethiol molecules; a sheet resistance of $4.4 \times 10^{10}\,\Omega$ was calculated. After immersing the device in a thiolated oligo(phenylene ethynylene) (OPE) solution to exchange alkanethiol molecules on the

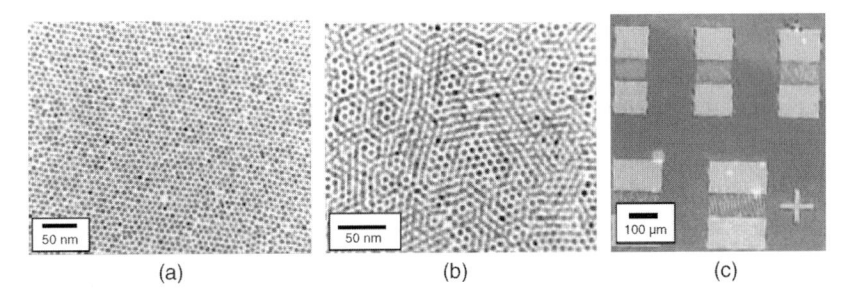

Figure 5.58 (a) TEM image of a nanoparticle monolayer printed onto a silicon nitride membrane window; (b) TEM image of a bilayer printed onto a silicon nitride membrane; (c) Optical image of a nanoparticle bilayer printed onto a lithographically fabricated electrode configuration. Reprinted with permission from Ref. [86]; © 2003, American Chemical Society.

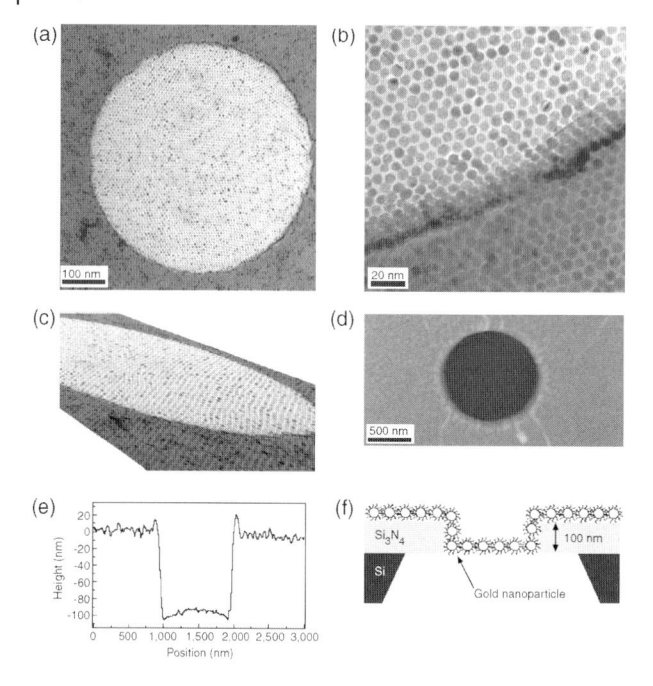

Figure 5.59 (a) TEM image of a free-standing array with a diameter of 250 nm; (b) A higher-magnification image of the membrane; (c) TEM image of a tilted membrane; (d) AFM height image of a free-standing membrane with a diameter of 550 nm; (e) Cross-section of the height image in panel (d); (f) Schematic diagram of the array configuration. Reprinted with permission from Ref. [87]; © 2007, Macmillan Publishers Ltd.

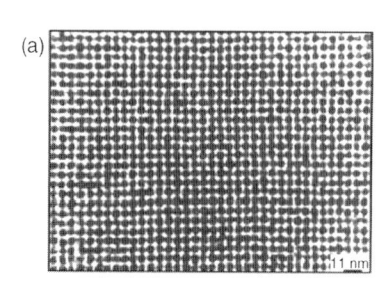

Figure 5.60 (a) TEM image of a 3-D aggregate consisting of 5.8 nm silver sulfide nanoparticles; (b) SEM image of superlattices of cobalt nanoparticles on HOPG; (c) Tilted SEM image showing the 3-D structures of the lattice in panel (a). Reprinted with permission from Ref. [88]; © 2003, American Chemical Society.

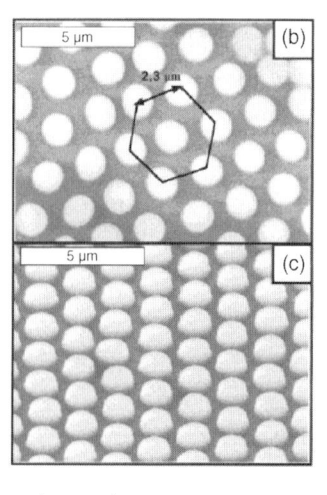

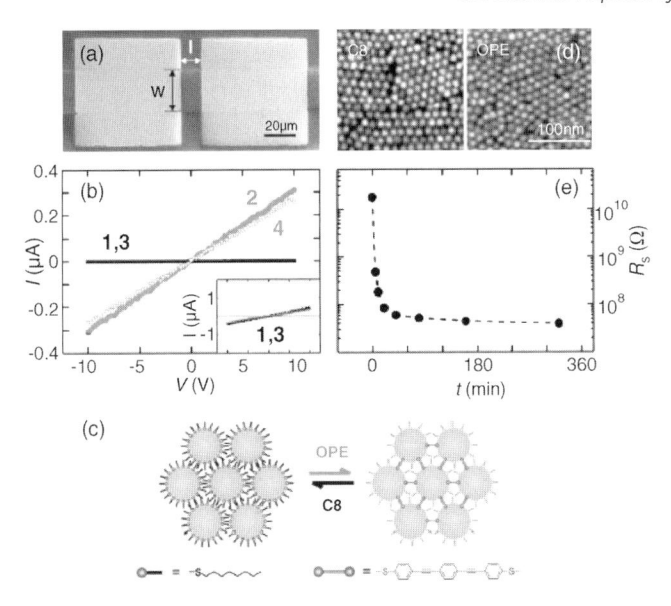

Figure 5.61 (a) Electron microscopy image of the investigated device; (b) I(V) characteristics of the device at different stages in the exchange experiments; (c) Scheme of the exchange reaction in the monolayer of the nanoparticles; (d) Electron microscopy images of the nanoparticle array before and after molecular exchange; (e) Resistance versus immersion time in the OPE measurement during the exchange reaction. Reprinted with permission from Ref. [89]; © 2007, Macmillan Publishers Ltd.

nanoparticles surface and their interlinking, the electrical behavior was re-measured. Due to interlinking with the OPE (a conjugated compound), the sheet resistance of the linked array was decreased by more than two orders of magnitude, at $6.3 \times 10^7\,\Omega$ (Figure 5.61b, curve 2). The formation of molecular junctions was reversible, as it could be proved in the electrical transport measurements (Figure 5.61b, curves 3/4).

A further step towards switchable molecular electronic circuits is the light-controlled conductivity of 2-D lattices of gold nanoparticles bridged by photochromic diarylethene molecules [90]. This was demonstrated by van der Molen *et al.*, who first prepared the device as described above [89] and then immersed it in a solution of "on" state switches so as to obtain an interlinked 2-D nanoparticle array (see Figure 5.62a and b). In this case, the nanoparticle array serves as a template for the switching molecules which form conductive bridges in-between.

The reversible conductance switching of the device was performed at room temperature, and started with an illumination with UV light (Figure 5.62c; $t < 0$). At $t = 0$, the UV light was deactivated and visible light activated. The conductance decreased rapidly, this being related to the switching molecule changing from the *on* to the *off* state. Repeating the switching by changing to UV light again caused a rapid increase in conductance, with four conductance switching cycles being demonstrated before the switching molecules decomposed. The decay in switching amplitude

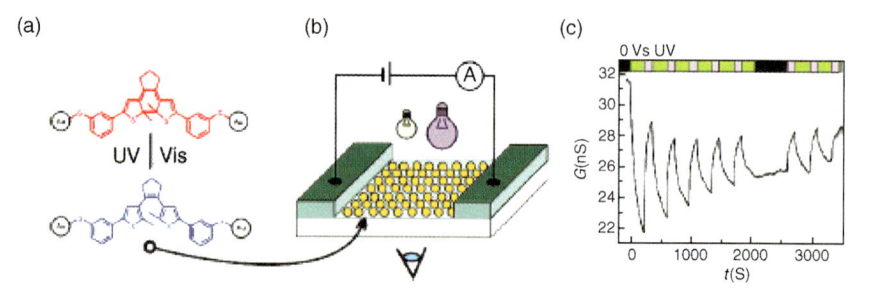

Figure 5.62 (a) Scheme of the switching molecule, a dithiolated diarylethene. Top: closed form ("on" state). Bottom: open form ("off" state); (b) Scheme of the measuring set-up for the light-induced conductance switching; (c) Diagram of the repeated conductance switching experiments, where conductance G is plotted versus illumination time. Reprinted with permission from Ref. [90]; © 2001, American Chemical Society.

was explained by the fact that, with every switching process, the number of switching molecules was decreased.

Temperature-dependent DC measurements on comparable samples with 2-D cluster linkage were investigated by Andres *et al.* [91], from which a Coulomb charging behavior and charging energy were obtained. From the Arrhenius relationship (as described above), the charging energy was calculated as $E_A = 97$ meV. The inter-particle resistance was revealed as 0.9 MΩ, from which a single-molecule resistance of 29 MΩ was obtained. This value was in good agreement with the prediction of 43 MΩ obtained from Hückel–MO calculations [62].

Another method for assembling nanoparticles in a 2-D array was reported by Koplin *et al.* [92], who used DNA to build monolayers of 15 nm gold nanoparticles on flat silicon substrates, and which had a high particle density to enable electronic exchange in-between. Both, direct current (DC) measurements and temperature-dependent impedance spectroscopy were performed to assess the electrical characterization of these layers, and both methods identified classical hopping transport, similar to 3-D nanoparticle structures. From the impedance data, an activation energy of 301 meV was calculated according to the Arrhenius relationship (Figure 5.63), whereas the transport properties were predominantly determined by disorder rather than by the properties of the DNA.

Beverly *et al.* [93] studied the temperature-dependent DC transport measurements on monolayers of self-assembled dodecanethiol-coated $\sim$7 nm silver nanoparticles as a function of particle size distribution-induced disorder. The superlattices disorder was adjusted by a stepwise variation of the particle size distribution. In the electrical transport measurements, six different monolayers of $\sim$7 nm silver nanoparticles, in which the size distribution was varied from 6.6% to 13.8%, were investigated at 300–10 K. Above $\sim$200 K, all films exhibited metallic conductivity, and below 200 K activated transport. However, between 30 and 100 K a second transition (T_{cross}) was observed that was based on the crossover from the simply activated transport to a

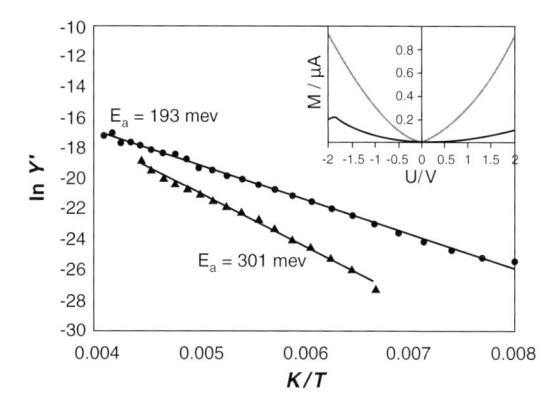

Figure 5.63 Arrhenius plots of samples prepared by specific (triangles) and nonspecific (dots) immobilization. The inset shows the corresponding $I(V)$ curves for specific (gray curve) and nonspecific (black curve) immobilization. Reprinted with permission from Ref. [92]; © 2009, American Chemical Society.

third regime of variable-range hopping. This "crossover" temperature, which was shown to depend basically on the superlattice disorder, correlated linearly with the particle size distribution width (see Figure 5.64).

These results imply that, at a size distribution of approximately 3%, it would be possible to observe a pure metallic conductivity at 0 K, which surely will be confirmed in future studies.

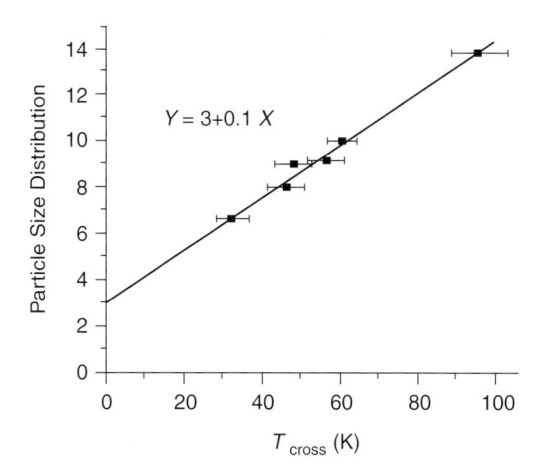

Figure 5.64 Plot of particle size distribution versus transition temperature (crossover point between activated transport mechanism and variable range hopping). Reproduced with kind permission from Ref. [93]; © 2002, American Chemical Society.

One step further to tune the electronic properties of a 2-D array of Ag nanoparticles with a size distribution of 7% was reported by Remacle et al. [94]. These authors discussed the experimental and computational results of temperature-dependent conductivity measurements as a function of size distribution, compression of the array, and the applied gate voltage. From the temperature-dependent source–drain measurements they obtained sigmoidal-shaped and nonlinear curves (Figure 5.65).

With increasing temperature, the electronic conduction of the array behaves more ohmic. This effect is characteristic for a transition from variable-range hopping at low temperatures to an activated behavior at high temperatures. In the computational results, the transition was observed over a larger temperature range compared to the experiment; this may be explained by a smaller size of array in the simulation, and means that 25- to 100-fold larger sizes of ordered domains in the experiments will lead to more dots, and thus to a higher density of states. Another important factor when comparing experiment with simulation is the coupling between the electrodes and the array. The magnitude of current flowing through the array is dependent on the coupling strength between the electrodes and the array, and also the coupling

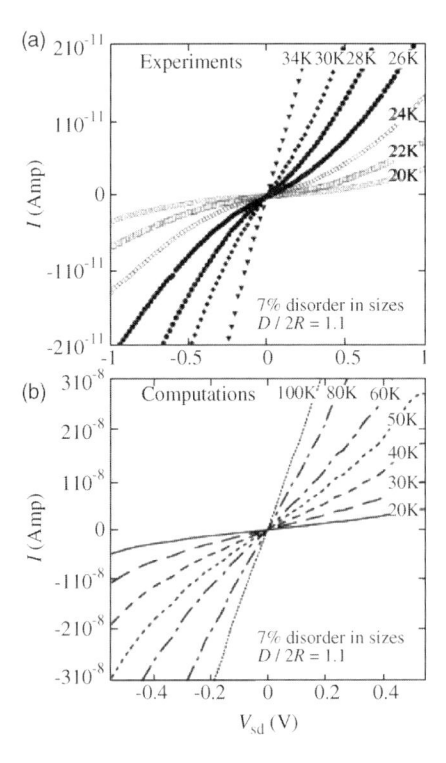

Figure 5.65 $I(V)$ characteristics of an array of 7 nm Ag particles. (a) Experimental characteristics and (b) calculated characteristics for an array consisting of 8911 dots with 7% size fluctuation and packing disorder. Reprinted with permission from Ref. [94]; © 2003, American Chemical Society.

between the dots. This can be tuned by varying the size and size distribution of the dots and by compressing the array. In addition, the gate voltage can be used to tune the electronic properties, although such a gating effect is also dependent on the compression of the array.

Greshnykh et al. [95] reported on DC transport measurements at different temperatures on cobalt–platinum nanoparticle monolayers in which the interparticle distance did not affect the activation energy. These authors synthesized cobalt–platinum particles with sizes of 4.4, 7.6 and 10.4 nm, and deposited them on a silicon oxide substrate with two gold electrodes. In order to probe the transport properties of the nanoparticle films (Figure 5.66a), DC measurements at different temperatures were performed. The initial result indicated that the electrical measurements showed no shape-induced effects. At room temperature, the detected $I(V)$ characteristics were point-symmetric and linear (Figure 5.66b), but at lower temperatures the $I(V)$ curves became nonlinear and showed a clear Coulomb blockade regime. From data fitting, it was possible to describe the characteristics for $T < 40$ K as electron-tunneling, and for $T > 40$ K as electron-hopping processes. In addition the activation energies, when calculated from the Arrhenius plot, showed that the smaller particles had higher activation energies, whereas the inter-particle distance did not affect the activation energy (Figure 5.66c). The film conductivity at room temperature scaled exponentially with the inter-particle distance. Another effect, namely an exponential decay of the activation energy at rising field strength, was also observed; this led to the conclusion that the activation energy was not equal to the charging energy.

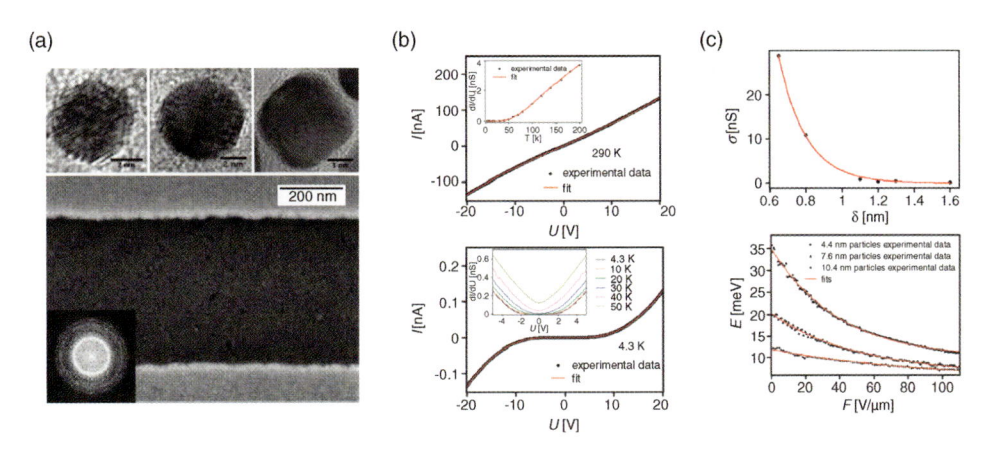

Figure 5.66 (a) HR-TEM images of the cobalt–platinum particles used. The larger particles have a cubic shape (top right). The lower image shows a monolayer of cobalt–platinum particles between two gold electrodes (the inset shows the obtained Fourier transformed pattern); (b) $I(V)$ characteristics of 7.6 nm particle monolayer at different temperatures; (c) Determined film conductivity versus interparticle distance (upper diagram) and field dependence of the activation energy (lower diagram). Reprinted with permission from Ref. [95]; © 2009, American Chemical Society.

5.2.5.2 Three-Dimensional Arrangements

Signatures of a transition from single-particle properties to 3-D charge transport have been reported by Pileni and coworkers [96], who performed tunneling spectroscopy studies at room temperature with 4.3 nm Ag nanoparticles stabilized in reverse micelles. The electrical transport measurements through single particles deposited on Au(111) substrates showed that, on increasing the applied voltage, the small capacitances of the double junction would be charged, while the detected current was close to zero because of the Coulomb blockade effect. Both, the $I(V)$ curve and dI/dV versus V plot (inset in Figure 5.67) clearly indicated a Coulomb blocking state – that is, a zero bias and the onset of conductivity through the nanoparticles above the threshold voltage. When the particles were deposited from solutions at higher concentrations, they arranged in a hexagonally ordered monolayer. The tunneling current through this

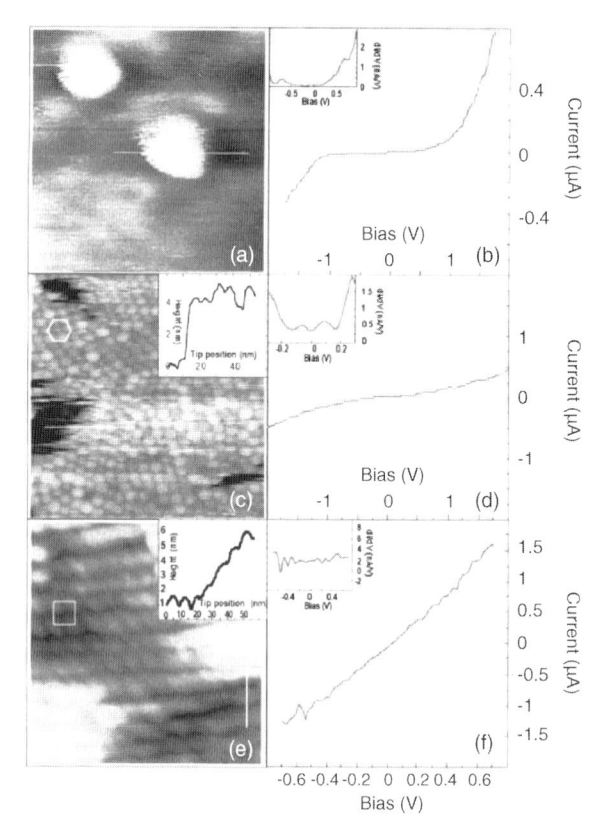

Figure 5.67 Left column: Constant-current mode STM image of isolated (a), self-organized in close-packed hexagonal network (c), and fcc structure (e) of silver nanoparticles deposited on an Au(111) substrate. Scan sizes:
(a) $17.1 \times 17.1 \, nm^2$, $V_t = -1 \, V$, $I_t = 1 \, nA$,
(c) $136 \times 136 \, nm^2$, $V_t = 2.5 \, V$, $I_t = 0.8 \, nA$,
(e) $143 \times 143 \, nm^2$, $V_t = -2.2 \, V$, $I_t = 0.72 \, nA$).
Right column: $I(V)$ curves and their derivatives in the inserts of isolated (b), self-organized in close-packed hexagonal network (d), and in fcc structure (f) of silver nanoparticles deposited on Au(111) substrate. Reproduced with permission from Ref. [96]; © 2000, Wiley-VCH Verlag GmbH & Co. KGaA.

layer showed a less-pronounced nonlinearity for low voltages, reflecting the fact that in 3-D systems an additional transport path for electrons appears between the adjacent nanoparticles. This caused the Coulomb gap between the individual particles to be smeared out, due to coupling of the electronic states in the close-packed system.

The electrical behavior of the 3-D system is also reflected in the electrical DC and AC response of compacts of ligand-stabilized nanoparticles [9]. As a common feature, at high temperatures (i.e., several tens of Kelvin below room temperature), the temperature-dependent DC and AC conductivities follow a simply activated behavior according to the Arrhenius relationship:

$$\sigma(T) = \sigma_0 + \exp\frac{-E_A}{k_B T},$$

where E_A, the activation energy, becomes temperature-dependent when the measuring temperature is decreased. This means that, down to very low temperatures, the conductivity follows the Variable-Range Hopping (VRH) expression proposed by Mott [97]:

$$\sigma(T) = \sigma_0 + \exp\left(\frac{-T_0}{T}\right)^{\gamma}$$

where $\gamma = 1$ $(d+1)$ in d dimensions. Although $\gamma = \frac{1}{4}$ $(d=3)$ might be expected from this general expression, $\gamma = \frac{1}{2}$ $(d=3)$ is predominantly observed in the case of the compacted metal cluster compounds. This temperature dependence, as well as the electric field-dependent conductivity, which reflects pronounced nonohmic behavior at strong electric fields, reveals a strong similarity to various heterogeneous materials, such as cermets, doped and amorphous semiconductors, or metal– and carbon–insulator composites. This behavior was carefully analyzed by van Staveren and Adriaanse [98–100], who applied different physical models of hopping conductivity, and concluded that the experimental data could best be fitted by a thermally activated stochastic multiple-site hopping process, whereas at high temperatures around room temperature, the nearest-neighbor hops predominated.

This means that at low temperatures, the number of charge carriers participating in the hopping process does not change with temperature. Instead, at high temperatures, where $k_B T$ becomes comparable to the charging energy E_c of the metal particle (note that E_c is determined by the total capacitance of the particle, and is therefore dependent on the particle size as well as on the inter-particle spacing, that is, the dot-to-dot distance), thermally excited extra charge carriers will participate in the hopping process. Thus, even at high temperatures, the activation energy reflects the energy required to transfer one electron from one electrically neutral particle to another, and may therefore be considered as the charge disproportionation energy. At low temperatures, the hopping transport would be expected to become zero if all particles were to become electrically neutral. As most reports on electrical conductivity reflect a residual conductivity at very low temperatures, it again becomes clear that packing, shape and size distribution prevent the localization sites from being identical (as noted at the start of this section).

In terms of a modified Anderson–Hubbard Hamiltonian [101, 102], orientational disorder and packing irregularities will lead to a distribution of the on-site Coulomb interaction, and of the interaction of electrons on different (at least neighboring) sites. This has been explicitly noted by Cuevas, Ortuno and Ruiz [103] who, in contrast to the Coulomb-gap model of Efros and Shklovskii [104], took three different states of charge of the mesoscopic particles into account – that is, neutral, positively charged, and negatively charged. The use of this model, VRH behavior, which dominates the electrical properties at low temperatures, can be fully explained.

In the high-temperature regime in many systems, nearest-neighbor hopping is predominant. Over this temperature range, it was shown that in 3-D systems of such cluster materials, the activation energy could be chemically tailored by means of bifunctional spacer molecules, which define the inter-particle spacing by the molecule length. This has been realized in the family of ligand-stabilized metal clusters for $Au_{55}(PPh_3)_{12}C_6$ and $Pd_{561}(phen)_{36}O_{200}$ by using bifunctional spacers with NH_2- or SH-groups on each terminus [105–107]. While the cluster size is kept constant, the length of the spacer molecules is increased such that an almost linear increase in the activation energy was observed. This can be explained by the decrease in junction capacitance C, which scales with $1/D$ (D = particle spacing), as long as the other geometric and dielectric parameters are held constant.

This simple relationship may be expected to be valid as long as the spacer molecules are not covalently bound to the cluster surface, and have no delocalized π-electron system along their backbone between the termini that is capable of supporting charge propagation. Hence, as soon as covalently linking species equipped with delocalized π-electrons enable inter-cluster electron transfer to occur, the activation energy drops, depending on the electronic structure of the molecule and its length. This reflects also that the activation energy observed is not identical to the charging energy of the individual clusters.

This relationship was controlled with eight different spacer molecules; the results are illustrated in Figure 5.68, in which the plot of activation energy versus cluster–cluster distance is represented.

Terril et al. characterized highly ordered 3-D systems, and showed that the electron-hopping conductivity depended on the activation energy of the electron transfer and the electron coupling term, β [108]. The term β represents the degree of participation of the ligands in the conduction process which control the dynamics of the electron core-to-core tunneling through the ligand shell and through the nonbonding contacts between ligand shells on adjacent nanoparticles. Terril et al. found, in the current–potential responses, characteristics with electron-hopping conductivity in which the electrons tunnel from Au core to Au core. With increasing chain length, the hopping rate was decreased and the activation energy increased, which is characteristic for a combination of a thermally activated electron transfer and a distance-dependent tunneling through the alkane coating of the nanoparticles. Therefore, the authors described the electron-tunneling barrier properties with the pre-exponential term of the Arrhenius relationship:

$$\sigma_{EL}(\delta, T) = \{\sigma_0 \exp[-\beta\delta]\} \exp[-E_A/RT].$$

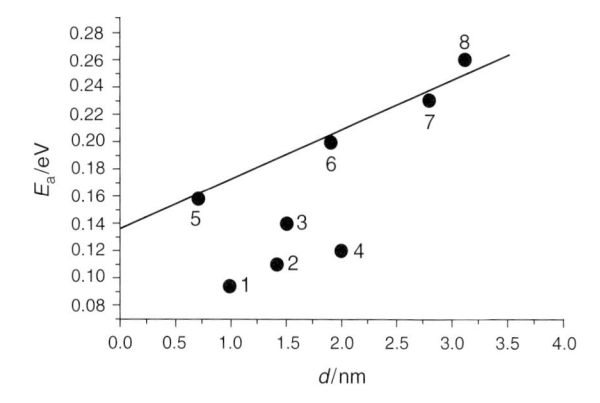

Figure 5.68 Diagram of activation energy measured in dependence of the cluster–cluster distance. Reproduced with permission from Ref. [107]; © 2003, Wiley-VCH Verlag GmbH & Co. KGaA.

In this equation, β is the electron-tunneling coefficient, δ is the average nanoparticle core edge–edge distance, E_A is the activation energy, and T is the temperature. The pre-exponential term ($\sigma_0 \exp[-\beta\delta]$) is the equivalent of an infinite-temperature electronic conductivity.

The temperature dependence of β for alkanethiolate molecules which connected 2.2 nm gold nanoparticles was studied by Wuelfing et al. [109], who concluded that larger activation barrier energies occurred at longer chain lengths. The differences in E_A could be avoided by plotting the Arrhenius intercepts against δ. Depending on the electronic structure of the ligand molecules, different β-values could be observed, ranging from 0.8 to 1.0 Å^{-1} for alkane ligands, and from 0.4 to 0.6 Å^{-1} for fully conjugated ligands [110].

Quinn et al. observed, for arrays of 1.6 nm gold nanoparticles, that the latter behaved as weakly coupled molecular solids comprising discrete nanoscale metallic islands, separated by insulating ligand barriers [111]. It was also found that the charge transport was dominated by the charging energy of the nanoparticle, the dielectric properties of the passivating ligands, the electrostatic coupling between neighboring cores, the inter-particle tunneling barrier resistance, and the dimensionality of the network of conducting paths.

On thin films of these clusters, Quinn et al. performed temperature-dependent DC measurements, from which charging energies between 106 meV < E_A < 112 meV could be derived, assuming an Arrhenius-like simple thermally activated behavior.

From simulations using the orthodox theory $E_A = e^2/2C_\Sigma$, the same authors calculated a total capacitance of $C_\Sigma = 0.75$ aF, with a mean activation energy of $E_A = 108$ meV. The total capacitance is given by the self-capacitance of an isolated nanoparticle (C_0) and the inter-particle capacitance arising from nearest-neighbor interactions (C_{nn}): $C_\Sigma = C_0 + C_{nn}$. In estimating the self-capacitance by simple electrostatics, by treating the nanoparticle as a conducting sphere of diameter d embedded in a dielectric of relative permittivity ε, the self-capacitance $C_0 = 2\pi\varepsilon_0\varepsilon d$ can thus be calculated to $C_0 \approx 0.25$ aF for $d = 1.65$ nm and $\varepsilon = 2.7$. As the value for the total capacitance exceeds the estimated nanoparticle self-capacitance, this suggests a

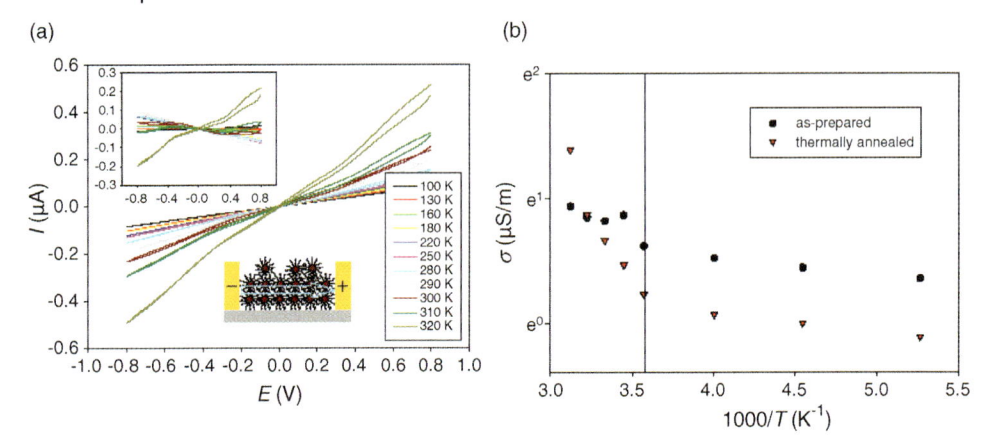

Figure 5.69 (a) $I(V)$ characteristics measured from a Au nanoparticle film at different temperatures after thermal annealing at 340 K for 3 h. The inset shows the difference spectra between as-prepared and annealed films; (b) Film conductivity measured as a function of temperature before and after annealing at 340 K. The vertical line at 280 K divides the Arrhenius profile (>280 K) from the tunneling profile (<280 K). Reprinted with permission from Ref. [112]; © 2009, American Institute of Physics.

substantial contribution from classical electrostatic inter-particle coupling from neighboring particles in the array.

Pradhan et al. observed a discrete charge transfer in nanoparticle solid films after thermal annealing, whereas before annealing only linear featureless current–potential characteristics could be measured [112]. The solid films were prepared by dropcasting a defined amount of 2.0 nm gold nanoparticles with octanethiol molecules as ligand shell on an interdigitated array electrode. The $I(V)$ characteristics measured from as-prepared films showed a linear behavior over the entire temperature range, whereas annealed films (340 K for 3 h) clearly showed Coulomb staircase features at temperatures >300 K that were attributed to a discrete charge transfer in the nanoparticle film (Figure 5.69a). This effect implied that a structural rearrangement of the nanoparticles had occurred. Another indicator for structural rearrangement is the higher conductivity of the as-prepared films compared to the lower conductivity of the annealed films (Figure 5.69b). Pradhan et al. concluded that the change in the $I(V)$ characteristics after annealing was mainly based on the increasing disorder of the organic matrix. The organic molecules favor gauche formation at elevated temperatures, which leads to a less-efficient electron tunneling, and therefore to a decrease in conductivity and a higher activation energy.

5.2.6
Concluding Remarks

In the preceding sections, it has been shown that the electrical properties of metal nanoparticles are strongly determined by the Coulomb charging energy. The respective phenomena can, in principle, be understood in terms of SET within the framework of the "orthodox theory of SET." Furthermore, for the smallest particles discussed here – that is, particles below 1–2 nm – additional effects due to the discrete nature of the energy states occur, and must be taken into account when discussing electrical transport phenomena and characteristic excitation energies [113].

To date, these effects still have not been studied experimentally or theoretically in great detail, and consequently most of the presently applied models are at least incomplete with respect to the huge variety of different control parameters employed in the chemical design of molecularly organized nanoparticles. In other words, the complex interplay of size and size distribution, of constitution, symmetry and conformation of the ligand molecules, of the state of charge of the particles, and of the embedding media and dielectric environment, is not yet understood to a sufficient degree. Hence, the electrical properties must be studied further to prove the reliability of the design strategies as described here for technological applications.

Whilst the application of chemically tailored metal nanoparticles in nanoelectronics might represent an alternative approach for the future, it remains an open question as to whether these nanoparticles – or other molecular objects, such as nanotubes or organic/organometallic molecules – will remain stable under the operating conditions of a high-performance electrical circuit. Moreover, the problem of inherent defects due to orientational disorders and isotope effects must be carefully considered.

Nevertheless, it is evident that, for any type of chemical approach, falling back onto the present-day paradigm of information exchange in logic elements or memory devices will require defect-tolerant computer architectures with suitable, intelligent software, as was proposed more than a decade ago [113]. On the other hand, alternative routes might be developed, which require more complex architectures and different principles of information exchange, where the burden with respect to redundancy, current density, performance speed – and especially of contact resistance and charge redistribution at the lead–molecule interface – are less relevant. Today, the solution of these problems represents one of the major challenges in chemistry.

References

1 Bardeen, J. and Brattain, W.H. (1948) *Phys. Rev.*, **74**, 230–231.

2 Okazaki, S. and Moers, J. (2005) *Nanoelectronics and Information Technology*, 2nd edn (ed R. Waser), Wiley-VCH, pp. 221–247.

3 Li, L., Gattass, R.R., Gershgoren, E., Hwang, H., and Fourkas, J.T. (2009) *Science*, **324**, 910–913.

4 Devoret, M.H. and Grabert, H. (1992) *Single Charge Tunneling Coulomb Blockade Phenomena in Nanostructures, NATO ASI Series*, vol. **294** (eds H. Grabertand M.H. Devoret), Plenum Press, New York, pp. 1–20.

5 Giaver, I. and Zeller, H.R. (1968) *Phys. Rev. Lett.*, **20**, 1504–1507.

6 Zeller, H.R. and Giaver, I. (1969) *Phys. Rev.*, **181**, 789–799.

7 Likharev, K.K. (1987) *IEEE Trans. Magn.*, **23**, 1142–1145.
8 Likharev, K.K. (1988) *IBM J. Res. Develop.*, **32**, 144–157.
9 Schön, G. and Simon, U. (1995) *Colloid Polym. Sci.*, **273**, 101–117.
10 Simon, U. (2004) *Nanoparticles - From Theory to Application* (ed. G. Schmid), Wiley-VCH, Weinheim, pp. 328–367.
11 Koplin, E. and Simon, U. (2007) *Metal Nanoparticles in Catalysis and Materials Science: The Issue of Size Control* (eds B. Corain, G. Schmid, and H. Toshima), Elsevier Science & Technology, pp. 107–128.
12 Schmid, G. and Simon, U. (2005) *Chem. Commun.*, 697–710.
13 Franke, M.E., Koplin, T.J., and Simon, U. (2006) *Small*, **2**, 36–50.
14 Fischler, M. and Simon, U. (2007) *Charge Migration in DNA: Perspectives from Physics, Chemistry, and Biology, NanoScience and Technology* (ed. T. Chakraborty), Springer-Verlag, Heidelberg, pp. 263–282.
15 Fischler, M., Homberger, M., and Simon, U. (2009) *Nanobioelectronics - for Electronics, Biology, and Medicine* (eds A. Offenhäusserand R. Rinaldi), Springer, New York, pp. 11–41.
16 Schmid, G., Brune, H., Ernst, H., Grünwald, W., Grünwald, A., Hofmann, H., Janich, P., Krug, H., Mayor, M., Rathgeber, W., Simon, U., Vogel, V., and Wyrwa, D. (2006) *Nanotechnology. Assessment and Perspectives*, Springer Verlag, Berlin.
17 Schmid, G. (1992) *Chem. Rev.*, **92**, 1709–1727.
18 Schmid, G. (1994) *Clusters and Colloids*, Wiley-VCH.
19 Feldheim, D.L. and Foss, C.A. (eds) (2002) *Metal Nanoparticles: Synthesis, Characterization and Application*, Marcel Dekker, New York.
20 Ashoori, R.C. (1996) *Nature*, **379**, 413–419.
21 Reed, M. (1993) *Sci. Am.*, **268**, 98–103.
22 Kastner, M.A. (1993) *Physics Today*, **46**, 24–31.
23 Feldheim, D.L. and Keating, C.D. (1998) *Chem. Soc. Rev.*, **27**, 1–12.
24 (a) Simon, U. (1998) *Adv. Mater.*, **10**, 1487–1492; (b) Simon, U. (1999) *Metal Clusters in Chemistry*, vol. **3**, Wiley-VCH, pp. 1342–1359.
25 Simon, U. and Schön, G. (2000) *Handbook of Nanostructured Materials and Nanotechnology*, vol. **3** (ed. H.S. Nalwa), Academic Press, p. 131.
26 (a) Simon, U., Schön, G., and Schmid, G. (1993) *Angew. Chem., Int. Ed. Engl.*, **32**, 250–254; (b) Schmid, G., Schön, G., and Simon, U. (1992) USA Patent No. 08/041. p. 239.
27 Schön, G. (1997) *Quantum Transport and Dissipation* (eds T. Dittrich et al..), Wiley-VCH, pp. 149–212.
28 Uchida, K. (2005) *Nanoelectronics and Information Technology*, 2nd edn (ed. R. Waser), Wiley-VCH, pp. 423–443.
29 Grabert, H. (1991) *Z. Phys. B*, **85**, 319–325.
30 Lafarge, P., Pothier, H., Williams, E.R., Esteve, D., Urbina, C., and Devoret, M.H. (1991) *Z. Phys. B*, **85**, 327–332.
31 Laaksonen, T., Ruiz, V., Liljeroth, P., and Quinn, B.M. (2008) *Chem. Soc. Rev.*, **37**, 1836–1846.
32 Murray, R.W. (2008) *Chem. Rev.*, **108**, 2688–2720.
33 Zabet-Khosousi, A. and Dhirani, A.-A. (2008) *Chem. Rev.*, **108**, 4072–4124.
34 Anselmetti, D., Richmont, D., Baratoff, A., Borer, G., Dreier, M., Bernasconi, M., and Güntherodt, H.-J. (1994) *Europhys. Lett.*, **25**, 297–302.
35 Dorogi, M., Gomez, J., Osifchin, R., Andres, R.P., and Reifenberger, R. (1995) *Phys. Rev. B*, **52**, 9071–9077.
36 Wang, C.-K., Fu, Y., and Luo, Y. (2001) *Phys. Chem. Chem. Phys.*, **3**, 5017–5023.
37 Schouteden, K., Vandamme, N., Janssens, E., Lievens, P., and Van Haesendonck, C. (2008) *Surf. Sci.*, **602**, 552–558.
38 van Kempen, H., Dubois, J.G.A., Gerritsen, J.W., and Schmid, G. (1995) *Physica B*, **204**, 51–56.
39 Dubois, J.G.A., Gerritsen, J.W., Shafranjuk, S.E., Boon, E.J.G., Schmid, G., and van Kempen, H. (1996) *Europhys. Lett.*, **33**, 279–284.
40 Averin, D.V. and Korotkov, A.N. (1990) *J. Low-Temp. Phys.*, **80**, 173–185.
41 Halperin, W.P. (1986) *Rev. Mod. Phys.*, **58**, 533–606.
42 Houbertz, R., Feigenspan, T., Mielke, F., Memmert, U., Hartmann, U., Simon, U.,

Schön, G., and Schmid, G. (1994) *Europhys. Lett.*, **28**, 641–646.
43 Chi, L.F., Hartig, M., Drechsler, T., Schaak, Th., Seidel, C., Fuchs, H., and Schmid, G. (1998) *Appl. Phys. A.*, **66**, S187–S190.
44 Zhang, U., Schmid, G., and Hartmann, U. (2003) *Nano Lett.*, **3**, 305–307.
45 Xu, L.-P. and Chen, S. (2009) *Chem. Phys. Lett.*, **468**, 222–226.
46 Gittins, D., Bethell, D., Schiffrin, D.J., and Nichols, R.J. (2000) *Nature*, **408**, 67–69.
47 Brousseau, L.C. III, Zhao, Q., Schultz, D.A., and Feldheim, D.L. (1998) *J. Am. Chem. Soc.*, **120**, 7645–7646.
48 Albrecht, T., Mertens, S.F.L., and Ulstrup, J. (2007) *J. Am. Chem. Soc.*, **129**, 9162–9167.
49 Klein, D.L., McEuen, O.L., Bowenkatari, J.E., Roth, R., and Alivisatos, A.P. (1996) *Appl. Phys. Lett.*, **68**, 2574–2576.
50 Klein, D.L., Roth, R., Lim, K.L.A., Alivisatos, A.P., and McEuen, P.L. (1997) *Nature*, **389**, 699–701.
51 Bezryadin, A., Dekker, C., and Schmid, G. (1997) *Appl. Phys. Lett.*, **71**, 1273–1275.
52 Chung, S.W., Ginger, D.S., Morales, M.W., Zhang, Z., Chandrasekhar, V., Ratner, M.A., and Mirkin, C.A. (2005) *Small*, **1**, 64–69.
53 Gasparian, V. and Simon, U. (1997) *Physica B*, **240**, 289–297.
54 Simon, U. and Gasparian, V. (1998) *Phys. Status Solidi B*, **205**, 223–227.
55 Landauer, R. (1979) *Philos. Mag.*, **21**, 863.
56 Landauer, R. (1984) *J. Phys. Cond. Mater.*, **1**, 8099.
57 Datta, S. (1995) *Electronic Transport in Mesoscopic Systems*, Cambridge University Press.
58 Simon, U. (1998) Struktur/Eigenschaftsbeziehungen und Strategien zur Steuerung des Ladungstranportes in nanoporösen Festkörpern. Habilitation Thesis, University of Essen.
59 Semrau, S., Schoeller, H., and Wenzel, W. (2005) *Phys. Rev. B*, **72**, 205443-1–205443-13.
60 Niemeyer, C.M. and Simon, U. (2005) *Eur. J. Inorg. Chem.*, 3641–3655.
61 Sato, T., Ahmed, H., Brown, D., and Johnson, B.F.H. (1997) *J. Appl. Phys.*, **82**, 696–701.

62 Samanta, M.P., Tian, W., Datta, S., Henderson, J.I., and Kubiak, C.P. (1996) *Phys. Rev. B*, **53**, R7626–R7629.
63 Schmid, G., Liu, Y.-P., Schumann, M., Raschke, T., and Radehaus, C. (2001) *Nano Lett.*, **1**, 405–407.
64 Lee, J.-H., Cheon, J., Lee, S.B., Chang, Y.-W., Kim, S.-I., and Yoo, K.-H. (2005) *J. Appl. Phys.*, **98**, 084315-1–084315-3.
65 Weiss, D.N., Brokmann, X., Calvet, L.E., Kastner, M.A., and Bawendi, M.G. (2006) *Appl. Phys. Lett.*, **88**, 143507-1–143507-3.
66 Bates, A.D., Callen, B.P., Cooper, J.M., Cosstick, R., Geary, C., Glidle, A., Jaeger, L., Pearson, J.L., Proupin-Pérez, M., Xu, C., and Cumming, D.R.S. (2006) *Nano Lett.*, **6**, 445–448.
67 Coskun, U.C., Mebrahtu, H., Huang, P.B., Huang, J., Sebba, D., Biasco, A., Makarovski, A., Lazarides, A., LaBean, T.H., and Finkelstein, G. (2008) *Appl. Phys. Lett.*, **93**, 123101-1–123101-3.
68 Blech, K., Noyong, M., Juillerat, F., Nakayama, T., Hofmann, H., and Simon, U. (2008) *J. Nanosci. Nanotechnol.*, **8**, 461–465.
69 Kouwenhoven, L.P. (1995) *Science*, **268**, 1440–1441.
70 Osterkamp, T.H., Fujisawa, T., van der Wiel, W.G., Ishibashi, K., Hijman, R.V., Tarucha, S., and Kouwenhoven, L.P. (1998) *Nature*, **395**, 873–876.
71 Blick, R.H., Hang, R.J., Weis, J., Pfannkuche, D., von Klitzing, K., and Erbel, K. (1995) *Phys. Rev. B*, **53**, 7899–7902.
72 Livermore, C., Crouch, C.H., Westervelt, R.M., Campman, K.L., and Gossard, A.C. (1996) *Science*, **274**, 1332–1335.
73 Remacle, F. and Levine, R.D. (2001) *ChemPhysChem*, **2**, 20–36.
74 Collier, C.P., Vossmeyer, T., and Heath, J.R. (1998) *Annu. Rev. Phys. Chem.*, **49**, 371–404.
75 Murray, C.B., Kagan, C.R., and Bawendi, M.G. (1995) *Science*, **270**, 1335–1338.
76 Sun, H.S., Murray, C.B., Weller, D., Folks, L., and Moser, A. (2000) *Science*, **287**, 1989–1992.
77 Collier, C.P., Saykally, R.J., Shiang, J.J., Henrichs, S.E., and Heath, J.R. (1997) *Science*, **277**, 1978–1981.

78 Alivisatos, A.P. (1996) *Science*, **271**, 933–937.
79 Murray, C.B., Norris, D.J., and Bawendi, M.G. (1993) *J. Am. Chem. Soc.*, **115**, 8706–8715.
80 Whetten, R.L., Khoury, J.T., Alvarez, M.M., Murthy, S., Vezmar, I., Wang, Z.L., Stephens, P.W., Cleveland, C.L., Luedtke, W.D., and Landman, U. (1996) *Adv. Mater.*, **8**, 428–433.
81 Taleb, A., Russier, V., Courty, A., and Pileni, M.P. (1999) *Phys. Rev. B*, **59**, 13350–13358.
82 Kiely, C.J., Fink, J., Brust, M., Bethell, D., and Shiffrin, D.J. (1998) *Nature*, **396**, 444–446.
83 Kiely, C.J., Fink, J., Zheng, J.G., Brust, M., Bethell, D., and Shiffrin, D.J. (2000) *Adv. Mater.*, **12**, 640–643.
84 Markovich, G., Collier, C.P., and Heath, J.M. (1998) *Phys. Rev. Lett.*, **80**, 3807–3810.
85 Janes, D.B., Kolagunta, V.R., Osifchin, R.G., Bielefeld, J.D., Andres, R.P., Henderson, J.I., and Kubiak, C.P. (1995) *Superlattice Microstruct.*, **18**, 275–282.
86 Santhanam, V., Liu, J., Agarwal, R., and Andres, R.P. (2003) *Langmuir*, **19**, 7881–7887.
87 Mueggenburg, K.E., Lin, X.-M., Goldsmith, R.H., and Jaeger, H.M. (2007) *Nat. Mater.*, **6**, 656–660.
88 Pileni, M.P. (2001) *J. Phys. Chem. B*, **105**, 3358–3371.
89 Liao, J., Bernard, L., Langer, M., Schönenberger, C., and Calame, M. (2006) *Adv. Mater.*, **18**, 2444–2447.
90 van der Molen, S.J., Liao, J., Kudernac, T., Agustsson, J.S., Bernard, L., Calame, M., van Wees, B.L., Feringa, B.L., and Schönenberger, C. (2009) *Nano Lett.*, **9**, 76–80.
91 Andres, R.P., Bielefeld, J.D., Henderson, J.I., Janes, D.B., Kolagunta, V.R., Kubiak, C.P., Mahoney, W.J., and Osifchin, R.G. (1996) *Science*, **273**, 1690–1693.
92 Koplin, E., Niemeyer, C.M., and Simon, U. (2006) *J. Mater. Chem.*, **16**, 1338–1344.
93 Beverly, K.C., Sampaio, J.F., and Heath, J.R. (2002) *J. Phys. Chem. B*, **106**, 2131–2135.
94 Remacle, F., Beverly, K.C., Heath, J.R., and Levine, R.D. (2003) *J. Phys. Chem. B*, **107**, 13892–13901.
95 Greshnykh, D., Frömsdorf, A., Weller, H., and Klinke, C. (2009) *Nano Lett.*, **9**, 473–478.
96 Taleb, A., Silly, F., Gusev, A.O., Charra, F., and Pileni, M.-P. (2000) *Adv. Mater.*, **12**, 633–637.
97 Mott, N.F. (1969) *Phil. Mag.*, **19**, 835–852.
98 van Staveren, M.P.J., Brom, H.B., and de Jongh, L.J. (1991) *Phys. Rep.*, **208**, 1–96.
99 Adriaanse, L.J. (1997) Charge carrier transport in metal/non-metal composites. PhD Thesis, University of Leiden, Netherlands.
100 de Jongh, I.J. (1994) *Physics and Chemistry of Metal Cluster Compounds*, Kluwer Academic Publishers, Dordrecht.
101 Anderson, P.W. (1958) *Phys. Rev.*, **109**, 1492–1505.
102 Hubbard, J. (1964) *Proc. Roy. Soc. A*, **277**, 237.
103 Cuevas, E., Ortuño, M., and Ruiz, J. (1993) *Phys. Rev. Lett.*, **12**, 1871–1874.
104 Efros, A.L. and Shklovskii, B.I. (1976) *Phys. Status Solidi B*, **76**, 475–485.
105 Simon, U., Flesch, R., Wiggers, H., Schön, G., and Schmid, G. (1998) *J. Mater. Chem.*, **8**, 517–518.
106 Simon, U. (2000) *Mater. Res. Soc. Symp. Proc.*, **581**, 77–82.
107 Torma, V., Vidoni, O., Simon, U., and Schmid, G. (2003) *Eur. J. Inorg. Chem.*, **6**, 1121–1127.
108 Terrill, R.H., Postlethwaite, T.A., Chen, C.H., Poon, C.D., Terzis, A., Chen, A., Hutchison, J.E., Clark, M.R., Wignall, G., Londono, J.D., Superfine, R., Falvo, M., Johnson, C.S. Jr, Samulski, E.T., and Murray, R.W. (1995) *J. Am. Chem. Soc.*, **117**, 12537–12548.
109 Wuelfing, W.P., Green, S.J., Cliffel, D.E., Pietron, J.J., and Murray, R.W. (2000) *J. Am. Chem. Soc.*, **122**, 11465–11472.
110 Wuelfing, W.P. and Murray, R.W. (2002) *J. Phys. Chem. B*, **106**, 3139–3145.
111 Quinn, A.J., Biancardo, M., Floyd, L., Belloni, M., Ashton, P.R., Preece, J.A., Bignozzi, C.A., and Redmond, G. (2005) *J. Mater. Chem.*, **15**, 4403–4407.
112 Pradhan, S., Kang, X., Mendoza, E., and Chen, S. (2009) *Appl. Phys. Lett.*, **94**, 042113-1–042113-3.
113 Heath, J.R., Kuekes, P., Snider, G., and Williams, S. (1998) *Science*, **280**, 1716–1721.

6
Semiconductor Quantum Dots for Analytical and Bioanalytical Applications
Ronit Freeman, Jian-Ping Xu, and Itamar Willner

6.1
Introduction

The synthesis and characterization of semiconductor nanoparticles or quantum dots (QDs) represent two of the major advances in material sciences during the past two decades. The photoexcitation of valence-band electrons into the conduction band of the semiconductor material yields electron–hole pairs that recombine and yield luminescence with energy corresponding to the band-gap of the semiconductor material. The theoretical understanding of the parameters controlling the luminescence features of QDs [1–4], and the experimental studies demonstrating the luminescence properties of numerous compositions of semiconductor materials exhibiting variable sizes and shapes [5–10], have led to the establishment of a new class of materials that finds broad applications in areas such as, energy conversion and storage [11–14], optoelectronic devices [15], sensor applications [16, 17], and photocatalysis [18–21]. The ingenious synthetic routes for preparing size-controlled semiconductor nanoparticles include the kinetics-controlled growth of monolayer- or thin films-protected QDs [22], or the synthesis of the QDs in nanoreactor systems such as, micelles [23], microemulsions [24, 25], or mesoporous inorganic supports [26, 27]. Further versatility in the synthesis of QDs was demonstrated with the preparation of core–shell semiconductor materials [28–35] or semiconductor materials exhibiting nanorod shapes [36–39] and even branched nanorod structures [6, 40–42]. The unique properties of QDs are represented by their photophysical and chemical properties:

- QDs exhibit high fluorescence quantum yields, and reveal a high stability against degradation and photobleaching [43].
- QDs reveal size-controlled luminescence features, due to the quantum confinement of the electrons in the particles. Thus, as the particles are smaller, the luminescence energies are blue-shifted to higher energies [1, 3, 44–46]. As control over the sizes of QDs was achieved by the different synthetic and separation methods, particles of the same material with luminescence properties covering a broad spectral region are available.

Nanoparticles: From Theory to Application. Edited by Günter Schmid
Copyright © 2010 WILEY-VCH Verlag GmbH & Co. KGaA, Weinheim
ISBN: 978-3-527-32589-4

- The fluorescence spectra of the QDs are usually characterized by narrow emission bands, exhibiting a large Stokes shift. These properties are important for coupling the emission functions of the QDs to other-sized QDs [47], or to organic fluorophores [48] (*vide infra*).
- Unlike molecular fluorophores, semiconductor QDs have broad absorbance bands, which allow the photoexcitation of different-sized QDs by a single wavelength.
- Finally, the arsenal of synthetic methodologies for the preparation of QDs leads to monolayer- or thin-film-protected nanoparticles with tailored functionalities (see Section 6.2). This enables the tethering of molecular or biomolecular units to the QDs [49], and also allows the electronic coupling between the photoexcited QDs and the tethered molecules or biomolecules.

Not surprisingly, QDs show great promise as photonic labels for analytical and bioanalytical applications. Specifically, the similar range of dimensions of QDs and biomolecules such as, enzymes, antigens/antibodies, protein receptors or nucleic acids suggest that coupling of the biomolecules to the QDs could yield hybrid materials that combine the photophysical properties of the nanoparticles, with the recognition or catalytic processes driven by the biomolecules. In fact, several coupled interactions within chemically functionalized QDs conjugates may occur, and these could be utilized for different analytical applications:

- The QDs may act as luminescent probes for recognition (sensing) events.
- Quenching of the luminescence of the QDs by an electron transfer (ET) mechanism, or by a fluorescence resonance energy transfer (FRET) route, could reflect the formation of a recognition pair, or may be used to follow the dynamics of chemical processes that occur on the modified QDs. In these systems, the sensing events or the respective biocatalytic transformations involve the association/dissociation of quencher/FRET acceptor units that electronically couple with the QDs.
- Light generated by chemical or biochemical processes (chemiluminescence or biochemiluminescence) may serve as the energy source for the photoexcitation of the QDs [50–52]. Thus, the labeling of sensing events with light-generating components may lead to the activation of the luminescence of the QDs.

The FRET process between donor and acceptor luminescent probes is a versatile spectroscopic tool to follow the intimate contact between molecular components. Hence, it also serves as a useful method to follow sensing events between recognition elements and their analytes [53–55].

The FRET process involves dipole–dipole interactions between a donor and acceptor pair, where the FRET probability decreases with distance by the sixth power. This causes the FRET efficiency, E, to be very sensitive to the distance, r, separating the donor–acceptor couple, (where R_o is the Förster radius):

$$E = \frac{R_o^6}{R_o^6 + r^6} \qquad (6.1)$$

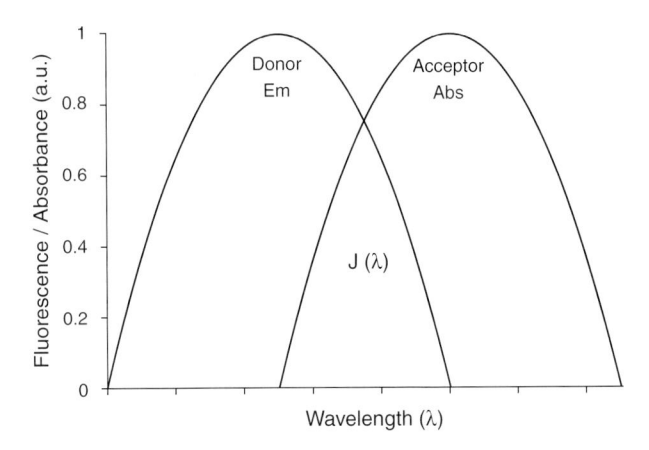

Figure 6.1 The overlap integral that controls the FRET process.

The Förster radius R_o is calculated via Eq. (6.2), where K^2 is a parameter that depends on the relative orientation of the dipoles and gains a value between 0 and 4 (for two randomly oriented dipoles the value of K^2 is 2/3). QY_D corresponds to the luminescence quantum yield of the donor fluorophore, n is the refractive index of the medium, and $J(\lambda)$ represents the overlap integral that provides a quantitative measure for the donor–acceptor spectral overlap (Figure 6.1). The overlap integral is calculated by Eq. (6.3), where $\varepsilon_A(\lambda)$ is the extinction coefficient of the acceptor and $F_D(\lambda)$ is the normalized emission spectrum.

$$R_0 = \sqrt[6]{\frac{8.8 \times 10^{-23} \kappa^2 QY_D J(\lambda)}{n^4}} \tag{6.2}$$

$$J(\lambda) = \int_0^\infty \varepsilon_A(\lambda) F_D(\lambda) \lambda^4 d\lambda \tag{6.3}$$

Typical Förster distances separate the donor and acceptor range between 2 and 8 nm. FRET signals are detectable up to about twice the Förster distance separating the donor–acceptor pair. If, however, the acceptor unit is tethered to n donor sites, the FRET efficiency increases and is given by Eq. (6.4):

$$E = \frac{nR_0^6}{nR_0^6 + r^6} \tag{6.4}$$

Substantial progress in the application of QDs for optical sensing and biosensing was accomplished in the past decade, and several reviews have summarized the advances in this field [56–58]. In this chapter, the aim is to emphasize the use of QDs for following molecular and biomolecular recognition events, and to probe the dynamics of chemical or biomolecular transformations by the application of modified hybrid QDs.

6.2
Water Solubilization and Functionalization of Quantum Dots with Biomolecules

Semiconductor QDs with superior photophysical features (i.e., high fluorescence quantum yields and stability against photobleaching) are usually synthesized from organometallic precursors, and are generally protected by a capping layer composed of organic ligands such, as trioctylphosphine (TOP) or trioctylphosphine oxide (TOPO) [59–62]. The resulting capped nanocrystals are hydrophobic and, therefore, insoluble in aqueous solvents, such that they cannot be directly applied in biological assays. Consequently, the organic capping ligands must either be exchanged or further functionalized with hydrophilic capping agents or, alternatively, aqueous synthetic methods for the production of QDs must be developed.

The introduction of QDs into aqueous media is usually accompanied by drastic decreases in the luminescence yields of the QDs. This effect presumably originates from the reaction of surface states with water, a process that yields surface traps for the conduction-band electrons [63]. As biorecognition events or biocatalytic transformations require aqueous environments for their reaction medium, it is imperative to preserve the luminescence properties of QDs in aqueous systems. Methods to stabilize the fluorescence properties of semiconductor QDs in aqueous media (Figure 6.2) have included surface passivation with protective layers, such as proteins [64, 65], as well as the coating of QDs with protective silicon oxide films [66, 67] or polymer films [43, 68, 69]. Alternatively, they can be coated with amphiphilic polymers, which have both a hydrophobic side chain that interacts with the organic capping layer of the QDs and a hydrophilic component, such as a poly(ethylene glycol) (PEG) backbone, for water solubility [70, 71]. Such water-soluble QDs may retain up to 55% of their quantum yields upon transfer to an aqueous medium.

Unfortunately, although these methods successfully preserve the QDs' photophysical properties, they will be endowed with thick stabilizing capping layers that hamper their photophysical functions. Typically, a thick passivation layer would create a barrier against the application of QDs in any FRET or ET quenching processes associated with sensing events or biocatalytic reactions. Thus, a delicate balance must be maintained between the nanostructure of the modifying capping layer associated with the QDs, and its effect on the photophysical features of the particles.

An alternative method for producing water-soluble QDs relies on the exchange of organic ligands linked to the QDs, such as TOPO or octadecylamine (ODA), with thiolated water-soluble ligands in a water–organic two-phase system. The most common thiolated molecules used to stabilize semiconductor QDs in aqueous media are thiolated aliphatic carboxylic acids, such as mercaptoacetic acid (MAA) [48, 65, 72], mercaptopropionic acid (MPA), or mercaptoundecanoic acid (MUA) (Figure 6.2). In order to achieve a higher stability, bidentate surface ligands composed of derivatives of dihydrolipoic acid (DHLA) [64, 73] or dithiothreitol (DTT) [74] have been used for the preparation of water-soluble CdSe/ZnS QDs. These bidentate ligands provide stable interactions with the QDs surfaces, owing to the chelating effect of the dithiol groups. Likewise, lipoic acid units

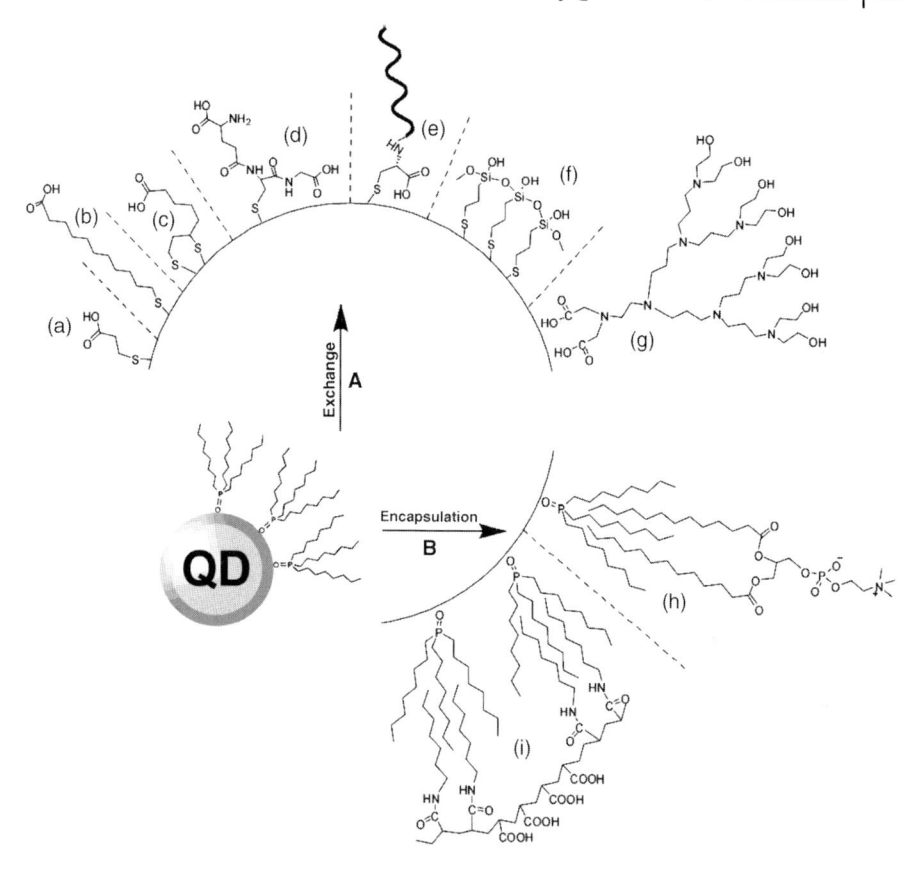

Figure 6.2 Modification of semiconductor quantum dots (QDs) with functional encapsulating layers for water solubilization and preservation of luminescence properties and/or secondary covalent modification of the surface with biomolecules. Path A: Exchange of the organic encapsulating layer with a water-soluble layer; (a–d) thiolated or dithiolated functional monolayers; (e) cysteine-terminated peptide; (f) thiolated siloxane; (g) carboxylic acid-functionalized dendron. Path B: Encapsulation of QDs stabilized with an organic encapsulating layer in functional bilayer films composed of (h) a phospholipid-encapsulating layer, and (i) a diblock copolymer.

tethered to PEG spacers that had been functionalized with linkable functionalities for biomolecules were used to modify the QDs [75, 76]. These modifiers eliminated any nonspecific adsorption processes, and also provided anchoring sites for any covalent immobilization of the biomolecules. Other thiol-containing materials, such as peptides with a polycysteine adhesive domain, were used to synthesize water-soluble QDs [77, 78]. Additionally, QDs were stabilized by exchanging ODA with water-soluble dendrons so as to yield water-soluble QDs, with quantum yields of 36% [79]. Interestingly, internally facing carboxylate groups on the dendrons retained a higher quantum yield in water compared to an internally facing thiol group [80].

Several methods for the synthesis of functionalized QDs directly in water have been reported [56, 81, 82]. When compared to organic syntheses, the aqueous synthetic approaches for QD preparation are simpler, reproducible and easily scaled-up, whilst the resultant QDs often exhibit a lower crystallinity or quantum yield. Different aqueous-based synthetic routes to prepare QDs that will retain a high fluorescence yield of the particles have been developed. Such routes are based on the reaction of metal salts (e.g., cadmium) and NaHTe or NaHSe in the presence of water-soluble molecular or macromolecular capping agents that kinetically protect the growth of the QDs by either covalent or ligand–ion interactions, so as to yield monolayer or thin-film-capped particles [83]. For example, nanocrystals such as CdSe [84, 85] and HgTe [86] were successfully prepared in the water phase using mercaptoethanol and thioglycerol as stabilizers, and the resultant QDs possessed high quantum yields (up to 40%). CdSe QDs with a narrow size distribution that revealed a high fluorescence quantum yield (25%) were also synthesized via a single-step procedure in water, using glutathione (GSH) as a stabilizing agent [87]. Similarly, glutathione-capped CdTe (GSH-CdTe) QDs with tunable fluorescence features (500–650 nm) and quantum yields as high as 45% were synthesized in aqueous media, and used to stain fixed cells [88]. The single-step synthesis of thiolated cyclodextrin-modified CdSe/CdS core–shell QDs resulted in water-soluble QDs with a quantum yield of 46% in water [89]. Aqueous syntheses of citrate-capped QDs [90, 91] and cysteine-modified QDs [92] have also been reported. Another aqueous synthesis of QDs, based on microwave irradiation under controlled temperature, allowed the preparation of a series of stable water-soluble MPA-capped CdTe QDs with tunable emission (505–733 nm) and high quantum yields (40–60%). This method was used also for the production of high-quality alloyed QDs [93, 94].

A secondary functionalization of the modified QDs with molecular or biomolecular recognition units is essential in order to develop hybrid QDs systems for sensing applications. For this, different methods were developed to tether molecular or biomolecular sensing functionalities to the QDs. Noncovalent interactions such as, electrostatic interactions have been used successfully to bioconjugate biomolecules to the QDs, while genetically engineered proteins that included a positively charged domain [64] or positively charged proteins (e.g., avidin [95] or papain [96]) were attached electrostatically to the negative surface of carboxylic acid-modified QDs. The protein–QD conjugates prepared in this way revealed a good water solubility and high fluorescence quantum yields (higher than for nonconjugated QDs), while the biomolecules retained their native activity. The covalent attachment of biomolecules to the functional capping layer protecting the QDs led to the production of stable biomolecular–QDs conjugates. For this, the most common methods have involved the coupling of primary amines (which were tethered to the biomolecules) to carboxylic acid residues on the capping layer associated with the nanoparticles. In an example of this, a DNA modified with a single primary amino group was linked to carboxylic acid-modified QDs [97]. A variety of antibodies that included free exposed thiol groups following reduction with DTT, were conjugated to QDs with free amine functionalities in their capping layer, by using heterobifunctional crosslinkers, such

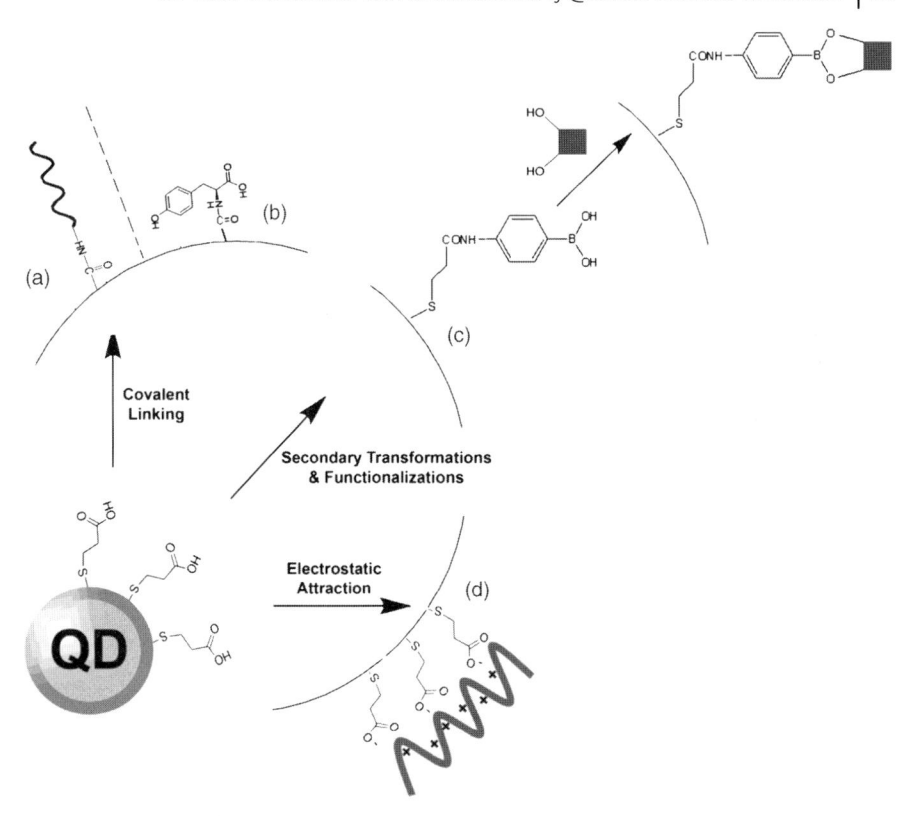

Figure 6.3 Modification of monolayer-functionalized QDs with molecular or macromolecular components. (a, b) Covalent attachment; (c) Attachment of ligand; (d) Electrostatic association of a polymer.

as succinimidyl 4-(*N*-maleimidomethyl)cyclohexane-1-carboxylate (SMCC). In addition, the functional groups ($-COOH$, $-NH_2$, $-SH$ or $-OH$) on a water-soluble QD surface can be further transformed into other functionalities that allow the secondary covalent attachment of molecules or biomolecules (Figure 6.3). For example, carboxylic acids on QDs have been converted to hydrazides, thereby allowing the secondary binding of carbohydrates [98]. The covalent tethering of phenylboronic acid ligands was also used to bind (via boronate esters) carbohydrates or *ortho*-dihydroxy benzene substrates [99]. Similarly, elaboration of the free hydroxyl groups of DTT-capped CdSe/ZnS QDs were transformed into imidazole-carbamates that enabled the subsequent tethering of DNA to the QDs surface. A further approach to conjugate biomolecules to QDs employed the polyhistidine tag (comprising six histidine residues) that binds carboxylic acid-functionalized QDs. Polyhistidine-tagged proteins [76, 100], antibodies [101], peptides [102], or DNA [103], were successfully coupled to QDs using this method.

6.3
Quantum Dot-Based Sensors

The chemical modification of semiconductor QDs with specific ligands or receptor units allows the development of optical sensors by the general mechanisms, as outlined in Figure 6.4. In this case, the analyte is labeled by a quencher unit that interacts with the QDs by FRET or ET mechanisms. The competitive binding of the analyte and the labeled analyte to the ligand (Figure 6.4, path a) or the equilibrium displacement of the label (Figure 6.4, path b) controls the resulting luminescence functions of the hybrid QDs. Alternatively, the association of the analyte with the ligand/receptor modifiers of the QDs may control the luminescence features of the QDs, thus leading to a quantitative detection of the analyte (Figure 6.4, path c).

6.3.1
Receptor- and Ligand-Functionalized QDs for Sensing

Chemical sensors based on the competitive binding of sugars to the boronic acid ligand have been reported. When 3-aminophenyl boronic acid was covalently linked to GSH-capped CdSe-ZnS QDs, the boronic acid ligand formed boronate esters with

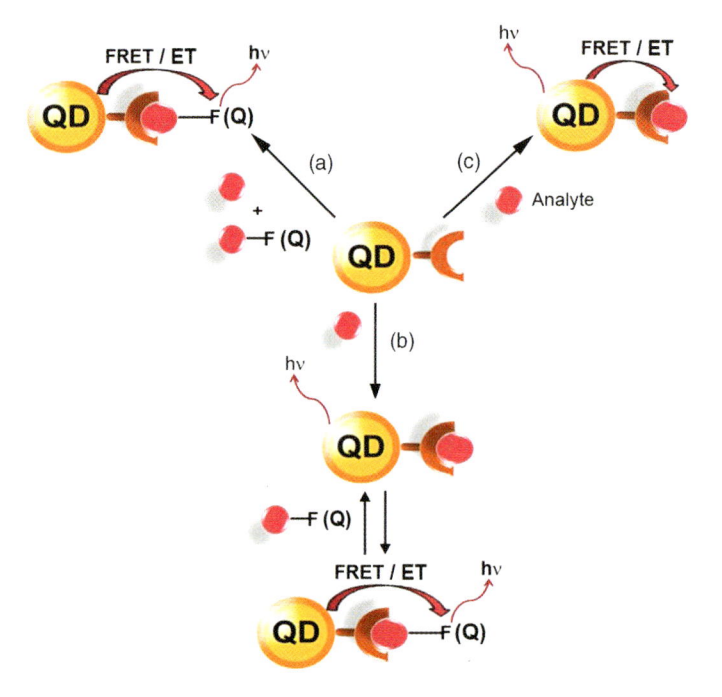

Figure 6.4 Different mechanisms for the application of QDs as optical sensors. Path (a): Competitive detection of an analyte using a labeled analyte and FRET/ET as transduction mechanisms. Path (b): Direct FRET/ET quenching of a labeled analyte. Path (c) Direct FRET/ET quenching by the analyte.

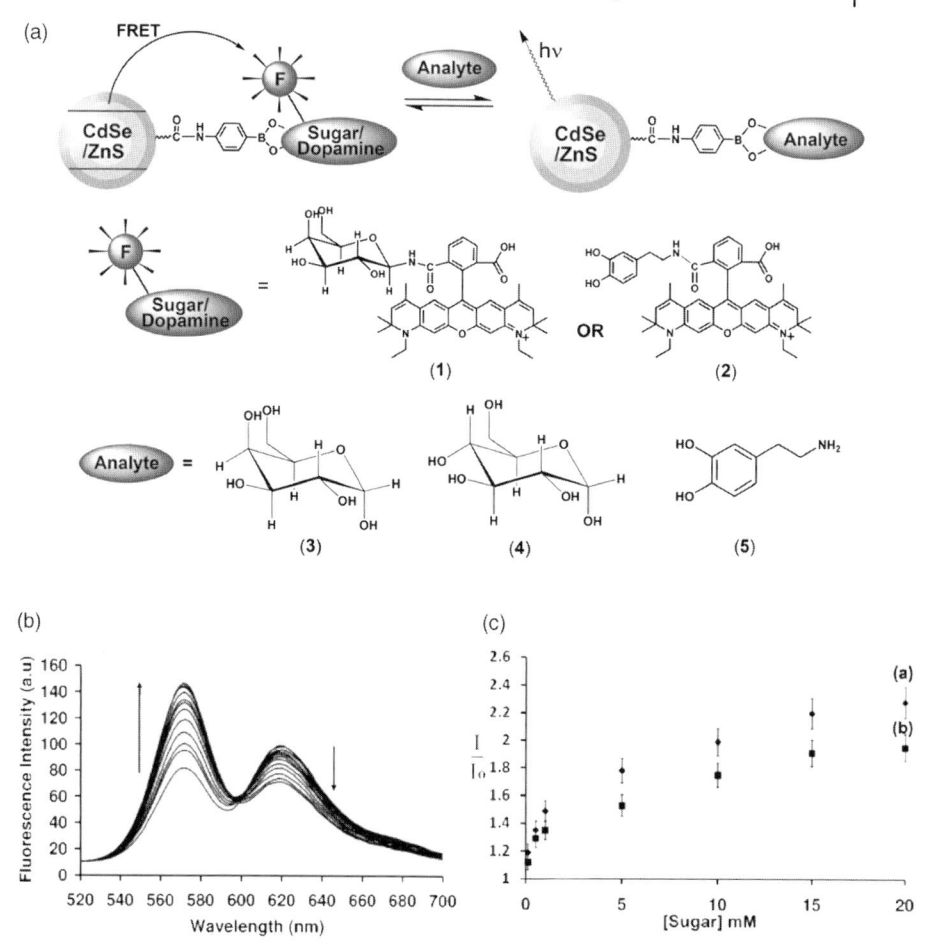

Figure 6.5 (a) Competitive analysis of monosaccharides or dopamine using fluorophore-labeled galactose or fluorophore-labeled dopamine, respectively; (b) Time-dependent fluorescence changes upon the interaction of the fluorophore-labeled galactose-functionalized QDs with galactose, 5 mM. The spectra were recorded at time intervals of 3 min; (c) Calibration curves corresponding to the competitive analysis of galactose (curve a) and glucose (curve b). (Reproduced with permission from [109]).

vicinal diols and, specifically, with saccharides [104–108] (Figure 6.5a). In this case, the dye-modified saccharide ATTO-590-labeled-galactose (**1**) was linked to the boronic acid ligand associated with the QDs, which resulted in the FRET from the QDs to the dye. In the presence of different monosaccharides, competitive binding to the QDs proceeded, which resulted in higher intensities of the luminescence of the QDs and lower fluorescence intensities of the acceptor dye. The system allowed the competitive optical assay of galactose (**3**), glucose (**4**), or dopamine (**5**) [109].

Different receptor-functionalized QDs were used for the analysis of different substrates, with cyclodextrins, crown-ethers and calixarenes being used as receptor units to modify the QDs. For example, β-cyclodextrin (β-CD) -modified CdSe/ZnS QDs have been used to develop both selective and chiroselective sensing platforms using either a competitive FRET assay or a direct ET quenching route. β-Cyclodextrin, a cyclic polysaccharide which consists of seven glucose units linked by 1–4 glycosidic bonds, is able to bind hydrophobic organic substrates to its hydrophobic cavity [110–114]. The binding properties of the CDs are controlled by the relative dimensions of their hydrophobic cavities; furthermore, the CDs are chiral receptors, and hence reveal chiroselective binding properties [115–120]. The β-CD was linked to CdSe/ZnS QDs modified by the boronic acid ligand (Figure 6.6a) to yield the receptor-functionalized QDs. Rhodamine B (a dye) was then incorporated into the receptor sites of β-CD which, upon excitation of the QDs, resulted in a decreased FRET emission of the dye

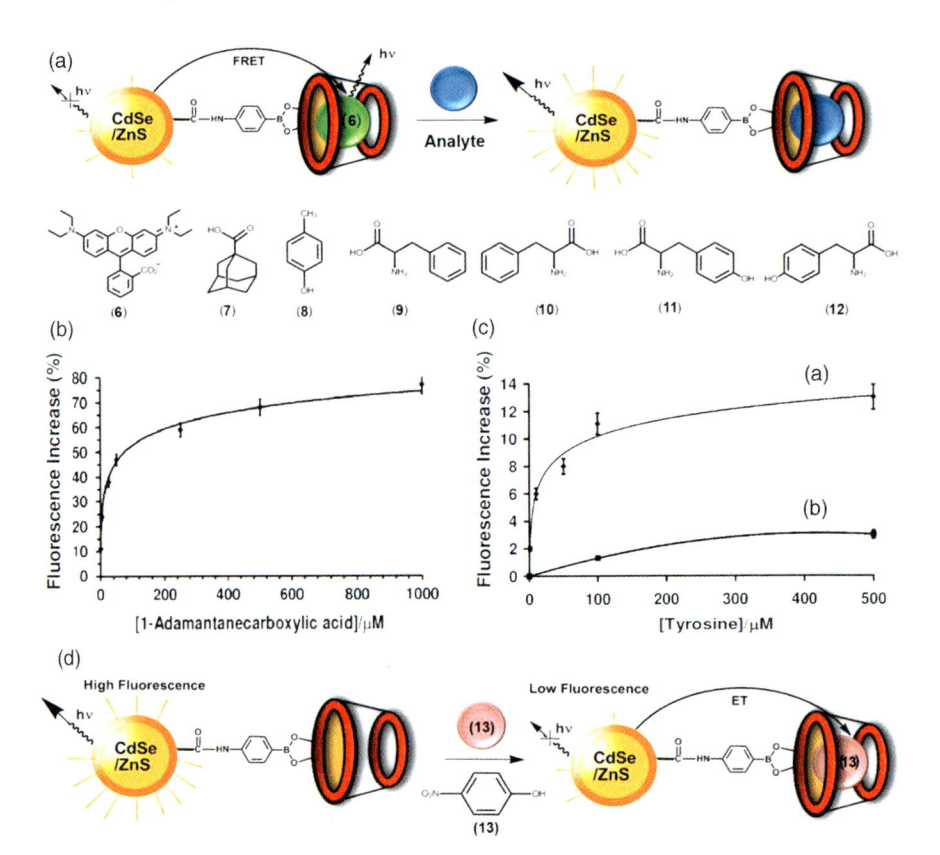

Figure 6.6 (a) Sensing of substrates by a competitive FRET assay using β-cyclodextrin-modified CdSe/ZnS QDs with receptor-bound rhodamine B (**6**); (b) Calibration curve corresponding to the analysis of variable concentrations of **7** by the rhodamine B/β-cyclodextrin-modified CdSe/ZnS QDs system; (c) Calibration curve corresponding to the analysis of variable concentrations of **11** (curve a) and **12** (curve b) by the rhodamine B/β-cyclodextrin-modified CdSe/ZnS QDs system; (d) Direct analysis of substrates by the β-cyclodextrin-modified CdSe/ZnS QDs, using an electron transfer quenching route. (Reproduced with permission from ref. [121]. Copyright 2009 American Chemical Society).

and a concomitant enhancement of the QDs' luminescence. The FRET between the QDs and rhodamine B was further used in the competitive analyses of adamantanecarboxylic acid (**7**) and *p*-hydroxytoluene (**8**). A chiroselective optical discrimination between D,L-phenylalanine (**9,10**) and D,L-tyrosine (**11, 12**) was demonstrated [121]. The fluorescence changes of the functionalized QDs, when analyzing different concentrations of adamantane carboxylic acid (**7**), are shown in Figure 6.6b, while luminescence changes monitored during the analysis of D- and L-tyrosine, revealing a chiroselective detection of the enantiomers, are shown in Figure 6.6c.

The β-CD-modified QDs were further used for the label-free analysis of *p*-nitrophenol (**13**) (see Figure 6.6d). In this case, the association of *p*-nitrophenol with the β-CD cavity resulted in an ET quenching of the QDs' luminescence, and a quantitative analysis of the guest substrate.

The modification of QDs with crown-ethers permitted an analysis of K^+ ions [47]. For this, the lipoic acid-tethered 15-crown-5-ether (**14**) was used to functionalize two different-sized CdSe/ZnS QDs (Figure 6.7a). The 15-crown-5 receptor is known to form a "sandwich"-type complex with the crown ether; accordingly, the addition of K^+-ions to the two different-sized QDs resulted in an interparticle association of the QDs through the bis-crown-ether-K^+ bridging complex (Figure 6.7b). A subsequent aggregation of the two different-sized QDs resulted in an energy transfer from the blue-emitting QDs to their red-emitting counterparts (Figure 6.7c). As the extent of interparticle aggregation was controlled by the concentration of K^+, the resultant FRET changes provided a quantitative signal for the K^+ ions. A similar macrocycle-modified QDs system was used for the analysis of Cd^{2+} using an ET mechanism [122], where the CdS:Mn/ZnS QDs were functionalized with 1,10-diaza-18-crown-6. As the amine-donor groups quenched the luminescence of the QDs by an ET mechanism, ligation of the Cd^{2+} ions to the macrocycle and its amine functionalities depleted the donor site, thus enhancing the luminescence of the QDs as Cd^{2+} concentration was raised.

Calixarenes are known to exhibit unique binding properties towards both ions (e.g., quaternary ammonium ions) and organic substrates [123–125]. When the calixarene receptor units were linked to TOPO-modified CdSe/ZnS QDs using hydrophobic interactions, the luminescence of the calixarene-functionalized QDs was quenched on association of the *N,N'*-dimethyl-4,4'-bipyridinium electron acceptor [126]. The tetrahexyl ether derivative of *p*-sulfonatocalix[4]arene receptor-functionalized QDs (Figure 6.8a) was used for the quantitative analysis of the acetylcholine (**15**) neurotransmitter (Figure 6.8b). As the concentration of the neurotransmitter was increased, the quenching of the QDs was raised [127] (Figure 6.8b). Whilst the luminescence of certain caly[6]arene-functionalized QDs was quenched by guest molecules acting as ET quenchers, the receptor sites were found to increase the luminescence intensities of the QDs. This phenomenon was attributed to a rigidification of the capping layer by the formation of receptor units, and an inhibition of the quenching of QDs by the solvent. By using this method it was possible to sense both methionine and phenylalanine [128]; likewise, *p*-sulfonatocalix[4]arene CdTe QDs conjugates were used to sense fenamithion and acetamipirid [129].

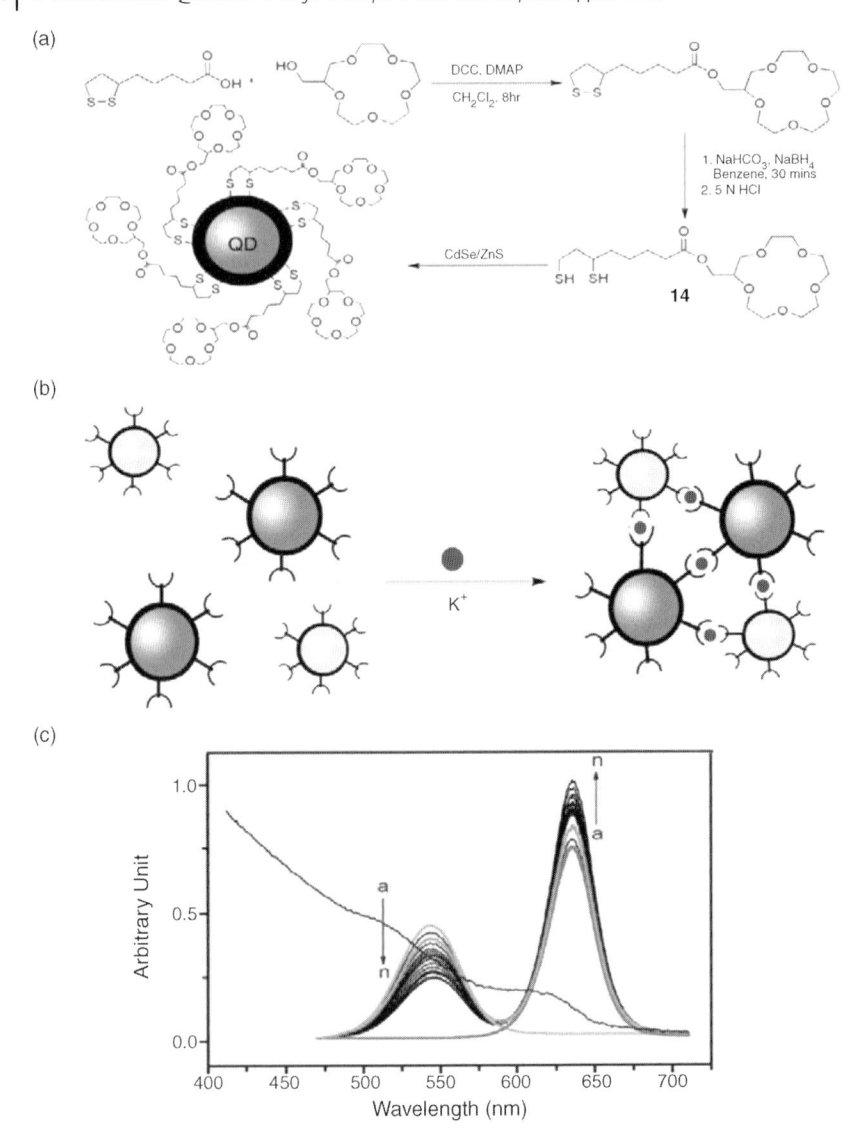

Figure 6.7 (a) Synthesis of 15-crown-5 and 15-crown-5-capped CdSe/ZnS QDs; (b) Aggregation of the 15-crown-5-modified two-sized CdSe/ZnS QDs upon the addition of K^+ ions; (c) Fluorescence titration spectra of the two-sized QDs upon addition of K^+ ions. (Reproduced with permission of the Royal Society of Chemistry from ref. [47]).

The capping layer of the QDs may act as ligand for the association of metal ions, such that an ET quenching of the QDs' luminescence will allow an analysis of the ions. For example, MAA-capped CdS QDs were shown capable of binding Hg^{2+} ions, whereby the fluorescence intensity of the QDs decreased linearly in the range 0.05×10^{-7} to 4.0×10^{-7} M of Hg^{2+}, thus enabling an analysis of the ions with a detection limit of 4.2×10^{-4} M. Different capping functionalities such as, MPA,

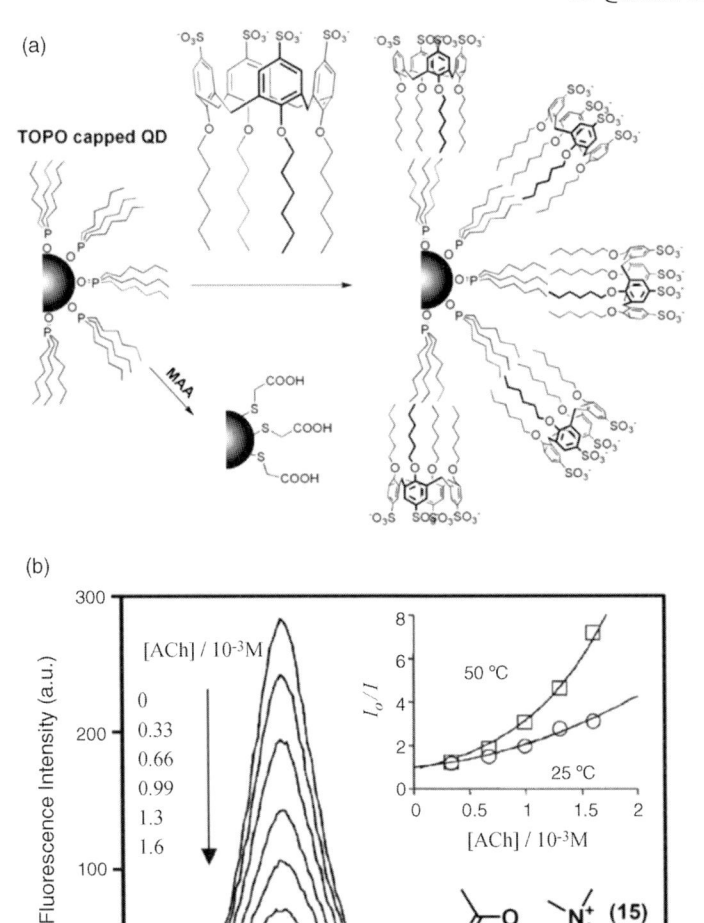

Figure 6.8 (a) Synthesis of TOPO-capped CdSe/ZnS QDs with tetrahexyl ether derivative of p-sulfonatocalix[4]arene receptor and MAA; (b) Quenching of the fluorescence emission of the modified CdSe/ZnS QDs upon the addition of different concentrations of acetylcholine. (Reproduced with permission from the Royal Society of Chemistry from ref. [127]).

MAA, GSH, mercaptosuccinic acid, thioglycolic acid or L-cysteine were used to bind different ions to a variety of QDs, leading to an ET quenching of the luminescence of the nanocrystals. However, by designing capping layers which consisted of peptides with tailored metal-binding capacities or Schiff-base ligands, enhanced sensitivities and selectivities could be demonstrated. The pentapeptide Gly-His-Leu-Leu-Cys was immobilized on CdS QDs, and the selective luminescence detection of Cu^{2+} or Ag^+ was demonstrated [130]. Similarly, CdS QDs modified with

Schiff-base functionalities revealed a specific sensing of Cu^{2+} and Fe^{3+} ions [131]. These solution-based optical sensors were further improved by the assembly of QDs on solid supports, and the optical detection of ions in complex mixtures. For example, mercaptosuccinic acid-functionalized CdTe QDs were deposited on quartz surfaces using a layer-by-layer (LbL) process that involved poly(dimethyldiallyl ammonium chloride) (PDDA) as an electrostatic "glue," with the resultant surfaces being used for the optical detection of Hg^{2+} [132].

It is well known that the capping layer which stabilizes the QDs may also interact with the analyte, thus affecting the luminescence properties of the semiconductor nanocrystals. Subsequently, oleic acid-coated CdSe QDs were synthesized and dispersed in chloroform, and the nitroaromatic explosives 2,4,6-trinitrotoluene (TNT), 2,4-dinitrotoluene (DNT), nitrobenzene (NB), 2,4-dinitrochlorobenzene were each intercalated between the hydrophobic layer surrounding the QDs. The quenching of the luminescence of the nanocrystals by the intercalated nitro-substituted analytes, via an ET mechanism, enabled a simultaneous analysis of the different explosives [133].

The adsorption of analytes onto the stabilizing capping layer may also affect the QDs' luminescence. Electrostatic interactions also facilitate the adsorption of substrates onto the QDs; for example, MPA-capped CdSe QDs were shown to adsorb edavarone by electrostatic interactions, which resulted in the blocking of nonradiative electron–hole recombination paths and an intensified luminescence of the QDs [134]. Similarly, the attraction of the negatively charged adenosine triphosphate by positively charged QDs affected the quenching of QDs; this system subsequently enabled the selective optical analysis of ATP in the presence of other nucleotides [135]. The fluorescence quenching of negatively charged MPA-functionalized CdSe/ZnS QDs by the positively charged herbicide N,N'-dimethyl-4,4'-bipyridinium by an ET mechanism was also demonstrated; the sensor system was further developed by coupling the receptor cucurbiturial that binds the bipyridinium salt to its cavity. The competitive deattachment of the bipyridinium salt from the QDs restored their luminescence, providing a positive fluorescence signal for sensing of the bipyridinium salt [136]. In analogy, a boronic acid-modified bipyridinium salt associated with positively charged QDs was used as a label for the competitive analysis of glucose [137]. While the adsorbed bipyridinium salt quenched the luminescence of the nanocrystals, the competitive formation of a glucose–boronate complex caused the bipyridinium units to detach from the QDs, leading to an intensified luminescence.

Both, purine and pyrimidine bases incorporated into nucleic acids, or in synthetic nucleic acid analogs, are able to bind specifically different metal ions [138]. For example, thymine yields a thymine–Hg^{2+}–thymine complex, while cytosine forms specifically a cytosine–Ag^+–cytosine complex. Such properties of the nucleic acids were utilized to develop QDs-based Hg^{2+}-ion and Ag^+-ion sensors, with different-sized QDs being implemented for the multiplexed analysis of Hg^{2+} and Ag^+ [139]. Two different-sized CdSe/ZnS QDs were modified with nucleic acids of specific ion-binding properties: blue-emitting QDs ($\lambda_{em} = 560$ nm) were functionalized with the thymine-rich nucleic acid (**16**) that binds Hg^{2+}-ions, while the red-emitting QDs ($\lambda_{em} = 620$ nm) were functionalized with the cytosine-rich nucleic acid (**17**) that associated with Ag^+-ions (Figure 6.9a). The formation of Hg^{2+}-modified complexes

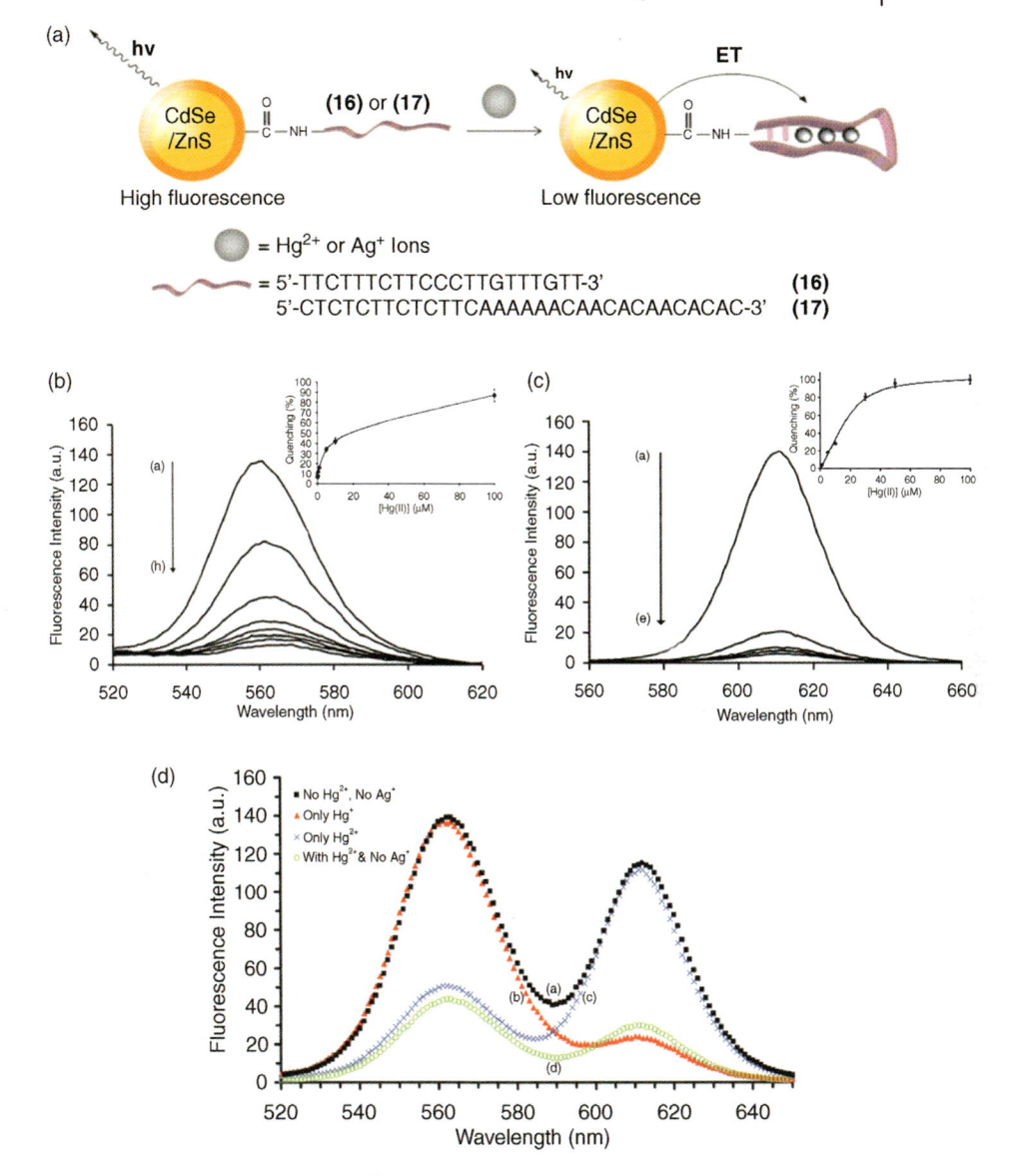

Figure 6.9 (a) Optical analysis of Hg^{2+}/Ag^+ ions by nucleic acid modified QDs; (b) Time-dependent luminescence spectra of **16**-QD^{560} upon interaction with 1×10^{-4} M Hg^{2+}. Inset: Calibration curve for various concentrations of Hg^{2+}; (c) Time-dependent luminescence spectra of **17**-QD^{620} upon interaction with 5×10^{-5} M Ag^+. Inset: Calibration curve for various concentrations of Ag^+; (d) Fluorescence changes of the **16**-QD^{560} and **17**-QD^{620} upon interaction with: (spectrum a) no Hg^{2+}, no Ag^+; (spectrum b) no Hg^{2+}, 30 μM Ag^+; (spectrum c) no Ag^+, 30 μM Hg^{2+}; (spectrum d) 30 μM Hg^{2+}, 30 μM Ag^+. (Reproduced with permission from ref. [139]).

on the (**16**)-modified QDs resulted in a quenching of the blue-emitting QDs (Figure 6.9b), whereas the association of Ag^+ with the (**17**)-modified QDs led to a quenching of the red-emitting QDs (Figure 6.9c). The respective QDs revealed selectivity, while other metal ions had no effect on the luminescence properties of the QDs. These QDs permitted an optical analysis of Hg^{2+} and Ag^+ with detection limits that corresponded to 10 nM (2 ppb) and 1 µM (200 ppb), respectively. A multiplexed analysis of the two ions was also accomplished by the functionalized QDs (Figure 6.9d) when, on mixing the two different-sized (**16**)- and (**17**)-functionalized QDs, a selective and simultaneous analysis of Hg^{2+} and Ag^+ ions was achieved by quenching of the respective QDs.

6.3.2
Functionalization of QDs with Chemically Reactive Units Participating in the Sensing

While the sensor system described in Section 6.3.1 included receptor or ligand sites for the selective association of a guest analyte, a variety of chemically modified QDs that use the capping layer as a chemically reactive interface can actively participate in the sensing process. In these systems, the analyte reacts with the capping layer, which leads, in turn, to changes in the fluorescence properties of the QDs that enables the quantitative optical detection of analytes. Semiconductor QDs functionalized with redox-active dyes or with redox-active capping components may cause changes in the photophysical properties of the QDs, due to bleaching of the dye by a redox process or by an electronic coupling of the redox-active capping component with the photoexcited electron–hole pair generated in the QDs.

For example, Nile-blue-functionalized CdSe/ZnS QDs were prepared and used as a hybrid material for optical sensing of the cofactor 1,4-dihydronicotinamide adenine dinucleotide (phosphate) (NAD(P)H). For this, water-soluble CdSe/CdS QDs were first produced by ligand exchange with MPA, after which the modified QDs were functionalized with a layer of bovine serum albumin (BSA) to which oxidized Nile blue was then covalently linked. The fluorescence of the QDs was quenched by oxidized Nile blue through a FRET quenching mechanism. In the presence of NADH, Nile blue acted as an electron mediator for the oxidation of the NAD(P)H cofactors. The reduced Nile blue units associated with the QDs lacked absorbance in the visible spectral region, and thus did not quench the QDs. As a result, reduction of the Nile blue capping layer by the NAD(P)H cofactors activated the fluorescence of the QDs (Figure 6.10a), which, in turn, enabled a quantitative optical detection of NADH (Figure 6.10b) [140]. This quantitative optical analysis of NADH using QDs enabled their use as luminescent probes to follow the activity of NAD^+-dependent enzymes, and their substrates (see Section 6.4.2).

In another example, phenyl boronic acid-functionalized CdSe/ZnS QDs were synthesized and modified by binding NAD^+ or NADH via the boronic esters, such that the vicinal diols were associated with the reduced/oxidized cofactors. Quenching of the QDs' luminescence by the NAD^+ cofactor, via an ET route, was shown to be substantially more efficient than quenching by the reduced cofactor, NADH. This difference in quenching features of NADH was used subsequently to optically sense

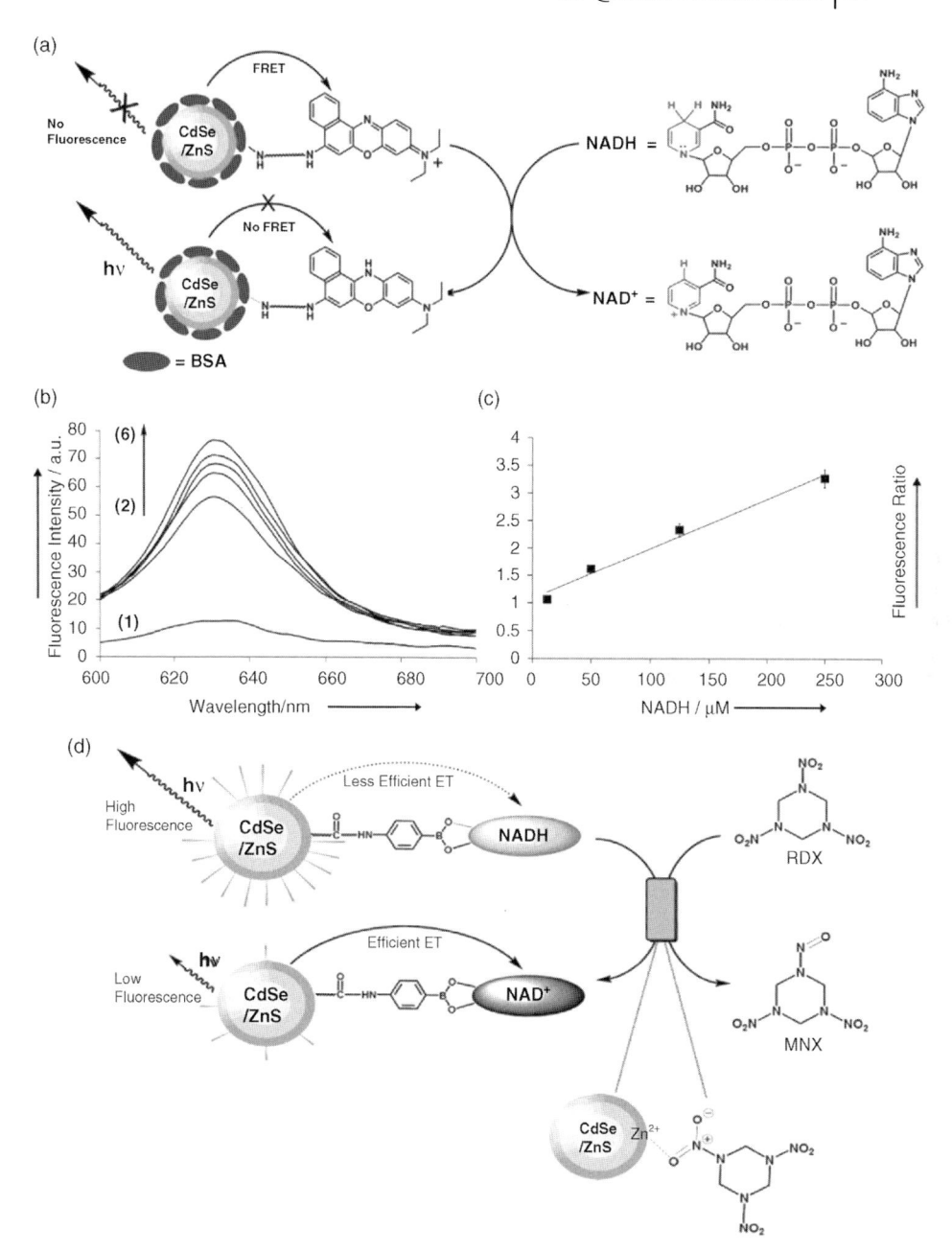

Figure 6.10 (a) Sensing of NADH by Nile blue-functionalized CdSe/ZnS QDs; (b) Time-dependent luminescence changes of the CdSe/ZnS QDs upon interaction of the QDs with NADH, 0.5 mM; (c) Derived calibration curve; (d) Sensing of the RDX explosive by the NADH-functionalized CdSe/ZnS QDs. (Reproduced with permission, parts (a) to (c) from ref. [140] and part (d) from ref. [141]).

the trinitrotriazine explosive, RDX [141]. In this case, the NADH coordinated to the boronic acid-functionalized CdSe/ZnS QDs reduced the explosive, while transforming the capping monolayer to the NAD$^+$ state; this effectively quenched the luminescence of the QDs (Figure 6.10c). These changes in the QDs fluorescence permitted the analysis of RDX with a detection limit corresponding to 1×10^{-10} M. Interestingly, in addition to the photophysical functions of the QDs for detecting RDX, a cooperative catalytic function of the QDs in the NADH-mediated reduction of RDX was also observed. Subsequently, the Zn^{2+}-ions present in the shell layer of the CdSe/ZnS QDs were found to coordinate as a Lewis acid to the nitro groups of RDX, thus polarizing the nitro group and enabling a hybrid transfer from NADH (see Figure 6.10c).

A nitric oxide (NO) QD-based sensor was developed via the NO-stimulated ligand substitution of a transition metal complex associated with CdSe/ZnS QDs [142]. The red-colored tris-(N-(dithiocarboxy) carcosine) iron(III)[Fe(DTCS)$_3$]$^{3-}$ was linked to ammonium-capped QDs by ionic interaction. As a consequence, the functionalized QDs reacted with NO by an ET process, followed by ligand substitution to yield the colorless paramagnetic bis(dithiocarbamato) nitrosyl iron(I) complex as a capping layer. This process triggered the luminescence of the QDs that enabled the detection of NO (Figure 6.11).

A QD-based sensor for carbon-centered radicals (alkyl or phenyl radicals) was developed by the use of 4-amino-2,2,6,6-tetramethylpiperidine oxide (4-amino-TEMPO)-functionalized CdSe QDs. The presence of a nitroxide moiety in the functional capping layer leads to quenching of the luminescence of the QDs. The nitroxide units of the capping layer react with the carbon-centered radical to yield the diamagnetic alkoxyamine-modified protecting layer. As the resultant units are unable to quench the luminescence of the QDs, this leads to an intensified luminescence upon interaction with reactive carbon-centered radicals.

QDs modified with carboxylic acid functionalities have revealed pH-controlled luminescence properties. Although the origin of this effect of protonation/deprotonation of the carboxylic acid is not fully understood, it is presumed that a neutralization of the cationic surface-states by the carboxylate units removes the electron traps, thus enhancing the fluorescence of the QDs. An interesting QDs-based pH sensor was developed by coupling CdSe nanocrystals with the pH-sensitive squaraine dye (Figure 6.12). The deprotonated dye (curve a in Figure 6.12) exhibits an appropriate adsorption to stimulate a FRET process between the QDs and the dye acceptor, while the protonated squaraine state (curve b) cannot act as a FRET acceptor. Thus, the pH-controlled ratio of (a)/(b) modulated the FRET intensity, while the ratio of the luminescence of the QD/squaraine fluorescence provided a quantitative measure of the solution's pH [58].

6.4
Biosensors

While the unique recognition properties of biomolecules (e.g., antigen–antibody, nucleic acid–DNA, enzyme–substrate, receptor–guest complexes) provide the basis

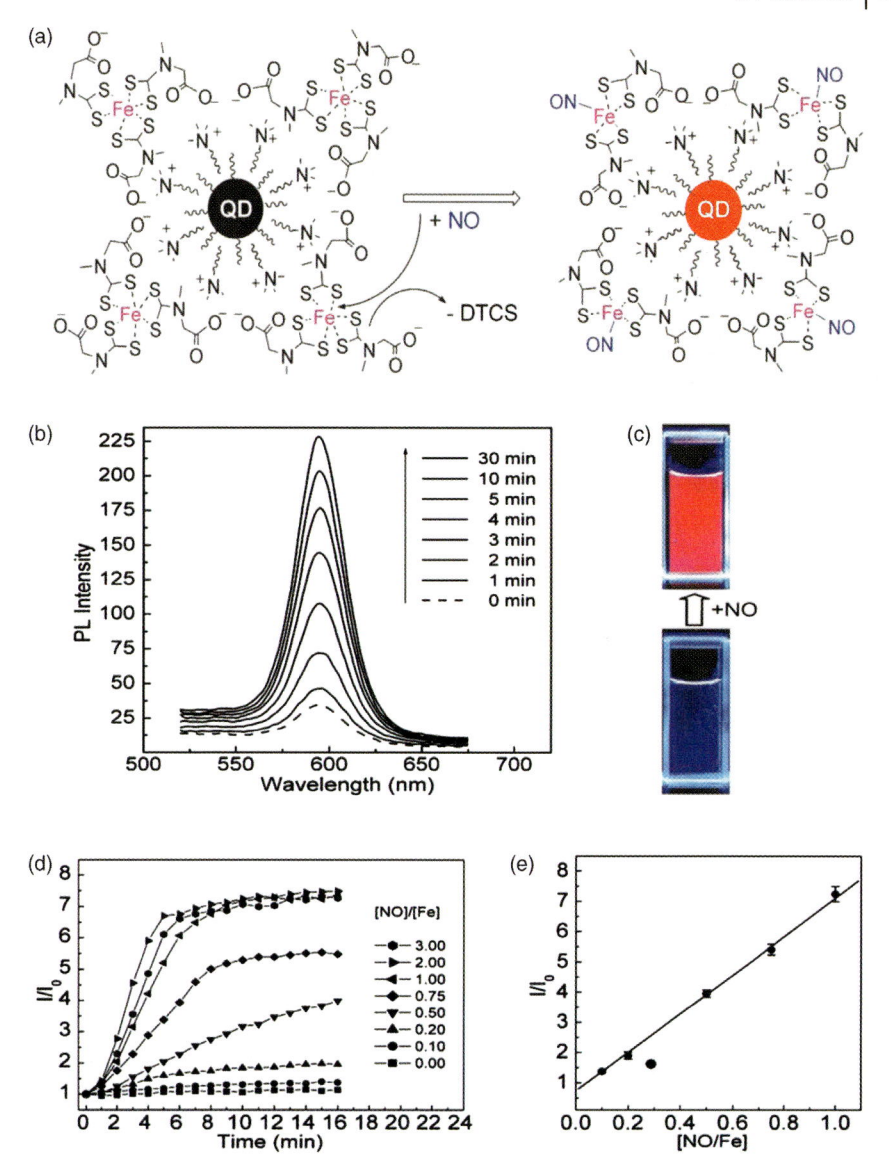

Figure 6.11 (a) The use of [Fe(DTCS)$_3$]$^{3-}$-functionalized QDs as optical sensor for NO. The reaction of NO with the capping layer triggers-on the luminescence of the QDs; (b) Time-dependent luminescence spectra of the modified QDs upon reaction with NO; (c) Luminescence features of the solution of the modified QDs upon their reaction with NO; (d) Time-dependent luminescence changes of the functionalized QDs upon reaction with variable concentrations of NO (expressed as [NO]/[Fe] ratios); (e) Derived calibration curve. (Reproduced with permission from the American Chemical Society Copyright 2009, from ref. [142]).

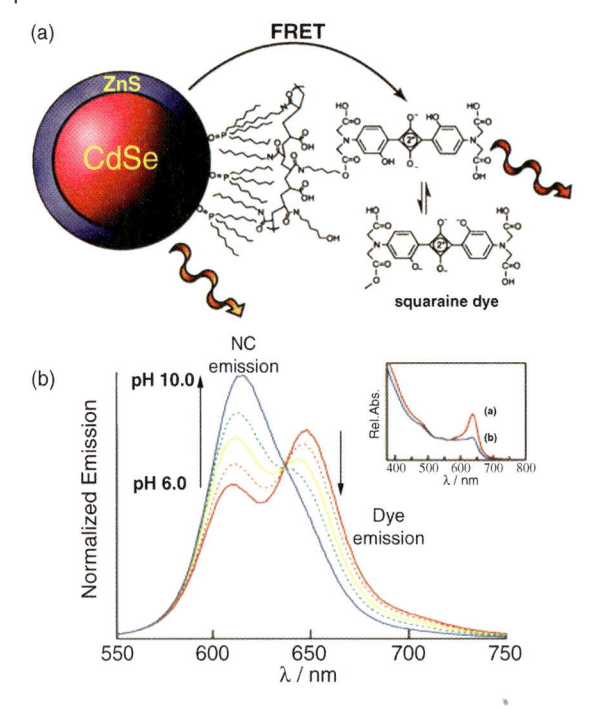

Figure 6.12 (a) Sensing of pH using the pH-sensitive squaraine dye conjugated to CdSe/ZnS QDs modified by an amphiphilic polymer associated with the hydrophobic nanocrystals; (b) Fluorescence spectra of the dye-modified QDs upon changing the pH. Inset: Absorption spectra of the squaraine dye at acidic (spectrum a) and basic (spectrum b) pH values. (Reproduced with permission from ref. [58]. Copyright 2006 American Chemical Society).

for the development of biosensors, nanotechnology provides nanoscale materials (e.g., nanoparticles, nanorods, nanotubes) of unique electronic and optical properties [5–10, 28–42]. Indeed, numerous studies have been implemented to investigate metallic nanoparticles or carbon nanotubes (CNTs) for the development of a variety of electronic or optical sensing platforms [143–145]. The unique size-controlled optical properties of semiconductor nanocrystals have paved the way for the development of new optical biosensing platforms, by integrating biomolecular recognition events with QDs. Besides the use of different-sized QDs as luminescence labels for the multiplexed analysis of biorecognition events, the activation of FRET or ET in biomolecule–QD hybrid systems provides a general means to develop new optical biosensing platforms. Specifically, the application of FRET or ET processes in QD–biomolecule conjugates can provide useful tools for monitoring the dynamics of biorecognition events or biocatalytic transformations. Furthermore, by incorporating QDs into living cells, effective new optical labels for imaging cell domains and for monitoring intracellular metabolic processes, can be envisaged.

6.4.1
Application of QDs for Probing Biorecognition Processes

The enzyme-linked immunosorbent assay (ELISA) provides a general sensing platform for the detection of antigen–antibody complexes on solid supports. Although, fluorescent dyes have often been used as labels to monitor the formation of immunocomplexes, the unique luminescent properties of QDs (high quantum efficiencies, optical stabilities, narrow emissions) support their use as an alternative approach to conventional organic dyes. Previously, QDs labeled with antibodies (Figure 6.13a) or nucleic acid-functionalized QDs (Figure 6.13b) have been used as functional QDs to monitor the formation of immune-complexes or DNA hybridization products, respectively. For example, when CdSe/ZnS QDs were functionalized with an avidin capping layer, biotinylated antibodies were seen to bind to the QDs through the protective capping layer. This led to the antibody-modified QDs being used as labels for a sandwich-type immunoassay in the detection of various toxins [95, 146].

The major advantage of QD–antibody bioconjugates is their ability to detect, simultaneously, a number of analytes. Indeed, different-sized CdSe/ZnS QDs, functionalized with the appropriate antibodies, were applied to a multiplexed sandwich-type fluoroimmunoassay detection of four different toxins, namely (cholera toxin (CT), ricin, shiga-like toxin 1 (SLT), and staphylococcal enterotoxin B (SEB) (Figure 6.14a). When the reaction was performed in single wells of a microtiter plate, in the presence of a mixture of all four QD–antibody conjugates as detecting units, the resultant fluorescence signal was shown to encode for the different toxins [147] (Figure 6.14b). In a further example, biotinylated denatured BSA-coated CdTe QDs of different sizes were used for the multiplexed simultaneous fluorescence determination of five different low-molecular-weight chemical drug residues (dexamethasone, medroxyprogesterone acetate, gentamicin, ceftiofur, and clonazepam) [148].

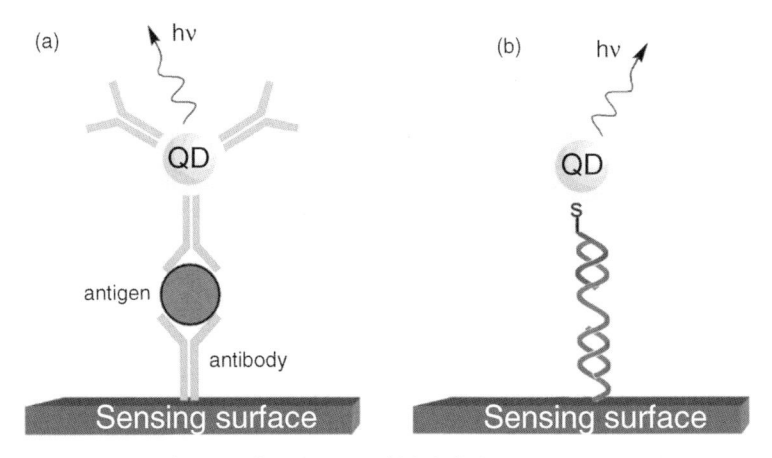

Figure 6.13 Application of QDs as optical labels for biorecognition events. (a) Formation of immune-complexes; (b) Probing DNA hybridization.

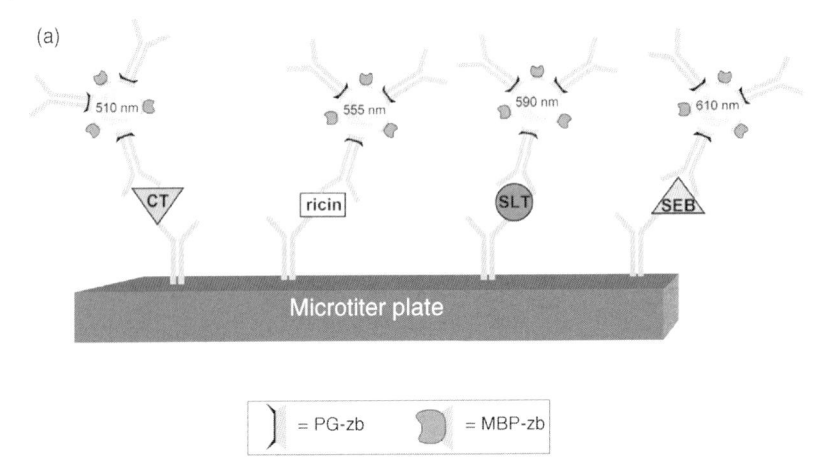

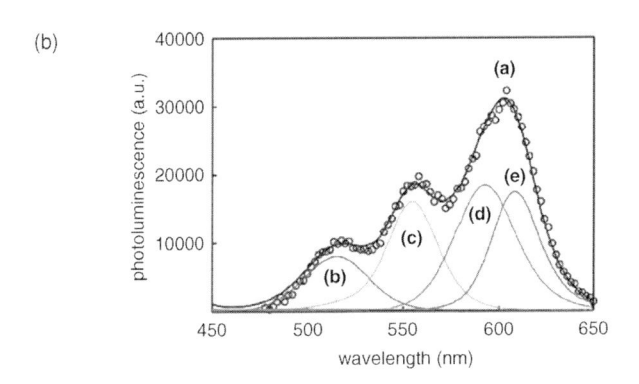

Figure 6.14 (a) The parallel optical analysis of antigens in a well array using the fluorescence of different-sized quantum dots; (B) Fluorescence spectrum observed for the four analytes (concentration $1\,\mu g\,ml^{-1}$) by the different-sized QDs (spectrum a), and deconvoluted spectra of individual toxins: (spectrum b) cholera toxin, (spectrum c) ricin, (spectrum d) shiga-like toxin 1, and (spectrum e) staphylococcal enterotoxin B. (Part (b) reproduced with permission from ref. [147]. Copyright 2004, American Chemical Society).

Conventional DNA/RNA fluorescent microarrays are based on the sandwich-type hybridization of target DNA/RNA between a capture probe attached to a surface, and a fluorophore-modified signaling label. In the past, many investigations have employed QDs as signaling labels for DNA/RNA hybridization on microarrays; examples include the use of QDs to detect single-nucleotide polymorphisms of the human oncogene p53, and the multiallele detection of hepatitis B and hepatitis C viruses in microarray configurations [149]. Different-sized CdSe/ZnS QDs modified with appropriate nucleic acids have also been used to probe hepatitis B and C genotypes in the presence of other human genes. Similarly, the perfectly matched sequence of the p53 gene was detected within minutes at room temperature in the presence of a background oligonucleotide mixture, which consisted of different single-nucleotide polymorphic sequences, with true-to-false signal ratios greater than 10 (under stringent buffer conditions).

DNA–QD conjugates have also been used as luminescent labels in fluorescence *in situ* hybridization (FISH) assays. The FISH assay is based on the denaturation of genomic DNA and the consecutive hybridization with a fluorescent-labeled DNA sequence, in order to visualize the presence or absence of specific DNA sequences in the chromosomes. For example, a QDs-based FISH labeling method was used to detect the Y chromosome in human sperm cells [74], and also the human metaphase chromosome [150]. The FISH technique was also used for the multiplex cellular detection of different mRNA targets [151], and to visualize a pUC18 plasmid inside the *Escherichia coli* HB101 bacterium [152].

FRET-based immunoassays that use QDs as the energy donors were developed by preparing QDs conjugated to fluorophore (quencher) -labeled analyte units. The competitive displacement of a fluorophore (quencher) by an analyte triggered a luminescence of the QDs, that in turn transduced the formation of the immune-complex with the target antigen. For example, CdSe/ZnS QDs were functionalized with an antibody fragment that selectively bound trinitrotoluene (TNT) [153]. The analogous substrate trinitrobenzene (TNB) was covalently linked to the quencher dye BHQ10 (**18**), and the quencher-labeled TNT analog was bound to the QD–antibody conjugate; this resulted in a quenching of the fluorescence of the QDs (Figure 6.15a). In the presence of the TNT analyte, the quencher-labeled TNB was competitively displaced from the antibody-modified QDs; this eliminated the FRET process

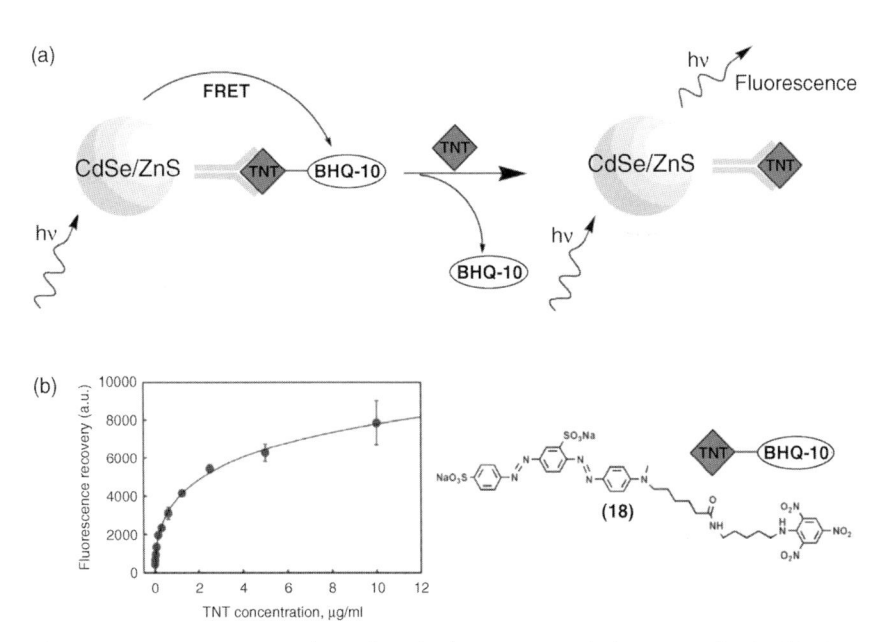

Figure 6.15 (a) Competitive analysis of TNT by the anti-TNT antibody associated with CdSe/ZnS QDs and using BHQ-10 as quencher; (b) Derived calibration curve for the analysis of TNT. (Reproduced with permission from ref. [153]. Copyright 2005 American Chemical Society).

between the QDs and the dye, and resulted in a switching-on of the fluorescence of the QDs (Figure 6.15b). In another example, a multiplexed competitive FRET-based fluoroimmunoassay for the simultaneous detection of three different antigens (*E. coli* 0571H7, *B. cereus*, and MS2-virus) was developed [154]. Different-sized CdSe/ZnS QDs, functionalized with the antibodies against the respective antigens, enabled the detection of all three antigens individually and simultaneously, via photoluminescence measurements in aqueous solution, or via the fluorescence imaging of conjugates trapped on the surface of a porous filter.

The FRET between luminescent QDs and dye molecules has been used to detect target DNAs. For example, the FRET between CdTe QDs and Cy3-labeled single-stranded DNA (ssDNA) probes, adsorbed onto the QD surface through a positively charged polymer capping layer, acting as an electrostatic linker, was used for the detection of DNA. In this case, the dye-labeled DNA probe would hybridize with the target DNA, thus reducing the FRET efficiency due to an increased distance between the QD and the dye [155]. In another example, DNA-modified CdSe/ZnS QDs were used to detect a dye (Alexa 594)-labeled complementary DNA probe [156] (Figure 6.16, route A). Hybridization of the dye-labeled DNA with a complementary DNA–QD conjugate brought the fluorophore into close proximity to the QD, and resulted in an efficient FRET process that encoded for the DNA probe. The competitive displacement of the labeled nucleic acid by the analyte decreased the FRET process, thus signaling a hybridization with the target DNA. Alternatively, the DNA-modified CdSe/ZnS QDs were used for the detection of an unlabeled complementary target DNA, in the presence of ethidium bromide, a specific double-stranded intercalator [97]. In the presence of the target DNA, a duplex DNA was formed; this led to the intercalation of ethidium bromide, with a resultant energy transfer from the QD to ethidium bromide (Figure 6.16, route B).

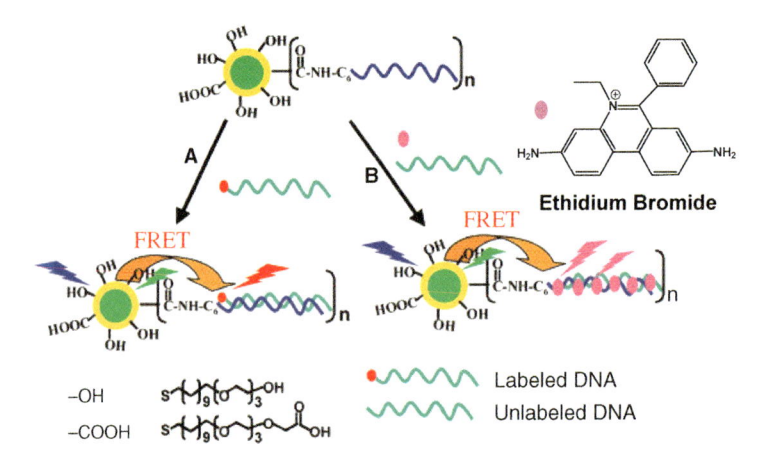

Figure 6.16 (a) Analysis of a DNA by a FRET process to a dye-functionalized complementary nucleic acid; (b) Analysis of DNA through a FRET process to ethidium bromide intercalated in the duplex DNA, as a result of hybridization with the complementary nucleic acid. (Reproduced with permission from ref. [156]. Copyright 2008 American Chemical Society).

The quenching of CdSe/ZnS–oligonucleotide conjugates by gold nanoparticles was used for the biosensing of DNA [157]. For this, a ssDNA was linked at its 5′-end to CdSe/ZnS QDs, and a complementary sequence was linked at its 3′ end to 1.4 nm gold nanoparticles. The formation of a duplex DNA via hybridization resulted in a significant quenching of the QDs emission by the gold nanoparticles. The addition of unlabeled oligonucleotides resulted in a recovery of the QDs emission, due to a separation of the QD–Au nanoparticle duplex structures.

The FRET quenching of CdSe/ZnS QDs was used to detect a target DNA in QD–molecular beacon (MB) conjugates [158–160]. In this case, a QD and a quencher molecule (Q) were tethered to the termini of a ssDNA oligonucleotide hairpin structure, which resulted in an effective proximity quenching of the QDs by the quenched units (Figure 6.17a). In the presence of a target DNA that was complementary to the single strand loop of the oligonucleotide hairpin structure, a hybridization to the loop region occurred which resulted in a separation of the stem duplex region. Subsequently, a spatial separation of the QDs from the quencher units restored the fluorescence of the QDs. For example, CdSe/ZnS QDs were linked to the 5′ end of a hairpin nucleic acid that included, at its 3′, end the quencher molecule 4-(4′-dimethylaminophenylazo)benzoic acid (Dabcyl), and used to detect single base mismatches [158]. Similarly, the multiplexed analysis of three different DNAs was accomplished using different-sized CdSe/ZnS QDs that had been functionalized with the appropriate hairpin structures, including the versatile black-hole quencher-2 (BHQ-2) unit [159]. Opening of the specific hairpin by the appropriate target led to a switch-on of the luminescence of the respective sized QDs (Figure 6.17b).

A three-component QD-based FRET cascade was also implemented to follow DNA hybridization [161]. In this case, the carboxylic acid-modified CdTe QDs were coated with the fluorescent conjugated polymer, poly[9,9-bis(3′-(N,N-bimethyl)-N-ethylammonium propyl)-2,7-fluorene-1,4-phenylene dibromide (PDFD) (19) (Figure 6.18a) that binds electrostatically to the negatively charged QDs. A nucleic acid probe, labeled with the dye IRD 700, was used to detect the analyte DNA. The duplex generated between the analyte DNA and the IRD 700-labeled nucleic acid was shown to bind by electrostatic interactions to the positively charged polymer-coated QDs, and this activated a two-step FRET cascade (Figure 6.18a). Photoexcitation of the PDFD polymer, that acted as an antenna ($\lambda_{em} = 440$ nm), resulted in the primary FRET process to the QDs that revealed a luminescence at 600 nm. The resulting QDs-stimulated emission activated the secondary FRET from the QDs to the dye, which was reflected by the fluorescence at 720 nm. As the probe dye-labeled nucleic acid was seen to bind strongly to the polymer only upon the formation of the duplex with the analyte DNA, the FRET cascade would be activated only in the presence of the analyte DNA (Figure 6.18b). This method enabled the analysis of DNA with a detection limit of 1 nM.

Aptamers are nucleic acid sequences that specifically bind proteins or low molecular-weight substrates. Aptamers are selected from a combinatorial library of 10^{15}–10^{16} DNAs, using the Systematic Evolution of Ligands by Exponential Enrichment Process (SELEX). Numerous aptamers that specifically bind proteins or low molecular-weight substrates have been elicited in recent years. Also, their recognition properties have been used extensively to develop electrochemical [162, 163] or optical

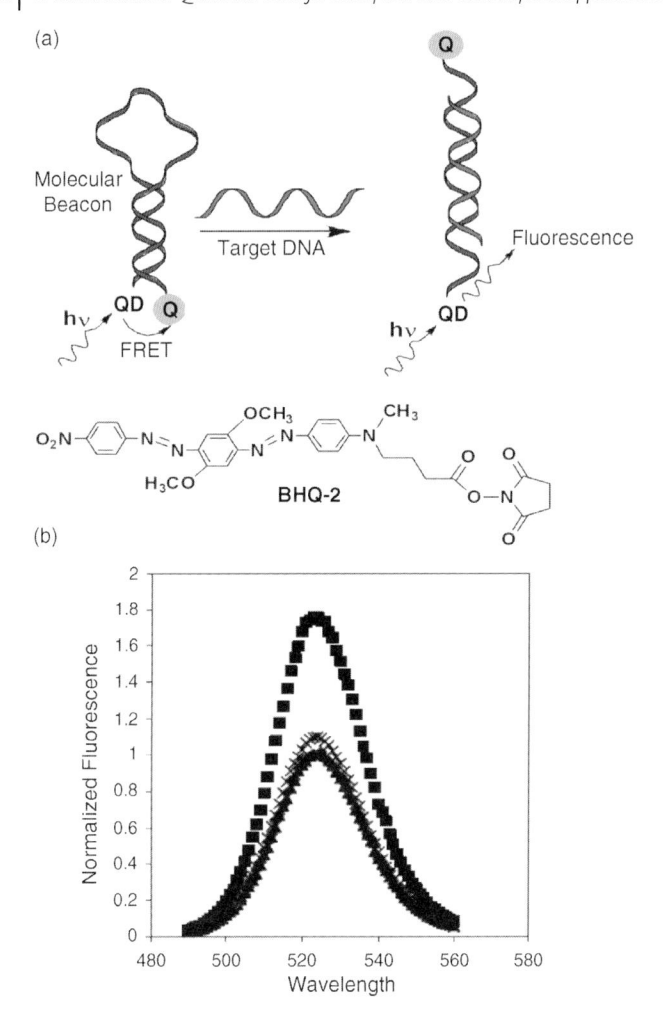

Figure 6.17 (a) Optical detection of DNA by a hairpin nucleic acid functionalized with QDs and quencher units Q; (b) Fluorescence intensities: ◆ without target DNA, ■ with target DNA, ▲ single-base mismatched target DNA (SMT), × noncomplementary target DNA (NST). (Reproduced with permission from IOP Publisher from ref. [159]).

[] sensor systems. QDs have also been used as optical labels for probing the formation of aptamer–protein complexes [166]. When the anti-thrombin aptamer was coupled to QDs, and the nucleic acid sequence hybridized with the complementary oligonucleotide–quencher conjugate (Figure 6.19). This resulted in the quenching of the fluorescence of the QDs in the QD–aptamer/quencher oligonucleotide duplex structure. In the presence of thrombin, the duplex was separated, whereupon the aptamer underwent a conformational change to the quadruplex structure that binds thrombin. Displacement of the quencher units from the blocked aptamer then activated the luminescence functions of the QDs. In a related study, thrombin was

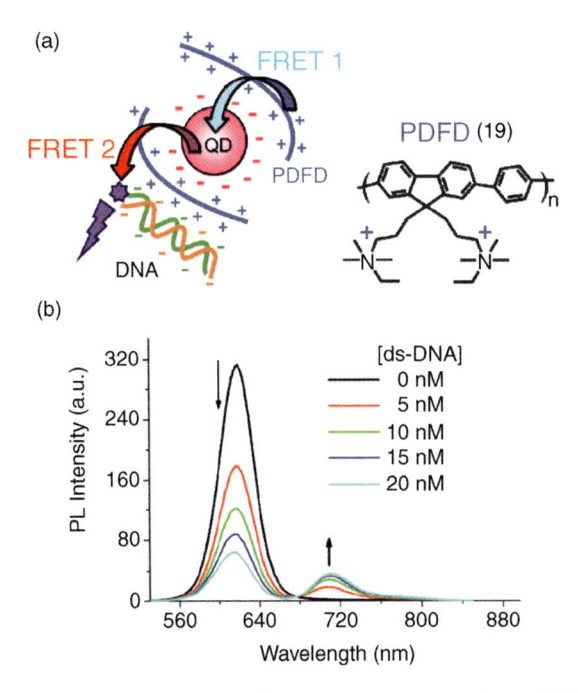

Figure 6.18 (a) Analysis of DNA by a two-step cascaded FRET using negatively charged QDs coated by the positively charged PDFD polymer chromophore that binds electrostatically the dye-labeled nucleic acid that hybridized with the target; (b) Fluorescence spectra observed upon analysis of different concentrations of the target DNA by the QDs/PDFD hybrid and the dye-functionalized nucleic acid as reporter. (Reproduced with permission from ref. [161]. Copyright 2009 American Chemical Society).

detected by an anti-thrombin aptamer conjugated to PbS QDs [167]. The binding of thrombin to aptamer-functionalized QDs led to a selective fluorescence quenching of the semiconductor nanocrystals, as the result of a charge-transfer process from thrombin to the QDs. This method not only enabled thrombin to be detected at concentrations down to 1×10^{-9} M, but also showed a high selectivity in the presence of high background concentrations of interfering proteins.

QD-based aptamer sensors for the detection of low-molecular-weight substrates have been similarly developed [168]. The aptamer against adenosine monophosphate

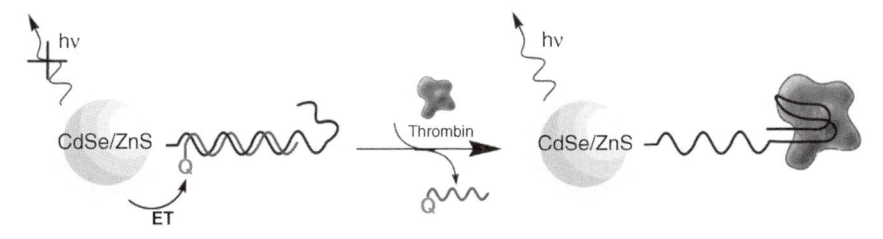

Figure 6.19 Optical analysis of thrombin by the formation of the aptamer–thrombin complex, and separation of the quencher-nucleic acid blocking unit.

(AMP) (**20**), bridged through a hybridization of the (**21**)-nucleic acid-functionalized CdSe/ZnS QDs and the (**22**)-nucleic acid-functionalized Au nanoparticles, led to QD–Au nanoparticle aggregates in which the fluorescence of the QDs was quenched by the gold nanoparticles (Figure 6.20a). In the presence of AMP, the aptamer–AMP

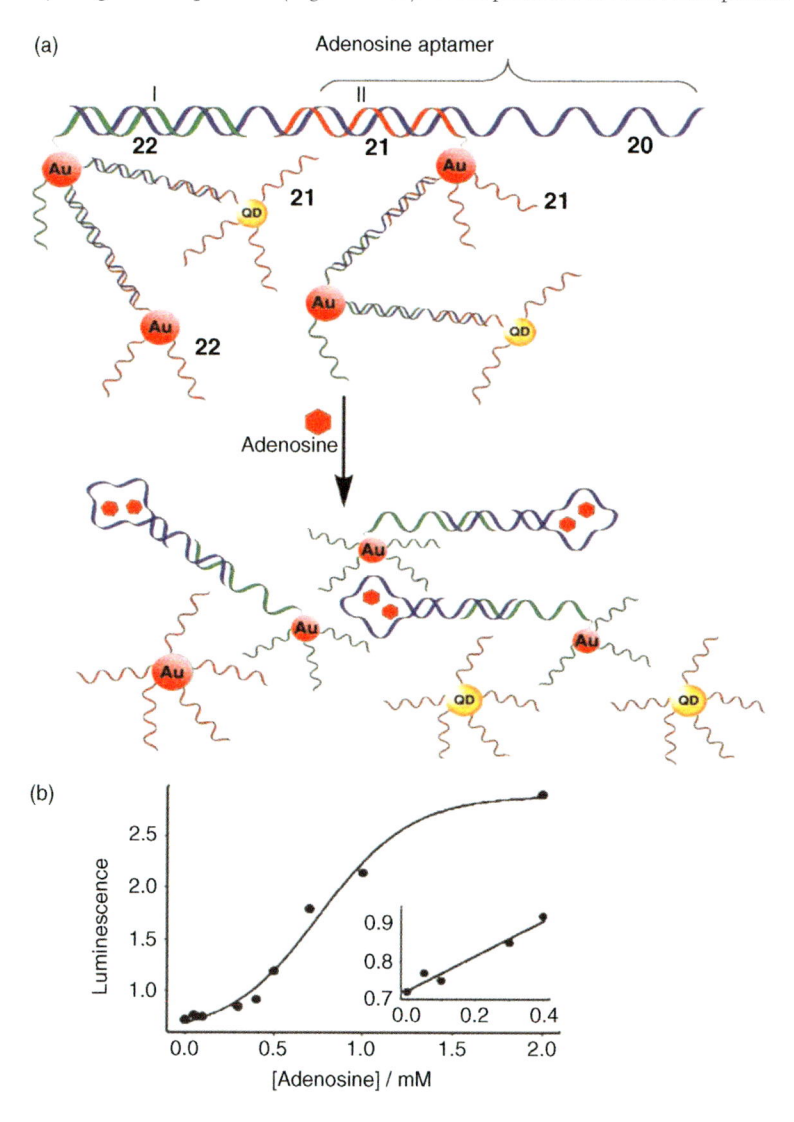

Figure 6.20 (a) Analysis of adenosine monophosphate (AMP) by fluorescence quenching of functionalized QDs by gold nanoparticles. The anti-adenosine monophosphate aptamer **20** bridges the CdSe QD functionalized with nucleic acid **21** and the nucleic acid **14**-modified with gold nanoparticles to form the respective aggregate. Upon analysis of AMP, the aggregate is separated, and the fluorescence of the QDs is switched on; (b) Fluorescence intensity changes as a function of the concentration of AMP. Inset: enlargement of the calibration curve in the region 0.0–0.4 mM AMP. (Reproduced with permission from ref. [168]. Copyright 2007 American Chemical Society).

complex was formed, which resulted in a separation of the QDs from the Au nanoparticles. As the distance between the QDs and the gold nanoparticles increased, the QDs regenerated their luminescent properties. Moreover, as separation of the aggregate was controlled by the AMP concentration, the resultant fluorescence would provide a quantitative signal for the analyte (Figure 6.20b). A similar system was applied to the analysis of cocaine [168].

QDs-based aptamer sensors were further developed by applying aptamer subunits that would self-assemble into supramolecular aptamer subunits–guest complexes in the presence of the analyte. CdSe/ZnS QDs were modified with a subunit of the anti-cocaine aptamer (**23**), while the second aptamer subunit was functionalized with an acceptor dye (**24**). In the presence of a cocaine analyte, supramolecular complex formation occurred between the aptamer subunits and cocaine, which led to the FRET process between the QDs and the dye acceptor (Figure 6.21). This permitted the detection of cocaine with a detection limit corresponding to 1×10^{-6} M [169]. The main advantage of this sensing platform over the use of an

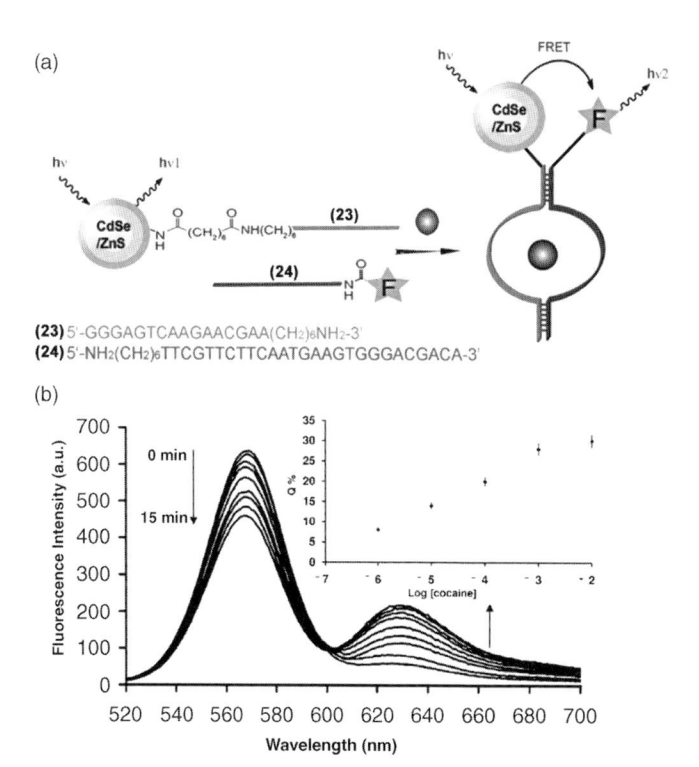

Figure 6.21 (a) QD-based optical sensing of cocaine by the formation of a cocaine–aptamer subunit supramolecular structure that stimulates a FRET process; (b) Time-dependent luminescence spectra of the system consisting of the **23**-functionalized QDs and Atto 590-modified **24**, upon interaction with cocaine, 1×10^{-3} M. Inset: Calibration curve corresponding to the luminescence quenching of the QDs upon interaction of the **23**-functionalized QDs and Atto 590-modified **24** with variable concentrations of cocaine. (Reproduced with permission of the Royal Society of Chemistry from ref. [169]).

intact aptamer sensing probe was reflected by the fact that any coincidental or partial folding of the intact aptamer was prevented, thus eliminating any background signals.

Besides the use of DNA–QDs conjugates for the analysis of target nucleic acids, QDs have also been implemented to follow the mechanical processes that occur on DNA scaffolds. The use of organized DNA scaffolds to activate "DNA machines" is a rapidly developing field [170–177], with a variety of such machines, including tweezers [178, 179], walkers [180, 181], gears [182] and others [183], having been developed. In this case, the QDs were used as an optical label to follow the dynamic walk-over ("walker") of a DNA strand on a DNA template [184] (Figure 6.22a). For this, CdSe/ZnS QDs were used as a solid support, while ATTO 590-functionalized nucleic acid (25) was applied as the moving DNA. The QDs were modified with the

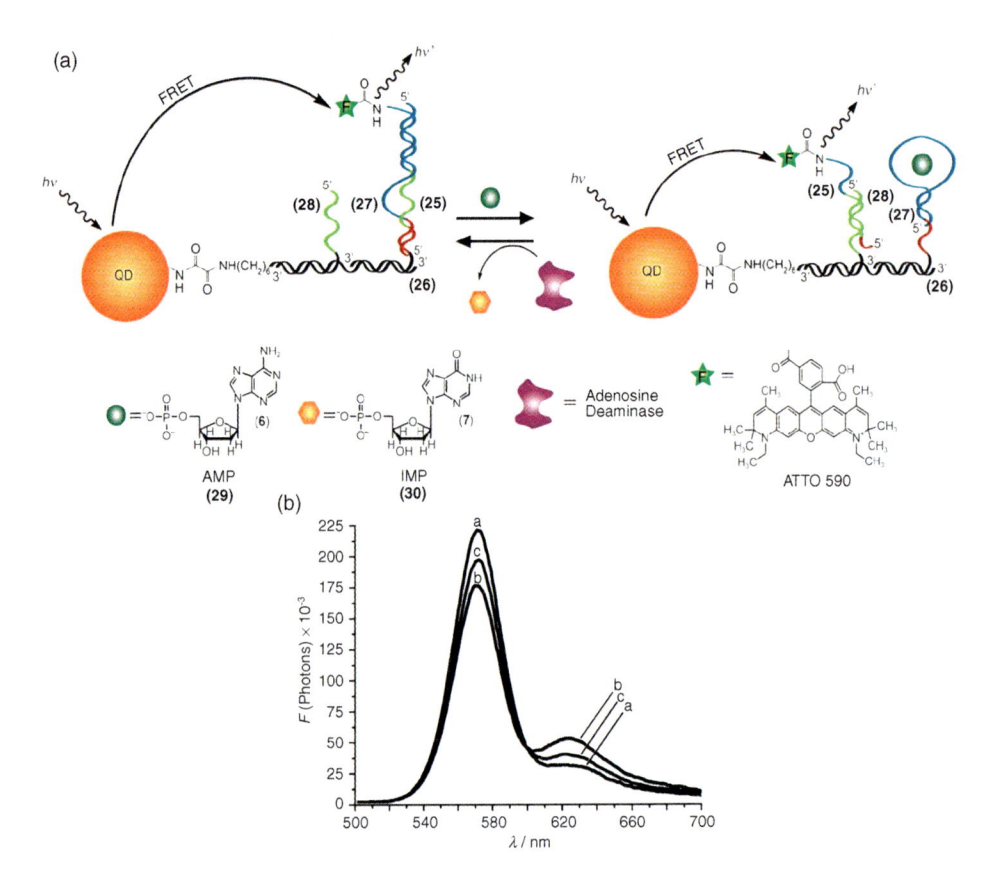

Figure 6.22 (a) Monitoring the switchable translocation of a dye-labeled nucleic acid on a DNA track associated with a CdSe/ZnS QD by FRET; (b) Fluorescence intensities corresponding to: (spectrum a) The dye-labeled nucleic acid 25 on foothold 27; (spectrum b) After treatment of the system with AMP and the translocation of B to foothold 28; (spectrum c) After treatment of the resulting system with adenosine deaminase, and translocation of 25 to foothold 27. (Reproduced with permission from Wiley-VCH from ref. [184]).

track DNA (26), while the two nucleic acids (27) and (28) were hybridized to the track DNA (26) and acted as two footholds. Foothold (27) included in its single-stranded domain the aptamer for the AMP (29) sequence, and this strand was blocked by the ATTO 590-modified nucleic acid (25), which served as the walking element in the system. Formation of the AMP–aptamer complex was favored energetically over hybridization with the blocker (25) that "walks-over" to the single-strand domain of foothold (28). The duplex between the walker and foothold (28) is energetically less favored compared to the blocked duplex between the walker and (27), while formation of the stabilized aptamer–AMP complex drives the formation of a less-stable duplex of the walker with foothold (28). Reaction of the system with adenosine deaminase transforms AMP to inosine monophosphate (IMP) (30), that lacks the binding affinity for the aptamer sequence. The release of IMP from the scaffold reactivates the translocation of the walker from foothold (28) to the energetically favored duplex structure on foothold (27). The programmed mechanical translocation of the walker unit on the DNA track was followed by the FRET process occurring between the CdSe QDs and the ATTO-590 acceptor dye. The distances separating the QDs and acceptor are controlled by the walkover process, the positioning of the walker on foothold (27) resulted in a less-efficient FRET process as compared to the FRET efficiency between the QDs and the walker positioned in close proximity to foothold (28) (Figure 6.22a). By the cyclic interaction of the QD–DNA walker-scaffold system with AMP or adenosine deaminase, the "walker" unit was reversibly cycled between footholds (27) and (28), resulting in efficient and less-efficient FRET processes, respectively (Figure 6.22b).

6.4.2
Probing Biocatalytic Transformations with QDs

The QD-stimulated FRET and ET quenching processes provide a means to follow the dynamics of chemical transformations and, specifically, of biocatalytic reactions. In fact, QDs and the respective FRET/ET processes have been used extensively to sense enzymes and/or to analyze their substrates. An early example [185] involved the detection of telomerase, a versatile marker for cancer cells. Typically, chromosomes are terminated by guanosine-rich nucleic acid segments that consist of constant repeat units, known as telomers. The telomers protect the genetic information in the chromosomes and their self-destruction and shortening provides the cell a signal to terminate cell proliferation [186, 187]. The enzyme telomerase, a ribonucleoprotein which catalyzes elongation of the telomer chains in parallel to their natural shortening, may accumulate in some cells. As a result, these cells will lack the triggering signal to terminate their life cycle, with the result that immortal and/or cancerous cells will be formed. In fact, high levels of telomerase have been detected in more than 95% of different cancer cells, such that the enzyme is now considered to be a versatile marker for the presence of cancer [188, 189]. When functionalized QDs were used to monitor telomerase activity (Figure 6.23a), the CdSe/ZnS QDs were modified with a thiolated nucleic acid (31) that was complementary to the RNA sequence embedded in telomerase, and would be recognized by the biocatalyst. In a

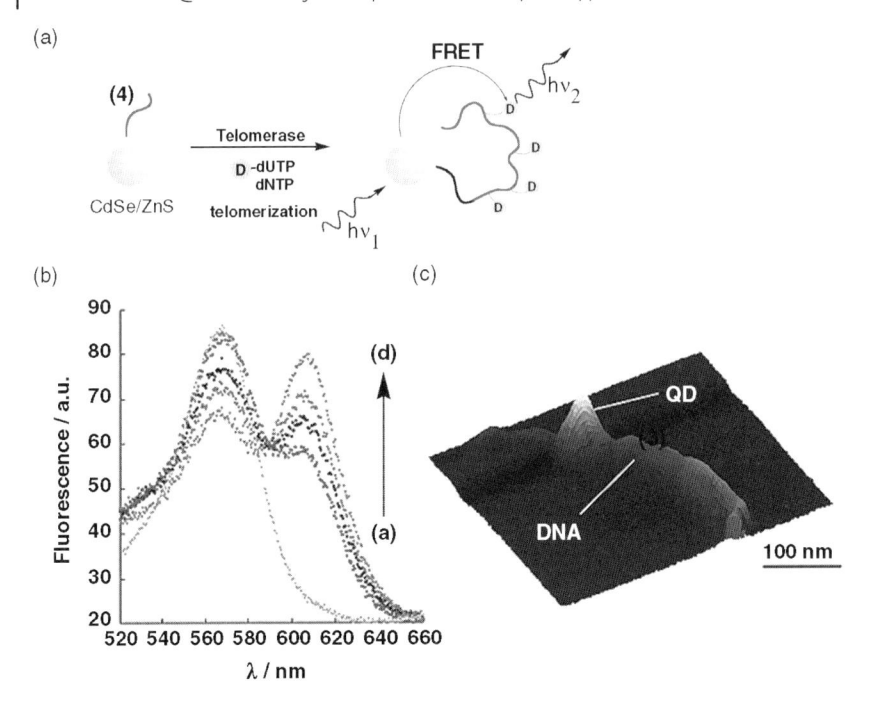

Figure 6.23 (a) Optical analysis of telomerase activity by the incorporation of Texas Red–dUTP into the telomeres associated with CdSe/ZnS QDs; (b) Time-dependent fluorescence changes upon telomerization of **31**-functionalized QDs in the presence of telomerase extracted from 10 000 HeLa cells, dNTPs (0.5 mM), and Texas Red dUTP (100 μM) at time intervals of (spectrum a) 0, (spectrum b) 10, (spectrum c) 30, and (spectrum d) 60 min; (c) AFM image of the telomeres generated on the QDs. (Reproduced with permission from ref. [185]. Copyright 2003 American Chemical Society).

HeLa cancer cell extract which included telomerase, the nucleotide mixture, dNTPs, and Texas Red-labeled dUTP, telomerization of the nucleic acid associated with the QDs proceeded such that the Texas Red-labeled nucleotide was incorporated into the elongated telomer chains, resulting in the FRET process (Figure 6.23b). The telomer chains generated on the QDs were also imaged at the molecular level using atomic force microscopy (AFM) (Figure 6.23c) [185]. A similar approach was employed to detect a target DNA (M13 phage), by using CdSe/ZnS QDs which had been modified with a nucleic acid that was complementary to a segment of the target DNA. When the QDs were interacted with M13 phage DNA, the resultant hybrid was replicated on the QDs in the presence of both the dNTPs mixture (which included the Texas Red-labeled dUTP) and polymerase. The subsequent FRET process that was stimulated between the QDs and the Texas Red dye served as a readout signal for the target DNA [185].

CdSe/ZnS QDs were also used to probe DNA hybridization and to follow the biocatalyzed scission of duplex DNA by DNase [190]. For this, the CdSe/ZnS QDs were functionalized with a nucleic acid primer (**32**), and then hybridized with the

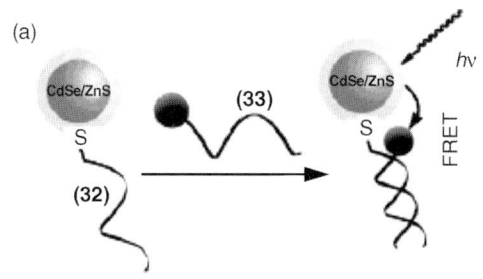

(a)

(32) : 5'-HS-(CH$_2$)$_5$GAATGCTTTAAGACACCTTTCAAGGT-3'
(33) : 5'-GAAAGGTGTCTTAAAGCATT-Texas Red-3'

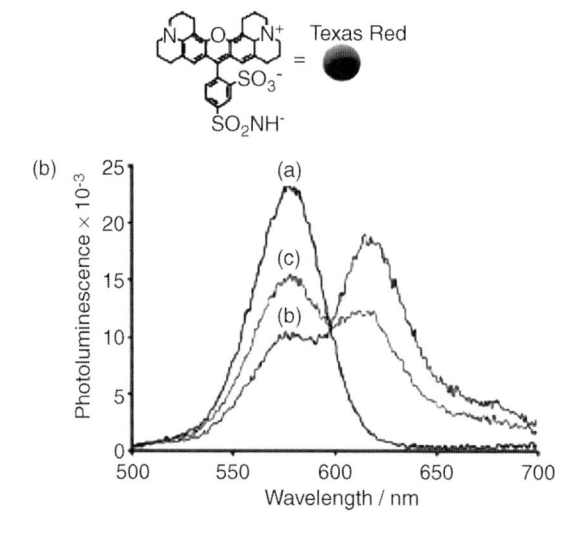

Figure 6.24 (a) Assembly of CdSe/ZnS and Texas Red-tethered duplex DNA; (b) The fluorescence spectra of (spectrum a) the **32**-functionalized CdSe/ZnS QDs; Spectrum b shows the DNA duplex **32/33** tethered to the QDs and the Texas Red chromophore; Spectrum c, after treatment with DNase I. (Reproduced with permission from Ref. [190]. Copyright 2005 American Chemical Society).

Texas Red-labeled complementary nucleic acid (**33**) (Figure 6.24a). The FRET process between the QDs and the Texas Red acceptor chromophore was used to probe the hybridization of DNA (Figure 6.24b, curve b). The reaction of the QD/Texas Red duplex with DNAse resulted in cleavage of the FRET acceptor and restoration of the luminescence of the QDs (Figure 6.24b, curve c). Unfortunately, the luminescence of the QDs was not fully restored, but this was attributed to a nonspecific adsorption of the Texas Red dye onto the QDs.

In a related study, CdSe/ZnS QDs modified with ssDNA–fluorescent dye conjugates were used to monitor the cleavage of ssDNA by micrococcal nuclease (MNase), with high sensitivity and specificity [191]. For this, the CdSe/ZnS QDs were first functionalized with streptavidin, after which a dye-modified, biotinylated, ssDNA was linked to the particles. In this configuration, the fluorescence of the QDs

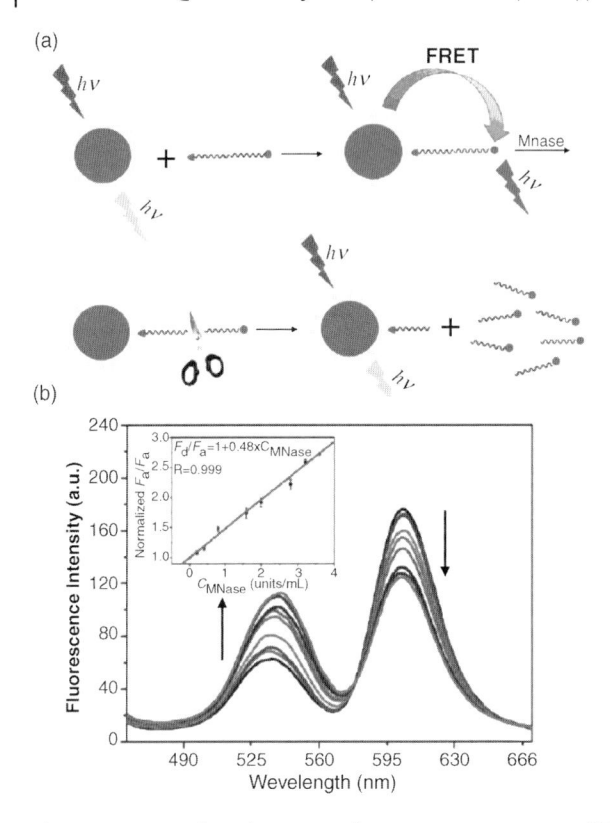

Figure 6.25 (a) Probing the activity of micrococcal nuclease (MNase) using QDs with a dye (6-carboxy-X-rhodamine, ROX)-modified nucleic acids; (b) Fluorescence spectra corresponding to the QDs/ROX-modified nucleic acid system upon treatment with different concentrations of MNase. As the concentration of MNase increases, the cleavage of the nucleic acid is enhanced, resulting in a lower FRET emission of ROX and higher luminescence of the QDs. Inset: Normalized calibration curve. (Reproduced with permission of the Royal Society of Chemistry from Ref. [191]).

was quenched via a FRET process. In the presence of MNase, the ssDNA was cleaved, thus releasing the dye acceptor from the QDs conjugates. The biocatalytic removal of the dye acceptor decreased the FRET process and intensified the luminesence of the QDs (Figure 6.25), which, in turn, enabled the MNase to be probed, with a detection limit corresponding to 0.06 units ml^{-1}.

Similarly, semiconductor QDs were integrated with proteins, such that the hybrid systems would permit the real-time analysis of catalytic transformations stimulated by the proteins [102, 192–194]. For example, the hydrolytic functions of a series of proteolytic enzymes were followed by the application of QD reporter units, using the FRET process as a readout mechanism. In this case, the CdSe QDs were modified with peptide sequences that were specific for different proteases, where the quencher units were tethered to the peptide termini. Within the QDs/fluorophore-modified hybrid assembly, the fluorescence of the QDs was quenched. Subsequent hydrolytic

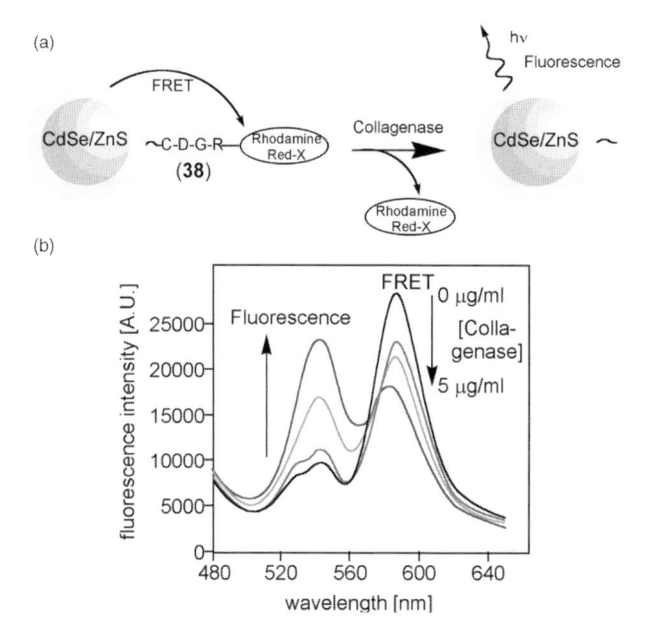

Figure 6.26 (a) CdSe/ZnS QDs for the optical analysis of the protease-mediated hydrolysis of the rhodamine red-X-functionalized peptide **38**; (b) Decrease in fluorescence of the dye and corresponding increase in fluorescence of the QDs on interaction with different concentrations of collagenase. (Reproduced with permission from Ref. [192]. Copyright 2006 American Chemical Society).

cleavage of the peptide resulted in a removal of the quencher units, which, in turn, restored the fluorescence generated by the QDs while decreasing the emission of the fluorophore. In an example of this process, collagenase was used to cleave the rhodamine Red-X dye-labeled peptide linked to CdSe/ZnS QDs [192] (Figure 6.26a). While the tethered dye quenched the fluorescence of the QDs, the hydrolytic scission of the dye, and its removal from the QD–peptide conjugate, led to a restoration of the fluorescence of the QDs (Figure 6.26b).

In another example, CdSe/ZnS QDs were used to follow the activity of β-lactamase [195] (Figure 6.27). For this, biotinylated lactam labeled with Cy5 dye was linked to the streptavidin-capped particles, resulting in FRET quenching of the QDs' luminescence by the Cy5-labeled lactam. Enzymatic cleavage of the lactam ring caused the Cy5 dye to be released from the QDs surface, while restoring the QDs emission.

The different biocatalytic transformations probed with QDs have led to the implementation of a FRET process to follow the reaction progress. The ET quenching of QDs represents an alternative mechanism for probing biocatalytic transformations; a typical example was the biocatalytic function of two enzymes, tyrosinase and thrombin, which were probed by using CdSe/ZnS QDs [196]. In this case, the CdSe/ZnS QDs were capped with a monolayer of methyl ester tyrosine (**34**). A subsequent tyrosinase-induced oxidation of tyrosine to dopa-quinone led to the generation of ET quencher units that suppressed the luminescence of the QDs (Figure 6.28a). The

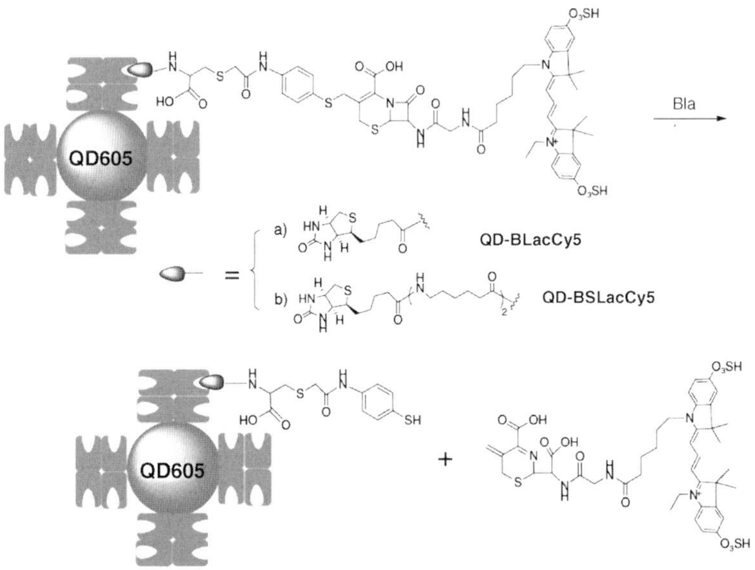

Figure 6.27 Optical detection of lactamase by modified-QDs. (Reproduced with permission from Ref. [193]).

depletion of fluorescence of the QDs, by interacting the functionalized particles with different concentrations of tyrosinase, was used to follow the tyrosinase activity via the time-dependent oxidation of L-DOPA residues (Figure 6.28b). Besides showing that QDs can be used to follow biocatalytic processes, the analysis of tyrosinase activity by using QDs has clear practical implication, as elevated levels of tyrosinase have been identified in the cells of melanoma. Hence, the optical detection of this biocatalytic biomarker using QDs might represent a useful diagnostic system in this situation.

The tyrosinase-stimulated oxidation of phenol residues was also employed to monitor the activity of thrombin, by using QDs [196]. Here, a tyrosine-terminated peptide (**35**) that included the specific sequence for cleavage by thrombin was linked to the QDs. Subsequent oxidation of the tyrosine residues by tyrosinase generated the o-quinone derivative of L-DOPA, which quenched the luminescence of the QD via ET (Figure 6.29a). The following thrombin-stimulated cleavage of the peptide, and removal of quinone quencher units from the QDs, caused a regeneration of the QD fluorescence (Figure 6.29b).

A similar approach was used for the optical detection of phenolic compounds and hydrogen peroxide, using CdTe QDs–enzyme hybrids. For this, mercaptosuccinic acid-capped CdTe QDs were first synthesized, after which the horseradish peroxidase-catalyzed oxidation of phenolic compounds by H_2O_2 to quinone products resulted in an efficient quenching of QD luminescence [197].

The different applications of QDs to monitor biocatalytic processes have required the specific modification of QDs with capping layers specific for target enzymes. The synthesis of QDs with a versatile modifier capable of analyzing a broad class of

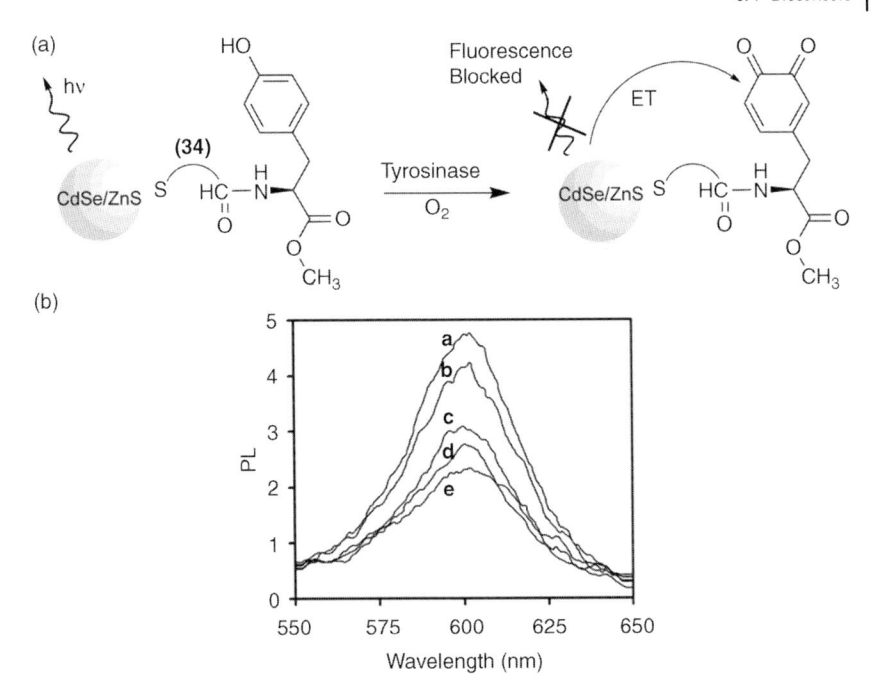

Figure 6.28 (a) Analysis of tyrosinase activity by the biocatalytic oxidation of the methyl ester tyrosine-functionalized CdSe/ZnS QDs to the dopaquinone derivative that results in the electron transfer quenching of the QDs; (B) Time-dependent fluorescence quenching of the QDs upon the tyrosinase-induced oxidation of the tyrosine-functionalized QDs: (spectrum a) 0, (spectrum b) 0.5, (spectrum c) 2, (spectrum d) 5 and (spectrum e) 10 min. (Reproduced with permission from Ref. [196]. Copyright 2006 American Chemical Society).

enzymes would clearly provide an advance in the bioanalytical applications of QDs. As many redox enzymes employ the common 1,4-nicotinamide adenine dinucleotide (phosphate) cofactor, $NAD(P)^+$, the use of appropriately functionalized QDs capable of analyzing the NAD(P)H cofactors could provide a generic method, not only for the analysis of $NAD(P)^+$-dependent enzymes, but also for the detection of their substrates. The synthesis of Nile blue-functionalized CdSe/ZnS QDs as a hybrid material that can optically sense NAD(P)H cofactors was described in Section 6.3.2. These NADH-sensitive QDs were used to follow not only NAD^+-dependent biocatalyzed transformations but also their substrates. As a model system, the QDs were applied to the analysis of ethanol in the presence of the NAD^+-dependent enzyme alcohol dehydrogenase (AlcDH) (Figure 6.30a). In this reaction, AlcDH catalyzes the oxidation of ethanol to acetaldehyde, with the concomitant reduction of NAD^+ to NADH; the resulting reduced cofactor is oxidized by quencher units associated with the QDs, and this results in an enhanced luminescence of the QDs (Figure 6.30b). The ethanol-mediated formation of NADH, in the presence of AlcDH, enabled the quantitative analysis of ethanol (Figure 6.30c). Moreover, as the concentration of ethanol was increased, the concentration of the resultant NADH was higher, leading to an intensified fluorescence of the QDs [140].

(a)

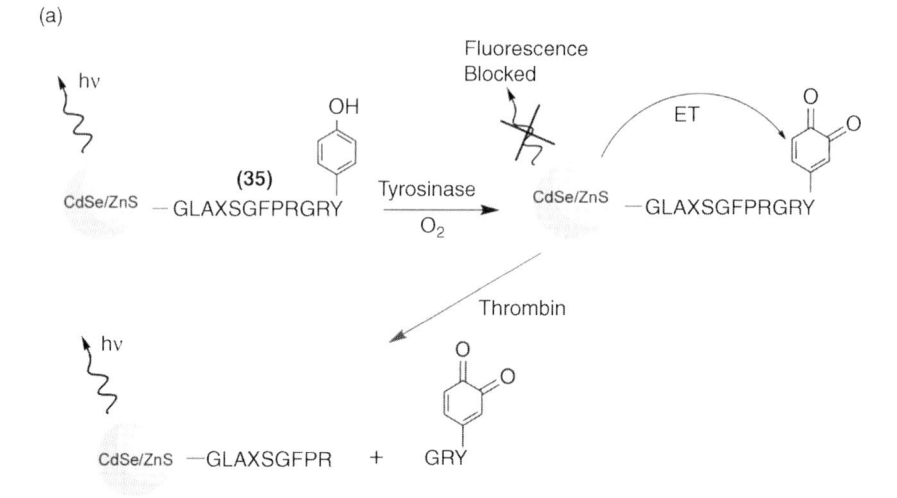

[Figure 6.29a scheme showing CdSe/ZnS QDs conjugated to peptide (35) GLAXSGFPRGRY with tyrosine-OH group; upon tyrosinase/O₂ oxidation to dopaquinone, fluorescence is blocked via ET; thrombin cleavage yields CdSe/ZnS–GLAXSGFPR + GRY and restores fluorescence]

N-Gly-Leu-Ala-Aib-Ser-Gly-Phe-Pro-Arg-Gly-Arg-Tyr-CONH₂

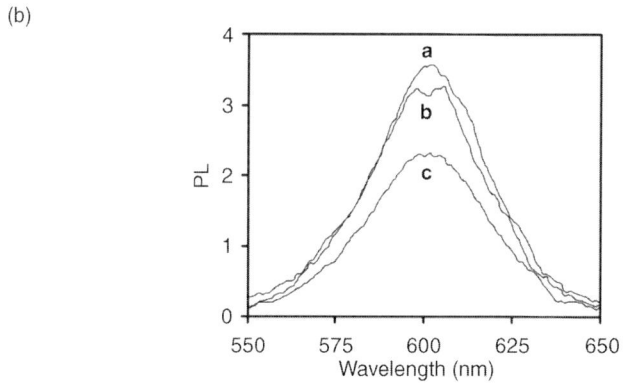

(b)

[PL spectra vs. Wavelength (nm) from 550 to 650, showing curves a, b, c]

Figure 6.29 (a) Sequential analysis of tyrosinase activity and thrombin activity by the tyrosinase-induced oxidation of the tyrosine-containing peptide **35** associated with the CdSe/ZnS QDs results in electron transfer quenching of the QDs, followed by the thrombin-induced cleavage of the dopaquinone-modified peptide to restore the QDs fluorescence; (b) (spectrum a) Fluorescence of QDs modified with **35** (spectrum b) after reaction of the QDs with tyrosinase for 10 min (spectrum c), and thereafter treatment of the dopaquinone-functionalized QDs with thrombin for 6 min (spectrum c). (Reproduced with permission from Ref. [196]. Copyright 2006 American Chemical Society).

In a further example, phenyl boronic acid-functionalized CdSe/ZnS QDs were used to bind NAD^+ or NADH via the boronic acid ligand to form boronate esters. The quenching of the QDs luminescence by the NAD^+ cofactor, via an ET quenching route, was substantially more efficient than was the quenching of QDs by NADH. This difference in quenching between NAD^+ and NADH enabled QDs to be used for the luminescence analysis of NAD^+-dependent enzymes and their substrates, such as the AlcDH/ethanol system [141] (Figure 6.31a). The biocatalytic reduction of the

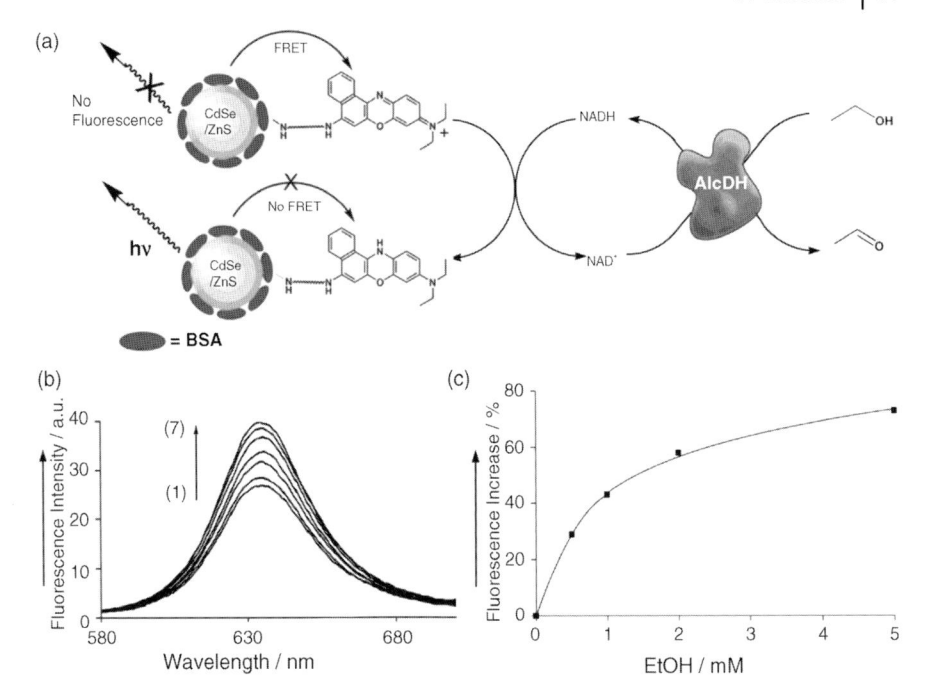

Figure 6.30 (a) Sensing of ethanol by Nile blue-functionalized CdSe/ZnS QDs; (b) Time-dependent fluorescence changes for the analysis of 1 mM ethanol by the functionalized QDs: (1) prior to the addition of ethanol; (2–7) after successive time intervals of 3 min; (c) Calibration curve corresponding to the optical analysis of different concentrations of ethanol by the functionalized QDs. (Reproduced with permission from Ref. [140]).

NAD^+ capping layer by ethanol, in the presence of AlcDH, yielded NADH-functionalized QDs in a process that switched-on the luminescence of the QDs (Figure 6.31b). This, in turn, enabled the quantitative analysis of different concentrations of the substrate, ethanol (Figure 6.31c). Consequently, as the concentration of ethanol was increased, the higher content of NADH associated with the QDs led to an intensification the luminescence of the QDs.

The use of Nile blue-functionalized QDs (Figure 6.30a) or NAD^+-modified QDs (Figure 6.31a) for the analysis of ethanol in the presence of AlcDH represents a generic approach, where the functionalized QDs can be applied to analyze any NAD^+-dependent enzyme, and its substrates.

A further bioanalytical application of QDs for following biocatalytic processes involved controlling the photophysical properties of QDs by means of enzyme-stimulated reactions. For example, the reaction of H_2O_2 with CdSe/ZnS QDs was found to result in a quenching of the luminescence properties of the QDs [198]. It was assumed that such quenching had originated from the formation of oxide traps for the conduction band electrons that would prevent radiative electron–hole recombination. Indeed, a detailed surface analysis of the reaction between CdTe QDs and H_2O_2 revealed the formation of Cd-oxide groups. Yet, whereas this phenomenon

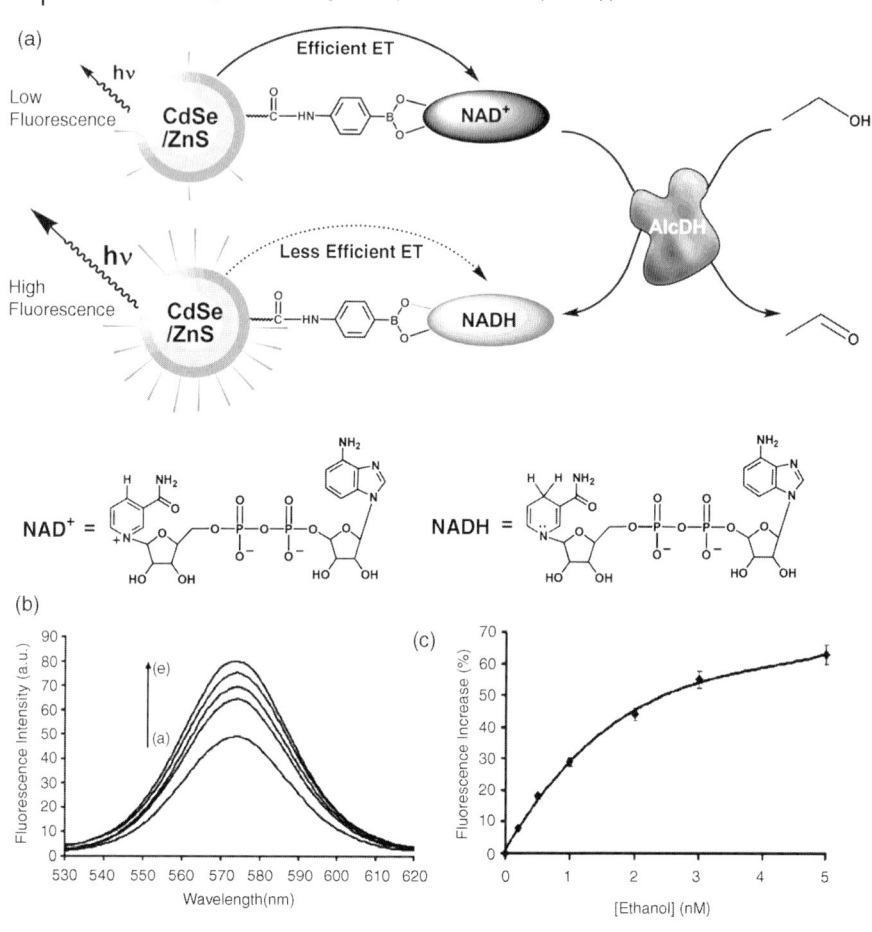

Figure 6.31 (a) The enzyme alcohol dehydrogenase (AlcDH) catalyzes the oxidation of ethanol to acetaldehyde, with the concomitant reduction of the NAD^+ units associated with the QDs to NADH; (b) Time-dependent fluorescence changes upon the interaction of the NAD^+-functionalized QDs with ethanol, 5 mM, in the presence of AlcDH, 5 units. The spectra were recorded at time intervals of 3 min; (c) Calibration curve corresponding to the fluorescence analysis of variable concentrations of ethanol by the NAD^+-functionalized QDs. (Reproduced with permission from Ref. [141]. Copyright 2009 American Chemical Society).

enabled the quantitative optical detection of H_2O_2, it also provided a general platform to monitor the activities of H_2O_2-generating oxidases, and their substrates.

For example, fluorescein-labeled avidin-functionalized CdSe/ZnS QDs were used for the optical analysis of glucose [198] (Figure 6.32a). In this case, biotinylated glucose oxidase was linked to the particles; the H_2O_2 generated by the biocatalyzed oxidation of glucose by O_2 then acted as a surface modifier of the QDs, resulting in a quenching of the QDs luminescence. Although the fluorescence of the fluorescein dye tethered to avidin was insensitive to H_2O_2, it served as an internal reference when probing the fluorescence changes of the QDs caused by H_2O_2. The quenching of the

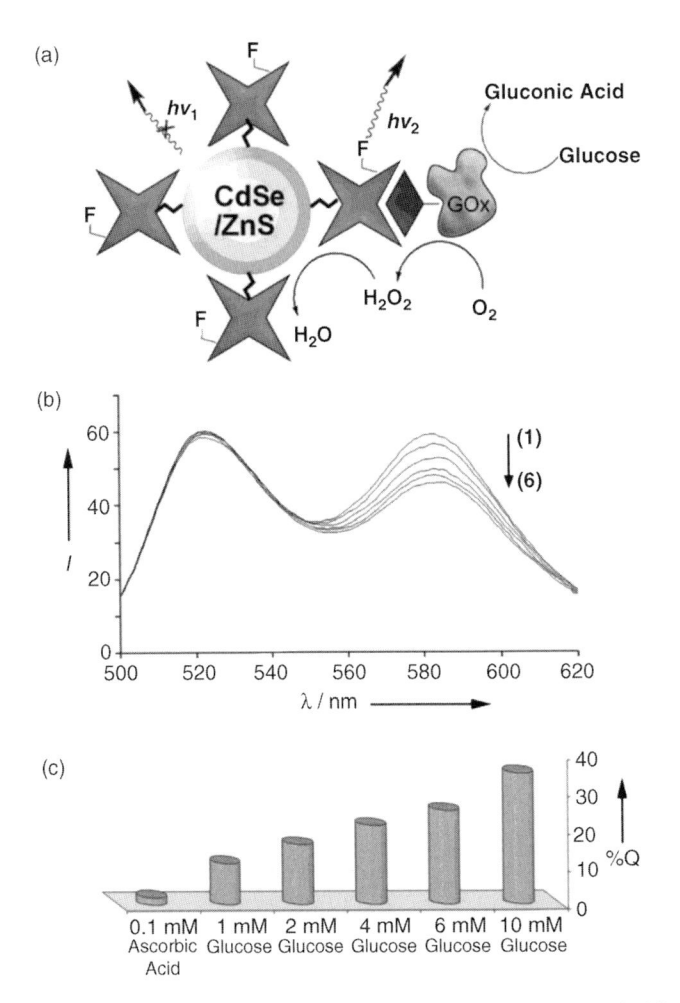

Figure 6.32 (a) Ratiometric analysis of glucose by biotinylated GOx associated with the fluorophore-labeled avidin–CdSe/ZnS QD conjugates; (b) Time-dependent ratiometric fluorescence analysis of glucose (5 mm) by GOx-modified QDs after 0, 2, 5, 10, 15, and 20 min; (c) Quenched fluorescence of the GOx-functionalized CdSe/ZnS QDs upon analyzing different concentrations of glucose for a fixed time interval of 10 min. (Reproduced with permission of Wiley-VCH from Ref. [198]).

QDs luminescence was enhanced when the concentration of glucose was elevated, such that the system permitted the quantitative ratiometric analysis of glucose (Figure 6.32b). The same QDs were used to follow the activity of acetylcholine esterase/choline oxidase, and to monitor the presence of acetylcholine oxidase inhibitors. Acetylcholine esterase catalyzes the hydrolysis of acetylcholine to yield choline, which subsequently is oxidized by O_2 to form betaine and H_2O_2 that, in turn, quenches the QD luminescence. The interaction of the biocatalytic cascade with the inhibitor neostigmine (a nerve gas simulator) inhibited the generation of H_2O_2 and eliminated the quenching of the QDs luminescence.

6.4.3
Probing Structural Perturbations of Proteins with QDs

The binding of substrates or inhibitors to enzymes, the association of ions to proteins, or the formation of protein–substrate or protein–protein complexes, involves conformational perturbations of the protein backbone. The sensitivity of the FRET process to the distance separating the donor–acceptor pair provides an effective spectroscopic tool to probe the conformational perturbations of proteins. This method was used to assemble a reagent-free, QD-based sensor for maltose [199]. In this case, the CdSe/ZnS QDs were functionalized with the maltose-binding protein (MBP) mutant, MBP41C, which includes a cysteine at a peristeric site that does not take part in the maltose-binding process. This residue was specifically labeled with the fluorescent dye, Cy3, which served as a FRET acceptor from the QD. Upon binding maltose, MBP undergoes conformational changes that alter the environment surrounding the dye, such that the fluorescence of Cy3 is altered in a concentration-dependent manner (Figure 6.33). These conformational changes result, in both, the enhanced FRET from the QDs to the dye, and an enhancement of the nonradiative decay of the dye.

Electron transfer was also used to follow the binding of maltose to MBP [200]. For this, MBP-coated CdSe QDs were functionalized with a ruthenium complex attached to the protein. In this configuration, the ruthenium was close to the QDs, enabling an efficient ET between the QDs and the ruthenium. On binding maltose, the MBP undergoes a conformational change that alters the ET distance between the QDs and the ruthenium complex, resulting in an increased luminescence of the QDs (Figure 6.34).

Similarly, solvation changes upon the association of a substrate with a protein-binding site could be probed by the ET quenching of a ruthenium(II) complex linked closely to the binding site on QDs that had been functionalized with the binding protein. For this, the CdSe QDs were functionalized with an intestinal fatty acid-binding protein that was further modified with the ruthenium(II) complex. Fluorescence changes which resulted from solvation differences on the binding of fatty acids to the protein were then used to probe the analytes [201].

In another example, QDs were applied to probe the structural perturbation of an enzyme during the substrate-binding process, with the fluorescence changes being

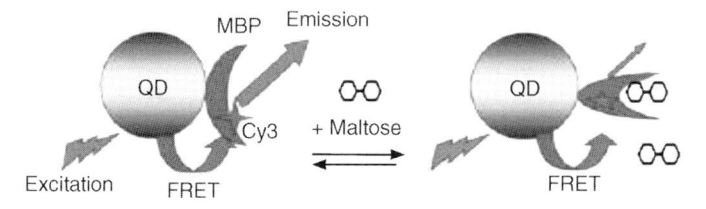

Figure 6.33 Control of the FRET process between CdSe and Cy3-functionalized maltose binding protein (MBP) upon association of maltose and induction of a conformational change in MBP. (Reproduced with permission of Wiley-VCH from Ref. [199]).

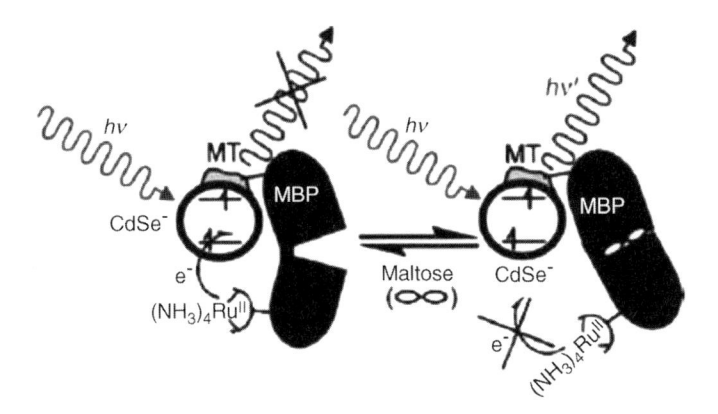

Figure 6.34 Controlling the degree of photoinduced electron transfer between CdSe and MBP functionalized with [Ru^{II}(phenanthroline)(NH$_3$)$_4$] upon association of maltose, resulting in a structural alteration of the electron transfer distances. (Reproduced with permission from Ref. [200]. Copyright 2005 American Chemical Society).

used for the optical sensing of paraoxon [202] (Figure 6.35). For this, organophosphorus hydrolase (OPH) was coupled to the CdSe/ZnS QDs through electrostatic interactions. In the presence of organophosphorus compounds (such as paraoxon), the secondary structure of OPH was changed, and this resulted in a quenching of the QDs photoluminescence. Typically, this system enabled the optical sensing of paraoxon with a detection limit of 10^{-8} M.

The FRET process that occurs between QDs and a dye-labeled substrate bound to the recognition pocket of a protein, may provide a general means for following the recognition events that occur between the proteins and their native substrates, by a displacement mechanism or a competitive assay. For example, when CdSe/ZnS QDs

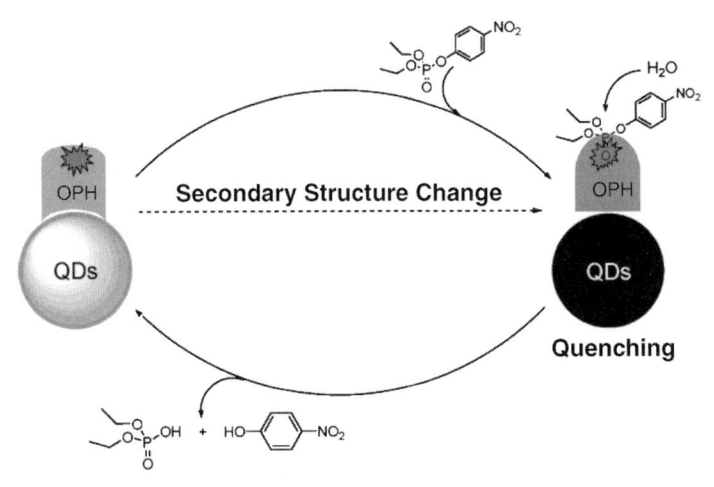

Figure 6.35 Probing the hydrolysis of paraoxon by the organophosphorus (OPH)-modified QDs. (Reproduced with permission from Ref. [202]. Copyright 2005 American Chemical Society).

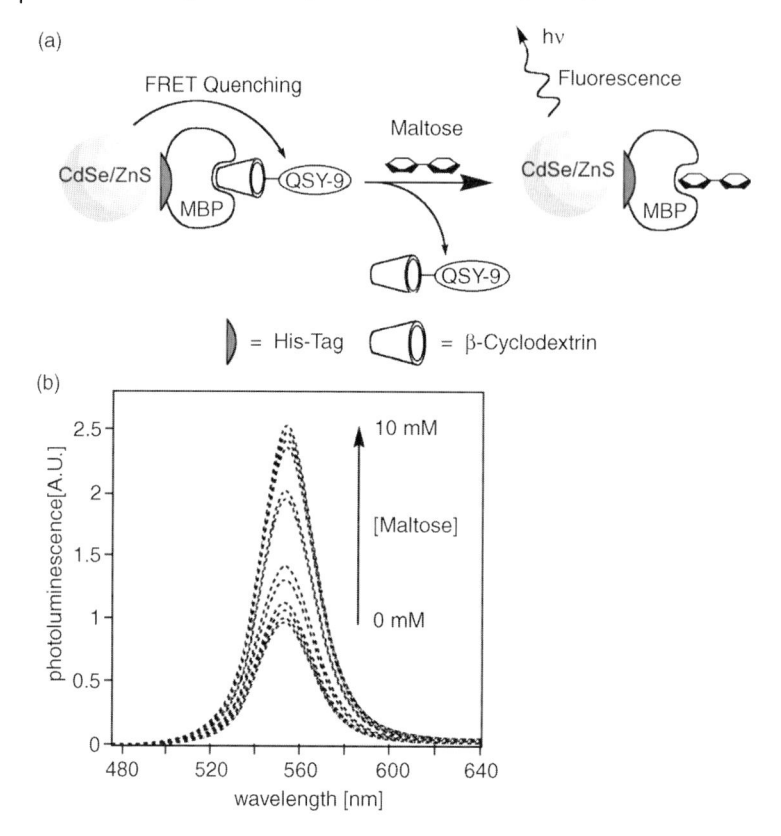

Figure 6.36 (a) Functionalized CdSe/ZnS QDs for the competitive assay of maltose using the maltose binding protein (MBP) as sensing material and β-cyclodextrin–QSY-9 dye conjugate β-CD-QSY-9 as FRET quencher; (b) Fluorescence changes of the MBP-functionalized QDs with increasing concentrations of maltose. (Reprinted by permission from Macmillan Publishers Ltd. Nature Materials Ref. [73]).

were linked to the MBP, the modified particles were interacted with the β-CD–QSY-9 dye conjugate, which resulted in the QD luminescence being quenched [73] (Figure 6.36a). However, displacement of the dye-labeled sugar from the protein on the addition of maltose resulted in a regeneration of QD luminescence (Figure 6.36b). This method enabled the development of a competitive QD-based sensor for maltose in solution.

Interactions between the HIV-1 regulatory protein, Rev, and the Rev responsive element (RRE) RNA, as well as the effect of inhibitors on these interactions, were monitored using a single-particle fluorescence-based method [203]. This biosensing method combined FRET and the colocalization of fluorophores as the means of detection. In the presence of an analyte, the dye-labeled recognition complex was assembled on the fluorescent QDs, and these were tunneled, under flow, into two detector channels that simultaneously analyzed the fluorescence of the QDs and the FRET emission of the dye. The two emissions were observed simultaneously only if

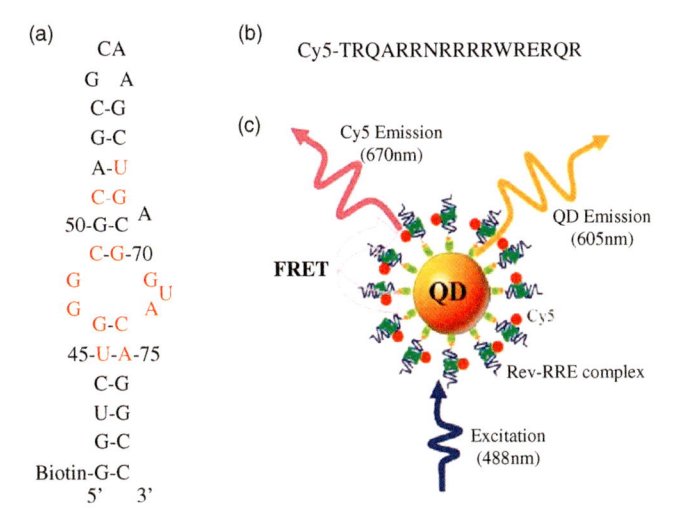

Figure 6.37 (a) Aptamer structure for the Rev regulatory protein; (b) Sequence of the Cy5-labeled Rev peptide; (c) Hybrid consisting of the aptamer-functionalized QDs for the FRET analysis of the Rev peptide. (Reprinted with permission from Ref. [203]. Copyright 2006 American Chemical Society).

colocalization had occurred. The system included biotin-modified RRE RNA, Cy5-labeled Rev peptide and streptavidin-conjugated QDs (Figure 6.37). The binding of Rev peptide to RRE formed a biotinylated Cy5-labeled Rev-peptide–RRE complex that assembled on the streptavidin-coated QDs. Fluorescence bursts were detected only by the two donor and acceptor detection channels when the interaction occurred. This method was also used to screen for inhibitors that eliminated the RNA–peptide interactions [203].

6.5
Intracellular Applications of QDs

To date, a variety of methods have been developed for the efficient delivery of QDs into cells [204, 205]. These include nonspecific endocytosis [206–209] or receptor-mediated endocytosis that involves QDs decorated with transfection reagents (peptides [210–215], proteins [65, 216–219], cationic liposomes [204], dendrimers [204], polymers [220, 221] or small molecules [222, 223]) that were used for the intracellular delivery of QDs. Physical techniques such as, electroporation [204, 224] or microinjection [204, 225, 226] have also been employed to deliver QDs into cells.

In the past, QDs have been used extensively for intracellular applications such as, cell imaging and the labeling of subcellular compartments, and these topics have been recently reviewed [201, 202]. The use of cell-incorporated QDs to probe drug delivery or to follow intracellular processes is, however, scarce, and the subject holds great promise for future developments. One major challenge would involve the use of

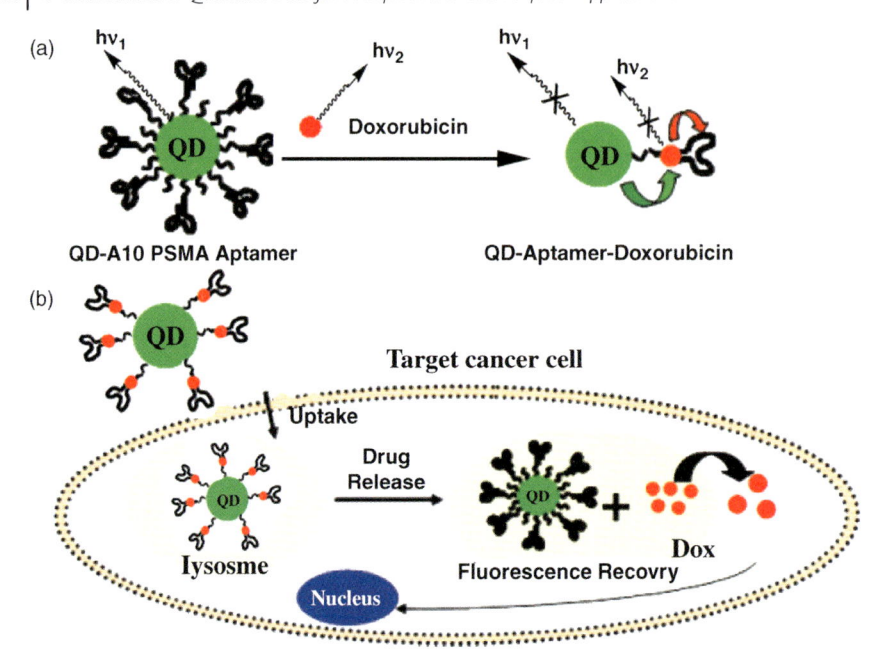

Figure 6.38 (a) Formation of the QDs–aptamer–doxorubicin complex; (b) Uptake of the QDs–aptamer–doxorubicin conjugates into target cancer cells, and the intracellular release of doxorubicin. (Reprinted with permission from Ref. [227]. Copyright 2007 American Chemical Society).

QDs as carriers of molecular or macromolecular "cargos" that could serve as therapeutic materials. The transport of drugs into the cells, and the release/localization of those drugs into the cell compartments, could be transduced by the photophysical properties of the QDs. For example [227], QDs have been modified with a RNA aptamer for the prostate-specific membrane antigen (PSMA), and the resulting bioconjugates interacted with doxorubicin, an anthracycline-based fluorescent anti-cancer drug that intercalates within the double-stranded CG sequences of the PSMA aptamer (Figure 6.38). The intercalation of doxorubicin resulted in the quenching of the QDs luminescence by a FRET process between the QDs and doxorubicin, and the fluorescence quenching of doxorubicin by the double-stranded RNA aptamer. The PSMA–doxorubicin–QDs were delivered into prostate cancer cells via endocytosis, and doxorubicin released in the cells; this resulted in the recovery of QDs luminescence and the fluorescence of the free doxorubicin. It also enabled sensing of the intracellular release of the therapeutic agent by activating the QDs luminescence, in addition to the simultaneous fluorescence localization and killing of the cancer cells.

QDs have also been used for the intracellular delivery of small interfering RNA (siRNA) for different genes, resulting in a synchronized treatment and imaging of the cells. The siRNAs prevent translation of the messenger RNA into proteins. Accordingly, when various QD–siRNA bioconjugates were synthesized and delivered into tumor cells, it resulted in an induced suppression of target genes such as the

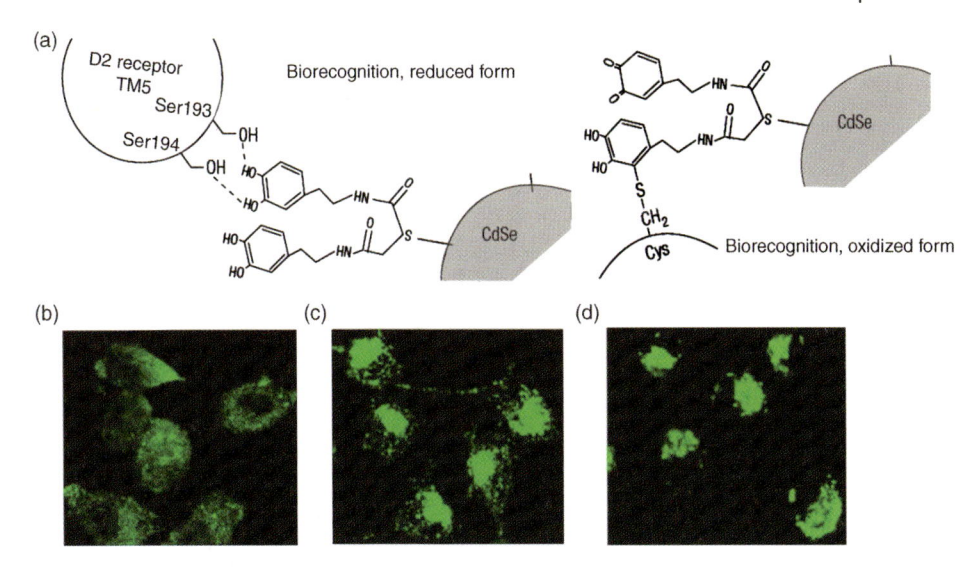

Figure 6.39 (a) Cell uptake mechanisms of dopamine-modified QDs in the reduced and oxidized forms; (b) Cells under oxidizing conditions labeled with dopamine-modified QDs; (c) Cells under normal conditions labeled with dopamine-modified QDs; (d) Cells under the most reducing conditions labeled with dopamine-modified QDs. (Reprinted by permission from Macmillan Publishers Ltd. Nature Materials Ref. [223]).

lamin A/C gene [228], the HER2 gene [229], and an enhanced synthesis of the green fluorescent protein (GFP) gene [229, 230].

One further challenge would be to use QDs as optical reporters capable of monitoring intracellular processes, such as cell metabolism. For example, CdSe/ZnS–dopamine conjugates were used to monitor the oxidative conditions within a cell as a measure of its metabolic state [223]. Dopamine, an electron donor, is oxidized to a dopaquinone derivative that serves as an electron acceptor. Dopamine plays a secondary role by enabling transport of the modified-QDs into the cells via the dopamine receptors (Figure 6.39). The fluorescence of the QDs may be altered, depending on the intracellular redox potential that determines the state of the neurotransmitter modifier. Under reducing conditions, fluorescence is observed only at the periphery of the cell (Figure 6.39d), but as the intracellular conditions become mildly oxidizing the luminescence of QDs is detected only in the membrane region that separated the nucleus of the cells from the cytoplasm (Figure 6.39c). Under highly oxidizing conditions, the QD luminescence is observed throughout the cell (Figure 6.39b), which implies that the QDs could be used to monitor the redox states of the different cell compartments.

QDs have also been used as fluorescence markers to follow intracellular metabolic pathways, with the system having been used to screen anticancer drugs [140]. When the CdSe/ZnS QDs were modified with the dye Nile blue, the chemically modified QDs could serve as optical labels for the sensing of NAD(P) cofactors (see Sections 6.3 and (6.4). As the NADH cofactor is formed upon activation of the intracellular

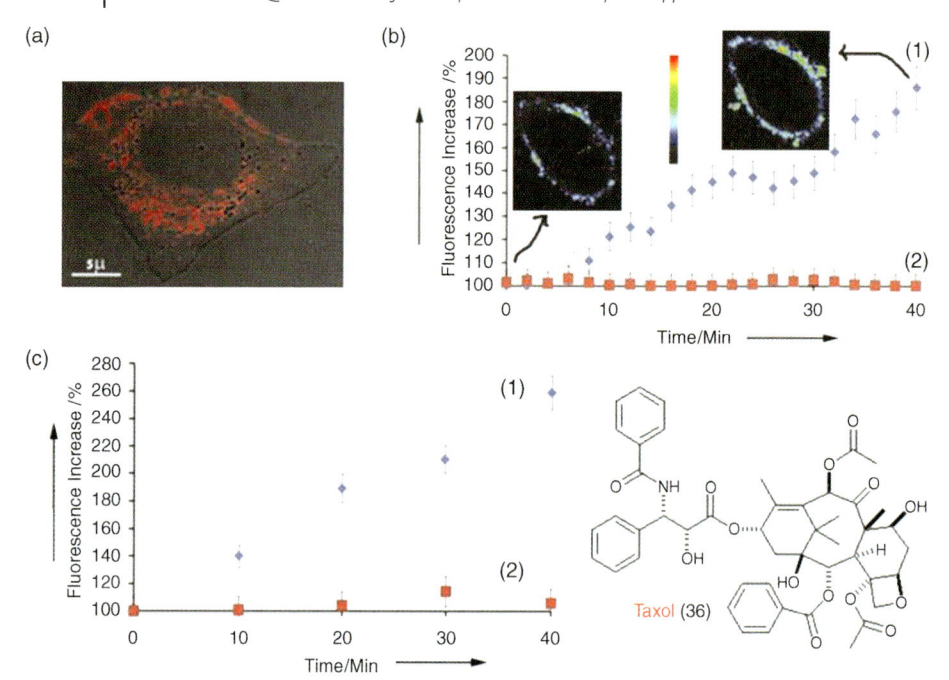

Figure 6.40 (a) Confocal fluorescence microscopy image of a HeLa cell that included incorporated Nile blue-functionalized QDs; (b) Curve 1: time-dependent fluorescence changes of HeLa cells that include the functionalized QDs upon interaction with 50 mM D-glucose. Curve 2: time-dependent fluorescence changes of HeLa cells that include the functionalized QDs upon interaction with 50 mM L-glucose. Each datum point corresponds to the analysis of 20 different cells. Inset: Fluorescence image of one representative cell before and after interaction with D-glucose; (c) Time-dependent fluorescence changes of HeLa cells that include functionalized QDs upon addition of 50 mM D-glucose to (curve 1) nontreated HeLa cells, and (curve 2) taxol-treated HeLa cells. Each of the datum points represents the averaging of the results obtained from 20 individual cells. (Reproduced with permission of Wiley-VCH from Ref. [140]).

metabolism, changes in the fluorescence of QDs incorporated into the cells can be used to probe the progress of the cell's metabolic pathways. When Nile blue-functionalized QDs were incorporated into HeLa cancer cells by electroporation (Figure 6.40a), the fluorescence image of a HeLa cell demonstrated the luminescence of QDs in the cytoplasm, and their delivery into the cells.

The HeLa cells with incorporated Nile blue-functionalized CdSe/ZnS QDs were subsequently grown under starvation conditions and then subjected to the addition of D-glucose (50 mM). The added glucose triggered the cell metabolism which, in turn, generated NADH that intensified the luminescence of the cells (Figure 6.40b, curve 1). The inset images show the luminescence of the cells prior to the addition of glucose, and following the activation of cell metabolism with D-glucose. Interestingly, L-glucose, which is not recognized by the cells, did not activate the cell metabolism (Figure 6.40b, curve 2). The successful monitoring of cellular metabolism by the

NADH-sensitive QDs was, in turn, used to screen anticancer drugs that affect cellular metabolism. As a model system, the effect of the anticancer drug taxol (**36**) on the metabolism of HeLa cancer cells was examined. The activation of intracellular metabolism by glucose addition was monitored in HeLa cancer cells with functionalized NADH-sensitive QDs, in the presence or absence of taxol treatment (Figure 6.40c, curves 1 and 2). The non-taxol-treated cells revealed a high luminescence, implying an effective intracellular metabolism, whereas the taxol-treated cells showed a very low luminescence, indicating that the anticancer drug had indeed inhibited the cellular metabolism. It appears that these functionalized QDs may well provide a versatile tool to screen for drugs that affect diverse intracellular metabolic pathways.

6.6 Conclusions and Perspectives

The unique photophysical properties of semiconductor QDs are, today, applied widely for both analytical and bioanalytical purposes. During the past few years, an impressive number of advances have been accomplished in the use of QDs as key materials for the development of molecular sensors and biosensors. The high fluorescence yields of QDs, their stability against photobleaching, and their size-controlled luminescence features support the use of QDs as superior materials for the development of optical sensors. The many chemical procedures available for the modification of QDs, with molecular functionalities or biomolecules, have paved the way to the synthesis of a variety of chemically modified or biomolecular-functionalized QD conjugates. These hybrid systems combine the unique optical properties of QDs with the specific recognition or catalytic functions of molecules and, indeed, numerous chemical sensors and biosensors based on QDs have been developed during the past few years. Particularly impressive are the advances in the development of QD-based biosensors; whereas early efforts employed QDs as passive optical labels for biorecognition events, later studies have incorporated the QDs as optical labels capable of probing the dynamics of biocatalytic transformations and conformational transitions in proteins. These fundamental studies have provided an important pillar in modern nanobiotechnology, and hold great promise for practical applications in the future. Specific among these are the size-controlled luminescence features of QDs that permit the excitation of different-sized QDs with the same excitation energy, to generate different luminescent particles from the same material, allowing the use of semiconductor QDs for the multiplexed parallel analysis of several analytes. Although multiplexed bioanalyses with QDs have been demonstrated, there are important practical implications ahead. It is possible to envisage different QD-based multiplexed assays for the rapid detection of different viral infections, for screening mutations that cause diseases or genetic disorders, and for the parallel analysis of toxins.

Whilst currently, the use of QDs as optical agents for biosensing events is at a mature stage of research, the use of QDs as optical probes for intracellular processes is in its infancy. Numerous recent studies have reported on the use of QDs as optical labels to image intracellular domains, yet this application appears to represent only the start of some challenging future applications. The incorporation of functionalized QDs into cells to probe intracellular biocatalytic pathways holds great promise for future nanomedicine, some likely examples being the rapid detection of cancer cells and of metabolically disordered cells, and/or for the screening of drugs.

During the past few years, some impressive progress has been made in the use of QDs as optical labels for chemical sensors. In this respect, the chemical modification of QDs with specific recognition ligands has led to the creation of chemosensors for ions, molecules, and macromolecules. The broad variety of specific binding ligands that has been developed during decades in the field of supramolecular chemistry will continue to provide unique opportunities to develop new chemosensors with important practical uses for monitoring environmental pollutants, for controlling food and water qualities, and for homeland security. However, advances in the synthesis of QDs with enhanced compositional complexities, such as core–shell or graded QDs, will lead to the introduction of new materials for the design of sensor devices that include cooperative functions of the core and shell ingredients (i.e., they will combine the optical properties of the core with the catalytic functions of the shell). Finally, the nanodimensions of QDs would enable their integration in the form of nanoscale sensor devices. Likewise, the optical detection of analytes with single QDs should be feasible.

The rapid progress that has been made in the analytical and bioanalytical applications of functionalized QD hybrid systems holds promise for exciting new results and practical sensing systems in the future. Clearly, the interdisciplinary activities of chemists, physicists, biologists and materials scientists alike must be combined to identify new areas for the development of these materials.

References

1 Alivisatos, A.P. (1996) *Science*, **271**, 933–937.
2 Alivisatos, A.P. (1996) *J. Phys. Chem.*, **100**, 13226–13239.
3 Nirmal, M. and Brus, L. (1999) *Acc. Chem. Res.*, **32**, 407–414.
4 Kuno, M., Lee, J.K., Dabbousi, B.O., Mikulec, F.V., and Bawendi, M.G. (1997) *J. Chem. Phys.*, **106**, 9869–9882.
5 Peng, X.G., Manna, L., Yang, W.D., Wickham, J., Scher, E., Kadavanich, A., and Alivisatos, A.P. (2000) *Nature*, **404**, 59–61.
6 Manna, L., Scher, E.C., and Alivisatos, A.P. (2000) *J. Am. Chem. Soc.*, **122**, 12700–12706.
7 Peng, Z.A. and Peng, X.G. (2001) *J. Am. Chem. Soc.*, **123**, 1389–1395.
8 Jun, Y.W., Lee, S.M., Kang, N.J., and Cheon, J. (2001) *J. Am. Chem. Soc.*, **123**, 5150–5151.
9 Jun, Y.W., Jung, Y.Y., and Cheon, J. (2002) *J. Am. Chem. Soc.*, **124**, 615–619.
10 Pinna, N., Weiss, K., Urban, J., and Pileni, M.P. (2001) *Adv. Mater.*, **13**, 261–264.
11 Ellingson, R.J., Beard, M.C., Johnson, J.C., Yu, P.R., Micic, O.I., Nozik, A.J., Shabaev, A., and Efros, A.L. (2005) *Nano Lett.*, **5**, 865–871.
12 Schaller, R.D. and Klimov, V.I. (2004) *Phys. Rev. Lett.*, **92**, 186601-1–186601-4.

13 Plass, R., Pelet, S., Krueger, J., Grätzel, M., and Bach, U. (2002) *J. Phys. Chem. B*, **106**, 7578–7580.
14 Nozik, A.J. (2002) *Physica E*, **14**, 115–120.
15 Kamat, P.V. (2007) *J. Phys. Chem. C*, **111**, 2834–2860.
16 Somers, R.C., Bawendi, M.G., and Nocera, D.G. (2007) *Chem. Soc. Rev.*, **36**, 579–591.
17 Walker, G.W., Sundar, V.C., Rudzinski, C.M., Wun, A.W., Bawendi, M.G., and Nocera, D.G. (2003) *Appl. Phys. Lett.*, **83**, 3555–3557.
18 Kamat, P.V. (1993) *Chem. Rev.*, **93**, 267–300.
19 Thompson, T.L. and Yates, J.T. (2005) *Top. Catal.*, **35**, 197–210.
20 Tachikawa, T., Fujitsuka, M., and Majima, T. (2007) *J. Phys. Chem. C*, **111**, 5259–5275.
21 Harris, C. and Kamat, P.V. (2009) *ACS Nano*, **3**, 682–690.
22 Yin, Y. and Alivisatos, A.P. (2005) *Nature*, **437**, 664–670.
23 Nakanishi, T., Ohtani, B., and Uosaki, K. (1998) *J. Phys. Chem. B*, **102**, 1571–1577.
24 Chen, D.L. and Gao, L. (2005) *Solid State Commun.*, **133**, 145–150.
25 Darbandi, M., Thomann, R., and Nann, T. (2005) *Chem. Mater.*, **17**, 5720–5725.
26 Kim, J., Kim, H.S., Lee, N., Kim, T., Kim, H., Yu, T., Song, I.C., Moon, W.K., and Hyeon, T. (2008) *Angew. Chem., Int. Ed.*, **47**, 8438–8441.
27 Kim, J., Lee, J.E., Lee, J., Yu, J.H., Kim, B.C., An, K., Hwang, Y., Shin, C.H., Park, J.G., Kim, J., and Hyeon, T. (2006) *J. Am. Chem. Soc.*, **128**, 688–689.
28 Talapin, D.V., Rogach, A.L., Kornowski, A., Haase, M., and Weller, H. (2001) *Nano Lett.*, **1**, 207–211.
29 Danek, M., Jensen, K.F., Murray, C.B., and Bawendi, M.G. (1996) *Chem. Mater.*, **8**, 173–180.
30 Rodriguez-Viejo, J., Jensen, K.F., Mattoussi, H., Michel, J., Dabbousi, B.O., and Bawendi, M.G. (1997) *Appl. Phys. Lett.*, **70**, 2132–2134.
31 Micic, O.I., Smith, B.B., and Nozik, A.J. (2000) *J. Phys. Chem. B*, **104**, 12149–12156.
32 Danek, M., Jensen, K.F., Murray, C.B., and Bawendi, M.G. (1994) *J. Cryst. Growth*, **145**, 714–720.
33 Dabbousi, B.O., Rodriguez-Viejo, J., Mikulec, F.V., Heine, J.R., Mattoussi, H., Ober, R., Jensen, K.F., and Bawendi, M.G. (1997) *J. Phys. Chem. B*, **101**, 9463–9475.
34 Hines, M.A. and Guyot-Sionnest, P. (1996) *J. Phys. Chem.*, **100**, 468–471.
35 Peng, X.G., Schlamp, M.C., Kadavanich, A.V., and Alivisatos, A.P. (1997) *J. Am. Chem. Soc.*, **119**, 7019–7029.
36 Kan, S., Mokari, T., Rothenberg, E., and Banin, U. (2003) *Nat. Mater.*, **2**, 155–158.
37 Yu, J.H., Joo, J., Park, H.M., Baik, S.I., Kim, Y.W., Kim, S.C., and Hyeon, T. (2005) *J. Am. Chem. Soc.*, **127**, 5662–5670.
38 Manna, L., Scher, E.C., Li, L.S., and Alivisatos, A.P. (2002) *J. Am. Chem. Soc.*, **124**, 7136–7145.
39 Pacholski, C., Kornowski, A., and Weller, H. (2002) *Angew. Chem., Int. Ed.*, **41**, 1188–1191.
40 Milliron, D.J., Hughes, S.M., Cui, Y., Manna, L., Li, J.B., Wang, L.W., and Alivisatos, A.P. (2004) *Nature*, **430**, 190–195.
41 Manna, L., Milliron, D.J., Meisel, A., Scher, E.C., and Alivisatos, A.P. (2003) *Nat. Mater.*, **2**, 382–385.
42 Wang, D.L. and Lieber, C.M. (2003) *Nat. Mater.*, **2**, 355–356.
43 Wu, X.Y., Liu, H.J., Liu, J.Q., Haley, K.N., Treadway, J.A., Larson, J.P., Ge, N.F., Peale, F., and Bruchez, M.P. (2003) *Nat. Biotechnol.*, **21**, 41–46.
44 Brus, L. (1991) *Appl. Phys. A: Mater. Sci. Process.*, **53**, 465–474.
45 Grieve, K., Mulvaney, P., and Grieser, F. (2000) *Curr. Opin. Colloid Interface Sci.*, **5**, 168–172.
46 Alivisatos, P. (2004) *Nat. Biotechnol.*, **22**, 47–52.
47 Chen, C.Y., Cheng, C.T., Lai, C.W., Wu, P.W., Wu, K.C., Chou, P.T., Chou, Y.H., and Chiu, H.T. (2006) *Chem. Commun.*, 263–265.
48 Willard, D.M., Carillo, L.L., Jung, J., and Van Orden, A. (2001) *Nano Lett.*, **1**, 469–474.

49 Pons, T., Uyeda, H.T., Medintz, I.L., and Mattoussi, H. (2006) *J. Phys. Chem. B*, **110**, 20308–20316.
50 So, M.K., Xu, C.J., Loening, A.M., Gambhir, S.S., and Rao, J.H. (2006) *Nat. Biotechnol.*, **24**, 339–343.
51 Yao, H.Q., Zhang, Y., Xiao, F., Xia, Z.Y., and Rao, J.H. (2007) *Angew. Chem., Int. Ed.*, **46**, 4346–4349.
52 Huang, X.Y., Li, L., Qian, H.F., Dong, C.Q., and Ren, J.C. (2006) *Angew. Chem., Int. Ed.*, **45**, 5140–5143.
53 Lakowicz, J.R. (2006) *Principles of Fluorescence Spectroscopy*, Springer, New York.
54 Miyawaki, A. (2003) *Dev. Cell*, **4**, 295–305.
55 Jares-Erijman, E.A., and Jovin, T.M. (2003) *Nat. Biotechnol.*, **21**, 1387–1395.
56 Medintz, I.L., Uyeda, H.T., Goldman, E.R., and Mattoussi, H. (2005) *Nat. Mater.*, **4**, 435–446.
57 Chan, W.C.W., Maxwell, D.J., Gao, X.H., Bailey, R.E., Han, M.Y., and Nie, S.M. (2002) *Curr. Opin. Biotechnol.*, **13**, 40–46.
58 Snee, P.T., Somers, R.C., Nair, G., Zimmer, J.P., Bawendi, M.G., and Nocera, D.G. (2006) *J. Am. Chem. Soc.*, **128**, 13320–13321.
59 Murray, C.B., Norris, D.J., and Bawendi, M.G. (1993) *J. Am. Chem. Soc.*, **115**, 8706–8715.
60 Peng, Z.A. and Peng, X.G. (2001) *J. Am. Chem. Soc.*, **123**, 183–184.
61 Qu, L.H., Peng, Z.A., and Peng, X.G. (2001) *Nano Lett.*, **1**, 333–337.
62 Qu, L.H. and Peng, X.G. (2002) *J. Am. Chem. Soc.*, **124**, 2049–2055.
63 Cordero, S.R., Carson, P.J., Estabrook, R.A., Strouse, G.F., and Buratto, S.K. (2000) *J. Phys. Chem. B*, **104**, 12137–12142.
64 Mattoussi, H., Mauro, J.M., Goldman, E.R., Anderson, G.P., Sundar, V.C., Mikulec, F.V., and Bawendi, M.G. (2000) *J. Am. Chem. Soc.*, **122**, 12142–12150.
65 Chan, W.C.W. and Nie, S.M. (1998) *Science*, **281**, 2016–2018.
66 Gerion, D., Pinaud, F., Williams, S.C., Parak, W.J., Zanchet, D., Weiss, S., and Alivisatos, A.P. (2001) *J. Phys. Chem. B*, **105**, 8861–8871.
67 Bruchez, M., Moronne, M., Gin, P., Weiss, S., and Alivisatos, A.P. (1998) *Science*, **281**, 2013–2016.
68 Pellegrino, T., Manna, L., Kudera, S., Liedl, T., Koktysh, D., Rogach, A.L., Keller, S., Radler, J., Natile, G., and Parak, W.J. (2004) *Nano Lett.*, **4**, 703–707.
69 Gao, X.H., Cui, Y.Y., Levenson, R.M., Chung, L.W.K., and Nie, S.M. (2004) *Nat. Biotechnol.*, **22**, 969–976.
70 Yu, W.W., Chang, E., Falkner, J.C., Zhang, J.Y., Al-Somali, A.M., Sayes, C.M., Johns, J., Drezek, R., and Colvin, V.L. (2007) *J. Am. Chem. Soc.*, **129**, 2871–2879.
71 Nikolic, M.S., Krack, M., Aleksandrovic, V., Kornowski, A., Forster, S., and Weller, H. (2006) *Angew. Chem., Int. Ed.*, **45**, 6577–6580.
72 Wang, S.P., Mamedova, N., Kotov, N.A., Chen, W., and Studer, J. (2002) *Nano Lett.*, **2**, 817–822.
73 Medintz, I.L., Clapp, A.R., Mattoussi, H., Goldman, E.R., Fisher, B., and Mauro, J.M. (2003) *Nat. Mater.*, **2**, 630–638.
74 Pathak, S., Choi, S.K., Arnheim, N., and Thompson, M.E. (2001) *J. Am. Chem. Soc.*, **123**, 4103–4104.
75 Uyeda, H.T., Medintz, I.L., Jaiswal, J.K., Simon, S.M., and Mattoussi, H. (2005) *J. Am. Chem. Soc.*, **127**, 3870–3878.
76 Liu, W., Howarth, M., Greytak, A.B., Zheng, Y., Nocera, D.G., Ting, A.Y., and Bawendi, M.G. (2008) *J. Am. Chem. Soc.*, **130**, 1274–1284.
77 Pinaud, F., King, D., Moore, H.P., and Weiss, S. (2004) *J. Am. Chem. Soc.*, **126**, 6115–6123.
78 Iyer, G., Pinaud, F., Tsay, J., and Weiss, S. (2007) *Small*, **3**, 793–798.
79 Liu, Y.C., Brandon, R., Cate, M., Peng, X.G., Stony, R., and Johnson, M. (2007) *Anal. Chem.*, **79**, 8796–8802.
80 Liu, Y.C., Kim, M., Wang, Y.J., Wang, Y.A., and Peng, X.G. (2006) *Langmuir*, **22**, 6341–6345.
81 Smith, A.M., Duan, H.W., Rhyner, M.N., Ruan, G., and Nie, S.M. (2006) *Phys. Chem. Chem. Phys.*, **8**, 3895–3903.
82 Pellegrino, T., Kudera, S., Liedl, T., Javier, A.M., Manna, L., and Parak, W.J. (2005) *Small*, **1**, 48–63.

83 Rogach, A.L., Kornowski, A., Gao, M.Y., Eychmüller, A., and Weller, H. (1999) *J. Phys. Chem. B*, **103**, 3065–3069.
84 Rogach, A.L., Katsikas, L., Kornowski, A., Su, D.S., Eychmüller, A., and Weller, H. (1996) *Ber. Bunsen. Phys. Chem.*, **100**, 1772–1778.
85 Gaponik, N., Talapin, D.V., Rogach, A.L., Hoppe, K., Shevchenko, E.V., Kornowski, A., Eychmüller, A., and Weller, H. (2002) *J. Phys. Chem. B*, **106**, 7177–7185.
86 Rogach, A., Kershaw, S., Burt, M., Harrison, M., Kornowski, A., Eychmüller, A., and Weller, H. (1999) *Adv. Mater.*, **11**, 552–555.
87 Baumle, M., Stamou, D., Segura, J.M., Hovius, R., and Vogel, H. (2004) *Langmuir*, **20**, 3828–3831.
88 Zheng, Y.G., Gao, S.J., and Ying, J.Y. (2007) *Adv. Mater.*, **19**, 376–380.
89 Palaniappan, K., Xue, C.H., Arumugam, G., Hackney, S.A., and Liu, J. (2006) *Chem. Mater.*, **18**, 1275–1280.
90 Rogach, A.L., Nagesha, D., Ostrander, J.W., Giersig, M., and Kotov, N.A. (2000) *Chem. Mater.*, **12**, 2676–2685.
91 Ma, Y., Yang, C., Li, N., and Yang, X.R. (2005) *Talanta*, **67**, 979–983.
92 Priyam, A., Chatterjee, A., Das, S.K., and Saha, A. (2005) *Res. Chem. Intermed.*, **31**, 691–702.
93 Li, L., Qian, H.F., and Ren, J.C. (2005) *Chem. Commun.*, 528–530.
94 Qian, H.F., Li, L., and Ren, J.C. (2005) *Mater. Res. Bull.*, **40**, 1726–1736.
95 Goldman, E.R., Balighian, E.D., Mattoussi, H., Kuno, M.K., Mauro, J.M., Tran, P.T., and Anderson, G.P. (2002) *J. Am. Chem. Soc.*, **124**, 6378–6382.
96 Lin, Z.B., Cui, S.X., Zhang, H., Chen, Q.D., Yang, B., Su, X.G., Zhang, J.H., and Jin, Q.H. (2003) *Anal. Biochem.*, **319**, 239–243.
97 Zhou, D.J., Piper, J.D., Abell, C., Klenerman, D., Kang, D.J., and Ying, L.M. (2005) *Chem. Commun.*, 4807–4809.
98 Pathak, S., Davidson, M.C., and Silva, G.A. (2007) *Nano Lett.*, **7**, 1839–1845.
99 Xing, Y., Chaudry, Q., Shen, C., Kong, K.Y., Zhau, H.E., Wchung, L., Petros, J.A., O'Regan, R.M., Yezhelyev, M.V., Simons, J.W., Wang, M.D., and Nie, S. (2007) *Nat. Protoc.*, **2**, 1152–1165.
100 Goldman, E.R., Medintz, I.L., Hayhurst, A., Anderson, G.P., Mauro, J.M., Iverson, B.L., Georgiou, G., and Mattoussi, H. (2005) *Anal. Chim. Acta*, **534**, 63–67.
101 Lao, U.L., Mulchandani, A., and Chen, W. (2006) *J. Am. Chem. Soc.*, **128**, 14756–14757.
102 Medintz, I.L., Clapp, A.R., Brunel, F.M., Tiefenbrunn, T., Uyeda, H.T., Chang, E.L., Deschamps, J.R., Dawson, P.E., and Mattoussi, H. (2006) *Nat. Mater.*, **5**, 581–589.
103 Medintz, I.L., Berti, L., Pons, T., Grimes, A.F., English, D.S., Alessandrini, A., Facci, P., and Mattoussi, H. (2007) *Nano Lett.*, **7**, 1741–1748.
104 James, T.D. (2005) Boronic Acid-Based Receptors and Sensors For Saccharides, in *Boronic Acids* (ed. D.G. Hall), Wiley-VCH, Weinheim, Ch. 12, pp. 441–479.
105 Lorand, J. and Edwards, J.O. (1959) *J. Org. Chem.*, **24**, 769–744.
106 James, T.D. and Shinkai, S. (2002) *Top. Curr. Res.*, **218**, 159–200.
107 deSilva, A.P., Gunaratne, H.Q.N., Gunnlaugsson, T., Huxley, A.J.M., McCoy, C.P., Rademacher, J.T., and Rice, T.E. (1997) *Chem. Rev.*, **97**, 1515–1566.
108 Eggert, H., Frederiksen, J., Morin, C., and Norrild, J.C. (1999) *J. Org. Chem.*, **64**, 3846–3852.
109 Freeman, R., Bahshi, L., Finder, T., Gill, R., and Willner, I. (2009) *Chem. Commun.*, 764–766.
110 Ueno, A. (1993) *Adv. Mater.*, **5**, 132–134.
111 Corradini, R., Dossena, A., Galaverna, G., Marchelli, R., Panagia, A., and Sartor, G. (1997) *J. Org. Chem.*, **62**, 6283–6289.
112 Haider, J.M. and Pikramenou, Z. (2005) *Chem. Soc. Rev.*, **34**, 120–132.
113 Furukawa, S., Mihara, H., and Ueno, A. (2003) *Macromol. Rapid Commun.*, **24**, 202–206.
114 Pagliari, S., Corradini, R., Galaverna, G., Sforza, S., Dossena, A., Montalti, M., Prodi, L., Zaccheroni, N., and Marchelli, R. (2004) *Chem. Eur. J.*, **10**, 2749–2758.

115 Stalcup, A.M. and Williams, K.L. (1992) *J. Liq. Chromatogr.*, **15**, 29–37.
116 Camilleri, P., Reid, C.A., and Manallack, D.T. (1994) *Chromatographia*, **38**, 771–775.
117 Risley, D.S. and Strege, M.A. (2000) *Anal. Chem.*, **72**, 1736–1739.
118 Schumacher, D.D., Mitchell, C.R., Xiao, T.L., Rozhkov, R.V., Larock, R.C., and Armstrong, D.W. (2003) *J. Chromatogr. A*, **1011**, 37–47.
119 Han, X., Zhong, Q., Yue, D., Della Ca, N., Larock, R.C., and Armstrong, D.W. (2005) *Chromatographia*, **61**, 205–211.
120 Sun, P., Wang, C., Armstrong, D.W., Peter, A., and Forro, E. (2006) *J. Liq. Chromatogr. Related Technol.*, **29**, 1847–1860.
121 Freeman, R., Finder, T., Bahshi, L., and Willner, I. (2009) *Nano Lett.*, **9**, 2073–2076.
122 Banerjee, S., Kara, S., and Santra, S. (2008) *Chem. Commun.*, 3037–3039.
123 Credi, A., Dumas, S., Silvi, S., Venturi, M., Arduini, A., Pochini, A., and Secchi, A. (2004) *J. Org. Chem.*, **69**, 5881–5887.
124 Beer, P.D. and Gale, P.A. (2001) *Angew. Chem., Int. Ed.*, **40**, 486–516.
125 Gale, P.A. (2003) *Coord. Chem. Rev.*, **240**, 191–221.
126 Gadenne, B., Yildiz, I., Amelia, M., Ciesa, F., Secchi, A., Arduini, A., Credi, A., and Raymo, F.M. (2008) *J. Mater. Chem.*, **18**, 2022–2027.
127 Jin, T., Fujii, F., Sakata, H., Tamura, M., and Kinjo, M. (2005) *Chem. Commun.*, 4300–4302.
128 Wang, X.Q., Wu, J.F., Li, F.Y., and Li, H.B. (2008) *Nanotechnology*, **19**, 205501–205508.
129 Qu, F.G., Zhou, X.F., Xu, J., Li, H.B., and Xie, G.Y. (2009) *Talanta*, **78**, 1359–1363.
130 Gattas-Asfura, K.A. and Leblanc, R.M. (2003) *Chem. Commun.*, 2684–2685.
131 Chen, Y.F. and Rosenzweig, Z. (2002) *Anal. Chem.*, **74**, 5132–5138.
132 Wang, C., Zhao, J., Wang, Y., Lou, N., Ma, Q., and Su, X.G. (2009) *Sens. Actuators, B*, **139**, 476–482.
133 Shi, G.H., Shang, Z.B., Wang, Y., Jin, W.J., and Zhang, T.C. (2008) *Spectrochim. Acta, Part A*, **70**, 247–252.
134 Liao, P., Yan, Z.Y., Xu, Z.J., and Sun, X. (2009) *Spectrochim. Acta, Part A*, **72**, 1066–1070.
135 Callan, J.F., Mulrooney, R.C., and Karnila, S. (2008) *J. Fluoresc.*, **18**, 1157–1161.
136 Yildiz, I., Tomasulo, M., and Raymo, F.M. (2006) *Proc. Natl Acad. Sci. USA*, **103**, 11457–11460.
137 Cordes, D.B., Gamsey, S., and Singaram, B. (2006) *Angew. Chem., Int. Ed.*, **45**, 3829–3832.
138 Clever, G.H., Kaul, C., and Carell, T. (2007) *Angew. Chem., Int. Ed.*, **46**, 6226–6236.
139 Freeman, R., Finder, T., and Willner, I. (2009) *Angew. Chem., Int. Ed.*, **48**, 7818–7821.
140 Freeman, R., Gill, R., Shweky, I., Kotler, M., Banin, U., and Willner, I. (2009) *Angew. Chem., Int. Ed.*, **48**, 309–313.
141 Freeman, R. and Willner, I. (2009) *Nano Lett.*, **9**, 322–326.
142 Wang, S.H., Han, M.Y., and Huang, D.J. (2009) *J. Am. Chem. Soc.*, **131**, 11692–11694.
143 Shipway, A.N., Katz, E., and Willner, I. (2000) *ChemPhysChem*, **1**, 18–52.
144 Baron, R., Willner, B., and Willner, I. (2007) *Chem. Commun.*, 323–332.
145 Katz, E. and Willner, I. (2004) *ChemPhysChem*, **5**, 1085–1104.
146 Goldman, E.R., Anderson, G.P., Tran, P.T., Mattoussi, H., Charles, P.T., and Mauro, J.M. (2002) *Anal. Chem.*, **74**, 841–847.
147 Goldman, E.R., Clapp, A.R., Anderson, G.P., Uyeda, H.T., Mauro, J.M., Medintz, I.L., and Mattoussi, H. (2004) *Anal. Chem.*, **76**, 684–688.
148 Peng, C.F., Li, Z.K., Zhu, Y.Y., Chen, W., Yuan, Y., Liu, L.Q., Li, Q.S., Xu, D.H., Qiao, R.R., Wang, L.B., Zhu, S.F., Jin, Z.Y., and Xu, C.L. (2009) *Biosens. Bioelectron.*, **24**, 3657–3662.
149 Gerion, D., Chen, F.Q., Kannan, B., Fu, A.H., Parak, W.J., Chen, D.J., Majumdar, A., and Alivisatos, A.P. (2003) *Anal. Chem.*, **75**, 4766–4772.
150 Xiao, Y. and Barker, P.E. (2004) *Nucleic Acids Res.*, **32**, e28.
151 Chan, P.M., Yuen, T., Ruf, F., Gonzalez-Maeso, J., and Sealfon, S.C. (2005) *Nucleic Acids Res.*, **33**, e161.

152 Wu, S.M., Zha, X., Zhang, Z.L., Xie, H.Y., Tian, Z.Q., Peng, J., Lu, Z.X., Pang, D.W., and Xie, Z.X. (2006) *ChemPhysChem*, **7**, 1062–1067.

153 Goldman, E.R., Medintz, I.L., Whitley, J.L., Hayhurst, A., Clapp, A.R., Uyeda, H.T., Deschamps, J.R., Lassman, M.E., and Mattoussi, H. (2005) *J. Am. Chem. Soc.*, **127**, 6744–6751.

154 Ramadurai, D., Norton, E., Hale, J., Garland, J.W., Stephenson, L.D., Stroscio, M.A., Sivananthan, S., and Kumar, A. (2008) *IET Nanobiotechnol.*, **2**, 47–53.

155 Peng, H., Zhang, L.J., Kjallman, T.H.M., Soeller, C., and Travas-Sejdic, J. (2007) *J. Am. Chem. Soc.*, **129**, 3048–3049.

156 Zhou, D.J., Ying, L.M., Hong, X., Hall, E.A., Abell, C., and Klenerman, D. (2008) *Langmuir*, **24**, 1659–1664.

157 Dyadyusha, L., Yin, H., Jaiswal, S., Brown, T., Baumberg, J.J., Booy, F.P., and Melvin, T. (2005) *Chem. Commun.*, 3201–3203.

158 Kim, J.H., Morikis, D., and Ozkan, M. (2004) *Sens. Actuators, B*, **102**, 315–319.

159 Kim, J.H., Chaudhary, S., and Ozkan, M. (2007) *Nanotechnology*, **18**, 195105–195111.

160 Cady, N.C., Strickland, A.D., and Batt, C.A. (2007) *Mol. Cell. Probes*, **21**, 116–124.

161 Jiang, G., Susha, A.S., Lutich, A.A., Stefani, F.D., Feldmann, J., and Rogach, A.L. (2009) *ACS Nano*, **3**, 4127–4131.

162 Hansen, J.A., Wang, J., Kawde, A.N., Xiang, Y., Gothelf, K.V., and Collins, G. (2006) *J. Am. Chem. Soc.*, **128**, 2228–2229.

163 Polsky, R., Gill, R., Kaganovsky, L., and Willner, I. (2006) *Anal. Chem.*, **78**, 2268–2271.

164 Pavlov, V., Xiao, Y., Shlyahovsky, B., and Willner, I. (2004) *J. Am. Chem. Soc.*, **126**, 11768–11769.

165 Liu, J.W. and Lu, Y. (2006) *Angew. Chem., Int. Ed.*, **45**, 90–94.

166 Levy, M., Cater, S.F., and Ellington, A.D. (2005) *ChemBioChem*, **6**, 2163–2166.

167 Choi, J.H., Chen, K.H., and Strano, M.S. (2006) *J. Am. Chem. Soc.*, **128**, 15584–15585.

168 Liu, J.W., Lee, J.H., and Lu, Y. (2007) *Anal. Chem.*, **79**, 4120–4125.

169 Freeman, R., Li, Y., Tel-Vered, R., Sharon, E., Elbaz, J., and Willner, I. (2009) *Analyst*, **134**, 653–656.

170 Mao, C.D., Sun, W.Q., Shen, Z.Y., and Seeman, N.C. (1999) *Nature*, **397**, 144–146.

171 Seeman, N.C. (2005) *Trends Biochem. Sci.*, **30**, 119–125.

172 Nutiu, R. and Li, Y.F. (2005) *Angew. Chem., Int. Ed.*, **44**, 5464–5467.

173 Beissenhirtz, M.K. and Willner, I. (2006) *Org. Biomol. Chem.*, **4**, 3392–3401.

174 Willner, I., Shlyahovsky, B., Zayats, M., and Willner, B. (2008) *Chem. Soc. Rev.*, **37**, 1153–1165.

175 Bath, J. and Turberfield, A.J. (2007) *Nat. Nanotechnol.*, **2**, 275–284.

176 Simmel, F.C. and Yurke, B. (2001) *Phys. Rev. E*, **63**, 041913.

177 Liedl, T., Sobey, T.L., and Simmel, F.C. (2007) *Nano Today*, **2**, 36–41.

178 Yurke, B., Turberfield, A.J., Mills, A.P., Simmel, F.C., and Neumann, J.L. (2000) *Nature*, **406**, 605–608.

179 Chen, Y., Wang, M.S., and Mao, C.D. (2004) *Angew. Chem., Int. Ed.*, **43**, 3554–3557.

180 Shin, J.S. and Pierce, N.A. (2004) *J. Am. Chem. Soc.*, **126**, 10834–10835.

181 Tian, Y., He, Y., Chen, Y., Yin, P., and Mao, C.D. (2005) *Angew. Chem., Int. Ed.*, **44**, 4355–4358.

182 Tian, Y. and Mao, C.D. (2004) *J. Am. Chem. Soc.*, **126**, 11410–11411.

183 Elbaz, J., Shlyahovsky, B., and Willner, I. (2008) *Chem. Commun.*, 1569–1571.

184 Elbaz, J., Tel-Vered, R. Freeman, R., Yildiz, H.B., and Willner, I.. (2009) *Angew. Chem., Int. Ed*, **48**, 133–137.

185 Patolsky, F., Gill, R., Weizmann, Y., Mokari, T., Banin, U., and Willner, I. (2003) *J. Am. Chem. Soc.*, **125**, 13918–13919.

186 Moyzis, R.K., Buckingham, J.M., Cram, L.S., Dani, M., Deaven, L.L., Jones, M.D., Meyne, J., Ratliff, R.L., and Wu, J.R. (1988) *Proc. Natl Acad. Sci. USA*, **85**, 6622–6626.

187 Bryan, T.M. and Cech, T.R. (1999) *Curr. Opin. Cell Biol.*, **11**, 318–324.

188 Kim, N.W., Piatyszek, M.A., Prowse, K.R., Harley, C.B., West, M.D., Ho, P.L.C., Coviello, G.M., Wright, W.E., Weinrich,

S.L., and Shay, J.W. (1994) *Science*, **266**, 2011–2015.

189 Shay, J.W. and Bacchetti, S. (1997) *Eur. J. Cancer*, **33**, 787–791.

190 Gill, R., Willner, I., Shweky, I., and Banin, U. (2005) *J. Phys. Chem. B*, **109**, 23715–23719.

191 Huang, S., Xiao, Q., He, Z.K., Liu, Y., Tinnefeld, P., Su, X.R., and Peng, X.N. (2008) *Chem. Commun.*, 5990–5992.

192 Shi, L.F., De Paoli, V., Rosenzweig, N., and Rosenzweig, Z. (2006) *J. Am. Chem. Soc.*, **128**, 10378–10379.

193 Chang, E., Miller, J.S., Sun, J.T., Yu, W.W., Colvin, V.L., Drezek, R., and West, J.L. (2005) *Biochem. Biophys. Res. Commun.*, **334**, 1317–1321.

194 Shi, L.F., Rosenzweig, N., and Rosenzweig, Z. (2007) *Anal. Chem.*, **79**, 208–214.

195 Xu, C.J., Xing, B.G., and Rao, H.H. (2006) *Biochem. Biophys. Res. Commun.*, **344**, 931–935.

196 Gill, R., Freeman, R., Xu, J.P., Willner, I., Winograd, S., Shweky, I., and Banin, U. (2006) *J. Am. Chem. Soc.*, **128**, 15376–15377.

197 Yuan, J., Guo, W., and Wang, E. (2008) *Anal. Chem.*, **80**, 1141–1145.

198 Gill, R., Bahshi, L., Freeman, R., and Willner, I. (2008) *Angew. Chem., Int. Ed.*, **47**, 1676–1679.

199 Medintz, I.L., Clapp, A.R., Melinger, J.S., Deschamps, J.R., and Mattoussi, H. (2005) *Adv. Mater.*, **17**, 2450–2455.

200 Sandros, M.G., Gao, D., and Benson, D.E. (2005) *J. Am. Chem. Soc.*, **127**, 12198–12199.

201 Aryal, B.P. and Benson, D.E. (2006) *J. Am. Chem. Soc.*, **128**, 15986–15987.

202 Ji, X.J., Zheng, J.Y., Xu, J.M., Rastogi, V.K., Cheng, T.C., DeFrank, J.J., and Leblanc, R.M. (2005) *J. Phys. Chem. B*, **109**, 3793–3799.

203 Zhang, C.Y. and Johnson, L.W. (2006) *J. Am. Chem. Soc.*, **128**, 5324–5325.

204 Derfus, A.M., Chan, W.C.W., and Bhatia, S.N. (2004) *Adv. Mater.*, **16**, 961–966.

205 Delehanty, J.B., Mattoussi, H., and Medintz, I.L. (2009) *Anal. Bioanal. Chem.*, **393**, 1091–1105.

206 Jaiswal, J.K., Mattoussi, H., Mauro, J.M., and Simon, S.M. (2003) *Nat. Biotechnol.*, **21**, 47–51.

207 Nabiev, I., Mitchell, S., Davies, A., Williams, Y., Kelleher, D., Moore, R., Gun'ko, Y.K., Byrne, S., Rakovich, Y.P., Donegan, J.F., Sukhanova, A., Conroy, J., Cottell, D., Gaponik, N., Rogach, A., and Volkov, Y. (2007 *Nano Lett.*, **7**, 3452–3461.

208 Pellegrino, T., Parak, W.J., Boudreau, R., Le Gros, M.A., Gerion, D., Alivisatos, A.P., and Larabell, C.A. (2003) *Differentiation*, **71**, 542–548.

209 Parak, W.J., Boudreau, R., Le Gros, M., Gerion, D., Zanchet, D., Micheel, C.M., Williams, S.C., Alivisatos, A.P., and Larabell, C. (2002) *Adv. Mater.*, **14**, 882–885.

210 Susumu, K., Uyeda, H.T., Medintz, I.L., Pons, T., Delehanty, J.B., and Mattoussi, H. (2007) *J. Am. Chem. Soc.*, **129**, 13987–13996.

211 Lewin, M., Carlesso, N., Tung, C.H., Tang, X.W., Cory, D., Scadden, D.T., and Weissleder, R. (2000) *Nat. Biotechnol.*, **18**, 410–414.

212 Rozenzhak, S.M., Kadakia, M.P., Caserta, T.M., Westbrook, T.R., Stone, M.O., and Naik, R.R. (2005) *Chem. Commun.*, 2217–2219.

213 Smith, B.R., Cheng, Z., De, A., Koh, A.L., Sinclair, R., and Gambhir, S.S. (2008) *Nano Lett.*, **8**, 2599–2606.

214 Biju, V., Muraleedharan, D., Nakayama, K., Shinohara, Y., Itoh, T., Baba, Y., and Ishikawa, M. (2007) *Langmuir*, **23**, 10254–10261.

215 Silver, J. and Ou, W. (2005) *Nano Lett.*, **5**, 1445–1449.

216 Qian, J., Yong, K.T., Roy, I., Ohulchanskyy, T.Y., Bergey, E.J., Lee, H.H., Tramposch, K.M., He, S.L., Maitra, A., and Prasad, P.N. (2007) *J. Phys. Chem. B*, **111**, 6969–6972.

217 Lidke, D.S., Nagy, P., Heintzmann, R., Arndt-Jovin, D.J., Post, J.N., Grecco, H.E., Jares-Erijman, E.A., and Jovin, T.M. (2004) *Nat. Biotechnol.*, **22**, 198–203.

218 Jaiswal, J.K., Goldman, E.R., Mattoussi, H., and Simon, S.M. (2004) *Nat. Methods*, **1**, 73–78.

219 Rajan, S.S., Liu, H.Y., and Vu, T.Q. (2008) *ACS Nano*, **2**, 1153–1166.
220 Gopalakrishnan, G., Danelon, C., Izewska, P., Prummer, M., Bolinger, P.Y., Geissbuhler, I., Demurtas, D., Dubochet, J., and Vogel, H. (2006) *Angew. Chem., Int. Ed.*, **45**, 5478–5483.
221 Duan, H.W. and Nie, S.M. (2007) *J. Am. Chem. Soc.*, **129**, 3333–3338.
222 Bharali, D.J., Lucey, D.W., Jayakumar, H., Pudavar, H.E., and Prasad, P.N. (2005) *J. Am. Chem. Soc.*, **127**, 11364–11371.
223 Clarke, S.J., Hollmann, C.A., Zhang, Z.J., Suffern, D., Bradforth, S.E., Dimitrijevic, N.M., Minarik, W.G., and Nadeau, J.L. (2006) *Nat. Mater.*, **5**, 409–417.
224 Chen, F.Q. and Gerion, D. (2004) *Nano Lett.*, **4**, 1827–1832.
225 Dubertret, B., Skourides, P., Norris, D.J., Noireaux, V., Brivanlou, A.H., and Libchaber, A. (2002) *Science*, **298**, 1759–1762.
226 Medintz, I.L., Pons, T., Delehanty, J.B., Susumu, K., Brunel, F.M., Dawson, P.E., and Mattoussi, H. (2008) *Bioconjugate Chem.*, **19**, 1785–1795.
227 Bagalkot, V., Zhang, L., Levy-Nissenbaum, E., Jon, S., Kantoff, P.W., Langer, R., and Farokhzad, O.C. (2007) *Nano Lett.*, **7**, 3065–3070.
228 Chen, A.A., Derfus, A.M., Khetani, S.R., and Bhatia, S.N. (2005) *Nucleic Acids Res.*, **33**, e190.
229 Derfus, A.M., Chen, A.A., Min, D.H., Ruoslahti, E., and Bhatia, S.N. (2007) *Bioconjugate Chem.*, **18**, 1391–1396.
230 Hoshino, A., Manabe, N., Fujioka, K., Hanada, S., Yasuhara, M., Kondo, A., and Yamamoto, K. (2008) *Nanotechnology*, **19**, 495102–495112.

7
Conclusions and Perspectives
Günter Schmid, on behalf of all the authors

Nanoparticles, whether they originate from semiconductors or from metals, are generally associated with the appearance of novel properties. The most far-reaching consequence of their small size is, without doubt, the change in their electronic properties. Next in importance are changes in their other properties, both physical and chemical. These changes are what ultimately constitute the value of nanosciences and nanotechnology. Novel properties become available without making new chemical compounds!

If particles are small enough, they become electronically comparable to atoms and molecules, and follow quantum mechanical rules instead of the laws of classical physics. That is why nanoparticles are referred to as "quantum dots" (QDs) or, sometimes, "artificial atoms." The reason for this behavior is the disappearance of the band structures and the formation of discrete energy levels, both of which are linked to increases in the already existing band gaps in the case of semiconductor particles, and the formation of band gaps in the case of metal nanoparticles. Two-dimensional arrays of QDs (quantum wells), as well as one-dimensional arrangements (quantum wires), will therefore reach a new dimension of quality with respect to possible future applications. The synthesis of all types of QDs is, therefore, a challenge for chemists and physicists alike. The easy availability of QDs is of decisive importance if nanoparticles are to play an industrial role, and in fact II–VI, III–V as well as Ib–VI semiconductor nanoparticles are routinely accessible by various wet-chemical methods, in gram quantities.

Even the most difficult step in all synthetic procedures – the achievement of monodispersity of the particles – can be controlled satisfactorily. This is true of both semiconductor and metal nanoparticles. Because metal nanoparticles have a long history, there has been plenty of time to develop perfect syntheses. One decisive finding opened up completely novel strategies for both semiconductor and metal nanoparticles, notably the use of protective ligand shells, which usually consist of organic molecules – amines, phosphines, phosphine oxides, thiols, oligomers and numerous other classes of compounds have been used in this role. Such coordinating molecules not only support the synthesis of monodisperse nanoparticles by their kinetic and size-limiting functions; they also render the particles soluble in solvents,

Nanoparticles: From Theory to Application. Edited by Günter Schmid
Copyright © 2010 WILEY-VCH Verlag GmbH & Co. KGaA, Weinheim
ISBN: 978-3-527-32589-4

depending on their chemical nature. In this case, traditional chemical procedures such as chromatography and crystallization can be used for purification. Semiconductor nanoparticles include the II–VI type, such as CdS, CdSe and CdTe, the III–V type, represented mainly by GaN, GaP, GaAs or InP and InAs, and the Ib–VI type, which are available as copper selenide and telluride nanoclusters, and as silver and gold selenide and telluride species. Recently, metal nanoparticles have also undergone a remarkable renaissance following a 100-years "sleep," and this is especially true of magnetic nanoparticles. Indeed, apart from providing a general improvement in our knowledge of magnetism, these offer unimagined application possibilities.

Although magnetic nanoparticles are currently being used in several industrial and medical applications, systematic research into the magnetic properties of nanoscale particles is ongoing. The reason for this is linked to a lack of high-quality samples, though recent improvements have been observed. Questions concerning magnetic phase transitions, such as that from a ferromagnetic to a superparamagnetic state, continue to attract considerable attention. Going one step further, fundamentally new results might be expected for highly ordered, dense magnetic systems. The alignment of the spins of neighboring particles relative to each other is of special interest for magnetic transport properties such as giant magnetic resistance (GMR). The possibility to prepare perfectly ordered domains of almost single-sized and well-defined particles of wide areas will result in a major leap forward in the exciting field of spintronics. Yet, other exciting developments in the use of magnetic nanoparticles in biology and medicine have already been realized. The so-called "liquid magnet hyperthermia" uses surface-modified magnetite nanoparticles which are attracted by tumor cells. Once located in the tumor cells, the particles are warmed by the application of a magnetic field, such that the tumor cells are killed.

The fascinating electronic "inner life" of nanoparticles is only of academic relevance if we fail to use them in nanoelectronics, optoelectronics, and other sophisticated techniques. One condition for such use is that we must learn how to organize nanoparticles not only in three dimensions but, even more importantly, also in two dimensions and one dimension. Whilst developments during the past decade have resulted in impressive progress, the self-assembly techniques of today have in particular been improved so much that not only semiconductors but also metal nanoparticles can be arranged in perfectly ordered assemblies of remarkable extensions. Some time ago, our first concepts regarding interparticle relationships were derived from these two-dimensional arrangements and, as it turned out, this type of "chemistry" between particles, among other things, determines the nature of the collective behavior. Yet, despite considerable progress having been made in self-organization, the main problem remains the transition from nano- to micro-dimensions – or even to macro-dimensions – these being the necessary steps to generate practical, working devices. Clearly, major efforts are still necessary to achieve these goals.

On the other hand, techniques have now been developed that even allow us to "write" with nanoparticles. Today, dip-pen lithography, accompanied by "nanoelectrochemical" modifications of surfaces, allow the fixation of nanoparticles in a final step. Perhaps these techniques will dominate over the principles of self-organization, as the links between nano, micro and macro, respectively, can more easily be controlled.

Much progress has been achieved in understanding charge transport phenomena through nanostructured materials. The electrical properties of nanoparticles are mainly determined by the Coulomb charging energy – a phenomenon that can, in principle, be understood in terms of single-electron tunneling (SET) within the framework of the "orthodox theory of SET". Moreover, in the case of the smallest particles, such as Au_{55}, additional effects due to the discrete nature of the energy states occur, and these must be taken into account when discussing electrical transport phenomena and characteristic excitation energies. This complicated interplay between the particles' size, their state of charge, their constitution and symmetry, the conformation of ligand molecules, and the dielectric environment – to name but a few – makes the physics of these structures extremely complicated. At least, the "orthodox theory" has to be extended for this size regime, but as the electron distribution will then no longer be a pure Fermi distribution, electrical capacitances cannot be discussed classically and even the tunneling time for electrons may be affected.

The fact that nanoparticles and biomaterials (such as enzymes, antibodies or nucleic acids) are of similar dimensions to nanoparticles means that the hybrid systems (which consist of both components) can become attractive nanoelements or building blocks of nanostructures and devices. Several functions and applications of biomaterial–nanoparticle hybrid systems, including analytical applications, signal-triggered electronic functions, nanostructures for circuitry, and the assembly of devices, are today of major importance. However, whilst some of these functions represent viable technologies, others are still at an embryonic stage that requires additional fundamental research. The analytical applications of nanoparticle–biomaterial systems have advanced tremendously during the past decade. Today, an understanding of the unique optical properties of nanoparticles, and the photophysics of coupled interparticle interactions, has allowed these particles to be used as optical labels for recognition events. Similarly, the catalytic properties of biomaterials have enabled biorecognition events to be amplified.

The interplay between nanoparticles and biological systems is of special relevance for semiconductor nanoparticles, known simply also as "quantum dots" (QDs). In recent years, these have emerged as ideal systems for molecular sensors and biosensors, based largely on their size-controlled luminescence. Yet, it is the wide variety of chemical functionalities with which QDs can be equipped that makes them ideal partners for different biosystems. In contrast to former passive optical labels, specifically functionalized QDs can operate as optical labels so as to observe the dynamics of biocatalytic transformations and conformational transitions of proteins. This development will surely open a wide variety of doors in modern nanobiotechnology.

Compared to the First Edition of *Nanoparticles*, this Second Edition of the book offers considerable progress in a host of different fields. Nonetheless, as worldwide research in this area of nanoscience and nanotechnology continues apace, it is to be expected that – in a few years time – further important developments will be reported, especially with regards to the application of techniques for biology and medicine.

Index

a

ab initio study of clusters   135
Ag–Au alloy nanoparticles   331
– TEM images   333
Ag nanoparticles, see silver nanoparticles
alcohol reduction process   220
arc discharge technique   241
Arrhenius equation   154
artificial atoms   372–379, 402, 435
atomic force microscopy (AFM)   316, 345, 355, 364ff, 440ff
Au_{13} clusters   330ff, 348
Au_{55} clusters   330ff, 347f, 351ff, 359, 364ff, 415
– AFM images   366
Auger processes   31ff
Au_{55} microcrystals   330
– SAXRD/WAXRD studies   332
– TEM image   333f
Au_{55} monolayers   415
Au nanoparticles, see gold nanoparticles (AuNPs)

b

band gap   28, 122ff
bioinspired techniques   231
biomedical imaging, contrast agents   239
biorecognition   475
biosensors   472ff, 503ff
blinking   29f
Bloch-functions   7
Bloch wall   272
blocking temperature   276
bottom-up strategy   219ff
bottom-up synthesis   58
box   21ff

c

carbon nanotubes (CNTs)   6ff, 18, 291ff, 474
catalysts   239
CdS colloid   74
CdSe dots   390
CdSe nanoclusters   74
CdSe nanocrystals   84f, 103, 252, 313ff, 323, 373, 379, 387f
– TEM images   314
CdSe nanoparticles   83
CdSe nanorods   314
CdSe quantum dots (QDs)   10, 23, 28, 30, 76, 81, 388, 464ff, 470ff, 474ff, 482ff, 491ff, 495ff
CdSe quantum rods (QRs)   385ff, 389
CdSe quantum well   87
CdSe–ZnS quantum rods (QRs)   389
CdS membranes   73
CdS nanocrystals   71ff, 75ff, 80ff, 322f, 379, 387. see also CdS nanoparticles
CdS nanoparticles   69, 73, 79, 315ff, 321ff
CdS nanorods   314
CdS quantum dots (QDs)   76, 460, 466, 470
CdTe   318
– nanocrystals   78ff, 315ff, 319ff, 322
– nanoparticles   318
– quantum dots (QDs)   460ff, 465, 468, 475, 478, 490ff, 493
CdTe–CdS–CdSe nanocrystals   88
CdTe–CdSe nanocrystals   88
CdTe–CdS nanocrystals   88
chalcogen-bridged copper clusters   128ff
chalcogen-bridged silver clusters   177
chemical switching   417
chemical vapor deposition (CVD)   241, 315
– electrospray organometallic   321
clusters   18ff

Nanoparticles: From Theory to Application. Edited by Günter Schmid
Copyright © 2010 WILEY-VCH Verlag GmbH & Co. KGaA, Weinheim
ISBN: 978-3-527-32589-4

c

cobalt nanoparticles   244, 246, 254, 279, 334f, 344ff, 361
– 1-D, TEM images   363
cobalt nanorods   253, 280, 343ff
– TEM/HR-TEM images   256
coercivity   272, 277, 279ff, 293
colloidal nanocrystals   27ff
colloidal quantum dots   26ff
colloidal syntheses   242
complementary metal-oxide semiconductor (CMOS)   41ff
Co nanocrystals   243, 245, 253, 260
Co nanodisks   253, 255, 280, 363
Co nanowires   253
conductivity channels   18
controlled decomposition   226
copper chalcogenide clusters   156, 157
copper selenide clusters   150ff, 161, 164, 395ff
copper selenide–selenolate clusters   156, 161ff
copper sulfide   139
copper sulfide clusters   152
copper telluride clusters   168, 170f, 173f, 396ff
copper-telluride-tellurolate clusters   172, 174f
copper tellurium cluster   169ff
$CoPt_3$ nanocrystals   243, 249ff, 334ff, 339ff
– HR-SEM images   336
– magnetic alloy nanocrystals   247–252
– SEM images   336
– TEM images   341ff
$CoPt_3$ nanocubes   280
$CoPt_3$ nanoparticles   248ff, 333
– TEM images   342ff
$CoPt_3$ nanowires   280
core–shell nanocrystals   83ff, 87ff, 102ff, 111ff, 116, 122ff, 124
core–shells   122
Coulomb attraction   22
Coulomb blockade (CB)   33ff, 405ff, 407f, 413, 415f, 420ff, 430ff, 445f,
Coulomb blocking   403
Coulomb gap   34, 38ff, 416f, 423ff, 428ff, 447ff
Coulomb interaction   21
cubes   228ff
cubo-octahedrons   229ff
Cu–S clusters   153
CuSe clusters   395
cylinders   228

d

1-D assemblies   421
data storage applications   281
1-D cluster wires   367
De Broglie relation   11
De Broglie wavelength   9, 13, 16, 18ff
decahedrons   229
demagnetization   272
density functional theory (DFT)   58f, 64, 131ff
density of states   12, 18ff
diblock copolymer   354ff
– TEM images   356
differential thermogravimetry (DTG)   397
digestive ripening   219
dip-pen nanolithography (DPN)   357ff, 362
1-D nanoparticle   362ff
3-D organization   328ff
1-D organized metal nanoparticle   347
3-D potential well   21
drug delivery agents   288, 290
drug-delivery systems   318
drug delivery techniques   291
2-D superlattices   339ff
3-D superlattices   344ff
dumb-bell-like nanoparticles   266ff

e

E_9 clusters   55
electrodeposition   241
electromagnetic radiation   228
electron-beam lithography (EBL)   25
electron diffraction   313
electron energy loss spectroscopy (EELS)   270
electron–hole pairs   21ff, 28, 31ff
exchange anisotropy   280
excitons   28
extended X-ray absorption fine structure (EXAFS)   79ff

f

Fe nanorods   280
FePt nanocrystals   258ff, 334ff, 351ff. *see also* FePt nanoparticles
– SEM images   335
– TEM image   335
FePt nanoparticles   257ff, 281ff, 333
Fermi energy   11
Fermi function   40ff
Fermi level   9, 35
Fermi velocity   11
Fermi wavenumber   11
ferrofluids   240ff

ferromagnetic nanocrystals   239
ferromagnets   258
FFT   331
fluorescence quantum yields   455
fluorescence resonance energy transfer
  (FRET)   456ff, 462ff, 472ff, 474, 477ff,
  483–485, 487ff, 496ff
Förster resonance energy transfer
  (FRET)   319ff, 323
Frank–Kasper polyhedra   170ff, 194
free electron gas   10
free electron model   10
full-shell clusters   222ff, 228ff
– TEM images   223

g

goethite nanocrystals   262ff
– TEM/HR-TEM image   263
gold nanocrystals   31, 32
gold nanoparticles (AuNPs)   338ff, 345, 347ff,
  350ff, 352, 354ff, 357ff, 416, 422, 424f, 434,
  436, 439, 441, 450ff, 482
– TEM images   354, 356
gold nanowires, TEM image   350ff

h

Hartree–Fock method   7
HeLa cancer cells   502ff
hematite nanocrystals   262ff
HgTe nanocrystals   78
high-resolution electron microscopy   322
high-resolution scanning electron microscopy
  (HR-SEM)   313, 334
high-resolution transmission electron
  microscopy (HR-TEM)   72ff, 89, 111,
  118f, 225, 244, 249, 251, 254, 267, 270,
  329, 445
hole   20ff
hollow magnetic nanocrystals   268–271
hollow nanoparticles   231
HOMO–LUMO gap   392, 395
hot-injection synthesis   81ff

i

Ib–VI nanoclusters   127
icosahedrons   228–230
III–V semiconductor nanocrystals   81, 101ff,
  105, 113
II–VI nanoparticles   69ff
II–VI semiconductor nanocrystals   76
InAs nanocrystals   103, 105–110, 373, 374
– STM measurements   375
inert gas condensation technique   241

InP nanocrystals   105–110, 317. see also InP
  nanoparticles
InP nanoparticles   321
intermetallic compounds   51
intermetalloid clusters   49, 58
isolobal concept   57
IV–VI semiconductor nanocrystals   81

k

Kirkendall effect   268, 270

l

Langmuir–Blodgett (LB) technique   315ff
– films   315
– ZnS–CdSe QD layers   315ff
laser irradiation   227
laser pyrolysis   241
LbL method   317ff
light-emitting diodes (LEDs)   116, 318ff, 381

m

maghemite nanocrystals   259ff
magnetic anisotropy   273
magnetic core–shell nanoparticles   265
magnetic drug delivery agents   286
magnetic hyperthermia   292ff
magnetic nanocrystals   243, 259–264
– hematite/ Wüstite/goethite
  nanocrystals   262ff
– magnetite nanocrystals   259ff
– metal ferrites nanocrystals   263
magnetic nanoparticles   239ff, 255–259, 266,
  288, 292, 300
– biomedical applications   282–301
– – drug delivery   285
– – gene delivery   290
– – magnetic resonance imaging (MRI)
  295ff
– – magnetic separation   291ff
– – tomographic imaging   297
– metal nanoparticles   271–281
– shape-controlled synthesis   252–255
magnetic resonance imaging (MRI)   283, 285,
  295ff, 297ff
magnetic susceptibility   273
magnetization   272
magnetofection   291
matrix assisted laser desorption/ionization-
  time-of-flight (MALDI-TOF) mass
  spectrometry   180, 185
metallic quantum dot (QD)   222
metal nanoparticles   31ff, 328, 367, 401
– stabilization   215–218

– – electrostatic stabilization 215, 217f
– – steric stabilization 215, 218
– three-dimensional organization 328–338
– two/one-dimensional structures 338ff
– – self-assembly 388ff, 348ff
– – structures 357–367
metalorganic chemical vapor deposition (MOCVD) 25, 315
metal-oxide-semiconductor field effect transistors (MOSFETs) 23
micelles 352ff
molecular beam epitaxy (MBE) 23, 101ff 241, 383
molecular crystals 311, 313ff
molecular dynamics (MD) simulations 347
multicomponent magnetic nanocrystals 264–271
– dumb-bell-like nanoparticles 266–268
– hollow magnetic nanocrystals 268–271
– magnetic core–shell nanoparticles 265
multiple quantum well 23

n

nanocrystals 9ff, 228ff
nano-onions 88
nanoparticles 76–81
– assemblies 435
– II–VI semiconductor nanocrystals 76
– III–V semiconductor nanoparticles 101–125
– organization 311
nanosensors 240ff
nanosticks 229
nanowires 229
Neel rotation 274
nido-E$_9$ cluster 57ff
noble metal nanoparticles 214
nonspherical particles 228

o

octahedrons 228ff
one-dimensional systems, *see* quantum wires
organometallic precursor molecules 227
Ostwald ripening 103, 219ff

p

particle-in-a-box approach 14ff, 17
particle-wave duality 6ff
PbSe nanocrystals 317
Peierls deformation 174
photochemical deposition 321
photoluminescence 73, 395ff
photolysis 227
photonic crystals 318

Plank constant 11
polyol method 220
pulsed laser deposition 241

q

quantum bits, *see* qubits
quantum computer 42
quantum confinement effects 15, 21, 110f, 111, 455
quantum dot quantum wells (QDQWs) 88ff
– CdS–CdSe–CdS 90
– CdS–HgS 89ff
– CdS–HgS–CdS 92
quantum dots (QDs) 3, 5–13, 18ff, 29ff, 38ff, 40ff, 87ff, 316ff, 322ff, 367ff, 371ff, 379ff, 383ff, 388, 390, 410ff, 455ff, 465ff, 472ff, 489ff, 498ff, 503ff
– Au nanoparticle 482
– CdSe QDs 10, 76, 317, 321f, 374
– CdSe/ZnS 497ff, 501ff
– CdS QDs 76
– cellular automata 41
– InAs QDs 377f, 380f
– lasers 32ff
– luminescence 493ff, 497ff
– optical properties 28ff
– semiconductor 10, 455ff, 462, 503
quantum Hall effect 16
quantum mechanics 6ff
quantum point contacts 18
quantum rods (QRs) 371ff, 389ff
quantum size effect 371, 395, 402
quantum wires 17ff 18
qubits 42

r

radiolysis 227
remanent magnetization 272
resonant tunneling transport 39
reverse micelles technique 241
rods 230

s

salt reduction 219–226
saturation magnetization 272
scanning electron microscopy (SEM) 223f, 334, 351ff, 383, 406, 422, 425, 429ff, 433, 435
scanning near-field optical microscopy (SNOM) 319
scanning transmission electron microscopy (STEM) 270
scanning tunneling microscopy (STM) 316, 363, 411ff, 416ff, 420ff, 446

scanning tunneling spectroscopy (STS)   386, 411ff, 415, 417ff
Schrödinger equation   6f, 11f, 14, 17, 21ff
selenide clusters   139
selenido-bridged gold clusters   209
selenido–selenolato-bridged copper clusters   138ff, 154
selenido–selenolato-bridged silver clusters   191
selenium-bridged   392
– copper clusters   136ff, 154, 180, 397
– gold clusters   208–210
– silver clusters   186–195
self-assembled monolayers (SAMs)   338, 367, 413, 416, 417
semiconducting nanocrystals, *see* quantum dots (QDs)
semiconductor nanocrystals   102ff, 110f, 312, 322ff, 372–379, 388. *see also* semiconductor nanoparticles
semiconductor nanoparticles   69, 311ff, 315, 328, 371ff
– II–VI nanoparticles   69ff
semiconductor quantum dots   10, 19–22, 455ff, 462
semiconductors   19ff
sensors   462ff
shape control   228–231
shape evolution   230
silver-chalcogenide clusters   183
silver nanoparticles   329ff, 438
– 2-D array of   444
silver selenide–selenolate clusters   157
single-crystal X-ray diffraction (XRD)   76, 311ff
single-electron devices   239
single-electron logic (SEL)   402
single electron memories   41
single-electron transfer   415
single-electron transistors   28, 36ff, 41ff, 82, 406ff, 429
single-electron tunneling (SET)   33, 35ff, 39, 41, 371, 373ff, 401ff, 409ff, 412, 414f, 418ff, 420ff
size-controlled luminescence   455
size-quantization   402
S-layers   355ff
small-angle scattering (SAS)   69
small-angle X-ray diffraction (SAXRD)   329–331
small-angle X-ray scattering (SAXS)   84, 251ff
sonochemical synthesis   242

specific absorption rate (SAR)   293ff
spherical nanoparticles   228
spherical resonators   318
spin casting   321
spin-coating   321
sputtering   241
stabilizers   240ff
stabilizing agents, *see* stabilizers
steric stabilizers   217
Stokes shift   29ff
Stranski–Krastanov regime   25
sulfur-bridged copper clusters   129ff, 139
sulfur-bridged silver clusters   178ff
superconductive quantum interference device (SQUID)   271
superlattices   311
super-paramagnetic iron oxide particles (SPIOs)   299
superstructures, *see* superlattices
surface plasmons   31

t

telluride–tellurolato-bridged silver clusters   199, 203, 206
tellurido–tellurolato-bridged copper clusters   166ff
tellurium-bridged copper clusters   165ff, 392
tellurium-bridged silver clusters   196–208
tetra-alkylammonium method   226
tetrahedrons   228ff
tetrel clusters   49
tetrel elements   50–56
thermogravimetric analysis (TGA)   397, 400
Tolman's cone angle   128
tomographic imaging   297
top-down techniques   219ff
TOP–TOPO method   74, 77
transmission electron microscopy (TEM)   29, 71, 73, 75ff, 108ff, 118, 248ff, 251ff, 255, 261, 265, 267, 269f, 313, 316, 329–333, 340, 344f, 347ff, 352, 359, 361ff, 377, 388f, 399, 439ff
tunneling transport   39
– resonant tunneling transport   39
– single-electron tunneling   39
tunnel junctions   34
TURBOMOLE   131, 152
two-dimensional arrangements   438

w

Wade–Mingos rules   57
Wade's rules   52, 54, 60ff
wavevector   11
wide-angle X-ray diffraction (WAXRD)   331

wires 228, 230
Wüstite nanocrystals 262ff

x

X-ray diffraction (XRD) 72ff, 79, 119ff, 225, 244, 247, 253, 258, 313, 316ff, 322, 398, 399
– powder XRD 81, 108ff, 119ff, 245
X-ray photoelectron spectroscopy (XPS) 92, 108, 117ff, 354, 397

z

zero-dimensional systems, *see* quantum dots (QDs)
Zintl clusters 50
Zintl ions 49–51, 56ff, 62
Zintl–Klemm–Busmann concept 50
Zintl phases 59
ZnSe nanocrystals 79
ZnS quantum dots (QDs) 464ff, 470ff, 474ff, 482ff, 491ff, 495ff